T0180839

Lecture Notes in Artificial Intelligence 10463

Subseries of Lecture Notes in Computer Science

More information about this series at http://www.springer.com/series/1244

YongAn Huang · Hao Wu
Honghai Liu · Zhouping Yin (Eds.)

Intelligent Robotics and Applications

10th International Conference, ICIRA 2017
Wuhan, China, August 16–18, 2017
Proceedings, Part II

 Springer

Editors
YongAn Huang
School of Mechanical Science
  and Engineering
Huazhong University of Science
  and Technology
Wuhan
China

Hao Wu
School of Mechanical Science
  and Engineering
Huazhong University of Science
  and Technology
Wuhan
China

Honghai Liu
Institute of Industrial Research
University of Portsmouth
Portsmouth
UK

Zhouping Yin
School of Mechanical Science
  and Engineering
Huazhong University of Science
  and Technology
Wuhan
China

ISSN 0302-9743   ISSN 1611-3349   (electronic)
Lecture Notes in Artificial Intelligence
ISBN 978-3-319-65291-7   ISBN 978-3-319-65292-4   (eBook)
DOI 10.1007/978-3-319-65292-4

Library of Congress Control Number: 2017948191

LNCS Sublibrary: SL7 – Artificial Intelligence

Printed on acid-free paper

This Springer imprint is published by Springer Nature
The registered company is Springer International Publishing AG
The registered company address is: Gewerbestrasse 11, 6330 Cham, Switzerland

Preface

The International Conference on Intelligent Robotics and Applications (ICIRA 2017) was held at Huazhong University of Science and Technology (HUST), Wuhan, China, during August 16–18, 2017. ICIRA 2017 was the 10th event of the conference series, which focuses on: (a) fundamental robotics research, including a wide spectrum ranging from the first industrial manipulator to Mars rovers, and from surgery robotics to cognitive robotics, etc.; and (b) industrial and real-world applications of robotics, which are the force driving the research further.

This volume of *Lecture Notes in Computer Science* contains the papers that were presented at ICIRA 2017. The regular papers in this volume were selected from more than 350 submissions covering various topics on scientific methods and industrial applications for intelligent robotics, such as soft and liquid-metal robotics, rehabilitation robotics, robotic dynamics and control, robot vision and application, robotic structure design and mechanism, robot learning, bio-inspired robotics, human–machine interaction, space robotics, mobile robotics, intelligent manufacturing and metrology, benchmarking and measuring service robots, real-world applications, and so on. Papers describing original works on abstractions, algorithms, theories, methodologies, and case studies are also included in this volume. Each submission was reviewed by at least two Program Committee members, with the assistance of external referees. The authors of the papers and the plenary and invited speakers come from the following countries and areas: Australia, Austria, China, Cyprus, Germany, Hong Kong, Japan, Korea, Singapore, Spain, Switzerland, UK, and USA.

We wish to thank all who made this conference possible: the authors of the submissions, the external referees (listed in the proceedings) for their scrupulous work, the six invited speakers for their excellent talks, the Advisory Committee for their guidance and advice, and the Program Committee and the Organizing Committee members for their rigorous and efficient work. Sincere thanks also go to the editors of the *Lecture Notes in Computer Science* series and Springer for their help in publishing this volume in a timely manner.

In addition, we greatly appreciate the following organizations for their support:

National Natural Science Foundation of China
School of Mechanical Science and Engineering, HUST
State Key Laboratory of Digital Manufacturing Equipment and Technology, China
State Key Laboratory of Robotic Technology and System, China
State Key Laboratory of Mechanical System and Vibration, China
State Key Laboratory of Robotics, China

August 2017

YongAn Huang
Hao Wu
Honghai Liu
Zhouping Yin

The original version of the book was revised:
For detailed information please see Erratum.
The Erratum to these chapters is available at
https://doi.org/10.1007/978-3-319-65292-4_78

Organization

Honorary Chair

Youlun Xiong   HUST, China

General Chair

Han Ding   HUST, China

General Co-chairs

Naoyuki Kubota	Tokyo Metropolitan University, Japan
Kok-Meng Lee	Georgia Institute of Technology, USA
Xiangyang Zhu	Shanghai Jiao Tong University, China

Program Co-chairs

YongAn Huang	HUST, China
Honghai Liu	University of Portsmouth, UK
Jinggang Yi	Rutgers University, USA

Advisory Committee Chairs

Jorge Angeles	McGill University, Canada
Tamio Arai	University of Tokyo, Japan
Hegao Cai	Harbin Institute of Technology, China
Xiang Chen	Windsor University, Canada
Toshio Fukuda	Nagoya University, Japan
Huosheng Hu	University of Essex, UK
Sabina Jesehke	RWTH Aachen University, Germany
Yinan Lai	National Natural Science Foundation of China, China
Jangmyung Lee	Pusan National University, Korea
Ming Li	National Natural Science Foundation of China, China
Peter Luh	University of Connecticut, USA
Zhongqin Lin	Shanghai Jiao Tong University, China
Xinyu Shao	HUST, China
Xiaobo Tan	Michigan State University, USA
Guobiao Wang	National Natural Science Foundation of China, China
Michael Wang	The Hong Kong University of Science and Technology, SAR China

Yang Wang   Georgia Institute of Technology, USA
Huayong Yang   Zhejiang University, China
Haibin Yu   Chinese Academy of Science, China

Organizing Committee Chairs

Feng Gao   Shanghai Jiao Tong University, China
Lei Ren   The University of Manchester, UK
Chunyi Su   Concordia University, Canada
Jeremy L. Wyatt   University of Birmingham, UK
Caihua Xiong   HUST, China
Jie Zhao   Harbin Institute of Technology, China

Organizing Committee Co-chairs

Tian Huang   Tianjin University, China
Youfu Li   City University of Hong Kong, SAR China
Hong Liu   Harbin Institute of Technology, China
Xuesong Mei   Xi'an Jiaotong University, China
Tianmiao Wang   Beihang University, China

Local Chairs

Kun Bai   HUST, China
Bo Tao   HUST, China
Hao Wu   HUST, China
Zhigang Wu   HUST, China
Wenlong Li   HUST, China

Technical Theme Committee

Gary Feng   City University of Hong Kong, SAR China
Ming Xie   Nanyang Technological University, Singapore

Financial Chair

Huan Zeng   HUST, China

Registration Chair

Jingrong Ge   HUST, China

General Secretariat

Hao Wu   HUST, China

Sponsoring Organizations

National Natural Science Foundation of China (NSFC), China
Huazhong University of Science and Technology (HUST), China
School of Mechanical Science and Engineering, HUST, China
University of Portsmouth, UK
State Key Laboratory of Digital Manufacturing Equipment and Technology, HUST, China
State Key Laboratory of Robotic Technology and System, Harbin Institute of Technology, China
State Key Laboratory of Mechanical System and Vibration, Shanghai Jiao Tong University, China
State Key Laboratory of Robotics, Shenyang Institute of Automation, China

Contents – Part II

Mechanism and Parallel Robotics

Machine and Robot Vision

Industrial Robot and Robot Manufacturing

An NC Code Based Machining Movement Simulation Method for a Parallel Robotic Machine

Xu Shen[1], Fugui Xie[1,2], Xin-Jun Liu[1,2(✉)], and Rafiq Ahmad[3]

[1] The State Key Laboratory of Tribology and Institute of Manufacturing Engineering, Department of Mechanical Engineering, Tsinghua University, Beijing 100084, China
x-che14@mails.tsinghua.edu.cn,
{xiefg,xinjunliu}@mail.tsinghua.edu.cn
[2] Beijing Key Lab of Precision/Ultra-Precision Manufacturing Equipments and Control, Tsinghua University, Beijing 100084, China
[3] Laboratory of Intelligent Manufacturing, Design and Automation (LIMDA), Mechanical Engineering Department, University of Alberta, Edmonton, AB T6G 2R3, Canada
rafiq.ahmad@ualberta.ca

Abstract. The virtual machine tool and Computer-Aided Manufacturing (CAM) simulation are widely adopted nowadays to lower the cost and save time. Although parallel robotic machines are becoming popular in industry due to its unique advantages in manufacturing application, few methods are available for its simulation. This paper presents a work achieved by combining conventional CAM analysis tool HSMWorks, Computer-Aided Design (CAD) software SolidWorks, and programming tools such as Python and MATLAB to realize the machining movement simulation of a parallel robotic machine. Firstly, an original NC code interpreter is compiled in Python that interprets G-code generated by HSMWorks. Then, necessary coordinate transformation and kinematic calculation are done by using MATLAB. Finally, driving data are imported into virtual machine tool in SolidWorks, and a complete motion simulation environment is then developed. The proposed method is a general approach, which can be upgraded and modified for the simulation of parallel robotic machines with any structure.

Keywords: Parallel robotic machine · NC code · Machining movement simulation · SolidWorks

1 Introduction

The concept of parallel mechanism machines is initially proposed by Pollard in 1938 [1] and then became an interesting topic both for research [2–9] and industrial applications. Since every chain is connected as a closed loop, errors in every chain are averaged and forces are distributed, which lead to higher accuracy [10], higher load capacity, and higher structure rigidity [11]. Due to its great potential and unique

© Springer International Publishing AG 2017
Y. Huang et al. (Eds.): ICIRA 2017, Part II, LNAI 10463, pp. 3–13, 2017.
DOI: 10.1007/978-3-319-65292-4_1

advantages, in cases where the complex faces were hard for the traditional serial machine tools to mill, it is a good choice for parallel robotic machines to finish rapid machining tasks [12–14].

Meanwhile, along with the fast development of the society, manufacturing industry are not very keen to use their resources (time and money) to build physical machine tools for performance test and optimization. The concept of Virtual Manufacturing (VM) is proposed as a system which can operate as a real machine to deal with models in the virtual world [15–19]. With this technology, researchers started to focus on how to better deal with the obstacles [20] and use visible decision-making method to generate safer too-paths [21]. Virtual Manufacturing and computer simulation system greatly avoided the unnecessary space waste in a real machine shop, saved money and work for building prototype and test, and increased the reliability and safety of the final product.

However, since the kinematics of parallel robot is sometimes more complex than traditional serial machine tools, it is troublesome to directly use manufacturing simulation software to simulate the working state of the parallel robotic machine. Therefore, a complete method for carrying out such simulation is in urgent need for manufacturing assessment.

A 5-degree-of-freedom (DOF) spatial parallel kinematic mechanism (PKM) with three limbs was proposed, and its mobility, singularity, and kinematics analysis were analyzed recently [22]. On this basis, the work in this paper is carried out. To achieve the aim of universality, the work here focuses on NC G-code interpreting to provide the robot with detailed machining data. Therefore, as long as there are G-code and its syntax documents, a complete manufacturing plan can be acquired no matter how the G-code is originally generated. After the complete process of "CAM Analysis— Interpreter—Kinematic Operator—Tool and Stock Setting", a machining movement simulation can be produced for further demands such as collision inspection and structure modification.

2   Computer-Aided Manufacturing (CAM) Analysis

The structure of a G-code program for 2D pocket milling generated by CAM software HSMWorks is presented and analyzed as follows. For lack of space, the program posted here contains only example of non-repetitive and motion-related commands.

The workpiece to be milled is presented in Fig. 1(a). The most important parameter to be set is the work coordinate system (WCS), which is crucial for the robotic machine tool presented in the following simulation. Post this milling process under Siemens SINUMERIK 810D postprocessor configuration, and the corresponding G-code and trajectory simulation can be seen in HSMWorks NC Editor (Fig. 1(b)). In real manufacturing cases, this G-code can be directly imported into the machine and start milling.

G-code:

N11 G90 G94

N12 G71

N14 G17

N15 G0 SUPA Z80 D0

N22 X-8.511 Y20.349

N24 Z86

N25 Z76

N26 G1 Z75.499 F500

N27 G17

N28 G3 X -8.498 Y20.331

Z75.203 I7.704 J5.559

......

N120 G2 X -78.169 Y23.55

I-151.12 J-64.574

......

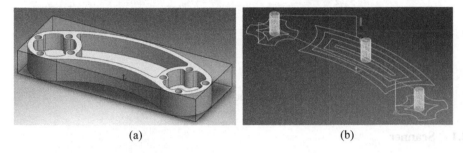

(a)   (b)

Fig. 1. Milling work setting: (a) workpiece; (b) CAM plan

According to SIEMENS SINUMERIK 840D sl/840Di sl/840D/840Di/810D Fundamentals [23], key commands in each block are demonstrated in Table 1.

Table 1. Some key commands in G-code program of this task

Commands	Meaning
G90	Absolute coordinate mode
G94	Feedrate per minute
G71	Metric dimensions (length [mm])
G17	XY-plane specification
G0	Rapid move
G1	Linear move
G2/G3	Circular interpolation, clockwise/counterclockwise

3   NC Code Interpreter Design

As a CAD software, SolidWorks is able to receive direct position coordinates rather than G-code commands. It is necessary to design an interpreter to interpret the G-code initially designed for serial machine tools into actual kinematics information.

Recognition method is illustrated in the flowchart as shown in Fig. 2. The interpreter consists of three units: Scanner, State Storage, and Interpolator.

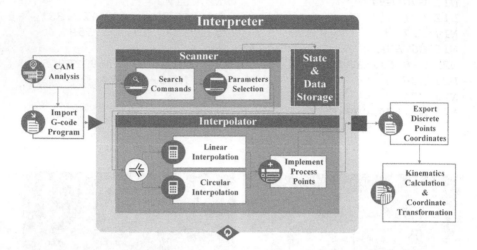

Fig. 2. Working principle of the interpreter

3.1 Scanner

The Scanner is used to read and analyze each G-code block, which contains key commands for tool's motion. By using Python "readlines" function, it is convenient to import each G-code block as a separate line to the interpreter and analyze its components. Commands such as G17/18/19 and G0/1/2/3 and numerical parameters following X/Y/Z/I/J/K are all major searching tasks.

3.2 State Storage

The State Storage contains various vital state markers which will define the working mode and all further operations. It receives information from Scanner and stores those states in some variables. As long as there is no other mode change appearing, these variables stored in the State Storage unit will keep the same, thus offering guidance for interpolation. Some of the relevant variables are Working Plane, Cutting Mode (rapid/linear/circular), Rotating Direction of the circular interpolation, Feedrate (machining speed), Step Size of the interpolation and so forth.

3.3 Interpolator

The Interpolator is the core unit of the interpreter. It will be utilized to interpolate processing points between the starting point and end point following a certain kind of motion. During the milling process, there will be four different motion modes—Rapid traverse motion (G0), Linear interpolation (G1), Circular interpolation clockwise (G2), and Circular interpolation counter-clockwise (G3).

Rapid Traverse Motion (G0): G-code program does not contain any speed definition for the rapid traverse. Because G0 only makes it faster than G1 to move tool to the working region, it is acceptable to set a certain larger G0 speed according to the feed rate for G1.

Linear Interpolation (G1): The program will supplement processing points linearly between the starting point and end point using particular step size, while defining the exact time at every point. As a result, the interpreter will export an extensive data matrix made up of four columns, where every row is a 4-dimensional array—(Time, X coordinate, Y coordinate, Z coordinate).

Circular Interpolation (G2/G3): According to mathematical theory, in 3D space, a parametric equation of a circle with radius r is given by

$$\begin{cases} x(\psi) = p_x + r \cos(\psi) \cdot a_x + r \sin(\psi) \cdot b_x \\ y(\psi) = p_y + r \cos(\psi) \cdot a_y + r \sin(\psi) \cdot b_y \\ z(\psi) = p_z + r \cos(\psi) \cdot a_z + r \sin(\psi) \cdot b_z \end{cases} \tag{1}$$

Where the center of the circle is $\mathbf{p_c} = (p_x, p_y, p_z)$, and two orthonormal vectors in the plane containing the circle are $\overrightarrow{\mathbf{a}} = (a_x, a_y, a_z)$ and $\overrightarrow{\mathbf{b}} = (b_x, b_y, b_z)$.

Table 2. Key data and parameters picked by the Scanner

Name & Type	Source command	Extraction result
Starting point coordinate	G0/1/2/3 X... Y... Z...	$\mathbf{p_s} = (x_s, y_s, z_s)$
Endpoint coordinate	G2/3 X... Y... Z...	$\mathbf{p_e} = (x_e, y_e, z_e)$
Center point offset	G2/3 I... J... K...	$\mathbf{c_o} = (i, j, k)$

As is demonstrated previously, key data can be acquired by the Scanner (Table 2): These parameters are further processed to fit in the mathematical theory in (1):

- Center of the circle: $\mathbf{p_c} = (p_x, p_y, p_z) = (x_s + i, y_s + j, z_s + k)$
- Two vectors in the plane containing the circle are $\overrightarrow{\mathbf{a}}_0 = (x_s - p_x, y_s - p_y, z_s - p_z)$, $\overrightarrow{\mathbf{c}}_0 = (x_e - p_x, y_e - p_y, z_e - p_z)$, then normalized to $\overrightarrow{\mathbf{a}} = \overrightarrow{\mathbf{a}_0}/|\overrightarrow{\mathbf{a}_0}|$, $\overrightarrow{\mathbf{c}} = \overrightarrow{\mathbf{c}_0}/|\overrightarrow{\mathbf{c}_0}|$
- Radius of circle is $r = \sqrt{i^2 + j^2 + k^2}$.

To finish the parametric Eq. (1), the vector $\vec{\mathbf{b}}$, which is orthonormal with $\vec{\mathbf{a}}$, is still in need. The normal vector of the circle plane is given by $\overrightarrow{\mathbf{n}} = \overrightarrow{\mathbf{a}} \times \overrightarrow{\mathbf{c}}$. So the vector $\vec{\mathbf{b}}$ can be calculated by $\overrightarrow{\mathbf{b_0}} = \overrightarrow{\mathbf{n}} \times \overrightarrow{\mathbf{a}}$, which is then normalized to $\vec{\mathbf{b}} = \overrightarrow{\mathbf{b_0}}/|\overrightarrow{\mathbf{b_0}}|$.

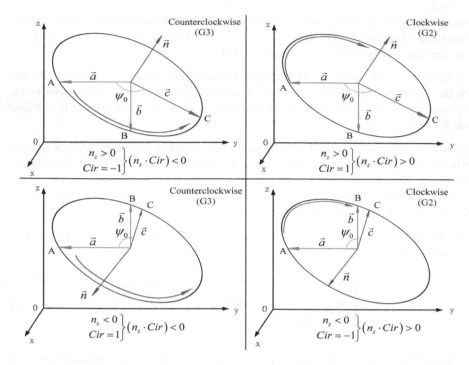

Fig. 3. All possible cases considering starting and end point positions and rotating directions

Considering the interpolating process, when $\psi = 0$, it is easy to find the relationship $(x(0), y(0), z(0)) = \vec{a_0}$, which means the start of the circular motion.

Define the angle between $\vec{a_0}$ and $\vec{c_0}$ to be ψ_0, then the angle the cutting tool needs to cover is ψ_0 or $2\pi - \psi_0$. The interpolator can start to create processing points from the starting point to the endpoint by changing ψ step by step under the restriction $|\psi| < \psi_0$ or $|\psi| < 2\pi - \psi_0$. The problem arises here: how to distinguish these two cases?

In the parametric Eq. (1), if ψ increases from 0 to ψ_0, the result $(x(\psi), y(\psi), z(\psi))$ will always move in the direction that passing a minor arc from point A to point B along the circle. In contrast, the effect of decreasing ψ from 0 to $-(2\pi - \psi_0)$ will step on the other way. So the location of point B will determine the effect of changing ψ. Since the vector $\vec{b}$ is derived from $\vec{n} \times \vec{a}$, the orientation of $\vec{n}$ (more specifically, the sign of n_z) will be a major focus in the end.

Figure 3 illustrates four possible situations during circular motion. For the convenience of programming and analysis, a marker "Cir" is used to describe expecting rotating direction, that is, G2 (clockwise) will lead to "$Cir = -1$" and G3 (counterclockwise) will result in "$Cir = 1$".

By investigating all four cases, it can be summarized that:

(I) $(n_z \times Cir) > 0$: ψ should increase from 0 to ψ_0.
(II) $(n_z \times Cir) < 0$: ψ should decrease from 0 to $-(2\pi - \psi_0)$.

Until now it is finally clear that how the Interpolator should control variable ψ to finish circular interpolation. By following the changing direction and the extremum restriction, it only needs to change ψ by a certain step size to supplement processing points between the starting point and end point.

It is worthwhile to make a further explanation that if only two parameters show in (I, J, K) but three in (X, Y, Z), it means that there will be circular motion on the plane indicated by the existing two coordinates in (I, J, K) while a linear move occurs perpendicular to it, thus resulting in a helix interpolation as a whole.

To conclude this part, in a complete G-code program, there will be multiple G0/G1/G2/G3 commands, so the interpolation methods described above will be reused for a significant number of times. Via the interpreter, the G-code program will finally be translated into a large data matrix ${}^{s}\mathbf{I}=\begin{bmatrix} \vec{t} & {}^{s}\mathbf{X} & {}^{s}\mathbf{Y} & {}^{s}\mathbf{Z} \end{bmatrix}$ in stock frame describing discrete points along its path (each line of $\begin{bmatrix} {}^{s}\mathbf{X} & {}^{s}\mathbf{Y} & {}^{s}\mathbf{Z} \end{bmatrix}$, Fig. 4) and their corresponding arriving time (column vector $\vec{t}$). It will be then processed as the input for parallel robotic machine inverse kinematic calculation.

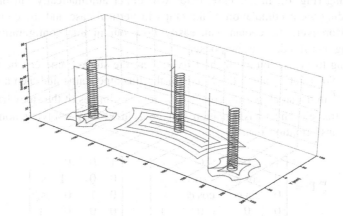

Fig. 4. Trajectory redrawn after interpretation

4 Machining Motion Simulation

The SPR-2(2UP-R-S) parallel robotic machine has been proposed in Ref. [22], the CAD model and kinematic scheme are shown in Fig. 5. From the inverse kinematics described in Ref. [22], it is clear that as long as there is a tool position coordinate, the length of five limbs will be available. However, the coordinate frames of the Interpreter and the Inverse Kinematic Operator (IKO) are different. The interpreter uses coordinate frame attached to stock while IKO uses frame attached to the machine base. So the combining of these two parts requires the transformation of the stock frame $\Re_{s}$ to the machine frame $\Re_{m}$.

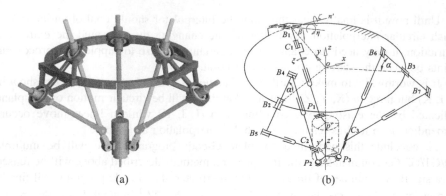

Fig. 5. The parallel robotic machine: (a) CAD model; (b) kinematic scheme

The force and position control of such a pocket milling job is vital for the final product's quality, where the chip removal will be a matter of concern with high priority. Therefore, the horizontal configuration is adopted for workpiece installation and machining (Fig. 6). In this case, chips will either automatically fall or be easily swept, avoiding the accumulation of the chips in vertical case and the corresponding problems. Moreover, the coolant can easily flow out in this configuration without accumulating and flooding in the pocket.

According to the definition of the two frames, the stock frame can be arrived by firstly a $\phi = 90°$ rotation along the X axis of the machine frame and then a translation $\begin{bmatrix} s_x & s_y & s_z \end{bmatrix}^T$ to the stock position (s_x, s_y, s_z are coordinates of the stock frame origin in machine frame). The transformation can be achieved by applying homogeneous coordinate transformation matrix

$$
{}^m_s\mathbf{T} = \begin{bmatrix} 1 & 0 & 0 & s_x \\ 0 & \cos\phi & -\sin\phi & s_y \\ 0 & \sin\phi & \cos\phi & s_z \\ 0 & 0 & 0 & 1 \end{bmatrix} = \begin{bmatrix} 1 & 0 & 0 & s_x \\ 0 & 0 & -1 & s_y \\ 0 & 1 & 0 & s_z \\ 0 & 0 & 0 & 1 \end{bmatrix} \tag{2}
$$

The coordinates in stock frame collected by the Interpreter will then be transformed to the machine frame by multiplying ${}^m_s\mathbf{T}$. This process is necessary before transporting data into the IKO.

$$
\begin{bmatrix} {}^m\mathbf{X} & {}^m\mathbf{Y} & {}^m\mathbf{Z} & 1 \end{bmatrix}^T = {}^m_s\mathbf{T} \cdot \begin{bmatrix} {}^s\mathbf{X} & {}^s\mathbf{Y} & {}^s\mathbf{Z} & 1 \end{bmatrix}^T \tag{3}
$$

Moreover, there are still two parameters required in IKO not defined— the azimuth angle and the tilt angle (φ, θ). Since this task is a 2D milling job, (φ, θ) are constants that won't change through the whole process. Input initial values (φ_0, θ_0) to generate two column vectors $\overrightarrow{\varphi} = \begin{bmatrix} \varphi_0 & \cdots & \varphi_0 \end{bmatrix}^T$ and $\overrightarrow{\theta} = \begin{bmatrix} \theta_0 & \cdots & \theta_0 \end{bmatrix}^T$, both of which have the same number of rows as $\vec{t}$. Here the machine tool is set to work under horizontal configuration, so $\varphi_0 = 90°$ and $\theta_0 = -90°$.

Fig. 6. Workpiece installation and machine tool application scene

Now all necessary inputs for IKO are ready as

$$^{m}\mathbf{I}=\left[\ \vec{\mathbf{t}}\quad ^{m}\mathbf{X}\quad ^{m}\mathbf{Y}\quad ^{m}\mathbf{Z}\quad \vec{\varphi}\quad \vec{\theta}\ \right] \qquad (4)$$

According to the algorithm in Ref. [22], $\mathbf{T} = [x, y, z]^{T}$ repeatedly picks three coordinates from every row in $\begin{bmatrix} ^{m}\mathbf{X} & ^{m}\mathbf{Y} & ^{m}\mathbf{Z} \end{bmatrix}$ and $\mathbf{R}(\varphi, \theta, \sigma)$ picks (φ, θ) from every row in $\vec{\varphi}$ and $\vec{\theta}$.

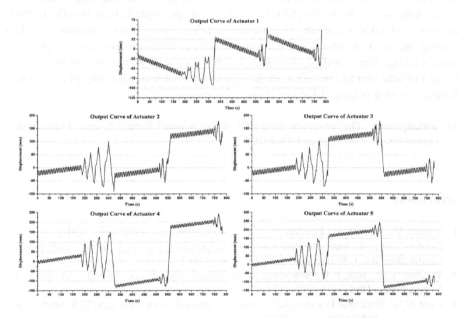

Fig. 7. Continuous output curve of five actuators L_1, L_2, L_3, L_4, and L_5

By defining the initial state of the machine tool and calculating all corresponding limb lengths as a datum, the displacement change that five actuators should generate will be determined. However, now they are only timeline-based discrete data showing the displacement change.

In SolidWorks, the feature "Motion Study" can define linear motors on actuators and import displacement data from files to offer instruction for motors' movement. It can also help users to plot discrete data points thus transforms discrete information into a continuous working curve (Fig. 7).

After finishing all setups and data import for actuators, click "Play" or export motion as a video file, the simulation is finally finished. Different camera views can be added to observe every desired detail of the machine tool motion. Moreover, various other analysis functions are provided by the software to provide the performance information of the robot.

5   Conclusion

In this paper, a machining motion simulation method is proposed for parallel robotic machines by utilizing G-code generated by CAM analysis and motion function in SolidWorks. There is no strict demand for CAM software or specific interface as long as manufacturing G-code for some machine tools and its supporting document are available. This feature significantly broadens the area where this method can be applied. In the part of the Interpreter design, Python programming language is employed because of its brevity in reading and processing information in each line of G-code text. Necessary interpolation in milling process is demonstrated with emphasis since G2 and G3 commands are hardest to be translated into motion data available for parallel kinematics. As for the IKO, the core solving principle is based on the result of previous work done in the lab. The last step requires some precise adjustment such as locating the stock position and choosing a proper machine tool configuration for manufacturing. The proposed simulation method can achieve the goal of closed-loop analysis to better detect the defects in the machine tool or manufacturing plan with high efficiency, which is crucial for fast design and prototyping.

Acknowledgments. This work was supported by the National Natural Science Foundation of China under Grants 51425501 and 51675290.

References

1. Pollard, V., Willard, L.: Position-controlling apparatus. US, US2286571 (1942)
2. Fichter, E.F.: A Stewart platform-based manipulator: general theory and practical construction. Int. J. Robot. Res. 5(2), 157–182 (1986)
3. Waldron, K.J., Hunt, K.H.: Series-parallel dualities in actively coordinated mechanisms. Int. J. Robot. Res. 10(5), 473–480 (1991)
4. Hunt, K.H.: Structural kinematics of in-parallel-actuated robot-arms. ASME J. Mech. Trans. Autom. Des. 105(4), 705–712 (1983)

5. Nanua, P., Waldron, K.J., Murthy, V.: Direct kinematic solution of a Stewart platform. IEEE Trans. Robot. Autom. **6**(4), 438–444 (1990)
6. Sugimoto, K.: Kinematic and dynamic analysis of parallel manipulators by means of motor algebra. ASME J. Mech. Trans. Autom. Des. **109**(1), 3–7 (1987)
7. Nguyen, C.C., Pooran, F.J.: Kinematic analysis and workspace determination of a 6 DOF CKCM robot end-effector. J. Mech. Work. Technol. **20**, 283–294 (1989)
8. Gosselin, C.: Determination of the workspace of 6-DOF parallel manipulators. ASME J. Mech. Des. **112**(3), 331–336 (1990)
9. Gosselin, C., Angeles, J.: Singularity analysis of closed-loop kinematic chains. IEEE Trans. Robot. Autom. **6**(3), 281–290 (1990)
10. Huang, Z., Kong, L.F., Fang, Y.F.: Theory and Control Mechanism for Parallel Robotics (1997)
11. Wang, Z., Wang, Z., Liu, W., Lei, Y.: A study on workspace, boundary workspace analysis and workpiece positioning for parallel machine tools. Mech. Mach. Theory **36**(5), 605–622 (2001)
12. Stewart, D.: A platform with six degrees of freedom. ARCHIVE Proc. Inst. Mech. Eng. 1847–1982 (vols. 1–196) **180**(1965), 371–386 (2013)
13. Kim, J., et al.: Performance analysis of parallel manipulator architectures for CNC machining applications. In: Proceedings of the IMECE Symposium on Machine Tools, Dallas (1997)
14. Ryu, S.-J., et al.: Eclipse: an overactuated parallel mechanism for rapid machining. In: Boër, C.R., Molinari-Tosatti, L., Smith, K.S. (eds.) Parallel Kinematic Machines. Springer, London (1999). doi:10.1007/978-1-4471-0885-6_32
15. Qingke, Y., Rujia, Z.: Virtual manufacturing system. China Mech. Eng. **4**, 10–12 (1995)
16. Iwata, K., et al.: Virtual manufacturing systems as advanced information infrastructure for integrating manufacturing resources and activities. CIRP Ann.-Manuf. Technol. **46**(1), 335–338 (1997)
17. Altintas, Y., et al.: Virtual machine tool. CIRP Ann. Manuf. Technol. **54**(2), 115–138 (2005)
18. Lin, W., Fu, J.: Modeling and application of virtual machine tool. In: International Conference on Artificial Reality and Telexistence–Workshops IEEE, pp. 16–19 (2006)
19. Carpenter, I.D., et al.: Virtual manufacturing. Manuf. Eng. **76**(S1), 113–116 (1997)
20. Ahmad, R., Plapper, P.: Safe and automated assembly process using vision assisted robot manipulator. Procedia CIRP **41**, 771–776 (2016)
21. Ahmad, R., Tichadou, S., Hascoet, J.-Y.: Generation of safe and intelligent tool-paths for multi-axis machine-tools in a dynamic 2D virtual environment. Int. J. Comput. Integr. Manuf. **29**(9), 982–995 (2016)
22. Xie, F., et al.: Mobility, singularity, and kinematics analyses of a novel spatial parallel mechanism. J. Mech. Robot. **8**(6), 061022 (2016)
23. Siemens Ltd.: Programming Manual of SINUMERIK 840D sl/840Di sl/840D/840Di/810D Fundamentals, 6FC5398-1BP10-2BA0, November 2006

Trajectory and Force Generation with Multi-constraints for Robotic Belt Grinding

Yangyang Mao, Huan Zhao[✉], Xin Zhao, and Han Ding

State Key Laboratory of Digital Manufacturing Equipment and Technology,
Huazhong University of Science and Technology, Wuhan 430074, Hubei,
People's Republic of China
huanzhao@hust.edu.cn

Abstract. In the robotic belt grinding process, the parameters such as robot feedrate and contact force play important roles on the material removal. In order to minimize the machining time and achieve high accuracy, a scheduling method of trajectory and force is proposed to maximize the robot feedrate. First, an optimization model with the constraints of joints speed, acceleration and robot feedrate, acceleration of robotic belt grinding is presented. Then, the feedrate scheduling problem is transferred to a linear programming problem, which can be solved efficiently. The contact force is also scheduled based on an empirical formula of grinding mechanism after solving the optimal feedrate. Finally, the simulation and experiment show the method can effectively achieve trajectory and force generation for robotic belt grinding with high efficiency and accuracy.

Keywords: Trajectory generation · Force generation · Robotic belt grinding

1 Introduction

To date, robotic belt grinding has been extensively applied to achieve high material removal rate in aeronautics industry, which can relieve hand grinders from their noisy work environment as well as for improving machining accuracy and product consistency [1]. Therefore, it is necessary to control the material removal process during belt grinding processes, especially in the complex workpiece surface machining (e.g., surfaces of turbine blades) [2]. However, the material removal is related to a variety of factors. It is revealed that the robot feedrate and the workpiece-wheel contact force are the two key process parameters as shown in Fig. 1. Therefore, the key problems of controlling the material removal amount of robotic belt grinding are to schedule the robot feedrate and contact force.

Song et al. [3] proposed the optimization model which is the minimum changes in robot feedrate and contact force at the next point to reduce the influence of the transition process caused by the control adjustment, through the predicted removal error at a grinding point. The method could effectively track the desired material removal, but the robot could not precisely track the desired parameter trajectories planned offline.

© Springer International Publishing AG 2017
Y. Huang et al. (Eds.): ICIRA 2017, Part II, LNAI 10463, pp. 14–23, 2017.
DOI: 10.1007/978-3-319-65292-4_2

Wu [4] employed a conservative strategy which could be used to solve the optimization problem by setting an objective function defined to minimize the difference between the actual material removal and the expected material removal. The robot feedrate and contact force to reach/pass the target were taken into consideration in the strategy. The method was the most direct and effective for the grinding of the complex geometries, but it required real-time measuring the amount of the complex blades so that the process was time-consuming and the production efficiency was not high. Ren and Kuhlenkötter [5] introduced a local process model where the force distribution in the contact area was determined via a finite element model constructed using a set of geometry and elasticity information on the intersection between the grinding belt and the workpiece. It is no doubt that this method increases the computational burden. Zhao et al. [6] developed an adaptive feedrate optimization model with the constraints of chord error, maximum feedrate, acceleration and jerk to minimize the machining time, but this method was only applicable to five-axis machining. Song et al. [7] presented a planning method for the control parameters of the grinding robot based on an adaptive modeling using the cooperative particle swarm optimization. It aimed to smooth the trajectories of the control parameters of the robot and shorten the response time in the transition process. However, the optimization goal was too complex which increased the computational load.

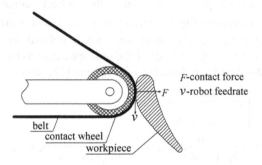

F-contact force
v-robot feedrate

belt
contact wheel
workpiece

Fig. 1. Ilustrition of the robot feedrate and contact force.

This paper presents a method of trajectory and force generation which aims to minimize the machining time. First, the optimization model simultaneously constrains the joints speed, acceleration and robot feedrate, acceleration. And the model is transferred to a linear optimization problem. Then, after solving the optimal robot feedrate, the contact force could be calculated based on an empirical approximate formula of grinding mechanism. Compared with the existing methods, the proposed method is efficient and can be applied in practical applications.

The rest of this paper is organized as follows. The tool path fitting into the B-spline with respect to parameter u is introduced in Sect. 2. Section 3 presents the scheduling algorithm with multi-constraints in detail. Simulation and experiment are provided in Sect. 4, followed by the conclusion in Sect. 5.

2   Tool Path Fitting into the B-Spline

The tool path of robotic belt grinding is commonly given by a sequence of discrete cut location (CL) points, which are generated according to chord error and scallop height [8, 9]. Each CL point consists of tool center position and orientation. The challenge in converting tool paths into axis motions is to command the tool to move at the specified feedrate. Because it is impractical to specify the axis positions at the motion controller loop closing frequency, interpolatory splines are used [10]. In addition, the tool tip and orientation locations are first fitted to splines independently to achieve geometric continuity while decoupling the relative changes in position and orientation of the cutter along the curved path [11]. Therefore, the tool center position and orientation of the spline path are defined as a function of parameter u.

$$
\begin{cases}
\boldsymbol{p}(u) = \displaystyle\sum_{i=0}^{n} N_{i,k}(u)\omega_i \boldsymbol{p}_i \Big/ \sum_{i=0}^{n} N_{i,k}(u)\omega_i \\[2mm]
\boldsymbol{o}(u) = \displaystyle\sum_{i=0}^{n} N_{i,k}(u)\omega_i \boldsymbol{o}_i \Big/ \sum_{i=0}^{n} N_{i,k}(u)\omega_i
\end{cases}
\tag{1}
$$

where $\boldsymbol{p}_i$ and $\boldsymbol{o}_i$ are control points, which are obtained by the post-processing of the tool path. $\boldsymbol{p}(u) = [p_x(u), p_y(u), p_z(u)]$ describes the wheel center curve. $\boldsymbol{o}(u) = [o_z(u), o_y(u), o_z(u)]$ represents the orientation axis curve, which are calculated by the control points. And if the orientation of the tool is defined by a unit vector, the orientation spline must lie on the surface of the unit sphere. ω_i is the weight coefficient. $N_{i,k}(u)$ is the k th B-spline basis function of k-degree [12], and is defined as

$$
\begin{aligned}
N_{i,0} &= \begin{cases} 0, & u_i \le u \le u_{i+1} \\ 1, & otherwise \end{cases} \\[2mm]
N_{i,k}(u) &= \frac{u - u_i}{u_{i+k} - u_i} N_{i,k-1}(u) + \frac{u_{i+k+1} - u}{u_{i+k+1} - u_{i+1}} N_{i+1,k-1}(u)
\end{aligned}
\tag{2}
$$

where $U = [u_0, u_1, \cdots, u_m]$ is the knot vector. And $x(u) = [\boldsymbol{p}(u), \boldsymbol{o}(u)]^{\mathrm{T}}$ can denote the pose of the robot.

3   Scheduling Algorithm for Trajectory and Force

3.1   Optimization Goal

The purpose of the scheduling method aims to minimize the machining time while taking joint speed, acceleration and robot feedrate, acceleration into consideration. Since $dt = |\boldsymbol{p}'| du/v(u)$, the optimization goal could be designed as follows:

$$
\min \quad \int_0^1 |\boldsymbol{p}'| du/v(u)
\tag{3}
$$

Because of solving the Eq. (3) is time-consuming, the optimization goal could be transferred to maximize the feedrate by discretizing the tool path with equal interval of parameter u. At last, the optimization goal could be rewritten as below:

$$\max \quad \sum_{i=0}^{n} \left(\frac{v(u_i)}{|p'(u_i)|} \right)^2 \tag{4}$$

3.2   Trajectory Generation

3.2.1   Joint Constraints

In the robotic machining process, joint constraints must be taken into consideration firstly. Therefore, the speed and the acceleration of six joints must not exceed their limits. Because the speed and acceleration of the six joints are done by interpolation of the joint values, they are taken into consideration in joint space. According to the conditions, the constraints could be evaluated as

$$\dot{\theta}_{min} \leq \dot{\theta}(u) \leq \dot{\theta}_{max} \tag{5}$$

$$\ddot{\theta}_{min} \leq \ddot{\theta}(u) \leq \ddot{\theta}_{max} \tag{6}$$

where $\dot{\theta}(u) = \left[\dot{\theta}_1(u), \dot{\theta}_2(u), \dot{\theta}_3(u), \dot{\theta}_4(u), \dot{\theta}_5(u), \dot{\theta}_6(u) \right]^T$ is the speed of six joints of the robot. And $\ddot{\theta}(u) = \left[\ddot{\theta}_1(u), \ddot{\theta}_2(u), \ddot{\theta}_3(u), \ddot{\theta}_4(u), \ddot{\theta}_5(u), \ddot{\theta}_6(u) \right]^T$ is the acceleration of six joints. $\dot{\theta}_{min}(\ddot{\theta}_{min})$ and $\dot{\theta}_{max}(\ddot{\theta}_{max})$ are the minimum and maximum values of the speed (acceleration) of six joints, respectively. And the limits of the speed and acceleration, which we can look up, are usually provided by the robot manufacturer.

Since $\dot{x}(u) = J\dot{\theta}(u)$, where $\dot{x}(u)$ denotes the derivatives of $x(u)$ with time, J is the jacobian matrix, Eqs. (5) and (6) can be described as

$$\dot{\theta}(u)^T = \frac{J^{-1}x'(u)^T}{|p'(u)|} v(u) \tag{7}$$

$$\ddot{\theta}(u)^T = J^{-1} \left[\left(\frac{1}{|p'(u)|^2} x''(u)^T - \frac{p'(u) \cdot p''(u)}{|p'(u)|^2} x'(u)^T \right) v^2(u) - \dot{J}\dot{\theta}(u) \right] \tag{8}$$

$x'(u)$ and $x''(u)$ are the derivatives of $x(u)$ with parameter u. $p'(u)$ and $p''(u)$ are the derivatives of $p(u)$ with parameter u.

3.2.2   Grinding Process Constraints

In the process of robotic belt grinding, the feedrate and acceleration of end-effector could be related to the speed and acceleration of the six joints through the jacobian. However, for the sake of convenience, the feedrate and acceleration of the end effector could be obtained by interpolation of the positions in base frame, which are collected in

Cartesian space. In addition, the limits of the feedrate and acceleration of the end effector are determined by the machining process experiences.

v_{lim} is the maximum robot feedrate, so the feedrate under this constraints is as following:

$$0 \le v(u_i) \le v_{\text{lim}} \tag{9}$$

The acceleration of robotic end-effector is obtained by the difference of the feedrate, which could be written as:

$$a(u_i) = \frac{v(u_{i+1}) - v(u_i)}{\Delta t} = \frac{v(u_{i+1})^2 - v(u_i)^2}{2|p'(u_i)|\Delta u} \tag{10}$$

where Δu is the interval of parameter u. When the maximum acceleration a_{lim} is given, the under constraints can be obtained.

$$-a_{\text{lim}} \le \frac{v(u_{i+1})^2 - v(u_i)^2}{2|p'(u_i)|\Delta u} \le a_{\text{lim}} \tag{11}$$

3.2.3 Detailed Algorithm

Because Eqs. (5), (6), (9) and (11) have established the linear constraints with respecting joint constraints and process constraints, the detailed algorithm is summarized by using the equations above.

Assuming that the parameter u is sampled with equal interval and using the Eqs. (7) and (9), the robot feedrate could be calculated as following:

$$v_r(u_i) = \min\left\{ \frac{|J\dot{\theta}(u_i)^{\text{T}}|}{|x'(u_i)^{\text{T}}|} |p'(u_i)|, v_{\text{lim}} \right\} \tag{12}$$

In the machining process, the robot feedrate $v(u_i)$ should be less than or equal to $v_r(u_i)$. At the beginning and the end of the path, the feedrates of the first and end point should be set to equal to a constant value, that is $v(0) = v_{first}, v(1) = v_{end}$. Using $v_i^2 (i = 0, \cdots n)$ as the design vector, the optimization model of robot feedrate is arranged as

$$\max \sum_{i=0}^{n} \frac{v_i^2}{|p'(u_i)|^2}$$

s. t.

$$\begin{cases} v_0 = v_{first}, v_n = v_{end} \\ 0 \le v_i^2 \le v_r^2(u_i) \\ -a_{\text{lim}} \le \frac{v(u_{i+1})^2 - v(u_i)^2}{2|p'(u_i)|\Delta u} \le a_{\text{lim}} \\ \ddot{\theta}(u)^{\text{T}} = J^{-1}\left[\left(\frac{1}{|p'(u)|^2}x''(u)^{\text{T}} - \frac{p'(u) \cdot p''(u)}{|p'(u)|^2}x'(u)^{\text{T}} \right)v^2(u) - \dot{J}\dot{\theta}(u) \right] \end{cases} \tag{13}$$

Note that Eq. (13) is a linear programming problem, so the scheduling feedrate could be solved effectively and easily.

3.3 Force Generation

Using the grinding mechanism, some empirical equations for modeling process are available and have been applied in practical systems [13]. For grinding a workpiece with a complex surface, an empirical approximate formula is given in [14] and [15] as

$$r = K \frac{(v_b)^\alpha}{(v)^\beta} (F)^\gamma \tag{14}$$

where r denotes the amount of the material removal, K is a constant that describes the grinding factors. v_b is the grinding belt velocity and usually could be set to a constant value. v is the robot feedrate and F is the contact force. α, β, γ are constant coefficients.

After solving the optimal robot feedrate from Eq. (13), the contact force could be calculated from Eq. (14) when r, K and v_b are known. Naturally, a maximum value of the contact force F_{lim} should be set because too high contact force would cause too much heat between the belt and the workpiece, which would burn the workpiece. At the beginning and the end of the path, the forces of the first and end point should be set to equal to a constant value, that is $F(0) = F_{first}$, $F(1) = F_{end}$. When the calculated contact force exceed the maximum value F_{lim}, the robot feedrate should be scheduled again from Eq. (13). While the robot feedrate and contact force could satisfy the all requirements finally, a cubic B-spline is used to fit the robot feedrate and contact force with respect to parameter u. Besides, after the contact force of robotic belt grinding is calculated and optimized, the position-based impedance control could be used for robotic time-varying force tracking, which would achieve the accurate robotic belt grinding force control [16, 17].

4 Simulation and Experiment

The proposed scheduling method is applied to a length of tool path based on robotic belt grinding as shown Fig. 2(a). And the tool path is described in the workpiece coordinate system. The parameters used in the scheduling method are list in Table 1. Figure 2(b) denotes the desire material removal. The robot feedrate and contact force could be scheduled according to the scheduling algorithm as shown Fig. 2(c). The acceleration of robot end-effector is described as shown Fig. 2(d). The speed and acceleration of six joints are given in Fig. 2(e–f).

We may know the scheduling feedrate and force are continuous from Fig. 1. The feedrate and acceleration of each joint, the robot feedrate, acceleration and contact force do not exceed their setting value, thus proving the effectiveness of the scheduling algorithm.

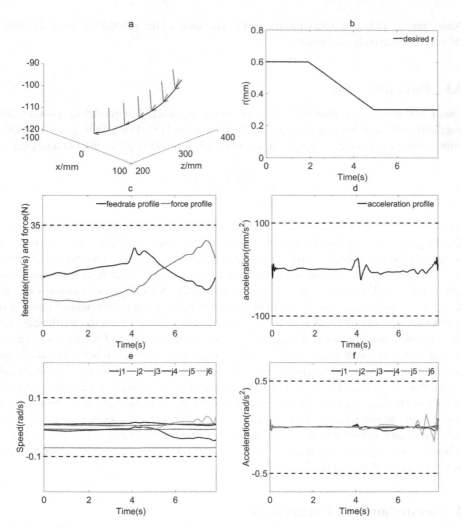

Fig. 2. Results of the scheduling method. (a) Tool path of robotic belt grinding. (b) Desired material removal. (c) Robot feedrate and contact force profile. (d) Robot acceleration profile. (e) The speed of six joints. (f) The acceleration of six joints.

The experiment has been conducted in the robotic belt grinding system to validate the proposed scheduling algorithm as shown Fig. 3. The robot is COMAU-Smart-NJ220 controlled by the C5G-Open which can acquire the robot feedrate. The force sensor is ATI omega160 which can measure the contact force accurately. The workpiece for the grinding experiments is a real airfoil blade made of titanium alloy with a free-form surface. The type of belt is GXK-51 P180 dedicated to grinding precise parts.

Table 1. Scheduling parameters for simulation

Parameters	Value
Parameter interval Δu	0.001
Interpolation period $T(\text{s})$	0.001
Maximum feedrate $v_{\lim}(\text{mm/s})$	35
Maximum force $F_{\lim}(\text{N})$	35
Feedrate of first point $v_{first}(\text{mm/s})$	17.4
Feedrate of end point $v_{end}(\text{mm/s})$	15.6
Force of first point $F_{first}(\text{N})$	9.1
Force of end point $F_{end}(\text{N})$	24.8
Maximum acceleration $a_{\lim}(\text{mm/s}^2)$	100
Minimum joints speed $\dot{\theta}_{\min}(\text{rad/s})$	$[-0.1]_{1\times6}^{\mathrm{T}}$
Maximum joints speed $\dot{\theta}_{\max}(\text{rad/s})$	$[0.1]_{1\times6}^{\mathrm{T}}$
Minimum joints acceleration $\ddot{\theta}_{\min}(\text{rad/s}^2)$	$[-1]_{1\times6}^{\mathrm{T}}$
Maximum joints acceleration $\ddot{\theta}_{\max}(\text{rad/s}^2)$	$[1]_{1\times6}^{\mathrm{T}}$
Belt velocity $v_b(\text{mm/s})$	25
Grinding factors constant K	1277.6
Constant coefficient α	−1.1107
Constant coefficient β	−0.5931
Constant coefficient γ	−0.7563

Fig. 3. Robotic belt grinding system.

The results of the experiment could be seen in Fig. 4. From the Fig. 4, we can know that the actual material removal is very close to the desired material removal. And the relative error is below 15%. This proves a higher grinding removal precision is obtained and the robot tracks the desired material removal more effectively after the optimal generation of the robot velocity and contact force.

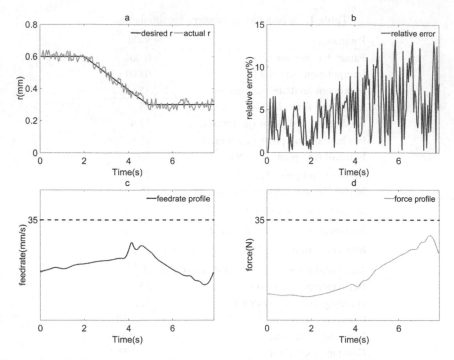

Fig. 4. Results of the experiment. (a) Desired material removal and actual one. (b) Relative error of the material removal. (c) Robot feedrate profile. (d) Contact force profile.

5 Conclusion

The trajectory and force for robotic belt grinding are scheduled effectively in this paper. First, the optimization method with respect to the constraints of joints speed, acceleration and robot feedrate, acceleration is proposed to maximize the feedrate in order to achieve high machining efficiency. Then, the model is transferred to a linear optimization problem which can be solved efficiently. After solving the optimal robot feedrate, the contact force could be calculated based on an empirical approximate formula of grinding mechanism. Finally, the simulation and experiment are conducted, which show the method can excellently achieve trajectory and force generation for robotic belt grinding with high efficiency and accuracy.

Acknowledgements. This work was supported by the National Natural Science Foundation of China under Grant Nos. 51535004, 51323009, 51375196 and 51405175.

References

1. Yun, H., Zhi, H.: Modern belt grinding technology and engineering applications (2009)
2. Wang, Y., Hou, B., Wang, F., et al.: A controllable material removal strategy considering force-geometry model of belt grinding processes. Int. J. Adv. Manufact. Technol., 1–11 (2016)

3. Song, Y., Yang, H., Lv, H.: Intelligent control for a robot belt grinding system. IEEE Trans. Control Syst. Technol. **21**(3), 716–724 (2013)
4. Wu, S.: Robotic conformance grinding modeling, control and optimization. Diss. Theses - Gradworks (4), 337 (2012)
5. Ren, X., Kuhlenkötter, B.: Real-time simulation and visualization of robotic belt grinding processes. Int. J. Adv. Manufact. Technol. **35**(11), 1090–1099 (2008)
6. Zhao, X., Zhao, H., Yang, J., Ding, H.: An adaptive feedrate scheduling method with multi-constraints for five-axis machine tools. In: Liu, H., Kubota, N., Zhu, X., Dillmann, R., Zhou, D. (eds.) ICIRA 2015. LNCS, vol. 9245, pp. 553–564. Springer, Cham (2015). doi:10. 1007/978-3-319-22876-1_48
7. Song, Y., Liang, W., Yang, Y.: A method for grinding removal control of a robot belt grinding system. J. Intell. Manufact. **23**(5), 1903–1913 (2012)
8. Choi, Y.K., Banerjee, A., Lee, J.W.: Tool path generation for free form surfaces using Bézier curves/surfaces. Comput. Ind. Eng. **52**(4), 486–501 (2007)
9. Choquet-Bruhat, Y.: Tool path generation and tolerance analysis for free-form surfaces. Int. J. Mach. Tools Manufact. **47**(3–4), 689–696 (2007)
10. Fleisig, R.V., Spence, A.D.: A constant feed and reduced angular acceleration interpolation algorithm for multi-axis machining. Comput.-Aided Des. **33**(1), 1–15 (2001)
11. Yuen, A., Zhang, K., Altintas, Y.: Smooth trajectory generation for five-axis machine tools. Int. J. Mach. Tools Manufact. **71**(8), 11–19 (2013)
12. Piegl, L., Tiller, W.: The NURBS Book. Springer, Heidelberg (1997). doi:10.1007/978-3-642-59223-2
13. Wang, J., Sun, Y., Gan, Z., Kazerounian, K.: Process modeling of flexible robotic grinding. In: Proceedings of the International Conference Control, Automation System (2003)
14. Hammann, G.: Modellierung des abtragaverhaltens elastischer robotergefuehrter schleifwerkzeuge. Ph.D. Dissertation, Institute for Control Engineering of Machine Tools and Manufacturing Units, University of Stuttgart, Stuttgart, Germany (1998)
15. Schueppstuhl, T.: Beitrag zum bandschleifen komplexer freiformgeometrien mit dem industrieroboter. Ph.D. Dissertation, Mechanical Engineering Department, Technical University of Dortmund, Dortmund, Germany (2003)
16. Xu, W., Cai, C., Yin, M., et al.: Time-varying force tracking in impedance control a case study for automatic cell manipulation. In: IEEE International Conference on Control and Automation. pp. 344–349. IEEE (2013)
17. Xie, Y., Sun, D., Liu, C., et al.: An adaptive impedance force control approach for robotic cell microinjection. In: IEEE/RSJ International Conference on Intelligent Robots and Systems, pp. 907–912. IEEE (2008)

Fractional-Order Integral Sliding Mode Controller for Biaxial Motion Control System

Xi Yu, Huan Zhao$^{(\boxtimes)}$, Xiangfei Li, and Han Ding

State Key Laboratory of Digital Manufacturing Equipment and Technology,
Huazhong University of Science and Technology, Wuhan 430074, Hubei,
People's Republic of China
huanzhao@hust.edu.cn

Abstract. Biaxial tables are widely applied in high performance motion control applications for high accuracy. Tracking error is one of the most significant indicators of machining precision, and tracking control is an effective means to eliminate the tracking error. In this paper, to attenuate the tracking error and reduce the chattering phenomenon in the control input simultaneously, a fractional-order integral sliding mode controller is proposed. Compared with the existing sliding mode controller, the proposed control law not only maintains the original robustness against variations but also reduces the tracking error effectively. At the same time, the overshoot can be weakened and the reaching law will converge to the sliding surface more rapidly. Experiments conducted on a biaxial table demonstrate that the proposed control scheme is easy to apply, the tracking error is smaller and the input chatter can be improved significantly compared to the integer SMC.

Keywords: Motion control · Fractional-order integral · Tracking error

1 Introduction

Although biaxial tables are widely used in high performance motion control applications for high accuracy, there are still errors in machining. Tracking error is regarded as one of the most significant indicators of machining precision. The effective approaches to improve machining accuracy contain tracking control methods, trajectory planning methods and so on.

To reduce the tracking error, there are a lot of researchers having developed several control algorithms over past few decades. Tomizuka [1] firstly proposed a zero phase error tracking controller (ZPETC) by canceling the stable dynamics of the servo drive to improve tracking accuracy. The main drawback of ZPETC is sophisticated, then a simple algorithm sliding mode control (SMC) was proposed by Altintas et al. [2], which has practical advantages in rapid tuning and implementation, but with severe chattering. Later, Sun [3] developed a new adaptive control approach to position synchronization of multiple motion axes, this new method guarantees asymptotic convergence to zero of both position and synchronization error. Barton and Alleyne [4]

© Springer International Publishing AG 2017
Y. Huang et al. (Eds.): ICIRA 2017, Part II, LNAI 10463, pp. 24–35, 2017.
DOI: 10.1007/978-3-319-65292-4_3

put forward a synthetic method for precision motion control by combining individual axis iterative learning control and cross-coupled iterative learning control into a single control input, which enhanced the precision motion control of the system through performance improvements in individual axis tracking. Because the effect of cross-coupled during high speed feed drives can not be ignored, a H∞ controller was designed by Yong et al. [5] to minimize the tracking error, which was augmented with integral action to achieve accurate tracking.

With the strong robustness against model uncertainties and disturbance rejection, SMC is one of the popular controllers in industrial applications all the time. However, because of the discontinuous nature of SMC, it will produce the chattering phenomenon and high-frequency oscillations in practice [6]. To counteract the chattering phenomenon in SMC, the fractional order sliding mode controller is received more and more attention. Delavari et al. [7] presented a fuzzy fractional order sliding mode controller for nonlinear systems to reduce the chattering phenomenon, and the fuzzy logical controller is used to replace the signum function at the reaching phase in the SMC. Zhang et al. [8] proposed a fractional order sliding-mode control for velocity control of permanent magnet synchronous motor, in which a fuzzy logic inference scheme is utilized to obtain the gain of switching control. Yin et al. [9] applied a fractional order sliding mode controller (FOSMC) in nonlinear systems to achieve extremum seeking. Tang et al. [10] designed a new fuzzy fractional order sliding mode controller for antilock braking system (ABS), this strategy can not only deal with the unsertainties but also track the desired slip faster than conventional SMC. In all of the above methods, it is clear to see that the robustness and chattering phenomenon can be effectively improved.

The original intention of this study is to maintain the robustness and decrease the chatter of SMC simultaneously. Thus, a fractional order integral sliding mode controller is proposed. The control law is tested on a biaxial table, and the performance of the proposed strategy is compared against the traditional SMC. Experiment results show that the tracking error is smaller and the chattering phenomenon is less than those of integer SMC.

2   Definition of Fractional Order Calculus

Fractional order calculus is a classical mathematical idea which allows to arbitrary order differentiation and integration, and it can be specified in terms of the fundamental operator $_aD_t^\alpha$ known as differ-integration operator [11],

$$f(t) = {_aD_t^\alpha} = \begin{cases} \frac{d^\alpha}{dt^\alpha}, & c > 0 \\ 1, & c = 0 \\ \int_a^t (d\tau)^\alpha, & c < 0 \end{cases} \tag{1}$$

where a and t are lower and upper limits, respectively, $\alpha \in R$ is the order of the fractional order operator. Among several basic definitions of arbitrary order differentiation and integration, the most popular two definitions are the Grunwald-Letnikov (G-L) and the Riemann-Liouville (R-L) definitions.

The α th-order Riemann-Liouville (R-L) fractional-order integration of continuous function $f(t)$ is given by,

$$_aI_t^\alpha f(t) = \frac{1}{\Gamma(\alpha)} \int_a^t \frac{f(\tau)}{(t-\tau)^{1-\alpha}} d\tau \qquad (2)$$

where $\alpha \in (0, 1)$, $\Gamma(\bullet)$ is the Gamma function and,

$$\Gamma(\alpha) = \int_0^\infty e^{-u} u^{\alpha-1} du \qquad (3)$$

The αth-order Riemann-Liouville (R-L) fractional-order derivative of continuous function $f(t)$ is defined as,

$$_aD_t^\alpha f(t) = \begin{cases} \frac{1}{\Gamma(m-\alpha)} \int_a^t \frac{f^{(m)}(\tau)}{(t-\tau)^{\alpha-m+1}} d\tau, & m-1<\alpha<m \\ \frac{d^m f(t)}{dt^m}, & \alpha = m \end{cases} \qquad (4)$$

where $m-1<\alpha\leq m$, $m \in N$, m is the minimum integer number in the value which are larger than α.

3   Controller Design

3.1   Dynamics of Biaxial Table Feed Drive

The dynamics of a single feed drive system of biaxial table shown as Fig. 1, which can be described as the following differential equation,

$$\frac{J}{K_aK_tR_g}\ddot{x}(t) + \frac{B}{K_aK_tR_g}\dot{x}(t) = u(t) - \frac{1}{K_aK_t}T_d(t) \qquad (5)$$

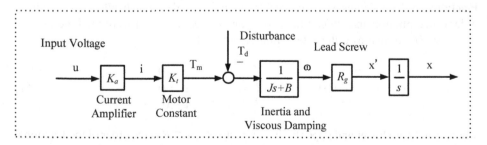

Fig. 1. A single feed drive system of the biaxial table

where J is the inertia and B is the viscous damping coefficient, both of them can be identified in advance. K_a is the current amplifier and K_t is the motor constant, R_g is the lead screw. $x(t)$, $\dot{x}(t)$ and $\ddot{x}(t)$ are the position, velocity and acceleration of the actuator, respectively. T_d is the external disturbance torque, u is the control input.

For simplicity, the coefficients of the dynamics can be rewritten as,

$$M = \frac{J}{K_a K_t R_g},\ C = \frac{B}{K_a K_t R_g},\ u_d = \frac{1}{K_a K_t} T_d(t) \tag{6}$$

So the dynamics differential equation can be expressed as,

$$M\ddot{x}(t) + C\dot{x}(t) = u(t) - u_d \tag{7}$$

In practice, our control objective is to drive the tracking error to zero asymptotically under any initial conditions. Because the exact knowledge of M, C and u_d are not known, there is only nominal or identified model can be available to design the controller, meanwhile these uncertainties would have a bad influence on robustness of system. Conventional SMC belongs to a class of nonlinear control strategies, which is robust to such uncertainties and time variations in the drive system [6]. So we can believe that fractional-order integral sliding mode controller is also robust to these uncertainties for it is a part of SMC essentially.

3.2 Design of Fractional Order Integral Sliding Mode Controller

The block diagram of fractional order sliding mode controller is shown in Fig. 2, the actual position is measured from an encoder, and the actual velocity and acceleration is estimated by utilizing the method of digital differentiation.

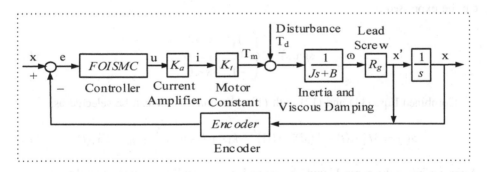

Fig. 2. Fractional-order sliding mode control scheme

Like designing a traditional SMC, there are also two fundamental steps in fractional-order sliding mode controller, one is the selection of a sliding surface and the other is the control law.

Since the oscillation induced by sliding surface may excite high frequency unmodeled dynamics of the system and damage the performance of the system, it is necessary to reduce the chattering phenomenon of the control input. To achieve such a goal of tracking accuracy, the fractional-order integral sliding surface is selected as,

$$s = \dot{e} + k_0 D_t^{-\alpha} e(t) \tag{8}$$

where k is a positive gain and is the obtainable and desired tracking bandwidth of the drive. $_0D_t^{-\alpha}$ is the function of fractional-order integral operator $(0 < \alpha < 1)$, e and $\dot{e}$ are the position and velocity tracking error of the drive system, respectively,

$$e = x_r - x_a \tag{9}$$

$$\dot{e} = \dot{x}_r - \dot{x}_a \tag{10}$$

Taking the derivative of s with respect to time yields,

$$\dot{s} = \ddot{e} + k\left(_0D_t^{-\alpha} e(t)\right)' \tag{11}$$

where $\ddot{e}$ is the tracking error of acceleration,

$$\ddot{e} = \ddot{x}_r - \ddot{x}_a = \ddot{x}_r - \frac{1}{M}(u(t) - u_d - C\dot{x}_a(t)) \tag{12}$$

Here, the reaching law is selected as,

$$\dot{s} = -k_1 s - k_2 \mathrm{sgn}(s) \tag{13}$$

where $k_1, k_2 \in R^+$ are sliding mode coefficients, sgn (s) is the sign function of s, which can be expressed as,

$$\mathrm{sgn}(s) = \begin{cases} 1, & s > 0 \\ 0, & s = 0 \\ -1, & s < 0 \end{cases} \tag{14}$$

Combined Eqs. (11) and (12) with (13), the control law can be selected as,

$$u(t) = M\left(\ddot{x}_r(t) + k\left(_0D_t^{-\alpha} e(t)\right)' + k_1 s + k_2 \mathrm{sgn}(s)\right) + u_d + C\dot{x}_a(t) \tag{15}$$

After designing the control input, the second step is stability analysis. Stability analysis has to satisfy the reaching condition of proposed fractional-order switching surface, which means wherever or whatever initial conditions state, the control output could drive initial states to switching surface. Here, the fundamental Lyapunov function is chosen as,

$$V = \frac{1}{2}s^2 \tag{16}$$

According to Eq. (16), the derivative of V can be expressed as:

$$\dot{V} = s\dot{s} \qquad (17)$$

For stability of nonlinear systems, the derivative of the Lyapunov function must be negative so as to fit the principle of conversation of energy.

By substituting s from Eq. (13) and sign function of s from (14), we can represent Eq. (17) as,

$$\dot{V} = -k_1 s^2 - k_2 s \, \text{sgn}(s) \begin{cases} = 0, & s = 0 \\ < 0, & s \neq 0 \end{cases} \qquad (18)$$

It is obvious that the Eq. (18) guarantees the asymptotic stability. So, next we will prove that the control law drives system to converge to switching surface in finite time. When the initial state satisfies $s(t_0) > 0$, we can represented Eq. (13) as,

$$\dot{s} = -k_1 s - k_2 \qquad (19)$$

Solving the Eq. (19), we can gain the solution of s,

$$s = \frac{-k_2 + [k_2 + k_1 s(t_0)]e^{-k_1(t-t_0)}}{k_1} \qquad (20)$$

When the Eq. (20) equals to zero, the system can converge to switching manifold, and the time can be expressed as,

$$t = -\frac{1}{k_1} \ln \frac{k_1}{k_2 + k_1 s(t_0)} + t_0 \qquad (21)$$

Similarly, when the initial state is $s(t_0) < 0$, the system can also converge to switching manifold when the time satisfies the following expression,

$$t = -\frac{1}{k_1} \ln \frac{k_1}{k_2 - k_1 s(t_0)} + t_0 \qquad (22)$$

So, once the time is longer than the following expression,

$$t > -\frac{1}{k_1} \ln \frac{k_1}{k_2 + k_1 |s(t_0)|} + t_0 \qquad (23)$$

the system will converge to switching surface at any initial state. From the Eq. (23), we can draw a conlusion that the converging time is associated with k_1 and k_2. What's more, we can also know that larger k_1 and smaller k_2 can make time shorter. In experiment, the parameters k_1 and k_2 can be determined by gradually increasing its value from zero to satisfactory tracking accuracy, and then we get the parameter of k_1 in fractional-order sliding mode controller is larger than that in traditional SMC and k_2 is smaller. For the parameter α, the tracking error will increase when it tends to be large, but when it becomes small, the chattering phenomenon will become serious, thus the selection of α should be a compromise between tracking accuracy and chattering elimination so as to get better performance.

4   Experiments Validation

The proposed fractional-order integral sliding mode controller is applied in a real time platform of biaxial table as shown in Fig. 3. A computer with matlabR2013a is used to achieve feedrate planing, interpolation and controller design, then the computer transmits the real-time information to dSPACE DS1103 controller platform by hardware fiber bus and software ControlDesk. Experiments are carried out in the current-control loop of Yaskawa AC servo motor system. For translational axes, motors will be coupled with a 10 mm/pitch lead screw, respectively. The dynamic parameters shown in Table 1 are from [12], whose identification method was proposed by Erkorkmaz [13].

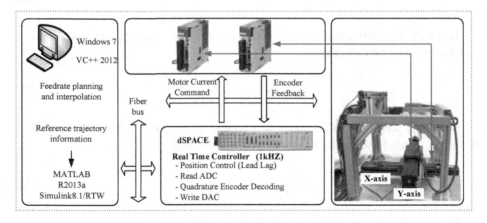

Fig. 3. Experimental platform of the biaxial table

Table 1. Dynamic parameters for biaxial table

Parameters	X-axis	Y-axis
m (Vs^2/m or Vs^2/rad)	0.1101	0.0271
c (Vs/m or Vs/rad)	0.4001	0.3871

The actual position is obtained from the encoder, and the actual velocity and acceleration are estimated by taking the derivative of the measured position from the linear encoder, so there may be noisy in velocity and acceleration. In this paper, we classify these uncertainties as external disturbance and give equivalent torque compensation, the relevant plants are given in Table 2.

Table 2. Disturbance parameters for biaxial table

Parameters	X-axis	Y-axis
u_d (if vel > 0)	0.11847	0.11881
u_d (if vel < 0)	−0.11466	−0.11151

In experiment, the parameters of X-axis and Y-axis are tuned separately and properly as shown in Table 3.

Table 3. Controller paremeters for biaxial table

Parameters	X-FOISMC	X-SMC	Y-FOISMC	Y-SMC
α	0.85	1	0.85	1
k	55	55	75	65
k_1	65	33	70	50
k_2	0.01	0.05	0.01	0.05

In order to facilitate the comparison of integer and fractional sliding mode controllers, the SMC strategy proposed in [2] and the method proposed in this paper are experimented simultaneously on biaxial table. The results of X-axis are shown in Fig. 4.

For a comparative analysis of the performance of integer and fractional-order controllers, we plot the control signals and tracking errors in the same picture in Fig. 5, respectively. From Fig. 5(a), it can be seen that the initial overshoots and chattering phenomenon of the proposed control signal are smaller than those of the integer SMC. In addition, it can be also seen from Fig. 5(b) that the tracking error of the presented method is smaller and smoother than that of the conventional SMC.

Machining accuracy includes tracking accuracy and contour accuracy, in this paper we only discuss the tracking error. Considering the machining accuracy, we must take contour error into account, contour error will be decoupled to each axis in joint space, which is essentially the error control of single axis, so we conduct experiments on Y axis, too. The results are shown in Figs. 6 and 7. From Fig. 7(a), we know that the chattering phenomenon decreases on a degree but the overshoot is still more than that of proposed controller, and the tracking error is more than that of integer SMC in

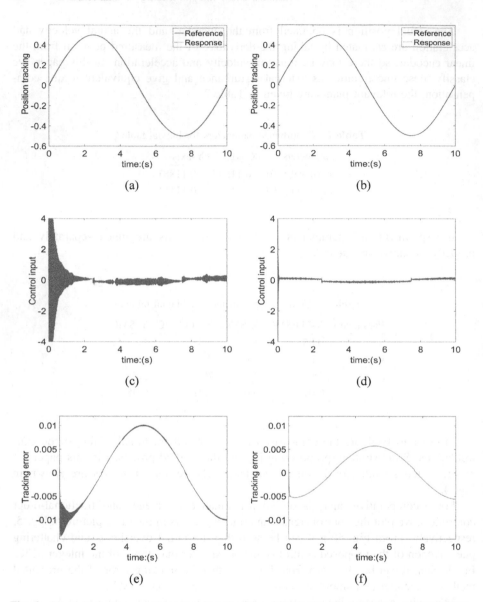

Fig. 4. Sinusoidal responses. (a), (c) and (e) Traditional SMC. (b), (d) and (f) Proposed FOISMC.

Fig. 7(b), we can make a conclusion that the fractional-order sliding mode controller obtains better performance in reducing overshoot and decreasing tracking error at the same time.

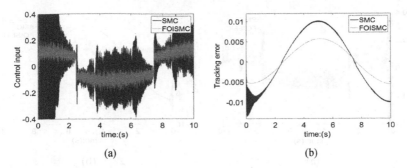

Fig. 5. Experimental results. (a) Control signal comparsion. (b) Tracking error comparsion.

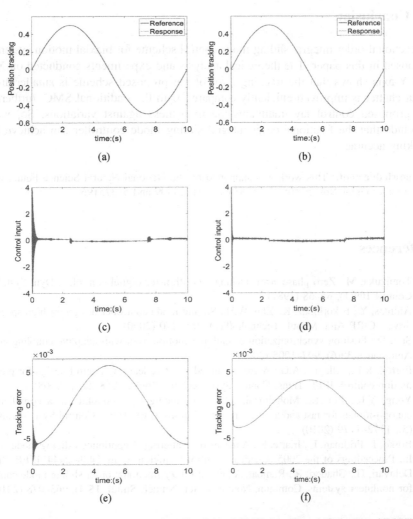

Fig. 6. Sinusoidal responses. (a), (c) and (e) Traditional SMC. (b), (d) and (f) Proposed FOISMC.

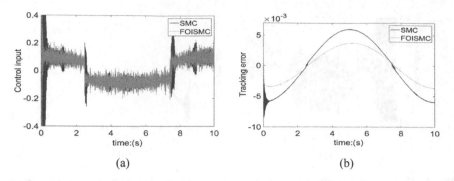

Fig. 7. Experimental results. (a) Control signal comparsion. (b) Tracking error comparsion.

5 Conclusion

A fractional-order integral sliding mode control scheme for biaxial motion system was proposed in this paper. The theoretical analysis and experiments conducted on X-axis and Y-axis shows that the tracking error of the proposed scheme is smaller and the input chatter is improved efficiently compared with the traditional SMC, furthermore, the proposed control law maintains the robustness against variations. So, we can conclude that the fractional-order integral sliding mode controller can achieve better tracking accuracy.

Acknowledgement. This work was supported by the National Natural Science Foundation of China under Grant Nos. 51405175, 51535004, 51323009 and 51375196.

References

1. Tomizuka, M.: Zero phase error tracking algorithm for digital control. J. Dyn. Syst. Meas. Control **109**(1), 65–68 (1987)
2. Altintas, Y., Erkorkmaz, K., Zhu, W.H.: Sliding mode controller design for high speed feed drives. CIRP Ann.-Manuf. Technol. **49**(1), 265–270 (2000)
3. Sun, D.: Position synchronization of multiple motion axes with adaptive coupling control. Automatica **39**(6), 997–1005 (2003)
4. Barton, K.L., Alleyne, A.G.: A cross-coupled iterative learning control design for precision motion control. IEEE Trans. Control Syst. Technol. **16**(6), 1218–1231 (2008)
5. Yong, Y.K., Liu, K., Moheimani, S.O.R.: Reducing cross-coupling in a compliant XY nanopositioner for fast and accurate raster scanning. IEEE Trans. Control Syst. Technol. **18**(5), 1172–1179 (2010)
6. Boiko, I., Fridman, L., Iriarte, R.: Analysis of chattering in continuous sliding mode control. In: Proceedings of the 2005 American Control Conference, pp. 2439–2444. IEEE (2005)
7. Delavari, H., Ghaderi, R., Ranjbar, A., et al.: Fuzzy fractional order sliding mode controller for nonlinear systems. Commun. Nonlinear Sci. Numer. Simul. **15**(4), 963–978 (2010)

8. Zhang, B.T., Pi, Y.G., Luo, Y.: Fractional order sliding-mode control based on parameters auto-tuning for velocity control of permanent magnet synchronous motor. ISA Trans. **51**(5), 649–656 (2012)

9. Yin, C., Chen, Y.Q., Zhong, S.: Fractional-order sliding mode based extremum seeking control of a class of nonlinear systems. Automatica **50**(12), 3173–3181 (2014)

10. Tang, Y., Zhang, X., Zhang, D., et al.: Fractional order sliding mode controller design for antilock braking systems. Neurocomputing **111**, 122–130 (2013)

11. Aghababa, M.P.: Finite-time chaos control and synchronization of fractional-order nonautonomous chaotic (hyperchaotic) systems using fractional nonsingular terminal sliding mode technique. Nonlinear Dyn. **69**(1), 247–261 (2012)

12. Li, X., Zhao, H., Zhao, X., et al.: Dual sliding mode contouring control with high accuracy contour error estimation for five-axis CNC machine tools. Int. J. Mach. Tools Manuf **108**, 74–82 (2016)

13. Erkorkmaz, K., Altintas, Y.: High speed CNC system design. Part II: modeling and identification of feed drives. Int. J. Mach. Tools Manuf. **41**(10), 1487–1509 (2001)

Pose Estimation with Mismatching Region Detection in Robot Bin Picking

Zhe Wang, Lei Jia, Lei Zhang, and Chungang Zhuang$^{(\boxtimes)}$

State Key Laboratory of Mechanical System and Vibration,
School of Mechanical Engineering, Shanghai Jiao Tong University,
Shanghai 200240, China
cgzhuang@sjtu.edu.cn

Abstract. 3D object detection and pose estimation based on 3D sensor have been widely studied for its applications in robotics. In this paper, we propose a new clustering strategy in Point Pair Feature (PPF) based 3D object detection and pose estimation framework to further improve the pose hypothesis result. Our main contribution is using Density Based Spatial Clustering of Applications with Noise (DBSCAN) and Principle Component Analysis (PCA) in PPF method. It was recently shown that point pair feature combined with a voting framework was able to obtain a fast and robust pose estimation result in heavily cluttered scenes with occlusions. However, this method may fail in the mismatching region caused by false features or features with insufficient information. Our experimental results show that the proposed method can detect mismatching region and false pose hypotheses in PPF method, which improves the performance in robot bin picking application.

Keywords: Pose estimation · Mismatching region detection · Point pair feature (PPF) · Robot bin picking

1 Introduction

In the last two decades, industrial robots have been widely used in various fields of manufacture to enable efficient automated production processes. Vision-guided industrial robots are one of key parts in the state-of-the-art manufacturing processes. One of the main tasks for industrial robot is picking object and placing it to the specific location. This task has a good solution in 2D placement case while in 3D scattered stacking case, as illustrated in Fig. 1, remains a challenge. Indeed, 3D object detection and pose estimation are significant for robot bin picking in 3D case. The goal is to locate some pickable object instances and determine their 6D poses, i.e. rotation matrix and translation matrix.

Various algorithms have been proposed for 3D object detection and pose estimation based on 2D images, range images or 3D point clouds. However, few of them are suited for the real-world industrial robot bin picking application for 3D scattered stacking case. There are some reasons: (a) the industrial objects are

© Springer International Publishing AG 2017
Y. Huang et al. (Eds.): ICIRA 2017, Part II, LNAI 10463, pp. 36–47, 2017.
DOI: 10.1007/978-3-319-65292-4_4

Fig. 1. Left: robot bin picking system. Middle: objects are stacked randomly in a bin. Right: scene point clouds from 3D sensor. The mismatching regions are marked with the red bounding box. (Color figure online)

generally texture-less objects with simple shape. Thus, traditional 2D/3D key point based object detection and pose estimation are not appropriate. (b) 3D scattered stacking objects form a heavy cluttered environment with occlusions. (c) 3D sensor may not obtain complete scene information because of reflection, transmission and absorption. (d) the industrial bin picking requires eliminating the wrong pose hypothesis. Therefore, no satisfactory general solution to this task exists yet.

In literatures, while several methods have been proposed to tackle with this problem based on different input data, we focus here on the approach from [1], which relies on point cloud data. This algorithm has been improved and extended by many authors [2–7]. Also, it has been used in various robot bin picking applications [5,8]. Recent studies [4] have shown its potential to deal with sensor noise, background clutters and occlusions, which is the major disadvantage of another popular pose estimation algorithm, LINEMOD [9,10]. Moreover, smarter sampling strategy [2–4] and GPU acceleration [6] enable PPF method to obtain a fast and robust pose estimation result in heavily cluttered scenes with occlusions. However, this method may fail in the mismatching region caused by false features or features with insufficient information. As shown in Fig. 1, the mismatching regions are marked with the red bounding boxes where the wrong pose may be evaluated a high score. Thus, robot may try to pick these high score wrong detected objects, which causes a decline in system robustness. Actually, even human may not able to identify the correct poses from these regions in point clouds. To tackle with this problem, we propose a new clustering strategy in Point Pair Feature (PPF) based 3D object detection and pose estimation framework to further improve the pose hypothesis result. Our main contribution is using Density Based Spatial Clustering of Applications with Noise (DBSCAN) [11,12] and Principle Component Analysis (PCA) in PPF method. Through further studying on the clustering step, we give a probable pose distribution perspective for mismatching region detection.

The remainder of this paper is organized as follows. Section 2 gives an extensive overview of various cutting-edge 3D object detection and pose estimation methods as well as their applications for robot bin picking. Our method for improved PPF method and mismatching region detection is demonstrated in Sect. 3. Experimental results are presented and discussed in Sect. 4 and the conclusion follows in Sect. 5.

2   Related Work

An extensive overview of the state-of-the-art approaches in 3D object detection and pose estimation is provided. This section also introduces their application in robot bin picking.

2.1   3D Object Detection and Pose Estimation

Many algorithms have been proposed for 3D object detection and pose estimation. We only focus on the recent work here and split them in several categories.

Point Wise Correspondence Based Methods. Iterative Closest Point (ICP) [13] is a classic point wise correspondence based method for point cloud registration, which can also be used for pose estimation. However, it requires a good initialization to avoid being trapped in a local minimum during optimization. Go-ICP [14] provides a global optimal registration by using global branch-and-bound (BnB) optimization in ICP. Since ICP enables a high accuracy point cloud registration, many algorithms combine ICP into a coarse-to-fine pose estimation framework.

Template Based Methods. Template based methods, which are widely used in 2D object detection, are extended in 3D case in recent years [9,10]. The templates store the different viewpoints of object appearances. An object is detected when a template matches the image and gives its 3D pose. LINEMOD, presented in [9,10], renders 3D object model to generate a large number of templates covering full view hemisphere. It combines 2D image gradients and 3D surface normals to build multiple modalities. Though such template based approaches can work accurately and quickly in practice, they still show robust problem in cluttered environment with occlusions.

Local Feature Based Methods. Local feature based methods, including point pair feature (PPF) [1], point feature histograms (PFH) [15] and 3D key point that are extended from traditional 2D key point like 3D-SURF [16], are robust to occlusions. A review can be found in [17]. These features are usually combined with other techniques like voting for object detection. However, they are usually computationally expensive and have difficulty in scenes with heavy clutters. In this paper, we focus one of the promising methods, PPF based method [1]. Several improvements and extensions of PPF based method have been proposed in recent years. [2–4] improve PPF method [1] in almost every stage by smarter subsampling strategy, smarter voting strategy and smarter pose evaluation strategy. [4] shows its potential to tackle with scene noise, cluttered environment with occlusions and loss of information caused by subsampling compared with original PPF method [1] and other state-of-the-art algorithms [3,9]. [5] discusses different point pair features for voting. [6] uses color point pair feature and gives its GPU implementation. RANSAC integrated PPF method is proposed in [7].

Learning Based Methods. With the rapid development of learning techniques, several learning based methods are proposed for this task. [18,19] use Iterative Hough Forest on local patches to detect and estimate 3D poses. [20–22] extend traditional convolutional neural network to 3D case for object detection and pose estimation. These recent studies show the potential to use the state-of-the-art learning techniques on this task.

2.2 3D Object Pose Estimation in Robot Bin Picking

Since most industrial objects are texture-less, LINEMOD [9,10] and PPF method [1,5,8] are popular in robot bin picking application. The first Amazon Picking Challenge [23,24] presents some state-of-the-art robot bin picking frameworks combined with various methods and tricks. SegICP, presented in [25], shows the potential to combine CNN-based object detection with ICP on this task.

3   Proposed Method

We describe here our contributions to improve the PPF method [1] more robust for robot bin picking. This paper focuses on PPF methods in pose clustering stage. Compared with traditional PPF method and its improvements, we give a probable pose distribution perspective to rethink about the PPF method and use it to detect mismatching region in the scene. This section introduces our improved PPF method first and then describes our clustering strategy in detail.

3.1   Improved PPF Method

We revisit PPF based object detection and pose estimation pipeline and describe our improved PPF method here. The improved PPF method can be divided into five stages.

Preprocessing the Model and Scene Data. Generally, the model data is CAD format and scene data is a point cloud without normals. In order to use PPF method, remeshing techniques are used to transform model and scene data to uniform dense point clouds with normals. We also use subsampling to accelerate the PPF method.

Model Training. Point pair feature is used to create a hash table for voting framework in this step. Every point pair on the model surface calculates the feature $\mathbf{F}_m(\mathbf{m}_i, \mathbf{m}_j)$ and the corresponding angle α_{ij}^{model}. The distance and angle in PPF are discretized in step d_{dist} and $d_{angle} = 2\pi/n_{angle}$. The indices of hash table is hashed PPF and the stored element is $(\mathbf{m}_i, \mathbf{m}_j, \alpha_{ij}^{model})$.

Voting. Voting framework is similar to the Generalized Hough Transform. Every local coordinates $(\mathbf{m}_r, \alpha)$ is presented in a two-dimensional accumulator array. α is discretized using d_{angle}. Here we subsample scene points as reference points. For each reference point s_r, the point pair feature $\mathbf{F}_s(s_r, s_j)$ and the corresponding angle α_{rj}^{scene} is calculated. Then corresponding $\mathbf{F}_m(\mathbf{m}_r, \mathbf{m}_j)$ is found and $\alpha = \alpha_{rj}^{model} - \alpha_{rj}^{scene}$. The accumulator position $(\mathbf{m}_r, \alpha)$ vote one. Compared to the original version, we compute vote value as

$$vote = \begin{cases} 1 - \lambda|n_1 \cdot n_2|, & 1 - \lambda|n_1 \cdot n_2| \geq v_{min} \\ v_{min}, & 1 - \lambda|n_1 \cdot n_2| < v_{min} \end{cases} \tag{1}$$

where $|n_1 \cdot n_2|$ denotes the absolute value of the dot product of two normals, λ is a weighting parameter in default set to 1 and v_{min} is the minimum boundary of the vote in default set to 0.1. The weighted vote will weaken the planar vote at some level. In addition, if the distance between two points is larger than the size of the model, it's unnecessary to process them because only point pairs belong to the same object shall be vote. After all points s_i are processed, the peak in the accumulator array are the corresponding local coordinate to s_r. Then, we obtain a set of possible object poses, which we called probable pose distribution in this paper.

Pose Clustering. In the original version of PPF method, the retrieved poses are sorted in descending order of the number of votes to make sure group most likely pose hypotheses first. Then it uses euclidean distance to cluster the proximal poses and the score of the cluster is the sum of the scores of the contained poses. Finally, it removes the isolated poses with low score and averages the clustered pose hypotheses to get the final pose hypotheses. In this paper, we use new clustering strategy combined with DBSCAN and PCA to further improve the pose hypotheses result. The following part will describe it in detail.

Post Processing the Pose Hypotheses. After the high score pose hypotheses are obtained, post processing is suggested to obtain higher accuracy of the detected poses. The pose hypothesis scores will be reevaluated to reject the wrong pose hypotheses and ICP algorithm will be used to get accuracy poses. One of the popular evaluation strategies is computing how many points on the model are detected on the scene point cloud. However, false pose hypotheses can get high evaluation scores in mismatching region, which will give wrong pose result to robot. Hence, we evaluate pose hypotheses in clustering stage first to detect the pose hypotheses in mismatching region.

3.2 Mismatching Region

Given a scene point cloud and a model point cloud, we define the mismatching region as region where wrong pose hypothesis can be evaluated a high score. For example, traditional evaluation strategies in the post processing stage, such as rerendering the depth image or computing area of the model surface detected

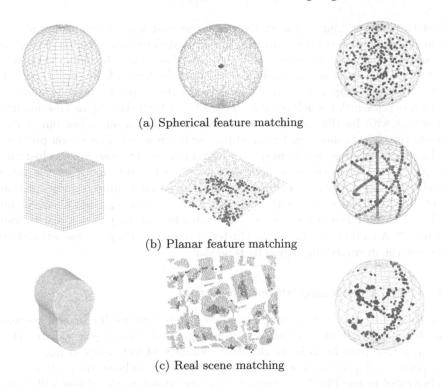

(a) Spherical feature matching

(b) Planar feature matching

(c) Real scene matching

Fig. 2. The probable pose distribution visualization on the mismatching region. Left: model point cloud. Middle: scene point cloud with pose positions visualization. Right: pose orientations visualization in normalized axis-angle space (Color figure online)

in the scene, will apparently evaluate a wrong pose hypothesis a high score in the mismatching region shown in Fig. 1. The mismatching region is caused by false features or features with insufficient information. This false features may appear when the scene features are similar to the model features, multiple model features couple or scene features couple with model features.

To vividly describe the cause of the mismatching region, we use PPF method in the scene with mismatching region and visualize the probable pose distribution, which is a set of high vote pose hypotheses after voting stage. The result is illustrated in Fig. 2. In the spherical feature matching, the pose positions gather at one point while the orientations randomly distributed in the normalized axis-angle space because the overlapping surface on the model can slide on the scene with small overlapping area change. Also, for the cube with six planar features, the pose positions can slide in a wide range on the planar scene while the orientations are constraint to be six spatial curves. And for real scene matching with the object in our robot bin picking system, it's obvious to see that the mismatching region contains a large pose position sliding area. Therefore, we know the mismatching region is the region where the overlapping surface on the model can slide on the scene with small overlapping area change. And we can detect the mismatching region by detecting the pose hypotheses sliding area in the probable pose distribution. In fact, the mismatching region highly depend on the shape of the detected objects and the way of stacking the detected objects. If the shape feature is simple, such as cube and cylinder, and the sensor can only scan the partial shape, the features cannot denote a specific pose because the pose hypotheses lack enough reliable constraints, which show up as false features or features with insufficient information. For example, in our robot bin picking system, the object, shown in Fig. 2, will cause the mismatching region problem.

To detect the pose in the mismatching region, we propose a new clustering strategy on the probable pose distribution. Since we only focus on the robot bin picking application, we believe that pose position is more significant than orientation, which may vary with symmetric structure but does not affect the bin picking. We use DBSCAN to cluster the probable pose by position distribution and use PCA to find the size of the clustered pose set. Then we use a threshold to detect the mismatching region by pose size.

3.3 DBSCAN: Density-Based Clustering

With the appropriate setting on the sampling step, here we list several assumptions for PPF method: (a) Every reference point on the object will return a high vote true pose hypothesis. (b) High visibility object sampled many reference points will gather plentys of high vote pose hypotheses on its true pose, as illustrated in the Fig. 2. (c) Scene feature dissimilar to the object will return a low vote wrong pose hypothesis. (d) High vote pose hypotheses, i.e. probable pose distribution, contain all visible object poses in the scene. (e) Mismatching region, where overlapping can slide, is a dense pose hypotheses cluster of arbitrary shape. Original PPF method uses greedy strategy in the pose clustering stage. It uses highest score pose hypotheses as a cluster center and cluster its nearest pose hypotheses. By repeating this loop, we get a descending order clustered pose hypotheses, which will be used to post processing. The cluster shape is similar to K-means, which limits the use of probable pose distribution.

Hence, we handle probable pose distribution using Density Based Spatial Clustering of Applications with Noise (DBSCAN) [11], which can discover clusters of arbitrary shape and also handle outliers effectively. DBSCAN algorithm defines cluster as a densely connected points by *eps* and *minpts*. *eps* is the maximum distance neighborhood for given point and *minpts* is the minimum number of points required in the *eps-neighborhood* of a point to form a cluster. This paper implements the DBSCAN algorithm in [11, 12] for clustering probable pose position. As we state before, only the position of the pose is used for clustering. The basic pipeline of DBSCAN is shown in Algorithm 1.

Figure 3 shows the result of clustered pose in the position space. Scene data is shown in Fig. 1. The high density clusters are detected and marked with different color while the isolated or low density points are detected as noise. Apparently, a clustered pose hypothesis with small size present a specific true pose.

Algorithm 1. DBSCAN(P,eps,$minpts$)

Input: probable pose set P, eps, $minpts$
Output: pose clusters C, pose cluster id I
1: mark all poses in P as unvisited, $C \leftarrow 0$, $I \leftarrow 0$
2: **for each** unvisited pose p in P **do**
3:      mark p as visited
4:      Neighbors $\leftarrow$ FindNeighbors(p,eps)
5:      **if** |Neighbors| $\leq$ $minpts$ **then**
6:          mark p as noise, $I(p) \leftarrow$ noise
7:      **else**
8:          $c \leftarrow$ next cluster
9:          add p to cluster c, $I(p) \leftarrow c$
10:         **for each** pose p' in Neighbors **do**
11:             **if** p' is unvisited **then**
12:                 mark p' as visited
13:                 pNeighbors $\leftarrow$ FindNeighbors(p',eps)
14:                 **if** |pNeighbors| $\geq$ $minpts$ **then**
15:                     Neighbors $\leftarrow$ Neighbors joined with pNeighbors
16:                 **end if**
17:                 **if** p' is not yet member of any cluster **then**
18:                     add p' to cluster c, $I(p') \leftarrow c$
19:                 **end if**
20:             **end if**
21:         **end for**
22:         add c to C
23:     **end if**
24: **end for**

3.4 Mismatching Region Detection

After we obtain clustered pose hypotheses by DBSCAN algorithm, we use Principle Component Analysis (PCA) to extract cluster features. PCA is a popular machine learning and statistics technique to extract the main directions of the dataset. In our approach, we first compute the center and main directions of the cluster, and then project the points in the cluster on the main directions to get the cluster size, which presents the overlapping sliding area. The sizes of all clusters are illustrated with 3D oriented bounding box in Fig. 3. Also, the mismatching region is detected by the sum of box size and marked with red color.

To obtain final pose hypotheses result, we use original greedy strategy in further clustering. The highest vote pose hypothesis in the cluster will be used to cluster its neighborhood and the final vote will be the sum of all neighbor votes. This cluster result will present the final pose hypothesis of the cluster. Then, we apply this method to every cluster as well as the mismatching region. Each cluster outputs only one good pose hypothesis with vote and the pose hypothesis from mismatching region will be marked as unreliable pose hypothesis. Finally, our approach only return high vote reliable pose hypotheses for the post processing stage.

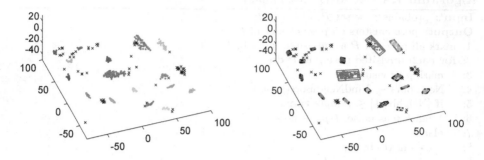

Fig. 3. Left: visualization of DBSCAN result. Right: visualization of PCA result. The oriented bounding boxes are drawn and the mismatching region are shown in red. The noise poses are marked as cross. (Color figure online)

4   Experimental Results

We evaluate our approach on the real robot bin picking application with mismatching region detection. Given a scene point cloud data, our algorithm not only detects the objects and estimates their poses, but also marks the poses in the mismatching region, as shown in Fig. 4. Multiple high score pose hypotheses are detected and wrong pose hypotheses with high evaluation score are marked with red color. Compared with the state-of-the-art PPF method, our approach improves the pose hypotheses result and enables robust 3D object detection and pose estimation in robot bin picking application. As we state in the Sect. 3, the mismatching region problem is caused by the shape of objects, the way of stacking objects and the capability of the sensor. If the mismatching region dosen't exist, our method works like the original version because DBSCAN works like original clustering. According to our analysis in Sect. 3, this method shall work in other object detection and pose estimation task.

Fig. 4. 3D object pose estimation with mismatching region detection on the real scene data for robot bin picking. The objects in mismatching region are marked with red color. (Color figure online)

The algorithm takes several parameters in DBSCAN and filter stage, which are related to sampling step and reference point selection strategy in PPF method. Also, since the clustering algorithm complexity stay same and the size of probable pose hypotheses is generally small, our approach dose not increase the much runtime compared with original version. However, this problem remains a challenge if the overlapping can not slide in some cases.

5   Conclusion

In this paper, we revisit the state-of-the-art PPF method and propose an improved PPF method with mismatching region detection for 3D object detection and pose estimation. We demonstrate the mismatching problem when using traditional PPF method for robot bin picking. To improve the pose hypothesis result, a new clustering strategy combined with DBSCAN and PCA is proposed. Also, we define the mismatching region and probable pose distribution to obtain more information in PPF method. Experiments on the real scene point clouds show that our approach provides the capability of mismatching region detection compared with original version.

We believe that PPF method still remain promising potential if we combine it with 2D image for 3D object detection and pose estimation problem. And the probable pose distribution has the potential to offer more constraints for other pose estimation techniques.

Acknowledgements. This work is partially supported by the National Natural Science Foundation of China (51375309). The support is gratefully acknowledged.

References

1. Drost, B., Ulrich, M., Navab, N., Ilic, S.: Model globally, match locally: efficient and robust 3D object recognition. In: 2010 IEEE Conference on Computer Vision and Pattern Recognition (CVPR), pp. 998–1005. IEEE (2010)
2. Pavel. Z.: Detection and localization of texture-less objects in RGB-D images (2015)
3. Birdal, T., Ilic, S.: Point pair features based object detection and pose estimation revisited. In: 2015 International Conference on 3D Vision (3DV), pp. 527–535. IEEE (2015)
4. Hinterstoisser, S., Lepetit, V., Rajkumar, N., Konolige, K.: Going further with point pair features. In: Leibe, B., Matas, J., Sebe, N., Welling, M. (eds.) ECCV 2016. LNCS, vol. 9907, pp. 834–848. Springer, Cham (2016). doi:10.1007/978-3-319-46487-9_51
5. Choi, C., Taguchi, Y., Tuzel, O., Liu, M.-Y., Ramalingam, S.: Voting-based pose estimation for robotic assembly using a 3D sensor. In: 2012 IEEE International Conference on Robotics and Automation (ICRA), pp. 1724–1731. IEEE (2012)
6. Choi, C., Christensen, H.I.: RGB-D object pose estimation in unstructured environments. Robot. Auton. Syst. **75**, 595–613 (2016)

7. Papazov, C., Burschka, D.: An efficient RANSAC for 3D object recognition in noisy and occluded scenes. In: Kimmel, R., Klette, R., Sugimoto, A. (eds.) ACCV 2010. LNCS, vol. 6492, pp. 135–148. Springer, Heidelberg (2011). doi:10.1007/978-3-642-19315-6_11

8. Abbeloos, W., Goedemé, T.: Point pair feature based object detection for random bin picking. In: 2016 13th Conference on Computer and Robot Vision (CRV), pp. 432–439. IEEE (2016)

9. Hinterstoisser, S., Cagniart, C., Ilic, S., Sturm, P., Navab, N., Fua, P., Lepetit, V.: Gradient response maps for real-time detection of textureless objects. IEEE Trans. Pattern Anal. Mach. Intell. **34**(5), 876–888 (2012)

10. Hinterstoisser, S., Lepetit, V., Ilic, S., Holzer, S., Bradski, G., Konolige, K., Navab, N.: Model based training, detection and pose estimation of texture-less 3D objects in heavily cluttered scenes. In: Lee, K.M., Matsushita, Y., Rehg, J.M., Hu, Z. (eds.) ACCV 2012. LNCS, vol. 7724, pp. 548–562. Springer, Heidelberg (2013). doi:10.1007/978-3-642-37331-2_42

11. Ester, M., Kriegel, H.-P., Sander, J., Xu, X., et al.: A density-based algorithm for discovering clusters in large spatial databases with noise. In: KDD 1996, pp. 226–231 (1996)

12. Kumar, K.M., Reddy, A.R.M.: A fast DBSCAN clustering algorithm by accelerating neighbor searching using groups method. Pattern Recogn. **58**, 39–48 (2016)

13. Besl, P.J., McKaY, N.D.: A method for registration of 3-D shapes. IEEE Trans. Pattern Anal. Mach. Intell. **14**(2), 239–256 (1992)

14. Yang, J., Li, H., Jia, Y.: Go-ICP: Solving 3D registration efficiently and globally optimally. In: Proceedings of the IEEE International Conference on Computer Vision, pp. 1457–1464 (2013)

15. Rusu, R.B., Blodow, N., Beetz, M.: Fast point feature histograms (FPFH) for 3D registration. In: IEEE International Conference on Robotics and Automation, ICRA 2009, pp. 3212–3217. IEEE (2009)

16. Knopp, J., Prasad, M., Willems, G., Timofte, R., Van Gool, L.: Hough transform and 3D SURF for robust three dimensional classification. In: Daniilidis, K., Maragos, P., Paragios, N. (eds.) ECCV 2010. LNCS, vol. 6316, pp. 589–602. Springer, Heidelberg (2010). doi:10.1007/978-3-642-15567-3_43

17. Guo, Y., Bennamoun, M., Sohel, F., Min, L., Wan, J.: 3D object recognition in cluttered scenes with local surface features: a survey. IEEE Trans. Pattern Anal. Mach. Intell. **36**(11), 2270–2287 (2014)

18. Kouskouridas, R., Tejani, A., Doumanoglou, A., Tang, D., Kim, T.-K.: Latent-class hough forests for 6 DoF object pose estimation. arXiv preprint arXiv:1602.01464 (2016)

19. Sahin, C., Kouskouridas, R., Kim, T.-K.: A learning-based variable size part extraction architecture for 6D object pose recovery in depth. arXiv preprint arXiv:1701.02166 (2017)

20. Maturana, D., Scherer, S.: VoxNet: a 3D convolutional neural network for real-time object recognition. In: 2015 IEEE/RSJ International Conference on Intelligent Robots and Systems (IROS), pp. 922–928. IEEE (2015)

21. Song, S., Xiao, J.: Deep sliding shapes for amodal 3D object detection in RGB-D images. In: Proceedings of the IEEE Conference on Computer Vision and Pattern Recognition, pp. 808–816 (2016)

22. Qi, C.R., Su, H., Nießner, M., Dai, A., Yan, M., Guibas, L.J.: Volumetric and multi-view CNNs for object classification on 3D data. In: Proceedings of the IEEE Conference on Computer Vision and Pattern Recognition, pp. 5648–5656 (2016)

23. Correll, N., Bekris, K.E., Berenson, D., Brock, O., Causo, A., Hauser, K., Okada, K., Rodriguez, A., Romano. J.M., Wurman, P.R.: Analysis and observations from the first Amazon picking challenge. IEEE Trans. Autom. Sci. Eng. (2016)
24. Zeng, A., Yu, K.-T., Song, S., Suo, D., Walker Jr., E., Rodriguez, A., Xiao, J.: Multi-view self-supervised deep learning for 6D pose estimation in the amazon picking challenge. arXiv preprint arXiv:1609.09475 (2016)
25. Wong, J.M., Kee, V., Le, T., Wagner, S., Mariottini, G.-L., Schneider, A., Hamilton, L., Chipalkatty, R., Hebert, M., Segicp, D.J., et al.: Integrated deep semantic segmentation and pose estimation. arXiv preprint arXiv:1703.01661 (2017)

A Five-Degree-of-Freedom Hybrid Manipulator for Machining of Complex Curved Surface

Yundou Xu[1,2], Jianhua Hu[1], Dongsheng Zhang[1], Jiantao Yao[1,2], and Yongsheng Zhao[1,2(✉)]

[1] Parallel Robot and Mechatronic System Laboratory of Hebei Province, Yanshan University, Qinhuangdao 066004, China
yszhao@ysu.edu.cn
[2] Key Laboratory of Advanced Forging & Stamping Technology and Science of Ministry of National Education, Yanshan University, Qinhuangdao 066004, China

Abstract. In this paper, a novel five-axis hybrid manipulator is constructed based on the 2-RPU&UPR parallel mechanism, which has less joints and all the axes of rotation are continuous. Then, the structure of the key components of the new five-axis hybrid manipulator is designed. Next, the analytic expression of the inverse position of the hybrid manipulator is established, based on which, the workpiece of the spherical surface is machined. All rotational degrees of freedom of this five-axis hybrid manipulator are of continuous axes, the analytical expression of the inverse position model can be obtained, so the trajectory planning and motion control can be realized easily. Moreover, the hybrid manipulator has less joints, which can guarantee high rigidity and high precision operation, so it has great application prospects.

Keywords: Hybrid manipulator · Parallel mechanism · Continuous axes · Inverse position

1 Introduction

In the high-precision machining of aerospace structural parts with large size, thin-walled and complex surfaces, the five-degree-of-freedom (DOF) serial-parallel hybrid robots have become more and more concerned by academia and industry [1, 2]. That is because the hybrid robot, as the compromise between parallel robots and series robots, has the characteristics of high rigidity, large workspace and high flexibility simultaneously [3, 4]. The five-DOF hybrid robots [5, 6] such as Tricept and Sprint Z3 head have been successfully applied in the field of aerospace processing, as well as automotive and aviation assembly, welding, drilling and so on.

The existing five-DOF hybrid robots can be classified into two categories [7]. The first category is composed of a three-DOF position parallel mechanism (3-UPS&UP [8], 2-UPS&UP [9], 2-UPR&SPR [10] parallel mechanism for example) plusing a 2-DOF rotating head attached to the moving platform, which has been exampled successfully by Tricept hybrid robot. Here, U, P, S and R denote the universal,

© Springer International Publishing AG 2017
Y. Huang et al. (Eds.): ICIRA 2017, Part II, LNAI 10463, pp. 48–58, 2017.
DOI: 10.1007/978-3-319-65292-4_5

prismatic, spherical and rotational joints, respectively. The second category is the integration of a three-DOF pose parallel mechanism (3-PRS [11], 3-RPS [12] parallel mechanism for example) and two long guideways, which has been exampled successfully by Sprint Z3 Head hybrid robot. However, these three-DOF parallel mechanisms above have a large number of joints, restricting the overall stiffness of the hybrid robot and operating accuracy. Based on a three-DOF parallel mechanism 2-RPU&UPR with less joints (the number of passive single-DOF joints is only 9 except the driven joints), a new five-DOF hybrid robot is constructed in this paper, which has the advantages of less joints, so it can meet the needs of high rigidity and high precision, and the application prospects is broad.

2   The Structure Composition of the Five-DOF Hybrid Manipulator

2.1   Principle Analysis of the Five-DOF Hybrid Robot Mechanism

As shown in Fig. 1, the robot consists of 2-RPU&UPR parallel mechanism, a head with one revolute joint, a mobile platform and the frame. The mechanism diagram is shown in Fig. 2. The 2-RPU&UPR mechanism has two rotational and one translational degrees of freedom, which are the rotation around the Y-axis on the fixed platform, the rotation of the x-axis on moving platform and the translation around of the Z-axis on the fixed platform. The rotation around the x-axis is used for the attitude adjustment in x-axis, while the rotation around the Y-axis are used to adjust a wide range of movement along X-axis. Therefore, a head with a revolute joint whose axis is parallel to the Y-axis, is connected to moving platform under the parallel mechanism to achieve attitude adjustment in y-direction, and a separate mobile platform along the Y-axis is installed on the workbench, so as to achieve the five-axis motions simultaneously.

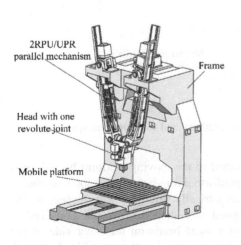

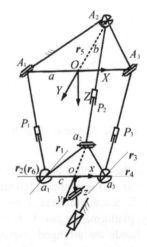

Fig. 1. Overall structure of 5-DOF hybrid robot   **Fig. 2.** Diagram of 5-DOF hybrid robot

2.2   Structural Design Principle

The Moving Platform

Three actuated branches and the head with one revolute joint are all connected to the moving platform, so two Hooke joints and two joints need to be designed on this moving platform, the structure model of the moving platform is designed as shown in Fig. 3.

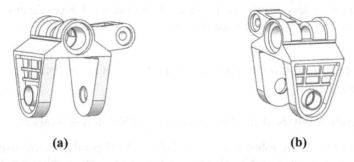

(a)   (b)

Fig. 3. Structure model of moving platform

In order to make the structure compact and reduce the distance between the motorized spindle and the moving platform, the driving motor of the head with one revolute joint is arranged at the front end of the moving platform. The relationship between the moving platform and the motorized spindle is shown in Fig. 4.

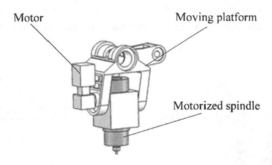

Fig. 4. Assembly diagram of the moving platform and the motorized spindle

Branches 1 and 2 are respectively connected to the moving platform by universal joints, branch 3 is connected to the moving platform through a revolute joint, as shown in Fig. 5. Since branches 1 and 2 are symmetrically arranged with respect to the moving platform, the two U joints are designed as T-shaped structures. Long end of two T heads are arranged coaxially. The two short heads on the other side of the T-shaped structures connect the branches in the same way, as shown in Fig. 6. Branch 3 is connected to the moving platform as shown in Fig. 7.

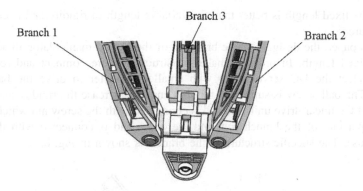

Fig. 5. Structure of connection for branches and moving platform

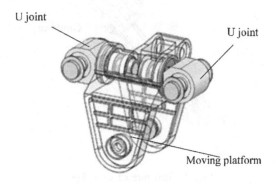

Fig. 6. Moving platform and U joints

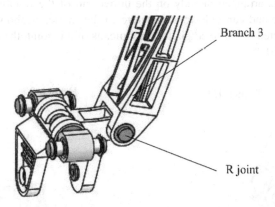

Fig. 7. Revolute joint and branch 3

The Design of Branches

The structure of the branches plays an important role in the performance of the whole hybrid manipulator, and its structure is divided into two types: fixed length and variable length. The fixed length is better than the variable length in rigidity and is easy to be manufactured.

In this paper, the design of three branches of the hybrid manipulator all adopts the form of fixed length. In order to make the structure more compact and reduce the quality, select the DC servo motor with smaller diameter to drive the ball screw rotating. The ball screw is supported by both ends to increase the rigidity and motion stability of the linear drive unit, one end is connected with the screw nut which is fixed at the upper end of the branch rod, and the other end is connected with the screw support base. The specific structure of the branch is shown in Fig. 8.

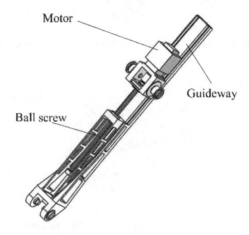

Fig. 8. Structure of branches

Branches 1 and 2 are connected to the fixed platform by means of revolute joints. The servo motor is arranged directly on the upper end of the R joint. The R joint is fixed on the frame and can only rotate relative to the frame, as shown in Fig. 9. The branch 3 is connected to the fixed platform by means of a U joint, the specific structure is shown in Fig. 10.

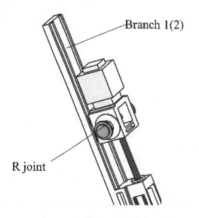

Fig. 9. Structure of R joint

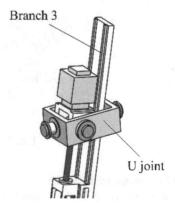

Fig. 10. Structure of U joint

2.3 Design of the Frame

The frame of the hybrid manipulator is a component which is connected with the worktable and the ground, whose stiffness and strength directly affect the accuracy and life of the manipulator. In this paper, the fixed platform is integrated with the frame, the overall structure uses welding parts. The specific structure of the frame is shown in Fig. 11.

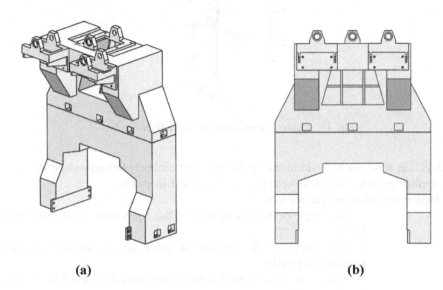

(a)   (b)

Fig. 11. Structure of the frame

3 Realization of Complex Surface Machining

3.1 Solution of Inverse Position

It can be seen from the literature [13] that the 2-RPU&UPR parallel mechanism has a translation and two rotations with continuous axes, so that the moving platform can obtain the pose through the following three transformations. Firstly rotates θ_1 around the Y-axis; then translate λ along the Z-axis; finally rotates θ_2 around the x-axis, so the mechanism can be equivalent to a serial mechanism composed of three joints of R, P and R. Consequently, the entire hybrid robot can be equivalent to a combination of a RPRR serial mechanism and a moving platform, as shown in Fig. 12, so the solving of the inverse position of the hybrid robot can be greatly simplified.

Next, we establish inverse position model of the hybrid robot. Since the hybrid robot consists of the 4-DOF serial mechanism RPRR and the 1-DOF mobile platform equivalently (the moving direction is along the y_0-direction), the orientation and the displacement in the x_0-axis and z_0-axis required for machining the workpiece are directly determined by the serial mechanism RPRR. While the displacement in the

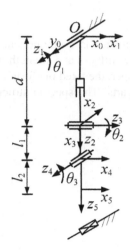

Fig. 12. Equivalent mechanism of the hybrid robot

y_0-direction needs to be determined by the tool and mobile platform, since the serial mechanism generates a part of coupling displacement in y-axis.

Define the following parameters:

θ_1 — The angle of the first R joint of the equivalent serial mechanism RPRR;

θ_2 — The angle of the second R joint of the equivalent serial mechanism RPRR;

θ_3 — The angle of the third R joint of the equivalent serial mechanism RPRR;

d_0 — The distance between the fixed platform and the moving platform in the initial state;

d_1 — The displacement of the P joint of the equivalent serial mechanism RPRR;

d_2 — The displacement of mobile platform in the $y0$-direction;

$[a_x \quad a_y \quad a_z]^T$ — The direction of the tool;

$[p_x \quad p_y \quad p_z]^T$ — The position of the end of the tool;

y_0 —The displacement of workpiece in the-direction.

According to the D-H method, the analytical expression of the inverse position solution can be obtained as follows:

$$\begin{cases} \theta_1 = \arctan[(-a_x l_2 + p_x)/(-a_z l_2 + p_z)] \\ \theta_3 = \arcsin(a_x c_1 - a_z s_1) \\ \theta_2 = \arcsin(-a_y/c_3) \\ d_1 = (-a_z l_2 + p_z - l_1 c_1 c_2)/c_1 - d_0 \\ d_2 = -a_y l_2 + a_y l_1/c_3 - y_0 \end{cases} \quad (1)$$

3.2   Examples: Spherical Surface Machining

As shown in Fig. 13, assuming that the locus of the workpiece to be machined is a sphere, the spherical radius is R, The radius of the circular trajectory being processed of the tool is r, so the distance between the center of the workpiece and the center of the circle being processed is $m = \sqrt{R^2 - r^2}$, and the distance between the center of the circle being processed and the $\{0\}$ coordinate system of the machine is h. the angle between the x-axis and the straight line connecting the center of the circle and the processing point is α.

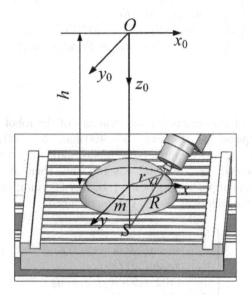

Fig. 13. Machining trajectory of workpiece

The position of the current processing point is:

$$(r\cos\alpha \quad -r\sin\alpha \quad h)^T, \ \alpha \in (0, 360°)$$

Here requires the normal of the tool and the tangent plane of the processing point are vertical, so the direction of the tool should be:

$$(-r\cos\alpha/R \quad r\sin\alpha/R \quad m/R)^T$$

That is:

$$(a_x \quad a_y \quad a_z) = (-r\cos\alpha/R \quad r\sin\alpha/R \quad m/R) \tag{2}$$

Since the workpiece moves only in the y-direction of the $\{0\}$ system on the mobile platform, the x and z coordinates of the tool and the workpiece in the $\{0\}$ system are the same at any time, then the x and z coordinates of the origin of {5} system in the {0} system are:

$$\begin{cases} p_x = r\cos\alpha \\ p_z = h \end{cases} \tag{3}$$

Substituting (2) and (3) into (1) yields:

$$\begin{cases} \theta_1 = \arctan[r(R+l_2)c\alpha/(Rh-ml_2)] \\ \theta_3 = \arcsin\left[\left(-rc\alpha c_1 - s1\sqrt{R^2-r^2}\right)/R\right] \\ \theta_2 = \arcsin[-rs\alpha/(Rc_3)] \\ d_1 = (Rh - ml_2 - Rl_1c_1c_2)/(Rc_1) - d_0 \\ d_2 = (-rs\alpha c_3 l_2 - rs\alpha l_1 + Rrs\alpha c_3)/(Rc_3) \end{cases} \tag{4}$$

3.3   Simulation

Given a specific set of parameters for the structure of the robot and the machining path of the workpiece, as follows: $a = 400\,\text{mm}$, $b = 300\,\text{mm}$, $c = 200\,\text{mm}$, $r_{10} = 824.62\,\text{mm}$, $r_{20} = 813.94\,\text{mm}$, $r_{30} = 824.62\,\text{mm}$, $l_1 = 50\,\text{mm}$, $l_2 = 60\,\text{mm}$, $r = 60\,\text{mm}$, $R = 100\,\text{mm}$, $m = 80\,\text{mm}$, $h = 150\,\text{mm}$. Then $d0$ can be calculated according to the structural parameters of the robot for 800 mm, substituting the given parameters into (4), the following results can be obtained:

$$\begin{cases} \theta_1 = \arctan(16c\alpha/17) \\ \theta_3 = \arcsin(-0.6c\alpha c_1 - 0.8s_1) \\ \theta_2 = \arcsin(-0.6s\alpha/c_3) \\ d_1 = (102 - 50c_1c_2)/c_1 - d_0 \\ d_2 = (36s\alpha c_3 - 30s\alpha - y_0c_3)/c_3 \end{cases}$$

For the (2-RPU&UPR) + R four-DOF mechanism, substitute the result of the inverse solution into the positive solution, we can get the direction and position of the tool's end. Figure 14 shows the track of the tool's direction. It can be seen from the figure that the orientation is exactly the same as that of the tool in processing the workpiece. The position of the workpiece is determined by that of the (2-RPU&UPR) + R four-DOF mechanism and the 1-DOF mobile platform, and the resulting position trajectory of workpiece is shown in Fig. 15. It can be seen from the figure that the track of the workpiece is a circle and its radius is 60, which is exactly the same as the pre-machined track of workpiece given earlier.

The previous analyses show that the hybrid robot can realize the linkage processing of the complex surface of the workpiece, and also verify that the inverse position model of the hybrid robot is completely correct.

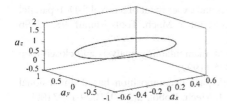

Fig. 14. Track of tool's orientation

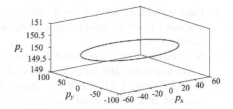

Fig. 15. Position of workpiece

4 Conclusion

A five-DOF hybrid manipulator is constructed based on the two-rotation-and-one-translation three-DOF parallel mechanism of 2-RPU&UPR. All the rotational axes of the robot are continuous and the robot has less number of passive joints. The key components are designed with reasonable structure, ensuring its high structural rigidity characteristics, so it can meet the needs of high precision operation.

The analytical expression of the position model of the robot is obtained, and the model is validated through the machining of a workpiece with spherical surface. The simulation results show that the robot can continuously adjust the attitude, so as to keep the normal of the tool and the normal of the machining surface coincident, and can finally finish the machining of the spherical surfaced workpiece.

Acknowledgment. This work was supported by the National Natural Science Foundation of China under Grant 51405425, and Key Basic Research Program of Hebei Province's Applied Basic Research Plan of China under Grant 15961805D.

References

1. Liu, N., Wu, J.: Kinematics and application of a hybrid industrial robot–Delta-RST. Sens. Transducers **169**, 186–192 (2014)
2. Terrier, M., Dugas, A., Hascoët, J.Y.: Qualification of parallel kinematics machines in high-speed milling on free form surfaces. Int. J. Mach. Tools Manuf. **44**(7–8), 865–877 (2004)
3. Hunt, K.H.: Structural kinematics of in-parallel-actuated robot-arms. J. Mech. Transmissions Autom. Des. **105**(4), 705–712 (1983)
4. Xie, F.G., Liu, X.J., Zhang, H., et al.: Design and experimental study of the SPKM165, a five-axis serial-parallel kinematic milling machine. Sci. China Technol. Sci. **54**(5), 1193–1205 (2011)
5. Wang, Y.Y., Huang, T., Zhao X.M., et al.: Finite element analysis and comparison of two hybrid robots-the tricept and the trivariant. In: IEEE/RSJ International Conference on Intelligent Robots and Systems, pp. 490–495 (2006)
6. Shi, J., Wang, Y., Zhang, G., et al.: Optimal design of 3-DOF PKM module for friction stir welding. Int. J. Adv. Manuf. Technol. **66**(9–12), 1879–1889 (2013)

7. Lian, B., Sun, T., Song, Y., et al.: Stiffness analysis and experiment of a novel 5-DoF parallel kinematic machine considering gravitational effects. Int. J. Mach. Tools Manuf. **95**, 82–96 (2015)
8. Siciliano, B.: The Tricept robot: inverse kinematics, manipulability analysis and closed-loop direct kinematics algorithm. Robotica **17**, 437–445 (1999)
9. Sun, T., Song, Y.M., Li, Y.G., Zhang, J.: Workspace decomposition based dimensional synthesis of a novel hybrid reconfigurable robot. J. Mech. Robot. **2**(3), 031009-1–031009-8 (2010)
10. Bi, Z.M., Jin, Y.: Kinematic modeling of Exechon parallel kinematic machine. Robot. Comput.-Integr. Manuf. **27**(1), 186–193 (2011)
11. Sun, T., Song, Y.M.: Dimensional synthesis of a 3-DOF parallel manipulator based on dimensionally homogeneous Jacobian matrix. Sci. China Technol. Sci. **53**(1), 168–174 (2010)
12. Schadlbauer, J., Walter, D.R., Husty, M.L.: The 3-RPS parallel manipulator from an algebraic viewpoint. Mech. Mach. Theory **75**(5), 161–176 (2014)
13. Xu, Y., Zhou, S., Yao, J., et al.: Rotational Axes Analysis of the 2-RPU/SPR 2R1T Parallel Mechanism, 2nd edn. Springer, Cham (2014)

A Dynamic Real-Time Motion Planning Method for Multi-robots with Collision Avoidance

Yonghong Zhang, Huan Zhao$^{(\boxtimes)}$, Congcong Ye, and Han Ding

State Key Laboratory of Digital Manufacturing Equipment and Technology
Huazhong, University of Science and Technology, Wuhan 430074, Hubei,
People's Republic of China
huanzhao@hust.edu.cn

Abstract. Collision avoidance is the major concern for the multi-robots operation. However, few literatures can generate a collision free path as well as a smooth motion profile at the same time. To solve this problem, this paper presents an integrated motion planning scheme for two manipulators working in a shared workspace. In this method, first, a collision free path is calculated in the path planning phase. Then, the smooth trajectory is generated by using a dynamic nonlinear filter. Both the path and the trajectory are calculated directly, thus the computation load is low and the approach can be applied in a real-time manner. Simulation results indicate that the proposed method is effective to realize collision avoidance for industrial robots working in a common workspace.

Keywords: Motion planning · Collision avoidance · Real-time · Trajectory generation

1 Introduction

Factory automation has been widely increasing through the use of industrial robots. Many industrial tasks are performed efficiently with the use of two manipulators systems, which increases productivity and reduces the amount of space needed in the factory. These systems execute mainly two types of operations: one type is that all manipulators cooperate for a given task, such as carrying an object; and another is that each manipulator accomplishes its own task independently in a common workspace, such as picking and placing. The basic requirement of fulfilling their tasks is that a robot must be able to move from its start point to goal and avoid possible collisions. A widely used technique is zone blocking method [1]: when one robot enters into the common workspace, a flag is activated to prevent other robots from entering the workspace. The method is not efficient because all robots can not work in the shared space at the same time.

Being a key challenge of robotics and automation engineering, path planning to avoid potential collisions has been studied extensively in the last two decades. Probabilistic Road Map [2, 3] and Rapid Random Trees [4, 5], as well as all their variants,

© Springer International Publishing AG 2017
Y. Huang et al. (Eds.): ICIRA 2017, Part II, LNAI 10463, pp. 59–69, 2017.
DOI: 10.1007/978-3-319-65292-4_6

are the most popular. Because such algorithms avoid the hard problem of building high-dimensional configuration space explicitly by sampling the configuration space.

The common characteristic of collision avoidance approaches in [2–5] is that obstacles are static. Two important factors limit the direct application of these collision avoidance approaches on robot coordination. The first factor is that a change in the environment can invalidate the previously planned path. The second factor is that they cannot respond to dynamic environment immediately. Hence, collision avoidance for multi robots working in the shared space is more complex and need specially treated.

Afaghani et al. [6–8] proposed a method to avoid collisions using time scheduling of the execution time of commands. If a collision were to occur, one robot could move while the other robots should stop and wait. The goal, however, cannot reach if the path is obstructed by other robots. To address this problem, Montaño A et al. [9, 10] proposed a coordination method in discretized coordination space [11]. The coordination is achieved through adjustment of movements of each robot along its original planned path. However, each robot must have information about the next movements of the other robots. Cheng [12] proposed a method to achieve coordination by adjusting the geometric paths of robots. Wang et al. [13] presents a collision avoidance method in configuration space. To reduce the computation time, only the first three joints of the robot are used for collision avoidance. Li et al. [14] presents a method tailored to path planning problems in changing environments where many moving obstacles crowed together. A roadmap is built using uniform random sampling method, and then A* algorithm is arranged to search for a feasible path. The aim of research in [12–14] is to find a collision-free path, which is a geometric representation of a plan to move from a start to a target. Motion planning has to produce an executable trajectory for a robot, and not merely a geometrical path. Chiddarwar and Babu [15] present a conflict free coordinated path planning for multiple robots in configuration-time space. The coordination strategy is moving to a safety position and waits there until conflict has been resolved. The smoothness of motion is not considered. Vannoy et al. [16] introduces an adaptive motion planning approach based on evolutionary computation. Although the smoothness of motion can be guaranteed, the trajectory is generated though iterative search process, which increases the calculation load.

This paper presents a dynamic motion planning for two industrial robots working in a shared workspace. In this method, a collision free path is obtained in path planning stage, and then the smooth trajectory is generated by employing a dynamic nonlinear filter. Unlike motion planning in discretized configuration space described in Refs. [13–15], the motion commands are generated by the dynamic nonlinear filter, which produces a trajectory with continuous velocity. Compared with Ref. [16],the path and trajectory are calculated directly instead of iteration, which decreases the computation load significantly.

The rest of this paper is organized as follows: Sect. 2 proposes the collision avoidance algorithm; Sect. 3 provides simulation results followed by the corresponding analysis; the paper is concluded in Sect. 4.

2 Dynamic Motion Planning for Robots with Collision Avoidance

Figure 1 is Architecture of the proposed method. The approach comprises of two major distinct phases, path planning and online trajectory generation. In path planning phase, SSV technique is used to detect the collision. When collision will occur, a new path is calculated immediately. In trajectory generation phase, a dynamic nonlinear filter is used to generate the smooth trajectory for each robot.

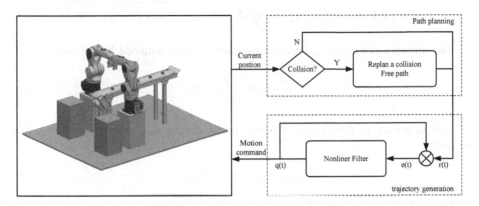

Fig. 1. Architecture of the proposed collision avoidance approach.

2.1 Path Planning

Modelling the manipulator plays an important role in collision avoidance. The model should be precise while being based on a simple mathematical representation for preventing high calculation cost. In this paper, SSV [17] method is used to determine detect collision. It is a safe and effective technique and easier to find the shortest distance, because cylinders and spheres are used to approximate the actual robot component geometry.

In SSV technique, each link is represented as a finite line $\vec{S}_n$, which is spanned through the two adjacent points $\mathbf{P}_n$ and $\mathbf{P}_{n+1}$, as shown in Fig. 2(a). Besides, a radius r_n is introduced which is bigger than the maximum distance between the line and the surface of the related robot component. As described above, $\vec{S}_n$ can be described as

$$\vec{S}_n(\mu) = \mathbf{P}_n + \mu(\mathbf{P}_{n+1} - \mathbf{P}_n) \tag{1}$$

where μ is a variable between 0 and 1.

Once the robot is modeled as shown in Fig. 2(b), the collision detection process between the robots will be realized in Cartesian coordinates. The first two components of robot are not considered because link1 is fixed and the movement range of link2 is small. As current joint angles are obtained from robot controller, $\vec{S}_n$ can be calculated using forward kinematics. As shown in Fig. 3, $\mathbf{P}_i = (X, Y, Z)$ is a point on the link $\vec{S}_n(\mu_n)$, $\mathbf{P}_j = (U, V, W)$ is a point on the link $\vec{S}_m(\mu_m)$

$$X = x_n + \mu_n(x_{n+1} - x_n), \ Y = y_n + \mu_n(y_{n+1} - y_n), \ Z = z_n + \mu_n(z_{n+1} - z_n)$$
$$U = x_m + \mu_m(x_{m+1} - x_m), \ V = y_m + \mu_m(y_{m+1} - y_m), \ W = z_m + \mu_m(z_{m+1} - z_m)$$
$$(2)$$

The distance between $\mathbf{P}_i$ and $\mathbf{P}_j$ can be calculated as follows

$$d = \|\mathbf{P}_i\mathbf{P}_j\| = \sqrt{f(\mu_n, \mu_m)} = \sqrt{(X - U)^2 + (Y - V)^2 + (Z - W)^2} \quad (3)$$

The minimal distance $d_{\min}$ between two links can be obtained by solving

$$\frac{\partial f}{\partial \mu_n} = 0, \ \frac{\partial f}{\partial \mu_m} = 0 \quad (4)$$

It is necessary to replan path for two robots if only $d_{\min}$ is less than the safety distance

$$d_{\min} < r_n + r_m = d_{safe} \quad (5)$$

For multiple robots path planning, prioritization scheme is a much more efficient technique [18, 19]. In this paper, the priority for each robot is assigned as follows:

(1) When one robot is completing a more important task than another robot, it has a higher priority;
(2) It has a higher priority if one robot is closer to its goal than another robot.

Figure 4 is an illustration of priority assignment. The priority of robot A is higher than robot B. Robot A can move along its previous path if there is no collision after robot B modifies its path. Otherwise both robots should modify their path to avoid collision. The modification only has to be good enough for a short period before the next cycle time since it will be corrected constantly. Based on this, a collision-free position of end-effector in the next cycle time can be defined as fellows

$$\mathbf{P}_{t+\Delta t} = \mathbf{P}_t + l \cdot \tau \quad (6)$$

where $\mathbf{P}_t$ is the current position of end-effector. τ, from $\mathbf{P}_i^{\min}$ to $\mathbf{P}_j^{\min}$, is the unit vector of line segment $\mathbf{P}_i^{\min}\mathbf{P}_j^{\min}$, as shown in Fig. 3. $\mathbf{P}_i^{\min}$ and $\mathbf{P}_j^{\min}$ are the points on their respective robot links with minimal distance to each other. l is the distance between $\mathbf{P}_{t+\Delta t}$ and $\mathbf{P}_t$, and it should larger than the difference between $d_{\min}$ and d_{safe}. Here, it is assumed that $l = 1.5|d_{safe} - d_{\min}|$. It is assumed that the pose of end-effector at $\mathbf{P}_{t+\Delta t}$ is the same as $\mathbf{P}_t$.

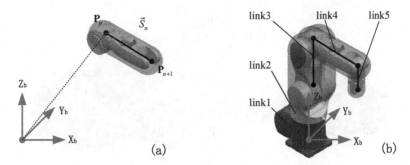

Fig. 2. Modelling a robot using SSV. (a) Description of a single robot link. (b) Description of a robot.

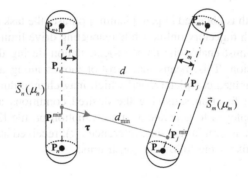

Fig. 3. Distance between two links.

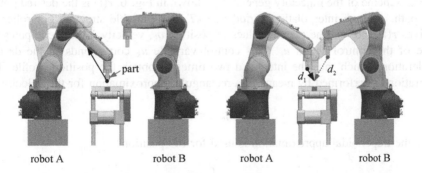

Fig. 4. An illustration of priority assignment. (a) Robot A holds a part. (b) Robot A is closer to its goal point.

As a result, the new path is moving the robot from $\mathbf{P}_t$ to goal $\mathbf{P}_{goal}$ via $\mathbf{P}_{t+\Delta t}$, as shown in Fig. 5.

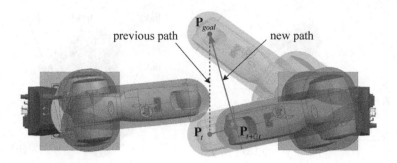

previous path

$\mathbf{P}_{goal}$

new path

$\mathbf{P}_t$ $\mathbf{P}_{t+\Delta t}$

Fig. 5. Illustration of path modification when collision occurs.

2.2 Online Trajectory Generation

As a collision free path is obtained in path planning phrase, the task of motion planning is generating a smooth trajectory online while respecting drive limits. Online trajectory generation methods must enable the path (re-)calculation during the robot motion in order to avoid collision. This means that a robot moves along a path that has not necessarily been computed completely, and which may change during the movement. Instead of defining a function satisfying the desired conditions as [16], a dynamic nonlinear filter is employed to generate a smooth motion profile [20]. This trajectory generator is able to transform in real-time reference $r(t)$, received as input, in a smooth output $q(t)$ which satisfies the following constraints:

$$\begin{aligned} |\dot{q}_k| &\le v_{\max} \\ |\ddot{q}_k| &\le a_{\max} \end{aligned} \tag{7}$$

The scheme of the trajectory generator is shown in Fig. 6. $r(t)$ is the desired potion of a path. At each interpolating period $t_k = kT_s$, the variable structure controller C_2 receives $r(t)$ and and the current values of position q_k, velocity $\dot{q}_k$, and computes the value of the control action u_k. This control variable u_k corresponds to the desired acceleration, which must be integrated two times to obtain the position profile. The integration is performed by means of a rectangular approximation for the velocity

$$\dot{q}_k = \dot{q}_{k-1} + T_s u_k \tag{8}$$

while the trapezoidal approximation is used for the position

$$q_k = q_{k-1} + \frac{T_s}{2}(\dot{q}_k + \dot{q}_{k-1}) \tag{9}$$

the control variable u_k is computed at each interpolating period $t_k = kT_s$ as

$$C_2 : \begin{cases} z_k = \frac{1}{T_s}\left(\frac{e_k}{T_s} + \frac{\dot{e}_k}{2}\right), \quad \dot{z}_k = \frac{\dot{e}_k}{T_s} \\ m = floor\left(\frac{1 + \sqrt{1 + 8|z_k|}}{2}\right) \\ \sigma_k = \dot{z}_k + \frac{z_k}{m} + \frac{m-1}{2}\,sign(z_k) \\ u_k = -a_{\max}\,\frac{1 + sign(\dot{q}_k sign(\sigma_k) + v_{\max} - T_s a_{\max}}{2}\,sat(\sigma_k) \end{cases} \tag{10}$$

where $e_k = (q_k - r_k)/a_{\max}$, $\dot{e}_k = \dot{q}_k/a_{\max}$ are the normalized tracking errors for position and velocity, z_k and $\dot{z}_k$ are the system state variables, $floor(\cdot)$ is the function 'integer part', $sign(\cdot)$ is the sign function, and $sat(\cdot)$ is a saturation function defined by

$$sat(x) = \begin{cases} -1, & x < -1 \\ x, & -1 \leq x \leq 1 \\ +1, & x > 1 \end{cases} \tag{11}$$

More details about the controller C_2 can be found in [20]. When the reference $r(t)$ is a step displacement, the output motion is perfectly equivalent to the trapezoidal trajectory.

One obvious advantage of the proposed collision avoidance algorithm is the smoothness of trajectory. Unlike motion planning in discretized configuration space as [13–15], motion commands are generated by the dynamic nonlinear filter, which produce a trajectory with continuous velocity. Another advantage of the proposed collision avoidance algorithm is its low calculation. The collisionfree path and the trajectory with smooth velocity profile are calculated directly rather than through an iterative search [16].

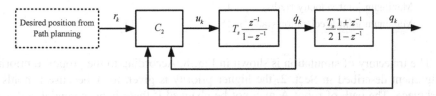

Fig. 6. Second order filter for online trajectory generator.

3 Simulation Validation

The proposed algorithm is evaluated using two KUKA robots (KR6R700), as shown in Fig. 7. In this environment, two robots are fixed on a board, with a distance between the central axes of the robots' bases 900 mm. The global coordinate system is located in the base of robot A. Each robot is controlled independently. The parameters used in the simulation are list in Table 1. In detail, robot A, holding a workpiece, moves from **P** to **P**$_A$, while robot B moves from **P**$_B$ to **P**. Apparently, it is highly possible that the collision of two manipulators occurs during the work process among the workspace.

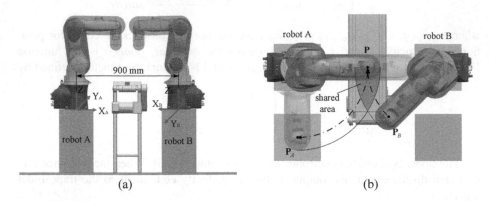

(a)   (b)

Fig. 7. Simulation system with two robots. (a) The snapshot of this simulation. (b) Initial and target position of robot A and B.

Table 1. Collision avoidance parameters used in the simulation

	Robot A	Robot B
Initial position	**P** (450, 0, 500)	**P**$_B$ (582, −318, 500)
Target position	**P**$_A$ (0, −450, 500)	**P** (450, 0, 500)
Maximum joint velocity (rad/s)	1	1
Maximum joint acceleration (rad/s^2)	15	15

The trajectory of simulation is shown in Fig. 8. According to the proposed priority assignment described in Sect. 2, the higher priority is given to A because it holds a workpiece. The path of robot A may not be changed if there is no potential collision possibility when path modification is performed for the robot B. As a result, the trajectory of robot A is changed slightly, while the trajectory of robot B is changed evidently. The distance between links is shown in Fig. 9. The motion profiles are shown in Fig. 10. The acceleration and velocity changed immediately to avoid collision when the distance between link4 of robot A and link4 of robot B is less than safety distance. It takes a little time to change movement direction of robot B. As a result, the distance between link4 of robot A and link4 of robot B decreases in an acceptable range even if it less than the safety distance. Though the movement of joint 6 has no influence on collision avoidance, it is responsible for the pose of the end-effector, which is adjusted when avoiding collisions. Because the collision-free path is planned locally in

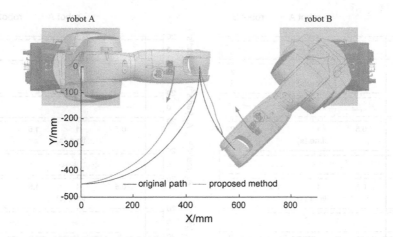

Fig. 8. Trajectory of simulation results.

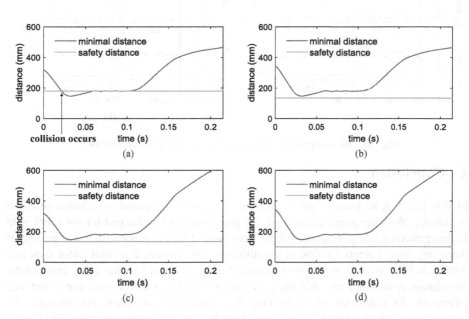

Fig. 9. Minimal distance. (a) Distance between link4 of robot A and link4 of robot B. (b) Distance between link4 of robot A and link5 of robot B. (c) Distance between link5 of robot A and link4 of robot B. (d) Distance between link5 of robot A and link5 of robot B.

time, there are still collisions in the nearly future. From Fig. 10, it is seen that the velocity profile is smooth and continuous. Besides, both the velocity and acceleration are under the constraint of maximum values.

Simulation results show that the presented method can realize collision avoidance effectively for industrial robots working in a common workspace.

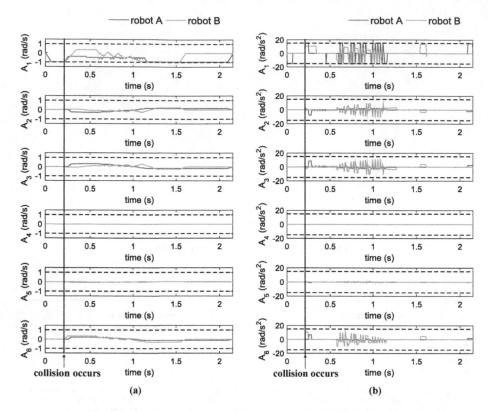

Fig. 10. Motion profiles. (a) Joint velocity. (b) Joint acceleration.

4   Conclusion

In this paper, a real-time motion planning method is proposed to achieve collision avoidance. With the proposed method, the path may not be changed for the robot with higher priority while path modification is performed for the robot with lower priority if there are some potential collision possibilities. Meanwhile, a second order dynamic nonlinear filter is used to generate smooth trajectory regarding the drive constraints. Simulation results indicate that the presented method can realize collision avoidance effectively for industrial robots working in a common workspace. Furthermore, the path as well as the trajectory is calculated directly, thus the computation load is low and the approach can be applied in real time.

Acknowledgement. This work was supported by the National Natural Science Foundation of China under Grant Nos. 51535004, 51323009, 51375196 and 51405175.

References

1. Zhou, J., Nagase, K., Kimura, S., et al.: Collision avoidance of two manipulators using rt-middleware. In: 2011 IEEE/SICE International Symposium on System Integration (SII), pp. 1031–1036. IEEE (2011)

2. Van Den Berg, J.P., Overmars, M.H.: Roadmap-based motion planning in dynamic environments. IEEE Trans. Robot. **21**(5), 885–897 (2005)
3. Al-Mutib, K., AlSulaiman, M., Emaduddin, M., et al.: D* lite based real-time multi-agent path planning in dynamic environments. In: 2011 Third International Conference on Computational Intelligence, Modelling and Simulation (CIMSiM), pp. 170–174. IEEE (2011)
4. Kuffner, J., LaValle, S.M.: RRT-Connect An efficient approach to single-query path planning. In: IEEE International Conference on Robotics and Automation, pp. 473–479 (2000)
5. LaValle, S.M., Kuffner, Jr., J.J.: Rapidly-exploring random trees: progress and prospects (2000)
6. Afaghani, A.Y., Aiyama, Y.: On-line collision avoidance between two robot manipulators using collision map and simple escaping method. In: 2013 IEEE/SICE International Symposium on System Integration (SII), pp. 105–110. IEEE (2013)
7. Afaghani, A.Y., Aiyama, Y.: On-line collision avoidance of two command-based industrial robotic arms using advanced collision map. Int. J. Robot. Mech. (JRM) **26**(3), 321–330 (2014)
8. Afaghani, A.Y., Aiyama, Y.: Advanced-collision-map-based on-line collision and deadlock avoidance between two robot manipulators with PTP commands. In: 2014 IEEE International Conference on Automation Science and Engineering (CASE), pp. 1244–1251. IEEE (2014)
9. Montaño, A., Suárez, R.: An online coordination algorithm for multirobot systems In: 2013 IEEE 18th Conference on Emerging Technologies & Factory Automation (ETFA), pp. 1–7. IEEE (2013)
10. Rodríguez, C., Montaño, A., Suárez, R.: Optimization of robot coordination using temporal synchronization. In: 2014 IEEE Emerging Technology and Factory Automation (ETFA), pp. 1–7. IEEE (2014)
11. Shin, Y., Bien, Z.: Collision-free trajectory planning for two robot arms. Robotica **7**(03), 205–212 (1989)
12. Cheng, X.: On-line collision-free path planning for service and assembly tasks by a two-arm robot. In: 1995 IEEE International Conference on Robotics and Automation, Proceedings, vol. 2, pp. 1523–1528. IEEE (1995)
13. Wang, S., Bao, J., Fu, Y.: Real-time motion planning for robot manipulators in unknown environments using infrared sensors. Robotica **25**(02), 201–211 (2007)
14. Li, Y., Liu, H., Ding, D.: Motion planning for robot manipulators among moving obstacles based on trajectory analysis and waiting strategy. In: SICE Annual Conference. IEEE, pp. 3020–3025 (2008)
15. Chiddarwar, S.S., Babu, N.R.: Conflict free coordinated path planning for multiple robots using a dynamic path modification sequence. Robot. Auton. Syst. **59**(7), 508–518 (2011)
16. Vannoy, J., Xiao, J.: Real-time adaptive motion planning (RAMP) of mobile manipulators in dynamic environments with unforeseen changes. IEEE Trans. Robot. **24**(5), 1199–1212 (2008)
17. Ennen, P., Ewert, D., Schilberg, D., et al.: Efficient collision avoidance for industrial manipulators with overlapping workspaces. Procedia CIRP **20**, 62–66 (2014)
18. Freund, E., Hoyer, H.: Real-time pathfinding in multirobot systems including obstacle avoidance. Int. J. Robot. Res. **7**(1), 42–70 (1988)
19. Warren, C.W.: Multiple robot path coordination using artificial potential fields. In: 1990 IEEE International Conference on Robotics and Automation, Proceedings, pp. 500–505. IEEE (1990)
20. Biagiotti, L., Melchiorri, C.: Trajectory Planning for Automatic Machines and Robots. Springer, Heidelberg (2008). doi:10.1007/978-3-540-85629-0

Reverse and Forward Post Processors
for a Robot Machining System

Fusaomi Nagata[1]($\boxtimes$), Yudai Okada[1], Takamasa Kusano[2],
and Keigo Watanabe[3]

[1] Tokyo University of Science, Yamaguchi, 1-1-1 Daigaku-Dori,
Sanyo-Onoda 756-0884, Japan
nagata@rs.tusy.ac.jp
[2] SOLIC Co. Ltd., 80-42 Yotsuyama-Machi, Omuta 836-0067, Japan
[3] Okayama University, 3-1-1 Tsushima-naka, Kita-ku, Okayama 700-8530, Japan

Abstract. This paper presents methods that can be widely and easily applied to data transformation process of machining robots. A reverse post processor is first introduced to regenerate original CLS (Cutter Location Sourse) data from post-processed NC (Numerical Control) data including variable axes codes. Then, a promising forward post processor is proposed to produce FANUC robot programs called LS data from CLS data. The proposed reverse and forward post processors allow an industrial machining robot to work based on the NC data that have been used for, e.g., a five axis NC machine tool with a tilting head.

Keywords: Reverse post processor · Forward post processor · CLS data · FANUC LS data · NC data · CAD/CAM · Industrial robot

1 Introduction

Many robot users demand that industrial robots can be controlled based on numerical control data called NC data as many kinds of NC machine tools, however, ordinary robot systems have their own not-standardized features. Up to now, many researchers have dedicated their efforts to enhance the relationship among CAD, CAM and robot programming. For example, Solvang et al. [1] considered a robot CAM for machining operations using an industrial robot. The system coped with the new machining standard called STEP-NC, which is a machine tool control language including geometric dimension and tolerance data for inspection. Andres et al. [2] addressed the inverse kinematics of a complex KUKA robotic system for milling works consisting of one robotic manipulator mounted on a linear axis and synchronized with a rotary table, then a functional post processor was implemented inside a CAM system to improve the communication between the CAM software and the manipulator. Also, many off-line programming methods have been also developed to reduce the robot programming load. However, many practical issues such as cable/hose tangling, robot configuration, collision, and reachability have not been sufficiently overcome yet.

© Springer International Publishing AG 2017
Y. Huang et al. (Eds.): ICIRA 2017, Part II, LNAI 10463, pp. 70–78, 2017.
DOI: 10.1007/978-3-319-65292-4_7

Chen and Sheng [3] discussed a new method to generate robot programs to deal with those problems. Besides, Zhan and Mao [4] studied the mechanism of the free-form surface polishing process by using an industrial robot and developed algorithms for robot programming and generation of NC code to realize robotic surface polishing. This research was based on a robot off-line programming system developed by a standard CAD/CAM system. Posada et al. [5] presented a novel approach to automatically generate robot programs based on variant CAD models and to enable online control to minimize the errors using a laser scanner sensor during deburring process.

The authors [6] have developed two type of robot machining systems. The first one could directly run based on CLS data without using any robot language and without conducting any complicated teaching tasks. The developed robotic CAM system enabled this function. On the other hand, the second one could run based on robot programs written in a robot language called the LS format. The developed forward post processor could automatically generate the robot program from CLS data without any teaching process [7]. These two developed systems have each own useful and serviceable feature. Recently, it is pointed out and expected by many robot users that industrial robot-based machining systems should be controlled based on NC data as many kinds of NC machine tools.

This paper presents methods that can be widely and easily applied to data transformation process of machining robots. A reverse post processor is first introduced to regenerate original CLS (Cutter Location Sourse) data from post-processed NC (Numerical Control) data including variable axes information. Then, a promising forward post processor is proposed to produce a robot program written in FANUC LS format from CLS data. The proposed reverse and forward post processors allow an industrial machining robot to work based on the NC data that have been used for, e.g., a five axis NC machine tool with a tilting head.

2   Reverse Post Processor

Generally speaking, the role of a post processor is to generate proper NC data for a target machine tool from CLS data made using the main processor of CAD/CAM. Post-processed NC data can be used only for the target NC machine tool, so that there is a need to reuse the NC data for other types of NC machine tools and/or industrial robots. Because there is often a circumstance in related manufactures, i.e., CAD model data and CLS data made in the past no longer remain to regenerate NC data due to the division of labor and process, but only NC data have been saved. That is the necessity of the reverse post processor.

In this section, a data converter called the reverse post processor is proposed to cope with the need. The reverse post processor enables to regenerate original CLS data from the post-processed NC data as shown in Fig. 1. Let's consider the conversion a bit more in detail. Although position components in an NC file are easily converted into a CLS file, orientation components should be noted

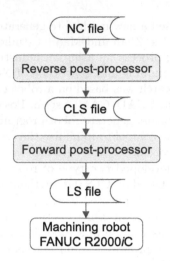

Fig. 1. The role of the proposed reverse and forward post processors.

because of the different data structure between two rotational angles in NC data and normal direction vectors in CLS data. In the proposed reverse post, a pair of two rotational angles B and C [deg.] around y- and z-axes in NC data is converted to a normal direction vector $\boldsymbol{n} = [n_x \ n_y \ n_z]^T$ in CLS data through

$$n_x = \sin\left(B \times \frac{\pi}{180}\right) \cos\left(C \times \frac{\pi}{180}\right) \tag{1}$$

$$n_y = \sin\left(B \times \frac{\pi}{180}\right) \sin\left(C \times \frac{\pi}{180}\right) \tag{2}$$

$$n_z = \cos\left(B \times \frac{\pi}{180}\right) \tag{3}$$

which, e.g., enables the conversion as

G01X29.314Y47.266Z − 76.915B − 38.248C − 20.494

⇓

GOTO/29.314, 47.266, −76.915, −0.5798851, 0.2167410, 0.7853385

Figure 2 illustrates the detailed conversion from a "G02" or "G03" code in NC data to a "CIRCLE" statement in CLS data, in which $\boldsymbol{X}_A = [X_A \ Y_A \ Z_A]^T$ and $\boldsymbol{X}_B = [X_B \ Y_B \ Z_B]^T$ are the position vectors on an arc's start and end points A, B, respectively. $\boldsymbol{X}_O = [X_O \ Y_O \ Z_O]^T$ is the center O of the arc. The one line including a G02 code in NC data forms the arc as shown in Fig. 2. "G02" is the code for a CW (clockwise) motion, in which the tool tip moves from the current point to the end point B along the arc with the center O. As can be seen, three statements consisting of "GOTO", "CIRCLE" and "GOTO" are generated from one "G02" code. Direction vector $[N_X \ N_Y \ N_Z]^T = [0 \ 0 \ 1]^T$ and $[N_X \ N_Y \ N_Z]^T = [0 \ 0 \ -1]^T$ correspond to "G02" and "G03", respectively. Through the above process, NC data can be reversely converted into original CLS data.

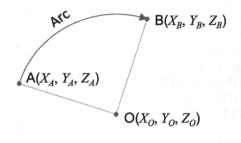

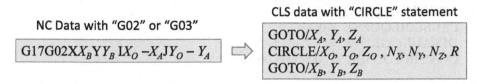

NC Data with "G02" or "G03"

$$G17G02XX_BYY_B IX_O - X_A JY_O - Y_A$$

CLS data with "CIRCLE" statement

$$GOTO/X_A, Y_A, Z_A$$
$$CIRCLE/X_O, Y_O, Z_O, N_X, N_Y, N_Z, R$$
$$GOTO/X_B, Y_B, Z_B$$

Fig. 2. Detailed reverse process for an arc entity.

3   Forward Post Processor for an Industrial Robot

3.1   Robot Program Written in FANUC LS Format

Fig. 3 illustrates the overall file structure of FUNUC robotic program whose file extension is given by "ls". The LS file is composed of header, program and data parts. First of all, the header part consists of the following statements.

```
/PROG fs
/ATTR
OWNER = MNEDITOR;
COMMENT = "Robot";
PROG_SIZE = 1030;
CREATE = DATE 16-06-12 TIME 23:33:50;
MODIFIED = DATE 16-06-12 TIME 23:33:50;
```

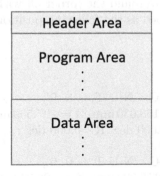

Fig. 3. FANUC robotic program (.ls) written in ASCII file format.

```
FILE_NAME = ;
VERSION = 0;
LINE_COUNT = 20;
MEMORY_SIZE = 1318;
PROTECT = READ_WRITE;
TCD: STACK_SIZE = 0,
TASK_PRIORITY = 50,
TIME_SLICE = 0,
BUSY_LAMP_OFF = 0,
ABORT_REQUEST = 0,
PAUSE_REQUEST = 0;
DEFAULT_GROUP = 1, *, *, *, *;
CONTROL_CODE = 00000000 00000000;
```

The program area is written as the following commands.

```
 1: JP[1] 100% CNT50;
 2: JP[2] 100% CNT50;
 3: LP[3] 250 mm/sec CNT50;
 4: LP[4] 250 mm/sec CNT50;
 5: LP[5] 250 mm/sec CNT50;
 6: LP[6] 250 mm/sec CNT50;
 7: LP[7] 250 mm/sec CNT50;
 8: LP[8] 250 mm/sec CNT50;
 9: LP[9] 250 mm/sec CNT50;
10: LP[10] 250 mm/sec CNT50;
11: LP[11] 250 mm/sec CNT50;
12: LP[12] 250 mm/sec CNT50;
                ⋮
```

Where "JP" means rapid motion like "G00" in NC file. "JP" is effective to rapidly manuever an end mill to the position near the initial cut for removing material. Also, "LP" means feed motion with a specified feed rate [mm/sec] like "G01". "CNT50" is added to round the corner off with a radius 50 mm.

The data area is described as the following position and orientation components in Cartesian space.

```
/POS
P[1] GP1:
 UF : 1, UT : 1, CONFIG : 'N U T, 0, 0, 0',
 X =−100.000 mm, Y =158.030 mm, Z =−97.559 mm,
 W =143.130 deg, P =0.000 deg, R =0.000 deg ;
P[2] GP1:
 UF : 1, UT : 1, CONFIG : 'N U T, 0, 0, 0',
 X =−100.000 mm, Y =156.992 mm, Z =−98.944 mm,
 W =180.000 deg, P =0.000 deg, R =0.000 deg ;
```

P[3] GP1:
UF : 1, UT : 1, CONFIG : 'N U T, 0, 0, 0',
X =−100.000 mm, Y =156.335 mm, Z =−99.672 mm,
W =180.000 deg, P =0.000 deg, R =0.000 deg ;
P[4] GP1:
UF : 1, UT : 1, CONFIG : 'N U T, 0, 0, 0',
X =−100.000 mm, Y =155.550 mm, Z =−100.258 mm,
W =180.000 deg, P =0.000 deg, R =0.000 deg ;
P[5] GP1:
UF : 1, UT : 1, CONFIG : 'N U T, 0, 0, 0',
X =−100.000 mm, Y =154.665 mm, Z =−100.679 mm,
W =180.000 deg, P =0.000 deg, R =0.000 deg ;

$$\vdots$$

Where X, Y and Z are the components of a position vector; W, P and R mean the rotational angles around x-, y- and z-axes as roll, pitch and yaw angles, respectively.

3.2 Forward Post Processor

The proposed post processor yields FANUC robot programs (.ls) from CLS data. As an example, CLS data are listed as

GOTO/0.025,0.000,0.000,0.00833333,0.00000000,0.99996527
GOTO/0.006,−0.024,0.000,−0.00185434,0.00812439,0.99996527
GOTO/0.025,0.000,0.000,−0.00833333,0.00000000,0.99996527
GOTO/−0.006,0.024,0.000,0.00185434,−0.00812433,0.99996527
GOTO/−0.025,0.000,0.000,0.00833333,0.00000000,0.99996527
GOTO/0.006,−0.034,0.000,−0.00213149,0.01141874,0.99993253

$$\vdots$$

Where one "GOTO" statement has position vector $\boldsymbol{x} = [x \ y \ z]^T$ and normal vector $\boldsymbol{n} = [n_x \ n_y \ n_z]^T$. Therefore, the rotation angles w, p and r [deg.] need to be calculated from the components of normal direction vector $\boldsymbol{n}$. In case of FANUC R2000iC, w and p are obtained by

$$w = -\text{atan2}(-n_y, -n_z) \qquad (4)$$

$$p = \text{asin}(n_x) \qquad (5)$$

The yaw angle r can be fixedly set to 0 because of the axis-symmetry of a ball-end mill attached to the robot's arm tip.

The proposed post processor only provides a rigorous data format conversion from CLS data to LS data. Although the kinematic problems about joints' movable ranges and robot singularity have not been considered yet, the users of the post processor have only to adequately set and use work coordinate systems according to target workpieces with various dimensions and shapes.

4 Limitation on Usage of LS File

The authors noticed a serious limitation on usage of LS file, i.e., LS file consisting of more than 9000 statements in program area can not be accepted by the RoboGuide and actual FANUC industrial robots. When a model with complex free-foamed surfaces is designed and main-processed using a 3D CAD/CAM, the outputted statements in the CLS file frequently exceeds 9000 lines. That is the reason why this limitation is a serious problem in order to realize an easy-to-use data interface between CAD/CAM systems and FANUC industrial robots.

To cope with this problem, the proposed post processor produces divided LS files consisting of less than 9000 statements in program area. For example, a CLS file (bspline.cls) consisting of far more than 9000 statements is firstly converted into a complete LS file (bspline.ls), then in addition, some divided LS files with less than 9000 statements are generated as "bspline1.ls", "bspline2.ls", $\cdots$, "bspline5.ls", i.e., the contents in bspline.ls are divided into five LS files as shown in Fig. 4. Due to this attentive function, a large sized CLS file can be smartly post-processed into divided LS files for FANUC industrial robots. At the same time as the conversion, furthermore, a list file named as bspline_list.ls is conveniently generated for executing long-time operation. Since the list file includes all names of divided LS files, all divided LS files can be executed in succession if the list file is selected. However, we strongly hope that this undesirable limitation will be improved by FANUC CORPORATION as soon as possible.

Finally, an experiment consisting of a reverse post process, forward post process and actual machining was conducted for evaluation using an impeller model as shown in Fig. 5. In the experiment, NC data including variable axes codes, e.x., for a five-axis NC machine tool with a tilting head were first generated from the model, then which were applied to the input of the reverse post process shown in Fig. 1, and furthermore, LS data outputted from the forward post processor were given to the FANUC industrial robot. It was verified as shown in Fig. 6 that the post processed LS data, i.e., FANUC robot program, were successfully available for the robot machining.

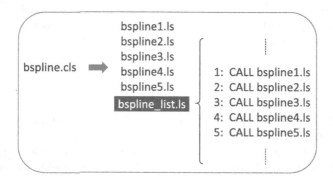

Fig. 4. The proposed post processor yields a list file (e.g., bspline list.ls) including all divided LS files for dealing with big-sized CLS data.

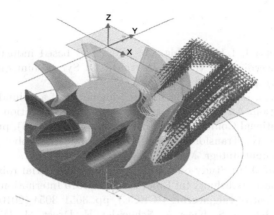

Fig. 5. An impeller model for evaluation. NC data generated from the model were applied to the input shown in Fig. 1.

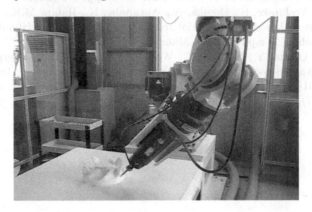

Fig. 6. Machining scene using the post-processed LS data (CLS data → LS data) after the reverse post process (NC data → CLS data).

5 Conclusions

This paper presented methods that can be widely and easily applied to data transformation process of machining robots. A reverse post processor was first introduced to regenerate original CLS (Cutter Location Sourse) data from post-processed NC (Numerical Control) data including variable axes codes. Then, a promising forward post processor was proposed to produce robot programs written in FANUC LS format from the CLS data. The proposed reverse and forward post processors allowed an industrial machining robot to work based on the NC data with variable axes information that had been used for, e.g., a five axis NC machine tool with a tilting head.

Acknowledgments. This work was supported by JSPS KAKENHI Grant Number JP16K06203.

References

1. Solvang, B., Refsahl, L.K., Sziebig, G.: STEP-NC based industrial robot CAM system. In: Proceedings of the 9th International Symposium on Robot Control (SYROCO 2009), pp. 361–366 (2009)
2. Andres, J., Gracia, L., Tornero, J.: Inverse kinematics of a redundant manipulator for cam integration. An industrial perspective of implementation. In: Proceedings of IEEE International Conference on Mechatronics (ICM 2009), pp. 1–6 (2009)
3. Chen, H., Sheng, W.: Transformative CAD based industrial robot program generation. Robot. Comput.-Integr. Manuf. **27**–5, 942–948 (2011)
4. Zhan, J.M., Mao, J.H.: Study on the algorithm for industrial robot programming in free-form surfaces polishing. In: Proceedings of 2010 International Conference on Electrical and Control Engineering (ICECE), pp. 3051–3054 (2010)
5. Posada, J.R.D., Kumar, S., Kuss, A., Schneider, U., Drust, M., Dietz, T., Verl, A.: Automatic programming and control for robotic deburring. In: Proceedings of 47th International Symposium on Robotics (ISR 2016), pp. 1–8 (2016)
6. Nagata, F., Yoshitake, S., Otsuka, A., Watanabe, K., Habib, M.K.: Development of CAM system based on industrial robotic servo controller without using robot language. Robot. Comput.-Integr. Manuf. **29**(2), 454–462 (2013)
7. Nagata, F., Okada, Y., Kusano, T., Watanabe, K.: CLS Data interpolation with spline curves and its post processing for generating a robot language. In: Proceedings of the 5th IIAE International Conference on Industrial Application Engineering (ICIAE 2017), pp. 358–363 (2017)

Research on Robot Grinding Technology Considering Removal Rate and Roughness

Shaobo Xie[✉], Shan Li, Bing Chen, and Junde Qi

Key Laboratory of Contemporary Design and Integrated Manufacturing
Technology of Ministry of Education, Northwestern Polytechnical University,
Xi'an 710072, China
xieshaobo@mail.nwpu.edu.cn

Abstract. In order to solve the problem of poor consistency and low processing efficiency of artificial grinding, this paper establishes a six-axis robot automatic grinding platform. In the first section of the paper, the qualitative relationship between the process parameters and the grinding quality could be learned from the single factor experiment, and the primitive range of the process parameter domain is obtained. On the basis, an orthogonal experiment is carried out, and a quantitative regression empirical model is established. Then the parameter sensitivity function is deduced based on the model. In regards to the grinding quality and stability constraints, the process parameter domain optimization is carried out. Finally, considering the influence of grinding attitude, the optimal attitude angle is found in the range of process parameters. The results show that the blade grinding system and the grinding scheme are effective.

Keywords: Robot grinding system · Grinding quality · Grinding angle · Optimized parameters domain · Turbine blades

1 Introduction

Turbine blades are one of the key components of Aero-engine and have a very important impact on the performance of an engine. At present, the turbine blade grinding is mainly rely on manual, the product pass rate of current is low and the consistency is poor, which directly reduces the blade production efficiency. However the unique multi-joint structure of the robot and the high processing flexibility make it very suitable for the grinding of the complex surface of the blade. The robot grinding product has high consistency and high machining precision, which can greatly reduce the labor cost and improve the automatic production efficiency [1].

At present, the research on grinding is mainly focused on the process system research or process parameters optimization. Tsaai and Huang [2] optimized the abrasive grain size, polishing pressure, tool speed and feed rate in the robot grinding system, and analyzed the influence of each parameter on surface roughness and grinding efficiency. Song et al. [3] studied the relationship between the material removal rate of the robot belt grinding system and the workpiece curvature, contact force, robot speed and belt speed, and established a robot grinding adaptive model. Sabourin et al. [4] introduced a method for grinding a turbine blade by a robot. By adjusting the process

© Springer International Publishing AG 2017
Y. Huang et al. (Eds.): ICIRA 2017, Part II, LNAI 10463, pp. 79–90, 2017.
DOI: 10.1007/978-3-319-65292-4_8

parameters, the surface roughness can be reduced to Ra = 0.1 μm. Ahn et al. [5] gives the formula for calculating the surface roughness, indicating that the surface roughness value is related to the tool speed, feed rate, grinding times and the thresholds that the abrasive can achieve. Tsai et al. [6] proposed a uniform material removal model for the constant force and constant speed grinding method. The results of experiments show that the method can reduce the third of the profile error.

In order to keep the grinding force constant, many scholars have done a lot of research. Pessoles and Tournier [7] proposed a 5-axis CNC grinding system based on grinding force and displacement control. Brecher et al. [8] developed a force control device and a control system for robot grinding. Pan and Zhang [9], aiming at the flexible grinding of industrial robots, created a grinding force control method has proven to obtain good surface quality.

Generally the existing force control system for the blade grinding process platform is complicated, and its control is relatively difficult and the scalability is also poor. Moreover, in current research, the parameters on the grinding process range are narrow and are mainly concerned with abrasive grain size, grinding force, grinding line speed and feed rate of four process parameters. The research about the influence of a grinding angle on a grinding surface quality is relatively limited.

In view of the above situation, this paper introduces the flexible adaptive force controller -ACF and constructs a flexible robot grinding system. Based on the system, the influence of process parameters including the attitude angle is analyzed and the optimization selection is observed. The first part introduces the composition of the flexible grinding system of the robot and the basic working principle of the flexible force controller; in the second part, the preliminary selection and optimization of process parameters are carried out, and the influence of grinding angle on grinding effect is analyzed, and the optimal parameters of grinding process and the optimal combination scheme are obtained. In the third part, the preliminary grinding experiments of the turbine blades are carried out, and the effectiveness of the flexible grinding system and the grinding scheme are verified by taking the surface roughness and the material removal rate as the consideration.

2   Robotic Flexible Grinding System

The flexible robotic grinding system is mainly built for the grinding process of the turbine blades. The blades are clamped on the special fixture of the table. The adaptive force controller and the grinding tool are connected to the end flange of the six-axis robot. Blade grinding is mainly done in the form of manual teaching or offline control. The main purpose is to make the material removal of the blade constant and to eliminate the milling lines in order to meet the requirements of surface accuracy and surface roughness. The main hardware of the grinding system consists of the robot body, the flexible force controller, the electric spindle, the grinding tool, the blade and the fixture as shown in Fig. 1.

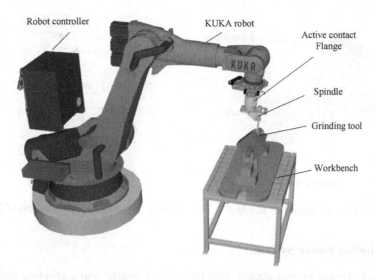

Fig. 1. The hardware of the robot grinding system

2.1 Robot Motion Control System

The robot motion control system consists of a KUKA KR210 robot body and robot control cabinet KRC2. The KUKA robot body is the main movement part of the grinding operation, which can be divided into three parts: 1. Robot base, which is used to fix the robot body; 2. The arms, which consist of six connecting rods by turning pairs; 3. Robot control cabinet, through which you can make the grinding tool to move in accordance with the program.

2.2 Adaptive Force Control System

Grinding force is a very important parameter in the grinding process, which has an important effect on the grinding quality and efficiency. Therefore, it is necessary to monitor and control the grinding force in the process accurately. This paper will introduce an adaptive force controller-ACF, which is a flange connected robot and grinding tool that can control the robot's end output to a constant force.

Flexible force control system first need to detect the grinding force, and then change the pressure from the proportional valve output to control the cylinder action, in order to achieve the adaptive compensation of the grinding movement. When a certain force is set, the amount of expansion and contraction of the ACF is detected, and the ACF control system controls the magnitude of the air pressure in the cylinder. The workflow is shown in the Figs. 2 and 3.

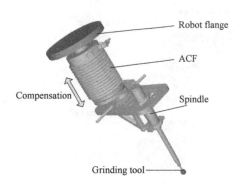

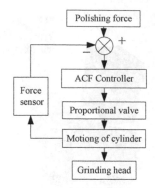

Fig. 2. The installation of the ACF   **Fig. 3.** The force control process of the ACF

2.3   Grinding Power System

The grinding power system consists of a high-speed spindle and a grinding tool that is held at the end of the motor spindle. The flexible force control system is installed at the end of the adaptive force controller to control the grinding speed and grinding force of the grinding tool. The most common grinding tools are rubber, leather, wool grinding heads, grinding wheels, sandpaper, etc. The base of a rubber grinding head generally uses a highly elastic material, which makes it flexible. Widely used in the field of aerospace, the rubber grinding head can properly compensate for the robot path error or bypass the processing error of the blade.

3   Robot Flexible Grinding Process

In order to study the relationship between the grinding process parameters and the surface quality of the blade, this section will discuss the selection of the appropriate abrasive and the initial process parameter field by single factor experiment. After that, the process parameters are selected in the grinding process parameter field for orthogonal experiment, and the measurement results and the grinding process parameters are analyzed by regression analysis. Then the sensitivity of the grinding process parameters has been analyzed, and the sensitive relationship between the surface removal rate, surface roughness and grinding parameters are obtained. The sensitivity of the process parameters is calculated by the sensitivity calculation of the sensitive process parameters, and the range of the sensitivity of the process parameters is obtained. Finally, the influence of the grinding attitude angle on grinding effect is analyzed by orthogonal experiment, and the validity of turbine blade is confirmed.

3.1   Grinding Power System

According to the experience manual and other papers [10, 11], the initial grinding process parameters range are robot feed rate $Vw = 10–70$ mm/s, grinding line speed $Vs = 5–25$ m/s, grinding force size $Fn = 10–50$ N.

Abrasive selection is an important process in conducting experimental research. In general, it depends on the characteristics of the workpiece, including the shape of the workpiece, size and material grinding performance. TC11 titanium alloy is difficult to be processed, in the grinding process, and the grinding force is relatively large and grinding temperature is relatively high, potentially causing adhesion and chilling phenomenon. Therefore, this paper chooses the grinding tool of silicon carbide material, which has the characteristics of high hardness and good grinding performance. In order to verify the effect of different size grinding head on the processing effect, the selected particle size was 240 #, 400 # and 600 # respectively.

3.2   Primary Experiment of Grinding Abrasive Based on Single Factor Experiment

Figure 4 shows the influence of the grinding characteristics and the grinding process parameters on the surface removal depth of the material.

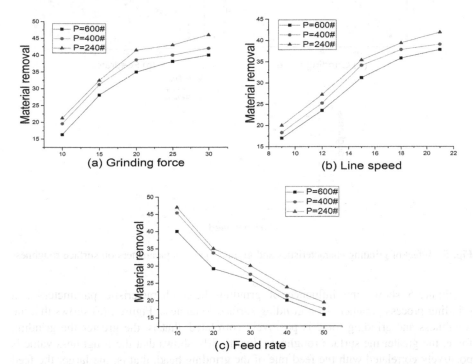

Fig. 4. Effect of grinding characteristics and grinding process parameters on depth of material removal

Figure 4(a) shows that the maximum removal depth increases with the grinding pressure. Figure 4(b) shows that the maximum removal depth of the material is positively related to the line speed, that is, the faster the grinding line speed, the larger the material removal depth. Figure 4(c) shows that the maximum removal depth is inversely related to the feed rate of the grinding head. The greater the feed rate, the lower the maximum removal depth of the material.

Based on the comprehensive analysis of three graphs, it can be found that the higher the particle size of the grinding head, the lowest depth of material removal, and the depth of material removal is inversely related to the abrasive grain size. The reason is that when the abrasive grain size is higher, the abrasive grain distribution density is bigger. In the case of the same process parameters, abrasive cutting depth is small, and elastic deformation will occur without material removal, so the material removal depth is relatively small.

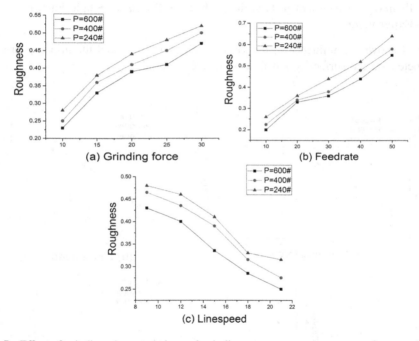

Fig. 5. Effect of grinding characteristics and grinding process parameters on surface roughness

Figure 5 shows the influence of grinding head characteristic parameters and grinding process parameters on grinding surface roughness. Figure 5(a) shows that the roughness and grinding force is positively correlated, that is, the greater the grinding force, the greater the surface roughness. Figure 5(b) shows that the roughness value is positively correlated with the feed rate of the grinding head, that is, the larger the feed rate is, the larger the surface roughness value is. As can be seen from Fig. 5(c), the roughness value is inversely related to the grinding line speed, that is, the larger the grinding line speed is, the smaller the surface roughness value is.

Based on these three images, compared with different grain size of the grinding head for the material surface roughness, it is convenient to find out that the higher the particle size of the grinding head is, the lower the surface roughness is. That means, the surface roughness of the material is inversely related to the abrasive grain size. The higher the abrasive grain size and the abrasive grain density is, the smaller the abrasive grain cutting depth in the same process parameters is. To sum up, considering the original definition of the surface roughness, it is easy to get the conclusion that the roughness value is relatively small.

The above-mentioned single factor experiments show that with the same grinding process parameters, the use of lower particle size of the grinding material paves the way to a higher depth of removal and a larger roughness. Taking into account the contour accuracy and roughness requirements of the blade shape, select the grinding tool with a particle size of 400 #. The initial range of other process parameters is: grinding head feed speed Vw = 10–40 mm/s, grinding head line speed Vs = 12–21 m/s, grinding force Fn = 10–30 N.

3.3 Effect of Grinding Process Parameters on Surface Quality Based on Orthogonal Experiment

In this section, the orthogonal experiment parameters are selected in the range of the grinding process parameters obtained by the single factor experiments, and the results of the measurement and the grinding process parameters are analyzed by regression analysis to establish the empirical formula of surface feature. Then, the sensitivity of the process parameters of the grinding surface is analyzed, and the sensitivity relation model between them is obtained. On this basis, the range of the sensitivity of the process parameters is obtained. These all support the optimization of the subsequent process parameter fields.

In order to establish the empirical formula, the multiple linear regression analysis method was used to fit the grinding process parameter data and the surface characteristic test data, and was established. As shown in Eq. (1)

$$
\begin{aligned}
R_a &= 0.4331 V_w^{0.273} F^{0.155} V_s^{-0.108} \\
H(0) &= 231.071 V_w^{-0.164} F^{0.173} V_s^{0.098}
\end{aligned}
\tag{1}
$$

According to the surface characteristic data, the relative sensitivity of the process parameters is defined: the relative sensitivity is the exponential part of the exponential empirical formula. The relative sensitivity of the surface features to the grinding parameters can be calculated as Eq. (2):

$$
\begin{aligned}
S_{R_a}^{\prime V_w} &= 0.273, S_{R_a}^{\prime F} = 0.155, S_{R_a}^{\prime V_s} = -0.108 \\
S_H^{\prime V_w} &= -0.164, S_H^{\prime F} = 0.173, S_H^{\prime V_s} = 0.098
\end{aligned}
\tag{2}
$$

In the range of experimental parameters, = 20 mm/s, = 15 m/s, F = 20N, surface roughness, surface material removal depth of the grinding head feed rate, grinding force, grinding head line speed absolute sensitivity model. As in Eq. (3):

$$S_{R_a}^{V_w} = 0.14V_w^{-0.727}, S_{R_a}^F = 0.114F^{-0.845}, S_{R_a}^{V_s} = -0.169V_s^{-1.108}$$
$$S_H^{V_w} = -82.97V_w^{-1.164}, S_H^F = 31.88F^{-0.827}, S_H^{V_s} = 23.26V_s^{-0.902}$$

(3)

From the relative sensitivity analysis, it can be seen that the surface roughness is the most sensitive to the change of the feed rate of the grinding head when the TC11 titanium alloy is grinding, and the material removal depth is the most sensitive to the grinding force.

According to Eq. (3), the absolute sensitivity curve of the surface roughness to the grinding process parameters is obtained, as is shown in Fig. 6(a–c).

As can be seen from Fig. 6(a), the absolute sensitivity of the [18N, 30N] in the grinding force interval is less than the [10N, 18N] interval. That is, when the grinding force changes from 18N to 30N, the surface roughness curve is smooth. Similarly, the Fig. 6(b) shows that when the feed rate changes from 20 mm/s to 40 mm/s, the surface roughness curve is relatively smooth. Figure 6(c) shows that when the grinding tool speed transforms from 16 m/s to 21 m/s, the surface roughness curve is relatively also smooth.

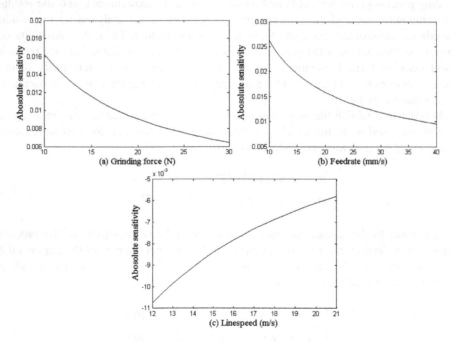

Fig. 6. The absolute sensitivity of surface roughness to grinding parameters

Similarly, the absolute sensitivity curve of the surface removal depth to the grinding parameters can be obtained. In the grinding force interval [18N, 30N] the absolute sensitivity value is less than [10N, 18N] interval, that means, when grinding force changes from 18N to 30N, the surface removal depth curve is relatively smooth.

The same, when the feed rate changes from 20 mm/s to 40 mm/s, the surface removal depth curve is relatively smooth. When the grinding line speed changes from 16 m/s to 21 m/s, the curve of the removal depth of the surface becomes relatively smooth.

3.4 Optimization of Grinding Process Parameters

Based on the single factor grinding process parameter experiment, the characteristic parameters of the grinding tool are firstly selected as follows: silicon carbide grinding head, particle size 400 #. In the optimization of the grinding parameters, it is necessary to consider not only whether the parameters are in the stable interval but also the requirements of the grinding surface characteristics and the sensitivity of the grinding parameters to the grinding surface characteristics. As shown in Table 1.

Table 1. Preferably grinding parameters of the process parameters interval

Category	Parameter range	Roughness range	Material removal
Feed rate (mm/s)	[20, 40]	<0.5	<40 μm
Grinding force (N)	[18, 30]	<0.5	<40 μm
Grinding line speed (m/s)	[16, 21]	<0.5	<40 μm

Based on the analysis above, this paper obtains the direction of the relatively stable grinding process parameters: grinding head feed speed Vw → 30 mm/s, grinding line speed Vs → 18 m/s, grinding force → 24N. Under these grinding conditions, with the use of 400 # silicon carbide grinding head, the TC11 alloy plate material was carried out to the grinding test. The surface roughness value was 0.26 μm and the single removal depth was 31 μm.

3.5 Research on Grinding Attitude Angle Based on Optimized Parameter Interval

It can be found that the quality of the surface of the workpiece would be significantly different when the grinding tool was tilted at a different grinding point from the surface of the workpiece. Different tilt angles can have a complex effect on the speed of the grinding head, the size of the contact area, the condition of the chip, and the wear. In order to study the relationship between the grinding attitude angle and the processing effect. The experiment parameters that affect the grinding quality of the robot are grinding force, grinding angle, grinding feed rate, grinding line speed, track spacing, etc. As shown in Fig. 7.

The surface roughness value of the sample after grinding was used as the response index, and the electric screw speed, the grinding force, the line spacing and the inclination angle were used as the four factors to carry out the 4-factor-3-level orthogonal test. During the grinding process, the surface was cooled with air, the depth of depression was 1 mm, and the feed rate was 1000 mm/min.

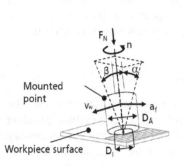

Fig. 7. Grinding process modeling

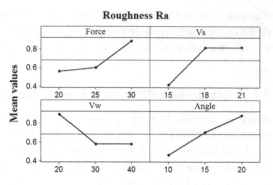

Fig. 8. Experimental results

As can be seen from Fig. 8, when the linear velocity is 21 m/s during the grinding process, the grinding force is 20 N, the feed rate is 30 mm/s, the inclination angle is 10°, the surface roughness can be minimized, with the uniform of the surface and no obvious texture.

4   Simulation and Experiment of Blade Grinding

4.1   Robot Grinding Platform

The turbine blade was selected as the experimental subject, whose material is titanium alloy TA 11. Its surface has a serious milling texture and the surface roughness value could reach above Ra0.9. According to processing requirements, the blade roughness after grinding must be less than 0.4. The hardware model and performance indicators of the construction process system are shown in the following Table 2.

Table 2. List of experimental hardware

Hardware	Model	Performance indicators
Robot	Kuka KR210-2	6 degrees of freedom; repeat positioning accuracy 0.06 mm
Force controller	FerRobotics ACF 110-10	Force accuracy 1N; Reaction time 4 ms
Electric spindle	NSK E4000	Max. speed 40000 min^{-1}; Max. output 1,200 W
Grinding tool	Artifex EK120	Tool diameter ⌀:−5 mm, chamfer

Turbine blade surface is a complex surface and its processing is particularly difficult. In the process of blade grinding, in order to make the grinding wheel axis and the blade surface method to maintain a certain position, it is nessary to adjust the attitude of the robot constantly. This process must rely on CAM software and robot offline programming software. Firstly, the blade CAD model is obtained and the grinding area is selected. Then, the grinding trajectory is set according to the process parameters. Finally, the robotic program is used to generate the robot executable program.

4.2   Quality Inspection of Blade

After grinding, the contour error of turbine blades was tested. At the time of testing, the specimen body was divided into several sections, and the precision of the blade shape in each section was about ±0.04 mm. In order to further verify the processing effect of the grinding process on the blade surface, the Infinite Focus G4 automatic zoom 3D surface measuring instrument was used to compare the microstructure before and after grinding. As shown in Fig. 9.

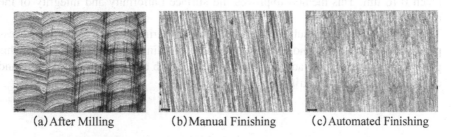

| (a) After Milling | (b) Manual Finishing | (c) Automated Finishing |

Fig. 9. Surface morphology under different grinding processes

Figure 10(a) shows more clearly of the different processing methods under the blade surface quality, after grinding the surface texture is basically eliminated. At the same time, in this paper a comparison with the robot grinding and artificial polishing after the surface roughness has completed.

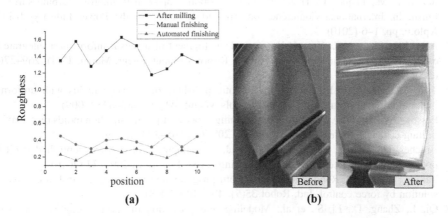

(a)   (b)

Fig. 10. (a) Roughness in three different grinding modes (b) The grinding quality of the blade after the preferred process parameters

After grinding, it can be seen from the Fig. 10(b) that the surface roughness of the blade has been significantly reduced. Robot grinding is better than manual grinding, and robots can have obvious advantages in efficiency.

5 Conclusions

1. This paper proposes a robot flexible grinding system which composed of robot and control cabinet, ACF flexible force controller, spindle, grinding tools and workbench. It is proven to be able to achieve automatic grinding and improve the grinding efficiency.
2. A grinding process is proposed based on the robot grinding platform, and a suitable grinding condition and process parameters are obtained. The texture left after the milling of the blade is substantially eliminated and the minimum roughness Ra can reach 0.16 μm. This method improves the surface uniformity and integrity of the blade grinding.
3. The turbine blade was subjected to grinding experiment with suitable process parameters. The results show that the flexible grinding system developed in this paper is suitable for the blade grinding process, which can provide an effective and quick way for the blade grinding.

References

1. Wu, S., Kazerounian, K., Gan, Z., et al.: Int. J. Adv. Manufact. Technol. **64**, 447 (2013). https://doi.org/10.1007/s00170-012-4030-6
2. Tsai, M.J., Huang, J.F.: Efficient automatic polishing process with a new compliant abrasive tool. Int. J. Adv. Manuf. Technol. **30**, 817–827 (2006)
3. Song, Y., Liang, W., Yang, Y.: A method for grinding removal control of a robot belt grinding system. J. Intell. Manuf. **23**(5), 1903–1913 (2012)
4. Sabourin, M., Paquet, F., Hazel, B., et al.: Robotic approach to improve turbine surface finish. In: International Conference on Applied Robotics for the Power Industry. IEEE Xplore, pp. 1–6 (2010)
5. Ahn, J.H., Shen, Y.F., Kim, H.Y., et al.: Development of a sensor information integrated expert system for optimizing die polishing. Robot. Comput. Integr. Manuf. **17**(4), 269–276 (2001)
6. Tsai, M.J., Huang, J.F., Kao, W.L.: Robotic polishing of precision molds with uniform material removal control. Int. J. Mach. Tools Manuf. **49**(11), 885–895 (2009)
7. Pessoles, X., Tournier, C.: Automatic polishing process of plastic injection molds on a 5-axis milling center. J. Mater. Process. Technol. **209**(7), 3665–3673 (2009)
8. Brecher, C., Tuecks, R., Zunke, R., et al.: Development of a force controlled orbital polishing head for free form surface finishing. Prod. Eng. **4**(2), 269–277 (2010)
9. Pan, Z., Zhang, H.: Robotic machining from programming to process control: a complete solution by force control. Ind. Robot **35**(5), 400–409 (2008)
10. Qi, J., Zhang, D., Li, S., et al.: Modeling and prediction of surface roughness in belt polishing based on artificial neural network. Proc. Inst. Mech. Eng. Part B: J. Eng. Manuf. 0954405416683737 (2017). https://doi.org/10.1177/0954405416683737
11. Huai, W., Tang, H., Shi, Y., et al.: Int. J. Adv. Manuf. Technol. **90**, 699 (2017). https://doi.org/10.1007/s00170-016-9397-3

Electromechanical Coupling Dynamic Model and Speed Response Characteristics of the Flexible Robotic Manipulator

Yufei Liu[1,2(✉)], Bin Zi[2], Xi Zhang[3], and Dezhang Xu[1]

[1] School of Mechanical and Automotive Engineering,
Anhui Polytechnic University, Wuhu 241000, China
Liuyufeiahpu@126.com
[2] School of Mechanical Engineering,
Hefei University of Technology, Hefei 230009, China
[3] Engineering Research and Training Center, Anhui Polytechnic University,
Wuhu 241000, China

Abstract. The flexible robotic manipulator, which has the advantages of lightweight, flexible operation and low energy consumption, is a typical electromechanical coupling system containing the driving system, transmission system and flexible manipulator. Considering the driving motor, transmission system and flexible manipulator as an integrated object, the electromechanical coupling dynamic model of the flexible robotic manipulator system (FRMS) was constructed based on the overall coupling relationship and electromechanical dynamics analysis approach. To reveal the electromechanical coupling mechanism of the FRMS, the speed response characteristics under electromechanical coupling effects are presented. The results indicate that the electromechanical coupling factors have significant impacts on the dynamic property of the FRMS, which are meaningful for the design and control of flexible robotic manipulators.

Keywords: Robotic · Flexible manipulator · Electromechanical coupling dynamic

1 Introductions

The robotic manipulators have widely application, such as mechanical processing, precision assembly, loading and spraying [1, 2]. Because the operation task is completed through the manipulator and its actuators, the structure and dynamic characteristics of the robotic manipulator have important effect on the performance of the whole system. As the modern robot technology developing to lower energy consumption, higher speed and higher precision, the robotic manipulator should be able to sustain the operation accuracy over a long period of time in the process of performing operations [3–5]. In order to realize the position precision of the actuators, traditional robotic manipulators are usually designed with rigid structure. However, the structural vibration during operations, start-stop and motion transformation, especially in high-speed operation, seriously reduces the operation accuracy and precision of the

© Springer International Publishing AG 2017
Y. Huang et al. (Eds.): ICIRA 2017, Part II, LNAI 10463, pp. 91–100, 2017.
DOI: 10.1007/978-3-319-65292-4_9

robotic manipulator [2]. Meanwhile, owing to the poor vibration controllability of the rigid manipulator, the structural vibration may persist for a long time which obviously limits the efficiency of the system. Besides, because the materials of the rigid manipulator are heavy, it will consume relatively more energy in high speed motion and wide range of space operation tasks [1, 6].

With the rapid development of modern manufacturing equipment, there has increasing attention been focused on the flexible manipulator which is a typical flexible structure [7]. The flexible manipulator can significantly reduce the weight of the structure while assure multi-function and has been widely used in the large lightweight structure, extending from the field of aeronautics and astronautics to the micro-operation robot and precision manufacturing with the development of modern industry and engineering applications [1, 8]. Due to the use of lightweight materials, however, the flexible manipulator usually has low stiffness and damping and exhibits residual vibrations during the operation. Therefore, the dynamic characteristics and control strategies for the flexible manipulator are the key problems for realizing the full advantages of its lightweight.

The flexible robotic manipulator is a typical complex electromechanical system which has complex electromechanical coupling relationship between the servo drives and the actuators [9]. For example, in the permanent magnet AC servo system, there is obvious coupling relationship between the mechanical parameters of the structure and the electromagnetic parameters of the driving system, whose concrete form can be divided into the coupling between the harmonic current in the armature circuit of the driving system and the transmission system, the coupling between the control parameters of the motor speed regulation system and the mechanics parameters of the servo system as well as the coupling between the servo system and the load system. These electromechanical coupling relationships of the flexible robotic manipulator system (FRMS) are easy to arouse the system vibration and affect the system dynamic characteristics [10]. Moreover, the systematic excitation caused by the coupling factors will be more prominent for high-speed and light structures, due to its low mode property and flexible factors, flexible robotic manipulator is easy to cause low resonance phenomenon which has obvious influence on the dynamic characteristics. However, most of the existing investigations considering the dynamic modeling and vibration control of the flexible robotic manipulator usually ignore the effect of the dynamic characteristics of driving system [11–13]. In order to accurately reflect the real dynamic characteristics of the FRMS, the electromechanical coupling factors should be considered. This paper considers the driving motor, transmission system and flexible manipulator as an integrated object, and based on the overall coupling relationship and electromechanical dynamics analysis approach, the electromechanical coupling dynamic model was constructed. To reflect the electromechanical coupling mechanism of the FRMS, the speed response characteristics of the FRMS under electromechanical coupling effects are revealed.

2 Electromechanical Coupling Dynamic Equation of the FRMS

Figure 1 shows the electromechanical coupling diagram of a permanent magnet synchronous motor (PMSM) - driven FRMS.

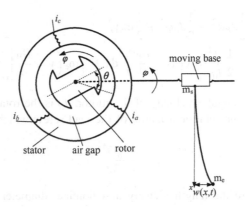

Fig. 1. Structure physical model of the FRMS

The AC servomotor is mainly composed of stator and rotor. When flowed through three-phase sinusoidal alternating current of 120 phase difference, the stator flux produces space rotating magnetic field whose rotating speed is associated with the frequency of the current. Then, through the driving torque produced by the interaction between stator magnetic field and rotor magnetic field, the electric energy is changed into the mechanical energy and can be used to drive the flexible manipulator.

The electromechanical coupling factors in the FRMS can be illustrated by the draging equation as shown in Eq. (1):

$$T - T_L = \frac{J}{9.55} \frac{dn}{dt} \tag{1}$$

where J is the moment of inertia, n is the motor output speed, T denotes the electromagnetic torque, T_L denotes the load torque. When the electromagnetic torque is larger than the load torque, the motor can drive the load. And with the increase of the load torque, the angle between rotor pole and stator rotating magnetic field increase, and the electromagnetic torque subsequently increase until equal to the load torque. Therefore, the stator current increases with the load increasing, and the armature reaction may cause the rise of the air gap flux as well as the stator counter electromotive force.

In this section, the electromechanical coupling dynamic equations of the FRMS, which considers driving motor, transmission system and flexible manipulator as a integrated object, will be constructed based on the electromechanical dynamics analysis approach. During the modeling, assumptions are made as: the iron core saturation, eddy current loss and hysteresis loss are ignored; the air gap is evenly distributed and the self-inductance as well as mutual-inductance of each phase winding is constant; the rotor does not have damping windings and nor the permanent magnet have damping effect while the form of counter electromotive force is sine curve; the flexible manipulator satisfies the Euler-Bernoulli beam theory and the shear and axial deformation are neglected.

The kinetic energy of the FRMS can be written as

$$
E_k = \frac{1}{2}J_R\varphi^2 + \frac{1}{2}J_b\varphi^2 + \frac{1}{2}m_s v_t^2 \\
+ \frac{1}{2}\int_0^L \rho A\left(v_t + \frac{\partial w}{\partial t}\right)^2 dx + \frac{1}{2}m_e\left(v_t + \frac{\partial w}{\partial t}\bigg|_{x=L}\right)^2
\tag{2}
$$

where φ indicates the angular velocity of the motor, J_R is the rotational inertia of the motor shaft and J_b is the rotational inertia of the lead screw which satisfies

$$
J_b = \frac{\pi\rho_b D^4 L_b}{32}
\tag{3}
$$

here L_b, ρ_b and D are the length, density and nominal diameter of the lead screw, respectively.

Based on the movement transitive relation, the relationship between the speed of the base v_t and the angular velocity of the motor φ can be expressed as

$$
v_t = \frac{D}{2}\varphi
\tag{4}
$$

Substituting Eq. (4) into Eq. (2), Eq. (2) can be simplified as

$$
E_k = \frac{1}{2}J_R\varphi^2 + \frac{1}{2}J_b\varphi^2 + \frac{1}{8}m_s D^2 \varphi^2 \\
+ \frac{1}{2}\int_0^L \rho A\left(\frac{D}{2}\varphi + \frac{\partial w}{\partial t}\right)^2 dx + \frac{1}{2}m_e\left[\frac{D}{2}\varphi + \frac{\partial w}{\partial t}\bigg|_{x=L}\right]^2
\tag{5}
$$

The elastic potential energy of the FRMS can be written as

$$
E_p = \frac{1}{2}\int_0^L EI\left(\frac{\partial^2 w}{\partial x^2}\right)^2 dx
\tag{6}
$$

Due to the electromechanical coupling effect between the motor and the flexible manipulator, the magnetic field energy should be taken into consideration. The magnetic field energy, produced by the stator current itself, the rotor permanent magnet and the interaction of the stator current and the flux linkage of the rotor, can be expressed as

$$
W_m = W_{m1} + W_{m2} + W_{m21}
\tag{7}
$$

The magnetic energy caused by the stator current can be represented as:

$$W_{m1} = \frac{1}{2}L_u i_u^2 + \frac{1}{2}L_v i_v^2 + \frac{1}{2}L_w i_w^2 + M i_u i_v + M i_v i_w + M i_u i_w \qquad (8)$$

where L_u, L_v, L_w express the self-inductance of the three-phase windings, M is the mutual-inductance and i_u, i_v, i_w are the current of the three-phase stator.

Owing to the slot number of each phase is large for the 60° phase distributed windings, the magnetic field energy, caused by the rotor permanent magnet, can be seen as a constant ($W_{m2} = C$).

The magnetic field energy, produced by the interaction of the stator current and the flux linkage of the rotor, can be shown as

$$W_{m21} = i_u \psi_f \cos\theta + i_v \psi_f \cos(\theta - \frac{2}{3}\pi) + i_w \psi_f \cos(\theta + \frac{2}{3}\pi) \qquad (9)$$

where θ and ψ_f are the position and flux linkage of the permanent magnet rotor, respectively.

From Eqs. (5)–(7), the Lagrange function of the flexible robot system can be obtained as

$$
\begin{aligned}
U &= E_k - E_p + W_m \\
&= \frac{1}{2}J_R\varphi^2 + \frac{1}{2}J_b\varphi^2 + \frac{1}{8}m_s D^2\varphi^2 \\
&\quad + \frac{1}{2}\int_0^L \rho A\left(\frac{D}{2}\varphi + \frac{\partial w}{\partial t}\right)^2 dx + \frac{1}{2}m_e\left[\frac{D}{2}\varphi + \frac{\partial w}{\partial t}\Big|_{x=L}\right]^2 - \frac{1}{2}\int_0^L EI\left(\frac{\partial^2 w}{\partial x^2}\right)^2 dx \\
&\quad + \frac{1}{2}L_u i_u^2 + \frac{1}{2}L_v i_v^2 + \frac{1}{2}L_w i_w^2 + M i_u i_v + M i_v i_w + M i_u i_w + C \\
&\quad + i_u \psi_f \cos\theta + i_v \psi_f \cos(\theta - \frac{2}{3}\pi) + i_w \psi_f \cos(\theta + \frac{2}{3}\pi)
\end{aligned}
\qquad (10)
$$

Combining Eq. (10), the Lagrange equation of FRMS can be expressed as [14]

$$\frac{d}{dt}\left(\frac{\partial U}{\partial \dot{r}_k}\right) - \frac{\partial U}{\partial r_k} + \frac{\partial F_h}{\partial \dot{r}_k} = Q_k \qquad (11)$$

here F_h is the dissipation functions and Q_k is the non-conservative generalized force.

For the FRMS, the dissipation functions include the dissipation of the mechanical system and electromagnetic system, which can be shown as

$$F_e = \frac{1}{2}R_u i_u^2 + \frac{1}{2}R_v i_v^2 + \frac{1}{2}R_w i_w^2 \qquad (12)$$

$$F_m = \frac{1}{2}(B_1 + B_e)\varphi^2 \qquad (13)$$

where R_u, R_v, R_w are the resistance of the motor three-phase windings, B_1 is the equivalent viscous damping coefficient of the driving system and B_e is the viscous damping coefficient of the motor. Then, the dissipation functions can be represented as

$$F_h = F_e + F_m = \frac{1}{2}R_u i_u^2 + \frac{1}{2}R_v i_v^2 + \frac{1}{2}R_w i_w^2 + \frac{1}{2}(B_1 + B_e)\varphi^2 \tag{14}$$

The non-conservative force is mainly considered as the frictional resistance between the base and the guide rail and can be expressed as

$$Q_i = -\mu m_s g \tag{15}$$

where μ is the friction coefficient between the base and the guide rail and g = 9.8 m/s^2.

Considering the FRMS is a electromechanical coupling system which contains the electromagnetic system and mechanical system, the Lagrange equation is converted into the Lagrange-Maxwell equations whose form is

$$\begin{cases} \dfrac{d}{dt}\left(\dfrac{\partial U}{\partial \dot{e}_i}\right) - \dfrac{\partial U}{\partial e_i} + \dfrac{\partial F_h}{\partial \dot{e}_i} = u_i \\ \dfrac{d}{dt}\left(\dfrac{\partial U}{\partial \dot{\lambda}_j}\right) - \dfrac{\partial U}{\partial \lambda_j} + \dfrac{\partial F_h}{\partial \dot{\lambda}_j} = Q_j \end{cases} \tag{16}$$

Through Eq. (16), the voltage equation of u, v and w stators winding can be respectively obtained as

$$u_1 = L_u \frac{di_u}{dt} + M\frac{di_v}{dt} + M\frac{di_w}{dt} - \varphi\psi_f \sin\theta + R_u i_u \tag{17}$$

$$u_2 = L_v \frac{di_v}{dt} + M\frac{di_u}{dt} + M\frac{di_w}{dt} - \varphi\psi_f \sin(\theta - \frac{2\pi}{3}) + R_v i_v \tag{18}$$

$$u_3 = L_w \frac{di_w}{dt} + M\frac{di_v}{dt} + M\frac{di_u}{dt} - \varphi\psi_f \sin(\theta + \frac{2\pi}{3}) + R_w i_w \tag{19}$$

Defining angular displacement of the motor as the generalized coordinate, According to Eq. (10), it can be obtained that

$$\left(J_R + J_b + \frac{1}{4}D^2 m_s + \frac{1}{4}D^2\rho AL + \frac{1}{4}D^2 m_e\right)\dot{\varphi} + \frac{D}{2}\int_0^L \rho A\frac{\partial^2 w}{\partial t^2}dx + \frac{D}{2}m_e \frac{\partial^2 w}{\partial t^2}\bigg|_{x=L}$$
$$= -i_u\psi_f \sin\theta - i_v\psi_f \sin(\theta - \frac{2}{3}\pi) - i_w\psi_f \sin(\theta + \frac{2}{3}\pi) - (B_1 + B_e)\varphi - \mu m_s g \tag{20}$$

Equation (20) reveals the coupling relationship between the angular velocity of the motor and the electromagnetic and mechanical structure parameters of the FRMS.

Based on this, the speed response characteristics of the FRMS under electromechanical coupling effects can be revealed and used to reflect the electromechanical coupling mechanism in the FRMS.

3　Speed Response Characteristics of the FRMS Under Electromechanical Coupling Effects

In order to analyze the electromechanical coupling characteristics of the PMSM speed, a simulation model, which is constructed in Matlab/Simulink, is used to solve Eq. (20) and the speed response characteristics of the FRMS. The structure of the electromechanical coupling simulation model is illustrated in Fig. 2. The main parameters of the FRMS in the electromechanical coupling dynamic simulation are shown in Table 1. The rotational inertia of the lead screw is $J_b = 1.643 \times 10^{-5}$ kg·m^2, which are much smaller than that of the shaft. For this, in the simulation analysis, the rotational inertia of the shaft is mainly considered.

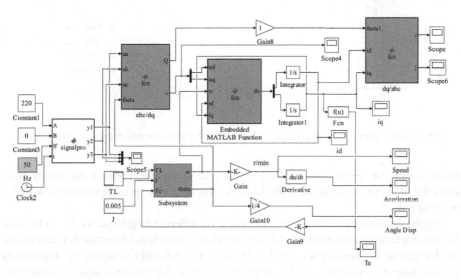

Fig. 2. Simulink model for speeds response of the FRMS under electromechanical coupling effects

The frequencies of the input voltages are defined as f = 20 Hz, 30 Hz, 40 Hz and 50 Hz. According to the relationship $\phi = 60f/p$, the steady state rotational speed of the PMSM can be got as 300r/min, 450r/min, 600r/min and 750r/min. However, as shown in Fig. 3, under the effect of electromechanical coupling, the speeds are not constant and exhibits certainly fluctuation in the startup process. Moreover, it can be

Table 1. Parameters of the system in the electromechanical coupling dynamic simulation

L	b	h	E	ρ	J_R
400 mm	45 mm	3 mm	25.24 GPa	2030 kg/m^3	0.005 kg·m^2
p	L_b	D	$ρ_b$	B_e	m_s
4	800 mm	10 mm	7850 kg/m^3	0.0005 kg·m^2/s	1 kg

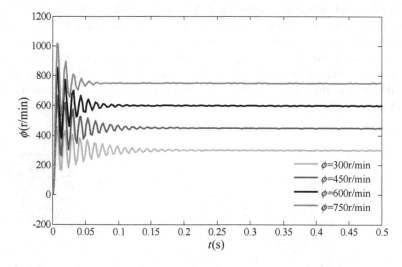

Fig. 3. The speed responses of the FRMS under the electromechanical coupling effects

obtained that the speeds are relatively stable and gradually approach the steady state when the time is more than 0.2 s. Besides, the fluctuation amplitude enhanced with the target speeds.

Figure 4 shows the fluctuation comparison of different target speeds. It can be seen that the fluctuation amplitudes of different rotational speeds exist certain differences but the fluctuation frequencies are similar. In order to uniformly compare the fluctuation level of different rotational speeds, the relatively speed fluctuation is defined as the ratio of the speed fluctuation peak and the target speed, which is expressed as

$$M_p = \frac{\phi_{\max} - \phi_\infty}{\phi_\infty} \times 100\% \qquad (21)$$

Through Eq. (21), the relatively speed fluctuations of 300r/min, 450r/min, 600r/min and 750r/min are calculated as 52.5%, 50.3556%, 42.85% and 35.8667%, which is shown in Table 2. It can be obtained that the relatively speed fluctuations are more obvious for the lower speeds.

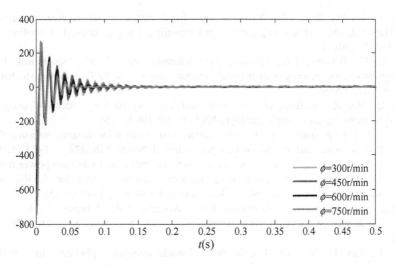

Fig. 4. The speed fluctuations of the FRMS under the electromechanical coupling effects

Table 2. Relatively speed fluctuations of different target speeds

ϕ(r/min)	300	450	600	750
M_p(%)	52.5	50.3556	42.8500	35.8667

4 Conclusions

Regarding the driving motor, transmission system and flexible manipulator as an integrated object, the electromechanical coupling dynamic model was established using the electromechanical dynamics analysis approach. Using the Matlab/Simulink, the dynamic simulation model is constructed to analyze the speed responses of the FRMS under electromechanical coupling effects. The results show that the electromechanical coupling effect have significant influence on the dynamic characteristic of the FRMS, the response speeds are not constant and exhibits certainly fluctuation in the startup process. Moreover, the fluctuation amplitude enhanced with the target speeds while the relatively speed fluctuations are more obvious for the lower speeds. The results are significant for the dynamic and control of flexible robotic manipulators.

Acknowledgements. This research work is supported by the Natural Science Research Project of Higher Education of Anhui Province (no. KJ2017A118), the Project funded by China Post-doctoral Science Foundation (no. 2017M612060) and the Research Starting Fund Project for Introduced Talents of Anhui Polytechnic University (no. 2016YQQ012). The authors sincerely thank the reviewers for their significant and constructive comments and suggestions.

References

1. Dwivedy, S.K., Eberhard, P.: Dynamic analysis of flexible manipulators, a literature review. Mech. Mach. Theory **41**, 749–777 (2006)

2. Liu, Y., Li, W., Wang, Y., Yang, X., En, L.: Coupling vibration characteristics of a translating flexible robot manipulator with harmonic driving motions. J. VibroEng. **17**(7), 3415–3427 (2015)
3. Kerem, G., Bradley, J.B., Edward, J.P.: Vibration control of a single-link flexible manipulator using an array of fiber optic curvature sensors and PZT actuators. Mechatronics **19**, 167–177 (2009)
4. Sen, Q., Bin, Z., Huafeng, D.: Dynamics and trajectory tracking control of cooperative multiple mobile cranes. Nonlinear Dyn. **83**(1–2), 89–108 (2016)
5. Bin, Z., Bin, Z.: A modified hybrid uncertain analysis method for dynamic response field of the LSOAAC with random and interval parameters. J. Sound Vib. **374**, 111–137 (2016)
6. Liu, Y., Li, W., Wang, Y., Yang, X., Ju, J.: Dynamic model and vibration power flow of a rigid-flexible coupling and harmonic-disturbance exciting system for flexible robotic manipulator with elastic joints. Shock Vib. Article ID 541057, 1–10 (2015)
7. Maria, A.N., Jorge, A.C., Ambrósio, L.M., Roseiro, A.A., Vasques, C.M.A.: Active vibration control of spatial flexible multibody systems. Multibody Syst. Dyn. **30**, 13–35 (2013)
8. Bin, Z., Jun, L., Sen, Q.: Localization, obstacle avoidance planning and control of cooperative cable parallel robots for multiple mobile cranes. Robot. Comput.-Integr. Manuf. **34**, 105–123 (2015)
9. Zhao, J.L., Yan, S.Z., Wu, J.N.: Analysis of parameter sensitivity of space manipulator with harmonic drive based on the revised response surface method. Acta Astronaut. **98**, 86–96 (2014)
10. Liu, Y., Li, W., Yang, X., Wang, Y., Fan, M., Ye, G.: Coupled dynamic model and vibration responses characteristic of a motor-driven flexible manipulator system. Mech. Sci. **6**(2), 235–244 (2015)
11. Mohsen, D., Nader, J., Zeyu, L., Darren, M.D.: An observerbased piezoelectric control of flexible Cartesian robot arms: theory and experiment. Control Eng. Pract. **12**, 1041–1053 (2004)
12. Qiu, Z.C.: Adaptive nonlinear vibration control of a Cartesian flexible manipulator driven by a ball screw mechanism. Mech. Syst. Sig. Proc. **30**, 248–266 (2012)
13. Wei, K.X., Meng, G., Zhou, S., Liu, J.W.: Vibration control of variable speed/acceleration rotating beams using smart materials. J. Sound Vib. **298**, 1150–1158 (2006)
14. Singiresu, S.R.: Mechanical Vibration, 4th edn. Pearson Education Inc., New York (2004)

Correction Algorithm of LIDAR Data for Mobile Robots

Wenzhi Bai, Gen Li[✉], and Liya Han

Huazhong University of Science and Technology,
Wuhan 430074, People's Republic of China
{wen,ligen_hust,hlyl993}@hust.edu.cn

Abstract. Laser range finder (LRF) or laser distance sensor (LDS), further referred to as LIDAR (light detection and ranging). LIDAR can obtain environmental point cloud data, while a robot can realize environmental sensing by adoption of the point cloud data generated and LIDAR-based SLAM (Simultaneous Localization And Mapping) algorithm. The precision of point clouds provided by the LIDAR determines that of environmental sensing of the LIDAR-based mobile robot. In this paper, a common correction algorithm has been proposed to correct the inaccuracy of measured point cloud data caused by mobile LIDAR, effectively improving the precision of point cloud data measured by the LIDAR under a mobile state. It also conducts mathematical derivation of the algorithm, presents simulation and real world experiments performed and verifies the necessity and effectiveness of the algorithm derived by experimental results in the paper.

Keywords: LIDAR · Data correction · Mobile robot · SLAM

1 Introduction

LIDAR were mostly applied in research-based robots in the past, featuring high scanning frequency and dense sampling points [1, 2]. At the same time, researchers ignored the missing data precision caused by LIDAR movement in a scan period due to the robot using LIDAR with slow moving speed. At present, LIDAR is developed towards low cost with the increasing demands of consumer robots [3]. Although low-cost LIDAR has low scanning frequency and sparse sampling point, demanding movement speed of the mobile robot has become requested. In this way, the precision of the measured point cloud of LIDAR in the moving state has begun to become crucial. It is worth noting that LIDAR supplier lack the power to achieve due to lack of trace data of the moving robots and the application party lack feature knowledge of LIDAR to solve the problems, since LIDAR data correction in the moving state situates in the intermediate zone between LIDAR supplier and the application party. Thereby, a data correction algorithm has been put forward in this paper in order to realize a more precise measurement for LIDAR.

The remainder of this paper is organized as follows. The second section and third section of the paper has introduced related research works and problem description as well as theoretical derivation of LIDAR data correction algorithm in moving state; In

© Springer International Publishing AG 2017
Y. Huang et al. (Eds.): ICIRA 2017, Part II, LNAI 10463, pp. 101–110, 2017.
DOI: 10.1007/978-3-319-65292-4_10

the fourth section, it has analyzed experiments and results as well as the contribution of the algorithm on the environment sensing of the robot; Conclusion has been conducted in the fifth section.

2 Related Work

In the early LIDAR application, laser scanning equipment with high scanning frequency was applied in detecting land-forms under the condition of airborne. With scanning frequency of LIDAR is high, flight speed in airborne is still fast. Thus, it's necessary to resolve the space attitude located by each measuring transient within the LIDAR scanning period, so as to obtain measured data of practical significance [4, 5]. Unlike other distance sensors such as sonars or IR sensors, an LiDAR is capable of fine angular and distance resolution, realtime behavior, and low false positive and negative rates. Similar to the application of airborne laser scanning equipment, LIDAR has been extensively used in the existing highly-efficient SLAM algorithm [6–8], serving numerous mobile robots. With the emergence of consumer robots, LIDAR has been developed in a low-cost plan with advantages of smaller size and lower consumption, scanning frequency, range and sampling points [3]. Although movement speed of the moving robot gets faster with decreasing scanning frequency of LIDAR, calculation is still conducted in scan [9] in the algorithms of mapping, localization and obstacle avoidance, which not only ignores the pose changes generated by movement in the measuring process of point cloud data in the scan, but also fails in presenting correction methods and realization means for data in the scan.

3 Correction Algorithm

3.1 Problem Describer

A common LIDAR application is presented in Fig. 1. The robot can sense the environment by the LIDAR carried. In general, LIDAR consists of fixed part and rotational part; the former is for fixing on the robot or other moving platform, while the latter is for an optical path realizing 360° environment scanning with utilization of rotated measuring units, so as to obtain environmental point cloud data of the whole plane. LIDAR is an important sensor for the application to sense the environment. The point cloud data obtained can construct a grid map through the SLAM algorithm. Further, the grid map is the basis of realizing navigation and obstacles avoidance of the robot. Also, cloud point data is important for matching with the existing map to realize the functions like localization. In this way, the precision of point cloud data will directly affect the realization of its dependent functions.

LIDAR is operated by scanning the whole plane environment with the optical path via rotating the rotational part to measure distances of the measured objects at a certain interval. Thus, measured point cloud data of the whole circle can be obtained. The whole data with time stamp can be offered to the robot for calculation. Two extreme cases greatly affecting the precision of LIDAR are proposed here. One is that we can

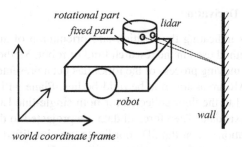

Fig. 1. Application of LIDAR

predict the point cloud data on its measured flat of the LIDAR are just multiple measurements of the same point when angular velocity of rotation of the rotational part of the LIDAR is consistent with that of robot in a reverse direction, as shown in case 1 of Fig. 2. Two is when the robot moves and scans a wall that is vertical to its moving direction, the robot moves rapidly from a distance far from the wall to near the wall. We can predict that the point clouds generated by the wall that is vertical to the moving direction will not vertical to the moving direction any more upon observing the point cloud data in a scanning period.

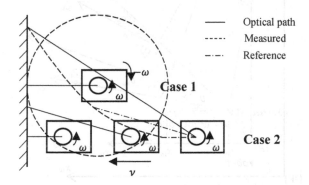

Fig. 2. Schematic diagram of extreme cases

We can obtain from the two cases that plane point cloud data given are close to the measured object with only LIDAR measuring errors, under the situation of rotation movement of the rotational part of LIDAR and no other relative movement, when the fixed part of LIDAR is static with respect to the measured object. However, when LIDAR is a moving state, the scanning movement within a scanning period contains not only the rotation movement of the rotational part of LIDAR, but also translation and rotation movements generated by the robot with fixed LIDAR with respect to the measured object. What we do in the paper is to consider these movements in moving state of LIDAR while correcting point cloud data of LIDAR in the scanning period, so that the point cloud data in the scanning plane can accurately present the measured object and improve the precision of LIDAR point cloud data.

3.2 Mathematical Derivation

Cartesian coordinates indicating the position and orientation of an object in the environment is a basic method used in the research field of robot. Method of the coordinates transformation in the moving process of the robot has been presented in [10]. The robot and all parts of LIDAR are assumed to be rigid bodies. Plane will be considered only. Plane data are parallel to the floor collected for both single-line LIDAR and multi-line LIDAR, ignoring Z axis data. Therefore, all data are projected to the OXY plane along Z axis for the calculation, when the 3D coordinate are simplified to a 2D one.

The world coordinate is defined as $\{W\}$ coordinate; The robot coordinate is defined as $\{R\}$ coordinate; The coordinate of the robot fixed with LIDAR is defined as $\{L\}$ coordinate; The robot coordinate is defined as $\{R'\}$ coordinate with Δt ($\Delta t < T$) time; The LIDAR coordinate is defined as $\{L'\}$ coordinate with Δt ($\Delta t < T$) time; P_1 and P_2 are the observing points in the world coordinate, of which, P_1 is the targeted point observed in the starting time, P_2 is the targeted point observed with Δt ($\Delta t < T$) time. Relative positions among each coordinate are presented in Fig. 3.

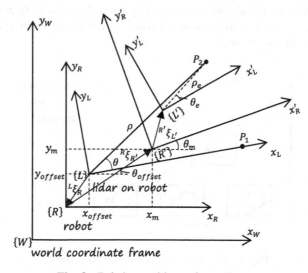

Fig. 3. Relative positions of coordinates

It can be seen from Fig. 3, $\{L\}$ coordinate is fixed on the $\{R\}$ coordinate due to the robot equipped with LIDAR, whose relative position is $^L\xi_R = \ominus^R\xi_L = \ominus\xi(x_{offset}, y_{offset}, \theta_{offset})$; $\{R'\}$ coordinate system is obtained by $\{R\}$ coordinate upon moving with Δt ($\Delta t < T$), whose relative position is regarded as $^R\xi_{R'} = \xi(x_m, y_m, \theta_m)$; $\{L'\}$ coordinate is obtained by $\{L\}$ coordinate upon moving with Δt ($\Delta t < T$); Thus, the relative position of $\{L'\}$ coordinate relating to $\{R'\}$ coordinate is $^{R'}\xi_{L'} = \xi(x_{offset}, y_{offset}, \theta_{offset})$ since LIDAR is fixed on the robot.

The conversion relation of coordinate:

$$^{L}\boldsymbol{P}_{2} = \left(^{L}\xi_{R} \oplus {}^{R}\xi_{R'} \oplus {}^{R'}\xi_{L'}\right) \cdot {}^{L'}\boldsymbol{P}_{2} \tag{1}$$

Where, each symbol stands for:

- $^{A}\boldsymbol{P}$ — the point P is described by coordinate vectors relative to frame $\{A\}$
- ξ — abstract representation of 3-dimensional Cartesian pose
- $^{A}\xi_{B}$ — abstract representation of 3-dimensional relative pose, frame $\{B\}$ with respect to frame $\{A\}$
- $\oplus$ — pose composition operator
- $\ominus$ — inverse of a pose (unary operator)
- $\cdot$ — transformation of a point by a relative pose, e.g. $\xi \cdot p$

From the above relations, it can obtain:

$$^{L}\xi_{R} = \ominus^{R}\xi_{L} \sim \begin{pmatrix} \cos\theta_{offset} & -\sin\theta_{offset} & x_{offset} \\ \sin\theta_{offset} & \cos\theta_{offset} & y_{offset} \\ 0 & 0 & 1 \end{pmatrix}^{-1} = {}^{R}\boldsymbol{T}_{L}^{-1} = {}^{L}\boldsymbol{T}_{R} \tag{2}$$

$$^{R}\xi_{R'} \sim \begin{pmatrix} \cos\theta_{m} & -\sin\theta_{m} & x_{m} \\ \sin\theta_{m} & \cos\theta_{m} & y_{m} \\ 0 & 0 & 1 \end{pmatrix} = {}^{R}\boldsymbol{T}_{R'} \tag{3}$$

$$^{R'}\xi_{L'} \sim \begin{pmatrix} \cos\theta_{offset} & -\sin\theta_{offset} & x_{offset} \\ \sin\theta_{offset} & \cos\theta_{offset} & y_{offset} \\ 0 & 0 & 1 \end{pmatrix} = {}^{R'}\boldsymbol{T}_{L'} \tag{4}$$

In the formula, we use the symbol $^{A}\boldsymbol{T}_{B}$ to denote that homogeneous transform representing frame $\{B\}$ with respect to frame $\{A\}$. Note that $^{A}\boldsymbol{T}_{B} = \left(^{B}\boldsymbol{T}_{A}\right)^{-1}$. The symbol $\sim$ is denoted that the two representations are equivalent.

$^{L}\boldsymbol{P}_{2}$ and $^{L'}\boldsymbol{P}_{2}$ are presented as $^{L}\widetilde{\boldsymbol{P}}_{2} = \left(^{L}x, {}^{L}y, 1\right)^{T}$ and $^{L'}\widetilde{\boldsymbol{P}}_{2} = \left(^{L'}x, {}^{L'}y, 1\right)^{T}$ in homogeneous form. That is,

$$^{L}\widetilde{\boldsymbol{P}}_{2} = {}^{L}\boldsymbol{T}_{R} \cdot {}^{R}\boldsymbol{T}_{R'} \cdot {}^{R'}\boldsymbol{T}_{L'} \cdot {}^{L'}\widetilde{\boldsymbol{P}}_{2} \tag{5}$$

(ρ_{e}, θ_{e}) is the $^{L'}\boldsymbol{P}_{2}$ observing data obtained by LIDAR, thus,

$$\begin{cases} ^{L'}x = \rho_{e} \cdot \cos\theta_{e} \\ ^{L'}y = \rho_{e} \cdot \sin\theta_{e} \end{cases} \tag{6}$$

Measured correction value of point $^{L}\boldsymbol{P}_{2}$ is (ρ, θ) with Δt $(\Delta t < T)$ time based on timestamp:

$$
\begin{cases}
\rho = \left| \sqrt{{}^L x^2 + {}^L y^2} \right| \\
\theta = \begin{cases}
\arctan \frac{{}^L y}{{}^L x} & ({}^L x > 0, \ {}^L y \in \mathrm{R}) \\
\arctan \frac{{}^L y}{{}^L x} + \pi & ({}^L x < 0, \ {}^L y \geq 0) \\
\arctan \frac{{}^L y}{{}^L x} - \pi & ({}^L x < 0, \ {}^L y < 0) \\
\frac{\pi}{2} & ({}^L x = 0, \ {}^L y < 0) \\
-\frac{\pi}{2} & ({}^L x = 0, \ {}^L y < 0)
\end{cases}
\end{cases}
\tag{7}
$$

A special situation should be noted, when it's ${}^L x = 0, {}^L y = 0$, θ has infinitely many solutions with $\rho = 0$. It has a practical significance in physics, namely, the measured value after Δt time is just the position of LIDAR at the moment represented by the timestamp. However, the measured value $\rho = 0$ is less than the minimum range of LIDAR, which will be ignored in the later use.

All data measured in a frame of LIDAR are converted into the coordinate of moment represented by the timestamp to complete the correction of data in that frame.

4 Experiments

In this section, it aimed to validate the necessity and effectiveness of the algorithm proposed in this paper through conducting simulation tests and real world experiment.

4.1 Simulation Tests

The LIDAR simulation parameters used in the simulation is conducted based on performance parameters of Hokuyo UTM-30LX and Inmotion ILD26TRI described in literature [1, 11] (Table 1):

Table 1. List of LIDAR simulation parameters

Type	Sample (point/scan)	Range (°)	Frequency (Hz)	Std dev (mm)
Hokuyo UTM-30LX	1440	360	40	30
Inmotion ILD26TRI	360	360	8	60

Linear velocity of the moving robot is set as $v = 1.5$ m/s and the angular velocity as $\omega = 2$ rad/s in normal speed, while linear velocity is set as $v = 6$ m/s and the angular velocity as $\omega = 5$ rad/s in high speed. Deviation generating between LIDAR installation center and rotational center of the robot has been considered in the translation speed occurring in the high-speed rotation during simulation.

The simulation is divided into three groups, including: measuring in a moving robot with installation of UTM-30LX LIDAR in normal speed, measuring in a moving robot with installation of ILD26TRI in normal speed and measuring in a moving robot with installation of UTM-30LX LIDAR in high speed; And moving status of each group is

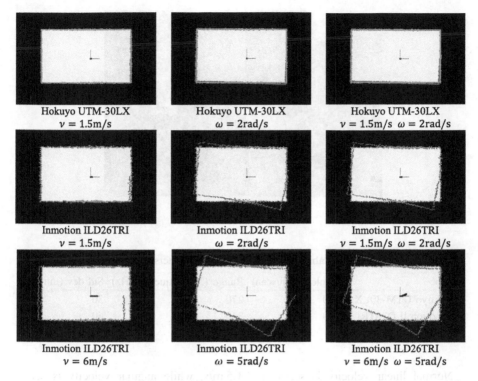

Fig. 4. Simulation result (Color figure online)

composed of translation, rotation and translation as well as rotation. Coordinate presented in Fig. 4 is the LIDAR coordinate at the moment of starting scanning. We can see from the simulation results, the black trace simulates the moving trace of a moving robot in the scanning process of a frame; the black and white areas simulate areas with obstacles and without obstacle in the environment, respectively. The red point cloud simulates a frame of point cloud data collected by LIDAR; while the green one is the point cloud data of a scan upon corrected by the algorithm proposed in this paper. Starting from the positive direction of X-axis, the scanning process is completed upon scanning a circle along the counterclockwise.

4.2 Real World Experiments

The experiment setup is composed of a mechanical arm carrying LIDAR, as shown in Fig. 5. Highly-precise movement can be realized and accurate movement locus can be obtained with the mechanical arm that can well restore the real scene of moving process of the moving robot.

The experiment is divided into two groups that robots are carried Hokuyo UTM-30LX and Inmotion ILD26TRI, respectively; each group consists of scanning plane point cloud in a static state, in the translation of normal speed and in the translation and rotation of normal speed. LIDAR parameters are presented in Table 2.

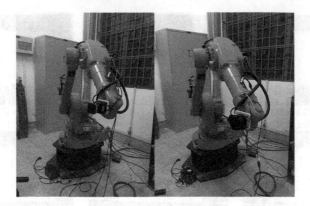

Fig. 5. Experiment setup

Table 2. List of LIDAR parameters

Type	Sample (point/scan)	Range (°)	Frequency (Hz)	Std dev (mm)
Hokuyo UTM-30LX	1080	270	40	30
Inmotion ILD26TRI	360	360	8	60

Normal linear velocity is set as $v = 1.5$ m/s, while angular velocity is set as $\omega = 2$ rad/s. Experiment results are shown in Fig. 6. The coordinate system shown in the figure is the LIDAR coordinate when starting scanning. The red point represents a scan of point cloud data collected by LIDAR; While the green one represents a scan of point cloud data upon correction. Starting from positive direction of X-axis, the scanning process is completed upon scanning a circle along the counterclockwise.

4.3 Discussion

We can see from the simulation and experiment results that the scanning frequency and movement speed of the robot are important factors affecting the measuring accuracy of point cloud. The LIDAR with low scanning frequency has a large pose change in the scanning process of a scan under the same moving state, presenting poor precision of its point cloud; The faster the movement speed, the larger influence will be exerted on the precision of the point cloud. Furthermore, rotation movement has a larger influence on the precision of cloud point compared to the translation movement. To some extent, although the correction algorithm changes the density distribution of the point cloud, but the correction results that point cloud accurately reflects the real environment to validate the necessity and effectiveness of LIDAR data correction algorithm in moving state.

Hence, we suggest that the correction algorithm is essential to the family service robot using low-cost LIDAR with low scanning frequency. Also, it shall pay attention

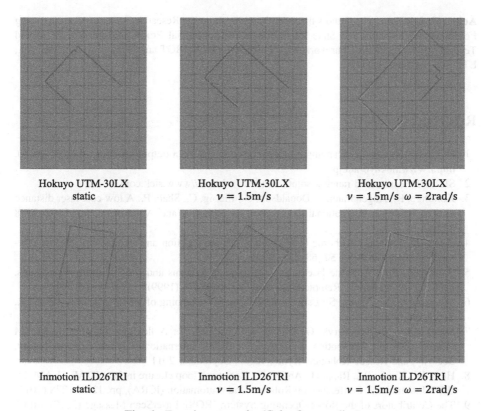

| Hokuyo UTM-30LX
static | Hokuyo UTM-30LX
$v = 1.5m/s$ | Hokuyo UTM-30LX
$v = 1.5m/s$ $\omega = 2rad/s$ |

| Inmotion ILD26TRI
static | Inmotion ILD26TRI
$v = 1.5m/s$ | Inmotion ILD26TRI
$v = 1.5m/s$ $\omega = 2rad/s$ |

Fig. 6. Experiment results (Color figure online)

to lower the rotational speed of such kind of robot in control. Moreover, the LIDAR shall be fixed around the rotational center of the robot to decrease offset, thus reducing high-speed translation caused by rotation.

5   Conclusions

In this paper, we have presented the necessity of correcting LIDAR measuring point cloud data in moving state and validated the effectiveness of the correction algorithm upon experiments. Considering pose changes of LIDAR moving platform in the scanning period, translation and rotation movement of the moving platform were overlaid with the rotational scanning movement in the algorithm, whose data can accurately reflect the real measured environment. The algorithm is suitable for processing point cloud data of 2D and 3D LIDAR flat scanning, especially the family service robot using low-cost LIDAR and the pursuit of rapid movement and guarantee no collision of the environment. The correction algorithm given in the paper can be popularized to all environmental sensing tasks using LIDAR, which can also provide algorithm references for sensor data pre-processing of the algorithm end and data encapsulation of LIDAR supplier, contributing to improving mapping, positioning, obstacle avoidance and other algorithms of robots using LIDAR data.

Acknowledgments. The authors thank National Engineering Research Center of Manufacturing Equipment Digitization and State Key Laboratory of Material Processing and Die & Mould Technology for supporting our work. Thank INMOTION ROBOT and Hokuyo Automation Co., LTD for providing LIDAR.

References

1. Hokuyo Automation: Scanning range finder, distance data output type for robotics (2017). http://www.hokuyo-aut.jp
2. SICK: Detection and ranging solutions (2017). https://www.sick.com
3. Konolige, K., Augenbraun, J., Donaldson, N., Fiebig, C., Shah, P.: A low-cost laser distance sensor. In: IEEE International Conference on Robotics and Automation, pp. 3002–3008 (2008)
4. Wehr, A., Lohr, U.: Airborne laser scanning-an introduction and overview. ISPRS J. Photogramm. Remote Sens. **54**, 68–82 (1999)
5. Baltsavias, E.P.: Airborne laser scanning: existing systems and firms and other resources. ISPRS J. Photogramm. Remote Sens. **54**(2–3), 164–198 (1999)
6. Montemerlo, M., Thrun, S.: Large-scale robotic 3-D mapping of urban structures. In: ISER, Singapore (2004)
7. Kohlbrecher, S., von Stryk, O., Meyer, J., Klingauf, U.: A flexible and scalable SLAM system with full 3D motion estimation. In: IEEE International Symposium on Safety, Security, and Rescue Robotics, Kyoto, Japan, September 2011
8. Hess, W., Kohler, D., Rapp, H., Andor, D.: Real-time loop closure in 2D LIDAR SLAM. In: IEEE International Conference on Robotics and Automation (ICRA), pp. 1271–1278 (2016)
9. The Contributors of the Robot Operating System (ROS): LaserScan Message (2017). http://docs.ros.org/api/sensor_msgs/html/msg/LaserScan.html
10. Corke, P.: Robotics, Vision and Control-Fundamental Algorithms in MATLAB, vol. 73. Springer, Heidelberg (2011). pp. 32–40
11. INMOTION ROBOT: 2D LiDAR product (2017). https://robot.imscv.com

Research and Application on Avoiding Twist Mechanism Based on Relative Rotation Platforms

Guobin Yang, Lubin Hang$^{(\boxtimes)}$, Jiuru Lu, Zhiyu Fu, Wentao Li,
and Liang Yu

College of Mechanical Engineering,
Shanghai University of Engineering Science, Shanghai 201620, China
hanglb@126.com

Abstract. Research on non-twist criterion of rope is the theoretical basis of the application of rope in engineering field. Aiming at the twist problem of ropes between two relative rotation platforms, this paper introduces a kind of avoiding twist mechanism and calculates its transmission ratio of the epicyclic gear train. Based on the twist model and mathematical description of rope, a non-twist criterion is analyzed and proposed. Then, this paper reveals that how the revolving speed ratio acts on rope twist by using the method of anti-rotation and non-twist criterion. Furthermore, the avoiding twist mechanism is extended to the case of three rotation platforms and its kinematic relation for avoiding twist is analyzed. Finally, the application of avoiding twist mechanism on optical imaging of rotation platforms is analyzed.

Keywords: Relative rotation platforms · Avoiding twist mechanism · Non-twist criterion · Optical imaging

1 Introduction

Energy or information transmissions between two relative rotation platforms are usually required, such as electrical energy, hydraulic media and optoelectronic information, etc. These transmissions are usually carried out by using pipeline or cable, which will lead to rope twist state, and limit the relative motion range of the two platforms.

To achieve energy or information transmission between the two platforms, the slip rings and rotary joints are the two means to avoid twist in continuous relative rotation of platforms.

Currently, the slip rings (see in Fig. 1) are used for transferring power and data from a stationary to a rotating structure [1, 2], in the field of the transmission of signal current. During transmission, the cables on both sides of the slip rings will not be twisted, and the structure of the slip rings is simple. However, the disadvantages could not be ignored. Due to the low electric level of the signal current and the variable contact resistance of the slip rings, there will be high interference in the signal transmission. In addition, Slip rings are prone to wear and fever, which may cause damage to the cables.

© Springer International Publishing AG 2017
Y. Huang et al. (Eds.): ICIRA 2017, Part II, LNAI 10463, pp. 111–123, 2017.
DOI: 10.1007/978-3-319-65292-4_11

Rotary joints (see in Fig. 2) are used to transfer liquid or gas media from stationary supply lines to rotating or moving machine parts. For the transmission of hydraulic energy [3–5], rotary joints possess the characteristics of good sealing property, compact structure, convenience for utilization and maintenance. But there exist weaknesses, including high cost and difficulty to manufacture, which limit its application.

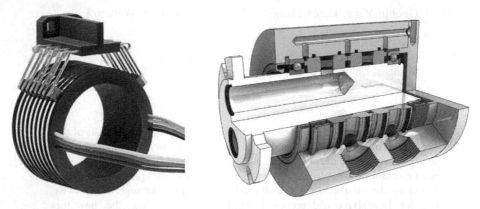

Fig. 1. The slip ring [6]   **Fig. 2.** Rotary joint [7]

From the above analysis, we can see that slip rings and rotary joints schemes have disadvantage, both of them can only be applied in a specific field. This paper aims at obtaining an avoiding twist mechanism, featuring wide range and low application cost, etc.

To solve the twist problem between two relative rotation platforms, an avoiding twist mechanism, which has little effect on energy or information transmission, is proposed in the patent [8]. But the essential relationship between the avoiding twist and the transmission ratio is not revealed in the reference. The structure and transmission of the mechanism are analyzed, furthermore, the motion model and equivalent model of the avoiding twist mechanism are established in this paper. By using the twist character of rope and ignoring the hysteretic and retard of rope twist, it is analyzed how the relative speed of the two platforms effect on the rope twist. Finally, the avoiding twist mechanism is researched further in the application of optical image transmission.

2 An Avoiding Twist Mechanism

An avoiding twist mechanism, which uses cables to transfer electrical energy between two relative rotation platforms, is proposed in [8]. The cables would not be twisted during the relative rotation for the mechanism's transmission ratio of two-to-one. The phenomenon of avoiding twist will be explained through the analysis and calculation of the mechanism. The other form of energy transmission mechanisms has the same structure principle.

2.1 Analysis of Avoiding Twist Mechanism

The model of the avoiding twist mechanism is shown in Fig. 3. The rotation platform 40 and the rotation roller 22, with the transmission ratio of two-to-one, are driven by gears. Thus, the cables, which one terminal connected on the stationary base and another terminal connected on the rotation platform, would avoid twist in the relative motion of two platforms.

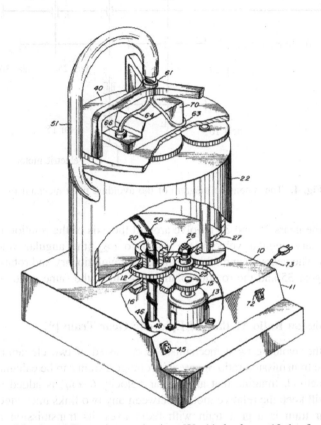

Fig. 3. The model of avoiding twist mechanism [8]. 11-the base; 12-the fixed gear; 13-the electric motor; 15-the drive gear; 22-the rotation roller; 40-the rotation platform; 46-the multicore cable; 51-the tubular channel

The schematic diagram of the avoiding twist mechanism shown in Fig. 4, in which the gears 15, 16, 25, 30, 33 and 35 are transmission gears and the gear 12 is fixed gear. The gears 12 and 30 have equal teeth number and the gears 33 and 35 also have equal teeth number. The electric motor drives the gear 15, which meshes with gear 16. Then the gear 16 drives the rotation roller 22 and leads to the revolution of the gears 25 and 30. The gear 25 meshes with the fixed gear 12 and be driven to rotate by its revolution. The gear 25 meshes with the gear 30 and the gear 30 drives the gear 33. The gear 33 meshes with the gear 35, which leads to the rotation of the rotation platform 40.

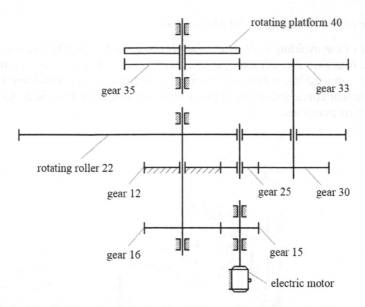

Fig. 4. The schematic diagram of the avoiding twist mechanism

In Fig. 4, the gears 25 and 30 revolve around the axis of the rotation roller 22 and their revolutionary angular velocity are equal to ω_{22} (the angular velocity of the rotation roller). The gears 30 and 33 have the same revolutionary and rotational angular velocity. The gear 35 and the rotation platform 40 have the same angular velocity.

2.2   Transmission Ratio Method of Epicyclic Gear Train [9]

From Fig. 4, the avoiding twist mechanism is consisted of two elementary epicyclic gear trains. The transmission ratio of any epicyclic gear train can be calculated by using anti-rotation method. Imagine that an angular velocity $(-\omega_H)$ is added to the whole gear train, it will keep the relative motion between any two links unchanged. Since the converted gear train is a gear train with fixed axes, its transmission ratio can be calculated as for an ordinary gear train. In the converted gear train, for example, the angular velocities of the gears m and n are $(\omega_m - \omega_H)$ and $(\omega_n - \omega_H)$, respectively, not ω_m and ω_n. They are denoted as ω_m^H and ω_n^H, respectively. The transmission ratio between the gears m and n in the converted gear train is denoted as i_{mn}^H. It can be derived in the same way as for a gear train with fixed axes as follows.

$$i_{mn}^H = \frac{\omega_m^H}{\omega_n^H} = \frac{\omega_m - \omega_H}{\omega_n - \omega_H} = \pm \frac{product\ of\ tooth\ numbers\ of\ all\ the\ driven\ gears}{product\ of\ tooth\ numbers\ of\ all\ the\ driving\ gears}$$

Train ratio i_{mn}^H is positive when gears m and n in the epicyclic gear train rotate in the same direction and negative when opposite.

2.3   Transmission Ratio Calculation of the Avoiding Twist Mechanism

The avoiding twist mechanism is consisted of two elementary epicyclic gear trains, and the rotating relationship between the rotation platform 40 and the rotation roller 22 should be analyzed.

Calculating the transmission ratio of the rotation platform 40 and the rotation roller 22 is the same as calculating the speed ratio of the gear 35 and the rotation roller 22. In order to facilitate the calculation, the rotation roller 22 is considered as the active gear by ignoring the existence of the electric motor, the gear 15 and the gear 16. The calculation process is as follow:

(1) Dividing the gear train into two subtrains
    The given combined gear train in the Fig. 4 can be divided into two subtrains. The gears 12, 25, 30 and the rotation roller 22 make up an elementary epicyclic gear train in which the rotation roller serves as the planet carrier only. The gears 33, 35 constitute the other epicyclic gear train. The gear 30 is fixed with the gear 33.

(2) Deriving the transmission ratio of subtrains independently
    The transmission ratio equations of the two subtrains are:

$$i_{(30)(12)}^{22} = \frac{\omega_{30}^{22}}{\omega_{12}^{22}} = \frac{\omega_{30} - \omega_{22}}{\omega_{12} - \omega_{22}} = \frac{z_{25}z_{12}}{z_{30}z_{25}} \tag{1}$$

$$i_{(33)(35)} = \frac{\omega_{33}^{22}}{\omega_{35}^{22}} = \frac{\omega_{33} - \omega_{22}}{\omega_{35} - \omega_{22}} = -\frac{z_{35}}{z_{33}} \tag{2}$$

From the above 2.1, we can know that

$$z_{12} = z_{30}, \; z_{33} = z_{35}$$

Then

$$i_{(30)(12)}^{22} = \frac{z_{25}z_{12}}{z_{30}z_{25}} = \frac{z_{12}}{z_{30}} = 1 \tag{3}$$

$$i_{(33)(35)} = -\frac{z_{35}}{z_{33}} = -1 \tag{4}$$

(3) Solving the transmission ratio of the combined gear train
    From Eqs. (1) and (3) we have:

$$i_{(30)(22)} = \frac{\omega_{30}}{\omega_{22}} = 1 - i_{(30)(12)}^{22} = 1 - \frac{z_{12}}{z_{30}} = 0$$

Then

$$\omega_{30} = 0$$

Because of the gear 33 is fixed with the gear 30. Thus

$$\omega_{33} = \omega_{30} = 0 \tag{5}$$

From Eqs. (2) and (4) we have:

$$\frac{\omega_{33} - \omega_{22}}{\omega_{35} - \omega_{22}} = -1 \tag{6}$$

Multiplying Eqs. (5) and (6) gives the result:

$$\omega_{35} = 2\omega_{22}$$

Thus

$$i_{(35)(22)} = \frac{\omega_{35}}{\omega_{22}} = 2$$

The rotating relationship between the rotation roller 22 and the gear 35 is proved.

3   The Twist Nature of Rope and the Principle of the Avoiding Twist Mechanism

It is proposed that rope would not twist when the transmission ratio equal to the two-to-one in [8]. But the essential relationship between the two-to-one ratio and the twist is not revealed. Also whether the other transmission ratio satisfies with the non-twist condition is not discussed. The following section will analyze the twist of the rope and give the intrinsic explanation on the above problems.

3.1   Definition of the Rope Twist State and Non-twist State

Suppose that the rope is composed of a set of line clusters with limited length. When all lines in the set of line clusters are straight lines and parallel to each other, it is indicated that the rope is in the non-twist state. While the two end faces of rope rotate relatively and each line of line clusters twist with helical, it is indicated that the rope is in the twist state, as shown in Fig. 5.

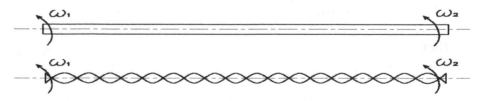

Fig. 5. The schematic diagram of the rope twist

3.2   The Mathematical Description of Twist and Non-twist Criterion

Suppose the length of the rope is l, the initial twist angle of the two end faces twisted in the same axial direction is α and β respectively. And at t time, the rotational angle is φ_1 and φ_2, the rotational angular velocity is $\omega_1(t)$ and $\omega_2(t)$; thus $\varphi_1 = \alpha + \int_0^t \omega_1(t)dt$; $\varphi_2 = \beta + \int_0^t \omega_2(t)dt$.

In per unit length, the cumulative relative twist angle ε is used to describe the degree of the rope twist, when the hysteretic and retard of rope twist are ignored, the degree of the twist:

$$\varepsilon = \frac{|\varphi_1 - \varphi_2|}{l} \tag{7}$$

In that, the value of ε is a function of time, which represents the degree of the twist of the rope at a time. And the greater its numerical value, the more obvious effect of the twist. While $\varepsilon = 0$, the mechanism is in the status of non-twist. But sometimes, the hysteretic and retard of rope twist cannot be ignored, then the value of ε is the average value actually. And it's usually different with the value ε of other location for the same rope.

When the conditions are satisfied that the hysteretic and retard of rope twist are ignored, a non-twist criterion is obtained.

Non-twist criterion is defined as follow:

According to the twist nature of rope, there is not twist on the rope when both the twist angular velocity of the two end faces and the direction along the axis of ropes are identical [10–12]. Otherwise, when these conditions are not be satisfied, there is cumulative twist on the rope.

3.3   The Equivalent Model of the Avoiding Twist Mechanism and Feature Analysis of the Transmission Ratio

The schematic model of the twist mechanism is shown in Fig. 6. In order to reveal the non-twist rule with using the method of reverse rotation, a common angular velocity is added on the overall system to make both the rotation roller and the rope stationary relative to ground, and at the same time, rotate the rotation platform and the base in reverse direction. Then an equivalent model can be obtained as shown in Fig. 7.

According to the non-twist criterion, the rope will not twist when the two rotating parts in the equivalent mechanism have same angular velocity and reverse direction. On the basis, adding a common angular velocity on the equivalent mechanism to make the mechanism back to the original motion form. In the other word, the base is fixed. From the above analysis, we can get the conclusion that the angular velocity of the rotation platform and the rotation roller is 2ω and ω respectively. Thus, the transmission ratio of two-to-one is the only correct solution for the avoiding twist mechanism.

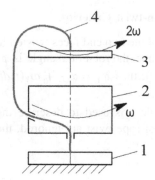

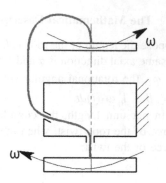

Fig. 6. The schematic model of the avoiding twist mechanism

Fig. 7. The equivalent model of a kind avoiding twist mechanism

3.4 The General Model of a Kind Avoiding Twist Mechanism and Non-twist Criterion

There are the fixed mechanical components in the above model. The general model of the avoiding twist mechanism is obtained when all the platforms are rotating as Fig. 8 shown.

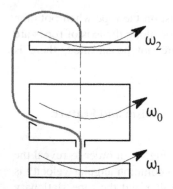

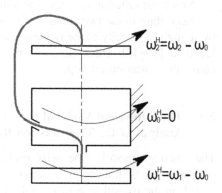

Fig. 8. The general model

Fig. 9. The equivalent general model

A common angular velocity $-\omega_0$ is added to make the rotation roller stationary relative to ground, shown in Fig. 9.

According to non-twist criterion, while there is no cumulative twist on the rope, it should satisfy with the condition:

$$\omega_1^H = -\omega_2^H \tag{8}$$

The minus in the expression (8) represents the directions of two angular velocities are opposite. That expression is equal to $\omega_1 - \omega_0 = -(\omega_2 - \omega_0)$, then $\omega_0 = \omega_2 + \omega_1/2$. Therefore, the relationship of angular velocity of the middle rotation platform in the common model is obtained. As any two parameters are known, the third parameter value can be determined.

4 Real-Time Transmission of Optical Imaging Based on the Avoiding Twist Mechanism

The transmission path of multi-beam light can be regarded as the rope arrangement in space. By using twist rule of rope, the problems of beam imaging could be explained. An optical imaging apparatus is proposed from using the avoiding twist mechanism in the patent [8]. This apparatus can realize that the image of the rotation platform has constant position and orientation on the projection screen with two-to-one transmission ratio. The structure of the apparatus is analyzed below and the optical transmission path of real-time image on the rotation platform is explained.

4.1 The Structure of the Optical Imaging Apparatus

The model of the optical imaging apparatus is shown in Fig. 10. The apparatus consists of the avoiding twist mechanism, a group of isosceles right triangular prism and an imaging screen.

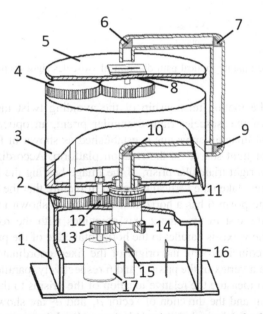

Fig. 10. The schematic diagram of the optical imaging apparatus. 1-the base; 2, 4, 8, 12, 13, 14-the transmission gear; 3-the rotation roller; 5-the rotation platform; 6, 7, 9, 10, 15-the isosceles right triangular prism; 11-the fixed gear; 16-the imaging screen; 17-the electric motor.

The Prisms 6, 7, 9, 10 fixed in the square tube and rotated synchronously with the rotation roller. At any time, the image of the rotation platform is projected on the imaging screen 16 through the prisms 6, 7, 9, 10, 15. Then the image realized transformation from the rotation platform to the stationary base. Thus, the image of the instruments and apparatus on the rotation platform shows on the imaging screen in real time.

4.2 Optical Path Analysis of the Optical Imaging Mechanism

According to the imaging principle of the isosceles right triangular prism, the virtual image is obtained by rotating the object in counterclockwise of 90°, in which the proportion and the relative relationship of the interior points, lines of the object are not change as shown in Fig. 11:

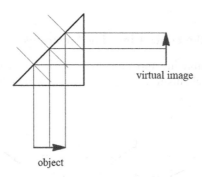

Fig. 11. The imaging optical path diagram of isosceles right triangular prism

According to the motion relationship of the avoiding twist mechanism and the imaging principle of the isosceles right triangular prism, an optical path equivalent model of the optical imaging apparatus is established as shown in Fig. 12:

There is a linear light source on the rotation platform. According to the imaging rule of the isosceles right triangular prism, if the image bearing the linear light source on the imaging screen, taken the vertical direction, it require that the linear light source on the surface of the prism 6 has a horizontal direction as shown in Fig. 12.

In the coordinate system xyz, the z axis coincides with the rotating axis of the rotation platform, the y axis is parallel to the horizontal edge of the prism 6. The end of linear light source coincides with the origin of the fixed coordinate system xyz. The vector a_1 and a_2 via a vertex of the prism 5 and 6 respectively, paralleling to the y axis, and they are used to measure the relative location of the prisms to the fixed coordinate system. The location and the direction of vector a_1 and a_2 are shown in Fig. 12.

At any time, suppose the included angle between the right-angle side on the surface of the prism 2 and the y axis of the coordinate system is ϕ_1, and the included angle between the direction of linear light source and the y axis of the fixed coordinate system is ϕ_2. When the apparatus in the initial position, $\phi_1 = \phi_2 = 0$, the imaging optical path in this status is shown in Fig. 13.

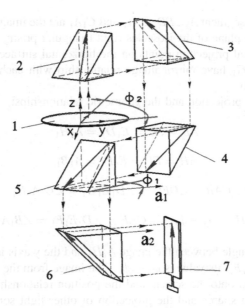

Fig. 12. The equivalent model of optical path. 1-the rotation platform; 2, 3, 4, 5, 6-isosceles right triangular prisms; 7-the imaging screen

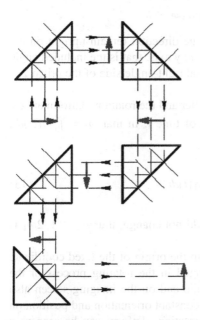

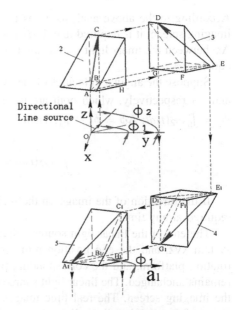

Fig. 13. The light path diagram at the initial position

Fig. 14. The transfer prism group

As shown in Fig. 14, the prism 2, 3, 4 and 5 formed the transfer prism group, which can achieve that the image of the linear light source transfer from the rotation platform to the base, at the same time, adjusting the new image parallel to the y axis.

The directed line segment AC, ED, E_1G_1 and C_1A_1 are the images of the linear light source on the incline plane of the isosceles right triangular prism, the points B, G, D_1 and H_1 are the vertical projection points on the horizontal surface of each prism; AC and ED, ED and E_1G_1 have mirror image relationship with each other respectively, E_1G_1 parallel to C_1A_1.

According to the projection and the geometric relationships:

$$AB = GE = E_1D_1 = A_1H_1 \tag{9}$$

$$AH = EF = E_1F_1 = A_1B_1 \tag{10}$$

$$\angle CAB = \angle DEG = \angle D_1E_1G_1 = \angle C_1A_1H_1 \tag{11}$$

$$\angle BAH = \phi_2 - \phi_1 = \angle GEF = \angle D_1E_1F_1 = \angle B_1A_1H_1 \tag{12}$$

The declination angle between the image A_1H_1 and the y axis is the included angle between A_1H_1 and A_1B_1, its value is $\phi_2 - \phi_1$. The image from the prism 5 through the prism 6 and projected onto the screen, and the position relationship between the projection of linear light source and the projection of other light source on the rotation platform would not be changed.

4.3 Optical Imaging Conditions of Constant Phase

According to the above analysis, to assure the image direction remain upright on the imaging screen, it is required that A_1H_1 parallel to the y axis, that is the included angle $\Delta\phi$ between the linear light source and the horizontal right-angle side of the prism 2 is equal to ϕ_1.

Suppose the angular velocities of the rotation roller and the rotation platform are ω_1 and ω_2 respectively, which both are the function of time t, in that $\phi_1 = \int_0^t \omega_1(t)dt$; $\phi_2 = \int_0^t \omega_2(t)dt$; $\Delta\phi = \phi_2 - \phi_1$. Then:

$$\int_0^t \omega_2(t)dt = 2\int_0^t \omega_1(t)dt \tag{13}$$

If the direction of the image on the screen would not change, thus $\omega_2 = 1/2\omega_1$ is required at any time.

The end of the linear light source coincides with the origin of the fixed coordinate system xyz, and the end position remains unchanged in the rotating process of the rotation platform, and the corresponding point of the end on the imaging screen also remains unchanged. The linear light source has the constant orientation and position on the imaging screen. The real time images of the rotation platform can be seen as a group of the countless linear light sources, the end of which coincides with the origin of the coordinate system xyz. That is, the images of the rotation platform can be shown on the imaging screen at real time, and the orientation and position will be unchanged. That could have certain practical significance in continuous observation and real-time monitoring of the instruments, workpiece and other apparatus on the rotation platform.

5   Conclusions

(1) With the twist nature of rope, the non-twist criterion is obtained. That is, there is no twist on the rope when both the twist angular velocity of the two end faces and the direction along the axis of rope are identical.
(2) By establishing an equivalent model, the rationality and correctness of the two-to-one transmission ratio of the avoiding twist mechanism are confirmed through the mathematical description. Further on, work situation of the avoiding twist mechanism is extended from two rotation platforms to three rotation platforms by using the rule analysis of rope twist.
(3) The correctness that the avoiding twist mechanism can be used to realize the constant phase imaging of rotation platform is presented.
(4) Non-twist criterion has important significance on the avoiding twist of cables, which can be applied in information communication and the imaging inspection fields.

Acknowledgments. This work is supported by the Shanghai Committee of Science and Technology Key Support Project (12510501100); Fund for Nature Foundation of Science (NFS51475050); Enterprise Project (E2-6203-16-10); Graduate Student Research Innovation Project of Shanghai University of Engineering Science: E3-0903-17-01215.

References

1. Zhou, X., Wang, Z.: Analysis on the main parameters of the precision conductive slip rings **27**(B11), 72–73 (2007). Proceedings of the Symposium on Metrology and Quality of the Metrology Technology Specialized Committee
2. Li, S.: Study on sliding friction and wear properties of silver based composite electrical contact materials. Hefei University of Technology (2009)
3. Wu, X., et al.: Study on the failure forms and the improvement ways of rotating union **3**, 18–23 (1997)
4. Wang, Y.: Development and research of multichannel high speed rotary joint. Guangdong University of Technology (2008)
5. Dong, H., et al.: Design of liquid rotary joint. J. Anhui Vocat. Coll. Electron. Inf. Technol. **11**(2), 23–25 (2012)
6. https://wqw.fabricast.com/fabricast-products/stock-slip-ring-assemblies/separate-rotor-and-brush-block-assemblies-1-50-inch-bore/
7. http://www.repack-s.com/applications/raccords-tournants.html
8. Adams, D.A.: Apparatus for providing energy communication between a moving and a stationary terminal. United States Patents: 810310.1971.06.22
9. Sun, H., et al.: Theory of Machines and Mechanisms. Higher Education Press, Beijing (2006)
10. Jiang, B.: Mathematics of String Figures. Dalian University of Technology Press, Dalian (2011)
11. Chiodo, M.: An Introduction to Braid Theory, 4 November 2005
12. Rooney, J.: Rotations of $4n\pi$ and the kinematic design of parallel manipulators. Open Mech. Eng. J. **4**, 86–92 (2010)

Optimal Motion Planning for Mobile Welding Robot

Gen Pan[1,2], Enguang Guan[2], Fan Yang[3(✉)], Anye Ren[1],
and Peng Gao[1]

[1] Institute of Aerospace System Engineering Shanghai, Shanghai,
People's Republic of China
[2] State Key Lab of Mechanical System and Vibration,
Shanghai Jiao Tong University, Shanghai, People's Republic of China
[3] Shanghai Aerospace Control Technology Institute,
Shanghai, People's Republic of China
Yangfan880109@163.com

Abstract. This paper focuses on the motion planning method for a novel mobile welding robot (MWR), based on the screw theory. The robot consists of a vehicle unit and a 5-DOF manipulator, which equipped a torch at the end of manipulator. In order to finish the welding task, the kinematic motion planning strategy is of great importance. As the traditional strategy which uses inverse kinematic and polynomial interpolation may cause a waste of computing time, the screw theory is chosen to improve the strategy. From the simulation and experiment results, it can be found that the optimal motion planning method is reliable and efficient.

Keywords: Mobile welding robot · Motion planning strategy · Screw theory

1 Introduction

Welding tasks involved in large unstructured equipment building such as ships and oil tankers are challenging and urge to be solved. As the working conditions are space constricted and dangerous, a novel mobile welding robot which can move on the slope or even vertical surfaces of huge equipments can be one potential solution. Recently, several research works have proved that automatic welding technology can improve the quality of welding seams signally [1–3]. Path planning of wheeled welding robots [4–6] and the algorithm research for trajectory planning [7–9] are also helpful for increasing the reliability and the efficiency of the MWR system.

In this paper, the MWR system has a novel combined mechanism for welding tasks. In our research team, Wu [10, 11] has made a detailed introduction about the MWR's structure in the published papers. The vehicle unit consists of three identical mobile adhesion parts and each of them comprises one wheel group, one magnetic adhesion unit and one lifting mechanism. The welding manipulator contains five rotational joints. The welding torch is held at the end of the manipulator.

Using the MWR as the controlled member, an optimal motion planning strategy is proposed. As the traditional strategy which uses inverse kinematic and polynomial

© Springer International Publishing AG 2017
Y. Huang et al. (Eds.): ICIRA 2017, Part II, LNAI 10463, pp. 124–134, 2017.
DOI: 10.1007/978-3-319-65292-4_12

interpolation [12, 13] may cause a waste of computing time, the screw theory model is used to improve the strategy performance. The structure of this paper is as follows: a brief introduction of the MWR is proposed in Sect. 2. The screw model of MWR and the motion planning strategy are presented in Sect. 3. Details about the simulation and experimental evaluation are demonstrated as the result and discussion and presented in Sect. 4. Conclusions are drawn in Sect. 5.

2 Mobile Welding Robot Design

The prototype of MWR is shown in Fig. 1. The MWR consists of one vehicle platform at the bottom and a 5-DOF manipulator arm mounted on the vehicle platform. The welding torch is clamped onto the manipulator. The wheel-foot mechanism of vehicle can move and cross obstacle flexibly adapting to large unstructured equipments, and three permanent-magnetic suction cups can adjust the air gap to control the adsorption force. The MWR has the mobility capability such as walking, climbing wall, crossing obstacles, turning and welding. To achieve a coordinate welding trajectory, need to study the motion planning algorithm insuring computational efficiency and path precision.

Fig. 1. Prototype of mobile welding robot

The main hardware structure of the control system of WMR is shown in Fig. 2. The control system is composed of an industrial personal computer, servo drivers, servo motors, stereo camera, sensor system, and software. IPC and servo drivers are configured as master-slave construction. Communication link between IPC and drivers or sensor is established by CAN bus. IPC is used for planning and sending mission commands. Servo drivers generate real-time control command of the motors, execute the motion planning and control algorithm, and drive the motors moving.

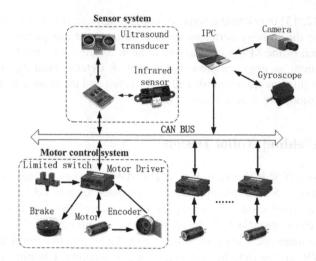

Fig. 2. The structure of hardware system

3   Motion Planning Strategy for MWR

3.1   Kinematical Modeling Based on Screw Theory

Different from the traditional kinematical modeling theory such as D-H theory, screw theory [14] is efficient in rotational and stretching DOF modeling. Under screw theory, the kinematical modeling can be described as follows:

Figure 3 shows the coordinate system of manipulator. If θ_i, $i = 1, 2, \ldots, 5$ do not equals to 0 at the same time, the revolving vectors of the rotation joints can be denoted as:

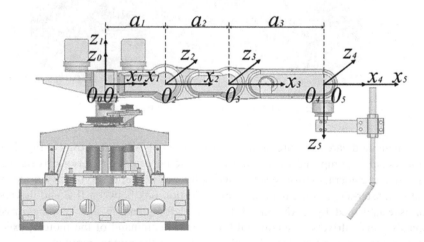

Fig. 3. The coordinate system of manipulator

$$\$_1 = (0 \quad 0 \quad 1; 0 \quad 0 \quad 0)$$

$$\$_2 = (-s_1 \quad c_1 \quad 0; 0 \quad 0 \quad a_1)$$

$$\$_3 = (-s_1 \quad c_1 \quad 0; a_2 c_1 s_2 \quad a_2 s_1 s_2 \quad a_1 + a_2 c_2)$$

$$\$_4 = (-s_1 \quad c_1 \quad 0; c_1(a_3 s_{23} + a_2 s_2) \quad s_1(a_3 s_{23} + a_2 s_2) \quad a_1 + a_3 c_{23} + a_2 c_2)$$

$$\$_5 = (-s_{23} c_1 \quad -s_{23} s_1 \quad -c_{23}; -s_1(a_1 c_{23} + a_2 c_3) \quad c_1(a_1 c_{23} + a_2 c_3) \quad 0)$$

$$\$_{wl} = (0 \quad 0 \quad 0; 0 \quad 1 \quad 0)$$

where $\theta_i (i = 1, 2, \ldots, 5)$ is the angle of the *i-th* joint, while $\$_i, i = 1, 2, \cdots, 5$ is the screw vector of the *i-th* rotation joint, and $\$_{wl}$ is the screw vector of the vehicle unit. c_i represents $\cos \theta_i$, and s_i represents $\sin \theta_i$, while c_{ij} and s_{ij} mean $\cos(\theta_i + \theta_j)$ and $\sin(\theta_i + \theta_j)$.

If the linear velocity of the vehicle unit is v_{wl}, and the rotational velocity of each rotation joint is ω_i, $i = 1, 2, \cdots, 5$, the screw vector at the end of the manipulator can be denoted as:

$$\$_m = v_{wl} \$_{wl} + \sum_{i=1}^{5} \omega_i \$_i = \begin{bmatrix} -s_1(\omega_2 + \omega_3 + \omega_4 + \omega_5) \\ c_1(\omega_2 + \omega_3 + \omega_4 + \omega_5) \\ \omega_1 \\ a_2 c_1 s_2 \omega_3 + c_1(a_3 s_{23} + a_2 s_2)(\omega_4 + \omega_5) \\ v_{wl} + a_2 s_1 s_2 \omega_3 + s_1(a_3 s_{23} + a_2 s_2)(\omega_4 + \omega_5) \\ a_1 \omega_2 + (a_1 + a_2 c_2)\omega_3 + (a_1 + a_3 c_{23} + a_2 c_2)(\omega_4 + \omega_5) \end{bmatrix} \quad (1) \quad (1)$$

For each joint, the motion ability is limited. That means $v_{wl\_max} \leq v_{wl} \leq v_{wl\_min}$ and $\omega_{i\_max} \leq \omega_i \leq \omega_{i\_min}$, $i = 1, 2, \cdots, 5$.

3.2   Linear Motion Modeling Based on Screw Theory

The linear trajectory which is the most common in the welding task is chosen as the motion target. Based on the screw theory, Eq. (1) can be rewritten as, $\$_m = (S_m; S_m^0)$, $S_m = \$_m(1:3)$, $S_m^0 = \$_m(4:6)$. $\$_m$ is a linear screw only when $S_m = 0$. Then $\omega_1 = 0$ and $\omega_2 + \omega_3 + \omega_4 + \omega_5 = 0$, we can get:

$$\$_m = \begin{bmatrix} 0 \\ 0 \\ 0 \\ a_2 c_1 s_2 \omega_3 + c_1(a_3 s_{23} + a_2 s_2)(\omega_4 + \omega_5) \\ v_{wl} + a_2 s_1 s_2 \omega_3 + s_1(a_3 s_{23} + a_2 s_2)(\omega_4 + \omega_5) \\ a_1 \omega_2 + (a_1 + a_2 c_2)\omega_3 + (a_1 + a_3 c_{23} + a_2 c_2)(\omega_4 + \omega_5) \end{bmatrix}$$

As the target trajectory is linear, the direction can be denoted as $\mathbf{S}_m^0 / \|\mathbf{S}_m^0\|$ and the velocity is $\|\mathbf{S}_m^0\|$, where $\|\bullet\|$ is the screw vector modulus. Then we can get:

$$\mathbf{S}_m^0 = \begin{bmatrix} a_2 c_1 s_2 \omega_3 + c_1 (a_3 s_{23} + a_2 s_2)(\omega_4 + \omega_5) \\ v_{wl} + a_2 s_1 s_2 \omega_3 + s_1 (a_3 s_{23} + a_2 s_2)(\omega_4 + \omega_5) \\ a_1 \omega_2 + (a_1 + a_2 c_2) \omega_3 + (a_1 + a_3 c_{23} + a_2 c_2)(\omega_4 + \omega_5) \end{bmatrix} \tag{2}$$

In Eq. (2), $\omega_1 = 0$, θ_1 is constant and θ_i, $i = 2, 3, \cdots, 5$ is variable, under $\omega_j \neq 0$, $j = 2, 3, \cdots, 5$. If the target velocity (v_x, v_y, v_z) is known, we can get:

$$\begin{cases} \omega_2 + \omega_3 + \omega_4 + \omega_5 = 0 \\ a_2 c_1 s_2 \omega_3 + c_1 (a_3 s_{23} + a_2 s_2)(\omega_4 + \omega_5) = v_x \\ v_{wl} + a_2 s_1 s_2 \omega_3 + s_1 (a_3 s_{23} + a_2 s_2)(\omega_4 + \omega_5) = v_y \\ a_1 \omega_2 + (a_1 + a_2 c_2) \omega_3 + (a_1 + a_3 c_{23} + a_2 c_2)(\omega_4 + \omega_5) = v_z \end{cases} \tag{3}$$

Equation (3) can be simplified by setting $\omega_{45} = \omega_4 + \omega_5$. Then it turns to:

$$\begin{cases} \omega_2 + \omega_3 + \omega_{45} = 0 \\ a_2 c_1 s_2 \omega_3 + c_1 (a_3 s_{23} + a_2 s_2) \omega_{45} = v_x \\ v_{wl} c_1 = v_y c_1 - v_x s_1 \\ a_2 c_2 \omega_3 + (a_3 c_{23} + a_2 c_2) \omega_{45} = v_z \end{cases}$$

The solution can be calculated as:

$$v_{wl} = v_y - v_x \tan(\theta_1)$$

$$\omega_3 = -\frac{v_x a_3 c_{23} + v_x a_2 c_2 - v_z c_1 (a_3 s_{23} + a_2 s_2)}{a_2 a_3 c_1 s_3}$$

$$\omega_{45} = \frac{v_x c_2 - v_z c_1 s_2}{a_3 c_1 s_3}$$

$$\omega_2 = -\omega_3 - \omega_{45}$$

By the previous velocity constraints, we can get:

$$\begin{cases} v_{wl\_Min} \leq v_{wl} \leq v_{wl\_Max} \\ \omega_{2\_Min} \leq \omega_2 \leq \omega_{2\_Max} \\ \omega_{3\_Min} \leq \omega_3 \leq \omega_{3\_Max} \\ \omega_{4\_Min} + \omega_{5\_Min} \leq \omega_{45} \leq \omega_{4\_Max} + \omega_{5\_Max} \end{cases}$$

Then

$$\begin{cases} v_{wl\_Min} \leq v_y - v_x \tan(\theta_1) \leq v_{wl\_Max} \\ \omega_{2\_Min} \leq -\omega_3 - \omega_{45} \leq \omega_{2\_Max} \\ a_2 a_3 c_1 s_3 \omega_{3\_Min} \leq -v_x(a_3 c_{23} + a_2 c_2) + v_z c_1 (a_3 s_{23} + a_2 s_2) \leq a_2 a_3 c_1 s_3 \omega_{3\_Max} \\ a_3 c_1 s_3 (\omega_{4\_Min} + \omega_{5\_Min}) \leq v_x c_2 - v_z c_1 s_2 \leq a_3 c_1 s_3 (\omega_{4\_Max} + \omega_{5\_Max}) \end{cases}$$

Finally

$$\begin{cases} (a_3 c_{23} + a_2 c_2)\omega_{45\_Min} + a_2 c_2 \omega_{3\_Min} \leq v_z \leq (a_3 c_{23} + a_2 c_2)\omega_{45\_Max} + a_2 c_2 \omega_{3\_Max} \\ a_2 c_1 s_2 \omega_{3\_Min} + c_1(a_3 s_{23} + a_2 s_2)\omega_{45\_Min} \leq v_x \leq c_1(a_3 s_{23} + a_2 s_2)\omega_{45\_Max} + a_2 c_1 s_2 \omega_{3\_Max} \\ a_2 c_1 s_3 \omega_{2\_Min} \leq v_x c_{23} - v_z c_1 s_{23} \leq a_2 c_1 s_3 \omega_{2\_Max} \\ v_{wl\_Min} \leq v_y - v_x \tan(\theta_1) \leq v_{wl\_Max} \end{cases}$$

$$(4)$$

where $\omega_{45\_Min} = \omega_{4\_Min} + \omega_{5\_Min}$, $\omega_{45\_Max} = \omega_{4\_Max} + \omega_{5\_Max}$.

By using Eq. (4), the proper rotational velocity can be found for each target velocity (v_x, v_y, v_z).

3.3   Optimal Motion Planning Strategy for MWR

In the traditional motion planning strategy based on D-H inverse kinematics theory, the points containing the position and pose information along the target trajectory are calculated one by one. The density of the points is depended on the accuracy requirement. All of the position and pose information between two points can be calculated by using polynomial interpolation. Therefore if the density of the points is too high, the inverse kinematics calculation will waste extra time; otherwise tracking error will be unacceptable, even may cause the system shock.

Under the previous screw theory model, if each joint angle is known and there is no velocity jump point, the modulated velocity processes will be smoothly and the instantaneous velocity direction at the end of the manipulator will keep along the target trajectory. In this ideal model, the joint angle, rotation velocity and the target velocity at the end of the manipulator can be related by an analytical expression. But in the discrete control mode, the joint angle is nonlinear to the velocity, and there is no analytical solution.

Considering the motion accuracy and instantaneity demands, an optimal motion planning strategy for MWR is designed. In this strategy, if the tracking error is under the threshold value, the screw model is chosen to finish the planning calculation. Otherwise, the D-H inverse kinematics model is chosen to get the tracking error back to the threshold value. The planning process is shown in Fig. 4.

The basic procedures of motion planning algorithm are exAApressed as follows:

(1) Initialize each joint angle and the linear velocity vector of MWR. If the position at the end of the manipulator is out of the work space, the controller will drive the MWR back to the work space.
(2) Initialize the step parameter, including step time, step linear distance and the maximum step joint angle. According to the MWR configuration, θ_2 and θ_3 are sensitive to the linearity of screw vector. Thus after ω_2 and ω_3 are calculated, the bigger one is chosen as the maximum step joint angle.
(3) According to the step parameters, calculate the joint angle, rotation velocity. Then drive the MWR motion.

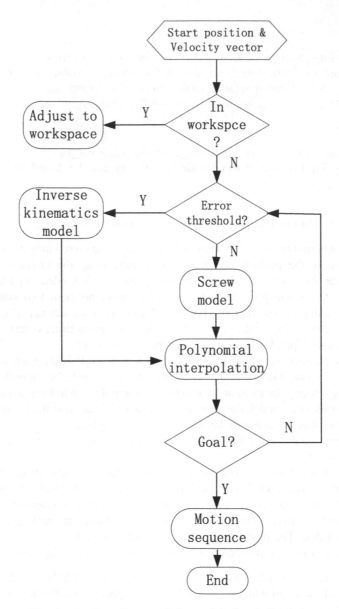

Fig. 4. The flow diagram of the motion planning strategy

(4) During the step motion, the rotation velocity constraints can forbid the velocity jump. Considering the beginning velocity, ending velocity and the motion accuracy demand, polynomial interpolation is used to calculate the joint angle and the rotation velocity.

(5) If the tracking error is under the threshold value, the screw model is chosen to calculate the step parameters. Otherwise, the D-H inverse kinematics model is chosen to get the tracking error back to the threshold value.

4 Simulation and Experiment

The linear trajectory motion planning of MWR was simulated on MATLAB, running on a desktop with a CPU Pentium E5200 Dual-Core, RAM 4 GB. The joint angle vector was initialized as $(0, -\pi/3, \pi/3, 0, 0)$; the target velocity was initialized as $(1, 0, 0)$, the velocity unit was mm/s; the simulation time duration was 50 s; the length of the linear trajectory is 50 mm. In the working space, the fixed X-Y-Z angle coordinate system is chosen to describe the terminal position of the manipulator. The angle coordinate values can be calculated by:

$$\beta = A \tan 2(-r_{31}, \sqrt{r_{11}^2 + r_{21}^2})$$
$$\alpha = A \tan 2(r_{21}/c\beta, r_{11}/c\beta)$$
$$\gamma = A \tan 2(r_{32}/c\beta, r_{33}/c\beta)$$

Where r_{ij} is the homogeneous transform matrix element of the tool frame, and $c\beta$ means $\cos \beta$.

In order to evaluate the modified motion planning strategy, the strategies only containing D-H inverse kinematics theory and screw theory are taken into the simulations at the same time.

The simulation results in Fig. 5 indicate that the linear motion process under D-H inverse kinematics theory is smooth and the tracking error is acceptable. The whole simulation time which does not contain the inverse operation of Jacobian matrix and polynomial interpolation is 4.8723 s.

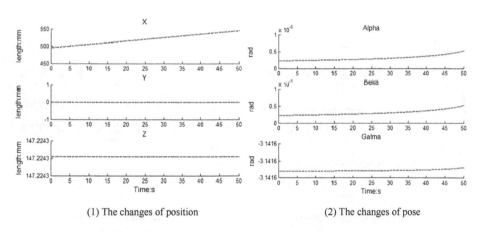

(1) The changes of position   (2) The changes of pose

Fig. 5. The simulation results under D-H inverse kinematics theory

The simulation results in Fig. 6 indicate that the linearity of motion under screw theory is worse than the inverse kinematics theory, especially in Z-axis. And the tracking error of posture becomes larger with the simulation time. But the whole simulation time is only 1.4029 s.

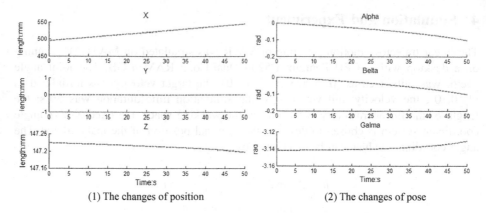

(1) The changes of position   (2) The changes of pose

Fig. 6. The simulation results under screw theory

The simulation results under the modified planning strategy are shown in Fig. 7. The threshold value of tracking error is initialized as 0.005 mm. If the tracking error is over the threshold value, the D-H inverse kinematics theory is activated. Thus the changes of position and pose are not continuous. It can be found that the amplitude of the change is limited, such as the amplitude of position change is under 0.0005 mm and the amplitude of pose change is under 0.01. These will not cause the system shock. The whole simulation time is 1.6356 s. It is a little worse than the screw theory, but is more efficient than the inverse kinematics theory. Therefore the optimal motion planning strategy is with a high accuracy and efficient.

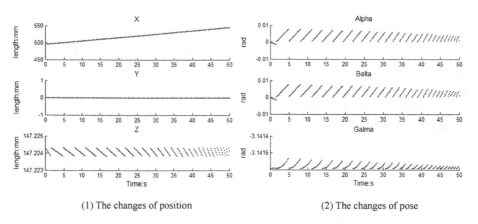

(1) The changes of position   (2) The changes of pose

Fig. 7. The simulation results under modified planning strategy

Through the optimization analysis of motion planning algorithm, the modified strategy is verified in the welding experiment. Figure 8(1) shows the welding process. The manipulator adjusts the position and posture to the ready state and then start welding. When finish welding, remove the torch from the workbench. Figure 8(2) shows the welding effect of the motion planning. The weld length is 50 mm, and result is uniform. It proves that the strategy is effective.

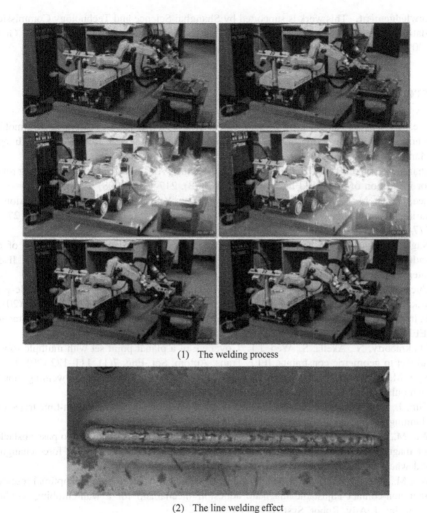

(1) The welding process

(2) The line welding effect

Fig. 8. The welding experiment of MWR

5 Conclusions

This paper has introduced a novel motion planning strategy for MWR. The mechanical structure and control system have been presented. To finish the linear welding task, the kinematic modeling and the motion modeling have been described in detail based on screw theory. Under the optimal motion planning strategy, the MWR is adaptive to the kinds of tracking error. Results from the simulation and experiments have demonstrated that the motion planning strategy is efficient, and it can be applied confidently in the domain of mobile welding robot systems. The the precision and the reliability will be increased with more sensors in the future.

Acknowledgement. This work is supported by Shanghai Science and Technology Commission Foundation under Grant No. 13DZ1108300, and National 863 plan of China under Grant No. 2009AAA042221.

References

1. Lee, D., Lee, S., Ku, N., Lim, C., Lee, K.Y., Kim, T.W., et al.: Development of a mobile robotic system for working in the double-hulled structure of a ship. Robot. Comput.-Integr. Manuf. **26**(1), 13–23 (2010)
2. Shang, J., Bridge, B., Sattar, T.P., Mondal, S., Brenner, A.: Development of a climbing robot for inspection of long weld lines. Ind. Robot **35**(3), 217–223 (2008)
3. Lee, D., Ku, N., Kim, T.W., Kim, J., Lee, K.Y., Son, Y.S.: Development and application of an intelligent welding robot system for shipbuilding. Robot. Comput.-Integr. Manuf. **27**(2), 377–388 (2011)
4. Nagarajan, U., Kantor, G., Hollis, R.L., (eds.): Trajectory planning and control of an underactuated dynamically stable single spherical wheeled mobile robot. In: IEEE International Conference on Robotics and Automation (2009)
5. Gueta, L.B., Chiba, R., Arai, T., Ueyama, T., Ota, J.: Practical point-to-point multiple-goal task realization in a robot arm with a rotating table. Adv. Robot. **25**(6–7), 717–738 (2011)
6. Toit, N.E.D., Burdick, J.W.: Robot motion planning in dynamic, uncertain environments. IEEE Trans. Robot. **28**(1), 101–115 (2012)
7. Chakraborty, N., Akella, S., Wen, J.T.: Coverage of a planar point set with multiple robots subject to geometric constraints. IEEE Trans. Autom. Sci. Eng. **7**(1), 111–122 (2010)
8. Ngo, M.D., Phuong, N.T., Duy, V.H., Kim, H.K.: Control of two wheeled welding mobile manipulator. Int. J. Adv. Robot. Syst. **4**(3) (2008)
9. Kim, J., Kim, S.R., Kim, S.J., Kim, D.H.: A practical approach for minimum-time trajectory planning for industrial robots. Ind. Robot **37**(1), 51–61 (2010)
10. Wu, M., Gao, X., Yan, W.X., Fu, Z., Zhao, Y., Chen, S.: New mechanism to pass obstacles for magnetic climbing robots with high payload, using only one motor for force-changing and wheel-lifting. Ind. Robot **38**(4), 372–380 (2011)
11. Wu, M., Pan, G., Zhang, T., Chen, S., Zhuang, F., Zhao, Y.Z.: Design and optimal research of a non-contact adjustable magnetic adhesion mechanism for a wall-climbing welding robot. Int. J. Adv. Robot. Syst. **10**(1), 1 (2013)
12. Zhang, T., Chen, S.B.: Optimal posture searching algorithm on mobile welding robot. J. Shanghai Jiaotong Univ. (Sci.) **19**(1), 84–87 (2014)
13. Zhang, T., Wu, M., Zhao, Y., Chen, X., Chen, S.: Optimal motion planning of mobile welding robot based on multivariable broken line seams. Int. J. Robot. Autom. **29**(2), 215–223 (2014)
14. Rodriguez-Leal, E., Dai, J.S., Pennock, G.R.: A study of the kinematics of the 5-RSP parallel mechanism using screw theory. In: Dai, J., Zoppi, M., Kong, X. (eds.) Advances in Reconfigurable Mechanisms and Robots I, pp. 355–369. Springer, London (2012). doi:10.1007/978-1-4471-4141-9_32

Off-Line Programmed Error Compensation of an Industrial Robot in Ship Hull Welding

Guanglei Wu[⊠], Delun Wang, and Huimin Dong

School of Mechanical Engineering, Dalian University of Technology,
Dalian 116024, China
{gwu,dlunwang,donghm}@dlut.edu.cn

Abstract. This paper presents the compliance modeling and error compensation for an industrial robot in the application of ship hull welding. The Cartesian stiffness matrix is derived using the virtual-spring approach, which takes the actuation and structural stiffness, arm gravity and external loads into account. Based on the developed stiffness model, a method to compensate the compliance error is introduced, being illustrated with an industrial robot along a welding trajectory. The results show that this compensation method can effectively improve the robot's operational accuracy, allowing the actual trajectory of the robot with auxiliary loads to coincide with the target one approximately.

Keywords: Industrial robot · Hull welding · Kinetostatics · Gravity effect · Error compensation

1 Introduction

As an automatic manufacturing device, industrial robots have been widely used and play an important role in many fields of industrial applications, thanks to their high repeatability accuracy and operational stability. Amongst the industrial applications, this type of serial manipulators are well adapted in welding industry [3], such as auto industry. Besides, the robots can also be used in ship-building industry, i.e., the large-scale hull welding, as displayed in Fig. 1. In this procedure, the welding trajectory is firstly detected by the vision system, where the data of the trajectory will be transferred to the control system for robot programming to accomplish the welding procedure. This means that the robot will conduct an off-line task.

The robots' motions are usually generated via the robotic controllers by virtue of the inverse kinematic model to compute the input signals for actuators corresponding to the desired end-effector position, where the compliance errors are ignored. On the other hand, as depicted in Fig. 1(c), under certain external load, particularly, the arm gravity and dynamic inertial forces, the kinematic control becomes non-applicable due to the elastic compliance errors caused by the limited strength of the robot components [12], namely, the actual trajectory

© Springer International Publishing AG 2017
Y. Huang et al. (Eds.): ICIRA 2017, Part II, LNAI 10463, pp. 135–146, 2017.
DOI: 10.1007/978-3-319-65292-4_13

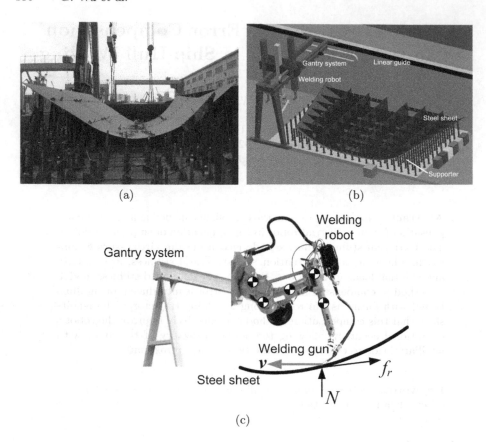

(a)   (b)

(c)

Fig. 1. The robot-based ship hull welding: (a) steel sheet on the supporter; (b) overall scheme; (c) reaction forces onto the welding gun.

will shift away from the desired path, resulting in the decreased product quality wherefrom high precision is needed in the applications.

The compliance error, i.e., geometric changes of the robot end-effector, can be compensated through the calibration [7,21], whereas, this technique is sometimes expensive. An economic way to handle the problem of error compensation is the modification of the robot control scheme [6,10] that defines the prescribed trajectory in Cartesian space: based on the error model, the loaded input trajectory is regenerated to achieve the coincidence between the output trajectory and the desired one, while input trajectory differs from the target one. The modification of the input trajectory is to be based on the compliance error model that calls for the computation of the stiffness matrix [8]. The stiffness matrix for the serial robotics was first derived by Salisbury [15], where only the actuation compliance described by one-dimensional linear springs was considered. In this approach, the derivation of the stiffness is on the basis of the assumption that the manipulator is in an unloaded equilibrium configuration. In practice, the external loads directly influence on the manipulator equilibrium configuration and may modify the stiffness properties [2,4]. As a consequence, the structural compliance and the robot geometry change due to external loads and gravity should be considered [11,13,14]. Thus, this work focuses on the compliance error modeling and compensation that is able to take into account the influence of the external and internal wenches and robot gravity onto the stiffness matrix and elastic deformations.

This paper deals with the stiffness modeling and compliance error compensation of an industrial robot in ship hull welding. The Cartesian stiffness matrix is computed in loaded configurations, where the actuation and structural stiffness is considered as well as the influence of the arm gravity and external loads. A method to modify the input trajectory is presented for error compensation. This method is numerically illustrated with the welding robot along a trajectory and the results show the effectiveness of the error compensation approach.

2   Industrial Welding Robot

Figure 2 shows the ABB IRB4600_60/205 robot [1] as the welding robot in this project. IRB 4600 series is ABB Robotics pioneer of the new sharp generation with enhanced and new capabilities, of which one industrial application is dedicated to welding.

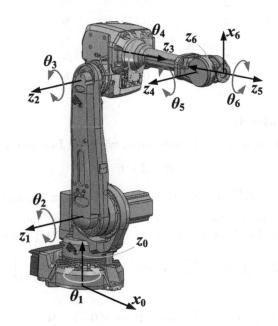

Fig. 2. The ABB IRB4600_60/205 robot and its coordinate systems.

2.1 Kinematics of the Industrial Robot

Following the Denavit–Hartenberg (D–H) convention [5], the Cartesian coordinate systems are established for each link of the robotic arm as shown in Fig. 2. Hereafter, let $\mathbf{i}$, $\mathbf{j}$ and $\mathbf{k}$ be the unit vectors along x-, y- and z-axis, respectively. The transformation matrix in the forward kinematics of the end-effector in the reference frame (x_0, y_0, z_0) is expressed as

$$^0\mathbf{A}_6 = \begin{bmatrix} \mathbf{R} & \mathbf{q} \\ \mathbf{0} & 1 \end{bmatrix} = \prod_{i=1}^{6} {}^{i-1}\mathbf{A}_i \quad \text{where} \quad {}^{i-1}\mathbf{A}_i = \begin{bmatrix} {}^{i-1}\mathbf{R}_i & {}^{i-1}\mathbf{q}_i \\ \mathbf{0} & 1 \end{bmatrix} \tag{1}$$

with

$$^{i-1}\mathbf{R}_i = \mathbf{R}_{z_{i-1}}(\theta_i)\mathbf{R}_{x_i}(\alpha_i) \tag{2a}$$

$$^{i-1}\mathbf{q}_i = \begin{bmatrix} a_i \cos\alpha_i & a_i \sin\alpha_i & d_i \end{bmatrix}^T \tag{2b}$$

where the D–H parameters are listed in Table 1. The inverse geometry problem of the six-axis robotics has been well documented in the literature [16].

Table 1. D-H parameters of the IRB4600_60/205 robot.

Joint i	α_i	a_i [m]	d_i [m]	θ_i
1	$\pi/2$	0.175	0.495	θ_1
2	0	1.075	0	θ_2
3	$\pi/2$	0.175	0	θ_3
4	$-\pi/2$	0	0.960	θ_4
5	$\pi/2$	0	0	θ_5
6	π	0	0.135	θ_6

2.2 Jacobian Matrix

The joint angular velocity can be calculated with the Jacobian matrix as below:

$$\dot{\boldsymbol{\theta}} = \mathbf{J}^{-1}\mathbf{t} \tag{3}$$

where $\dot{\boldsymbol{\theta}} = \begin{bmatrix} \dot{\theta}_1 & \dot{\theta}_2 & \dots & \dot{\theta}_6 \end{bmatrix}^T$ denotes the vector of the joint angular velocities, and $\mathbf{t} = \begin{bmatrix} \boldsymbol{\omega}^T & \dot{\mathbf{q}}^T \end{bmatrix}^T$ stands for the end-effector twist, $\boldsymbol{\omega}$ being the angular velocities. Moreover, $\mathbf{J}$ is the Jacobian matrix of the robotic arm [17], namely,

$$\mathbf{J} = \begin{bmatrix} \mathbf{j}_1 & \mathbf{j}_2 & \dots & \mathbf{j}_6 \end{bmatrix} \quad \text{where} \quad \mathbf{j}_i = \begin{bmatrix} \mathbf{z}_{i-1} \\ \mathbf{p}_{i-1} \times \mathbf{z}_{i-1} \end{bmatrix} \tag{4}$$

with

$$\mathbf{z}_{i-1} = \mathbf{R}_{i-1}\mathbf{k}; \quad \mathbf{p}_{i-1} = \mathbf{q}_{i-1} - \mathbf{q} \tag{5}$$

where $\mathbf{R}_{i-1}$ and $\mathbf{q}_{i-1}$ denote the rotation matrix and position vector of the transformation matrix from the reference coordinate system to the $(i-1)$th coordinate system, respectively, which can be extracted from $\prod_{i=0}^{i-1} {}^{i-1}\mathbf{A}_i$ in Eq. (1).

2.3   Robot Dynamics

The dynamic behavior of the robot under the load $\mathbf{f}_e$ during the welding can be described as

$$\mathbf{M}\delta\ddot{\mathbf{t}} + \mathbf{C}\delta\dot{\mathbf{t}} + \mathbf{K}\delta\mathbf{t} = \mathbf{f}_e \tag{6}$$

where $\mathbf{M}$ is 6-dimensional mass matrix that represents the global behavior of the robot in terms of natural frequencies, $\mathbf{C}$ is 6-dimensional damping matrix, and $\mathbf{K}$ is 6-dimensional Cartesian stiffness matrix of the robot under the external loading $\mathbf{f}_e$ [18]. Moreover, $\delta\mathbf{t}$, $\delta\dot{\mathbf{t}}$ and $\delta\ddot{\mathbf{t}}$ are the instantaneous dynamic displacement, velocity and acceleration of the tool end-point, respectively.

The mass matrix $\mathbf{M}$ in Eq. (6) can be solved based on the dynamic equation of motion (EOM) [9] of a serial robot as below:

$$\mathbf{M}(\boldsymbol{\theta})\ddot{\boldsymbol{\theta}} + \mathbf{v}(\boldsymbol{\theta}, \dot{\boldsymbol{\theta}}) + \mathbf{g}(\boldsymbol{\theta}) = \boldsymbol{\tau} \tag{7}$$

where $\boldsymbol{\theta}$, $\dot{\boldsymbol{\theta}}$, $\ddot{\boldsymbol{\theta}}$ are 6-dimensional vectors of generalized joint angles, angular velocities, and angular accelerations, respectively. $\mathbf{M}(\boldsymbol{\theta})$ is the 6×6 general inertial matrix, $\mathbf{v}(\boldsymbol{\theta}, \dot{\boldsymbol{\theta}})$ is a 6-dimensional vector representing the Coriolis and centrifugal forces, and $\mathbf{g}(\boldsymbol{\theta})$ is a 6-dimensional vectors accounting for the force due to gravity. Moreover, $\boldsymbol{\tau}$ is the vector of general external forces applied at joints.

3   Compliance Error Modeling and Compensation

In order to compensate the positioning errors of the robot end-effector for high-quality welding, the compliance errors caused by the external payloads and robot gravity should be calculated precisely, which calls for the computation of the stiffness matrix of the robot.

3.1   Elastostatic Modeling

In this work, the virtual spring approach [13] is adopted to derive the stiffness matrix, based on the screw coordinates [19, 20]. Figure 3 shows the VJM model of the robotic arm, where $\mathbf{g}_j$, $j = 1, 2, \ldots, 8$, stand for the gravity and $\mathbf{f}_e$ denotes the external loads.

Let $\boldsymbol{\theta}$, $\boldsymbol{\theta}'$ be the original and the deformed angular displacements of the joints, respectively, in accordance with the principle of virtual work, the work of the auxiliary loads is equal to the work of internal forces $\boldsymbol{\tau}_\theta$ [13], namely,

$$\sum(\mathbf{g}_j^T \delta\mathbf{t}_j) + \mathbf{f}_e^T \delta\mathbf{t} = \boldsymbol{\tau}_\theta^T(\boldsymbol{\theta}' - \boldsymbol{\theta}) \tag{8}$$

where the virtual displacements $\delta\mathbf{t}_j$ and $\delta\mathbf{t}$ can be computed from the linearized geometrical model derived from $\delta\mathbf{t}_j = \mathbf{J}_j(\boldsymbol{\theta}'-\boldsymbol{\theta})$ and $\delta\mathbf{t} = \mathbf{J}_\theta(\boldsymbol{\theta}'-\boldsymbol{\theta})$, respectively, $\mathbf{J}_j$ and $\mathbf{J}_\theta$ being the Jacobians, namely,

$$\mathbf{J}_\theta = \begin{bmatrix} \mathbf{j}_1 \ \mathbf{j}_2 \ \mathbf{J}_U \ \mathbf{j}_3 \ \mathbf{j}_4 \ \mathbf{J}_F \ \mathbf{j}_5 \ \mathbf{j}_6 \end{bmatrix} \in \mathbb{R}^{6\times 18} \tag{9}$$

$$\mathbf{J}_j = \mathbf{J}_\theta(:, 1{:}k) \tag{10}$$

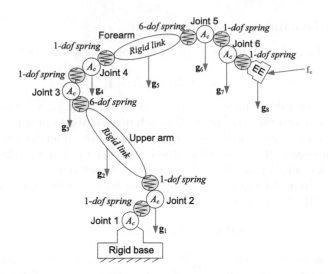

Fig. 3. Virtual-spring model of the industrial robot with auxiliary loads, where A_c and EE stand for the actuator and end-effector, respectively.

where $\mathbf{J}_\theta(:, 1:k)$ stands for the first k columns in $\mathbf{J}_\theta$ and k stands for the number of the mobilities of the virtual springs from the base to $\mathbf{g}_j$. Moreover, $\mathbf{J}_U$ and $\mathbf{J}_F$, respectively, relate the link deflections of the upper arm and forearm to the end-effector, expressed as

$$\mathbf{J}_U = \begin{bmatrix} \mathbf{x}_1 & \mathbf{y}_1 & \mathbf{z}_1 & \mathbf{0} & \mathbf{0} & \mathbf{0} \\ \mathbf{q}_2 \times \mathbf{x}_1 & \mathbf{q}_2 \times \mathbf{y}_1 & \mathbf{q}_2 \times \mathbf{z}_1 & \mathbf{x}_1 & \mathbf{y}_1 & \mathbf{z}_1 \end{bmatrix} \tag{11a}$$

$$\mathbf{J}_F = \begin{bmatrix} \mathbf{z}_3 & \mathbf{x}_3 & \mathbf{y}_3 & \mathbf{0} & \mathbf{0} & \mathbf{0} \\ \mathbf{q}_4 \times \mathbf{z}_3 & \mathbf{q}_4 \times \mathbf{x}_3 & \mathbf{q}_4 \times \mathbf{y}_3 & \mathbf{z}_3 & \mathbf{x}_3 & \mathbf{y}_3 \end{bmatrix} \tag{11b}$$

Equation (8) is rewritten as

$$\sum (\mathbf{g}_j^T \mathbf{J}_j(\boldsymbol{\theta}' - \boldsymbol{\theta})) + \mathbf{f}_e^T \mathbf{J}_\theta(\boldsymbol{\theta}' - \boldsymbol{\theta}) = \boldsymbol{\tau}_\theta^T(\boldsymbol{\theta}' - \boldsymbol{\theta}) \tag{12}$$

sequentially, the force equilibrium equation for the robotic arm is derived as

$$\boldsymbol{\tau}_\theta = \sum (\mathbf{J}_j^T \mathbf{g}_j) + \mathbf{J}_\theta^T \mathbf{f}_e = \mathbf{J}_g^T \mathbf{G} + \mathbf{J}_\theta^T \mathbf{F} \tag{13}$$

with

$$\mathbf{J}_g^T = \begin{bmatrix} \mathbf{J}_1 \ \mathbf{J}_2 \ldots \mathbf{J}_8 \end{bmatrix} \in \mathbb{R}^{6 \times 48}; \quad \mathbf{G} = \begin{bmatrix} \mathbf{g}_1^T \ \mathbf{g}_2^T \ \cdots \ \mathbf{g}_8^T \end{bmatrix}^T \in \mathbb{R}^{48} \tag{14}$$

With the linear force-deflection relation, the equilibrium condition is written as

$$\mathbf{J}_g^T \mathbf{G} + \mathbf{J}_\theta^T \mathbf{f}_e = \mathbf{K}_\theta(\boldsymbol{\theta}_i' - \boldsymbol{\theta}_i) \tag{15}$$

with the stiffness matrix $\mathbf{K}_\theta$ in the joint space expressed as

$$\mathbf{K}_\theta = \mathrm{diag} \begin{bmatrix} K_{\mathrm{act},1} \ K_{\mathrm{act},2} \ \mathbf{K}_U \ K_{\mathrm{act},3} \ K_{\mathrm{act},4} \ \mathbf{K}_F \ K_{\mathrm{act},5} \ K_{\mathrm{act},6} \end{bmatrix} \tag{16}$$

where $K_{\text{act},i}$ is the actuation stiffness, and $\mathbf{K}_U$ and $\mathbf{K}_F$, calculated by the Euler-Bernoulli beam model, represent the 6×6 stiffness matrices of the upper arm and forearm, respectively.

To compute the stiffness matrix of the loaded mode, let assume that a neighborhood of the loaded configuration, where the external loads and the joint displacement are supposed to be incremented by some small values $\delta \mathbf{f}_e$ and $\delta \boldsymbol{\theta}$, also satisfies the equilibrium conditions, leading to

$$(\mathbf{J}_g + \delta \mathbf{J}_g)^T \mathbf{G} + (\mathbf{J}_\theta + \delta \mathbf{J}_\theta)^T (\mathbf{f}_e + \delta \mathbf{f}_e) = \mathbf{K}_\theta (\boldsymbol{\theta}' - \boldsymbol{\theta} + \delta \boldsymbol{\theta}) \tag{17}$$

with the linearized kinematic constraint:

$$\delta \mathbf{t} = \mathbf{J}_\theta \delta \boldsymbol{\theta} \tag{18}$$

Upon the removal of the unchanged equilibrium condition of Eq. (15) from Eq. (17), after linearization, one obtains

$$\mathbf{H}_g^T \otimes \mathbf{G} \delta \boldsymbol{\theta} + \mathbf{J}_\theta^T \delta \mathbf{f}_e + \mathbf{H}_\theta^T \otimes \mathbf{f}_e \delta \boldsymbol{\theta} = \mathbf{K}_\theta \delta \boldsymbol{\theta} \tag{19}$$

where the symbol $\otimes$ represents the Kronecker product between matrices, and $\mathbf{H}_g = \partial \mathbf{J}_g / \partial \boldsymbol{\theta}$, $\mathbf{H}_\theta = \partial \mathbf{J}_\theta / \partial \boldsymbol{\theta}$, namely, the Hessian matrices. Combing Eqs. (18) and (19), the kineto-static model of the industrial robot is reduced to

$$\begin{bmatrix} \mathbf{0} & \mathbf{J}_\theta \\ \mathbf{J}_\theta^T & \mathbf{K}_E - \mathbf{K}_\theta \end{bmatrix} \begin{bmatrix} \delta \mathbf{f}_e \\ \delta \boldsymbol{\theta} \end{bmatrix} = \begin{bmatrix} \delta \mathbf{t} \\ \mathbf{0} \end{bmatrix} \tag{20}$$

with

$$\mathbf{K}_E = \mathbf{H}_g^T \otimes \mathbf{G} + \mathbf{H}_\theta^T \otimes \mathbf{F} \tag{21}$$

From the force-deformation equation $\delta \mathbf{f}_e = \mathbf{K} \delta \mathbf{t}$ of the end-efector, the Cartesian stiffness matrix $\mathbf{K}$ of the robot is calculated as,

$$\mathbf{K} = \left(\mathbf{J}_\theta \left(\mathbf{K}_\theta - \mathbf{K}_E \right)^{-1} \mathbf{J}_\theta^T \right)^{-1} \tag{22}$$

3.2   Error Compensation Procedure

The motions of the industrial robots are usually generated according to the inverse kinematics, from which the input signals of the actuators $\boldsymbol{\theta}$ corresponding to the desired end-effector position $\mathbf{p}$ are computed. However, with the external loads $\mathbf{f}_e$ exerted to the end-effector, the kinematic control becomes non-applicable due to the compliance error $\delta \mathbf{t}$ of the end-effector, thus, the actual position is computed from the stiffness model

$$\mathbf{p}' = \mathbf{p} + \delta \mathbf{t} \quad \text{where} \quad \delta \mathbf{t} = \mathbf{K}^{-1} (\mathbf{f}_e + \mathbf{J}^{-T} \boldsymbol{\tau}) \tag{23}$$

here, $\mathbf{f}_e + \mathbf{J}^{-T} \boldsymbol{\tau}$ stands for the combination of the force and inertial forces. In order to make the end-effector be located in the desired position $\mathbf{p}$, the compliance error between $\mathbf{p}$ and $\mathbf{p}'$, under the loads $\mathbf{f}_e$, should be compensated. Let

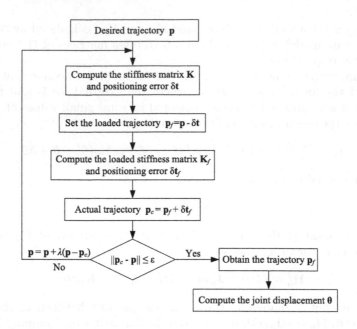

Fig. 4. Error compensation procedure for the welding robot.

suppose the modified end-effector position to be $\mathbf{p}_f = \mathbf{p} - \delta\mathbf{t}$, under the same loads $\mathbf{f}_e$, the actual position after compensation $\mathbf{p}_c$ should be in the neighborhood of the desired position, namely,

$$\mathbf{p}_c = \mathbf{p}_f + \delta\mathbf{t}_f \approx \mathbf{p} \quad \text{where} \quad \delta\mathbf{t}_f = \mathbf{K}_f^{-1}(\mathbf{f}_e + \mathbf{J}_f^{-T}\boldsymbol{\tau}_f) \tag{24}$$

where $\mathbf{K}_f$ is the stiffness matrix evaluated at the deflected position, and $\mathbf{J}_f$ and $\boldsymbol{\tau}_f$ are the Jacobian and joint torque at the actual positions, respectively. As a consequence, the modified end-effector location $\mathbf{p}_f$ can be calculated from the following iterative procedure,

$$\mathbf{p}'_c = \mathbf{p} + \lambda(\mathbf{p} - \mathbf{p}_c) \tag{25}$$

where the prime term corresponds to the next iteration, and $\lambda = \Delta_t/\|\mathbf{p} - \mathbf{p}_c\|$ is the scalar parameter to achieve the convergence, Δ_t being the maximum magnitude among the elements in $\mathbf{p} - \mathbf{p}_c$. This iterative method, presented in Fig. 4, will stop until $\|\mathbf{p} - \mathbf{p}_c\| \leq \epsilon$, where ϵ is an acceptable tolerance. Using this procedure to modify the reference trajectory in the robotic control, it is possible to compensate compliance errors to follow accurately the desired trajectory.

4   Case Study of Error Compensation

The previously presented stiffness modeling and error compensation procedure are illustrated with the welding robot under study, namely, the ABB IRB 4600 robot, along a trajectory as displayed in Fig. 5 together with the motion profiles of the end-effector. Based upon the FEA based evaluation, the upper arm and forearm can be treated as rigid links, compared to the actuation stiffness as listed in Table 2. Other detailed technical parameters and specifications of the robot can be found from the user manual [1], and the external wrench applied to the robot end-effector is supposed to be pure forces $\mathbf{f}_e = \begin{bmatrix} 200 & -200 & -200 \end{bmatrix}^T$, which is considered to be constant. Here, the requirement on the welding accuracy is higher than 0.5 mm.

Table 2. Joint stiffness values of the welding robot (unit: $[10^6 \cdot \text{Nm/rad}]$).

$K_{act,1}$	$K_{act,2}$	$K_{act,3}$	$K_{act,4}$	$K_{act,5}$	$K_{act,6}$
0.237	3.32	2.79	0.486	0.521	0.38

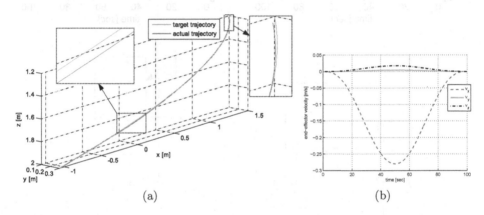

(a)   (b)

Fig. 5. The motion profiles of the welding robot: (a) desired and actual trajectories without error compensation; (b) end-effector velocities.

Figure 6 shows the comparison of the positioning errors and trajectories before and after compensation. It is seen that the maximum positioning error can be reduced to 0.01 mm when the robot tracks the modified trajectory, meaning that the positioning error after compensation can be ignored as it is much smaller than the acceptable error. Compared to the positioning error without compensation, the robot accuracy improves around 98%, which makes the actual trajectory after compensation coincident with the desired trajectory approximately, as shown in Fig. 6(b). The comparison reveals that the approach of error compensation can effectively improve the operational accuracy of the robot, also applicable to other industrial applications, such as milling.

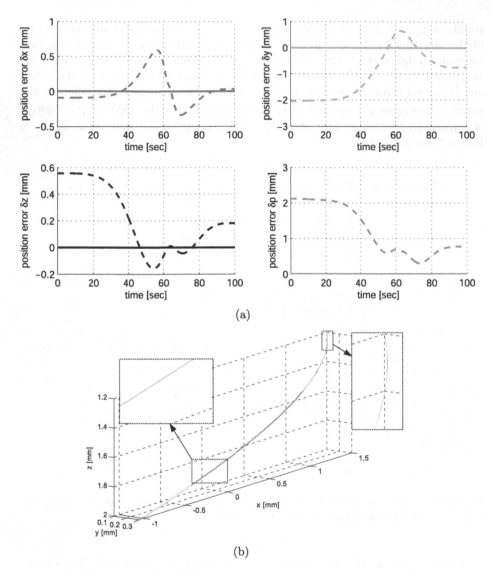

(a)

(b)

Fig. 6. The positioning errors and trajectory after compensation: (a) positioning errors (dashed line–before compensation; solid line–after compensation); (b) comparison of target and actual trajectories.

5 Conclusions

This paper deals with the kinetostatic modeling and compliance error compensation for an industrial robot in the ship hull welding. Besides the actuation and structural stiffness, the Cartesian stiffness matrix of the robot is derived with the consideration of the arm gravity and the external loads in the welding process as well as the inertial forces. An iterative error compensation method is introduced and is numerically illustrated along a welding trajectory. The results show that the error compensation procedure can effectively improve the operational precision to make the manipulator track the desired trajectory within acceptable positioning errors. The proposed approach implies that the manipulator accuracy can be effectively improved when the control strategy is based on the combination of the kinematic, kinetostatic and dynamic models, which is applicable to other industrial applications, such as machining.

Acknowledgement. The reported work is partly supported by the National Science and Technology Major Project (No. 2013ZX04003041-6), the Fundamental Research Funds for the Central Universities (No. DUT16RC(3)068), and the Liaoning Province STI major projects (No. 2015106007).

References

1. ABB IRB 4600 Product specification. http://new.abb.com/products/robotics/en/industrial-robots/irb-4600
2. Alici, G., Shirinzadeh, B.: Enhanced stiffness modeling, identification and characterization for robot manipulators. IEEE Trans. Robot. **21**(4), 554–564 (2005)
3. Bolmsjö, G., Olsson, M., Cederberg, P.: Robotic arc welding-trends and developments for higher autonomy. Ind. Robot **29**(2), 98–104 (2002)
4. Chen, S.F., Kao, I.: Conservative congruence transformation for joint and cartesian stiffness matrices of robotic hands and fingers. Int. J. Robot. Res. **19**, 835–847 (2000)
5. Denavit, J., Hartenberg, R.: A kinematic notation for lower-pair mechanisms based on matrices. ASME J. Appl. Mech. **22**, 215–221 (1955)
6. Dépincé, P., Hasco, J.Y.: Active integration of tool deflection effects in end milling. Part 2. Compensation of tool deflection. Int. J. Mach. Tools Manuf. **46**(9), 945–956 (2006)
7. Eastwood, S.J., Webb, P.: A gravitational deflection compensation strategy for HPKMs. Robot. Comput.–Integr. Manuf. **26**(6), 694–702 (2010)
8. Huang, T., Zhao, X., Whitehouse, D.J.: Stiffness estimation of a tripod-based parallel kinematic machine. IEEE Trans. Robot. Autom. **18**(1), 50–58 (2002)
9. Jalon, J.G.D., Bayo, E.: Kinematic and Dynamic Simulation of Multibody Systems: The Real Time Challenge. Springer, New York (2011)
10. Klimchik, A., Pashkevich, A., Chablat, D., Hovland, G.: Compliance error compensation technique for parallel robots composed of non-perfect serial chains. Robot. Comput.–Integr. Manuf. **29**(2), 385–393 (2013)
11. Kövecses, J., Angeles, J.: The stiffness matrix in elastically articulated rigid-body systems. Multibody Syst. Dyn. **18**(2), 169–184 (2007)
12. Meggiolaro, M.A., Dubowsky, S., Mavroidis, C.: Geometric and elastic error calibration of a high accuracy patient positioning system. Mech. Mach. Theory **40**(4), 415–427 (2005)
13. Pashkevich, A., Klimchik, A., Chablat, D.: Enhanced stiffness modeling of manipulators with passive joints. Mech. Mach. Theory **46**(5), 662–679 (2011)
14. Quennouelle, C., Gosselin, C.M.: Stiffness matrix of compliant parallel mechanisms. In: Lenarčič, J., Wenger, P. (eds.) Advances in Robot Kinematics: Analysis and Design, pp. 331–341. Springer, Netherlands (2008). doi:10.1007/978-1-4020-8600-7_35

15. Salisbury, J.: Active stiffness control of a manipulator in cartesian coordinates. In: 19th IEEE Conference on Decision and Control Symposium on Adaptive Processess, vol. 19, pp. 95–100 (1980)
16. Siciliano, B., Khatib, O.: Springer Handbook of Robotics. Springer, Heidelberg (2016). doi:10.1007/978-3-319-32552-1
17. Tsai, L.: Robot Analysis: The Mechanics of Serial and Parallel Manipulators. Wiley, Hoboken (1999)
18. Wittbrodt, E., Adamiec-Wójcik, I., Wojciech, S.: Dynamics of Flexible Multibody Systems. Springer, Heidelberg (2006). doi:10.1007/978-3-540-32352-5
19. Wu, G., Bai, S., Kepler, J.: Mobile platform center shift in spherical parallel manipulators with flexible limbs. Mech. Mach. Theory **75**, 12–26 (2014)
20. Wu, G., Guo, S., Bai, S.: Compliance modeling and error compensation of a 3-parallelogram lightweight robotic arm. In: Bai, S., Ceccarelli, M. (eds.) Recent Advances in Mechanism Design for Robotics. MMS, vol. 33, pp. 325–336. Springer, Cham (2015). doi:10.1007/978-3-319-18126-4_31
21. Zhuang, H.: Self-calibration of parallel mechanisms with a case study on Stewart platforms. IEEE Trans. Robot. Autom. **13**(3), 387–397 (1997)

Study and Experiment on Positioning Error of SCARA Robot Caused by Joint Clearance

Changyu Xu[1], Huimin Dong[1(✉)], Shangkun Xu[1], Yu Wu[1],
and Chenggang Wang[2]

[1] Dalian University of Technology, Dalian 116024, China
donghm@dlut.edu.cn
[2] Dalian Yunming Automation Technology Company, Dalian 116000, China

Abstract. Joint clearance is a main cause of generating positioning error in industry robot and the low repeatability and uncertainty make it difficult to compensate. This paper mainly focus on SCARA robot and study the effect of revolute joint clearance in positioning accuracy. First, the positioning error model in XOY plane related to joint clearance is set up and analysis to obtain the extreme error value is made. Then a combined idea to study the joint clearance is presented. From this way, the seemingly random clearance error vector can be predicted by some rough deciding factors. Therefore, if the distribution pattern of clearance vector and the rough decided value are obtained, a compensation is possible. Finally, the Ball-Bar was applied to carry out experiment for invalidation and further correction. From the experiment, some patterns are obtained and show that the clearance is indeed rough decided by some factors.

Keywords: SCARA robot · Joint clearance · Positioning error · Ball-Bar

1 Introduction

As robot becomes more and more important in industry automation field, the investigation of its accuracy and characteristics become popular among researchers. A very crucial aspect for an industry robot is the positioning error which can be influenced by many factors, particularly, the joint clearance. For most mechanisms, joint clearance is inevitable and main sources of it usually from the manufacturing error, wearing, and bearing gap and deformation. Unlike existing structural error by manufacturing or compliance which can be compensated, the low repeatability of joint clearance made it difficult to compensate.

Some researchers focused on the kinematics when exist joint clearance which include assessment of mechanism accuracy, the repeatability and the mathematic model of end positioning error etc. There are mainly two possible approaches to the kinematic problems: deterministic way which determines the contact point position in joint clearance by supposed contact force model, and stochastic approach that handles the problem from a probabilistic way.

In order to obtain the mathematic model of end positioning error, Innocenti [1] had proposed a method based on the principle of the virtual work to get the relation of the joint contact configuration and the resulting displacement. Then Stefano and Vincenzo [5]

© Springer International Publishing AG 2017
Y. Huang et al. (Eds.): ICIRA 2017, Part II, LNAI 10463, pp. 147–156, 2017.
DOI: 10.1007/978-3-319-65292-4_14

modified this method. Some scholars considered the joint clearance as an additional massless link, like the work of Ting et al. [6] and Tsai and Lai [7], which used geometry method to get the maximum deviation region. For some simple specific robot mechanisms, some researchers just used coordinate transform to get the final error model caused by joint clearance. In the study of Nicolas and his fellows [3], a system way of displaying the end effector error by coordinate transformation had been proposed.

For investigating the kinematic effects, most scholars focused on the maximum and minimum positioning error of end effector so as to give a maximum allowable accuracy guarantee [3, 6, 7]. While some studied the working path deviation when joint clearance exist. They try to reduce and minimize the deviations from the target locations by properly selecting the working path when considering the clearance in the model [2, 4]. This study mainly focus on SCARA robot and investigates the effects of joint clearance in positioning error of XOY plane. First set up the positioning error model, then propose an idea of synthesizing deterministic way and stochastic approach. After that, the experiment is carried out for validation and correction. Finally, some conclusions are made.

2   Setting up Positioning Error Model

In general, SCARA robot has three revolute pairs and one prismatic pair and axes of the three revolute joints are in parallel with each other. Because of the compact structure and parallel axes feature, SCARA manipulator can realize high accuracy motion, which makes it widely used in electronic assembly field. However, this special structure feature also limits the workspace of SCARA to plane working only. Figure 1 is the 3-D model of YAMAHA YK600XG SCARA robot, the experimental object of this study provided by Dalian YUNMING Automation Technology Company. The mechanism diagram is shown as Fig. 2.

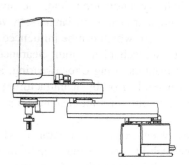

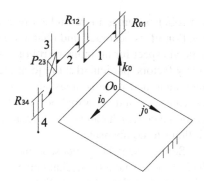

Fig. 1. 3-D Model of YAMAHA SCARA robot   **Fig. 2.** The mechanism diagram

Joint R_{01}, R_{12}, R_{34} are revolute joint and the axes of them are parallel with axis Z. Joint P_{23} is a prismatic joint along Z axis. Combinating motion of joint $1(R_{01})$ and $2(R_{12})$ realizes the positioning in XOY plane and motion of joint $3(P_{23})$ realizes movement in Z direction. Revolute joint 4 is used for assembly work like screwing.

From the mechanism diagram, we can obtain that the end-effector position is depend on joint 1, 2 and 3. For simplifying the analysis, take the base coordinate as coordinate system for all components. Denote the origin of base coordinate system O_0 as the intersection point of joint R_{01}'s axis and the bottom installing plane. And denote k_0 as the direction of joint R_{01}'s axis, define i_0 as the direction of arm 1 when rotation angle of joint 1 is zero. Shown as Fig. 2. Therefore, the projection of end-effector point Q in XOY plane is shown as Fig. 3. Denote the rotation angle of joint 1 as θ_1 and rotation angle of joint 2 as θ_2. Positive rotation direction is in the anti-clockwise. The position movement of Q in Z direction is the moving parameter of joint 3. Denote P_3 as the moving parameter of joint 3 and h_0 is distance between the bottom installing plane and the extreme position of joint 3, shown as Fig. 4. L_1, L_2 are the axes distance of joint 1, 2 and joint 2, 4 respectively, which is just the length of arm1 and arm2.

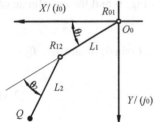

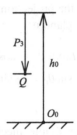

Fig. 3. Projection of Q in XOY plane   **Fig. 4.** Position of Q in Z direction

So, the coordinate of Q based on O_0 is:

$$Q = (L_1 \cos \theta_1 + L_2 \cos(\theta_1 + \theta_2), L_1 \sin \theta_1 + L_2 \sin(\theta_1 + \theta_2), P_3 + h_0) \qquad (2.1)$$

According to the parameters of the provided YAMAHA YK600XG SCARA robot, it have, $L_1 = 300$ mm, $L_2 = 300$ mm, $h_0 = 137.6$ mm, joint motion scope, P_3: $(-200$ mm, $0)$, θ_1: $(-130°, 130°)$, θ_2: $(-145°, 145°)$. Then the coordinate of Q's projection on XOY plane can be simplified as:

$$Q_p = (L(\cos \theta_1 + \cos(\theta_1 + \theta_2)), L(\sin \theta_1 + \sin(\theta_1 + \theta_2))) \qquad (2.2)$$

For simplifying the joint clearance model and considering the structure characteristic and main affecting factors, two reasonable assumptions are made. First, assume that the revolute joints of the 3-DOF manipulator only exist circular clearance and a continuity contact between shaft and hole. Second, for the study purpose is to investigate the positioning error of XOY plane and do not consider the orientation error, the parallel axes assumption is made between the shaft and hole of R-joint.

So, the joint clearance of joint 1, 2 can be simplified as a plan vector, whose magnitude is a constant $\varepsilon = R - r$ (radius difference between shaft and hole). Uncertain direction θ_ε are denoted as the angle between i_0 of base coordinate system O_0 and the clearance error vector (Positive angle is anti-clockwise). Show as Fig. 5. ε_i and $\theta_{\varepsilon i}$ present the joint clearance magnitude and direction angle of joint $i(i = 1, 2)$.

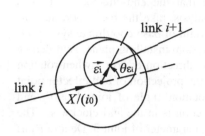

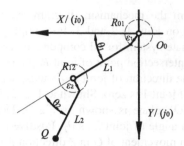

Fig. 5. Joint clearance error vector   **Fig. 6.** Projection of Q in XOY (with clearance)

Ideal state of Q in XOY plane is shown as Fig. 3. When introducing the joint clearance, the projection of Q in XOY plane is changed to Fig. 6 and the coordinate of Q in XOY plane is:

$$Q'_p = (L(\cos\theta_1 + \cos(\theta_1 + \theta_2)) + \varepsilon_1\cos\theta_{\varepsilon_1} + \varepsilon_2\cos\theta_{\varepsilon_2}, L(\sin\theta_1 + \sin(\theta_1 + \theta_2))$$
$$+ \varepsilon_1\sin\theta_{\varepsilon_1} + \varepsilon_2\sin\theta_{\varepsilon_2})$$

$$(2.3)$$

For the simple mechanism structure of SCARA robot, the positioning error of end-effector Q in XOY plane can be obtained by direct subtraction.

$$\Delta Q_p = Q'_p - Q_p = (\varepsilon_1\cos\theta_{\varepsilon_1} + \varepsilon_2\cos\theta_{\varepsilon_2}, \varepsilon_1\sin\theta_{\varepsilon_1} + \varepsilon_2\sin\theta_{\varepsilon_2}) \qquad (2.4)$$

ε_1, ε_2 are constant value, $\theta\varepsilon_1$ and $\theta\varepsilon_2$ are random uncertain quantities. From (2.4), the extreme value of end-effector positioning error can be obtained from analytical way. $\theta\varepsilon_1$ and $\theta\varepsilon_2$ are independent variables. The extreme value of X, Y direction positioning error are:

$$|\varepsilon_x|_{max} = \varepsilon_1 + \varepsilon_2\big|_{for\ \cos\theta_{\varepsilon_1},\cos\theta_{\varepsilon_2}=1}, \quad |\varepsilon_x|_{min} = 0\big|_{for\ \cos\theta_{\varepsilon_1},\cos\theta_{\varepsilon_2}=0}$$
$$|\varepsilon_y|_{max} = \varepsilon_1 + \varepsilon_2\big|_{for\ \sin\theta_{\varepsilon_1},\sin\theta_{\varepsilon_2}=1}, \quad |\varepsilon_y|_{min} = 0\big|_{for\ \sin\theta_{\varepsilon_1},\sin\theta_{\varepsilon_2}=0}$$

$$(2.5)$$

Define the distance error $d^2 = \varepsilon_x^2 + \varepsilon_y^2$, then:

$$d^2 = \varepsilon_1^2 + \varepsilon_2^2 + 2\varepsilon_1\varepsilon_2\cos(\theta_{\varepsilon_1} - \theta_{\varepsilon_2}) \qquad (2.6)$$

So, the extreme value of distance error d is:

$$d_{max} = \varepsilon_1 + \varepsilon_2\big|_{for\ \cos(\theta_{\varepsilon_1}-\theta_{\varepsilon_2})=1}, \quad d_{min} = |\varepsilon_1 - \varepsilon_2|\big|_{for\ \cos(\theta_{\varepsilon_1}-\theta_{\varepsilon_2})=-1} \qquad (2.5)$$

The analysis above only consider the extreme value and take $\theta\varepsilon_1$ and $\theta\varepsilon_2$ totally random and have ignored possible influence factors.

3   Combined Idea for Revolute Joint Clearance

The clearance vector direction of revolute joint can be influenced by many factors, and some factors are even uncertain itself. So complete deterministic way is hard to obtain. Meanwhile, total stochastic approach will ignore the main external influence factors, like load force, link gravity, inertia effect etc. A combined way is introduced in this section: first, use the main influence factors to determine a contact point of shaft and hole; then according to this point direction build a deviation distribution for the influence of subtle uncertain factors, such as friction, impact, vibration etc. Specific possible deviation distribution type need be obtained from the experiment.

3.1   Main External Influence Analysis to Clearance

For SCARA robot, the link gravity is along Z-axis. If not consider the orientation error of end-effector, the force along Z axis has no impact in the positioning error of XOY plane. And when robot do certain assembly task, the state of robot motion must be static in order to finish the assembly work. So no centrifugal force exist. Therefore, the mainly external influence factor may be the inertia effect which is related to the rotate direction of each joint. Show as Fig. 7. Then from the arm rotation direction the contact point can be obtained so as to obtain $\theta \varepsilon_i$, denoted as θ_b. Therefore, when arm rotate from the same side to a position, the joint clearance effect will be the same ideally. While approaching a position from different rotate direction, the effect of joint clearance will be different. Validation of this inference will be shown in the experiment result.

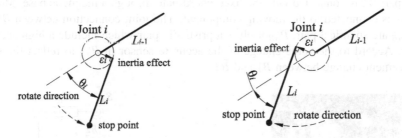

Fig. 7. Inertia effect for revolute joint clearance, left: anti-clockwise, right: clockwise

The discussion above ignored the effect of other subtle uncertain factors. In fact, some disturbance may be generated around the obtained contact point before. A way to consider these uncertain factors is to introduce a deviation distribution above the obtained point. Make average of the deviation distribution θ_b and the standard deviation σ to stand for those uncertain disturbances. Specific distribution pattern should be obtained from the experiment.

3.2   Monte Carlo Method to Obtain Positioning Error

Monte Carlo simulation is a random number simulation method supported by basic mathematic theory, i.e. the law of large number and central-limit theory. In Monte Carlo method, suppose function $Y = f(X_1, X_2, \ldots, X_n)$, and the probability distribution of variable $X_1, X_2, \ldots, X_n$ are already known, a group of value $x_1, x_2, \ldots, x_n$ of $X_1, X_2, \ldots, X_n$ is generated using a random number generator by sampling. And then obtain the value of Y according to the function relation $y_i = f(x_{1i}, x_{2i}, \ldots, x_{ni})$. Repeat the sampling procedure a number of times ($i = 1, 2, \ldots, m$), then a large group of sample $y_1, y_2, \ldots, y_m$ can be obtained. As long as the sampling time m is large enough, the numerical characteristic of the sample can represent real one of Y accurately.

For certain SCARA robot, the clearance error magnitude ε_1, ε_2 in Eq. (2.4) are constant, which can be assured from the manufacturer or measured by the experiment. Equations (2.4) and (2.6) give the relation function of ε_x, ε_y, d based on $\theta\varepsilon_1$, $\theta\varepsilon_2$. And the probability distribution of $\theta\varepsilon_1$, $\theta\varepsilon_2$ can be obtained from the way presented in Sect. 3.1. Then we can apply the Monte Carlo method described above to Eqs. (2.4) and (2.6) and obtain the distribution of positioning error as well as some numerical characteristics.

4   Experiment by Ball-Bar

In this experiment, the Ball-Bar is applied to measure the joint clearance effects to positioning error. Ball-Bar is an error measurement instrument and usually applied in CNC machine to measure the positioning error of the machine caused by factors like machine geometry error, wear and control system error etc. shown as Fig. 8. It has three main parts, fixed part $B2$, center rod P_{AB} and the moving part $B1$, shown as Fig. 9. Fixed part $B2$ is connected with the fixed coordinate through a magnetic base. Moving part $B1$ is connected to the moving component. The joint connection between $B1$, $B2$ and P_{AB} are spherical pair. P_{AB} itself is a prismatic pair which include a high accurate sensor. Accord to the value change of the accurate sensor in P_{AB} to reflect the small displacement change between $B1$ and $B2$.

Fig. 8. Photo of Ball-Bar measuring

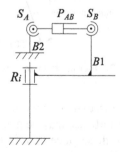

Fig. 9. Mechanism of Ball-Bar measuring

Not limited in CNC, Ball-Bar can also be adapted to measure the revolute joint error when referring to the SCARA robot. There are two purposes of using the Ball-Bar, one is to find out whether the combined idea above is adaptable and the other is to study the pattern of the deviation as well as the positioning performance of joint clearance.

Experiment mechanism diagram for joint 1 is shown as Fig. 10. And Fig. 11 is the experiment photo. For fixing the fixed part $B2$ of the Ball-Bar in the axis of joint 1 accurately, fixture is designed, shown as Fig. 11.

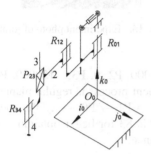

Fig. 10. Experimental diagram for joint 1   **Fig. 11.** Experiment photo of joint 1

For fixture limitation, only 7 points is measured, P0: −330000 pulses, P1: −270000, P2: −210000, P3: −150000, P4: −90000, P5: 30000, P6: 130000. The unit of the measure position is in pulse for the convenience of the control way in the servo motor in joint 1. Measure point shown as Fig. 12. The experiment process is first regular planned as a cycle P0 → P1 → P2 → P3 → P4 → P5 → P6 → P5 → P4 → ... then a random measuring way from point to point is measured.

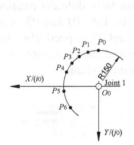

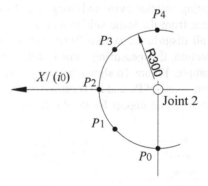

Fig. 12. Experiment procedure for joint 1   **Fig. 13.** Experiment procedure for joint 2

Experiment mechanism diagram for joint 2 is shown as Fig. 14 and Fig. 15 is the experiment photo. Fixture is shown in Fig. 15.

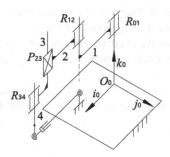

Fig. 14. Experimental diagram for joint 2   **Fig. 15.** Experiment photo of joint 2

5 position is measured, P0: 300000 pulses, P1: 150000, P2: 0, P3: −150000, P4: −300000. Measure point shown as Fig. 13. The experiment process is regular planned as a cycle P0 → P1 → P2 → P3 → P4 → P3 → P2 → P1 → P0 → ... the measuring times for each point in the same rotate direction is 83 for better obtaining the uncertainty pattern and other features of the joint clearance.

5   Experiment Data and Analysis

When measuring the joint clearance error, which is a random error, the fixed deviation of the fixture and the robot structure error generate no impact to the measurement. Although each point the value of Ball-Bar is different due to fixed error, the disturbance feature of each point is not influenced by the system fixed error for the reason that in the same position the fixed structure error has the same value. Below are the experiment data and some analysis.

For joint 1, considering position P1, P2, P3, P4 and P5 in regular motion when rotating anticlockwise and clockwise, the experiment data shows that when manipulator rotate from the same side to a position, the measuring value is relatively stable with a small disturbance within 0.001. While approaching a position from different rotation direction, the measuring value is different for all measure points. Take P1 and P3 as an example. Figure 16 shows the plot of measuring value of P1 and P3 respectively. Different value of P1 and P3 is caused by the fix error of the fixture and robot structure error, which has no impact for the disturbance measurement for certain position.

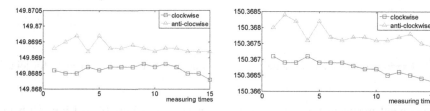

Fig. 16. Measuring value of P1(L), P3(R); red represent clockwise, black represent anticlockwise (Color figure online)

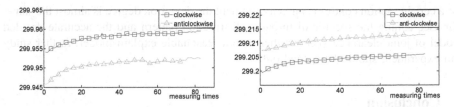

Fig. 17. Value of P1(L) and P3(R), red represent clockwise, black represent anticlockwise (R_{12}) (Color figure online)

The result verifies the inference made in Sect. 3.1 and shows that the seemingly random joint clearance vector indeed rough decided by some factors. And some deviation also exist whose pattern will be roughly show in the measurement of joint 2 when repeat the measurement a lot of times. The rough decided law can also be verified in the experimental data of joint 2.

For joint 2, consider position P1, P2, P3 in Fig. 14 which has two direction approaches. P0 and P4 only have one approach direction. For verification, Fig. 17 shows the plot of measuring value of P1 and P3 from different approaching direction respectively. The result once again verifies that the seemingly random joint clearance vector indeed rough decided by some factors.

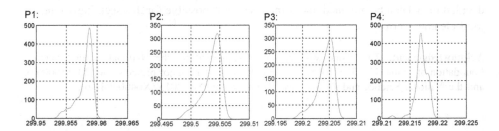

Fig. 18. Probability density function of measuring value in Pi in clockwise rotation.

For the same approaching direction, the experiment shows that uncertainty deviation indeed exist and the uncertainty pattern of different positions in same rotation direction has similar features. Shown as Fig. 18. The figure is the probability density function of the measuring value of position Pi from the data, having smoothed by kernel estimation method. The approach direction is clockwise.

From Fig. 18, we can found the disturbance pattern are almost similar, one peak density in the distribution and then plummet as the value away from the center value. The pattern is close to a normal distribution but a little different from ideal normal distribution. Possible reason is that so many small uncertain factors affect the joint clearance, like friction, compact, vibration, elasticity, etc. and according to the central-limit theory, the distribution of the disturbance tend to a normal distribution. And from the data, it shows that the standard deviation of P1, P2, P3 and P4 are 0.0013 mm, 0.0016 mm, 0.0016 mm and 0.0011 mm respectively, which shows that for a certain motion condition and environment the deviation degree is almost the same. For ascertaining the specific disturbance distribution pattern and the accurate decided model of joint clearance vector rather than a reasonable explain model, further study and experiment should be carried out.

6 Conclusion

This paper first simplified the revolute joint clearance and set up the positioning error model of SCARA robot. According to the error model, some simple analysis to obtain the extreme error value is made, which take clearance direction $\theta \varepsilon_1$ and $\theta \varepsilon_2$ totally random and have ignored possible influence factors for joint clearance. Then a combined idea is presented for considering main external influence factors and uncertain factors separately. Finally, experiments scheme and result is shown to verify whether joint clearance vector is totally random or roughly decided by some factors. The result show that the joint clearance is indeed exist and it is not totally random which means compensation is possible. And the experiment also show similar pattern for the uncertainty of joint clearance, from these patterns, a norm distribution may be adaptable for the deviation distribution of clearance vector direction.

Above combined idea is just giving a method to illustrate the feasibility of separating the rough deciding factors and uncertain factors. For further study, the rough decided model of the joint clearance should be investigated accurately and if the rough decided model can be obtained, the compensation is possible for this seemingly random joint clearance error.

Acknowledgement. This work was financially supported by the National Natural Science Foundation of China (No. 51375065), Liaoning Province STI major projects (No. 2015106007), and the National Science and Technology Major Project (No. 2013ZX04003041-6).

References

1. Innocenti, C.: Kinematic clearance sensitivity analysis of spatial structures with revolute joints. In: ASME, Design Engineering Technical Conference, Paper DETC98/DAC-8679 (1999)
2. Lai, M.-C., Chan, K.-Y.: Minimal deviation paths for manipulators with joint clearances. In: ASME, Mechanisms and Robotics Conference, p. V05BT08A058 (2014)
3. Nicolas, B., Philippe, C., Stéphane, C.: The kinematic sensitivity of robotic manipulators to joint clearances. In: Computers and Information in Engineering Conference (2010)
4. Erkaya, S.: Trajectory optimization of a walking mechanism having revolute joints with clearance using ANFIS approach. Nonlinear Dyn. **71**(1–2), 75–91 (2013)
5. Stefano, V., Vincenzo, P.C.: A new technique for clearance influence analysis in spatial mechanisms. ASME J. Mech. Rob. **127**(3), 446–455 (2002). Montreal, Quebec, Canada
6. Ting, K.L., Zhu, J.M., Watkins, D.: The effects of joint clearance on position and orientation deviation of linkages and manipulators. Mech. Mach. Theory **35**(3), 391–401 (2000)
7. Tsai, M.J., Lai, T.H.: Kinematic sensitivity analysis of linkage with joint clearance based on transmission quality. Mech. Mach. Theory **39**(11), 1189–1206 (2004)

Real-Time Normal Measurement and Error Compensation of Curved Aircraft Surface Based on On-line Thickness Measurement

Yuan Yuan[✉], Qingzhen Bi, Limin Zhu, and Han Ding

Shanghai Jiao Tong University, 800 DongChuan Rd., Shanghai 200240, China
yyaizhen@sjtu.edu.cn

Abstract. Large aerospace thin-walled workpieces easily give rise to random deformation in clamping and machining processes, in which the real-time monitoring of wall thickness needs a high-quality normal technology. A high-precision on-line surface normal measurement and a real-time compensation strategy are developed in this paper. Firstly, the deviation between the actual normal vector and the spindle direction of curved workpiece surface is calculated from the data measured by four eddy current displacement sensors which are installed at the front end of the spindle; then, the deviation is converted into the compensation of each axis via homogeneous coordinate transformation and post-processing of tools. Meanwhile online compensation results get finished on the move. A simulated and experimental platform is established on an A-C five-axis machine tool in order to measure deviations of sensors at each point position and record the result of compensation. The application of this method in practical engineering can greatly improve the efficiency of measurement and control.

Keywords: Aerospace thin-walled workpieces · Normal error · Real-time measurement and compensation · Kinematics transformation · Post-processing

1 Introduction

In current industrial production, the normal direction plays a practical significance. In the process of drillings and specific millings, an accurate normal vector needs to be obtained in order to realize higher machining precision. Moreover, in the measurement of thickness of parts by the ultrasonic, the right normal ensures the accuracy and stability of thickness data.

Current computing methods about random normal of curved surface are generally divided into Cross Product method, Quadric Surface Fitting, NURBS curve method, Triangulation Algorithm, etc. [1]. The cross product could be easily calculated by getting points coordinates of measuring curved surface but it has a low accuracy; the surface fitting is complex and has low applicability of more curved surface though it is a high-precision method; curve and triangulation have to collect a large amount of data which is hard to apply to the real-time measurement.

The original version of this chapter was revised: the acknowledgement section was added. The erratum to this chapter is available at https://doi.org/10.1007/978-3-319-65292-4_78

Y. Huang et al. (Eds.): ICIRA 2017, Part II, LNAI 10463, pp. 157–170, 2017.
DOI: 10.1007/978-3-319-65292-4_15

Moreover, in aviation industrials, normal measurement and error compensation is applied to drilling and riveting fields and it is divided into three main types, off-line programming without adjustment, off-line programming in site adjustment and online programming in site adjustment.

The first sets an example that a normal off-time calibration of the drilling hole which has a poor accuracy [2]; the second has an application on automatic riveting that off-time method realizes normal compensation but cannot match workpieces with complex deformation [3]. Within the limit of application background, most companies and researchers have conducted a series of research of normal error compensation more from this aspect. For example, Lin Minqing proposed *The Surface-Normal Measurement Method and Micro-Adjustment Mechanism* [4] in his article that used Gauss equations of several points to solve the current normal vector of every drilling hole and make a micro-adjustment at the same position. Similar applications has existed in Yuan's paper [5] where related articles mainly selected many points on the large curved surface as objects of studies and obtained off-line coordinates of all points so as to reach the fixed position and realize the independent adjustment without interference; the last can realize an online normal measurement but real-time continuous motion compensation has not been completed such as robotic drilling of Electroimpact Inc. (EI) company [6] and mirror milling equipment for aircraft skin of M.torres in Spain.

However, most large aerospace thin-walled workpieces like skins and aircraft panels easily give rise to random deformation in clamping and machining processes, which leads to changeable surface curvature and unknown difficulties of off-line programming. On one hand, the reason lies in the processing error of stocks preparation, on the other hand, it derives from stress release in the milling. What's more, in the surface processing of large aerospace thin-walled workpieces, instead of simple points positions, traditional normal error compensation methods are unable to satisfy rapid milling of these parts. Thus, a simpler and quicker method of normal adjustment and compensation is required.

This paper proposes a new normal method which realize a combination of online measurement and real-time compensation. It is able to acquire real-time normal direction with complex surface by reading points information in the continuous motion of tools and implement an adaptive adjustment of errors with the NC numerical control system. The final aim lies in real-time machining processing of complex curved surface to further improve processing efficiency and machining precision and solve a series of industrial problems caused by normal.

2   Real-Time Normal Measurement and Error Compensation

2.1   *Real-Time* of Numerical Control System

Compared with traditional fixed-point measurement and off-line programming, the real-time normal error compensation is able to be completed within a period of point position, which experiences a process of sensors scanning to implementation movement of tools. The finally result is to realize follow-up processes and dynamic online compensation. Detailed data flow chart is shown as Fig. 1.

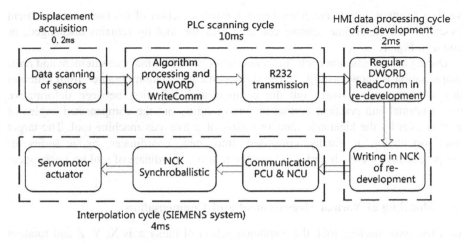

Interpolation cycle (SIEMENS system)
4ms

Fig. 1. Sequence flow chart of measurement system

From above, it is easily noted that an entire cycle of normal error compensation needs only about 10 ms on account of a series of parallel structures from data processing and transmission to actuator implementation, in which machine tools can finish all actions of measurement and compensation. Assume that feed speed of tools is 2000 mm/min and the machine motion just has a movement about 0.333 mm within a cycle of normal error compensation. It is so tiny that the deviation is negligible and the real-time measurement and compensation of normal direction is accurate and efficient in the processing of parts. From above, to completely achieve the real-time of fast compensation, computation time deriving from algorithm complexity and response mechanism will decide the efficiency of real-time dynamic compensation of each point position. Thus, a new strategy is shown as follows.

2.2 Measurement and Compensation Strategy

To acquire random deformation of large and complex curved aircraft parts, the online normal error compensation with an online measurement needs to adjust all axes of a

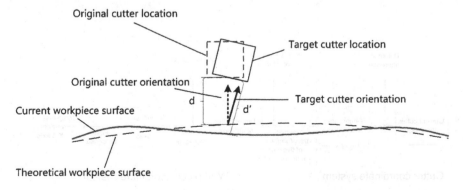

Fig. 2. The deformation of curved surface and comparison of original and target normal

five-axis machine tool to reach the target normal direction of thickness-measurement device. At the same time relative coordinate of the tool tip remains unchanged, as shown in Fig. 1.

During a compensation cycle, the main flow chart of normal measurement and error compensation is shown as Fig. 2. The displacement deviation obtained from different eddy current sensors works out the compensated deflection between deformative curved normal and practical normal to further figure out the compensated angles of rotation axes by the kinematic transformation of a five-axis machine tool. The target cutter orientation is finally converted into new coordinates by a series of post-processing of tools in order to keep the relative coordinate of tool tip unchanged.

2.3   Modeling of Normal Measurement and Compensation

For a five-axis machine tool, the combined action of linear axis X, Y, Z and rotation axis like A-C or A-B determines the relative position of the measuring head in tools. On certain large curved randomly surface, data of real-time displacement is measured from four eddy current displacement sensors and it is supposed to be exactly right normal direction of this point when all four readings are the same. If there are deviations in any two sensors especially for diagonal ones, these deviations have to be converted into the compensation of all axes, as shown in Fig. 3. In the cutter coordinate system, four sensors are distributed on a circle with a radius of R, in which sensor 1 and 3 are located on the X_T axis while sensor 2 and 4 are located on the Y_T axis. Their measured values are respectively l1, l2, l3, l4. On account of the tiny range of sensors (about 4 mm) in view of their distribution radius, the angle of rotation by the deviation, α, β can be calculated approximately as follows,

$$\alpha \approx \sin \alpha = \frac{l_1 - l_3}{\sqrt{(l_1 - l_3)^2 + 4R^2}} \tag{1}$$

$$\beta \approx \sin \beta = \frac{l_2 - l_4}{\sqrt{(l_2 - l_4)^2 + 4R^2}} \tag{2}$$

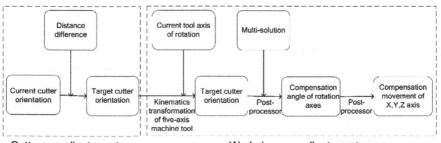

Fig. 3. Online normal error compensation with normal measurement

With the combination of original plane composed of displacement sensors and target plane of measuring points, as shown in Fig. 4, the normal vector of original plane $\overrightarrow{N} = \overrightarrow{V}_T$ is firstly rotated α angle around X_T axis and then β angle around Y_T axis to form a new normal vector $\overrightarrow{N'} = \overrightarrow{V'}_T$. This target vector has nothing to do with the order of the axis of rotation selected because of the tiny compensation angle.

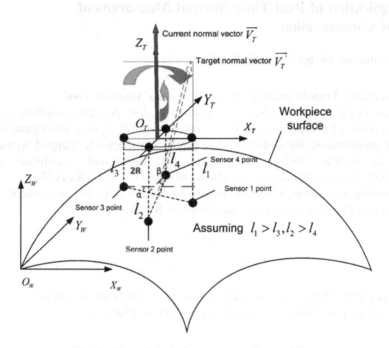

Fig. 4. Normal deviation compensation modeling

The current original normal vector in the cutter coordinate system, $O_T X_T Y_T Z_T$, $\overrightarrow{V_T} = [0 \ \ 0 \ \ 1]^T$, is presumptively converted into the target normal vector in this coordinate system by two rotation. The target vector $\overrightarrow{V'_T}$ is as follows,

$$V'_T = \begin{bmatrix} R(X, \beta) & 0 \\ 0 & 1 \end{bmatrix} \begin{bmatrix} R(Y, \alpha) & 0 \\ 0 & 1 \end{bmatrix} V_T \tag{3}$$

In the kinematic transformation, it is known that $R(Y, \alpha)$ and $R(X, \beta)$ is respectively regarded as a rotational transfer matrix which is rotated α and β around Y axis and X axis,

$$\begin{bmatrix} R(Y, \alpha) & 0 \\ 0 & 1 \end{bmatrix} = \begin{bmatrix} \cos \alpha & 0 & \sin \alpha & 0 \\ 0 & 1 & 0 & 0 \\ -\sin \alpha & 0 & \cos \alpha & 0 \\ 0 & 0 & 0 & 1 \end{bmatrix}, \begin{bmatrix} R(X, \beta) & 0 \\ 0 & 1 \end{bmatrix} = \begin{bmatrix} 1 & 0 & 0 & 0 \\ 0 & \cos \beta & -\sin \beta & 0 \\ 0 & \sin \beta & \cos \beta & 0 \\ 0 & 0 & 0 & 1 \end{bmatrix}$$

According to formula (2), the target vector can be obtained as follows,

$$\overrightarrow{V'_T} = [\sin \alpha \quad -\cos \alpha \sin \beta \quad \cos \alpha \cos \beta]^T \tag{4}$$

3 Application of Real-Time Normal Measurement and Compensation

3.1 Acquiring Target Normal Direction

The Kinematic Transformation of A-C Five-Axis Machine Tool
As shown in Fig. 6, to describe the movement of A-C dual-table machine tool like Fig. 5, a series of coordinate systems are established [3]. After a homogenous coordinate transformation, the cutter location point is respectively mapped to machine coordinate system (MCS) $O_M X_M Y_M Z_M$ [7] from tool coordinate system (TCS) $O_T X_T Y_T Z_T$ and workpiece coordinate system (WCS) $O_W X_W Y_W Z_W$.

According to relevant kinematic transformation of A-C five-axis machine tool, there are two transformation matrix equations as follows,

$$
\begin{aligned}
{}^M P &= {}^M_T T {}^T P \\
{}^M P &= {}^M_A T {}^A_C T {}^C_W T {}^W P
\end{aligned}
\tag{5}
$$

Furthermore, ${}^Y_X T$ shows a transformation from X coordinate system to Y coordinate system and 20 pose variables are included as follows (Table 1),

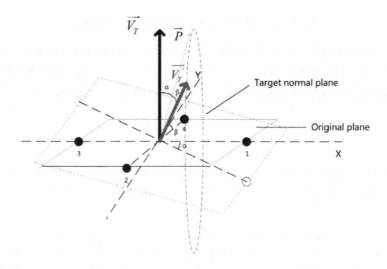

Fig. 5. Compensation angle of the normal vector

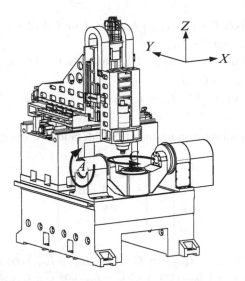

Fig. 6. A-C Dual-table machine tool

Table 1. Pose variables of all axes

Pose position and orientation	$_T^M T$		$_A^M T$	$_C^A T$	$_W^C T$
Translation and Rotation	(X, Y, Z)		A	C	$(L_{cwx}, L_{cwy}, L_{cwz})$
	$(L_{acx}, L_{acy}, L_{acz})$	$(L_{cwx}, L_{cwy}, L_{cwz})$	$(L_{max}, L_{may}, L_{maz})$	$(L_{acx}, L_{acy}, L_{acz})$	

Consequently, the cutter location in WCS can be expressed as follows,

$$
\begin{aligned}
{}^W P &= (x, y, z)^T = {}_W^C T^{-1} {}_C^A T^{-1} {}_A^M T^{-1} {}_T^M T^T P = {}^W T_T [0 \quad 0 \quad 0 \quad 1]^T \\
{}^W V &= (i, j, k)^T = {}_W^C T^{-1} {}_C^A T^{-1} {}_A^M T^{-1} {}_T^M T^T V = {}^W T_T [0 \quad 0 \quad 1 \quad 0]^T
\end{aligned}
\tag{6}
$$

Similarly, an integrated kinematic transformation of A-C five-axis machine tool is shown,

$$
\begin{aligned}
\begin{bmatrix} {}^W V & {}^W P \end{bmatrix} &= F_{TW}(X, Y, Z, A, C) \begin{bmatrix} {}^T V & {}^T P \end{bmatrix} \\
&= \begin{bmatrix}
\cos C & -\cos A \sin C & \sin A \sin C & M_x \\
\sin C & \cos A \cos C & -\sin A \cos C & M_y \\
0 & \sin A & \cos A & M_z \\
0 & 0 & 0 & 1
\end{bmatrix} \begin{bmatrix} {}^T V & {}^T P \end{bmatrix}
\end{aligned}
\tag{7}
$$

$$M_x = (X + L_{acx} + L_{cwx})\cos C - (Y + L_{acy} + L_{cwy})\cos A \sin C + (Z + L_{acz} + L_{cwz})\sin A \sin C$$
$$\quad - L_{acx}\cos C + L_{acy}\sin C - L_{cwx}$$
$$M_y = (X + L_{acx} + L_{cwx})\sin C + (Y + L_{acy} + L_{cwy})\cos A \cos C - (Z + L_{acz} + L_{cwz})\sin A \cos C$$
$$\quad - L_{acx}\sin C - L_{acy}\cos C - L_{cwy}$$
$$M_z = (Y + L_{acy} + L_{cwy})\sin A + (Z + L_{acz} + L_{cwz})\cos A - L_{acz} - L_{cwz}$$

In general, the transformation formula from MCS to WCS is expressed as follows,

$$\begin{bmatrix} i \\ j \\ k \\ 0 \end{bmatrix} = \begin{bmatrix} \sin A \sin C \\ -\sin A \cos C \\ \cos A \\ 0 \end{bmatrix} \tag{8}$$

$$\begin{bmatrix} x \\ y \\ z \\ 1 \end{bmatrix} = \begin{bmatrix} \begin{pmatrix} (X + L_{acx} + L_{cwx})\cos C - (Y + L_{acy} + L_{cwy})\cos A \sin C \\ + (Z + L_{acz} + L_{cwz})\sin A \sin C - L_{acx}\cos C + L_{acy}\sin C - L_{cwx} \end{pmatrix} \\ \begin{pmatrix} (X + L_{acx} + L_{cwx})\sin C + (Y + L_{acy} + L_{cwy})\cos A \cos C \\ -(Z + L_{acz} + L_{cwz})\sin A \cos C - L_{acx}\sin C - L_{acy}\cos C - L_{cwy} \end{pmatrix} \\ (Y + L_{acy} + L_{cwy})\sin A + (Z + L_{acz} + L_{cwz})\cos A - L_{acz} - L_{cwz} \\ 1 \end{bmatrix} \tag{9}$$

Target Normal Vector of WCS

The assumed target normal vector in TCS is replaced with $\overrightarrow{V_T'}$ in Eq. 4 from $\overrightarrow{V_T} = \begin{bmatrix} 0 & 0 & 1 \end{bmatrix}^T$

$$\overrightarrow{V_T'} = {}^T\overrightarrow{V'} = \begin{bmatrix} \sin \alpha & -\cos \alpha \sin \beta & \cos \alpha \cos \beta \end{bmatrix}^T$$

According to Eq. 7, the target normal vector in WCS is shown as follows,

$${}^W\overrightarrow{V} = [l, m, n]^T$$
$$l = \sin(C)\sin(A)\cos(\alpha)\cos(\beta) + \sin(C)\sin(\beta)\cos(A)\cos(\alpha) + \sin(\alpha)\cos(C)$$
$$m = -\cos(C)\sin(A)\cos(\alpha)\cos(\beta) - \cos(C)\sin(\beta)\cos(A)\cos(\alpha) + \sin(C)\sin(\alpha) \tag{10}$$
$$n = (-\sin(A)\sin(\beta) + \cos(A)\cos(\beta))\cos(\alpha)$$

Finally, ${}^W\overrightarrow{V}$ is unitized as follows,

$$\overrightarrow{\xi} = \left(\frac{l}{\sqrt{l^2 + m^2 + n^2}}, \frac{m}{\sqrt{l^2 + m^2 + n^2}}, \frac{n}{\sqrt{l^2 + m^2 + n^2}}\right)^T = (i, j, k)^T \tag{11}$$

3.2 Post-processing and Compensation Strategy

Note that a new normal vector $(i, j, k)^T$ presents the ideal position. By inversely calculating the (x, y, z, i, j, k) by Eqs. 8 and 9 in Sect. 3.1, the compensation of all axes is respectively computed. As a result, a new position of rotation axes A and C can be calculated by Eq. 8 at the first. Moreover, other new liner axes positions can also be worked out.

Nevertheless, according to the Eq. 8, the inverse value of A' or C' is possible to exist in multiple solution. To allow for all condition, a special discussion of multiple solution in five-axis machine tool is expressed as follows:

Firstly, the value of k should be considered as follows,

1. When $k = 1$,

$$A' = 0$$
$$C' = C \tag{12}$$

2. When $k \neq 1$,

$$
\begin{cases}
A_1 = \arccos(k) \\
C_1 = a \tan 2(\frac{i}{\sin A}, \frac{j}{-\sin A})
\end{cases}
\tag{13}
$$

Or

$$
\begin{cases}
A_2 = -\arccos(k) \\
C_2 = a \tan 2(\frac{i}{\sin A}, \frac{j}{-\sin A})
\end{cases}
\tag{14}
$$

Then, $a \tan 2(y, x)$ represents the azimuth between the origin and point (x, y) which the range is defined with $(-\pi, \pi]$. And it not only depends on the value of $\arctan(\frac{y}{x})$, but also is determined where the point (x, y) is located.

Consequently, there are four kinds of solutions in $(C - 2\pi, C + 2\pi]$. To get specific and unique solutions of A' and C', a minimal path of rotation [6, 7] is applied as follows,

(1) Current rotation position (A, C);
(2) Possible target position (A_1, C_1), (A_2, C_2), (A_3, C_3), (A_4, C_4);
(3) Each A_1, A_2, A_3, A_4 is respectively compared with original A and the minimal deviation of one or several is reserved, which is the target A';
(4) Retain the unique one of minimal deviation with original C in step 3). That is the target C'.

Furthermore, a reverse post-processing in Eq. 9 completed and new target line axis shows as follows,

$$X' = (x + L_{cwx}) \cos C' + (y + L_{cwy}) \sin C' - L_{cwx}$$
$$Y' = -(x + L_{cwx}) \cos A' \sin C' + (y + L_{cwy}) \cos A' \cos C' + (z + L_{acz} + L_{cwz}) \sin A'$$
$$+ L_{acy} \cos A' - L_{acy} - L_{cwy}$$
$$Z' = (x + L_{cwx}) \sin A' \sin C' - (y + L_{cwy}) \sin A' \cos C' + (z + L_{acz} + L_{cwz}) \cos A'$$
$$- L_{acy} \sin A' - L_{acz} - L_{cwz}$$

$$(15)$$

In the end, within a compensation cycle, a new group of NC code X', Y', Z', A', C' is worked out and the machine tool is able to realize a dynamic real-time compensation

$$\Delta X = X' - X, \quad \Delta Y = Y' - Y, \quad \Delta Z = Z' - Z, \quad \Delta A = A' - A, \quad \Delta C = C' - C$$

4 Case Study

4.1 Experiment Setup

To verify the feasibility and effectiveness of real-time normal measurement and compensation, an experiment is carried out on a 5-axis of A-C dual-table machine tool. An aerospace thin-walled curved plate ($\Phi2250$) is used as a model to realize online normal error compensation when four displacement sensors display the same values within a margin of error. The experiment platform is established in Fig. 7. A high-precision MICRO-EPSILON eddy current displacement sensor with a precision of 0.1% is chosen.

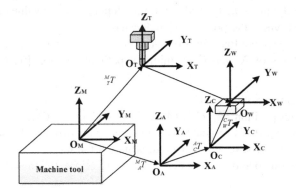

Fig. 7. Coordinate system chains of A-C dual-table machine tool

The model of curved plate and motion is displayed in Fig. 8. The measurement equipment starts from No. 0 position and experiences a series of point position No. 1, No. 2 and so on. A rectilinear motion is completed between two point positions and the motion span is set as 100 mm with a feed rate of 1000 mm/min. At the moment, it is seen from Fig. 9 that the series of original cutter normal vector N_1', N_2' have a deviation with the real normal vector N_2, N_3. Record values of four sensors before and after compensation and so Table 2 is shown as follows,

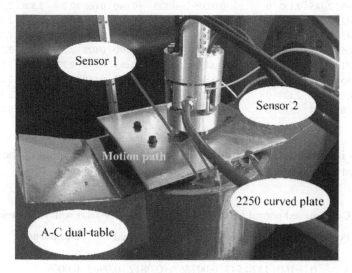

Fig. 8. Experiment platform of normal compensation verification

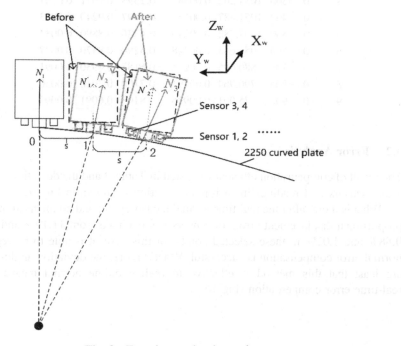

Fig. 9. Experiment planning path

Table 2. Experimental data of all point positions

Displayed value of sensors (before)/mm				Compensation of machine tool					Displayed value of sensors (after)/mm			
l_1	l_2	l_3	l_4	ΔX/mm	ΔY/mm	ΔZ/mm	ΔA/°	ΔC/°	l_1'	l_2'	l_3'	l_4'
1	2.436	2.418	2.225	2.149	0	0.00069	−0.542	−0.073	0.098	2.312	2.323	2.335	2.340
2	2.378	2.398	2.245	2.132	0	0.00039	−0.300	−0.040	0.068	2.296	2.308	2.312	2.315
3	2.576	2.552	2.305	2.207	0	0.00181	−0.654	−0.087	0.132	2.343	2.295	2.321	2.319
4	2.317	2.345	2.241	2.196	0	0.00047	−0.076	−0.010	0.051	2.270	2.313	2.281	2.305
5	2.343	2.305	2.217	2.233	0	−0.00126	−0.599	−0.080	0.022	2.292	2.319	2.290	2.276
6	2.340	2.371	2.139	2.067	0	−0.00014	−0.621	−0.083	0.080	2.278	2.334	2.250	2.354
7	2.354	2.395	2.180	2.141	0	0.00081	−0.387	−0.052	0.089	2.267	2.269	2.291	2.301
8	2.489	2.428	2.192	2.138	0	0.00072	0.869	−0.116	0.124	2.278	2.286	2.302	2.273
9	2.584	2.520	2.193	2.180	0	−0.00023	−1.305	−0.175	0.142	2.297	2.307	2.292	2.293

Table 3 is shown as a comparison between real normal direction of the plate and compensated normal vector by error compensation above.

Table 3. Compensated normal vector of each point and deviation with real normal vector

Point	Real normal vector			Compensated normal vector			Deviation/(°)
1	0	−100	1120.547	0.00736	−0.0812	0.9965	0.0069
2	0	−200	1107.079	0.00422	−0.1772	0.9920	0.0046
3	0	−300	1084.262	0.00543	−0.2593	0.9641	0.0101
4	0	−400	1051.487	0.00152	−0.3497	0.9240	0.0023
5	0	−500	1007.782	0.00249	−0.4386	0.8890	0.0034
6	0	−600	951.643	0.00348	−0.5290	0.8370	0.0037
7	0	−700	880.695	0.00075	−0.6187	0.7930	0.0091
8	0	−800	790.964	0.00491	−0.7100	0.7010	0.0050
9	0	−900	675.001	0.00975	−0.8103	0.6091	0.0097

4.2   Error Analysis

The result of compensation efficiency is listed in Table 2 and the deviation of sensors 1, 3 and sensors 2, 4 readings in each point position is shown in Fig. 9.

What is more, after the real-time normal measurement and compensation algorithm, proportional displacement errors of sensors has declined from 11.19% and 12.10% to 0.88% and 1.02% in these selected points. In this case, it is able to be regraded that normal error compensation is successful. What's more, the deviation angles in Table 3 are least that this method is effective to realize online normal measurement and real-time error compensation (Fig 10).

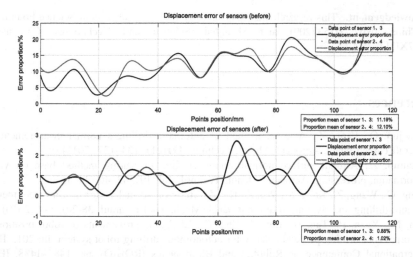

Fig. 10. Error proportion of displacement sensors before and after compensation

5 Conclusion

To acquire real-time normal direction of large aerospace thin-walled surface and reach the right position in time, this paper proposes an efficient method for automatic normal measurement and error compensation of cutter bit to adapt any curved surface with a random deformation. Moreover, the integrated design skillfully turns the deviation of displacement value from sensors into angles in order to realize rapid response and improve the manufacturing efficiency. After an experimental verification, this compensation method is valid and the adjusted normal error is less than 0.01°. And this method simplifies the measurement and compensation processing and is exactly efficient to let the cutter bit reach the right normal with a micron class accuracy. The contributions of this paper are listed as follows:

1. The algorithm in the paper converts skillfully the normal error into the feedbacks of eddy sensors values so as to avoid complicated solution processes of real normal vector. In fact, the axis motion of tools just needs a series of increment value instead of coordinates while the new normal vector in Table 3 is measured by current eddy sensors values in the case study. Meanwhile, the method greatly improves computational speed of the CNC system, timeliness and validity of machine tools in order to provide favorable conditions to achieve real-time compensation.
2. During the dynamic measurement process, the algorithm in the paper is real and effective where displacement measurements of each measurement point are compensated in place, that is, each point reaches real normal direction. By constantly changing the machine feed rate, the compensation effect is the same effect and effective so that the real-time requirement of dynamic compensation is achieved. The real-time dynamic compensation of normal error fills the blank of the dynamic monitoring for the process of large aerospace thin-walled workpieces. At the same time, it lays the foundation of obtaining real-time parts thickness data and other lots of related technologies and opens up a new application field.

Acknowledgement. This research was sponsored by the National Natural Science Foundation of China [No. U1537209] and Shanghai Science and Technology Commission [No.17XD1422500].

References

1. Shaojun, Q., Fang, Z.: A novel method of automatic drilling's normal measurement for aircraft curved panel. Aeronaut. Manuf. Technol. **491**(21), 134–137 (2015)
2. Weimin, Z.: Aircraft surface normal vector on line measuring method base on visual information processing. Aeronaut. Manuf. Technol. **489**(19), 43–46 (2015)
3. Ying, G., Wang, Z., Kang, Y., Wu, Z., Hu, Z.: Study on normal vector measurement method in auto-drilling and riveting of air-craft panel. Mach. Tool Hydraul. **38**(20), 1–4 (2010)
4. Lin, M., Yuan, P., Tan, H., Liu, Y., Zhu, Q., Li, Y.: Improvements of robot positioning accuracy and drilling perpendicularity for autonomous drilling robot system. In: 2015 IEEE International Conference on Robotics and Biomimetics (ROBIO), pp. 1483–1488. IEEE, December 2015
5. Yuan, P., Wang, Q., Wang, T., Wang, C., Song, B.: Surface normal measurement in the end effector of a drilling robot for aviation. In: 2014 IEEE International Conference on Robotics and Automation (ICRA), pp. 4481–4486. IEEE May 2014
6. DeVlieg, R.: Robotic trailing edge flap drilling system (No. 2009-01-3244). SAE Technical Paper (2009)
7. Huang, N., Jin, Y., Bi, Q., Wang, Y.: Integrated post-processor for 5-axis machine tools with geometric errors compensation. Int. J. Mach. Tools Manuf. **94**, 65–73 (2015)
8. Munlin, M., Makhanov, S.S., Bohez, E.L.: Optimization of rotations of a five-axis milling machine near stationary points. Comput.-Aided Des. **36**(12), 1117–1128 (2004)
9. Makhanov, S.S., Munlin, M.: Optimal sequencing of rotation angles for five-axis machining. Int. J. Adv. Manuf. Technol. **35**(1–2), 41–54 (2007)

Feasibility of the Bi-Directional Scanning Method in Acceleration/deceleration Feedrate Scheduling for CNC Machining

Jie Huang, Xu Du, and Li-Min Zhu[✉]

State Key Laboratory of Mechanical System and Vibration,
School of Mechanical Engineering, Shanghai Jiao Tong University,
Shanghai 200240, People's Republic of China
{thk2dth,xinxiangyangguang,zhulm}@sjtu.edu.cn

Abstract. The Acceleration/Deceleration (AD) feedrate scheduling is widely used to plan the feedrate for CNC machining. Since the toolpath in CNC machining consists of enormous blocks (e.g., G01 or G02 block), the AD scheduling will result in lots of successive feedrate profiles. The feedrate at the junction of the adjacent profiles can be discontinuous, which will saturate the actuator and deteriorate the machining performance. The Bi-Directional Scanning Method (BDSM) is used to make the profiles overall continuous. To alleviate the computational burden, the BDSM is usually applied within a look-ahead buffer. When the buffer is filled with feedrate profiles, the BDSM updates the feedrates in the buffer. Conventional works believe that the BDSM with look-ahead (BDSMLA) will only increase the feedrate in each updating and is always feasible. We find that the BDSM can, however, decrease the feedrates in the buffer. The feedrate decrease will result in overall feedrate discontinuity and make the BDSM infeasible. We also propose a tweak method which can guarantee the feasibility of the BDSMLA. Simulation reveals the feedrate decrease in the BDSMLA and verifies the effectiveness of the tweak method.

Keywords: Feedrate scheduling · Bi-directional scanning · Look-ahead buffer · CNC machining

1 Introduction

The productivity of Computer Numerical Control (CNC) machining is highly related to the feedrate along the toolpath. Therefore, the time-minimum feedrate scheduling for CNC machining has received a significant amount of attention in the literature. Various techniques have been proposed to schedule the feedrate for CNC machining. Due to the real-time requirement of CNC systems, the AD scheduling method is one of the most widely-used scheduling methods for CNC machining [1,2]. The AD feedrate scheduling is usually applied to each block of the toolpath, which will result in a series of feedrate profiles.

© Springer International Publishing AG 2017
Y. Huang et al. (Eds.): ICIRA 2017, Part II, LNAI 10463, pp. 171–183, 2017.
DOI: 10.1007/978-3-319-65292-4_16

Though each feedrate profile is internally continuous, the continuity at the junction of the adjacent profiles is not assured. The feedrate discontinuity at the junction will saturate the actuator and deteriorate the machining performance. To guarantee the smoothness of the overall feedrate profile, the scheduling methods usually involve a BDSM. After bi-directional scanning, the overall feedrate profile obtained by the AD feedrate scheduling is continuous and *kinematically time-minimum*.

The BDSM was first used to plan time-minimum trajectories for manipulators and robots to follow specified paths. Bobrow et al. [3] and Slotine and Yang [4] employed a method similar to BDSM to find the switching points where the torque of an actuator switches from one bound to another. The switching point method along with the BDSM was later used to schedule the feedrate for NC machining by Timar and Farouki [5] and Dong and Stori [6]. For switching point method, the feasibility and optimality of the BDSM had been proved. However, the BDSM in switching point method cannot be directly used to schedule feedrate for CNC machining due to the following reasons [1]. (1) While the trajectory in the switching method is mostly planned off-line, the feedrate for CNC machining should be scheduled on-line due to the real-time requirement of CNC systems. (2) The toolpath in CNC machining consists of numerous line or arc blocks. The toolpath composition makes the task to identify all the switching points impractical. (3) The feedrate for CNC machining may be overridden on-line by the user, which will make the trajectory planned in the switching point method invalid.

For CNC machining, the AD feedrate scheduling method is widely adopted due to its robustness and computational efficiency. To make the overall feedrate smooth, the AD feedrate scheduling method also utilizes the BDSM. To alleviate the computational burden for CNC systems, the BDSM is usually applied within a FIFO (First-In-First-Out) look-ahead buffer. Whenever the head of the buffer is fed with a new scheduling block from the interpreter, the BDSM is invoked immediately to update the feedrate in the buffer. Then, the tail of the buffer will be sent to the interpolator to generate motion commands [7]. The BDSM in the AD feedrate scheduling noticeably different from that in the switching point method in the following two ways. (1) To schedule or update the feedrate, the BDSM is used only once. (2) The BDSM is employed within the look-ahead buffer rather than along the whole path. Therefore, the feasibility of the BDSM in the AD feedrate scheduling method needs further studies.

Though the BDSM is widely used in the AD feedrate scheduling, its feasibility is not fully studied [8]. Luo et al. [9] used a doubly linked list as a look-ahead buffer to schedule the feedrate. The BDSM was applied within the list. Zhao et al. [10] detailed the BDSM and applied it within a look-ahead window. Dong [11] proposed a generic look-ahead feedrate plan (GLAFP) scheme for AD feedrate scheduling. A lot of properties of GLAFP were also revealed. To ensure the continuity of the overall feedrate profile, Lee et al. [12] proposed a backward scanning process to handle successive short blocks in priority. No look-ahead buffer was involved in their method. To make the overall feedrate profile smooth,

Fan et al. [13] proposed a backtracking algorithm. The prevalent open source motion controller for CNC milling, Grbl[1], also employs the BDSM within a look-ahead buffer to optimize the feedrate [14]. However, none of them fully addresses the feasibility of BDSM in AD feedrate scheduling.

In this paper, the feasibility of the BDSM without look-ahead is elaborated firstly. Then, the feasibility of the BDSMLA is discussed. A tweak method which can ensure the feasibility of the BDSMLA is also proposed.

2   BDSM Without Look-Ahead

In this section, the necessity of the BDSM is discussed first. Then the procedure of the BDSM is detailed. Finally, the feasibility of the BDSM without look-ahead is elaborated.

2.1   Necessity of the BDSM

To connect boundary feedrates, the AD feedrate scheduling uses four types of AD profiles: ACCeleration (ACC), DECeleration (DEC), ACC+DEC and ACC+(Constant Feed) (CF)+DEC [12,13,15], as shown in Fig. 1. To transit the feedrate, if the traversal length is short, an ACC or a DEC profile will be adopted; otherwise, an ACC+CF+DEC, or an ACC+DEC profile will be adopted. These types of profiles are also widely adopted to plan the trajectories in joint space for robots and manipulators [16–18]. However, trajectory planning in joint space usually involves no BDSM. The joint boundary velocities are always feasible since the joint velocity can be positive as well as negative [16]. Therefore, the velocity at the junction of the adjacent joint profiles is continuous and BDSM is unnecessary.

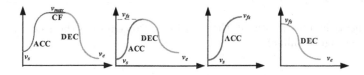

Fig. 1. Four types of the AD feedrate profiles.

The feedrate for CNC machining is the tool tip velocity in the workpiece coordinate which is usually non-negative. The non-negative property of the feedrate in CNC machining makes the BDSM indispensable. As mentioned by Lee et al. [12] and Fan et al. [13], short traversal distance will result in an AD profile with a decreased boundary feedrate. If the boundary feedrate is decreased, the feedrate at the junction of two adjacent profiles will be discontinuous. To make the feedrate profile overall continuous, the normally adopted BDSM is indispensable.

---

[1] Grbl is written in optimized C that run on a straight Arduino. It has gotten 2149 stars and 1381 forks on GitHub by May 2017.

2.2   Procedure of the BDSM

If no look-ahead buffer is involved, the BDSM is applied to the overall path. The overall start and end feedrates are both set to zero. Suppose there are N feedrate profiles to be scheduled, which means the BDSM should determine $N+1$ boundary feedrate v_i. Initially, the boundary feedrate v_i is limited by the nominal feedrate v_i^n which is usually determined by the kinematic and geometric constraints [12,15]. The determined boundary feedrates cannot exceed their nominal values. Let v_i^b denote the boundary feedrate determined by the backward process and v_i^f denote the boundary feedrate determined by the forward process. Term the feedrate profile transiting feedrate from v_{i-1} to v_i as the ith feedrate profile, the given traversal distance of the ith feedrate profile as s_i. Let $f(t, v_s, v_e, s, T)$ denote an ACC profile which transits the feedrate from v_s to v_e ($v_e > v_s$) with a traversal distance s and a traversal time T. The BDSM consists of two sequential processes: a backward process which scans the feedrate profiles from the end to the start, and a forward process which scans the feedrate profile from the start to the end as schematically illustrated in Fig. 2 and detailed in Fig. 3 [9,10]. The purple line in Fig. 2 representing the final feedrate profile is used to indicate the existence of a feasible feedrate profile. The feasible profile can be any profile types.

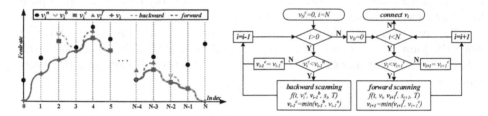

Fig. 2. Schematic illustration of the BDSM. (Color figure online)   **Fig. 3.** Procedure of the BDSM.

2.3   Feasibility of the BDSM Without Look-Ahead

To analyze the feasibility of the BDSM, two properties of ACC profile are given first. Property 1 has been proved in [15]. Property 2 is proved in Appendix A.

Property 1. *Suppose there exists an ACC profile $f(t, v_s, v_e, s, T)$. It must exist an ACC+DEC or ACC+CF+DEC profile transiting the feedrate from v_s to v_e' with the same traversal distance s if $v_e' \in [v_s, v_e]$.*

Property 2. *For symmetrical kinematic constraints, $s = (v_s + v_e) T/2$, regardless of the AD functions used.*

Property 1 ensures the feasibility of the BDSM. By observing the BDSM procedure, one can easily find that: (a) backward scanning only decreases the start boundary feedrate, and (b) forward scanning only decreases the end boundary feedrate. The feedrate decrease impacts the adjacent feedrate profile. In backward scanning, the impact will propagate backward and influence the previous profile. In forward scanning, the impact will propagate forward and influence the next profile. Since the overall start feedrate is zero, the first profile needs no backward scanning. Therefore, the overall start feedrate will keep unchanged. The overall end feedrate will keep unchanged, analogously. After bi-directional scanning, all the boundary feedrates are determined. Based on Property 1, the determined boundary feedrates are all feasible when constructing the final feedrate profiles.

3   BDSM with Look-Ahead

If applied to the overall toolpath which consists of enormous blocks, the BDSM will become computationally intensive. To alleviate the computational burden, the so-called look-ahead buffer is used. In this section, the mechanism of the look-ahead buffer is introduced first. Then the feasibility of the BDSMLA is discussed. Finally, to guarantee the feasibility, a tweak method is proposed.

3.1   Mechanism of the Look-Ahead Buffer

The look-ahead buffer bridges the interpreter with the interpolator. If a look-ahead buffer is used, the Numerical Control Kernel (NCK) of a CNC system can be schematically illustrated by the diagram in Fig. 4. The trajectory planning task will read a few blocks to the look-ahead buffer and schedule the feedrate within the buffer before interpolation. After scheduling, the tail of the buffer will be taken away by the interpolator to generate motion commands. The new block retrieved from the interpreter is then appended to the head of the buffer. The end boundary feedrate for the new block is always set to zero. The start feedrate of the current tail remains unchanged. All the intermediate boundary feedrate are reset to their nominal values v_i^n. The feedrate is then re-scheduled with the BDSM. Note that the start feedrate of the current tail is identical to the end feedrate of the profile previous taken away by the interpolator. To keep the feedrate continuous, the re-scheduling should never change the start feedrate of the current tail.

The BDSM will update the boundary feedrates and only determine one boundary feedrate every time. In the jth BDSM, only the jth boundary feedrate v_j is to be determined. Note that the $(j-1)^{th}$ boundary feedrate v_{j-1} has been determined in the previous BDSM. All the other boundary feedrates after the jth BDSM are undetermined and will be updated in the following BDSM. Denote the ith boundary feedrate after the jth BDSM $v_{i,j}$. The feedrate scheduling with look-ahead is illustrated with a look-ahead buffer of size 4, as depicted in Fig. 6.

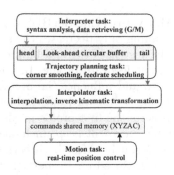

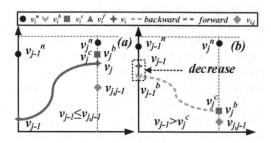

Fig. 4. NCK of CNC systems.   **Fig. 5.** Two situations when $v_{j-1} > 0$.

After the buffer initially filled by the first four blocks retrieved from the interpreter, the first BDSM is invoked. Without loss of generality, suppose the types of the four feedrate profiles after the first BDSM are ACC, ACC+CF+DEC, ACC+DEC and DEC, respectively. After the first BDSM, v_1 is determined, and $v_{2,3,4}$ are undetermined. The tail of the buffer, i.e., the first feedrate profile (of type ACC), is then taken away by the interpolator. As the 5th block fed in, the second BDSM is invoked. In the second BDSM, v_1 remains unchanged, and $v_{2,3,4}$ are set to their nominal values and will be updated. After the second BDSM, v_2 is determined, and the tail, i.e., the second feedrate profile (of type ACC+CF+DEC), is taken away. As new block continues coming, the BDSM is re-invoked every time. If there is no new block, the scheduling is completed.

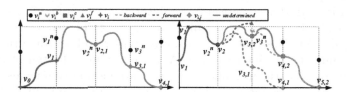

Fig. 6. Feedrate scheduling with a look-ahead buffer of size 4.

It should be noted that there are some tricks to accelerate the execution of the BDSMLA. For example, if the feedrate $v_{i,j} = v_i^n$, which means $v_{i,j}$ cannot be further increased, the boundary feedrates from v_{j-1} to v_i are actually all determined (see $v_{2,1}$ in Fig. 6). Thus, the profiles from the buffer tail to the ith profile need no updating, and the BDSM can be directly applied to the rest of the buffer. This trick is also used in Grbl. Moreover, if the size of the look-ahead buffer is large, the parallel BDSM [8] can accelerate the scanning process.

3.2   Feasibility of the BDSMLA

The only different situation of the BDSM when a look-ahead buffer is involved is that the overall start feedrate is non-zero. We find that the non-zero overall start feedrate can make the feasibility of the BDSM problematic. While the first profile without look-ahead needs no backward scanning, the buffer tail with a non-zero start feedrate may need a backward scanning. The backward scanning can decrease the start feedrate of the buffer tail, which will make the overall feedrate discontinuous.

Suppose the jth BDSM is invoked to determine the jth boundary feedrate v_j. As aforementioned, all the intermediate boundary feedrates have been reset to their nominal values. For the buffer tail, there are two situations, as listed in Fig. 5. In situation (a) where $v_j^c \geq v_{j-1}$, the tail needs no backward scanning, and $v_j \in [v_{j,j-1}, v_j^c]$ is determined by the forward scanning. In situation (b) where $v_j^c < v_{j-1}$, the tail needs a backward scanning. However, an ACC profile transiting the feedrate from v_j^c to v_{j-1} cannot be guaranteed based on Property 1. The backward scanning may result in a start feedrate v_{j-1}^b smaller than v_{j-1}, which contradicts the requirement that the start feedrate cannot be changed. The backward will not decrease the start feedrate of the buffer under the following assertion.

Assertion 1. *Suppose there exists an ACC profile $f(t, v_s, v_e, s, T)$. It must exist a profile transiting the feedrate from v_s' to v_e with the same traversal distance s if $v_s' \in [v_s, v_e]$.*

Contrary to conventional works that took the assertion for granted [10, 11, 13], we find that the validity of the assertion is related to the function used for the AD feedrate profile. For simplicity, suppose the kinematic constrains are symmetric. For ACC profiles, the traversal distance can be calculated according to Property 2.

If d is monotonically decreasing w.r.t. v_s, the assertion will hold. For acceleration-bounded AD profile, d can be calculated according to (1), where a_m is the specified maximum acceleration. Fixed v_e, d is monotonically decreasing w.r.t. v_s when $v_s \in [0, v_e]$. Thus, the assertion holds, and the BDSMLA is feasible.

$$d = \left(v_e^2 - v_s^2\right)/2a_m \tag{1}$$

However, the monotonically decreasing relationship is not assured for a general AD feedrate profile. To illustrate, take a jerk-bounded AD profile with the normally adopted values for example,

$$v_m = 200\,\mathrm{mm/s}, a_m = 1000\,\mathrm{mm/s^2}, j_m = 10000\,\mathrm{mm/s^3}, t_s = 0.5\,\mathrm{ms}.$$

where v_m, j_m and t_s are specified maximum feedrate, specified maximum jerk and interpolation period, respectively. Let v_s and v_e both range from 0 to v_m, and calculate the traversal distance d of a DEC or an ACC profile transiting the feedrate from v_s to v_e. When $v_s \leq v_e$, distance of an ACC profile is calculated; otherwise, distance of a DEC profile is calculated. The traversal distance is calculated according to Property 2 and is given in Fig. 7. Fixed $v_e = v_m$, the distance is not monotonically decreasing when v_s increases, as shown in Fig. 7(a). Denote the maximum traversal distance d_{max}. For an ACC profile, the start

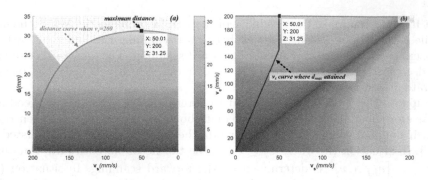

Fig. 7. Traversal distance transiting from v_s to v_e: (a) distance w.r.t. v_s when $v_e = 200$; (b) v_s where d_{max} of an ACC profile is attained.

feedrate v_s where d_{max} attained w.r.t. v_e is shown in Fig. 7(b). The start feedrates except for the trivial one when $v_e = 0$ are all greater than zero, which indicates the traversal distance for an ACC profile is normally not monotonically decreasing w.r.t. the start feedrate. Therefore, the assertion fails, and the feasibility of BDSMLA is not assured for jerk-bounded AD profile. Moreover, since such monotonically decreasing relationship does not exist for the jounce-bounded and trigonometric-jerk AD profiles [8], the BDSMLA for them is not necessarily feasible either.

3.3   Tweak of the BDSM

In this section, we tweak the traditional BDSM and make it feasible for general AD feedrate profiles. As mentioned in Sect. 2.3, the impact of boundary feedrate decrease introduced in forward scanning will propagate forward. Since the end feedrate of the buffer is still zero, the impact in forward scanning can be resolved. Therefore, if the decrease of the start feedrate in backward scanning is circumvented, the feasibility of the BDSMLA can be achieved.

To assure the feasibility, we tweak the BDSM locally when the backward scanning decreases the start feedrate of the buffer. Suppose the backward scanning in the jth BDSM decreases the start feedrate of the buffer v_{j-1}, the tweak procedure is detailed in Fig. 8(a). Without loss of generality, take the example as shown in Fig. 8(b) to illustrate the procedure. The jth backward scanning decreases the $(j-1)^{th}$ boundary feedrate from v_{j-1} to v_{j-1}^c. Thus, keep v_{j-1} unchanged and let $v_j = v_{j,j-1}$. The jth profile is not updated in the jth BDSM. For the $(j+1)^{th}$ profile, we have $v_{j+1}^c < v_j$. Scanning backward from v_{j+1}^c obtains v_j^a. Since $v_j^a < v_j$, let $v_{j+1,j} = v_{j+1,j-1}$. For the $(j+2)^{th}$ profile, since $v_{j+2}^c > v_{j+1,j}$, the tweak is completed and continue the forward scanning from the $(j+1)^{th}$ boundary feedrate.

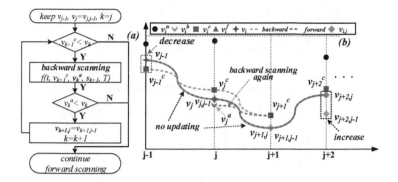

Fig. 8. Tweak of the BDSM: (a) procedure of the tweak; (b) a tweak example.

4 Simulation

The BDSM without look-ahead is widely used in the literature, and its feasibility has been sufficiently validated by simulations and experiments [8–10]. For succinctness, in this section, we skip the BDSM without look-ahead and discuss the feasibility of the BDSMLA.

The butterfly curve[2] in Fig. 9 is used to demonstrate: (1) the BDSMLA can decrease the start feedrate of the buffer; (2) the proposed tweak method of the BDSM is effective. The jerk-bounded AD feedrate profile and the parameters in Sect. 3.2 are used for the simulation.

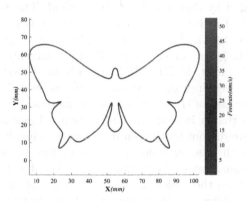

Fig. 9. Contour of the butterfly curve.

Firstly, 1050 points are sampled on the parametric curve with a chord error $\delta = 0.02$ mm. The nominal feedrate v_i^n of each sampled point is determined

---

[2] The curve data is available at https://github.com/thk2dth/CNCLite/tree/master/ Kernel/NurbsData.

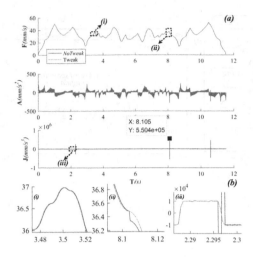

Fig. 10. Feedrate profiles.

according to (3), where θ is the corner angle, and $\varepsilon = \delta/2$ is a specified deviation error [19].

$$v_i^n = \min\left(\sqrt{\frac{q\varepsilon a_m}{1-q}}, v_m\right), q = \sqrt{\frac{1+\cos\theta}{2}} \qquad (2)$$

Then, the BDSMs with and without the proposed tweak method are applied within a size-50 look-ahead buffer. After obtaining the feedrate, the acceleration and jerk are evaluated by digital differential method. The results are given in Fig. 10. As shown in Fig. 10(b), the feedrate without tweak is discontinuous, and the jerk without tweak shows several peaks. The maximum jerk value is almost 55 times of the given value, which will surely saturate the actuator. This phenomenon indicates that the BDSMLA actually decreased the start feedrate of the buffer. After tweaking, the feedrate profile is overall continuous, which validates the effectiveness of the proposed tweak method.

Besides, we find that a look-ahead buffer with a large size (e.g., 100) can eliminate the feedrate discontinuity even if no tweak is involved. The reason for this result is that with a larger buffer size a boundary feedrate is more likely to approach its nominal value. As mentioned in Sect. 3.1, the BDSM will not update the profiles from the buffer tail to the feedrate identical to the nominal value. However, it is difficult to determine a sufficiently large size of the buffer to make the BDSM feasible for different toolpaths. Moreover, the buffer size in a controller is usually limited due to the real-time requirement. The proposed tweak method can guarantee the feasibility regardless of the buffer size and will benefit the AD feedrate scheduling for CNC machining.

5   Conclusions

To meet the real-time requirement of CNC systems, the BDSM with AD feedrate scheduling is usually adopted to update the feedrates within a look-ahead buffer for CNC machining. It is believed that the BDSMLA in each updating only increases the feedrate and is always feasible. By analyzing the properties of the ACC feedrate profile, we find that the BDSMLA may decrease the feedrate and can be infeasible. There are two main contributions in this paper.

- We find that the BDSMLA can be infeasible if the traversal distance of an ACC profile is not monotonically decreasing w.r.t. its start feedrate.
- A tweak method which can guarantee the feasibility of the BDSMLA is proposed.

Simulation has demonstrated the feedrate decrease in BDSMLA and verified the effectiveness of the proposed tweak method. The tweak method can benefit the feedrate scheduling with look-ahead for CNC machining. If the look-ahead buffer size is limited, the tweak method will be of great help for improving machining quality.

Acknowledgment. This work was partially supported by the National Natural Science Foundation of China under grant No. 51325502, and the Science & Technology Commission of Shanghai Municipality under grant No. 15550722300.

A   Proof of Property 2

Suppose an ACC feedrate profile $f(t), t \in [0, T]$ transits the feedrate from $f(0) = v_s$ to $f(T) = v_e$ with a traversal time T. Since the kinematic constraints are symmetric, the first derivative $f'(t)$ is always symmetric about $t = T/2$. Based on Dirichlet's conditions, we can represent $f'(t)$ by the following Fourier series, where b_i is the Fourier coefficients.

$$f'(t) = \frac{b_0}{2} + \sum_{i=1}^{\infty} b_i \cos \frac{4i\pi (t - T/2)}{T}, t \in [0, T] \tag{3}$$

Integrating $f'(t)$ yields,

$$f(t) = \frac{b_0 t}{2} + \sum_{i=1}^{\infty} \frac{b_i T}{4i\pi} \sin \frac{4i\pi (t - T/2)}{T} + c_0, t \in [0, T] \tag{4}$$

where c_0 is the integration constant. According to the boundary feedrates, we have

$$c_0 = v_s, b_0 = \frac{2 (v_e - v_s)}{T} \tag{5}$$

Substituting (5) into (4) and integrating $f(t)$ obtain the distance function,

$$s(t) = \frac{b_0 t^2}{4} - \sum_{i=1}^{\infty} b_i \left(\frac{T}{4i\pi} \right)^2 \cos \frac{4i\pi (t - T/2)}{T} + c_0 t + c_1, t \in [0, T] \tag{6}$$

where c_1 is the integration constant. Therefore, the traversal distance can be evaluated as

$$s = s(T) - s(0) = \frac{v_s + v_e}{2} T \tag{7}$$

References

1. Suh, S.H., Kang, S.K., Chung, D.H., Stroud, I.: Theory and Design of CNC Systems. Springer Series in Advanced Manufacturing. Springer, London (2008). doi:10.1007/978-1-84800-336-1

2. Erkorkmaz, K., Altintas, Y.: High speed CNC system design. Part I: jerk limited trajectory generation and quintic spline interpolation. Int. J. Mach. Tools Manuf. **41**(9), 1323–1345 (2001)

3. Bobrow, J., Dubowsky, S., Gibson, J.: Time-optimal control of robotic manipulators along specified paths. Int. J. Robot. Res. **4**(3), 3–17 (1985)

4. Slotine, J.J.E., Yang, H.S.: Improving the efficiency of time-optimal path-following algorithms. IEEE Trans. Robot. Autom. **5**(1), 118–124 (1989)

5. Timar, S.D., Farouki, R.T.: Time-optimal traversal of curved paths by Cartesian CNC machines under both constant and speed-dependent axis acceleration bounds. Robot. Comput.-Integr. Manuf. **23**(5), 563–579 (2007)

6. Dong, J., Stori, J.A.: Optimal feed-rate scheduling for high-speed contouring. J. Manuf. Sci. Eng. **129**(1), 63 (2007)

7. Schuett, T.: A Closer Look At Look-Ahead (1996). http://www.mmsonline.com/articles/a-closer-look-at-look-ahead. Accessed 17 Feb 2017

8. Huang, J., Du, X., Zhu, L.: Parallel acceleration/deceleration feedrate scheduling for CNC machine tools based on bi-directional scanning technique. Proc. Inst. Mech. Eng. Part B J. Eng. Manuf. (2017). doi:10.1177/0954405417706997

9. Luo, F., Zhou, Y., Yin, J.: A universal velocity profile generation approach for high-speed machining of small line segments with look-ahead. Int. J. Adv. Manuf. Technol. **35**(5–6), 505–518 (2007)

10. Zhao, H., Zhu, L., Ding, H.: A real-time look-ahead interpolation methodology with curvature-continuous B-spline transition scheme for CNC machining of short line segments. Int. J. Mach. Tools Manuf. **65**, 88–98 (2013)

11. Dong, J.: Research on key technologies for reconfigurable CNC controller. Ph.D. thesis, Tianjin University (2010)

12. Lee, A.C., Lin, M.T., Pan, Y.R., Lin, W.Y.: The feedrate scheduling of NURBS interpolator for CNC machine tools. CAD Comput. Aided Des. **43**(6), 612–628 (2011)

13. Fan, W., Gao, X.S., Yan, W., Yuan, C.M.: Interpolation of parametric CNC machining path under confined jounce. Int. J. Adv. Manuf. Technol. **62**(5–8), 719–739 (2012)

14. Jeon, S.S.: grbl/planner (2016). https://github.com/grbl/grbl/blob/master/grbl/planner.c. Accessed 17 Feb 2017

15. Huang, J., Zhu, L.M.M.: Feedrate scheduling for interpolation of parametric tool path using the sine series representation of jerk profile. Proc. Inst. Mech. Eng. Part B J. Eng. Manuf. doi:10.1177/0954405416629588

16. Kröger, T.: On-Line Trajectory Generation in Robotic Systems. Springer Tracts in Advanced Robotics, vol. 58. Springer, Heidelberg (2010)

17. Siciliano, B., Sciavicco, L., Villani, L., Oriolo, G.: Robotics: Modelling, Planning and Control. Advanced Textbooks in Control and Signal Processing. Springer, London (2009)
18. Ezair, B., Tassa, T., Shiller, Z.: Planning high order trajectories with general initial and final conditions and asymmetric bounds. Int. J. Robot. Res. **33**(6), 898–916 (2014)
19. Jeon, S.S.: Improving Grbl: cornering algorithm (2011). https://onehossshay. wordpress.com/2011/09/24/improving_grbl_cornering_algorithm/. Accessed 04 Apr 2017

A Feed-Direction Stiffness Based Trajectory Optimization Method for a Milling Robot

Gang Xiong, Ye Ding, and LiMin Zhu$^{(\boxtimes)}$

State Key Laboratory of Mechanical System and Vibration,
School of Mechanical Engineering, Shanghai Jiao Tong University,
Shanghai 200240, People's Republic of China
{xionggang,y.ding,zhulm}@sjtu.edu.cn

Abstract. The post-processing process for an industrial robot in milling applications suffers from a redundancy problem when converting a 5-axis tool path to the corresponding 6-axis robot trajectory. This paper proposes a feed-direction stiffness based index to optimize the redundant freedom of the robot after identifying its stiffness model. At each cutter location point, the stiffness of the robot machining system along the feed direction is maximized, and an optimal robot configuration is obtained. The optimized robot trajectory via the proposed method has an advantage of improving the machining stability and production efficiency. Experiments verify the validity of the method.

Keywords: Robot machining · Feed-direction stiffness · Stiffness identification · Trajectory optimization

1 Introduction

The traditional application areas for industrial robot involve tasks that require good repeatability but not necessarily good accuracy such as handling, assembly, painting and welding. Compared with CNC machine tools, industrial robots have the advantages of wider workspace, lower cost and greater flexibility, thus can offer an efficient solution for large and complex shaped product manufacturing in aerospace industry [1]. Robot machining has become a significant trend in manufacturing industry, and many researches have been conducted in this area [2].

However, industrial robots are still rarely used in milling applications in industry. Except for the inherent low absolute accuracy and stiffness characteristics of industrial robots, the lack of a standard post-processing software is the main obstacle. This paper aims to solve this problem. In milling application, the robot cutting trajectory is usually converted from cutter location (CL) data generated by a 5-axis milling module in a commercial CAD/CAM software package [3]. Nevertheless, a milling task only requires five degrees of freedom (DOFs), three of which are used to locate the tool center point (TCP) and the rest two

© Springer International Publishing AG 2017
Y. Huang et al. (Eds.): ICIRA 2017, Part II, LNAI 10463, pp. 184–195, 2017.
DOI: 10.1007/978-3-319-65292-4_17

are used to determine the tool axis direction, while a standard commercial industrial robot usually has 6 DOFs. Using a 6-axis industrial robot to perform 5-axis milling tasks will result in one redundant DOF, which is the rotation about the milling spindle axis. This is a common situation because that the spindle axis is often mounted not to be aligned with the last joint axis to improve the dexterity and manipulability of the robot [4].

Until recently, a lot of researchers have been working on optimizing the redundant DOF and introducing new methodologies into the post-processor of a cutting robot. Xiao et al. [4] propose an optimization method which simultaneously optimizes the robot singularity criteria, joint limit criteria and obstacle avoidance criteria via one-dimensional search. Laurent et al. [5] and Huo et al. [6] use the Moore-Penrose pseudo inverse with an optimization term to solve the redundancy problem considering a set of performance criteria of the robot including joint limits, singularity avoidance and so on. These researches mainly concentrate on the geometrical and kinematical constraints of the robot, while neglect the machining performance in milling process. In milling operations, insufficient stiffness of the robot will cause static deformation of the TCP and vibrations of the robot structure [7], which will degrade the machining quality significantly. Therefore, some stiffness-based performance indexes have been proposed to optimize the redundant DOF in machining applications. Angeles [8] suggests the norm of the Cartesian stiffness matrix would be a plausible candidate. While Guo et al. [9] point out that the norm makes no physical sense as the Cartesian stiffness matrix has entries with disparate physical units, and they propose a performance index in terms of the determinant of the translational compliance sub-matrix (TCSM). However, this performance index just describes the overall stiffness of the robot at a certain posture, and doesn't pay enough attention to the anisotropy of the Cartesian stiffness [10]. In milling applications, the stiffness of the TCP along certain directions is usually paid special attention. Peng et al. [11] suggest that the feed-direction stiffness should be maximized in their 7-axis 5-linkage machine tool, which guarantees a high machining efficiency via adopting a larger feed rate. This concept can be extended to robot milling, in which the process parameters such as feed rate and cut of depth are always conservative due to the low stiffness of the robot [12]. In this paper, a feed-direction stiffness based performance index is proposed to optimize the redundant DOF for robot milling. This index takes the anisotropic force ellipsoid of the TCP into account and will improve the production efficiency of the robot machining system.

The remainder of this paper is organized as follows. In Sect. 2, the mathematic formula of the redundant problem is derived. In Sect. 3, the stiffness model of a MOTOMAN MH80 industrial robot is identified. In Sect. 4, a new performance index for optimizing the redundant DOF is proposed. In Sect. 5, Experiments have been done to verify the effectiveness of the proposed method. The paper is concluded in Sect. 6.

2   The Redundancy in Robot Milling

The most important part of post-processing for robot machining is to determine the joint space, i.e. the set of revolute joint variables $\theta_i (i = 1, \ldots, 6)$, from the CL data $CL(x, y, z, i, j, k)$, where (x, y, z) is the coordinate of the TCP, while (i, j, k) is a unit vector representing the tool axis direction. Practically, the CL data is generated in a standard CAD/CAM software such as NX, the positions and orientations are all described in the workpiece frame. When the CL data $CL(x, y, z, i, j, k)$ is obtained, the posture of the end effector (EE) of the robot with respect to the workpiece frame can be determined, and it can be represented by a six-dimension vector, namely $(x, y, z, \alpha, \beta, \gamma)$, where α, β and γ are the z-y-z type Euler angles. Knowing the posture of the EE, the robot joint angles can be finally calculated via inverse kinematics.

Let a frame attached to the cutter with its origin at the TCP and its axis along the tool axis is described by a six-dimension vector $(x, y, z, \alpha, \beta, \gamma)$ with respect to the workpiece frame, and the corresponding homogeneous transformation matrix $^{w}\mathbf{T}_t$ will be

$$^{w}\mathbf{T}_t = rot(z, \alpha)rot(y, \beta)rot(z, \gamma)trans(x, y, z) \tag{1}$$

where rot and $trans$ are rotation and translation transformations. So to perform a milling task, the CL data $CL(x, y, z, i, j, k)$ and $^{w}\mathbf{T}_t$ have to satisfy

$$\begin{bmatrix} i & x \\ j & y \\ k & z \\ 0 & 1 \end{bmatrix} = {^{w}\mathbf{T}_t} \begin{bmatrix} 0 & 0 \\ 0 & 0 \\ -1 & 0 \\ 0 & 1 \end{bmatrix} \tag{2}$$

It can be deduced from (1) and (2) that

$$\begin{bmatrix} i & j & k \end{bmatrix}^T = \begin{bmatrix} -cos\alpha sin\beta & -sin\beta sin\alpha & cos\beta \end{bmatrix}^T \tag{3}$$

Equation (3) shows that given a CL point $CL(x, y, z, i, j, k)$, five components of the six-dimension vector $(x, y, z, \alpha, \beta, \gamma)$ can be determined, i.e. (x, y, z, α, β), while the third Euler angle γ is arbitrary, so that infinite robot configurations can be chosen to produce the specified CL. The redundant third Euler angle γ should be optimized.

The post-processing problem can be finally formulated as

$$\theta(CL, \gamma) = f(CL(x, y, z, i, j, k), \gamma) \tag{4}$$

where $\theta = [\theta_1, \ldots, \theta_6]$ represents the robot configuration, and $f(\cdot)$ denotes the inverse kinematics of the robot. For a standard six revolute joints industrial robot with the last three joint axis lines intersecting at a common point, analytical inverse kinematic solutions can be easily deduced [13].

To conclude, the key of the post-processing problem for a milling robot is to find a proper milling performance related index H to optimize the third Euler angle γ, and determine a unique optimal set of θ, i.e. the robot configurations along the milling path for a given set of CL data.

3   Stiffness Model and Its Identification

A robot deforms under external forces because of insufficient stiffness, thus the stiffness characteristic of the robot has a significant influence on the milling performance. The optimizing index should consider the stiffness model of the robot. Generally speaking, the robot stiffness mainly depends on the torsional stiffness of the gearbox and the drive shaft of each joint, the links are assumed to be infinitely stiff. Therefore, each joint of the robot is usually modeled as a linear torsion spring, and a constant diagonal matrix with each diagonal term defining the stiffness of a joint is used to represent the robot joint space stiffness characteristic [14, 15], i.e.

$$\mathbf{K}_\theta = diag([K_{\theta_1}, K_{\theta_2}, K_{\theta_3}, K_{\theta_4}, K_{\theta_5}, K_{\theta_6}]) \tag{5}$$

The Cartesian Stiffness of the EE is

$$\mathbf{K}_x = \mathbf{J}_F^{-T}(\mathbf{K}_\theta - \mathbf{K}_c)\mathbf{J}_x^{-1} \tag{6}$$

where $\mathbf{J}_x$ and $\mathbf{J}_F$ are the Jacobian matrixes which satisfy $d\mathbf{x} = \mathbf{J}_x d\theta$ and $\tau = \mathbf{J}_F^T \mathbf{F}$, $\mathbf{F}$ denotes the six-dimension wrench vector applied at a certain point on the EE, τ denotes the torques on the joints caused by $\mathbf{F}$, $d\theta$ denotes the elastic deformations of the robot joints caused by τ, and $d\mathbf{x}$ denotes the six-dimension Cartesian deformations of the EE and it is measured at a point different from the force bearing point. $\mathbf{K}_c = [\frac{\partial \mathbf{J}^T}{\partial \theta_1}\mathbf{F}, \frac{\partial \mathbf{J}^T}{\partial \theta_2}\mathbf{F}, \frac{\partial \mathbf{J}^T}{\partial \theta_3}\mathbf{F}, \frac{\partial \mathbf{J}^T}{\partial \theta_4}\mathbf{F}, \frac{\partial \mathbf{J}^T}{\partial \theta_5}\mathbf{F}, \frac{\partial \mathbf{J}^T}{\partial \theta_6}\mathbf{F}]$ is a complementary term [16], and practically, the contribution of the complementary term to the total robot deformation is very small under normal robot payload [17], thus it can be omitted, i.e.

$$\mathbf{K}_x \approx \mathbf{J}_F^{-T}\mathbf{K}_\theta\mathbf{J}_x^{-1} \tag{7}$$

Finally, the relationship between $d\mathbf{x}$ and $\mathbf{F}$ can be formulated as

$$d\mathbf{x} = \mathbf{K}_x^{-1}\mathbf{F} = \mathbf{C}\mathbf{F} = \mathbf{J}_x\mathbf{K}_\theta^{-1}\mathbf{J}_F^T\mathbf{F} \tag{8}$$

Equation (8) can be unfolded as

$$d\mathbf{x} = \begin{bmatrix} \mathbf{J}_{x11}\sum_{i=1}^{n}\mathbf{J}_{Fi1}F_i & \cdots & \mathbf{J}_{x16}\sum_{i=1}^{n}\mathbf{J}_{Fi6}F_i \\ \vdots & \ddots & \vdots \\ \mathbf{J}_{x61}\sum_{i=1}^{n}\mathbf{J}_{Fi1}F_i & \cdots & \mathbf{J}_{x66}\sum_{i=1}^{n}\mathbf{J}_{Fi6}F_i \end{bmatrix}\begin{bmatrix} \frac{1}{K_{\theta_1}} \\ \vdots \\ \frac{1}{K_{\theta_6}} \end{bmatrix} = \mathbf{Ak} \tag{9}$$

where F_i is the ith component of $\mathbf{F}$ and $\mathbf{J}_{xij}$ or $\mathbf{J}_{Fij}$ denotes the ith row jth column element of the matrix $\mathbf{J}_x$ or $\mathbf{J}_F$, $i, j = 1, \ldots, 6$.

To identify the stiffness coefficients $\mathbf{K}_\theta$ (i.e. $\mathbf{k}$ in (9)), it usually involves measuring a set of $\mathbf{F}_i$ and $d\mathbf{x}_i$ at several robot configurations, $i = 1, \ldots, m$, and constructing the following equation systems:

$$[d\mathbf{x}_1, \cdots, d\mathbf{x}_m]^T = [\mathbf{A}_1, \cdots, \mathbf{A}_m]^T\mathbf{k} \tag{10}$$

Equation (10) is simplified as

$$y = Mk \tag{11}$$

Then the stiffness coefficients can be identified by the linear least square method as

$$k = M^+y \tag{12}$$

The robot used in this paper is a Motoman MH80 industrial robot from Yaskawa, it's maximal payload is 80 kg and the repeated positioning accuracy is ±0.07 mm. The schematic diagram of the robot is shown in Fig. 1, and the D-H parameters are given in Table 1. To identify the stiffness model of the robot via the above-mentioned method, an experiment sytem is etablished and shown in Fig. 2.

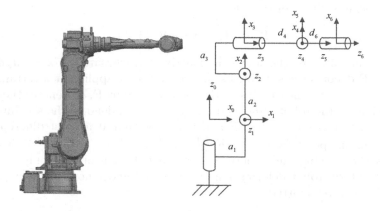

Fig. 1. Schematic diagram of Motoman MH80 industrial robot

Table 1. The D-H parameters of the Motoman MH80 industrial robot

i	α_i	a_i	d_i	θ_i
1	$\pi/2$	145	0	θ_1
2	0	870	0	$\theta_2 + \pi/2$
3	$\pi/2$	210	0	θ_3
4	$-\pi/2$	0	1025	θ_4
5	$\pi/2$	0	0	θ_5
6	0	0	175	θ_6

In the experiment, 70 robot configurations are randomly generated. At each configuration, the robot is loaded by a spring, and the force applied on the EE is measured by an Omega160 force sensor. The EE frame posture and its deformation are measured by a Leica AT906 Laser tracker via three magnetic holders and a spherically mounted reflector (SMR). The forces and deformations are all transformed to the robot base frame. After the measurements, 50 configurations are randomly chosen for parameter identification, and the remaining 20 configurations are used for verification. The joint stiffness are finally identified and listed in Table 2.

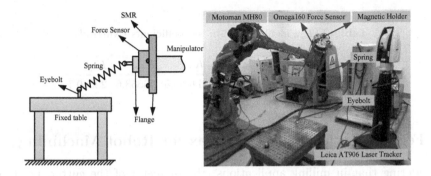

Fig. 2. Experiment setup for stiffness model identification

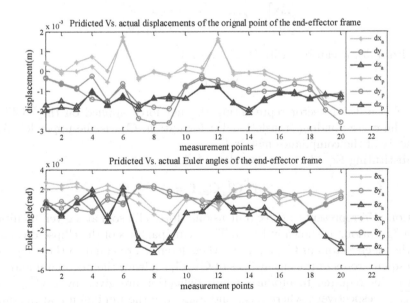

Fig. 3. Predicted deformations and the actual deformations of the EE

Figure 3 shows the model predicted deformations based on the identified joint stiffness and the actual measured deformations for the 20 verification robot configurations. In this figure, the subscript "a" denotes the actual deformation while the subscript "p" denotes the predicted deformation, dx, dy and dz represent the displacements of the original point of the EE frame while δ_x, δ_y and δ_z represent three rotations about x, y and z axis. It can be seen that the deformation predicted via the identified parameters is fairly precise and the modeling error is limited in a narrow band, which verifies the correctness of the parameter identification process.

Table 2. The identified stiffness coefficients (Nm/rad)

K_{θ_1}	K_{θ_2}	K_{θ_3}	K_{θ_4}	K_{θ_5}	K_{θ_6}
481488.5	1394618.1	677034.5	64258.5	46653.6	22060.1

4   Feed-Direction Stiffness Index for Robot Machining

Considering that in milling applications, the diameter of the cutter, i.e. lever arm, is relatively small, and the spindle speed is high, the moment exerted by the workpiece on the EE and the corresponding rotation deformation are negligible [18]. Thus Eq. (8) can be unfolded as

$$\begin{bmatrix} \Delta x \\ 0 \end{bmatrix} = \mathbf{CF} = \begin{bmatrix} \mathbf{C}_{xx} & \mathbf{C}_{xr} \\ \mathbf{C}_{xr} & \mathbf{C}_{rr} \end{bmatrix} \begin{bmatrix} \mathbf{f} \\ 0 \end{bmatrix} \tag{13}$$

From Eq. (13), it can be easily obtained that

$$\Delta \mathbf{x} = \mathbf{C}_{xx}\mathbf{f} \tag{14}$$

where $\mathbf{f}$ is a 3×1 vector representing the net force applied on the TCP and $\Delta \mathbf{x}$ is the corresponding translational deformation, $\mathbf{C}_{xx}$ is the upper left 3×3 submatrix of the compliance matrix $\mathbf{C}$.

Substituting Eq. (14) into $\Delta \mathbf{x}^T \Delta \mathbf{x} = 1$, i.e.

$$\mathbf{f}^T \mathbf{C}_{xx}^T \mathbf{C}_{xx} \mathbf{f} = 1 \tag{15}$$

Equation (15) represents a three dimensional force ellipsoid at a given manipulator configuration, as shown in Fig. 4. The principal axes of the ellipsoid coincide with the eigenvectors of $\mathbf{C}_{xx}^T \mathbf{C}_{xx}$, and their lengths are equal to the reciprocals of the square roots of the eigenvalues of $\mathbf{C}_{xx}^T \mathbf{C}_{xx}$. Hence the maximum and minimum forces required to produce a unit deflection are given by $1/\sqrt{\lambda_{min}}$ and $1/\sqrt{\lambda_{max}}$, respectively, where λ_{min} and λ_{max} are the minimum and maximum eigenvalues of $\mathbf{C}_{xx}^T \mathbf{C}_{xx}$. A vector from the center of the force ellipsoid to any point on the surface of the force ellipsoid reflects the stiffness performance of robot machining system along the direction of the vector.

As mentioned before, in order to improve the production efficiency, the stiffness along the cutting feed direction is required high enough to improve the feed rate during the machining process. Figure 5 shows the force ellipsoid at a certain

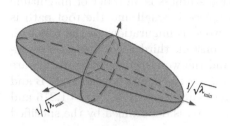

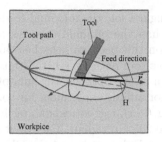

Fig. 4. Three dimensional force ellipsoid   **Fig. 5.** The force ellipsoid at a CL point

cutter location along a tool path, it is obvious that the length of the axis along the tool feed direction $\mathbf{r}$ of the ellipsoid H reflects the stiffness of the machining robot along the feed direction. H is determined by the feed direction $\mathbf{r}$ and the robot configuration θ, thus H can be used as an optimizing index to eliminate the redundant freedom of the milling robot, which will guarantee a higher production efficiency. Therefore, a redundancy elimination method is proposed as following:

$$\max_{\gamma} \quad H(\theta(CL, \gamma), \mathbf{r})$$
$$s.t. \quad \theta_{min} \leq \theta \leq \theta_{max} \tag{16}$$

where θ_{min} and θ_{max} are the lower and upper limits of the joint variables. Note that γ is an arbitrary variable, and the nonlinear optimal problem (16) is defined on $\mathbb{R}$. The problem is numerically solvable by conventional optimization methods, such as the optimization toolbox in MATLAB.

Therefore, the whole post-processing process for the robot milling can be summarized as following: First, the CL data for the specified workpiece is generated by a commercial CAD/CAM software package. Then, five components of the six-dimension vector that describe the posture of the cutting tool are determined. Finally, the last component of the posture vector, i.e. the third Euler angle γ, is optimized based on Eq. (16), and the corresponding optimal robot configuration can be calculated by analytical kinematic inverse solution. Through these steps, the redundancy is eliminated and a unique robot trajectory will be generated. The robot trajectory will have the advantage of higher machining efficiency.

5   Experiment

To verify the effectiveness of the proposed method, for simplicity, a face milling tool path is first generated in the NX 8.5 software. Then the tool path is converted to the robot trajectory via Eqs. (3) and (4). To show the influence of the redundant third Euler angle γ on the feed-direction stiffness index, the H values are calculated with different γ values at a CL point and along a tool path. As shown in Fig. 6, the feed-direction stiffness of the robot varies with the redundant

Euler angle a lot, the maximal feed-direction stiffness is an order of magnitude higher than the minimal one. As the workpiece is small and the tool path is similar on this occasion, the best and the worst configurations of the robot on the whole tool path vary little, thus three constant third Euler angles in representation of the optimal, the suboptimal and the worst robot configuration are chosen for milling experiment for comparison, i.e. the points labeled as A, B and C in Fig. 6 (a), the corresponding Euler angles are $\gamma_A = 1.531$, $\gamma_B = 1.216$ and $\gamma_C = -0.2052$ in radian. The robot configuration is determined by the specified redundant Euler angle and the CL point.

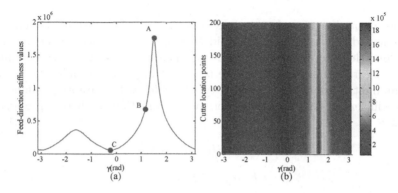

Fig. 6. Feed-direction stiffness varies with γ (a) at a CL point and (b) along a tool path

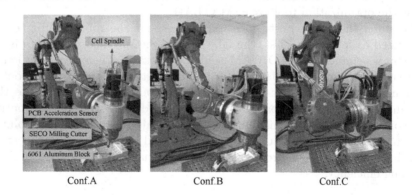

Fig. 7. Robot configurations corresponding to the three specified third Euler angles

The experiment setup is shown in Fig. 7 Conf. A. Within the robot milling system, a SECO milling cutter with a diameter of 10 mm is mounted on a CELL EBS-120g-24000r spindle for cutting. The spindle has a rated power of 7.5 KW and can provide a toque as much as 6 Nm, and it is mounted on the mounting flange of the Motoman MH80 industrial robot. A 6061 aluminum block with a dimension of 260 mm × 110 mm×80 mm is used as the cutting workpiece and it is fixed on a metallic workbench. The acceleration signal during the milling process is monitored by a PCB 352A25 acceleration sensor attached on the spindle. In the experiment, the robot is first commanded to mill the workpiece under the feed rate of 150 mm/min, 300 mm/min and 600 mm/min with the three previously specified poses as shown in Fig. 7 (Conf. A, Conf. B and Conf. C are corresponding to the three specified Euler angles γ_A, γ_B and γ_C), the acceleration signals are collected and they can help evaluate the machining stability. By analysis of the acceleration signals some conclusions can be obtained. During the whole experiment process, the axial depth of cut is 1.5 mm, the radial depth of cut is 4 mm and the rotational speed of the cutter is 5000 r/min.

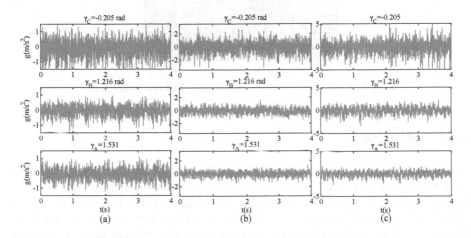

Fig. 8. Accelerations with feedrate (a)150 mm/min (b)300 mm/min (c)600 mm/min

Figure 8 shows the acceleration signals during the machining process under different robot configurations and different feed rates. By comparing the acceleration signals under different feed rates and the same robot configuration, it can be found that when the feed rate increases, the acceleration amplitude becomes higher. For example, when machining with Conf. C, the acceleration amplitude is in the range of ±1.5 g under feed rate 150 mm/min, and it becomes ±3 g and ±5 g under feed rate 300 mm/min and 600 mm/min. On the other hand, it can be easily observed that when machining under the same feed rate, the acceleration amplitude decreases progressively when using robot configuration C, B and A. That is to say, a higher feed-direction stiffness can efficiently improve the machining stability, in other words, using the proposed optimization method to generate the robot milling trajectory, a higher feed rate can be chosen and the production efficiency can be improved, which verifies the effectiveness of the proposed optimizing index and the redundancy elimination method.

To further verify the proposed method, three different areas of a plane are machined with the three poses under the feed rate of 300 mm/min, and the machining quality is observed. The machined plane is shown in Fig. 9. It can be easily found that the area machined via the optimized robot configuration A has the best surface quality while the area machined via the worst robot configuration C has the worst surface quality, which is caused by insufficient feed-direction stiffness of the machining robot.

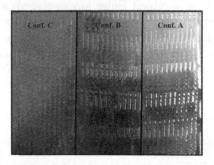

Fig. 9. Areas machined via different robot configurations

6   Conclusion

The post-processing of a milling robot involves converting the 5-axis CL data from a commercial CAD/CAM software to the corresponding 6-axis robot trajectory, which will result in a redundant freedom in the robot system. As the stiffness characteristic of a robot is influenced by the robot configuration, a feed-direction stiffness index is proposed to optimize the redundant freedom after identifying the stiffness model of the robot. Via maximizing the feed-direction stiffness of the robot, the redundant freedom of the robot is eliminated. According to the analysis of the acceleration signals and the machining quality in the experiment, it can be concluded that stronger the stiffness of the robot machining system along feed-direction is, better machining stability and quality is. Thus the post-processing technique for a milling robot proposed in this paper is effective and can enhance the machining performance.

References

1. Leali, F., Vergnano, A., Pini, F., Pellicciari, M., Berselli, G.: A workcell calibration method for enhancing accuracy in robot machining of aerospace parts. Int. J. Adv. Manufact. Technol. **85**(1–4), 47–55 (2016)
2. Lehmann, C., Pellicciari, M., Drust, M., Gunnink, J.W.: Machining with industrial robots: the COMET project approach. Commun. Comput. Inf. Sci. **371**, 27–36 (2013)
3. Chen, Y., Dong, F.: Robot machining: recent development and future research issues. Int. J. Adv. Manufact. Technol. **66**(9–12), 1489–1497 (2013)

4. Xiao, W., Huan, J.: Redundancy and optimization of a 6R robot for five-axis milling applications: singularity, joint limits and collision. Prod. Eng. **6**(3), 287–296 (2012)
5. Sabourin, L., Subrin, K., Cousturier, R., Gogu, G., Mezouar, Y.: Redundancy-based optimization approach to optimize robotic cell behaviour: application to robotic machining. Ind. Robot: Int. J. **42**(2), 167–178 (2015)
6. Huo, L., Baron, L.: The self-adaptation of weights for joint-limits and singularity avoidances of functionally redundant robotic-task. Robot. Comput.-Integr. Manuf. **27**(2), 367–376 (2011)
7. Schneider, U., Ansaloni, M., Drust, M., Leali, F., Verl, A.: Experimental investigation of sources of error in robot machining. In: Robotics in Smart Manufacturing, pp. 14–26 (2013)
8. Angeles, J.: On the nature of the Cartesian Stiffness Matrix. Ingeniería mecánica, tecnología desarrollo **3**(5), 163–170 (2010)
9. Guo, Y., Dong, H., Ke, Y.: Stiffness-oriented posture optimization in robotic machining applications. Robot. Comput.-Integr. Manuf. **35**, 69–76 (2015)
10. Lin, Y., Zhao, H., Ding, H.: Posture optimization methodology of 6R industrial robots for machining using performance evaluation indexes. Robot. Comput.-Integr. Manuf. **48**, 59–72 (2017). (April 2016)
11. Peng, F.Y., Yan, R., Chen, W., Yang, J.Z., Li, B.: Anisotropic force ellipsoid based multi-axis motion optimization of machine tools. Chin. J. Mech. Eng. **25**(5), 960–967 (2012)
12. Zhang, H., Pan, Z.: Robotic machining: material removal rate control with a flexible manipulator. In: 2008 IEEE Conference on Robotics, Automation and Mechatronics, pp. 30–35. IEEE, September 2008
13. Tsai, L.W.: Robot analysis: the mechanics of serial and parallel manipulators. Wiley, Hoboken (1999)
14. Wang, J., Zhang, H., Fuhlbrigge, T.: Improving machining accuracy with robot deformation compensation. In: 2009 IEEE/RSJ International Conference on Intelligent Robots and Systems, pp. 3826–3831. IEEE, October 2009
15. Dumas, C., Caro, S., Cherif, M., Garnier, S., Furet, B.: Joint stiffness identification of industrial serial robots. Robotica **30**(04), 649–659 (2012)
16. Chen, S.F., Kao, I.: Conservative congruence transformation for joint and cartesian stiffness matrices of robotic hands and fingers. Int. J. Robot. Res. **19**(9), 835–847 (2000)
17. Alici, G., Shirinzadeh, B.: Enhanced stiffness modeling, identification and characterization for robot manipulators. IEEE Trans. Robot. **21**(4), 554–564 (2005)
18. Zargarbashi, S., Khan, W., Angeles, J.: Posture optimization in robot-assisted machining operations. Mech. Mach. Theory **51**, 74–86 (2012)

Mechanism and Parallel Robotics

Kinematic Analysis and Performance Evaluation of a Redundantly Actuated Hybrid Manipulator

Lingmin Xu[1], Qiaohong Chen[2(✉)], Leiying He[1], and Qinchuan Li[1]

[1] Mechatronic Institute, Zhejiang Sci-Tech University, Hangzhou,
Zhejiang Province 310018, People's Republic of China
xulingmin1993@163.com, heleiying@163.com,
lqchuan@zstu.edu.cn
[2] School of Information, Zhejiang Sci-Tech University, Hangzhou,
Zhejiang Province 310018, People's Republic of China
chen_lisa@zstu.edu.cn

Abstract. This paper deals with the kinematic analysis and performance evaluation of a 2-URPR-UPR (U, R, P standing for universal, revolute and prismatic joint, respectively) redundantly actuated hybrid manipulator. First, the kinematic analysis of the proposed manipulator is presented, including mobility analysis, inverse kinematics, and singular analysis. Then, the reciprocal of the condition number based on a dimensionally homogeneous Jacobian matrix is used to evaluate the dexterity by changing the operating height and radius ratios of the mechanism separately. Finally, the global conditioning index is carried out to describe global dexterity performance with different radius ratios.

Keywords: Redundantly actuated hybrid manipulator · Kinematic analysis · Performance evaluation

1 Introduction

Due to the merits in terms of eliminating singularity, improving stiffness and dexterity characteristics [1–5] comparing with the non-redundantly actuated parallel manipulators (PMs), the PMs with actuation redundancy have attracted considerable attention in recent decades [4–9]. The actuation redundancy of PMs includes two categories [4, 5]: (1) replacement of passive joints with active joints, and (2) addition of active limbs without changing the mobility of the PM.

Kinematic analysis and performance evaluation of the redundantly actuated PMs have been a research hotspot, and much process has been carried out [1–3, 9–16]. Kim et al., made many contributions about redundantly actuated PMs [3, 9–11]. Particularly, they investigated the stiffness enhancement and antagonistic stiffness of the redundantly actuated PMs. Wu et al., proposed a novel 3-degree-of-freedom (DOF) planar parallel manipulator with redundantly actuated and analyzed the performances including workspace, singularity, dexterity and stiffness [1, 12]. Wang et al., proposed a novel type of 3-DOF redundantly parallel tool head with modified spherical joint,

© Springer International Publishing AG 2017
Y. Huang et al. (Eds.): ICIRA 2017, Part II, LNAI 10463, pp. 199–211, 2017.
DOI: 10.1007/978-3-319-65292-4_18

which largely increases the workspace [13]. Dai et al., made relevant contributions about ankle rehabilitation and proposed redundantly actuated PMs with high performance [2, 14]. Fang and his colleagues also proposed redundantly actuated PMs for ankle rehabilitation, and discussed their kinematics and performances [15, 16].

In this paper, a 3-DOF redundantly actuated 2-URPR-UPR hybrid manipulator is presented. Here, the notations U, R, P denote universal, revolute and prismatic joints, respectively. The actuation redundancy is implemented by replacing the passive U joint in URPR limb with active joint, which is the serial subsystem. Meanwhile, the serial module presented here is used to achieve the required DOF of the manipulator.

The paper is organized as follows. Section 2 introduces the 2-URPR-UPR hybrid manipulator. The mobility analysis is performed using screw theory in Sect. 3. The inverse kinematic and velocity are presented in Sects. 4 and 5, respectively. With the Jacobian matrix, the singularity analysis of the robot is discussed in Sect. 6. In Sect. 7, the dexterity analysis of the PM is performed in detail based on the dimensionally homogeneous Jacobian matrix, and the relationship between the GCI and the different radius ratio is discussed. Finally, conclusions are presented in Sect. 8.

2   Description of the 2-URPR-UPR Hybrid Manipulator

A prototype of 2-URPR-UPR hybrid manipulator with actuation redundancy is shown in Fig. 1. The moving platform is connected to the fixed base through three limbs, two limbs of which have identical kinematic chains URPR, and the third one is UPR limb. Limb 1 and 2 are coplanar and limited to within the plane $\varPi_1$, which is always perpendicular to the plane $\varPi_2$. The plane $\varPi_2$ represents the symmetrical plane to which the limb 3 belongs. For the limb 1 and 2, the two revolute axes of the U joint are limited in the plane $\varPi_1$ and $\varPi_2$, respectively. The revolute axes of the first R joints in limb 1 and 2 are parallel to each other. Meanwhile, they are parallel to axes of the R joints connected on the moving platform. The P joints in limb 1 and 2 are perpendicular to the revolute

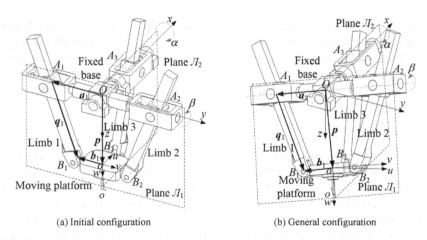

(a) Initial configuration   (b) General configuration

Fig. 1.   Schematics of the 2-URPR-UPR hybrid manipulator with actuation redundancy.

axes of the R joints. For the limb 3, the first revolute axis of the U joint is coaxial with that of the U joint in limb 1 and 2. The second revolute axis of the H joint is parallel to the revolute axis of the R joint connected the moving platform, which is perpendicular to the P joint in limb 3. The proposed hybrid manipulator is actuated by four actuators, three linear actuators in P joints and one rotary actuator in the U joint of limb 1 and 2, respectively. The center of the U joint in limb 1 and 2 is denoted by O. Let A_1 and A_2 denote the central points of the first R joints in limb 1 and 2, while the central point of the U joint in limb 3 is denoted by A_3. The centers of the second R joints in limb 1 and 2 are denoted by B_1 and B_2, and the centers of the R joint in limb 3 is denoted by B_3.

The coordinate frames are established as shown in Fig. 1. A fixed reference coordinate frame $O\text{-}xyz$ is attached to the fixed base and the origin O is at the midpoint of line A_1A_2. Let x-axis always point in the direction of OA_3, y-axis point along OA_2 in the initial configuration, and the z-axis point downward vertically. A moving platform coordinate frame $o\text{-}uvw$ is attached on the moving platform, as shown in Fig. 1. Let u-axis point in the direction of oB_3 and v-axis point along oB_2. The w-axis is acting downward vertically with respect to the moving platform. The architectural parameters of 2-URPR-UPR hybrid manipulator with actuation redundancy are defined as follows: $OB_1 = OB_2 = OB_3 = e_1$, $OA_1 = OA_2 = OA_3 = e_2$ and $oo' = H$.

3   Mobility Analysis

Mobility analysis based on the screw theory [17] is essential for determining the motion of a manipulator. In screw theory, a unit screw $\boldsymbol{S}$ is defined as

$$\boldsymbol{S} = (\boldsymbol{s}; \boldsymbol{s}_0) = (\boldsymbol{s}; \boldsymbol{r} \times \boldsymbol{s} + h\boldsymbol{s}), \tag{1}$$

where $\boldsymbol{s}$ is a unit vector along the direction of the screw axis, $\boldsymbol{r}$ is the position vector of any point on the screw axis, and h is the pitch.

A screw $\boldsymbol{S}^r = (\boldsymbol{s}_r; \boldsymbol{s}_{0r})$ that is reciprocal to a set of screws, $\boldsymbol{S}_1, \boldsymbol{S}_2, \ldots, \boldsymbol{S}_n$, is defined as

$$\boldsymbol{S}_i \circ \boldsymbol{S}^r = \boldsymbol{s}_i \cdot \boldsymbol{s}_{0r} + \boldsymbol{s}_r \cdot \boldsymbol{s}_{0i} = 0 \quad i=1, 2, \ldots, n, \tag{2}$$

where $\circ$ and $\boldsymbol{S}_i$ represent the reciprocal product and the ith screw in the screw set, respectively. A screw is called a twist when it is used to represent an instantaneous motion of a rigid body. A wrench screw is used to represent a force or a coaxial couple acting on a rigid body. In the following analysis, $\boldsymbol{S}_{ij}$ denotes the unit twist associated with the jth kinematic joint of the ith limb, and $\boldsymbol{S}_{ij}^r$ denotes the jth unit constraint wrench acting on the moving platform by the ith limb.

With respect to the fixed coordinate frame $O\text{-}xyz$, the position vectors of $B_i (i = 1, 2, 3)$ are defined as $[x_{B_1} \quad y_{B_1} \quad z_{B_1}]^{\text{T}}$, $[x_{B_2} \quad y_{B_2} \quad z_{B_2}]^{\text{T}}$ and $[x_{B_3} \quad y_{B_3} \quad z_{B_3}]^{\text{T}}$, respectively. The rotation angle around x-axis and A_1A_2 are denoted as α and β, thus the position vectors of A_i $(i = 1, 2, 3)$ can be defined as $[0 \quad -e_2c_\alpha \quad -e_2s_\alpha]^{\text{T}}$, $[0 \quad e_2c_\alpha \quad e_2s_\alpha]^{\text{T}}$ and

$[e_2 \quad 0 \quad 0]^T$, respectively, where s and c denote sin and cosine functions, respectively. As shown in Fig. 1(b), the twist system of limb 1 can be written as

$$
\begin{cases}
\$_{11} = (1 \quad 0 \quad 0; \quad 0 \quad 0 \quad 0) \\
\$_{12} = (0 \quad c_\alpha \quad s_\alpha; \quad 0 \quad 0 \quad 0) \\
\$_{13} = (c_\beta \quad s_\alpha s_\beta \quad -c_\alpha s_\beta; \quad e_2 s_\beta \quad -e_2 s_\alpha c_\beta \quad e_2 c_\alpha c_\beta) \\
\$_{14} = (0 \quad 0 \quad 0; \quad x_{B_1} \quad y_{B_1} + e_2 c_\alpha \quad z_{B_1} + e_2 s_\alpha) \\
\$_{15} = (c_\beta \quad s_\alpha s_\beta \quad -c_\alpha s_\beta; \quad -s_\beta(y_{B_1} c_\alpha + z_{B_1} s_\alpha) \quad z_{B_1} c_\beta + x_{B_1} c_\alpha s_\beta \quad x_{B_1} s_\alpha s_\beta - y_{B_1} c_\beta)
\end{cases}
\tag{3}
$$

Based on the reciprocal screw theory, the wrench system of limb 1 can be calculated by using Eq. (2)

$$
\$_{11}^r = (c_\beta \quad s_\alpha s_\beta \quad -c_\alpha s_\beta; \quad 0 \quad 0 \quad 0),
\tag{4}
$$

where $\$_{11}^r$ represents a constraint force in the direction of u-axis.

Similarly, the wrench systems of limb 2 and 3 are given by

$$
\$_{21}^r = (c_\beta \quad s_\alpha s_\beta \quad -c_\alpha s_\beta; \quad 0 \quad 0 \quad 0),
\tag{5a}
$$

and

$$
\begin{cases}
\$_{31}^r = (0 \quad 0 \quad 0; \quad 0 \quad -s_\alpha \quad c_\alpha) \\
\$_{32}^r = (0 \quad c_\alpha \quad s_\alpha; \quad 0 \quad 0 \quad 0)
\end{cases},
\tag{5b}
$$

where $\$_{22}^r$ is the same as $\$_{12}^r$. $\$_{31}^r$ represents a constraint couple perpendicular to the U joint in the limb 3, and $\$_{32}^r$ is a constraint force parallel to the second revolute axis of the U joint in limb 3.

Thus, using reciprocal screw theory, the moving platform twist system of the redundant 2-URPR-UPR hybrid manipulator can be obtained as shown

$$
\begin{cases}
\$_1^{pm} = (1 \quad 0 \quad 0; \quad 0 \quad 0 \quad 0) \\
\$_2^{pm} = (0 \quad c_\alpha \quad s_\alpha; \quad 0 \quad 0 \quad 0) \\
\$_3^{pm} = (0 \quad 0 \quad 0; \quad s_\beta \quad -s_\alpha c_\beta \quad c_\alpha c_\beta)
\end{cases}.
\tag{6}
$$

Equation (6) shows that the 2-URPR-UPR hybrid manipulator has three DOFs, including two rotations around x-axis and $A_1 A_2$, and one translation along w-axis.

4   Inverse Kinematics

The inverse kinematics of the 2-URPR-UPR hybrid manipulator involves the calculations of the actuated joint parameters (θ, q_1, q_2, q_3) given the position and orientation parameters of the moving platform (α, β, z).

The rotation matrix between the moving coordinate frame o-uvw and the fixed coordinate frame O-xyz can be written as

$$
{}^{O}\mathbf{R}_o = \begin{bmatrix} c_\beta & 0 & s_\beta \\ s_\alpha s_\beta & c_\alpha & -s_\alpha c_\beta \\ -c_\alpha s_\beta & s_\alpha & c_\alpha c_\beta \end{bmatrix}. \tag{7}
$$

As shown in Fig. 1, the position vectors A_iB_i ($i = 1, 2, 3$) with respect to the fixed coordinate frame are defined as q_i, and vector $p = [z\tan_\beta/c_\alpha \quad -z\tan_\alpha \quad z]^T$ denotes the position vector of the point o with respect to the fixed coordinate frame O-xyz, where tan denotes the tangent function. The position vectors oA_i and oB_i ($i = 1, 2, 3$) with respect to the fixed coordinate frame are denoted as a_i and b_i, respectively. Thus we can obtain

$$
\begin{cases} a_1 = [0 \quad -e_2 c_\alpha \quad -e_2 s_\alpha]^T \\ a_2 = [0 \quad e_2 c_\alpha \quad e_2 s_\alpha]^T \\ a_3 = [e_2 \quad 0 \quad 0]^T \end{cases}, \tag{8a}
$$

and

$$
\begin{cases} b_1 = {}^{O}\mathbf{R}_o[0 \quad -e_1 \quad 0]^T \\ b_2 = {}^{O}\mathbf{R}_o[0 \quad e_1 \quad 0]^T \\ b_3 = {}^{O}\mathbf{R}_o[e_1 \quad 0 \quad 0]^T \end{cases}. \tag{8b}
$$

As shown in Fig. 1, the position vectors q_i ($i = 1, 2, 3$) can be written in the following form

$$
q_i = p - a_i + b_i. \tag{9}
$$

Thus, we can obtain the explicit expressions of q_i ($i = 1, 2, 3$) through Eqs. (7–9)

$$
\begin{cases} q_1 = [z\tan_\beta/c_\alpha \quad e_2 c_\alpha - e_1 c_\alpha - z\tan_\alpha \quad z + e_2 s_\alpha - e_1 s_\alpha]^T \\ q_2 = [z\tan_\beta/c_\alpha \quad e_1 c_\alpha - e_2 c_\alpha - z\tan_\alpha \quad z + e_1 s_\alpha - e_2 s_\alpha]^T \\ q_3 = [z\tan_\beta/c_\alpha + e_1 c_\beta - e_2 \quad e_1 s_\alpha s_\beta - z\tan_\alpha \quad z - e_1 c_\alpha s_\beta]^T \end{cases}, \tag{10}
$$

Since the angle θ is equal to α, the inverse kinematics of the redundant 2-URPR-UPR hybrid manipulator can be written as

$$
\begin{cases} \theta = \alpha \\ q_1 = \sqrt{(z\tan_\beta/c_\alpha)^2 + (e_2 c_\alpha - e_1 c_\alpha - z\tan_\alpha)^2 + (z + e_2 s_\alpha - e_1 s_\alpha)^2} \\ q_2 = \sqrt{(z\tan_\beta/c_\alpha)^2 + (e_1 c_\alpha - e_2 c_\alpha - z\tan_\alpha)^2 + (z + e_1 s_\alpha - e_2 s_\alpha)^2} \\ q_3 = \sqrt{(z\tan_\beta/c_\alpha + e_1 c_\beta - e_2)^2 + (e_1 s_\alpha s_\beta - z\tan_\alpha)^2 + (z - e_1 c_\alpha s_\beta)^2} \end{cases}. \tag{11}
$$

5 Velocity Analysis

The Jacobian matrix represents the mapping between the rates of the actuated joints $\dot{q} = \begin{pmatrix} \dot{\theta} & \dot{q}_1 & \dot{q}_2 & \dot{q}_3 \end{pmatrix}^{\mathrm{T}}$ and the velocities of the moving platform $\dot{X} = \begin{pmatrix} \dot{\alpha} & \dot{\beta} & \dot{z} \end{pmatrix}^{\mathrm{T}}$. By taking the derivative of Eq. (11) with respect to time leads to

$$J_q\dot{q} = J_x\dot{X}, \tag{12}$$

where $\boldsymbol{J}_q = \begin{bmatrix} J_{q_{11}} & 0 & 0 & 0 \\ 0 & J_{q_{22}} & 0 & 0 \\ 0 & 0 & J_{q_{33}} & 0 \\ 0 & 0 & 0 & J_{q_{44}} \end{bmatrix}$, $\boldsymbol{J}_x = \begin{bmatrix} J_{x_{11}} & J_{x_{12}} & J_{x_{13}} \\ J_{x_{21}} & J_{x_{22}} & J_{x_{23}} \\ J_{x_{31}} & J_{x_{32}} & J_{x_{33}} \\ J_{x_{41}} & J_{x_{42}} & J_{x_{43}} \end{bmatrix}$, $J_{q_{11}} = 1$, $J_{q_{22}} = q_1$,

$J_{q_{33}} = q_2$, $J_{q_{44}} = q_3$, $J_{x_{11}} = 1$, $J_{x_{12}} = 0$, $J_{x_{13}} = 0$, $J_{x_{21}} = z^2 s_\alpha / \left(c_\alpha^3 c_\beta^2 \right)$, $J_{x_{22}} = z^2 s_\beta / \left(c_\alpha^2 c_\beta^3 \right)$, $J_{x_{23}} = z / \left((s_\alpha^2 - 1)(s_\beta^2 - 1) \right)$, $J_{x_{31}} = z^2 s_\alpha / \left(c_\alpha^3 c_\beta^2 \right)$, $J_{x_{32}} = z^2 s_\beta / \left(c_\alpha^2 c_\beta^3 \right)$, $J_{x_{33}} = z / \left((s_\alpha^2 - 1)(s_\beta^2 - 1) \right)$, $J_{x_{41}} = z s_\alpha (z - e_2 c_\alpha c_\beta s_\beta) / \left(c_\alpha^3 c_\beta^2 \right)$, $J_{x_{42}} = (z s_\beta - e_2 z c_\alpha c_\beta + e_1 e_2 c_\alpha s_\beta c_\beta) / \left(c_\alpha^2 c_\beta^3 \right)$, $J_{x_{43}} = (z - e_2 c_\alpha s_\beta c_\beta) / \left(c_\alpha^2 c_\beta^2 \right)$.

Thus, the velocity equation of the 2-URPR-UPR hybrid manipulator can be expressed as follows

$$\dot{q} = J_q^{-1} J_x \dot{X} = J\dot{X}, \tag{13}$$

where $J = J_q^{-1} J_x$ is a 4×3 Jacobian matrix.

6 Velocity Analysis

Singularity analysis is a fundamental issue for the kinematic analysis. The mechanism becomes uncontrollable in singular configurations and the kinematic performance will thus deteriorate. Usually, the kinematic singularities can be divided into following three types: the forward kinematic singularity, the inverse kinematic singularity and the combined singularity [18].

Based on Ref. [4, 15], forward kinematic singularity occurs when $\left| J_x^{\mathrm{T}} J_x \right| = 0$ and $\left| J_q \right| \neq 0$, where

$$J_x^{\mathrm{T}} J_x = \begin{bmatrix} J_{x_{11}}^2 + J_{x_{21}}^2 + J_{x_{31}}^2 + J_{x_{41}}^2 & J_{x_{21}} J_{x_{22}} + J_{x_{31}} J_{x_{32}} + J_{x_{41}} J_{x_{42}} & J_{x_{21}} J_{x_{23}} + J_{x_{31}} J_{x_{33}} + J_{x_{41}} J_{x_{43}} \\ J_{x_{21}} J_{x_{22}} + J_{x_{31}} J_{x_{32}} + J_{x_{41}} J_{x_{42}} & J_{x_{22}}^2 + J_{x_{32}}^2 + J_{x_{42}}^2 & J_{x_{22}} J_{x_{23}} + J_{x_{32}} J_{x_{33}} + J_{x_{42}} J_{x_{43}} \\ J_{x_{21}} J_{x_{23}} + J_{x_{31}} J_{x_{33}} + J_{x_{41}} J_{x_{43}} & J_{x_{22}} J_{x_{23}} + J_{x_{32}} J_{x_{33}} + J_{x_{42}} J_{x_{43}} & J_{x_{23}}^2 + J_{x_{33}}^2 + J_{x_{43}}^2 \end{bmatrix}. \tag{14}$$

From Eq. (14), $\left| J_x^{\mathrm{T}} J_x \right| = 0$ holds when

$$\begin{cases} J_{x_{11}} = 0 \\ J_{x_{21}} = 0 \\ J_{x_{31}} = 0 \\ J_{x_{41}} = 0 \end{cases}, \text{ or } \begin{cases} J_{x_{22}} = 0 \\ J_{x_{32}} = 0 \\ J_{x_{42}} = 0 \end{cases}, \text{ or } \begin{cases} J_{x_{23}} = 0 \\ J_{x_{33}} = 0 \\ J_{x_{43}} = 0 \end{cases}. \tag{15}$$

Through calculating these three conditions respectively, and then we can obtain

$$\begin{cases} \beta = 0 \\ z = 0 \end{cases}. \tag{16}$$

However, the condition $z = 0$ in Eq. (16) cannot occur in a real application. So the 2-URPR-UPR hybrid manipulator has no forward kinematic singularity.

When $|J_q| = 0$ and $|J_x^T J_x| \neq 0$, inverse kinematic singularity occurs. Since J_q is a diagonal matrix and $J_{q_{11}} = 1$, $|J_q| = 0$ occurs when any of $J_{q_{22}}$, $J_{q_{33}}$ or $J_{q_{44}}$ is equal to zero.

If $J_{q_{22}} = 0$, we have

$$\begin{cases} e_1 = e_2 \\ z = 0 \end{cases}. \tag{17}$$

Similar to the Eq. (16), the condition $z = 0$ in Eq. (17) cannot occur in a real application. This situation also happens on $J_{q_{33}} = 0$ and $J_{q_{44}} = 0$. Thus the 2-URPR-UPR hybrid manipulator has no inverse kinematic singularity.

Since the combined singularity occurs when $|J_x^T J_x| = 0$ and $|J_q| = 0$ simultaneously, we can easily deduce that the proposed manipulator has no combined singularity based on the above analysis. Consequently, the 2-URPR-UPR redundant hybrid manipulator has no kinematic singularity.

7 Dexterity Analysis

As mentioned in Sect. 2, the 2-URPR-URP hybrid manipulator is actuated by three linear actuators and one rotary actuator. Meanwhile, this robot has combined types of mobility, i.e., two rotational DOFs and one translational DOF, which will cause the inconsistency of the units of the elements in the Jacobian matrix and erroneous interpretations. So it is necessary to obtain a new Jacobian matrix in which the dimensions of every elements are the same physical units. Considering the method proposed by Angeles et al. [19, 20], here the length of moving platform, e_1, is used as characteristic/natural link lengths L_q and L_x to generate a new homogeneous Jacobian matrix. The function of the characteristic lengths is to homogenize the original matrix. The homogeneous matrices J_{qh} and J_{xh} are then formulated as follows

$$J_{qh} = J_q \text{diag}\left(\tfrac{1}{L_q} \quad 1 \quad 1 \quad 1 \right), \tag{18a}$$

and

$$J_{xh} = J_x \text{diag}\left(\tfrac{1}{L_x} \quad \tfrac{1}{L_x} \quad 1\right). \tag{18b}$$

Thus, the homogeneous Jacobian matrix J_h can be obtained as

$$J_h = J_{qh}^{-1} J_{xh}. \tag{19}$$

7.1 Local Dexterity Index

Dexterity is defined as the capability of a robot to arbitrarily change its pose, or apply forces and torques in arbitrary directions [21], which is configuration-dependent. The condition number of the Jacobian matrix is an important local dexterity index that is used to describe the kinematic performance of a mechanism. And the condition number c of the homogeneous Jacobian matrix is given by

$$c = \|J_h\| \cdot \|J_h^{-1}\|, \tag{20}$$

where the notation $\|\bullet\|$ is defined as the 2-norm of the matrix.

The condition number can be used to illustrate the magnification between relative input errors and relative output error [22]. Condition number ranges from one to infinite. The closer the condition number is to unity, the better the kinematic performance of this configuration will be.

Here, we use the reciprocal of the condition number, i.e., $C = 1/c$ ($C \in [0, 1]$) in this paper. To obtain the same orientation capability, the rotation ranges are set as follows: $-40° \le \alpha \le 40°$, $-40° \le \beta \le 40°$. For 2-URPR-UPR hybrid manipulator, the operating height z affects the distribution of condition number. Based on the architecture parameters $e_1 = e_2 = 120$ mm and the length of tool head $H = 40$ mm, the distribution variations of C in three-dimensional workspace is shown in Fig. 2, which is symmetrical about the plane $y = 0$. Note that the values in the z-axis is the height of the end-point with tool head. As shown in Fig. 2, the orientation workspace of is symmetrical with respect to angle $\alpha = 0$.

As shown in Fig. 2, the 2-URPR-UPR hybrid manipulator with different operating heights have different performance distributions. However, the situation that the architecture parameters of fixed base and moving platform are the same ($e_1 = e_2$) rarely occurs in practice. Based on the situation that $z = 1.5e_2$, Fig. 3 shows the distributions of C for the proposed mechanism with different radius ratios (e_1/e_2). From the Fig. 3, we can clearly find that a bigger radius ratio produces a better dexterity of the hybrid manipulator.

7.2 Global Conditioning Index

Although the local dexterity index mentioned above shows the distribution of the kinematic performance characteristic, it can hardly evaluate the global dexterity performance of the mechanism within the attainable workspace. The study of the global dexterity performance is also essential for evaluation of the mechanism. To describe the mechanism performance better, the global conditioning index (GCI) proposed by

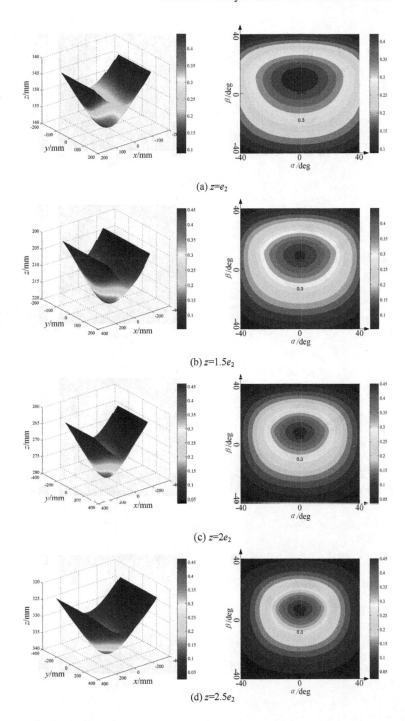

(a) $z=e_2$

(b) $z=1.5e_2$

(c) $z=2e_2$

(d) $z=2.5e_2$

Fig. 2. Distributions of the reciprocal of the condition number for the 2-URPR-UPR redundant hybrid manipulator with different operating height.

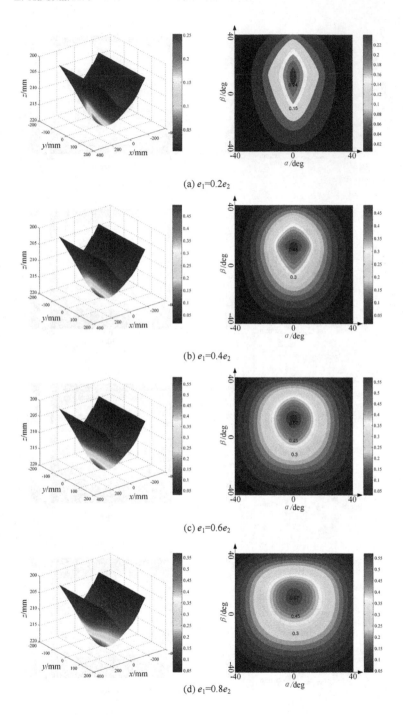

(a) $e_1 = 0.2e_2$

(b) $e_1 = 0.4e_2$

(c) $e_1 = 0.6e_2$

(d) $e_1 = 0.8e_2$

Fig. 3. Distributions of the reciprocal of the condition number for the 2-URPR-UPR redundant hybrid manipulator with different radius ratios.

Gosselin et al. [23] is used to evaluate the global dexterity characteristic in this study, which is defined as

$$\eta = \frac{\int_W C dW}{W}. \qquad (21)$$

Apparently, the purpose of calculating the GCI is to obtain the average performance over the whole reachable workspace (W). GCI ranges from zero to unity. And the closer the GCI is to unity, the better the global dexterity the mechanism has. By changing the radius ratio (e_1/e_2) of the 2-URPR-UPR hybrid manipulator, the value of GCI can be calculated and the tendency of variation is shown in Fig. 4. The tendency shows that when the radius ratio ranges zero to 0.8, the bigger radius ratio produces the better GCI of the 2-URPR-UPR hybrid manipulator, and when radius ratio is bigger than 0.8, the value of GCI is almost unchanged.

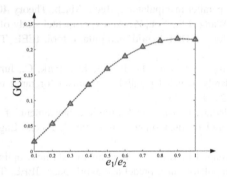

Fig. 4. GCI of the 2-URPR-UPR hybrid manipulator versus different radius ratios.

8   Conclusions

The kinematic analysis and performance evaluation of a 2-URPR-UPR hybrid manipulator with actuation redundancy are presented. Mobility analysis shows that the proposed robot has one translational DOF and two rotational DOFs. Based on the inverse analysis and Jacobian matrix, it is found that the 2-URPR-UPR hybrid manipulator has no singularity. The dimension of the Jacobian matrix is then homogenized by using the characteristic length. Local and global indices of dexterity based on the homogenized Jacobian matrix are obtained, including the reciprocal of the condition number and GCI.

Acknowledgments. The work is supported by the National Natural Science Foundation of China (NSFC) under Grant 51275479 and 51525504 and Natural Science Foundation of Zhejiang Province under Grant LZ14E050005.

References

1. Wang, J.S., Wu, J., Li, T.M., Liu, X.J.: Workspace and singularity analysis of a 3-DOF planar parallel manipulator with actuation redundancy. Robotica **27**, 51–57 (2009)
2. Saglia, J.A., Dai, J.S., Caldwell, D.G.: Geometry and kinematic analysis of a redundantly actuated parallel mechanism that eliminates singularities and improves dexterity. J. Mech. Des. **130**, 124501 (2008)
3. Kim, S.H., Jeon, D., Shin, H.P., In, W., Kim, J.: Design and analysis of decoupled parallel mechanism with redundant actuator. Int. J. Precis. Eng. Man. **10**, 93–99 (2009)
4. Merlet, J.P.: Redundant parallel manipulators. Lab. Robot. Autom. **8**, 17–24 (1996)
5. Kim, S.: Operational quality analysis of parallel manipulators with actuation redundancy. In: IEEE International Conference on Robotics and Automation, pp. 2651–2655 (1997)
6. Firmani, F., Podhorodeski, R.P.: Force-unconstrained poses for a redundantly-actuated planar parallel manipulator. Mech. Mach. Theory **39**, 459–476 (2004)
7. Nokleby, S.B., Fisher, R., Podhorodeski, R.P., Firmani, F.: Force capabilities of redundantly-actuated parallel manipulators. Mech. Mach. Theory **40**, 578–599 (2005)
8. Wang, L.P., Wu, I., Wang, J.S., You, Z.: An experimental study of a redundantly actuated parallel manipulator for a 5-DOF hybrid manipulator tool. IEEE. Trans. Mechatronics. **14**, 72–81 (2009)
9. Kim, J., Park, F.C., Ryu, S.J., Kim, J., Hwang, J.C., Park, C., Iurascu, C.C.: Design and analysis of a redundantly actuated parallel mechanism for rapid machining. IEEE Trans. Robot. Autom. **17**, 423–434 (2001)
10. Shin, H., Kim, S., Jeong, J., Kim, J.: Stiffness enhancement of a redundantly actuated parallel manipulator tool by dual support rims. Int. J. Precis. Eng. Man. **13**, 1539–1547 (2012)
11. Shin, H., Lee, S., Jeong, I.J., Kim, J.: Antagonistic stiffness optimization of redundantly actuated parallel manipulators in a predefined workspace. IEEE. Trans. Mech. **18**, 1161–1169 (2013)
12. Wu, J., Wang, J.S., Wang, L.P., Li, T.M.: Dexterity and stiffness analysis of a three-degree-of-freedom planar parallel manipulator with actuation redundancy. P. I. Mech. Eng. C-J. Mec. **221**, 961–969 (2007)
13. Wang, D., Fan, R., Chen, W.: Performance enhancement of a three-degree-of-freedom parallel tool head via actuation redundancy. Mech. Mach. Theory **71**, 142–162 (2014)
14. Saglia, J.A., Tsagarakis, N.G., Dai, J.S., Caldwell, D.G.: A high performance redundantly actuated parallel mechanism for ankle rehabilitation. Int. J. Rob. Res. **28**, 1216–1227 (2009)
15. Wang, C.Z., Fang, Y.F., Guo, S., Chen, Y.Q.: Design and kinematical performance analysis of a 3-RUS/RRR redundantly actuated parallel mechanism for ankle rehabilitation. J. Mech. Robot. **5**, 041003 (2013)
16. Wang, C.Z., Fang, Y.F., Guo, S., Zhou, C.C.: Design and kinematic analysis of redundantly actuated parallel mechanisms for ankle rehabilitation. Robotica **33**, 366–384 (2015)
17. Huang, Z., Li, Q.C.: Type synthesis of symmetrical lower-mobility parallel mechanisms using the constraint-synthesis method. Int. J. Rob. Res. **22**, 59–79 (2003)
18. Gosselin, C.M., Angeles, J.: Singularity analysis of closed-loop kinematic chains. IEEE Trans. Robot. Autom. **6**, 281–290 (1990)
19. Ma, O., Angeles, J.: Optimum architecture design of platform manipulators. In: Proceedings of the 5th International Conference on Advanced Robotics, vol. 2, pp. 1130–1135 (1991)
20. Angeles, J.: The design of isotropic manipulator architectures in the presence of redundancies. Int. J. Rob. Res. **11**, 196–201 (1992)

21. Stoughton, R.S., Arai, T.: A modified Stewart platform manipulator with improved dexterity. IEEE Trans. Robot. Autom. **9**, 166–173 (1993)
22. Merlet, J.P.: Jacobian, manipulability, condition number, and accuracy of parallel robots. J. Mech. Des. **128**, 199–206 (2006)
23. Gosselin, C., Angeles, J.: A global performance index for the kinematic optimization of robotic manipulators. J. Mech. Des. **113**, 220–226 (1991)

Topology Optimization of the Active Arms for a High-Speed Parallel Robot Based on Variable Height Method

Qizhi Meng[1], Fugui Xie[1,2], and Xin-Jun Liu[1,2(✉)]

[1] Department of Mechanical Engineering,
The State Key Laboratory of Tribology & Institute of Manufacturing
Engineering, Tsinghua University, Beijing 100084, China
xinjunliu@mail.tsinghua.edu.cn
[2] Beijing Key Lab of Precision/Ultra-Precision Manufacturing Equipments
and Control, Tsinghua University, Beijing 100084, China

Abstract. This paper presents a solution for the topology optimization of the active arms for high-speed parallel robots. The guide-weight method is introduced into the topology optimization of continuum structures as a numerically iterative criterion. A new hypothetical material interpolation scheme is established as the theoretical foundation of the proposed variable height method. Based on the guide-weight and variable height methods, an efficient and intuitive topology optimization algorithm for flexible manufacturing which includes subtractive and additive manufacturing is put forward. The procedure of topology optimization algorithm for flexible manufacturing is described in detail. Two typical numerical examples of minimum compliance under the weight constraint are tested. In order to improve the static stiffness and dynamic response performance of the high-speed parallel robots, the presented approach is finally applied to optimize the topology of a parallel robot's active arms.

Keywords: Topology optimization · Guide-weight method · Variable height method · Flexible manufacturing · Parallel robot

1 Introduction

The increasing demand for structure parts to be cost-efficient and high-performance, gives new challenges to mechanical designers especially in the field of high-speed parallel robots. As an optimal design method, topology optimization will find out the best material layout for given design objectives and constraints with respect to design variables under predetermined boundary conditions. Since the pioneer paper by Bendsøe and Kikuchi [1], topology optimization has long been a focus of academic researches and been widely used in industrial designs.

To realize topology optimization of continuum structures, two critical issues need to be considered. One is the modeling strategy for the physical material. The other is the solving strategy for a computing model. The modeling strategy has been investigated extensively, and various methods have been proposed, i.e., homogenization method (HM) [2], simplified isotropic material with penalization (SIMP) [3], evolutionary

© Springer International Publishing AG 2017
Y. Huang et al. (Eds.): ICIRA 2017, Part II, LNAI 10463, pp. 212–224, 2017.
DOI: 10.1007/978-3-319-65292-4_19

structural optimization (ESO) [4], level set method [5], and independent continuous mapping method (ICM) [6]. The solving strategy can be summarized into three categories including the optimality criteria method (OC) [7], the mathematical programming method (MP) [8] and the heuristic methods [9]. From the aspect of traditional manufacturing, there is a defect that, in some cases, the topology optimization results cannot be realized in practical application because of microstructure [10]. Therefore, the topology optimization is usually implemented combined with additive manufacturing [11]. Inspired by power law [12], the variable height method is proposed to establish the physical model and avoid the existences of the intermediate density elements or microstructures to realize flexible manufacturing. And the so-called flexible manufacturing, which includes additive manufacturing, subtractive manufacturing and even equivalent manufacturing, can be realized due to the fact that only the basis material's height needs to be changed in this method. In the field of topology optimization, the guide-weight method [13] is initially used for the optimal design of antenna structures. Because of its simple form and fast convergence, the guide-weight method is applied as an OC method in this paper.

The existing topology optimization application objects are generally concentrated on structural components [14], compliant mechanisms [15], heat transfer systems [16], and etc. To the authors' knowledge, there are few researches which focus on the topology optimization of the key parts of spatial parallel mechanisms. We will do some work within this area.

This paper is organized as follows. Combined with the guide-weight and variable height methods, the mathematical principle of topology optimization algorithm for flexible manufacturing is deduced and applied in Sect. 2. In Sect. 3.1, the formulation of the topology problem for minimum compliance under the weight constraint is stated. With that, the sensitivity analysis and the iterative criterion are briefly derived, while the optimization procedure is described in the middle part of Sect. 3. Two classical numerical examples are presented to prove the validity of the topology optimization algorithm in Sect. 3.5. Then, as an application, topology optimization of a high-speed parallel robot's active arms is carried out in Sect. 4. The conclusion is presented in the last section.

2 Topology Optimization Algorithm

The topology optimization algorithm for flexible manufacturing consists of two key parts, i.e., the numerical solution criterion and physical model construction strategy. Their realizations are investigated as follows.

2.1 Guide-Weight Method

General formulation of an optimization problem with single objective can be written as

$$\begin{cases} \text{find}: & X = [x_1, \quad x_2, \quad \cdots, \quad x_N]^T \in R^N, \\ \text{min}: & f(X), \\ \text{subject to}: & g(X) \leq 0, \\ & x_i^L \leq x_i \leq x_i^U \qquad i = 1, 2, \cdots, N, \end{cases} \tag{1}$$

where $X = [x_1, x_2, \cdots, x_N]^T$ is an N-dimensional vector of the design variables; $f(X)$ and $g(X)$ are the objective and constraint functions respectively. To deal with this problem, the Lagrange function is constructed first

$$L = f(X) + \lambda g(X), \tag{2}$$

where λ is the Lagrange multiplier of $g(X)$. Based on the Kuhn-Tucker condition, in the optimal solution X^*, the following formulas must be satisfied:

$$\begin{cases} (1)\ \frac{\partial L}{\partial x_i} = \frac{\partial f}{\partial x_i} + \lambda \frac{\partial g}{\partial x_i} \begin{cases} \leq 0 & \text{if } x_i = x_i^U \\ = 0 & \text{if } x_i^L < x_i < x_i^U \\ \geq 0 & \text{if } x_i = x_i^L \end{cases}, \\ (2)\ \lambda \geq 0, \\ (3)\ \lambda g(X) = 0 \text{ and } \begin{cases} \lambda = 0 & \text{if } g(X) < 0 \\ \lambda \geq 0 & \text{if } g(X) = 0 \end{cases}, \\ (4)\ g(X) \leq 0, \\ (5)\ x_i^L \leq x_i \leq x_i^U \qquad i = 1, 2, \cdots, N. \end{cases} \tag{3}$$

According to the first equation of Eq. (3), when $x_i \in \left(x_i^L, x_i^U\right)$ the following equation can be obtained:

$$-\frac{\frac{\partial g}{\partial x_i}}{\frac{\partial f}{\partial x_i}} = \frac{-x_i \frac{\partial g}{\partial x_i}}{x_i \frac{\partial f}{\partial x_i}} = \frac{1}{\lambda}. \tag{4}$$

The above equation is the central formula of the guide-weight method, based on the idea of numerical iteration and convergence, it leads to

$$(x_i)^{k+1} = \alpha \left(\frac{-x_i \frac{\partial f}{\partial x_i}}{\lambda \frac{\partial g}{\partial x_i}} \right)^k + (1 - \alpha)(x_i)^k, \tag{5}$$

where k is the iteration step and α is employed as the step factor. For the specific issues, the Lagrange multiplier λ can be derived as a solvable expression.

Let

$$G_i = -x_i \frac{\partial f}{\partial x_i}, \tag{6}$$

$$H_i = \frac{\partial g}{\partial x_i}, \tag{7}$$

where G_i is the guide weight of x_i, H_i is the proportional weight of x_i. Take the bounds of x_i into consideration, it can be obtained as

$$x_i^{(k+1)} = \begin{cases} x_i^U & \text{if } x_i \geq x_i^U, \\ \alpha \left(\frac{G_i}{\lambda H_i} \right)^{(k)} + (1 - \alpha) x_i^{(k)} & \text{if } x_i^L < x_i < x_i^U \quad i = 1, 2, \cdots, N, \\ x_i^L & \text{if } x_i \leq x_i^L. \end{cases} \tag{8}$$

The above equation is the iterative expression of the guide-weight method for the optimization problem, i.e., Eq. (1).

2.2   Variable Height Method

Combined with the finite element method (FEM), the continuous design variable h is introduced as the relative height of the element. Inspired by the SIMP material interpolation scheme, we establish the following artificial formula between the height and physical property by the power law:

$$E(x) = f(h,p)E_0 = h^{p-1}E_0, \tag{9}$$

where $E(x)$ is the newest Young's modulus of the material, $f(h, p)$ is the interpolation function, p is the penalization factor. In the plane stress field, we have:

$$k_i = \int_{V_i} B^T DB dV, \tag{10}$$

where B is a constant strain matrix, D is the elasticity matrix and V is the volume of the element. In a 2D problem, we have

$$D = \frac{E}{1 - u^2} \begin{bmatrix} 1 & u & 0 \\ u & 1 & 0 \\ 0 & 0 & 1 - u/2 \end{bmatrix} \quad \text{and} \quad V = sh, \tag{11}$$

Where u is the Poisson's ratio, s is the floorage of the element. Substituting Eqs. (9) and (11) into Eq. (10) leads to

$$k_i = h_i^p k_{i0}, \tag{12}$$

where k_i and k_{io} are the current and initial stiffness matrixes, respectively; and h_i is the relative height of the ith element.

For intuition, the illustration of modeling strategies is shown in Fig. 1, from which we can see that p and h were used as design variables in the SIMP method and the variable height method, respectively.

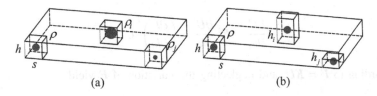

(a)   (b)

Fig. 1. Modeling strategies: (a) the SIMP method; (b) the variable height method

3 Compliance Optimization

3.1 Form of the Compliance Optimization

Optimization problems of minimum compliance under the weight constraint can be expressed as

$$
\begin{cases}
\text{find}: & \mathbf{H} = [h_1, \ h_2, \ \cdots, \ h_N]^T \in R^N, \\
\text{min}: & C(\mathbf{H}), \\
\text{subject to}: & M \leq fM_0, \\
& 0 < h_i^L \leq h_i \leq h_i^U \quad i = 1, 2, \cdots, N.
\end{cases}
\tag{13}
$$

In this problem, compared with the general expression in Eq. (1), $f(\mathbf{H}) = C(\mathbf{H})$ denotes the mean compliance of the structure, whereas $g(\mathbf{H}) = M - fM_0$ denotes the weight constraint, where f is the weight fraction. The design variable T is the relative thickness of the element.

To make Eq. (13) solvable, we should find the explicit expression of $C(\mathbf{H})$ and $M(\mathbf{H})$. The mean compliance can be expressed as

$$
C = \mathbf{F}^T \mathbf{U} = \mathbf{U}^T \mathbf{K} \mathbf{U} = \sum_{i=1}^{N} \mathbf{u}_i^T \mathbf{k}_i \mathbf{u}_i,
\tag{14}
$$

where $\mathbf{F}$ is the load vector; $\mathbf{U}$ is the displacement vector; $\mathbf{K}$ is the total stiffness matrix; $\mathbf{u}_i$ is the displacement vector of the ith element. Based on VHM, substituting Eq. (9) into Eq. (14), one can obtain

$$
C = \sum_{i=1}^{N} h_i^p \mathbf{u}_i^T \mathbf{k}_{io} \mathbf{u}_i.
\tag{15}
$$

The weight of the design domain can be derived as follows

$$
M = \sum_{i=1}^{N} \rho s h_i.
\tag{16}
$$

3.2 Sensitivity Analysis

According to Eq. (15), we can get the sensitivity of $C(\mathbf{H})$ as follows:

$$
\frac{\partial C}{\partial h_i} = \mathbf{F}^T \frac{\partial \mathbf{U}}{\partial h_i} + \left(\frac{\partial \mathbf{F}}{\partial h_i}\right)^T \mathbf{U}.
\tag{17}
$$

According to $\mathbf{F} = \mathbf{K}\mathbf{U}$, and neglecting the variation of $\mathbf{F}$ yields

$$
\frac{\partial \mathbf{F}}{\partial h_i} = \frac{\partial \mathbf{K}}{\partial h_i} \mathbf{U} + \mathbf{K} \frac{\partial \mathbf{U}}{\partial h_i} = \mathbf{0}.
\tag{18}
$$

Substituting Eq. (18) into Eq. (17), one can obtain:

$$\frac{\partial C}{\partial h_i} = -F^T K^{-1} \frac{\partial K}{\partial h_i} U = -U^T \frac{\partial K}{\partial h_i} U = -\sum_{j=1}^{N} u_j^T \frac{\partial k_j}{\partial h_i} u_j. \tag{19}$$

The sensitivity of the object function $C(H)$ and the constraint function $M(H)$ related to h_i can be derived as

$$\frac{\partial C}{\partial h_i} = -p h_i^{p-1} u_i^T k_{io} u_i, \tag{20}$$

$$\frac{\partial M}{\partial h_i} = v_i. \tag{21}$$

3.3   Iterative Criterion

Based on the two expressions defined in Eqs. (6) and (7), we can get

$$G_i = -\rho_i \frac{\partial C}{\partial t_i} = p h_i^p u_i^T k_{io} u_i, \tag{22}$$

$$H_i = \frac{\partial M}{\partial h_i} = \rho s. \tag{23}$$

Thus, as in Eq. (8), the iteration formula can be written as

$$h_i^{(k+1)} = \begin{cases} 1 & \text{if} \quad h_i \geq h_i^U, \\ \alpha \left(\frac{p h_i^p u_i^T k_{io} u_i}{\lambda s_i} \right)^{(k)} + (1-\alpha) h_i^{(k)} & \text{if} \quad h_i^L < h_i < h_i^U, \\ h_i^L & \text{if} \quad h_i \leq h_i^L, \end{cases} \tag{24}$$

where λ can be obtained as

$$\lambda = \frac{\sum_{i=1}^{N} p h_i^p u_i^T k_{io} u_i}{f M_0}. \tag{25}$$

Based on Eqs. (24) and (25), the optimization problem in Eq. (13) can be solved. Of note is that, considering Eqs. (4), (23) and (25), one obtains

$$M_i = h_i s = h_i H_i = \frac{G_i}{\lambda} = \frac{G_i}{G} f M_0. \tag{26}$$

The above equation is the distribution criterion of weight, and its physical meaning is that, the weight and the guide weight of each element should be proportional in the optimum structure.

3.4   The Topology Optimization Procedure

On the basis of the above principles, we design a set of algorithm programs which are operated in the MATLAB environment. Major steps are illustrated in a flowchart as shown in Fig. 2, detailed descriptions can be outlined as follows:

Flow 1: *Parameters initialization*. The indispensable parameters are defined, such as penalty factor p, step factor α, volume fraction f and error value limit ε.

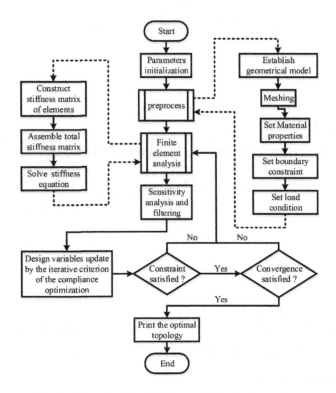

Fig. 2. Flowchart for topology optimization procedure.

Flow 2: *Preprocess*. A geometrical model is established as the design domain in advance. Then, the continuous design domain is dispersed as different square elements. Finally, material properties, boundary constraint and load condition are assigned.

Flow 3: *Finite element analysis*. In this step, three subflows are included. Taking the planar four-node isoparametric element into consideration, the stiffness matrix of elements can be constructed according to Eq. (10). Then, the total stiffness matrix is assembled and nodal displacements can be obtained by solving stiffness equation.

Flow 4: *Sensitivity analysis and height filtering*. The sensitivities can be obtained by computing the objective, constraint functions and their derivative. Height filtering is necessary to avoid the checkboard and mesh dependency.

This paper employs an averaging strategy to bring about the filtering, as shown in Fig. 3, we firstly calculate the height of each element, then, we assign the average value of four heights to their common node, finally the average height of four nodes are assigned to their common element. To confirm the validity of the averaging strategy, a classical filtering example with loading and support conditions is displayed in Fig. 4(a). The optimal results are shown in Fig. 4(b), (c) and (d), which indicate that the topologies are consistent with different mesh sizes.

Flow 5: *__Design variables update with the AFM__*. The design variables are updated by the iterative criterion of the compliance optimization.

Flow 6: *__Constraint and convergence satisfied__*? The flows 2–5 should be cycled until the weight constraint and the convergence criterion are satisfied.

Flow 7: *__Print the optimal topology__*. If the weight constraint and the convergence criterion are satisfied, the final optimal result can be printed.

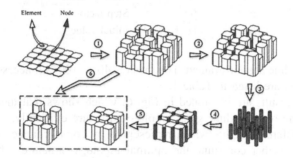

Fig. 3. Diagrammatic sketch of the averaging strategy.

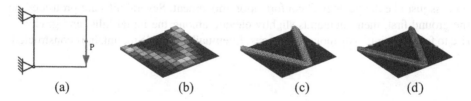

(a)   (b)   (c)   (d)

Fig. 4. A filtering example: (a) working condition, and (b), (c), (d) topology results obtained with different mesh sizes 10 × 10, 40 × 40 and 70 × 70.

3.5   Numerical Examples

To verify the effectiveness of the proposed method, the numerical implementation is demonstrated with two examples. The weight fraction f which means utilization factor of raw materials is chosen as 30%.

(1) Simply supported beam problem

As shown in Fig. 5, the design domain is a 0.2 m × 0.5 m rectangle. The load P is 1500 N, located in the middle of the top boundary line, and the left and right points of

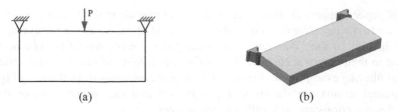

Fig. 5. Simply supported beam problem: (a) working condition; (b) design domain

Table 1. Parameters for topology optimization of simply supported beam problem

Parameters	Values	Meanings
E_0	2.06×10^{11}	Young's module
u	0.3	Poisson's ratio
p	3	Penalty factor
α	0.5	Step factor
h_0	[1, 1, …, 1]	Initial value

the top boundary line are constrained. The mesh is 40×100. All necessary parameters during calculation are listed in Table 1.

The optimal results are illustrated in Fig. 6, which shows that the topology optimization algorithm can produce great topology structure efficiently. In Fig. 6(a), the elements with different heights which represented by colors changed gradually from blue to yellow together constitute the optimal topology structure. The convergence process and the changes of topology structures in the iterative process are provided in Fig. 6(b). The red convergence curve demonstrates that the optimal solution can be obtained within 30 iterations. More interesting, the changes of topology in the iterative process just like the geological conformation movement. Several hills are protruded on the ground first, then, highlands slightly elevate among the initial hills, and gradually rise to form continuously mountain ranges. Eventually, a stable mountain is constructed.

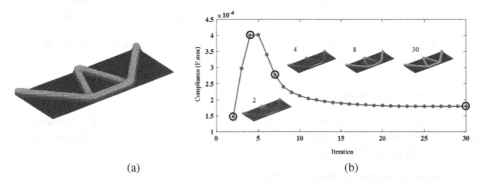

Fig. 6. Optimal results: (a) optimal topology; (b) convergence process (Color figure online)

(2) Cantilever beam problem

As shown in Fig. 7, the load P is located in the middle of the right boundary line, and the left boundary line is constrained. Other parameters are the same as that in example (1). The optimal results are illustrated in Fig. 8. The optimal solution can be obtained within 40 iterations.

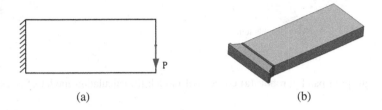

Fig. 7. Cantilever beam problem: (a) working condition; (b) design domain

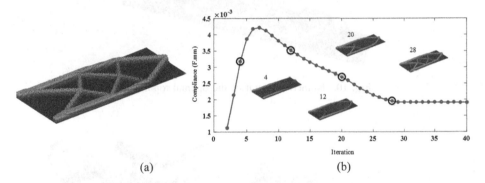

Fig. 8. Optimal results: (a) optimal topology; (b) convergence process

4 Application

The architecture of a high-speed parallel robot is shown in Fig. 9(a), which is composed of a base, four limbs (I, II, III and IV) and a mobile platform with an end-effector. The four limbs are connected with the base and mobile platform, collectively these components make up the spatial closed-loop mechanism.

The output motions of this robot include three translations and one rotation about the vertical direction, which is in great need in industries to realize high-speed pick-and-place manipulation [17]. In order to improve its performance in terms of speed, acceleration and dynamic response, topology optimization of the active arms is our primary strategy. The calculation model of the active arms is presented in Fig. 9(b). and the design domain is a 0.05 m × 0.35 m rectangle.

The optimal topology of the active arms is shown in Fig. 10. The continuous material distribution between the force location and the constraint is considered as the

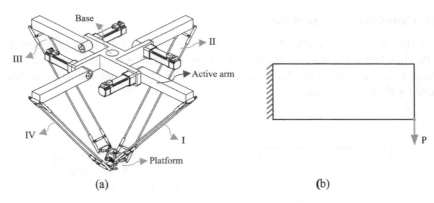

(a)   (b)

Fig. 9. High-speed parallel robot: (a) conceptual model; (b) calculation model of active arms

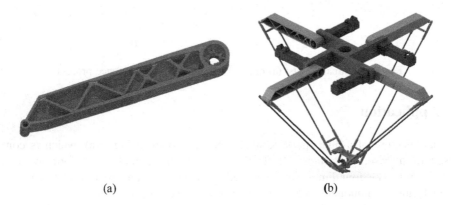

Fig. 10. Numerical result of the optimal topology.

(a)   (b)

Fig. 11. CAD models: (a) active arm; (b) assembly of the complete robot.

optimal transmission path of force. Based on the numerical information of the topology, the corresponding CAD model is established in Fig. 11(a). By the aid of flexible manufacturing processes such as 3D printing, milling, casting and so on, the active arms can be processed. Then, the complete robot, which is expected to have good dynamic performance, can be assembled Fig. 11(b).

5 Conclusions

This paper deals with the topology optimization of the active arms for a high-speed parallel robot. The key parts of the optimization algorithm consist of the introduced guide-weight method and the proposed variable height method. The variable height method provides an attractive alternative for material interpolation scheme of the topology optimization problems. The best material layout of the optimal topology is realized by the height variations of the elements. Therefore, the optimal structure can be processed via flexible manufacturing. In addition to theoretical analysis, the procedure and the relevant flowchart are described in detail. Two classical numerical examples of minimum compliance under the weight constraint are tested, then the fast computational efficiency and good processing adaptability have been confirmed. As a result, the topology of the active arms is derived in MATLAB environment and the corresponding CAD model is then built. As a summary, the design of the main structure part for a high-speed parallel robot has been completed.

Acknowledgment. This work was supported by the National Natural Science Foundation of China under Grant 51425501 and by Beijing Municipal Science & Technology Commission under Grant Z171100000817007.

References

1. Bendsøe, M.P., Kikuchi, N.: Generating optimal topologies in structural design using a homogenization method. Comput. Methods Appl. Mech. Eng. **71**(2), 197–224 (1988)
2. Suzuki, K., Kikuchi, N.: A homogenization method for shape and topology optimization. Crit. Care Med. **93**(3), 291–318 (1991)
3. Bendsøe, M.P., Sigmund, O.: Material interpolation schemes in topology optimization. Arch. Appl. Mech. **69**(9), 635–654 (1999)
4. Sun, X.F., Yang, J., Xie, Y.M., et al.: Topology optimization of composite structure using bi-directional evolutionary structural optimization method. Procedia Eng. **14**(2259), 2980–2985 (2011)
5. Mei, Y., Wang, X.: A level set method for structural topology optimization and its applications. Comput. Methods Appl. Mech. Eng. **35**(7), 415–441 (2004)
6. Sui, Y., Du, J., Guo, Y.: Independent continuous mapping for topological optimization of frame structures. Acta. Mech. Sin. **22**(6), 611 (2006)
7. Yin, L., Yang, W.: Optimality criteria method for topology optimization under multiple constraints. Comput. Struct. **79**(20–21), 1839–1850 (2001)
8. Zillober, C.: A globally convergent version of the method of moving asymptotes. Struct. Optim. **6**(3), 166–174 (1993)
9. Smith, O.D.S.: Topology optimization of trusses with local stability constraints and multiple loading conditions—a heuristic approach. Struct. Optim. **13**(2–3), 155–166 (1997)
10. Liu, S., Li, Q., Chen, W., et al.: An identification method for enclosed voids restriction in manufacturability design for additive manufacturing structures. Front. Mech. Eng. **10**(2), 126–137 (2015)
11. Zegard, T., Paulino, G.H.: Bridging topology optimization and additive manufacturing. Struct. Multi. Optim. **53**(1), 175–192 (2016)

12. Rietz, A.: Sufficiency of a finite exponent in SIMP (power law) methods. Struct. Multi. Optim. **21**(2), 159–163 (2001)
13. Chen, S., Wei, Q., Huang, J.: Meaning and rationality of guide-weight criterion for structural optimization. Chin. J. Solid Mech. **34**(6), 628–638 (2013)
14. Li, B., Hong, J., Liu, Z.: A novel topology optimization method of welded box-beam structures motivated by low-carbon manufacturing concerns. J. Cleaner Prod. **142**, 2792–2803 (2017)
15. Jin, M., Zhang, X.: A new topology optimization method for planar compliant parallel mechanisms. Mech. Mach. Theor. **95**, 42–58 (2016)
16. Dbouk, T.: A review about the engineering design of optimal heat transfer systems using topology optimization. Appl. Thermal Eng. **112**(2017), 841–854 (2016)
17. Xie, F.G., Liu, X.J.: Design and development of a high-speed and high-rotation robot with four identical arms and a single platform. J. Mech. Robot. **7**(4), 041015 (2015)

Stiffness Analysis of a Variable Stiffness Joint Using a Leaf Spring

Lijin Fang$^{(\boxtimes)}$ and Yan Wang(iD)

Northeastern University, Shenyang 110819, China
ljfang@mail.neu.edu.cn, yan_wang0221@163.com

Abstract. A variable stiffness joint using a leaf spring is designed to ensure physical safety. The joint stiffness is often controlled by changing the effective length of the leaf spring. The stiffness model based on the small-deformation theory cannot solve large deflection problems of the leaf spring caused by larger joint deflected angles. The elliptic integral solution is considered to be the most accurate method for analyzing large deflections of beams. In this paper, a stiffness analysis model based on the elliptic integral solution is proposed. The joint stiffness property is analyzed. The simulation results show the joint stiffness is nonmonotonic and strong nonlinear. A stiffness simplified model is presented by nonlinear curve fitting for application simplicity. The experiment is carried out to verify the stiffness property and the stiffness analysis model.

Keywords: Variable stiffness joint · Stiffness analysis · Large deflection

1 Introduction

Variable stiffness joints are designed and implemented in recent years to solve the safety problems of physical human-robot interaction. The power input and output of a mechanically variable stiffness joint are connected by elastic elements, i.e., springs that can absorb impact shocks and store energy. The nonlinear stiffness is controlled by mechanical reconfiguration which can change the force-displacement relationships of springs. The joint stiffness of a variable stiffness joint using a leaf spring is mostly controlled by varying the effective length of the leaf spring. By this way, the stiffness has been controlled in VSJ [1], MIA [2], ADEA [3], VSR-joint [4] and MeRIA [5].

The bending deformation of the leaf spring is simplified based on the small-deformation theory in order to calculate the stiffness easily. As in Refs. [1, 3], the deflection of the leaf spring is in a linear relationship with the force, and the horizontal displacement is ignored. This stiffness model is easy to use, but it is limited because the large deflection of the leaf spring is not taken into account when the joint deflected angles are larger in practical application. Therefore, the stiffness model for solving the large deflection problem is required to analyze the stiffness property of the variable stiffness joint using a leaf spring. The nonlinearity associated with the large deflection problem poses modeling and computational challenges, and complicates the stiffness

© Springer International Publishing AG 2017
Y. Huang et al. (Eds.): ICIRA 2017, Part II, LNAI 10463, pp. 225–237, 2017.
DOI: 10.1007/978-3-319-65292-4_20

analysis [6, 7]. The elliptic integral solution is often considered to be the most accurate method for modeling large deflections of beams that are so thin and flexible that the effects of axial elongation and shear are negligible [8]. Alkhaldi and Abu-Alshaikh [7] presented a novel approach to obtain the closed-form solutions of large deflection of beams subjected to inclined non-conservative load and tip-end moment. Zhang and Chen derived a comprehensive elliptic integral solution to the large deflection problems of thin beams in compliant mechanisms, based on which the cross-spring pivot was precisely modeled [8, 9]. The joint stiffness can be accurately analyzed based on the elliptic integral, but the calculation procedure is more complicated than that of the small-deformation theory. A simple method is needed to calculate the stiffness of the variable stiffness joint accurately.

In this paper, a stiffness analysis model based on the elliptic integral solution is proposed for solving the stiffness of a variable stiffness joint using a leaf spring. The joint stiffness property is analyzed. For calculating the stiffness easily, a stiffness simplified model is presented by nonlinear curve fitting. An experiment is carried out to verify the stiffness property and the stiffness model.

2 Principle of a Variable Stiffness Joint Using a Leaf Spring

The schematic of the variable stiffness joint using a leaf spring is shown in Fig. 1, which is studied in this paper to analyze the joint stiffness property. Both the joint arm and the rotation carrier could rotate around the central shaft. One end of the leaf spring is fixed to the rotation carrier, and the other end is acted on by the external force point which is fixed on the joint arm. The slider slides along the leaf spring to change the effective length L, which is the distance between the slider and the external force point. The slider can move from the left of the central shaft to the right. The deflected angle is θ_j when the joint arm is subjected to the external moment M_j.

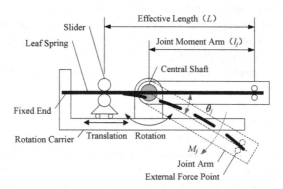

Fig. 1. Variable stiffness joint using a leaf spring in its initial position (solid lines) and a deflected position (dashed lines)

For studying the effect of the effective length L in the initial position on the joint stiffness k_j, k_j is analyzed based on the small-deformation theory. The variable stiffness schematic based on the small-deformation theory is shown in Fig. 2. When M_j is applied to the joint, the external force F is exerted on the leaf spring via the external force point B. The friction is ignored. And the distance between the slider O' and the fixed end A is unchanged during the joint deflection. The joint moment arm l_j is the distance between B and the central shaft O.

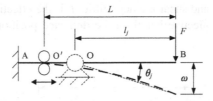

Fig. 2. Variable stiffness schematic based on the small-deformation theory

According to the solution presented in Ref. [1], θ_j is given as $\theta_j \approx \tan\theta_j = \omega/l_j$, and F is given as $F = 3EIl_j\theta_j/L^3$. Knowing that $M_j = Fl_j$, the joint stiffness k_j based on the small-deformation theory is given as

$$k_j = \frac{dM_j}{d\theta_j} = \frac{3EIl_j^2}{L^3} \tag{1}$$

Knowing from Eq. (1), k_j is only affected by L. The stiffness property is illustrated by a case. The constant parameters of the variable stiffness joint are: $E = 206$ GPa, $I = 2.25$ mm^4, $l_j = 195$ mm. The stiffness k_j in the joint's initial position for different L is shown in Fig. 3. The longer L is, the smaller k_j is. The stiffness model based on the small-deformation theory can solve the stiffness in the joint's initial position, but it cannot reflect the stiffness changes when the joint deflected angles are larger.

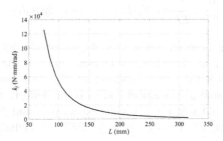

Fig. 3. Stiffness k_j in joint's initial position for different L

3   Stiffness Analysis Model of a Variable Stiffness Joint Using a Leaf Spring

3.1   Stiffness Analysis Model Based on Elliptic Integral Solution

In this section, the stiffness analysis model based on the elliptic integral solution to large deflection problems of cantilever beams is proposed to study the effect of the joint deflected angle θ_j on the joint stiffness k_j. Figure 4 shows the variable stiffness schematic based on the large deflection. The bending deflection is only considered in this model. Axial elongation and shear are negligible. L is the effective length in the joint's initial position, and the effective length in the deflected position is represented by l.

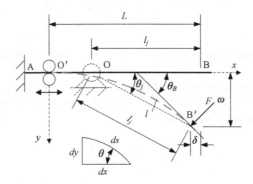

Fig. 4. The variable stiffness schematic based on large deflection in its initial position (black solid lines) and a deflected position (red dashed lines) (Color figure online)

According to the Bernoulli-Euler beam theory, the bending moment at an arbitrary point P (x, y) on the deflected leaf spring is proportional to the curvature at that point [10], that is

$$\frac{d\theta}{ds} = \frac{M}{EI} = \frac{F}{EI}[\cos\theta_B(L - \delta - x) + \sin\theta_B(\omega - y)] \tag{2}$$

where M is the bending moment, $d\theta/ds$ is the curvature, θ is the angular deflection of the leaf spring, s is the arc length of deflection curve of the leaf spring, x is the horizontal coordinate, y is the vertical coordinate, θ_B is the angular deflection at the point B, δ is the horizontal displacement and ω is the deflection at the point B.

Differentiating both sides of Eq. (2) with respect to s, and then integrating it with the boundary conditions, i.e., $\theta = \theta_B$, $d\theta/ds = M/EI = 0$. Defining $\alpha^2 = F/EI$, Eq. (2) can be rewritten as

$$\frac{d\theta}{ds} = \alpha\sqrt{2\sin(\theta_B - \theta)} \tag{3}$$

Note that

$$\begin{cases} \frac{dy}{ds} = \sin\theta \\ \frac{dx}{ds} = \cos\theta \end{cases} \tag{4}$$

Substituting Eq. (4) into Eq. (3) and integrating them result in

$$\alpha l = \int_0^{\theta_B} \frac{d\theta}{\sqrt{2\sin(\theta_B - \theta)}} \tag{5}$$

$$\alpha\omega = \int_0^{\theta_B} \frac{\sin\theta d\theta}{\sqrt{2\sin(\theta_B - \theta)}} \tag{6}$$

$$\alpha(l - \delta) = \int_0^{\theta_B} \frac{\cos\theta d\theta}{\sqrt{2\sin(\theta_B - \theta)}} \tag{7}$$

The elliptic integral solution is used to solve Eqs. (5–7). The incomplete elliptic integrals of the first and second kinds are respectively defined as [10]

$$F(k, \varphi) = \int_0^{\varphi} \frac{dt}{\sqrt{1 - k^2 \sin^2 t}} \text{ and } E(k, \varphi) = \int_0^{\varphi} \sqrt{1 - k^2 \sin^2 t} \, dt$$

where, φ is called the amplitude and k ($-1 \leq k \leq 1$) is called the modulus. When $\varphi = \pi/2$, they become the complete elliptic integrals of the first and second kinds and are respectively defined as $F(k)$ and $E(k)$. Equations (5–7) are rewritten as

$$\alpha l = f \tag{8}$$

$$\alpha\omega = -2e \cos\theta_B + f \cos\theta_B + 2k \sin^{\frac{3}{2}}\theta_B \tag{9}$$

$$\alpha(L - \delta) = 2e \sin\theta_B - f \sin\theta_B + 2k \cos\theta_B \sin^{\frac{1}{2}}\theta_B \tag{10}$$

where, $f = F(k) - F(k, \varphi)$, $e = E(k) - E(k, \varphi)$, $k = 1/\sqrt{2}$ and $\varphi = \arcsin\sqrt{1 - \sin\theta_B}$. Because l_j is constant when the joint is deflected, ω and δ are expressed as

$$\begin{cases} \omega = l_j \sin\theta_j \\ \delta = l_j(1 - \cos\theta_j) \end{cases} \tag{11}$$

An index β is defined as $\beta = \alpha\omega/\alpha(L - \delta)$ from Eqs. (9) and (10). The joint deflected angle θ_j is derived from Eqs. (9–11) as

$$\theta_j = \arccos\left[\frac{(L - l_j)\beta}{l_j\sqrt{1 + \beta^2}}\right] - \arctan\left(\frac{1}{\beta}\right) \tag{12}$$

Knowing θ_j, α can be calculated by Eqs. (9) and (11) as $\alpha = \alpha\omega/\omega$. The effective length l in the deflected position is derived from Eq. (8) as $l = f/\alpha$. The external moment M_j is expressed as

$$M_j = \alpha^2 EI \cos(\theta_B - \theta_j) l_j \tag{13}$$

Because both M_j and θ_j are related with θ_B as the intermediate variable, the joint stiffness k_j is calculated by Eqs. (12) and (13) as

$$k_j = \frac{dM_j}{d\theta_j} = \frac{dM_j}{d\theta_B} \bigg/ \frac{d\theta_j}{d\theta_B} \tag{14}$$

Knowing that $0 \leq \theta \leq \theta_B$, the range of θ_B is derived from Eq. (3) as $0 \leq \theta_B \leq \pi/2$, which is considered in the stiffness analysis model.

3.2 Analysis of Stiffness Property

The parameters in Sect. 2 are used in the stiffness analysis model to analyze the stiffness property. Define ζ as the effective length factor, which is the ratio of the effective length L in the initial position to the joint moment arm l_j, i.e., $\zeta = L/l_j$. The effective length l in the deflected position is nondimensionalized as the length change index λ, i.e., $\lambda = l/L$. The relationship between λ and θ_j is shown in Fig. 5.

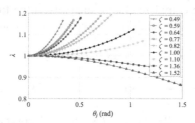

Fig. 5. Plots of length change index λ versus deflected angle θ_j for different ζ

Figure 6 shows the different deflected shapes of the leaf spring for different deflected angles θ_j to observe the deflected configurations. Knowing the central shaft O at x = 0, the black link OB represents the joint moment arm. The curve O'B represents the deflected shape of the leaf spring. The change of l is different when L is different.

Figure 7 shows the relationship between M_j and θ_j for different ζ. Figure 8 shows the relationship between k_j and θ_j for different ζ. The stiffness k_j dose not always increase or decrease with the increase of θ_j. Compared with Fig. 5, k_j is related to l. When l gets longer, k_j decreases; however, when l gets shorter, k_j increases. It is found that k_j decreases drastically when $L \leq l_j$, but k_j changes slowly when $L > l_j$.

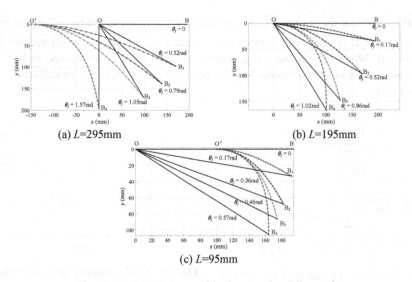

Fig. 6. Deflected shape of leaf spring for different θ_j

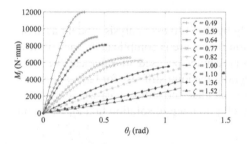

Fig. 7. Plots of M_j versus θ_j for different ζ **Fig. 8.** Plots of k_j versus θ_j for different ζ

The joint stiffness properties are analyzed as follows: (1) the stiffness changes are nonmonotonic; (2) the stiffness is affected by the changes of the effective length of the leaf spring; (3) the stiffness is strong nonlinear when the effective length is relatively shorter.

4   Stiffness Simplified Model for Large Deflection

4.1   Calculation Procedure of Stiffness Analysis Model

Although θ_j can be calculated by θ_B, θ_B could not be calculated by θ_j, as observed from Eq. (12). Therefore, the variables (e.g., l, M_j, k_j) could be directly calculated by θ_B rather than by θ_j. However, θ_j rather than θ_B can be measured in practical application. When θ_j is known, $\overline{\theta_B}$, a trial value of θ_B, can be found by a search procedure. Figure 9 shows the calculation procedure of the stiffness analysis model. Taking the accuracy,

$|\overline{\theta_j} - \theta_j| \leq 10^{-4}$ as an example, $\overline{\theta_j}$, a trial value of θ_j, is calculated by setting an initial value, $\overline{\theta_B} = 0$. If the accuracy requirement is not met, 10^{-5} is added to $\overline{\theta_B}$, and $\overline{\theta_j}$ is recalculated. When the accuracy requirement is satisfied, k_j is computed at last. The error could be reduced by improving the accuracy. However, the procedure is very complicated.

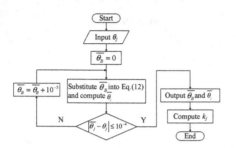

Fig. 9. Calculation procedure of the stiffness analysis model

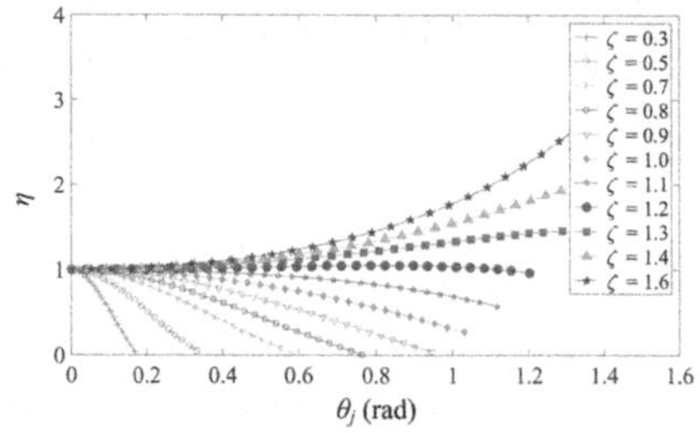

Fig. 10. Nondimensional stiffness coefficient η for different ζ

4.2 Stiffness Simplified Model

The joint stiffness can be calculated accurately by the stiffness analysis model based on the elliptic integral solution, but the calculation procedure is complicated. The stiffness model based on the small-deformation theory cannot be used to calculate the joint stiffness during the joint deflection. In this section, the joint stiffness for large deflection is simplified by the nonlinear curve fitting to calculate the stiffness easily.

The relationship between k_j based on the elliptic integral solution and k_{js} based on the small-deformation theory is given as

$$k_j = \eta k_{js} \tag{15}$$

where, η is the nondimensional stiffness coefficient, $\eta = f(\theta_j, \zeta)$.

The stiffness coefficient η for different effective length factors ζ is shown in Fig. 10. The value of η is related to θ_j and ζ. First, applying the Least Square Polynomial Fitting Method, a Cubic Polynomial of θ_j is used here to fit η. The fitting curve of η can be presented in following equation

$$\eta = p_1\theta_j^3 + p_2\theta_j^2 + p_3\theta_j + p_4 \tag{16}$$

where, the values of p_1, p_2, p_3 and p_4 are determined by the each value of ζ.

The values of p_1, p_2, p_3 and p_4 for different effective length factors ζ are shown in Table 1. It is found that the values of p_1 and p_2 change obviously when $0.2 \leq \zeta < 0.8$. To improve the fitting correlation coefficient R^2, the range of ζ is divided into two parts, i.e., $0.2 \leq \zeta < 0.8$ and $0.8 \leq \zeta \leq 1.6$. Applying the Exponential Model, a Double

Table 1. The values of p_1, p_2, p_3 and p_4 for different ζ

	p_1	p_2	p_3	p_4
$\zeta = 1.6$	0.840047	−0.45101	0.417623	0.96537
$\zeta = 1.4$	0.036675	0.540218	−0.03251	1.003143
$\zeta = 1.2$	−0.35246	0.486979	−1.0393	1.006613
$\zeta = 1.0$	0.039417	−0.72669	0.000779	1.000035
$\zeta = 0.8$	2.13218	−3.40232	0.060352	0.998698
$\zeta = 0.6$	11.52068	−10.1892	0.090969	0.999163
$\zeta = 0.4$	69.07511	−33.5874	0.169772	0.999113
$\zeta = 0.2$	870.9202	−182.697	0.43305	0.998876

Exponential Model of ζ is used to fit p_1, p_2, p_3 and p_4 when $0.2 \leq \zeta < 0.8$. The fitting curve of p_1, p_2, p_3 and p_4 can be presented in following equations

$$p_{11} = 47994.63e^{-23.0184\zeta} + 2316.338e^{-8.93315\zeta} \tag{17}$$

$$p_{21} = -2562.69e^{-16.7665\zeta} - 280.623e^{-5.54755\zeta} \tag{18}$$

$$p_{31} = 1.325318e^{-8.50993\zeta} + 0.268132e^{-1.93587\zeta} \tag{19}$$

$$p_{41} = -2562.69e^{-16.7665\zeta} - 280.623e^{-5.54755} \tag{20}$$

Applying the Least Square Polynomial Fitting Method, a Quartic Polynomial of ζ is used to fit p_1, p_2, p_3 and p_4 when $0.8 \leq \zeta \leq 1.6$. The fitting curve of p_1, p_2, p_3 and p_4 can be presented in following equations

$$p_{12} = 10.05243\zeta^4 - 61.6123\zeta^3 + 144.6535\zeta^2 - 149.89\zeta + 56.82624 \tag{21}$$

$$p_{22} = -3.74534\zeta^4 + 22.23417\zeta^3 - 62.0405\zeta^2 + 81.697\zeta \quad 38.8612 \tag{22}$$

$$p_{32} = 0.106359\zeta^4 + 3.938932\zeta^3 - 12.9999\zeta^2 + 13.18879\zeta - 4.2354 \tag{23}$$

$$p_{42} = -0.31162\zeta^4 + 1.082235\zeta^3 - 1.30561\zeta^2 + 0.636523\zeta + 0.898663 \tag{24}$$

The fitting values of p_1, p_2, p_3 and p_4 are shown in Fig. 11.

The joint stiffness k_{jcf} based on the simplified model can be computed by Eq. (15). The calculation procedure of the stiffness simplified model is shown in Fig. 12.

The error ε between k_{jcf} based on the simplified model and k_j based on the analysis model is given as Eq. (25). The error ε for different effective length factors ζ is shown in Fig. 13. The range of ε is ± 0.05 in a certain range of θ_j.

$$\varepsilon = \frac{k_j - k_{jcf}}{k_j} \tag{25}$$

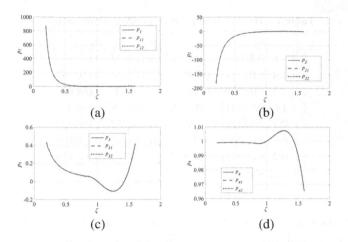

Fig. 11. Fitting values of p_1, p_2, p_3 and p_4

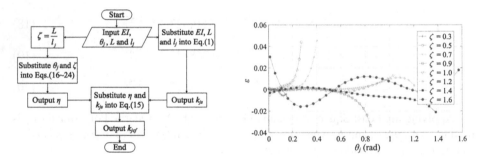

Fig. 12. Calculation procedure of the stiffness simplified model

Fig. 13. Error ε for different ζ

5 Experiment

According to the variable stiffness schematic shown in Fig. 1, an experiment was conducted to show the stiffness property and verify the stiffness analysis model. Figure 14 shows the experimental setup. The parameters in Sect. 2 are used in the experiment. The joint arm rotates around the central shaft by pulling the rope which connects the tension sensor and the external force point. The tension value measured by the tension sensor is shown on the display instrument. The deflected angle of the joint arm measured by the angular sensor is sent to the PC. The slider is manually adjusted to change the effective length of the leaf spring. The effective length in the deflected position can be read by the scale on the leaf spring.

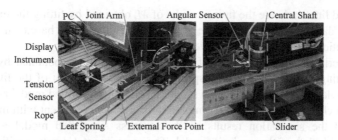

Fig. 14. Experimental setup

Four effective lengths in the joint's initial position, i.e., $L = 95$ mm, $L = 145$ mm, $L = 195$ mm and $L = 245$ mm, are selected in the experiment. The comparison of the simulation results based on the elliptic integral solution with the experimental results of the effective length l in the deflected position is shown in Fig. 15. The dot indicates the experimental data of l. The dashed lines indicate the simulation results.

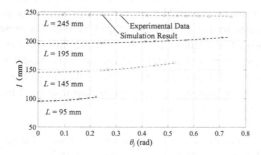

Fig. 15. Comparison of simulation results with experimental results of l

The experimental data of the external moment M_j can be calculated by three parameters, i.e., the measured tension, the deflected angle and the joint moment arm. Figure 16 shows the comparison of the simulation results based on the elliptic integral solution with the experimental results of M_j. The dot indicates the experimental data of

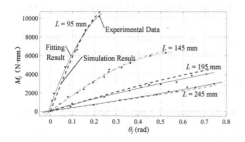

Fig. 16. Comparison of simulation results with experimental results of M_j

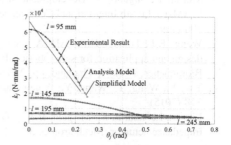

Fig. 17. Comparison of simulation results with experimental fitting results of k_j

M_j. The solid lines indicate the fitting results of M_j obtained by fitting the experimental datum. The dashed lines indicate the simulation results of M_j based on the elliptic integral solution. The simulation curves are close to the fitting curves.

The experimental fitting results of the joint stiffness k_j are obtained by differentiating the fitting results of M_j with respect to θ_j. The comparison of the fitting results with the simulation results of the analysis model and the simplified model are shown in Fig. 17. The changing trends of the simulation results are consistent with that of fitting results. And the simulation results of the stiffness simplified model are in good agreement with that of the analysis model. The experimental results verified the joint stiffness property and the validity of the stiffness analysis model.

6 Conclusions

A stiffness analysis model based on the elliptic integral solution to large deflection problems of cantilever beams is proposed to analyze the joint stiffness of the variable stiffness joint using a leaf spring. The simulation results show the joint stiffness is affected by the large deflection of the leaf spring. And the changes of the stiffness are nonmonotonic and strong nonlinear. However, the calculation procedure of the analysis model is too complicated to use. A stiffness simplified model is presented by nonlinear curve fitting to improve the stiffness accuracy and simplify the calculation procedure. The error between the simplified model and the analysis model is in an acceptable range of ± 0.05. The experiment is carried out to verify the stiffness property, and the experimental results are consistent with the simulation results of the stiffness analysis model and the stiffness simplified model.

Acknowledgment. The authors gratefully acknowledge the financial support from the National Natural Science Foundation of China under Grant No. 51575092.

References

1. Choi, J., Hong, S., et al.: A robot joint with variable stiffness using leaf springs. IEEE Trans. Robot. **27**(2), 229–238 (2011)
2. Morita, T., Sugano, S.: Development of an anthropomorphic force-controlled manipulator WAM-10. In: 8th International Conference on Advanced Robotics, Monterey, CA, pp. 701–706 (1997)
3. Wang, R.J., Huang, H.P.: ADEA—Active variable stiffness differential elastic actuator: design and application for safe robotics. In: IEEE International Conference on Robotics and Biomimetics, Phuken, Thailand, pp. 2768–2773 (2011)
4. Tao, Y., Wang, T., Wang, Y.: A new variable stiffness robot joint. Ind. Robot **42**(4), 371–378 (2015)
5. Liu, L., Leonhardt, S., Misgeld, B.J.E.: Design and control of a mechanical rotary variable impedance actuator. Mechatronics **39**, 226–236 (2016)

6. Tari, H., Kinzel, G.L., Mendelsohn, D.A.: Cartesian and piecewise parametric large deflection solutions of tip point loaded Euler-Bernoulli cantilever beams. Int. J. Mech. Sci. **100**, 216–225 (2015)
7. Alkhaldi, H.S., Abu-Alshaikh, I., et al.: Closed-form solution of large deflection of a spring-hinged beam subjected to non-conservative force and tip end moment. Eur. J. Mech. A/Solids **47**, 271–279 (2014)
8. Zhang, A., Chen, G.: A comprehensive elliptic integral solution to the large deflection problems of thin beams in compliant mechanisms. J. Mech. Robot. **5**(2), 021006 (2013)
9. Zhang, A., Chen, G., Jia, J.: Large deflection modeling of cross-spring pivots based on comprehensive elliptic integral solution. Jixie Gongcheng Xuebao **50**(11), 80–85 (2014)
10. Howell, L.L.: Compliant Mechanisms. Wiley-Interscience, New York (2001)

Design of a Series Variable Stiffness Joint Based on Antagonistic Principle

Shipeng Cui, Yiwei Liu$^{(\boxtimes)}$, Yongjun Sun, and Hong Liu

State Key Laboratory of Robotics and System, Harbin Institute of Technology,
Harbin 150001, China
lyw@hit.edu.cn

Abstract. In order to achieve a safe human-robot interaction environment, a variable stiffness joint (VSJ) intended to apply in elbow of robot arm is introduced. The configuration of the VSJ is converted from antagonistic to series by employing a mirror pair of nonlinear elastic transmissions (NETs) which are based on compliant four-bar mechanism (CFM). In this paper, the preliminary mechanical design and the working principle details of the VSJ are presented together with systematic design process of the NET. Simulation results validate the feasibility of the design of the VSJ and the NET.

Keywords: Variable stiffness joint · Antagonistic · Series · Nonlinear elastic transmission · Compliant four-bar mechanism

1 Introduction

With the increasing interactions between humans and robots, applications of compliant design have been a trend in robotic field. As a solution of mechanically compliant design, VSJ has attracted much attention.

Although several types of VSJ have been proposed, the methods for implementing variable stiffness can be classified into two main approaches. The first approach employs antagonistic configuration inspired by the musculoskeletal system. In antagonistic configuration, two motors actuate antagonistically a link through nonlinear elastic transmissions placed between the motors and the link. The position and torque of the joint are simultaneously controlled by two motors. This type of VSJ has a simple but large structure and the range of the joint output is limited by the structure. VSA-II [1] and BAVS [2] are the representative examples for this approach.

Alternatively, in the series configuration, the variable stiffness actuator is composed of a main motor and an auxiliary motor. Position and torque of the joint are controlled by the main motor and the joint stiffness is regulated by the auxiliary motor. This type of VSJ has a compact and integrated but complex structure. In most of the series VSJ such as VS-Joint [3] and FSJ [4], the stiffness is adjusted by changing the pretension of the elastic transmission. However, some designs rely on the lever mechanism concept. AwAs-II [5], CompAct-VSA [6] and vsaUT-II [7] are good examples of this approach.

In this paper, a series VSJ based on antagonistic principle is proposed, designed and modeled. The main novelty of this paper is focused on the mechanical structure design of the VSJ. This paper is organized as follows. Section 2 describes the general working

Y. Huang et al. (Eds.): ICIRA 2017, Part II, LNAI 10463, pp. 238–249, 2017.
DOI: 10.1007/978-3-319-65292-4_21

principle of an antagonistic VSJ and the series VSJ based on it. Section 3 presents the design of the NET based on CFM. Section 4 elaborates the mechanical design and the working principle details of the VSJ together with the simulation results. Finally, conclusions and future works are discussed in Sect. 5.

2 Series VSJ Based on Antagonistic Principle

In this section, combining the advantages of the series VSJ and the antagonistic VSJ, a series VSJ based on antagonistic principle is proposed.

The antagonistic variable stiffness can be classified into two types: unidirectional and bidirectional. In the simple unidirectional antagonistic design, each motor is connected to the elastic transmission which can be only driven in one direction. This means that the maximum joint torque is decided by the torque capability of one motor although two motors actuate the link. In the bidirectional antagonistic design, the elastic transmissions are used to provide a bidirectional coupling between the motors and the link. Both motors can push and pull on the joint, thus the load capacity of the joint increases to the sum of torques which can be provided by the two motors. This paper only employs the bidirectional antagonistic design. In Fig. 1a, the schematic of the bidirectional antagonistic VSJ is presented.

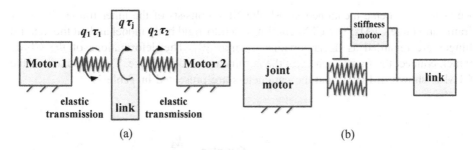

Fig. 1. Schematic of the bidirectional antagonistic VSJ

As shown in Fig. 1a, q_1 and q_2 are the angular positions of motor 1 and motor 2 respectively, and q_j is the link angular position. Assuming that the elastic transmissions have the same torque function with regard to angular displacement $M(\cdot)$, namely $\tau_{\{1,2\}} = M(\varphi_{\{1,2\}}) = M(q_{\{1,2\}} \pm q_j)$, the total torque τ_j applied to the joint is:

$$\tau_j = \tau_1 - \tau_2 \tag{1}$$

The joint stiffness k_j can be calculated as:

$$k_j = \frac{\partial \tau_1}{\partial q_j} - \frac{\partial \tau_2}{\partial q_j} \tag{2}$$

Equation 2 indicates that if the linear elastic transmissions ($M(\varphi_{\{1,2\}}) = k_e(\varphi_{\{1,2\}})$) are used in the antagonistic VSJ, the joint stiffness is constant ($k_j = 2k_e$). Therefore, in order to achieve a VSJ, it is necessary to use NETs.

The working principle of the bidirectional antagonistic VSJ can be understood as: the identical directional motion of the motors drives the link and the opposite directional motion of the motors adjust the stiffness of the joint. Considering series configuration has more compact structure, the antagonistic VSJ can be converted into the series one equivalently and the converted schematic is shown in Fig. 1b. As shown in Fig. 1b, two elastic transmissions can be regarded as an entirety. The joint motor drives the link and the stiffness motor adjusts the stiffness of the joint by changing the stiffness of the elastic transmissions entirely.

3   NET Design

In this section, the NET is designed as a cubic torsion spring. The main novelty of the NET is the kinematic structure design. Compared with the existing nonlinear torsion spring, this design method is simple and the mechanical structure is compact.

3.1   Geometry of the CFM

Refer to the scheme mentioned in [8], the NET consists of the inner frame, the outer frame and three CFMs. The CFM employs pseudo-rigid body model and all the flexural hinges are regarded as linear torsional springs [9]. The deformations of the CFMs which connect the frames generates the nonlinear reaction torque. The schematic of the CFM is shown in Fig. 2, and the parameters are presented in Table 1.

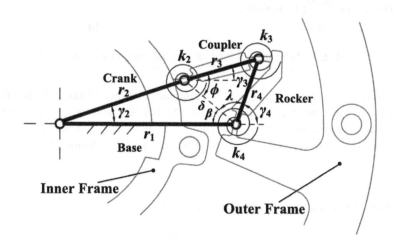

Fig. 2. Schematic of the CFM

Given the lengths of the four bars, the geometry of the CFM can be described by γ_2.

$$\delta = \sqrt{r_1^2 + r_2^2 - 2r_1 r_2 \cos \gamma_2} \tag{3}$$

$$\beta = \arccos \frac{r_1^2 + \delta^2 - r_2^2}{2r_1 \delta} \tag{4}$$

$$\lambda = \arccos \frac{r_4^2 + \delta^2 - r_3^2}{2r_4 \delta} \tag{5}$$

$$\gamma_3 = \phi - \beta \tag{6}$$

$$\gamma_4 = \pi - \lambda - \beta \tag{7}$$

Table 1. Symbols used for schematic description of the CFM

Symbol	Description	Unit
γ_{20}	Initial angle between base and crank without loading	rad
γ_2	Angle between base and crank	rad
γ_3	Angle between base and coupler	rad
γ_4	Angle between extension line of base and rocker	rad
β	Angle between base and connecting line of hinge 2 and hinge 3	rad
λ	Angle between rocker and connecting line of hinge 2 and hinge 3	rad
ϕ	Angle between coupler and connecting line of hinge 2 and hinge 3	rad
r_1, r_2, r_3, r_4	Base length, Crank length, Coupler length, Rocker length	mm
δ	Length between hinge 2 and hinge 3	mm
k_2, k_3, k_4	Hinge 2 stiffness, Hinge 3 stiffness, Hinge 4 stiffness	Nm/rad

The angular displacements of the hinges with respect to the unloaded configuration can be defined as:

$$\psi_1 = \gamma_2 - \gamma_{20} \tag{8}$$

$$\psi_2 = \gamma_2 - \gamma_{20} - \gamma_3 + \gamma_{30} \tag{9}$$

$$\psi_3 = \gamma_4 - \gamma_{40} - \gamma_3 + \gamma_{30} \tag{10}$$

$$\psi_4 = \gamma_4 - \gamma_{40} \tag{11}$$

where γ_{30} and γ_{40} can be calculated by posing $\gamma_2 = \gamma_{20}$ in Eqs. 3–7. Therefore, the torques of the hinges can be calculated as:

$$T_2 = k_2\psi_2 \tag{12}$$

$$T_3 = k_3\psi_3 \tag{13}$$

$$T_4 = k_4\psi_4 \tag{14}$$

By applying the principle of the virtual works, T_1 can be calculated as:

$$T_1 = T_2 - (T_2 + T_3)h_{32} + (T_3 + T_4)h_{42} \tag{11}$$

where

$$h_{32} = \frac{r_2 \sin(\gamma_4 - \gamma_2)}{r_3 \sin(\gamma_3 - \gamma_4)} \tag{12}$$

$$h_{42} = \frac{r_2 \sin(\gamma_3 - \gamma_2)}{r_4 \sin(\gamma_3 - \gamma_4)} \tag{13}$$

Equation 11 gives the relationship between T_1 and ψ_1. Figure 2 indicates ψ_1 is equal to the angular displacement between the inner frame and the outer frame.

3.2   Optimal Design of the NET

The parameters of the NET can be obtained by means of an optimization process. Defining the desired torque function with regard to angular displacement φ:

$$\tau_d = 428.26\varphi^3 + 132.21\varphi^2 + 47.18\varphi \tag{14}$$

Recalling that ψ_1 is equal to φ and the NET has three CFMs, the desired torque function of each CFM can be defined as:

$$T_{1d}(\psi_1) = \frac{\tau_d}{3} = 142.75\psi_1^3 + 44.07\psi_1^2 + 15.73\psi_1 \tag{15}$$

Defining the target function of the optimization process:

$$J = \int_{\psi_{1\min}}^{\psi_{1\max}} [T_1(\psi_1) - T_{1d}(\psi_1)]^2 d\psi_1 \tag{16}$$

where $T_1(\psi_1)$ can be expressed by means of Eq. 3–13. γ_{20} is posing $\pi/6$ for simplifying optimization process and assuming the maximum passive displacement $\varphi_{\max} = 20°$. Defining $\Lambda = [r_1, r_2, r_3, r_4, k_2, k_3, k_4]$, the optimization process is searching for the optimal value of Λ which minimizes the target function:

$$\min_\Lambda J : \Lambda_{\min} < \Lambda < \Lambda_{\max} \tag{17}$$

where $\Lambda_{\min}$ and $\Lambda_{\max}$ are suitable bounds of Λ to avoid unreasonable solutions. The optimization process can be implemented in MATLAB by means of the *lsqnonlin* algorithm which can solve nonlinear least-squares problems. The optimal parameters of the CFM are listed in Table 2.

Table 2. Optimal parameters of the CFM

Parameter	Value	Unit
r_1	20	mm
r_2	15.4	mm
r_3	8.5	mm
r_4	8	mm
k_2	1.3	Nm/rad
k_3	1.8	Nm/rad
k_4	2	Nm/rad

The torque of the optimized NET is plotted together with the desired torque $\tau_d(\varphi)$ in Fig. 3. As shown in Fig. 3, the optimized torque curve is generally coincident with the desired torque curve. The influence on the joint torque of the VSJ brought by the discrepancy between the optimized torque of the NET and the desired torque will be discussed in Sect. 4.

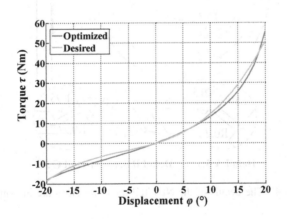

Fig. 3. Torque curve of the optimized NET

3.3 Model Design of the NET

Assuming the rotational center of the hinge locates in the middle point of the beam, the stiffness of the hinge can be calculated by employing the simple beam model:

$$I_i = \frac{h_i^3 b}{12} \tag{18}$$

$$k_i = \frac{EI_i}{l_i} \tag{19}$$

$$l_i \geq 5h_i \tag{20}$$

where E is the Young's modulus of the material, b is the thickness of the NET and I_i is the moment of inertia of the hinge. h_i and l_i are the height and the length of the i-th hinge respectively. The thickness of the NET is assumed constant ($b = 5$ mm) and choosing Aluminum 7075-T6 whose Young's modulus is 71.7 GPa as the material, then the details of the NET is shown in Fig. 4.

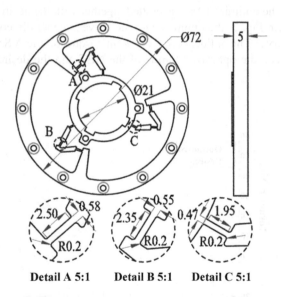

Detail A 5:1   **Detail B 5:1**   **Detail C 5:1**

Fig. 4. The details of the NET (all the dimensions are in [mm])

4   VSJ Design and Simulation Results

The intention to apply the designed VSJ in elbow yields some requirements:

- To be extremely compact and highly integrated
- To imitate the performance of a human elbow, reasonable maximum torque appears to be about 60 Nm and the range of stiffness appears to be as wide as possible
- To keep friction as low as possible

In order to satisfy the requirements above, the designed VSJ employs the series configuration mentioned in Sect. 2. The CAD assembly of the VSJ which is similar with the FSJ of DLR is shown in Fig. 5a.

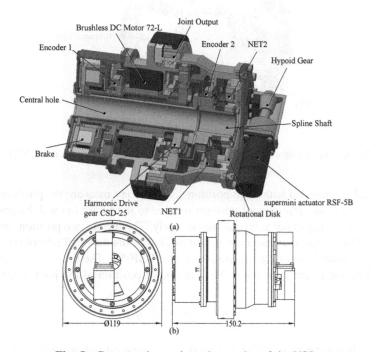

Fig. 5. Cross section and maximum size of the VSJ.

The VSJ adopts the design of central hole for wiring and employs the brushless DC motor 72-L with brake developed by Harbin Institute of Technology. The spline shaft is fixed to the Harmonic Drive gear CSD-25 (ratio $n_{main} = 160{:}1$). A mirror pair of NETs lay concentric and the inner frames of them are driven by spline shaft. The outer frame of one NET is fixed to the joint output by screw and the other NET's outer frame is attached to the rotational disk by spline. The joint output, the rotational disk installed with hypoid gear (ratio $n_h = 45{:}1$) and the supermini actuator RSF-5B constitute the stiffness adjusting setup. The supermini actuator RSF-5B integrated a gearbox (ratio $n_r = 100{:}1$), a brake and an encoder is provided by Harmonic Drive. The two NETs are shown in Fig. 6 and they are regarded as an entirety. Stiffness adjusting is implemented by rotating the outer frames of NETs and the whole elastic transmission has a certain stiffness by rotating the outer frames of the NETs. The sizes of two NETs' outer frames are different because of the restriction of the VSJ structure, but they have the same torque function with regard to angular displacement between the inner frame and the outer frame.

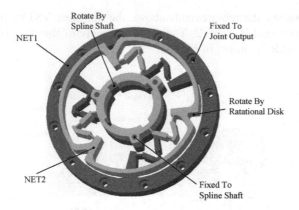

Fig. 6. The whole elastic transmission constituted by a mirror pair of NETs

The VSJ is equipped with three position sensors, they measure the position of joint motor axis (q_{M1}), the joint output position relative to joint base (q) and the position of the stiffness adjusting motor axis (q_{M2}) respectively. The former two position sensors are 18-bit absolute magnetic encoder developed by Harbin Institute of Technology. The last sensor is the attachment of the supermini actuator RSF-5B. The three sensors provide a full identification of the joint state. The gear output position of the joint motor is

$$\theta_1 = n_{main}q_{M1} = \frac{1}{160}q_{M1} \tag{21}$$

With n_r being the ratio of RSF-5B gearbox, the gear output position of the stiffness adjusting motor is

$$\theta_2 = n_r n_h q_{M2} = \frac{1}{4500}q_{M2} \tag{22}$$

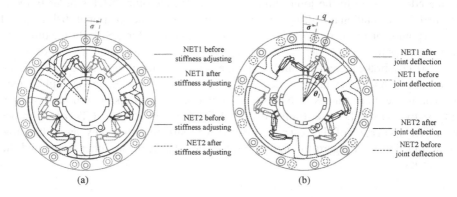

(a)   (b)

Fig. 7. Deformations of the CETs

When the stiffness adjusting setup works, the outer frames of the NETs rotate the same stiffness adjusting angle σ contrarily because the two NETs have the same torque characteristic. As shown in Fig. 7a, the stiffness adjusting angle σ is given as:

$$\sigma = \frac{1}{2}\theta_2 \tag{23}$$

As shown in Fig. 7b, the passive joint deflection φ_j can be expressed as:

$$\varphi_j = q - \sigma - \theta_1 = q - \frac{1}{2}\theta_2 - \theta_1 \tag{24}$$

Therefore, the total displacement of the NETs is given as:

$$\varphi_{\{1,2\}} = \sigma \pm \varphi_j \tag{25}$$

According to Eq. 1, the joint output torque τ_j can be calculated as:

$$\tau_j = \tau_1 - \tau_2 = M(\sigma + \varphi_j) - M(\sigma - \varphi_j) \tag{26}$$

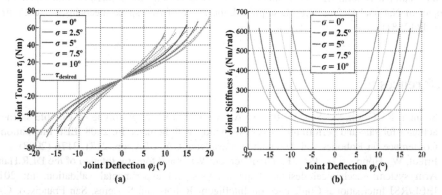

Fig. 8. Torque and stiffness curves of the optimized VSJ with different stiffness adjusting angle

The VSJ equipped with the desired NETs and the optimized NETs designed in Sect. 3.3 are called desired VSJ and designed VSJ respectively. Figure 8a shows the joint torques of the designed VSJ for different stiffness adjusting angles of $\sigma = [0, 2.5, 5, 7.5, 10]°$ and the dashed lines represent correspondingly the desired joint torques. It can be seen that the torque curves of the designed VSJ is moderate at small joint deflection but steep at large joint deflection. This is because that the optimized torque curve of the NET is below at the small angular displacement but above at the large angular displacement as shown in Fig. 3. The joint stiffness curves of the designed VSJ are shown in Fig. 8b.

Although the range of the joint stiffness (111.30 Nm/rad–660.18 Nm/rad) is narrower than the range of the FSJ, the maximum torque of the designed VSJ (73.47 Nm) is larger. Comparing to the FSJ based on the compression spring and the cam-roller mechanism, the use of the NETs not only makes the VSJ manufactured and assembled easily but also keeps the friction low.

5   Conclusions

The design of the series VSJ based on antagonistic principle is presented. The NET employing the CFM results in not only convenience to manufacture and assembly but also low friction compared to the FSJ based on the compression spring and the cam-roller mechanism. The VSJ is fully integrated with two motors, three position sensors and a mirror pair of the NETs. The central hole of the VSJ is convenient for wiring. The designed VSJ has a highly compact structure and can be easily served as a module to build up robot arm.

Future works will focus on the prototype manufacture of the NET and the VSJ together with experimental verification of them.

Acknowledgements. This work is supported by the Foundation for Innovative Research Groups of the National Natural Science Foundation of China (Grant No. 51521003) and Open Fund Project of National Defense Key Discipline Laboratory of Aerospace mechanism and Control (No. 1112881).

References

1. Schiavi, R., Grioli, G., Sen, S., Bicchi, A.: VSA-II: a novel prototype of variable stiffness actuator for safe and performing robots interacting with humans. In: 2008 IEEE International Conference on Robotics and Automation, Pasadena, CA, USA, pp. 2171–2176 (2008)
2. Friedl, W., Höppner, H., Petit, F., Hirzinger, G.: Wrist and forearm rotation of the DLR Hand Arm system: mechanical design, shape analysis and experimental validation. In: 2011 IEEE/RSJ International Conference on Intelligent Robots and Systems, San Francisco, CA, USA, pp. 1836–1842 (2011)
3. Wolf, S., Hirzinger, G.: A new variable stiffness design: matching requirements of the next robot generation. In: 2008 IEEE International Conference on Robotics and Automation, Pasadena, CA, USA, pp. 1741–1746 (2008)
4. Wolf, S., Eiberger, O., Hirzinger, G.: The DLR FSJ: energy based design of a variable stiffness joint. In: 2011 IEEE International Conference on Robotics and Automation, Shanghai, China, pp. 5082–5089 (2011)
5. Jafari, A., Tsagarakis, N.G., Caldwell, D.G.: AwAS-II: a new actuator with adjustable stiffness based on the novel principle of adaptable pivot point and variable lever ratio. In: 2011 IEEE International Conference on Robotics and Automation, Shanghai, China, pp. 4638–4643 (2011)
6. Tsagarakis, N.G., Sardellitti, I., Caldwell, D.G.: A new variable stiffness actuator (CompAct-VSA): design and modelling. In: 2011 IEEE/RSJ International Conference on Intelligent Robots and Systems, San Francisco, CA, USA, pp. 378–383 (2011)

7. Groothuis, S.S., Rusticelli, G., Zucchelli, A., Stramigioli, S., Carloni, R.: The variable stiffness actuator vsaUT-II: mechanical design, modeling, and identification. IEEE/ASME Trans. Mech. **19**(2), 589–597 (2014)
8. Palli, G., Melchiorri, C., Berselli, G., Vassura, G.: Design and modeling of variable stiffness joints based on compliant flexures. Montreal, Quebec, Canada, pp. 1069–1078 (2010)
9. Lobontiu, N.: Corner-filleted flexure hinges. J. Mech. Des. Trans. ASME **123**, 346–352 (2001)

Two-Degree-of-Freedom Mechanisms Design Based on Parasitic Motion Maximization

Zhenyang Zhuo, Yunjiang Lou$^{(\boxtimes)}$, Bin Liao, and Mingliang Wang

Shenzhen Engineering Lab for Intelligent Cordless Ultrasonic Imaging Technology
in Medicine, School of Mechatronics Engineering and Automation,
Harbin Institute of Technology, Shenzhen, HIT Campus,
University Town of Shenzhen, Xili, Nanshan, Shenzhen, China
louyj@hit.edu.cn

Abstract. Parasitic motion is widely considered as an important generic property of mechanisms, which is always believed as a drawback of lower mobility parallel mechanisms. Many researchers have tried to remove or reduce the parasitic motion by optimization or designing new structural layouts. However, in some applications, the parasitic motion can enable lower degree-of-freedom (DoF) mechanisms to complete a higher DoF motion task. In this paper, a new method is proposed to utilize the parasitic motion. By the example of a five-bar mechanism and a two-link mechanism, the optimization method is used to maximize the parasitic motion so that the mechanisms can manipulate objects with a large orientation range. This optimization based mechanism design concept will lead to lower cost, lower complexity of kinematics and control.

Keywords: Parasitic motion · Optimization · Mechanism design

1 Introduction

Parasitic motion is a kind of movement in a non-independent direction. The motions in unspecified motion coordinates were first defined as the parasitic motion by Carretero *et al.* [1,2]. The parasitic motion is a common phenomenon in mechanisms and a generic property of lower mobility parallel mechanisms, *e.g.*, the 3-RPS mechanism [3], the omni wrist [4] and the 3-CUP parallel mechanism [5]. Researchers generally regard the parasitic motion as a drawback of lower mobility parallel mechanisms, which will bring about more difficulties in control. After years of research, there have been many methods used to remove or reduce the parasitic motion. Carretero *et al.* used kinematic optimization to minimize the parasitic motion in order to obtain the pure motion in 2-DoF rotation and 1-DoF translation. Li *et al.* [6,7] analyzed a family of [PP]S parallel mechanisms and took the 3-PRS parallel mechanism as an example to reveal the relationship between structural parameters and parasitic motion, and then showed the necessary structural condition for a 3-PRS parallel mechanism without parasitic motion. Liu *et al.* [8] simplified the descriptions of parallel mechanisms to reduce the impact of parasitic motion on the kinematic model.

© Springer International Publishing AG 2017
Y. Huang et al. (Eds.): ICIRA 2017, Part II, LNAI 10463, pp. 250–260, 2017.
DOI: 10.1007/978-3-319-65292-4_22

Wu *et al.* [9] studied the quotient kinematics machines and proposed that two different modules make up mechanism with higher DoF, which considered as a new way of dealing with the parasitic motion.

Though parasitic motion is widely regarded unwanted, it is not always disadvantageous in any circumstances. In some applications, parasitic motion can be utilized to make a lower-DoF mechanism complete some operation that require higher DoF. Five-bar mechanisms are the simplest parallel robot. Because the structure is simple and its control is easy, five-bar mechanism has been widely used in industry. In the past few decades, many researchers have analyzed the five-bar mechanism, including workspace [10,11] and stiffness [12], kinematics and singularity analysis [13] and optimal design [14]. In packaging industry, a general task is to pick disordered objects from a conveyer and place them in a box with a specified orientation. Under normal circumstances, the planar five-bar mechanism cannot complete this type of tasks because it is a 2-DoF mechanism. However, if the parasitic motion in rotation can be maximized by optimization, a 2-DoF five-bar parallel manipulator would be able to accomplish the task in a given range as the V3 robot [15] to a certain degree. Based on the design, adding an air cylinder to provide translation in Z-direction, the five-bar mechanism can accomplish the general 4-DoF pick-and-place task in many applications. It is generally considered that the parasitic motion is property of parallel mechanisms. However, the parasitic motion also exists in the lower-DoF serial mechanism. Therefore, the optimization method for the parasitic motion of the parallel mechanism can also be used in the serial mechanism, as a result of which the mechanism can also accomplish some tasks that require higher-DoF mechanism. In this paper, the planar five-bar parallel mechanism and the two-link serial mechanism are taken as examples to show an optimization based new way in utilizing parasitic motion. This implementation leads to lower cost, simpler kinematic structure and control of robots.

The organization of this paper is as follows. In Sect. 2, we analyze the five-bar mechanism to show the existence of the parasitic motion. Then an optimization method is used to maximize the parasitic motion to achieve our target. In Sect. 3, we use the method to complete the analysis and optimization of the two-link mechanism. In Sect. 4, a conclusion is drawn.

2   Parasitic Motion Maximization for a Five-Bar Mechanism

2.1   Mechanism Configuration and Parasitic Motion

A 2-DoF planar five-bar mechanism is the simplest parallel mechanism in structure. Planar five-bar mechanism is also rich in configuration. In this paper, the mechanism being analyzed is a 5R planar five-bar mechanism as shown in Fig. 1, where θ_1 and θ_2 measured from X-direction in counterclockwise are input angles of two revolute joints in point A and E, and ϕ_1 and ϕ_2 are output orientation of link BC and CD. The coordinates of the points in Fig. 1 are expressed as $[x_i \ y_i]^T$, where i=A, B, C, D, E.

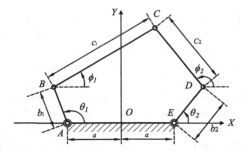

Fig. 1. Schematic representation of a five-bar mechanism

A complete description of the position and orientation of the output bar of five-bar mechanism can be $[x_c \; y_c \; \phi_1]^T$. According to Fig. 2, the orientation angle ϕ_1 changes with X and Y coordinates, where $a = 0.25\,\text{m}$, $b_1 = 0.5\,\text{m}$, $b_2 = 0.5\,\text{m}$, $c_1 = 0.5\,\text{m}$, $c_2 = 0.5\,\text{m}$. However, since the five-bar mechanism is a 2-DoF mechanism, it implies that only two variables can be specified independently, and the third one is dependent on the two variables, which is called parasitic motion.

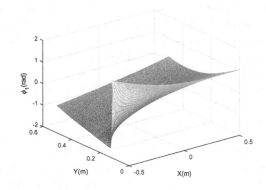

Fig. 2. Parasitic motion of a five-bar mechanism

2.2   Parasitic Motion Maximization

In this subsection, we show a new way to deal with the parasitic motion of five-bar mechanism. In those three motion coordinates, the orientation coordinate is treated as parasitic motion. For the application, it requires to maximize the parasitic motion to adjust the orientation of the target objects. More details are as follows.

Workspace and Working Process. It is not only positions but also orientations that needs controlling in planar pick and place applications using five-bar mechanisms. Figure 3 shows the working process.

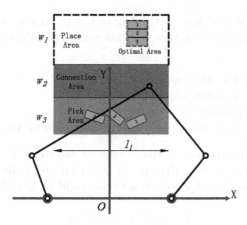

Fig. 3. Pick-and-place operation and workspace of the five-bar mechanism

Objects come along the X-direction with Y-coordinate y_o and orientation ϕ_o in the Pick Area. They are expected to be put in the Optimal Area of the Place Area with a certain coordinate $[x_{pl}\ y_{pl}]^T$ and the fixed desired orientation ϕ_d. Considering the connection of the conveyor belts in two directions and the actual size of the objects, we set up the Connection Area between the Place Area and Pick Area.

At first, we give a definition of the workspace of the five-bar mechanism. As shown in the Fig. 3, the whole workspace consists of three parts described above, which are symbolled as W_{pi}, W_{co} and W_{pl}, respectively. Obviously, the workspace of mechanism $W_{wo} = W_{pi} + W_{co} + W_{pl}$.

Now, we show how the mechanism works and determine where we pick the object and where we place it. In practical applications, given a visual system, coordinates and orientation of the objects are measured. Firstly, we need to determine where the five-bar mechanism picks the object up. And after the placement, the orientation of the objects will be adjusted with a certain angle. This angle is defined as $\phi_\Delta = \phi_d - \phi_o$. Then we find out where to do the picking operation. The picking orientation of the end-effector is

$$\phi_{pi} = \phi_{pl} - \phi_\Delta = \phi_{pl} + \phi_o - \phi_d \qquad (1)$$

Its picking Y-coordinate is $y_{pi} = y_o$. ϕ_{pl} is the placing orientation coordinate of end-effector in point $[x_{pl}\ y_{pl}]^T$. Since ϕ_{pl} is known, we can calculate ϕ_{pi} by Eq. (1). That means we have already obtained the $[y_{pi}\ \phi_{pi}]$, which are two coordinate parameters of end-effector in the point where the object is picked up. Further more we can easily calculate the x_{pi} by the inverse kinematics. Here we can find the object is picked up at point $[x_{pi}\ y_{pi}]^T$ with orientation ϕ_o, if we place the object at the point $[x_{pl}\ y_{pl}]^T$, the object will have the orientation ϕ_d.

Relevant Design Variables. According to the arrangement of five-bar mechanism and the requirement of task, the potential optimization parameters are shown below:

(1) a: the half length of the base bar;
(2) $b_i, i = 1, 2$: the actuated bar lengths;
(3) $c_i, i = 1, 2$: the passive bar lengths;
(4) $x_i, y_i, i = 1, 2, 3$: the coordinate of the objects placed in the Optimal Area.

The parameters with regard to the length of bars can be defined as $X = [a \; b_1 \; b_2 \; c_1 \; c_2]$. In this paper, the optimization target is the orientation angle of the end-effector, so the objective functions would be based on the measure:

$$\eta \equiv \eta(X, x_c, y_c) = \phi_1 \tag{2}$$

where $[x_c \; y_c] \in W_{re}$ is the point of reachable workspace W_{re}.

Objective Function and Constraints. Through the previous analysis, it is possible to complete the pick-and-place operation by optimal design. Now, we show a new optimization to take advantage of parasitic motion. Since the picking operation is done in the Pick Area, the placement operation is done in the Place Area, and we hope that the difference in the orientation angle of the two areas to be as much as possible, which means that error angle adjusted by the mechanism is greater. The objective function can be defined as follows:

$$
\begin{aligned}
f = \min_{x_C} \{ \max_{y_C} \; [\eta(X, x_c, y_c)] \} \\
- \max\{\eta(X, x_1, y_1), \eta(X, x_2, y_2), \eta(X, x_3, y_3)\} \\
+ \min\{\eta(X, x_1, y_1), \eta(X, x_2, y_2), \eta(X, x_3, y_3)\} \\
- \max_{x_C} \{ \min_{y_C} \; [\eta(X, x_c, y_c)] \}
\end{aligned}
\tag{3}
$$

where $[x_c \; y_c] \in W_{pi}$ and $[x_i, y_i], i = 1, 2, 3$ is in the Optimal Area of the Place Area. Obviously, the value of the function ought to be as large as possible.

Next we define some constraints about mechanism performance and optimization goal. In this paper, we use the inverse condition number of the Jacobian matrix to evaluate the flexibility of the mechanism, which is defined as:

$$\kappa = \frac{\sigma_{min}(J)}{\sigma_{max}(J)} \tag{4}$$

The inverse condition number, which is bounded by $[0, 1]$, is widely used as a evaluation of the distance to the singular configuration. $\kappa = 0$ indicates that the mechanism is singular and $\kappa = 1$ indicates that the mechanism is isotropic. In order to guarantee that the mechanism have a satisfactory performance, we give the constraint

$$\kappa_{min} \geq 0.1 \tag{5}$$

for all $[x_c \ y_c] \in W_{re}$. In order to narrow the range of the search when optimization, we limit the size of the mechanism:

$$\begin{cases} a + b_1 + c_1 \leq 1 \\ a + b_2 + c_2 \leq 1 \\ [0.001 \ 0.001 \ 0.001 \ 0.001 \ 0.001]^T \leq X \leq [1.000 \ 1.000 \ 1.000 \ 1.000 \ 1.000]^T \end{cases}$$

$$(6)$$

Considering the size of the motor, we give a constraint of the distance between the workspace and motor:

$$\min(y_C) \geq 0.2 \tag{7}$$

Because the initial orientation angles of the objects from the conveyor belt may be positive or negative, in order to ensure that the positive and negative adjustment of the angle is close, we define the following constraint:

$$\begin{aligned} \Delta = |(\min_{x_C}\{\max_{y_C} [\eta(X, x_c, y_c)]\} \\ - \max\{\eta(X, x_1, y_1), \eta(X, x_2, y_2), \eta(X, x_3, y_3)\}) \\ - (\min\{\eta(X, x_1, y_1), \eta(X, x_2, y_2), \eta(X, x_3, y_3)\} \\ - \max_{x_C}\{\min_{y_C} [\eta(X, x_c, y_c)]\})| \leq 2° \end{aligned} \tag{8}$$

Through the above analysis, we can obtain the following optimization problem:

$$\max : f(X)$$

$$\begin{aligned} s.t. \quad & \kappa_{min} \geq 0.1, \\ & a + b_1 + c_1 \leq 1, \\ & a + b_2 + c_2 \leq 1, \\ & [0.001 \ 0.001 \ 0.001 \ 0.001 \ 0.001]^T \leq X \\ & \leq [1.000 \ 1.000 \ 1.000 \ 1.000 \ 1.000]^T, \\ & \min(y_C) \geq 0.2, \\ & \Delta \leq 2°. \end{aligned} \tag{9}$$

Optimization and Results Analysis. For better modeling and optimization, we define the shape and size of the workspace. The workspace W_{wo} is defined as a rectangle whose side lengths are $l_1 = 0.250$ m, $w_1 = 0.160$ m, $w_2 = 0.100$ m and $w_3 = 0.120$ m. And the distance of the three objects in the Optimal Area is defined as $d = 0.025$ m, which means $y_2 - y_1 = y_3 - y_2 = 0.025$ m. For the optimization problem, it is difficult to obtain the analytical expression for the objective function. Worse still, the position of workspace and Optimal Area are all uncertain. So we need to choose the optimal position in the process of the optimization. Here, we use heuristic search algorithm to tackle it. In the computer science field, genetic algorithm (GA) is a method of searching the optimal solution by simulating the natural evolutionary process. Genetic algorithm is characterized by good convergence, less computation time and high robustness. In this paper, we use GA as the optimization tool. In order to prevent the local optimum solution, we increase the mutation rate 0.015 per two hundred generations.

The optimization results are shown in Fig. 4. Figure 5 shows contour plot of parasitic motion of end-effector in the Pick Area. It can be seen from the Fig. 5 that the parasitic motion along the X axis in the Pick Area changes significantly.

The optimal parameters are $X = [0.005\ 0.725\ 0.248\ 0.220\ 0.741]$ and $f = 1.125$ rad. The positive adjustment of the angle is 0.547 rad (31.33°), the negative adjustment of the angle is 0.578 rad (33.15°) at least. It means that the mechanism can rearrange a object with nearly ±30° angle error at least. If the objects always come through a guide device, the results can satisfy the vast majority of pick-and-place tasks.

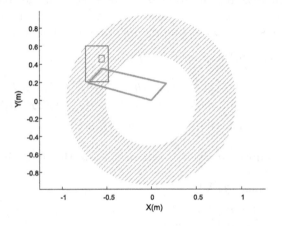

Fig. 4. Optimal workspace of the five-bar mechanism

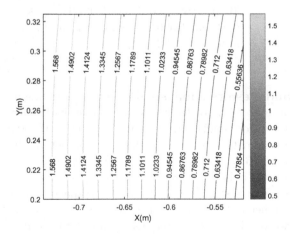

Fig. 5. Parasitic motion in W_{pi} of the five-bar mechanism

3   Parasitic Motion Maximization for a Two-Link Mechanism

3.1   Mechanism Configuration and Parasitic Motion

Similar to the planar five-bar mechanism, a 2-DoF planar two-link mechanism is the simplest serial mechanism in structure. Figure 6 shows the coordinate system, OD and DC are two active links. θ_1 measured from the X-direction in counterclockwise is one input angles of the revolute joint in point O. And θ_2 measured from the OD in counterclockwise is another input angles of the revolute joint in point D. ϕ_1 measured from the X-direction in counterclockwise is the output orientation of the link DC. The coordinates of the points in Fig. 6 are expressed as $[x_i \ y_i]^T$, where $i = $ C, D.

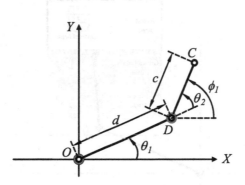

Fig. 6. Schematic representation of a two-link mechanism

A complete description of the position and orientation of the output link of two-link mechanism can be $[x_c \ y_c \ \phi_1]^T$. According to Fig. 7, the orientation angle ϕ_1 also changes with the X and Y coordinates, where $c = 0.5$, $d = 0.5$. Similar to the five-bar mechanism, two-link mechanism is a 2-DoF mechanism, which implies that only two variables can be specified independently, and the third one is parasitic motion.

3.2   Parasitic Motion Maximization and Results Analysis

Since the objects and the task is similar to the five-bar mechanism, we omit the similar key points. Figure 8 shows the process of planar pick-and-place application. The workspace and working process are the same as that of the five-bar mechanism.

According to the arrangement of two-link mechanism and the requirements for task, the potential optimization parameters consist of lengths of two links and the coordinates of the objects placed in the Optimal Area. The parameters

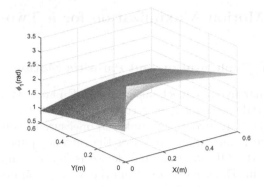

Fig. 7. Parasitic motion of a two-link mechanism

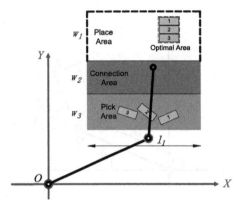

Fig. 8. Pick-and-place operation and workspace of the two-link mechanism

about the length of bars can be defined as $X = [c\ d]$. And the objective functions are based on the measure:

$$\eta \equiv \eta(X, x_c, y_c) = \phi_1 \tag{10}$$

where $[x_c\ y_c] \in W_{re}$ is the point of reachable workspace W_{re}.

The objective function and constraints are similar to the five-bar mechanism. So we can obtain the following optimization problem for the two-link mechanism:

$$\max : f(X)$$

$$s.t. \quad \kappa_{min} \geq 0.1,$$
$$c + d \leq 1,$$
$$[0.001\ 0.001]^T \leq X \leq [1.000\ 1.000]^T, \tag{11}$$
$$\min(y_C) \geq 0.2,$$
$$\Delta \leq 2°.$$

And we also use the GA as the optimization tool. The optimization results are shown in Fig. 9. Figure 10 shows contour plot of parasitic motion of end-effector

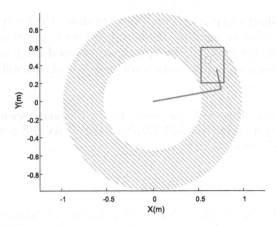

Fig. 9. Optimal workspace of the two-link mechanism

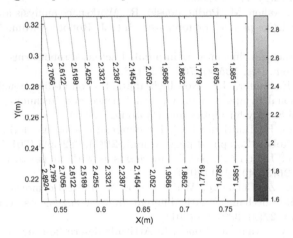

Fig. 10. Parasitic motion in W_{pi} of the two-link mechanism

in the Pick Area. It can be seen from the Fig. 10 that the parasitic motion along the X axis in the Pick Area changes significantly.

The optimal parameters are $X = [0.223\ 0.760]$ and $f = 1.120$ rad. The positive adjustment of the angle is 0.562 rad (32.20°), the negative adjustment of the angle is 0.558 rad (31.94°). It means the mechanism can rearrange a object with nearly ±30° angle error at least. If the objects always come through a guide device, the results can satisfy the vast majority of pick-and-place tasks.

4   Conclusions

It is generally considered that the parasitic motion is a drawback of parallel mechanisms. In the paper, we show that lower mobility serial mechanisms also have the parasitic motion. And we take a five-bar mechanism and a two-link mechanism as examples to propose a new way to show that by utilizing parasitic motion it is achievable to manipulate the objects with an orientation range of approximate ±0.52 rad (±30°). The optimization based mechanism design can deal with the parasitic motion to make some pick-and-place applications cheaper and easier.

Acknowledgements. This work was supported by Shenzhen Science and Technology Program, partially from No. JCYJ20150731105134064 and partially from No. JSGG201506011537 23042.

References

1. Carretero, J., Podhorodeski, R., Nahon, M., Gosselin, C.: Kinematic analysis and optimization of a new three degree-of-freedom spatial parallel manipulator. J. Mech. Des. **122**(1), 17–24 (2000)
2. Carretero, J., Nahon, M., Podhorodeski, R.: Workspace analysis and optimization of a novel 3-DOF parallel manipulator. Int. J. Robot. Autom. **15**(4), 178–188 (2000)
3. Hunt, K.: Structural kinematics of in-parallel-actuated robot-arms. ASME J. Mech. Transm. Autom. Des. **105**(4), 705–712 (1983)
4. Rosheim, M.E., Sauter, G.F.: New high-angulation omni-directional sensor mount. In: International Symposium on Optical Science and Technology, pp. 163–174. International Society for Optics and Photonics (2002)
5. Cuan-Urquizo, E., Rodriguez-Leal, E.: Kinematic analysis of the 3-cup parallel mechanism. Robot. Comput.-Integr. Manuf. **29**(5), 382–395 (2013)
6. Li, Q., Hervé, J.M.: 1T2R parallel mechanisms without parasitic motion. IEEE Trans. Rob. **26**(3), 401–410 (2010)
7. Li, Q., Chen, Z., Chen, Q., Wu, C., Hu, X.: Parasitic motion comparison of 3-PRS parallel mechanism with different limb arrangements. Robot. Comput.-Integr. Manuf. **27**(2), 389–396 (2011)
8. Liu, X., Wu, C., Wang, J., Bonev, I.: Attitude description method of [pp] s type parallel robotic mechanisms. Jixie Gongcheng Xuebao/Chin. J. Mech. Eng. **44**(10), 19–23 (2008)
9. Wu, Y., Wang, H., Li, Z.: Quotient kinematics machines: concept, analysis, and synthesis. J. Mech. Robot. **3**(4), 041004 (2011)
10. Shuanglin, Z., Huijun, Z., Weizhong, G., Yanan, Y.: Study on the flexible workspace of plane closed-loop five-bar mechanism. Chin. J. Mech. Eng. **36**(11), 10–15 (2000)
11. Joubair, A., Slamani, M., Bonev, I.A.: Kinematic calibration of a five-bar planar parallel robot using all working modes. Robot. Comput.-Integr. Manuf. **29**(4), 15–25 (2013)
12. Gao, F., Liu, X., Gruver, W.A.: Performance evaluation of two-degree-of-freedom planar parallel robots. Mech. Mach. Theory **33**(6), 661–668 (1998)
13. Liu, X.J., Wang, J., Pritschow, G.: Kinematics, singularity and workspace of planar 5R symmetrical parallel mechanisms. Mech. Mach. Theory **41**(2), 145–169 (2006)
14. Lou, Y., Zhang, Y., Huang, R., Li, Z.: Integrated structure and control design for a flexible planar manipulator. Intell. Robot. Appl. **7101**, 260–269 (2011)
15. Liao, B., Lou, Y.: Optimal kinematic design of a new 3-DOF planar parallel manipulator for pick-and-place applications. In: 2012 International Conference on Mechatronics and Automation (ICMA), pp. 892–897. IEEE (2012)

Novel Design of a Family of Legged Mobile Lander

Rongfu Lin and Weizhong Guo[✉]

State Key Laboratory of Mechanical Systems and Vibration,
School of Mechanical Engineering, Shanghai Jiao Tong University,
Shanghai 200240, China
{rongfulin, wzguo}@sjtu.edu.cn

Abstract. During extraterrestrial planet exploration programs, autonomous robots are deployed using separate landers. In this paper, a concept of a novel legged robot is introduced which has inbuilt the features of lander and rover, including landing and walking capabilities as well as being deployable, orientation adjusted and terrain adaptable. Firstly, motion characteristics of the novel legged robot mapping its functions are extracted, which can be divided into global and local motion characteristics. Secondly, structures of legs are designed according to the extracted motion characteristics, mainly composed of upper and lower parts. Finally, numerous structures of legged mobile landers are obtained and presented by assembling the same or different structures of legs.

Keywords: Legged mobile lander · Novel design · Parallel mechanism

1 Introduction

Some astronomical bodies, such as Moon or Mars, are of explorative scientific significance to many nations. The lander and the rover play important roles for extraterrestrial planet exploration [1]. At present, there have been different types of landers and rovers, which are launched together in the above mentioned exploratory missions.

As for landers: Surveyor 1 [2] was the first lunar soft-lander in the unmanned Surveyor program of the National Aeronautics and Space Administration (NASA, United States). Luna 16 [3] was the first robotic probe to land on the Moon and return a sample of lunar soil to Earth. As for Apollo Program, the structural subsystem, design requirements, configuration description, design and manufacturing problems, and design verification of the lunar module were discussed in reference [4]. There are some other significant lunar landers, such as EuroMoon 2000 [5] designed by the European space agency, ELENE-B [6] designed by Japan, Altair lunar lander [7], Chang'e 3 [8] and so on. One conceptual design of a lunar lander was proposed in [9]. In addition, some Mars landers, such as Viking 1 lander [10] and Phoenix lander [11] were proposed. As for rovers: Many wheeled rovers were proposed and studied, such as Sojourner rover [12], Spirit and Opportunity rovers of the Mars Rover mission [13], Jade Rabbit rover [14], Shrimp and SOLERO rovers[15, 16], and ExoMars-E rover [17]. Most of the existing rovers with wheels have the limited capability to traverse tough terrains and have problems with the orientation adjustment of the body.

© Springer International Publishing AG 2017
Y. Huang et al. (Eds.): ICIRA 2017, Part II, LNAI 10463, pp. 261–272, 2017.
DOI: 10.1007/978-3-319-65292-4_23

Particularly, some conceptual robots with a wheel-legged mobile system are designed to expand their maneuvering range, enhancing their exploration capability, such as ATHLETE [18, 19] and Spider-bot [20]. However, the functions of lander and rover are still separated, which means: the lander is immovable while the rover cannot be used for landing.

Up to now, in these missions, a lander with a rover are launched together, and the rover is on the upper module. Being immovable after landing, the most important task of the lander is to help the rover step on the surface. Then, the rover accomplishes the exploration work. These devices increase the cost of mission, such as manufacture and launching. This mode for exploratory mission also has some other limitations [21–23]. As is known, during the landing, the engine of lander will blow away rocks and debris, so in order to protect the base station, the landing site has to be a few hundred meters to several kilometers away depending upon the scenario. This will bring some challenges for astronauts and other equipments to transport between the landing site and the base station. Because the lander is immovable, it requires considerable time and energy for transportation between the landing site and the base station. Furthermore, it leads to limit the exploration range of the rover because it has to receive energy or other aids from the lander after extravehicular activity. The rover and astronauts cannot go to a far away place from the landing site. In addition, as the requirement of the exploratory mission gets more complex, novel legged mobile landers can be a solution to the above issues or limitations. The aim of this paper is to investigate the novel design of a legged mobile lander combining capabilities of lander and rover inspired by the configurations of existing landers [2–11] and the experience of walking robot from our laboratory [24].

2   Overall Concept for the Legged Mobile Lander

The schematic of the legged mobile lander is presented in the middle of Fig. 1, which consists of the body and four legs. The structure of each leg can be divided into two parts, the upper part and the lower part. The upper part is a parallel mechanism and the lower part has the terrain adaptability mechanism connected with a foot pad. Furthermore, the upper part has three limbs: the main backbone and the two sightly thinner auxiliary backbones. The energy absorber such as a metal honeycomb can be installed into the limbs of the upper part of the legs and the linkage between the middle platform and the foot pad. During landing, the main bumper backbone will absorb vertical impact shock and the auxiliary bumper backbone will absorb horizontal impact shock.

For facilitating the conceptual idea, the design concept is explained in 3 phases, as shown in Fig. 1. In the beginning phase, Phase 1, integrated motion characteristics [25] mapping multi-functions of the mobile lander are extracted. Considering the complexity of these functions, the corresponding motion characteristics are divided into two categories to design the legs, the *global motion characteristics*, attached to the body to describe its motions and the *local motion characteristics* distributed on the leg describing those of the respective leg. Next, in Phase 2, the structure of the legs for the mobile lander is designed to satisfy the motion characteristics mapping all the mentioned functions. In Phase 3, legged mobile landers are achieved by assembling the same or different legs designed in the first phase.

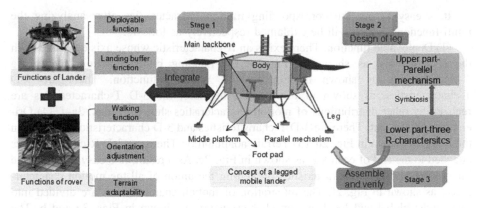

Fig. 1. Concept of constructing a legged mobile lander

3   Motion Characteristics of the Legged Mobile Lander

3.1   Motion Characteristics of the Mobile Lander

In order to represent the motion characteristics intuitively and vividly, some symbols are listed in Table 1. Note that $\tilde{R}$ regards a movable R-characteristic satisfying the theorem of the rotating axis movement in the 2D plane or the 3D space [25]. If there exists a movable R-characteristic, the rigid body rotates around any axis parallel to that of the R-characteristic.

Table 1. Symbols definition and explanation

Element	Symbol	Motion characteristic
	$\tilde{R}$	R-characteristic which is movable[25]
	R	R-characteristic which is not movable
	T	T-characteristic
	2T	2 dimensional(2-D) T-characteristics
	3T	3 dimensional(3-D) T-characteristics

It is easy to get the corresponding motion characteristics after analyzing the multi-functions, which will be explained respectively as follows:

(1) Deployable function: There exists an R-characteristic whose axis coincides with the specific axis, as shown in Fig. 2a. (2) Landing buffer function: there is a T-characteristic, as shown in Fig. 2b. (3) Walking function: There are 3-D T-characteristics, as shown in Fig. 2c. Furthermore, the 3-D T-characteristics are realized by other distributions of motion characteristics shown in Figs. 3a–f. (4) Orientation adjustment: There are 3-D T-characteristics and 3-D characteristics attached on the body, as shown in Fig. 2d. (5) Terrain adaptability: There are 3-D R-characteristics attached to the end of each leg, as shown in Fig. 2e. As a result, it is easy to obtain the integration of motion characteristics by getting the union of all the motion characteristics, as shown in Fig. 2f. The integration of motion characteristics are divided into two kinds: global and local motion characteristics, as shown in Figs. 3a and b. The motion characteristics of the body are regarded as global, determined by all legs while those of each leg are called local motion characteristics.

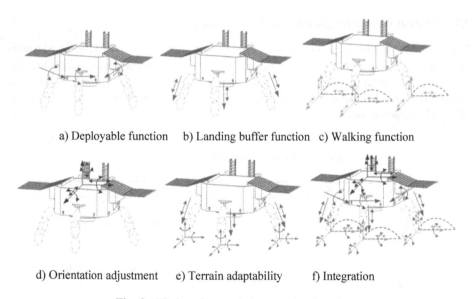

a) Deployable function   b) Landing buffer function   c) Walking function

d) Orientation adjustment   e) Terrain adaptability   f) Integration

Fig. 2. Motion characteristics mapping functions

3.2   Determination of the Motion Characteristics of the Legs

The global motion characteristics are six dimensional which include 3-D T-characteristics and 3-D R-characteristics. The leg of the mobile lander consists of two parts, an upper part and a lower part. The lower part owns the 3-D R-characteristics designed by a spherical joint, as shown in Fig. 4c. The upper part is regarded as a parallel mechanism whose moving platform allows for the rest of the motion characteristics except the 3-D R-characteristics, as shown in Fig. 4d. Among these motion characteristics, the 3-D T-characteristics are realized by other distributions of motion characteristics

shown in Figs. 3a–f. In addition because of the existence of other two motion characteristics (one T-characteristic and one R-characteristic) three kinds of motion characteristics for upper part of the leg are easily obtained and are represented as $(RR)_OT$, RTR
and CT (C stands for cylindrical motion characteristics combining an R-characteristic
and a T-characteristic with the same direction, and $()_O$, meaning that the axes of the
R-characteristics intersect at point O) which are presented in Figs. 3a–c, respectively.

According to the scope of this paper, the mobile lander on the basis of the first kind
of union of motion characteristic $((RR)_OT)$ is designed and implemented. Furthermore
other kinds can be designed by the same method.

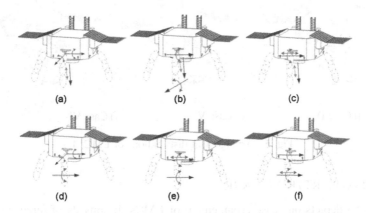

Fig. 3. Motion characteristics mapping walking function

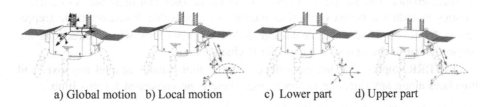

a) Global motion b) Local motion   c) Lower part   d) Upper part

Fig. 4. Motion characteristics of the legged mobile lander

4 Leg Design of the Legged Mobile Lander

4.1 Determination of Motion Characteristics for Limbs of Legs (LMC)

Based on the motion characteristic movement theorem and intersection rules [25] of the
G_F set theory, there exist six cases of arrangements of motion characteristics of limbs
according to the motion characteristic of the end-effector, i.e. $(RR)_OT$, as shown in
Fig. 5a, b, respectively.

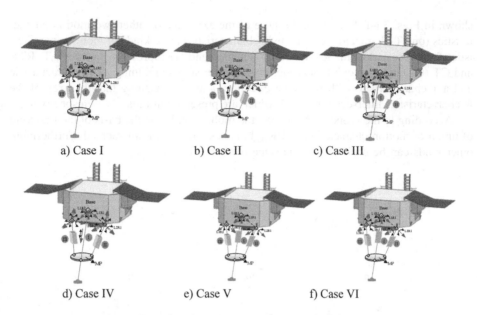

a) Case I   b) Case II   c) Case III

d) Case IV   e) Case V   f) Case VI

Fig. 5. Arrangement distributions of LMC

Case I: $(RR)_O T \cap RTT\tilde{R} \cap TTT\tilde{R}\tilde{R}$

Figure 5a depicts one case arrangement of LMCs. It consists of three limbs with different motion characteristics. The first limb's motion characteristic is $(RR)_O T$, the second one is $RTT\tilde{R}$ and the third one is $TTT\tilde{R}\tilde{R}$. $(RR)_O T$ refers to the axes of two R-characteristics, intersecting at a point O; $RTT\tilde{R}$ denotes that there are four characteristics including a nonmovable R-characteristic, a movable R-characteristic (represented by $\tilde{R}$, as listed in Table 1) and 2-D T-characteristics whose directions are perpendicular to the axis of the movable R-characteristic.

$TTT\tilde{R}\tilde{R}$ implies that the end-effector of the limb rotates around any axis and translates along any direction in space since there is no constraint for this limb.

Case II: $(RR)_O T \cap TTT\tilde{R}\tilde{R} \cap TTT\tilde{R}\tilde{R}$

Figure 5b shows one case arrangement of LMCs. It consists of three limbs with three kinds of motion characteristics. The first kind of motion characteristic is $(RR)_O T$, the second one is $TTT\tilde{R}\tilde{R}$ and the third one is $TTT\tilde{R}\tilde{R}$.

$TTT\tilde{R}\tilde{R}$ denotes that there are 3-D T-characteristics and two movable R-characteristics.

It is noted that: the planes spanned by the two R-characteristics of the first and second limbs are parallel to each other in this case.

Case III: $(RR)_O T \cap (RR)_O TT\tilde{R} \cap TTT\tilde{R}\tilde{R}$

Figure 5c shows one case arrangement of LMCs which consists of three limbs with three kinds of motion characteristics $(RR)_O T, (RR)_O TT\tilde{R}$ and $TTT\tilde{R}\tilde{R}$.

$(RR)_0 T T \tilde{R}$ indicates that there are two nonmovable R-characteristics, an R-characteristic with movable property and two T-characteristics whose directions are perpendicular to the axis of the movable R-characteristic.

It is noted that the axis of $L_1 R_1$ coincides with that of $L_2 R_1$. The axis of $L_1 R_2$ is parallel to that of $L_2 R_2$. The direction of T-characteristic of the first limb is parallel to the plane of the 2-D T-characteristics of limb 2 and the third limb is located at any position between base and moving platform.

Case IV: $(RR)_0 T \cap TTT\tilde{R}\tilde{R} \cap TTT\tilde{R}\tilde{R}$

Figure 5d shows one case arrangement of LMCs, which consists of three limbs with two kinds of motion characteristics. The first kind of motion characteristic is $(RR)_0 T$, the second one is $TTT\tilde{R}\tilde{R}$.

It is noted that: there is no special geometrical condition of arrangement in this case.

Case V: $RTT\tilde{R} \cap RTT\tilde{R} \cap RTT\tilde{R}$

Figure 5e shows one case arrangement of LMCs, which consists of three limbs with only one kind of motion characteristics.

It is noted that the arrangement condition is special. The axis of $L_1 R_1$ coincides with that of $L_2 R_1$. The axis of $L_3 R_1$ is parallel to that of $L_1 R_1$. The axes of $L_3 R_2$, $L_3 R_2$, $L_3 R_2$, is parallel to each other and the plane spanned by 2-D T-characteristics of each limb is be parallel to each other at the same time.

Case VI: $RTT\tilde{R} \cap RTT\tilde{R} \cap (RR)_0 TT\tilde{R}$

Figure 5f shows one case arrangement of LMCs which consists of three limbs with different kinds of motion characteristics.

The geometrical conditions of arrangement of three limbs are the same as that of case V.

4.2 Limb Design for Legs of the Mobile Lander

There are five kinds of motion characteristics of limbs to design the legs of the mobile lander, i.e., $(RR)_0 T$, $RTT\tilde{R}$, $(RR)_0 TT\tilde{R}$, $TTT\tilde{R}\tilde{R}$, $TTT\tilde{R}\tilde{R}$. The structures of limbs corresponding to these five motion characteristics are displayed in this section according to the G_F set theory [25]. Only some schematic diagrams of practical limbs are depicted in this section. (1) $(RR)_0 T$: there are two types of limbs are synthesized: $(RR)_0 P$ and UP. Some simple joints, such as the revolute(R), prismatic(P) joints, and some combined joints such as cylinder(C), spherical(S), universal(U) joints, pure-translation universal joint(U^*) and the parallelogram joint(Pa) are used in this paper. (2) $RTT\tilde{R}$: Numerous configurations with motion characteristic $RTT\tilde{R}$ are enumerated as listed in Table 2. What's more $(RR)_0 TT\tilde{R}$, $TTT\tilde{R}\tilde{R}$, $TTT\tilde{R}\tilde{R}$ are shown in Tables 3, 4 and 5.

Table 2. Configurations for a limb with RTT$\tilde{\text{R}}$ motion characteristic

Joint	Limbs			
R P	RPPR	RPRP	RRPP	RPRR
	RRPR	RRRP	RRRR	
R P Pa	RPaPR	RPaRP	RPPaR	RPRPa
	RRPPa	RRPPa	RPaRR	RRRPa
C R P Pa	CPR	CRP	CRR	CPaR
	CRPa	CRPa		

Table 3. Configurations for a limb with $(\text{RR})_\text{O}$TT$\tilde{\text{R}}$ motion characteristic

Joint	Limbs		
R P	$(\text{RR})_\text{O}$PPR	$(\text{RR})_\text{O}$PRP	$(\text{RR})_\text{O}$RPR
	$(\text{RR})_\text{O}$RPP	$(\text{RR})_\text{O}$PRR	$(\text{RR})_\text{O}$RRP
R P Pa	$(\text{RR})_\text{O}$PaPR	$(\text{RR})_\text{O}$PaRP	$(\text{RR})_\text{O}$PaRR
	$(\text{RR})_\text{O}$PPaR	$(\text{RR})_\text{O}$PRPa	$(\text{RR})_\text{O}$RRPa
U R P Pa	UPPR	UPRP	URPP
	UPRR	URPR	URRP
	URRR	URPaR	UPaRR
U S P R	SPP	SPR	SRP
	SRR	SPaP	SPaR

Table 4. Typical configurations for a limb with TTT$\tilde{\text{R}}\tilde{\text{R}}$ motion characteristic

Joint	Limbs		
P R	PPPRR	PPRPR	RPPPR
	RRPPP	PR(RRR)$_\text{pl}$	PR(RPR)$_\text{pl}$
	PR(PRR)$_\text{pl}$	PR(RRP)$_\text{pl}$	RR(RRR)$_\text{pl}$
	RR(PRR)$_\text{pl}$	RR(RPR)$_\text{pl}$	RR(RRP)$_\text{pl}$
P U* R	PU*RR	U*RPR	RPU*R
R U U	URU	RUU	UUR
C R U	CRU	CUR	URC
	UCR	RCU	RUC
P C U	PCU	PUC	CPU
	CUP	UPC	UCP
P U U	UPU	PUU	UUP

4.3   Structure Design for Leg of the Mobile Lander

The structure of the legs is synthesized by assembling three corresponding limbs (listed in Tables 2, 3, 4 and 5) under the six cases of arrangement conditions (as shown in Fig. 6). These typical cases of configurations of legs are listed in Table 6 corresponding to their distributions.

Table 5. Typical configurations for a limb with $TTT\tilde{R}\tilde{R}\tilde{R}$ motion characteristic

Joint	Limbs			
P R	$PPP(RRR)_O$	$RPP(RRR)_O$	$PRR(RRR)_O$	$RRP(RRR)_O$
R P_R R	$PPP(RP_RR)_O$	$RPP(RP_RR)_O$	$PRR(RP_RR)_O$	$RRP(RP_RR)_O$
Pa P_R R	$PaPaPa(RRR)_O$	$PaPaP(RRR)_O$	$PaPaP(RP_RR)_O$	$PaPaPa(RP_RR)_O$
P U^* R Pa	$PU^*(RRR)_O$	$U^*Pa(RRR)_O$	$UR(RP_RR)_O$	$PU^*(RP_RR)_O$
	$U^*Pa(RP_RR)_O$	$UP(RRR)_O$	$UP(RP_RR)_O$	$UR(RRR)_O$
R U S	RUS	RSU SUR	USR SRU	URS
R Pa S S	RSS	SRS	UPaS	PaUS

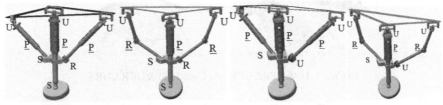

a) Case I: UP&UPR&UPS b) Case I: UP&URR&URS c) Case II: UP&UPU&UPS d) Case II: UP&URU&URS

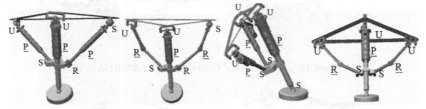

e) Case III: UP&SPR&UPS f) Case III: UP&SRR&URS g) Case IV: 2-UPS&UP   h) Case IV: 2-URS&UP

Fig. 6. Some typical legs with $(RR)_O T$ motion characteristic

Table 6. Typical Configurations of legs with $(RR)_O T$ motion characteristic

Case	Configurations of legs		
Case I	UP&URR&UPS	UP&URR&PSS	UP&UPR&PSS
	UP&URR&URS	UP&UPR&RSS	**UP&UPR&UPS**
	UP&URU&URS	UP&URU&RSS	**UP&UPU&UPS**
Case III	**UP&SPR&UPS**	UP&SPR&PSS	**UP&SRR&URS**
	UP&SRR&UPS	**UP&SRR&PSS**	**UP&SRR&RSS**
Case IV	**2-UPS&UP**	**2-URS&UP**	2-U*PS&UP
	2-PSS&UP	**2-RSS&UP**	2-SRU&UP
Case V	**3-URR**	3-CRR 3-CPR	**3-UPR**
Case VI	**2-URR&SRR**	2-CRR&SRR	**2-UPR&SPR**

Figure 6a and b synthesize the case I as shown in Fig. 5, composed of three kinds of limbs with $(RR)_O T, RTT\tilde{R}$ and $TTT\tilde{R}\tilde{R}\tilde{R}$ motion characteristics listed in Tables 2 and 5, respectively. Figure 6c and d synthesize the case II. Figure 6e and f depicts the case III. Figure 6g and h have the details about the case IV. In particular, Fig. 6i and j bring the case V.

5 Some Typical Legged Mobile Landers

A mobile lander is constructed by connecting the body with four legs, listed in Table 6. The structures of mobile landers are classified into two kinds in terms of types of legs: one with the same structure of legs, the other one with different structures of legs.

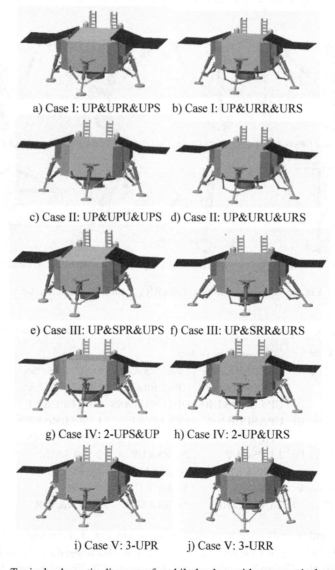

a) Case I: UP&UPR&UPS   b) Case I: UP&URR&URS

c) Case II: UP&UPU&UPS   d) Case II: UP&URU&URS

e) Case III: UP&SPR&UPS   f) Case III: UP&SRR&URS

g) Case IV: 2-UPS&UP   h) Case IV: 2-UP&URS

i) Case V: 3-UPR   j) Case V: 3-URR

Fig. 7. Typical schematic diagram of mobile landers with symmetrical structures

A large amount of mobile landers with all structurally symmetric legs are obtained by assembling a body with the same structures of legs chosen from Table 6. Consequently, some configurations of the legged mobile landers with symmetrical structures are depicted in Fig. 7.

6 Conclusion

The novel concept of the legged mobile lander combining features and capabilities of lander and rover is introduced in this paper. The legged mobile lander has some advantages detailed as following: 1, It expands the maneuvering range and enhances the capability of exploration. 2, It has more capability and less volume than conventional landers and rovers. Ultimately, the cost of mission, such as manufacturing and launching can be reduced greatly. 3, In comparison with wheeled rovers, the mobile lander walks across obstacles more efficiently and effectively.

The systematic procedure containing three stages and the corresponding method for the novel design of the legged mobile lander are proposed. Following the procedure, numerous legged mobile landers are achieved and presented. Some architectures of the proposed legs of the robot have the potential for other practical industrial applications, such as light machining tools for deburring, polishing, and grinding of curved surfaces. The key method proposed in this paper provide reference for the structure design of walking robots. In addition, wheels can be added at the end of the legs of the legged mobile lander to be a wheel-legged mobile lander.

Acknowledgments. The author thanks the partial financial supports under the projects from the National Natural Science Foundation of China (Grant No. 51323005, Nos. 51335007, U1613208), the National Basic Research Program of China, (Grant No. 2013CB035501), and the research project of State Key Laboratory of Mechanical System and Vibration (Grant No. NSVZD201608).

References

1. Donahue, B.B., Caplin, G., Smith, D.B., Behrens, J.W., Maulsby, C.: Lunar lander concepts for human exploration. J. Spacecr. Rocket. **45**, 383–393 (2008)
2. Hapke, B.: Surveyor I and Luna IX pictures and the Lunar soil. Icarus **6**, 254–269 (1967)
3. Williams, R.J., Gibson, E.K.: The origin and stability of lunar goethite, hematite and magnetite. Earth Planet. Sci. Lett. **17**, 84–88 (1972)
4. Weiss, S.P.: Apollo experience report: lunar module structural subsystem (1973)
5. Parkinson, R.: The use of system models in the EuroMoon spacecraft design. Acta Astronaut. **44**, 437–443 (1999)
6. Okada, T., Sasaki, S., Sugihara, T., Saiki, K., Akiyama, H., Ohtake, M., et al.: Lander and rover exploration on the lunar surface: a study for SELENE-B mission. Adv. Space Res. **37**, 88–92 (2006)
7. Prinzell III, L.J., Kramer, L.J., Norman, R.M., Arthur III, J.J., Williams, S.P., Shelton, K.J., et al.: Synthetic and enhanced vision system for altair lunar lander (2009)

8. Wu, W., Yu, D.: Key technologies in the Chang'E-3 soft-landing project. J. Deep Space Explor. **1**, 105–109 (2014)
9. Peng, H., Liu, J., Zhang, Z.: Conceptual design of a lunar lander. Spacecr. Eng. **1**, 006 (2008)
10. Ringrose, T., Towner, M., Zarnecki, J.: Convective vortices on Mars: a reanalysis of Viking Lander 2 meteorological data, sols 1–60. Icarus **163**, 78–87 (2003)
11. Heet, T.L., Arvidson, R., Cull, S., Mellon, M., Seelos, K.: Geomorphic and geologic settings of the Phoenix Lander mission landing site. J. Geophys. Res.: Planet. **114** (2009)
12. Iagnemma, K., Shibly, H., Rzepniewski, A., Dubowsky, S.: Planning and control algorithms for enhanced rough-terrain rover mobility. In: International Symposium on Artificial Intelligence Robotics & Automation in Space (2001)
13. Lindemann, R.: Mars exploration rover mobility development-mechanical mobility hardware design, development, and testing. Robot. Autom. Mag. IEEE **13**, 19–26 (2006)
14. Chuankai, L., Baofeng, W., Jia, W.: Integrated INS and vision based orientation determination and positioning of CE-3 lunar rover. J. Spacec. TT&C Tech. **33**, 250–257 (2014)
15. Bertrand, R., Lamon, P., Michaud, S., Schiele, A., Siegwart, R.: The SOLERO rover for regional exploration of planetary surfaces (2003)
16. Estier, T., Crausaz, Y., Merminod, B., Lauria, M., Piguet, R., Siegwart, R.: An innovative space rover with extended climbing abilities. Space Robot. **36**, 333–339 (2000)
17. Baglioni, P., Elfving, A., Ravera, F.: The ExoMars rover - overview of phase B1 results (2008)
18. Wilcox, B.H.: ATHLETE: a mobility and manipulation system for the moon. In: Aerospace Conference, pp. 1–10 (2007)
19. Wilcox, B.H.: ATHLETE: an option for mobile lunar landers. In: Aerospace Conference, pp. 1–8 (2008)
20. Laird, D., Raptis, I.A., Price, J.: Design and validation of a centimeter-scale robot collective. In: IEEE International Conference on Systems, Man and Cybernetics, pp. 918–923 (2014)
21. Liang, L., Zhang, Z., Guo, L., Yang, C., Zeng, Y., Li, M., Ye, P.: Mobile lunar lander crewed lunar exploration missions. Manned Spacefl. **21**(5), 472–478 (2015)
22. Renzhang, Z., Hongfang, W., Xiaoguang, W., et al.: Advances in the Soviet/Russian EVA spaceflight. Manned Spacefl. **1**(15), 25–45 (2009)
23. Birkenstaedt, B.M., Hopkins, J., Kutter, B.F., et al.: Lunar lander configurations incorporating accessibility, mobility, and centaur cryogenic propulion experience. In: AIAA Space Conference, vol. 7284, pp. 6–9
24. Yang, P., Gao, F.: Leg kinematic analysis and prototype experiments of walking-operating multifunctional hexapod robot. Proc. Inst. Mech. Eng. Part C: J. Mech. Eng. Sci. **228**, 2217–2232 (2014)
25. Gao, F., Li, W., Zhao, X., Jin, Z., Zhao, H.: New kinematic structures for 2-, 3-, 4-, and 5-DOF parallel manipulator designs. Mech. Mach. Theory **37**, 1395–1411 (2002)

Designing of a Passive Knee-Assisting Exoskeleton for Weight-Bearing

Bo Yuan[1(✉)], Bo Li[2], Yong Chen[3], Bilian Tan[4], Min Jiang[1],
Shuai Tang[1], Yi Wei[2], Zhijie Wang[1], Bin Ma[5], and Ju Huang[5]

[1] Department of Machinery and Electrical Engineering,
Logistical Engineering University, Chongqing, China
ramboyuanbo@aliyun.com
[2] Department Petroleum Supply Engineering,
Logistical Engineering University, Chongqing, China
[3] College of Aerospace Science and Engineering,
National University of Defence Technology, Changsha, China
[4] College of Arts and Science, New York University, New York, USA
[5] Mechanical Engineering College,
Chongqing Industrial and Commercial University, Chongqing, China

Abstract. Weight-bearing exoskeleton can effectively help the wearer to bear heavier burden, while assisting his ambulation. However, current researches in this field are relatively scarce on the passive weight-bearing exoskeleton. This research aims to design an unpowered knee-assisting exoskeleton used for weight-bearing, which can store human metabolic energy and assist human locomotion, through utilization of Teflon string, pulley, and compression spring. Biomechanical analysis of human weight-bearing locomotion shows that knee-flexion angle can be used to identify level walking or ascending movement of a person, where the largest sagittal flexion angle in level walking does not exceed 60°. Hence, the contour of pulley used for winding string is designed to be eccentric, in which the assisting torque varies nonlinearly according to the knee-flexion angle. Through mechanical modeling of eccentric pulley, we predict the assisting torque of the device and compare it with actual experimental statistics. Results show that such passive exoskeleton exhibits multi-stage nonlinear assisting augmentation under different knee-flexion angles; with least possible knee assistance during level walking, and remarkable assistance during climbing.

Keywords: Weight-bearing · Knee-assisting · Unpowered · Ascending locomotion · Eccentric pulley

1 Preface

Exoskeleton is a wearable mobile device that can amplify the strength or save the energy consumption of an individual in motion, through the use of external power supply or human power [1, 2]. In recent years, exoskeleton technology has been used in various fields such as load-carrying, rehabilitation of paraplegics [3], human locomotion, object-lifting, and industrial manufacturing. The main goal of load-carrying

© Springer International Publishing AG 2017
Y. Huang et al. (Eds.): ICIRA 2017, Part II, LNAI 10463, pp. 273–285, 2017.
DOI: 10.1007/978-3-319-65292-4_24

exoskeleton is to enhance strength, reduce fatigue, as well as protect musculature damage of a healthy wearer [4]. The basic functions of such device are for weight support, gait assistance, while ensuring the wearer's flexibility in undesirable road conditions, and retaining long-lasting working time.

The Berkley's BLEEX [5, 6], Harvard's Soft Exosuit [7, 8], Japan's Hal [9], and France's Hercule [10] are some typical active powered exoskeletons made to decrease fatigue and increase productivity. Though these machines greatly improved the wearer's capabilities of the natural human skeletal structure, there are some limitations like bulky size, high power consumption (e.g. the power sources can only sustain one exoskeleton for no longer than a few hours), complicated systems, expensive research and production costs, have constrained the machine's applications and promotions. In contrast to powered exoskeleton, passive exoskeleton has simpler structures, lower production costs, and requires no electric power at all, thus making it more practical and promotable than active exoskeletons.

Lockheed Martin's Fortis is a representative example of passive exoskeleton [11]. It employs linkage mechanism and dead-center principle, which gives the Fortis a higher productivity than ever before. This unpowered framework allows the wearer to effortlessly lift heavy tools. Unfortunately, due the design's lack of strength-amplification feature, Fortis is more useful in standing position, and not for outdoors weight-bearing activities.

In fact, not only can powered exoskeleton provide weight support for the wearer, but also can passive exoskeleton. For instance, Cameron University designs an ankle exoskeleton using techniques like strings, stretch springs, and ratchet [12, 13]. The passive ankle elastic mechanism can be used to produce ankle exoskeleton that enables the wearer to walk with minimal actuation. Such kind of purely passive exoskeleton can reduce the walker's metabolic energy by up to 7%.

Over centuries of evolution, human beings have already adapted well-coordinated mobility. During walking, dead-centre principle explains why the knee stands upright; pendulum effect is observed when the walker's legs swing freely without bearing weight of the torso on sagittal plane. This manner of walking permits humans to walk on flat roads at higher efficiency with lower energy consumption [14]. Therefore, the unpowered exoskeleton developed by Cameron University marks the beginning of humans' attempt to optimize human locomotion through exoskeleton.

Cameron University focuses on the development of ankle exoskeleton used for gait assistance, which is somewhat different from weight-bearing exoskeleton. As a person ascends (climbing uphill or walking upstairs), his quadriceps are constantly doing work, thus, offering knee torque for its extension. Such torque overcomes gravitational potential energy exerted on the walker's own body weight and his loads. If the human body ascends while bearing excessive weight, it would not take long before he drains out his energy in the lower limbs, resulting in exercise-induced fatigue and even exercise-induced exhaustion.

Hence, this exoskeleton not only need to provide effective weight-bearing support for the wearer both in standing position and in motion, but also to provide additional knee torque when ascending, in order to reduce output power of prime mover muscles (e.g. quadriceps), and ultimately to prevent weariness from overworking the leg muscles. As a result, our team will concentrate on passive knee joint exoskeleton development. We will use springs to collect the wearer's metabolic energy, provide added knee-assisting torque while he walks upwards, hoping to put this purely passive weight-bearing exoskeleton technology to use.

2 Human Weight-Bearing Gait Biomechanics Analysis

In order for a purely passive exoskeleton to provide supplementary strength for the wearer, it must obtain energy from the wearer first. To simplify the structure's complexity, the best way for the knee-assisting device to store energy should also come from the knee joint movement. When the human body is walking (on flat road, uphill, or downhill), the knee joint movement involves both extension and flexion of the legs. For every one step upward, the leg stepping forward first will experience leg flexion and then leg extension when it brings the body upwards. Thigh experiences fatigue mainly because when the leg extends to prepare for rising, the quadriceps need to overcome gravitational potential energy to do work. If an assisting device is present, it can store the energy that would otherwise be used for leg flexion when moving upward. This device works in that the hamstring tendon uses more than needed energy to pull the strings of the device when the wearer bends leg for rising, the stored energy is later released when the wearer rises. The goal of this process is to distribute the wearer's metabolic energy between leg flexion and extension, in order to prolong work period throughout the step going upward. Even though this approach cannot reduce the total amount of work required, it could reduce the maximum burden on the wearer, hence increases the total time of weight-bearing ambulation that the wearer can endure.

Meanwhile, the human body also experiences leg extension and flexion in level walking to avoid the legs from hitting the ground when moving forward. Here, the knee-assisting device should reduce the effect of weight assistance and energy storage, or it would interfere with normal human ambulation and cause unnecessary energy consumption of the wearer. Therefore, our device needs to function differently between level walking and climbing up, to provide visible assistance for the wearer when going upward, and minimize intervention in the wearer's natural marching on level ground. As a purely passive device with no electrical monitors, the best alternative to distinguish level walking and climbing up movements is by looking at the knee joint flexion angle, where the angle is considerably bigger in level walking than in climbing up.

To find the most suitable knee joint flexion angle to distinguish level walking and ascending locomotion, we conducted a "maximum knee-flexion angle" statistical test for the human body under different conditions. During the experiment, we placed the camera axis perpendicular to wearer's sagittal plane, and videotaped the wearer's movements where the lower limbs stay in the center of the video. We marked the measuring points on the thigh and calf to precisely measure the respective flexion angles. Figure 1 below displays pictures from a previous experiment.

During this experiment, the measurable value is the maximum knee-flexion angle, and factors that could affect experimental results include wearer's gait and whether he carries any load while in motion. (See Table 1 for details.) To derive data of significant statistical meaning, we conducted the same experiment on 5 individual samples, in which every sample was tested for 10 times under each different experimental

Fig. 1. Maximum knee-flexion angle statistical test for level walking

environment. Eventually, we obtained the statistical mean and standard error of maximum knee-flexion angle for every individual sample and under different experiment conditions. The experiment results shown in Table 1 are the average level of maximum knee-flexion angle of all samples, with the standard error among different individuals following behind.

Table 1. Maximum knee-flexion angle statistical test for various locomotion

Locomotion		0 kg	10 kg	30 kg
Level walking	Stroll (1.36 m/s)	$59.5 \pm 1.9°$	$59.9 \pm 1.8°$	$64.8 \pm 1.3°$
	Speed walk (1.59 m/s)	$62.9 \pm 2.1°$	$63.5 \pm 2.0°$	$67.9 \pm 1.5°$
	Jog (2.55 m/s)	$77.4 \pm 2.1°$	$78.3 \pm 1.8°$	$81.4 \pm 2.4°$
Step ascending	Bump (10 cm)	$67.0 \pm 1.9°$	$68.6 \pm 1.7°$	$72.8 \pm 1.4°$
	Stair (20 cm)	$80.9 \pm 1.4°$	$81.1 \pm 1.6°$	$89.6 \pm 2.2°$
	Step (30 cm)	$98.8 \pm 1.7°$	$101.6 \pm 1.7°$	$112.0 \pm 2.1°$
	Bleacher (40 cm)	$109.6 \pm 2.0°$	$112.0 \pm 1.7°$	$118.3 \pm 2.1°$

From Table 1 above, we can see that, on average, human level walking flexion angle is $59.5 \pm 1.9°$. This range is correspondent to the CGA's flexion angle of $58.2°$. In contrast, human ascending flexion angle is about $67.0–118.3°$. For example, Fig. 2 below shows a sample's maximum flexion angles when climbing up different altitudes.

Fig. 2. Maximum knee-flexion angle at different heights

As the inclination altitude increases, the human knee-flexion angle can become greater than $118.3°$. Under normal conditions, such manner of walking would no longer be single average strides, but of steep-slope climbing. Therefore, knee joint weight-assisting device should be designed in which it provides most effective weight-assistance at flexion angle of 60–120°, and lowest possible weight-assistance at flexion angle of 0–60°.

3   Passive Knee-Assisting Device Design

3.1   Primary Design Objectives

a. Elastic potential energy storage during knee flexion; torque assistance during extension;
b. Insignificant torque assistance during level walking, and apparent torque assistance during ascending, determined by flexion angle.

3.2   Structural Design

One applicable design is to use pulley, Teflon strings, and compression springs to actualize our exoskeleton's passive knee-assisting function. Our exoskeleton is worn along outer human leg, with one rod placed on the outer thigh and another on the outer calf, and knee-joint components located on the same height as the wearer's natural knee-joint. (As shown in the dotted lines in Fig. 3a). Figure 3b displays a profile of our exoskeleton, essential components include the shank exo-caput (1) (the exoskeleton component which connects the shank rod to the knee joint), thigh exo-caput (2), compression spring (3), Teflon string (4) with lead tails (401, 402) on both ends, slider block (5), rotation shaft (6), and thigh rod (7). Among these, both the shank exo-caput and thigh exo-caput are equipped with a channel (101, 201) where the Teflon string penetrates, and ends of the channel are the string outlets (102, 202) in which the string gets out from exo-caput's channel. Specifically, the string outlet on the thigh exo-caput is named as upper outlet (202), and the corresponding string outlet on the shank side is named as lower outlet (202). The partial pulley of the shank exo-caput is also equipped with string grooves (103). In the exo-thigh sector, the compression spring (3) sits between the slider block (5) and thigh exo-caput (2). As Teflon string passes the inside of compression spring, it then penetrates through the string channels and outlets of both exo-caputs, and eventually fastens the lead tail onto the end surface of the shank exo-caput.

During leg flexion, the shank and thigh exo-caput revolves around the rotation shaft respectively. Because total length of the Teflon string remains constant, the length of string between two exo-caputs continue to expands during leg flexion, and gradually coils up inside the string grooves. Consequently, the string segment located inside the thigh rod shrinks, causing it to tag along the slider block and press against the compression spring, and ultimately enables the device to store energy when the leg flexes. In fact, the shank exo-caput is an incomplete pulley, and one side of the string tails fastens on this incomplete pulley, thus forming a winding device. Such design allows the energy-storing spring to be built in the thigh rod of the exoskeleton, which in return

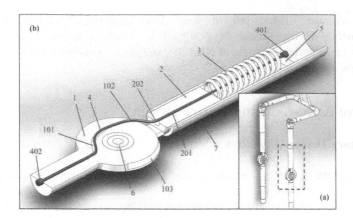

Fig. 3. Passive knee-assisting structure illustration

reduces knee-joint complexity and, increases space utilization. What is more meaningful is that the relative motion between the string and the pulley's surface is purely rolling. They won't slide against each other during locomotion, which makes the device more durable.

In leg extension, the compression spring brings the Teflon string back into the thigh rod, and induces assisting torque to shank exo-caput. Accordingly, this assisting torque (T_a) is determined by the spring force (F_s) and the force lever (h_l), where force lever is the distance between the center of rotation shaft and tangent line of the eccentric curve, shown as blue dashed line in Fig. 4a. Essentially, for the unpowered knee-assisting device to have least productivity in level walking and maximum productivity in ascending locomotion, the assisting torque (T_a) must be a nonlinear function of the knee flexion angle θ. More concretely, the function increases monotonically with the increasing of θ. However, when $\theta < 60°$, the incremental slope is relatively small, and the slope increases when $\theta < 60°$. A nonlinear assisting torque function can be achieved by using a nonlinear spring, varying the deformation of spring (Δs) or the arm of string force (h_l) nonlinearly from change in θ. Indeed, we can use nonlinear spring to obtain desired assisting behavior, but doing so would increase the level of difficulty in design and manufacturing. In contrast, to change the force lever (h_l) (Fig. 4a, blue dashed line segment) and string extension (s_d) (Fig. 4a, green dotted line segment), would only require making the pulley contour radius (R_k) to be an eccentric pulley that changes with the knee flexion angle θ. Such newly designed apparatus is simple and reliable, and easier to modify the mechanical characteristic for different knee-assisting devices. For this reason, our research will utilize eccentric pulley to actualize nonlinear knee-assistance.

It must be noted that, when pulley contour curve becomes eccentric, the center of curvature for any points of the pulley contour do not necessarily match with the rotation center of the knee-joint. Thus, the distance (R_k) between one edge point and the rotation center, does not necessarily equal to the distance (h_l) between the tangent line crossing such edge point and the rotation center. One example of such situation can be seen in Fig. 4b. The geometric shape of the eccentric pulley can be determined by

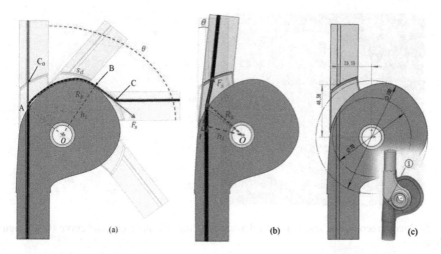

Fig. 4. Eccentric pulley parameter definition (Color figure online)

the contour's included angle (α) and contour radius (R_k). However, in order for our derivations to have actual meanings, we should set knee-joint leg flexion angle θ as the independent variable when deriving the string extension length (s_d) or string force lever (h_l). The offset distance between the upper outlet (Fig. 3b, 202) and rotation shaft, as well as the eccentric contour figure, all contribute to the nonlinear relation between flexion angle θ and the contour's included angle α, resulting in $\theta \neq \alpha$.

The outline of the eccentric pulley employed in this research is based on two concentric circles with different diameters (70 mm and 100 mm, respectively) and smoothed crossover with spline curves, as shown in Fig. 4c. Concurrently, for the knee-joint exoskeleton to retain angle-restricting function, we offset axis of the shank and thigh rods against the knee-joint rotation center, as well as set up restricting blocks on both sides of the eccentric pulley. When the exoskeleton is in upright position, shank and thigh exo-caputs' restricting blocks touch each other (Pointed as ① in Fig. 4c), thus avoiding knee-joint to bend forward, which could damage the human knee joints. Such design will also cause the upper outlet offsetting certain distance in regard to the rotation center of knee-joint (horizontal offset 29.59 mm, vertical offset 46.38 mm).

3.3 Mechanical Modeling

In order to analytically calculate the torque of such knee-assisting device, we need to obtain the analytical formula of the eccentric spline contour curve first. By using Matlab's image recognition, the numerical solution of eccentric curve (shown in Fig. 5a) can be extracted from the cutaway view of exoskeleton's knee-joint, which are plotted in Solidworks. Thus, the analytical formula of the eccentric contour curve can be derived and replotted in a polar coordinate, which can be found as a blue thick line in Fig. 5b, by using the polynomial fitting to the previous numerical solution. Those

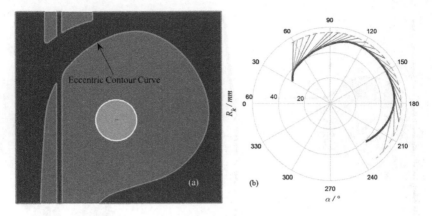

Fig. 5. Image recognition and tangent calculation of the eccentric contour curve (Color figure online)

colorful straight segments around the contour curve are the tangent lines with different contact points on that curve, which are calculated by linear fitting the slope of its differential part. Each segment's length represents the impending Teflon string between two exo-caputs, and the other ending of the segment implies the outlets of the thigh exo-caput.

According to the offset position of the upper outlet as compared to the rotation center, the rotating trajectory of the outlet can be predicted with the motion of knee flexion, which is shown as a red dotted arc in Fig. 5b. By merging formulae of such trajectory with those of different tangent lines, the variation of the string's stretching length (s_d) and the position of upper outlet (shown as C point in Fig. 4a) can be derived against the contour's included angle (α). The angle between the upper outlet's present (C point) and original position (C_0 point in Fig. 4a) indicates the flexion angle of knee joint (θ), which further gives the nonlinear relationship between θ and α shown in Fig. 6. It should to be noted that the original contour's included angle (α_0) is nonzero, which can be confirmed in Fig. 5b, hence Fig. 6 annotates a reference line ($\alpha - \alpha_0$) for a better observation of nonlinearity.

The arm of the string's force (h_l) can be calculated by the tangent line of the eccentric contour at the string's contact point (for instance, point B in Fig. 4a). In Cartesian coordinates, the distance from the rotation center (x_0, y_0) to the tangent line $Ax + By + C = 0$ is obtained as the following formula:

$$h_l = \frac{Ax_o + By_o + C}{\sqrt{A^2 + B^2}} \tag{1}$$

Hence, we retain the analytical result about the force lever of string (h_l) corresponding to the knee-flexion angle, which is shown in Fig. 7a. Obviously, the arm of force keeps increasing nonlinearly when $\theta < 60°$, while maintaining the level around 50 mm when $\theta < 60°$.

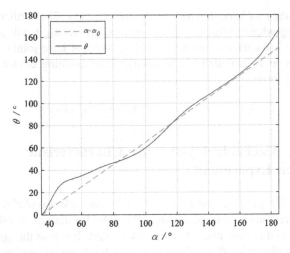

Fig. 6. The nonlinear relationship between knee flexion angle θ and contour's included angle α

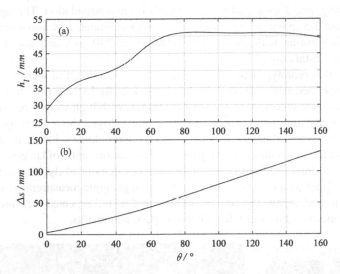

Fig. 7. Variations of string's force lever (a) and the deformation of spring (b)

In addition, the deformation of spring Δs can be obtained by the stretch length of string (s_d) between two outlets when the knee joint flexes by θ (shown as $\overset{\frown}{AC}$ Fig. 4a), subtracting its original stretch length (s_{d0}) when the knee keeps upright, $\Delta s = s_d - s_{d0}$. The length of curve $\overset{\frown}{AC}$ is composed of the arc length ($s_{\overset{\frown}{AB}} = \int_0^\alpha R_k(\varphi)d\varphi$) where Teflon string coils around the groove on shank exo-caput, and the distance of the impended straight string ($\overline{BC}$). Whereas the original stretch length of string (s_{d0}) can be obtained by calculating the distance of $\overline{AC_0}$. Result shows the deformation of spring Δs caused by winding the string around such eccentric pulley has a highly linear increase with the flexion angle of knee θ, which is illustrated in Fig. 7b. Furthermore, when we use a linear spring whose stiffness is k and preload sets as s_0, it will give the string a stretch force $F_s = k(\Delta s(\theta) + s_0)$ with the bending angle of knee-joint equals θ. Finally, the extension torque T_a provided by the passive knee-assisting device can be derived by the folllowing formula:

$$T_a(\theta) = F_s \cdot h_l = k \cdot (\Delta s(\theta) + s_0) \cdot h_l(\theta) \tag{2}$$

4   Distinctive Mechanical Assistance Characteristic Verification Experiment

We use Nylon 3D printing technology to fabricate the exoskeleton's knee-joint designed in the previous section; aluminum-alloy tube for the hip, shank rod, and thigh rod components; and carbon fiber sheet for backboard. Because the spring mentioned above needs to be placed inside the thigh rod, its outer diameter need to be less than the rod's inner diameter (24 mm), and the preloaded spring length needs to be less than the rod length (240 mm). Meanwhile, the exoskeleton needs to provide appropriate torque during climbing, and does not affect the wearer's normal ambulation. Through repeated trials, we chose 2.5 mm wire diameter, 23 mm outer diameter, 230 mm length, 20 coils of compression spring with approximate stiffness of 240 g/m, and about 17 mm of preload in our installation.

Based on this prototype exoskeleton, we implement amount of weight-bearing and knee-assisting experiments, respectively (shown in Fig. 8). In the former test, we proved that the exoskeleton can carry up to 70 kg of weight, with high weight-bearing productivity. In knee-assisting torque measurements, we employed the tension dynamometer (Fig. 8b), to measure the amount of force used at different knee-joint leg flexion angles, with the arm of loading points remaining fixed to the rotation center of knee-joint. We then used high-definition camera to record both the measurement of dynamometer and the bending knee-joint with uniform speed. Lastly, through angle measurement software, to measure the forced applied in every 5° interval, and with 5 repeated measurements, we obtained the mean and standard deviation of experiment results.

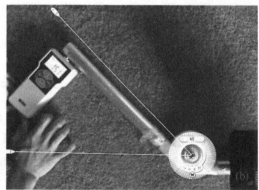

Fig. 8. Weight-bearing (a) and knee-assisting (b) experiments based on the prototype

In Fig. 9, the blue curve shows the analytical result of knee-assisting torque which is calculated according to Eq. 2, while the red curve indicates the mean value of experimental results with its error bars represent the maximum and minimum samples for interval bending angles. It can be observed that the analytical result matches the experimental result well. Overall, the assisting torque increases monotonically with the increase of the flexion angle. Specifically, the knee joint with eccentric contour exhibits multi-stage nonlinear assisting augmentation under different ranges of flexion angle. When $\theta \leq 50^{\circ}$, the torque increases slowest with θ, due to the short lever of string's stretch force; when $50^{\circ} \leq \theta \leq 80^{\circ}$, the torque rises up rapidly (steepest increasing slope among all range), mainly because the arm of force and deformation of spring expand simultaneously; when $\theta \geq 80^{\circ}$, the increasing momentum of the torque slows down again, since the arm of force does not keep increasing any more, but maintains around the maximum level about 50 mm.

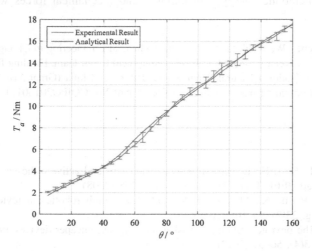

Fig. 9. The comparison between analytical and experiment results of the knee-assisting torque (Color figure online)

5   Conclusion

In our biomechanics research for human weight-bearing ambulation, we found that the knee joints flexion angle exhibits great differences between level walking and climbing. Normally, knee joint flexion angle does not exceed 60° during level walking, while knee joint flexion angle ranges from 65.8°–118.9° during climbing. With added weight on the walker, knee joint flexion angle increases even at the same altitude as when no extra weight was added. We decided to include pulley, springs, and strings in building a passive weight-assisting and energy saving knee-joint exoskeleton. Pulley has an eccentric contour, its curvature changes along with knee-joint flexion. Through theoretical analysis and calculations, we have derived deformation and force lever of the spring. Through experiment on the relation between knee-joint moment and flexion angle, we performed regression analysis on spring stiffness and preload. Results show that the actual spring stiffness is equivalent to the hypothesized value, and the analytical and experimental characteristics of knee-assisting moment match well. The device exhibits subtle assistance when flexion angle is less than 60°, while the assisting level gradually arises, which can grow up to 18 Nm, when the flexion angle further increases.

Nonetheless, we still face several limitations. Firstly, in the weight-bearing walking analysis, we tested for joint flexion angle, when we should also take into account thigh muscle activity during climbing. Another issue is the knee-joint's eccentric contour, in which we only have one applicable method so far, to obtain optimal assisting results, we need to think of better alternatives. Lastly, the mechanical verification experiment is not sufficient to prove that our passive knee-assisting device can reduce muscles' activities during weight-bearing ascending locomotion, and improve human metabolic productivity. Thus, an integrated experiment by simultaneously recording the muscle's EMG signal, metabolic energy consumption, and mechanical forces will be further pursued.

Acknowledgement. We thank the College of Mechanical Engineering in Chongqing Industrial and Commercial University for lending BioPac equipment to our team. Funding for this research were provided by National Natural Science Foundations of China (Grant Nos. 11504427 and 51505494) and Chongqing patent analysis project (Grant No. CQIPO2016011).

References

1. Dollar, A.M., Herr, H.: Lower extremity exoskeletons and active orthoses: challenges and state-of-the-art. IEEE Trans. Robot. **24**(1), 144–158 (2008)
2. Slavka, V., Patrik, K., Marcel, J.: Wearable lower limb robotics: a review. Biocybern. Biomed. Eng. **33**(2), 96–105 (2013)
3. Mertz, L.: The next generation of exoskeletons: lighter, cheaper devices are in the works. IEEE Pulse **3**(4), 56–61 (2012)
4. Walsh, C.J., Pasch, K., Herr, H.: An autonomous, underactuated exoskeleton for load-carrying augmentation. In: 2006 IEEE/RSJ International Conference on Intelligent Robots and Systems, pp. 1410–1415 (2006)
5. Zoss, A.B., Kazerooni, H., Chu, A.: Biomechanical design of the Berkeley lower extremity exoskeleton (BLEEX). IEEE/ASME Trans. Mechatron. **11**(2), 128–138 (2006)
6. Justin, G., Ryan, S., Kazerooni, H.: Control and system identification for the Berkeley lower extremity exoskeleton (BLEEX). Adv. Robot. **20**(9), 989–1014 (2006)
7. Asbeck, A.T., Schmidt, K., Galiana, I., et al.: Multi-joint soft exosuit for gait assistance. In: 2015 IEEE International Conference on Robotics and Automation, Seattle, Washington (2015)
8. Wehner, M., Quinlivan, B., Aubin, P., et al.: A lightweight soft exosuit for gait assistance. In: IEEE International Conference on Robotics and Automation (2013)
9. Kawamoto, H., Lee, S., Kanbe, S., et al.: Power assit method for HAL3 using EMG based feedback controller. In: IEEE International Conference on Systems (2003)
10. Hercule. France (2017). http://www.rb3d.com/en/exo/
11. Young, A., Ferris, D.: State-of-the-art and future directions for robotic lower limb exoskeletons. IEEE Trans. Neural Syst. Rehabil. Eng. (2016)

12. Collins, S.H., Wiggin, M.B., Sawicki, G.S.: Reducing the energy cost of human walking using an unpowered exoskeleton. Nature **522**(7555), 212–215 (2015)
13. Wiggin, M.B., Sawicki, G.S., Collins, S.H.: An exoskeleton using controlled energy storage and release to aid ankle propulsion. In: 2011 IEEE International Conference on Rehabilitation Robotics (2011)
14. Kuo, A.D., Donelan, J.M., Ruina, A.: Energetic consequences of walking like an inverted pendulum: step-to-step transitions. Exerc. Sport Sci. Rev. **33**(2), 88–97 (2005)

Development of HIT Humanoid Robot

Baoshi Cao, Yikun Gu, Kui Sun$^{(\boxtimes)}$, Minghe Jin, and Hong Liu

State Key Laboratory of Robotics and System, Harbin Institute of Technology,
Harbin 150001, China
cbob.robot@gmail.com, sun.kui@163.com

Abstract. This paper gives an overview on HIT humanoid robot, which is developed as a research platform for replicating human in special environments. The system has two upper extremities, each upper extremity includes a 7-DOF humanoid arm and a 15-DOF dexterous hand as end-effector. Two humanoid arms and dexterous hand are combined with a 2-DOF dynamic torso and a 3-DOF binocular head to form the humanoid upper body. A wheeled mobile robot is employed to provide mobility. In this paper, we describe the design specification and give an overview on mechanical design of each subsystem. Additionally, we give a short introduction to the hardware architecture and software of HIT humanoid robot.

Keywords: HIT humanoid robot · Dual-arm system · Dexterous hand · Robotic

1 Introduction

Humanoid robot has emerged as a significant potential research field and has been identified as critical for future robotic applications. Humanoid robots can be used in many practical applications to release human being from repetitive work and dangerous work environments. Humanoid robot is also a good example to gain deeper understanding on human body motion. In addition, humanoid robots drive the development of highly integrated mechatronic components and robust control methods [1]. Therefore, humanoid robots should have anthropomorphic appearance and human capabilities as its design principle.

In humanoid robot design, there are two design sequences: "from up to down" and "from down to up". "From up to down" means humanoid arm is designed first, then humanoid torso is designed according to existing upper body, leg is complete in the end, while "from down to up" is on the contrary. Advantage of first sequence is defining specific design requirements at the initial stages of lower body and torso design, such as load capacity, workspace, etc. Another sequence can provides suggestions for upper body design.

Most biped robots are design in "from up to down" sequence, such as WABIAN - the first biped robot [2, 3], HRP series robot [4–6] form AIST, and the milestone robot ASIMO series designed by HONDA company [7]. Those robot mainly focus on dynamically walk and locate capability, often their allowed payload is quite small and manipulation skill is usually relatively poor. In contrary, some humanoid robot focus on manipulation ability of upper body, such as PR-2 robot [8] developed by Willow

© Springer International Publishing AG 2017
Y. Huang et al. (Eds.): ICIRA 2017, Part II, LNAI 10463, pp. 286–297, 2017.
DOI: 10.1007/978-3-319-65292-4_25

Garage, NASA's Robotnaut 1 [9] and Robotnaut 2 [10], female appearance robot AILA [11] developed by DFKI Robotics Innovation Center and DLR's Justin. Those systems all have a wheeled robot as their initial lower body to provide mobility so that they can decrease system complexity and receive a good balance while manipulation.

Based on our previous work experience, we have designed a dual-arm humanoid robot with a wheeled mobile robot. The total of 49 degrees of freedom of upper body can be controlled in one central controller and joint controller on each motion unit. HIT humanoid robot can be considered as a new member of the latter category.

This paper is organized as follows. Section 2 introduces design specifications. Mechanical design is detailed shown in Sect. 3. Hardware and software are separately described in Sects. 4 and 5. A summary is given in Sect. 6.

2 Design Specifications

HIT humanoid robot, as shown in Fig. 1, is designed with "from up to down" sequence and consists of an anthropomorphic upper body which is mounted on a wheeled mobile robot. The upper body carries a 3-DOF binocular head [12, 13], two extremities, each of them consists a 7-DOF humanoid arm [14, 15] and a 15-DOF five-fingered dexterous hand [16], and a torso with two joints [17]. The wheeled mobile robot is employed as a first solution to provide mobility to robot when main focus of design was on the stable operation of upper body, future developments will concentrate on a new lower limbs of mobility.

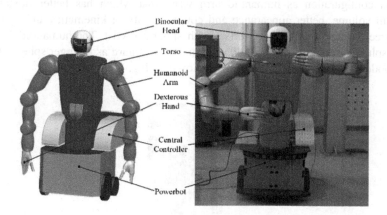

Fig. 1. Overview of HIT humanoid robot system

Table 1 gives an overview of the 51 actuated DOF, and more detail is presented in the following sections.

Table 1. System overview

Sub-system	Head	Arms	Hands	Torso	Mobile robot	Total
DOF	3	7 × 2	15 × 2	2	2	51

The initial specifications for design of the robot is to be a versatile platform for research on replicating human in special environments. Therefore, it should have an anthropomorphic kinematic configuration for research on bi-manual grasping in addition of a human-like appearance. For the mechanical design the following requirements have been taken into account: The system should be able to reach objects on the floor as well as objects on a table of about 2 m high. To achieve stability operation of manipulators, the center of gravity of upper limbs should be adjustable with regard to the mobile robot. Two six-axis force-torque sensors are combined between humanoid arm and five-fingered dexterous hand to perceive force of Cartesian space. The center-controller box, which consists of center-controller board, communication board, power board, is located on the mobile robot. A battery is mounted on mobile robot to supply power for the whole system, make the system could run without cable.

3 Mechanical Design

3.1 Humanoid Dual-Arms

Like most humanoid robot, humanoid dual-arm system is the key subsystem for operation. HIT humanoid arm is designed to achieve human-like manipulative capabilities and approximately size of a human arm. Similar to the motions of human arm, the degree of freedom of HIT humanoid arm is equivalent to seven revolutions: a three degrees of freedoms (roll-pitch-roll) combination as humanoid arm shoulder; another spherical configuration as humanoid arm wrist that which has better dexterously, minimum volume, better appearance and controllability of kinematics and dynamics; one degree of freedom elbow non-offset with spherical wrist. The humanoid arm with S-R-S (spherical-roll-spherical) configuration can be regard as a bigger spherical joint. CAD model and actual humanoid arm is shown in Fig. 2.

Fig. 2. CAD model and actual humanoid robot arm

The humanoid arm joints are accomplished with three similar modular construction with just little difference in size of components. All joints and links are trade-off between mass and stiffness, the humanoid arm is about 1.1 m from the base to the end connector and total weigh less than 18.5 kg. The payload capability is over 10 kg when it fully expands in terrestrial gravitation condition. The acceleration of each joint is over 1 rad/s2 and arm speed is over 0.5 m/s.

Each modular joint is composed of mechanics, multisensory systems, electronics and controller, mainly components is shown in Fig. 3.

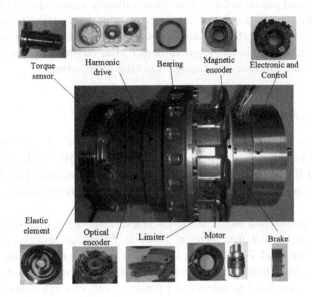

Fig. 3. The critical component selected in the integrated SEA

The mechanic transmission chain is start from brushless DC motor, through harmonic reducer and torque sensor, to the joint output. A high efficiency and low power consumption fail-safe brake is fixed on the other side of motor. In addition, mechanical limit is designed on joint output to keep different revolution range, a large hole along the joint axis is developed to pass cables. Multisensory systems is introduced in each joint to interact with environment as similar functionalities to human arm, all sensors are listed in Table 2.

Table 2. Sensors requirement of one humanoid arm joint

Sensor	Quantity	Principle
Motor position sensor	3	Hall effect
Motor position sensor	1	Magnetic
Joint position sensor	1	Optical
Joint torque sensor	1	Strain gauge
Temperature sensor	1	Thermistor

Magnetic encoder rotary wheel is fixed to motor shaft and sensor-board is fixed to joint shell, then the motor relative angle position is gained while motor runs. Optical encoder is integrated between joint output shaft and joint shell, a circle of joint' rotation

(360°) is divided into 2^{15}. In the original version humanoid arm joint, a two full bridges torque sensor with cross-shaped mechanic frame is designed to measure the output torque. The linearity of joint torque sensor is 0.3% and accuracy is 16-bit. Four limiters is designed to avoid plastic deformation. Then a discoid elastic element torque sensor with larger flexibility is developed to decrease joint passive impedance and more forgiving shock tolerance [18]. And also a few of new elastic elements are analyzed in [19]. Temperature sensor is employed to detect joint environment temperature, it is used to estimate the joint stays in suitable working temperature.

3.2   Five-Fingered Dexterous Hand

The requirements for interacting with human being and tools is the starting point in dexterous robot hands' design. Based on the experience of HIT/DLR Hand I [20], the HIT/DLR Hand II was designed to be stronger and more compact. The fingertip force of each finger increases from 7 N to 10 N with its size and weight decrease to 2/3 of its former version.

To grasp dexterous and firmly, HIT/DLR Hand II, shown in Fig. 4, is composed of an independent palm and five identical modular fingers to imitate appearance and functions of human hand. Four modular fingers are parallelly mounted on one side, ups and downs like human hand while the last identical finger is mounted on the other side as thumb. Each finger has 3 actuators and 1 passive joint. Through harmonic drives and timing belts, the transmission from two identical motors, which are parallel mounted on the bottom of finger base, is delivered to a differential gear to realize two axis rotation of finger base. The least motor is used to drive middle and distal phalanx through harmonic drive and high speed timing belt by means of a wire mechanism. All the actuators and electronics are integrated in the finger body and palm. The finger and palm envelop is designed to show a nice appearance, good grasping and well protection, It is also a skeleton part of the hand. It is proved that HIT/DLR Hand II is advantage in operating area by aid of simulation.

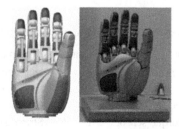

Fig. 4. CAD model and actual HIT/DLR Hand II

The HIT/DLR Hand II has a variety of sensors, such as position sensor and force sensor, in each finger to increase its precision ability. The sensors are detailed shown in Table 3.

Each finger motor has a digital hall sensor to collect motor position information, the signal is used to drive motor. Joint position sensor and joint torque, which are specially designed to reduce volume of a finger, locate in output axis of each transmission, they are indispensable to complete compliance manipulation tasks. A tiny 6 dimensional force/torque sensor is employed in each fingertip, it can get force and toque information of fingertip, that is quite useful in interactive tasks. A tactile sensor [21], which is based on the shape and available space of fingertip, is attached on fingertip surface. It can provide more detailed measurements of contact between finger and target object in interactive tasks.

Table 3. Sensors requirements of HIT/DLR Hand II

Sensor	Quantity	Principle
Motor position sensor	3 × 5	Hall effect
Joint position sensor	3 × 5	Potentiometer(2)/magnetic(1)
Joint torque sensor	3 × 5	Strain gauge
Fingertip force/torque sensor	1 × 5	Strain gauge
Temperature sensor	1	Thermistor
Tactile sensor	1 × 5	Piezoresistive

3.3 Binocular Head

The primary function of robot head is to percept environment and robot itself (especially robot arms and hands). Two generation of binocular heads were designed to collect environment information of robot. They both consist of a two video cameras system to imitate human eyes and a 3-DOF neck to move the visual system (Fig. 5).

Fig. 5. CAD model, actual binocular head and envelop

In the first version, neck base joint is designed to achieve pitching and swing movement by using bevel gear differential mechanism. An individual joint is used to roll the head. To reduce transmission gap in bevel gear differential mechanism, a special adjust ring is introduced in the second neck. A super flat BLDC motor with a tiny harmonic driver is employed instead of a super flat BLDC motor with planetary reducer to avoid backlash. Envelop of head is divided into two parts, lower part is fixed on camera brace and upper part is bolted to lower part, so the cover can move with video cameras system.

3.4 Humanoid Torso

To build a humanoid robot, dynamic torso is advantaged in extending workspace of humanoid arms and visual range of binocular head. It also verified that dynamic torso can lower energy consumption of torso rigidly to 73.5% in same walking task. HIT humanoid torso is based on prior work about the other subsystems (binocular head, humanoid arm and dexterous hand) of humanoid robot. To achieve a good balance, a PowerBot mobile robot is employed as lower body in initial stage. As a subsystem of wheel humanoid robot, a two degrees of freedom humanoid torso with pitch and yaw configuration is designed to act and move similar as human torso (Fig. 6).

Fig. 6. CAD model and actual humanoid torso

Each torso joint consists of mechanics, multisensory systems, electronics and controller. It weighs 10.9 kg with the capability of handling 741 nm payload. Similar to arm joint, brushless DC motor is used as actuator in torso joint, through synchronous belt and harmonic reducer, the motion is transmitted to joint output shaft. Motor shaft and output shaft parallel but not intersect is arranged in torso joint to reduce rotational axial dimension of torso joint. A new fail-safe brake is designed and installed. Rotation disk is fixed on motor shaft, wear-resisting friction disks is compacted by pressing disk when power off, the motor shaft is free of axial motion (Fig. 7).

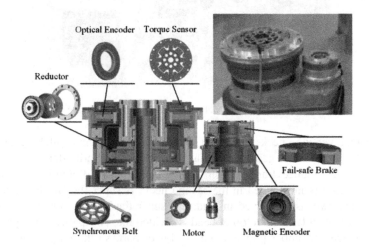

Fig. 7. Humanoid torso joint

A variety of sensors similar to arm joint are introduced in torso joint to enhance perception of torso, the sensors are detail shown in Table 4.

Table 4. Sensors requirement of one humanoid torso joint

Sensor	Quantity	Principle
Torque sensor	1	Strain gauge
Joint position sensor	1	Optical
Motor position sensor	1	Magnetic
Motor position sensor	3	Hall effect
Current sensor	2	Resistance drop
Temperature sensor	1	Thermistor
Electrify detection sensors	2	Electrical flow

Torque sensor, joint position sensor, motor position sensor and temperature sensor are the same function as arm joint. To ensure motor torque accurately, two current sensors are mounted on joint circuit board. Electrify detection sensors are installed inside joint fail-safe brake to reduce brake current after opening brake effectively, it can save energy and reduce heat accumulating.

4   Hardware Architecture of HIT Humanoid Robot

4.1   Overview [22]

HIT humanoid robot is a highly integrated robotic system, there are 49 degrees of freedom as foresaid. Each degree of freedom includes a motor as actuator and various of sensors to percept environment, all subsystems are connected to the central controller via PPSeCo, a custom high-speed serial communication bus. The architecture of HIT humanoid robot controllers is compose of real-time part and non-real-time part, as shown in Fig. 8. Those two parts communicate with each other through Ethernet.

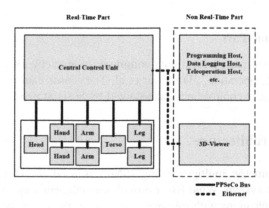

Fig. 8. Overview of the hardware architecture of controllers for humanoid robotics [22]

4.2   Central Control Unit [22, 23]

The central control unit, which located in central controller box, is developed to achieve real-time control. The real-time control is divided into Cartesian space control and joint space control, as shown in Fig. 9. They run parallel in different cores, and share same memory module to communicate with each other in cooperating tasks.

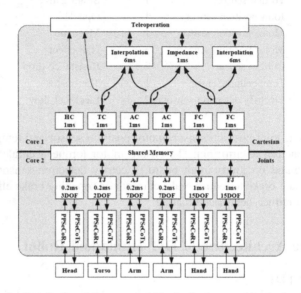

Fig. 9. Detail of real-time control blocks for humanoid robotics [22]

4.3   Joint Controller [15]

The joint controller is used to control the motor, collect sensors information, and transmit data with central control et al. The joint controller structure is detailed shown in Fig. 10, FPGA with a 32 bit Nios software core microprocessor is used as main controller.

4.4   Non-real-Time Part

For the non-real-time part, a local data communications network for distributed processing such as tele-operation, 3D viewer et al. The sensor data of robotic systems is broadcasted by central control unit and received by hosts at need.

5   Software Architecture

VxWorks for symmetric multiprocessing (SMP) is used in central control unit as real-time operation system. Wind River provides an efficient way to develop real-time and embedded applications with minimal intrusion on the target system.

Simulink provides a graphical user interface (GUI) for building models as block diagrams and a comprehensive block library of sinks, sources, linear and nonlinear components, and connectors. Real-Time Workshop technology generates C or C++ source code and executable for algorithms that are modeled graphically in the Simulink environment. A hardware-in-the-loop (HIL) testing is also available for tuning and monitoring the generated code by using Simulink blocks and built-in analysis capabilities (Fig. 11).

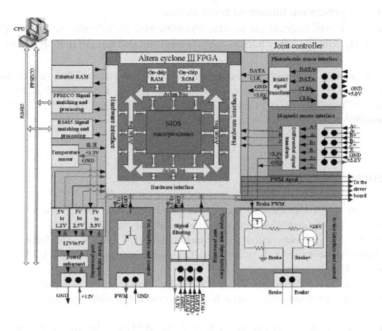

Fig. 10. Structure of the joint controller [15]

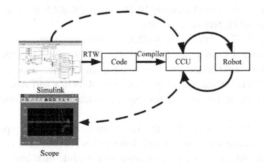

Fig. 11. Overview of a hardware-in-loop setup [22]

6   Conclusion

In this paper, we gave an overview on HIT humanoid Robot, a torque-controlled research platform, which has evolved from a seven degrees of freedom humanoid arm. HIT humanoid robot's design specifications, mechanical and mechatronics (include electronics, sensors and hardware architecture), software architecture were described. In retrospect, HIT humanoid robot is available platform for research on human working environment. Knowledge from working with HIT humanoid robot will be very beneficial for next generation humanoid robot design.

Future work will concentrate inverse dynamic and impedance based manipulation, the whole body control, safety human-robot interaction and human in loop tele-operation.

Acknowledgements. This work was supported in part by the National Program on Key Basic Research Project 973 Program under Grant 2013CB733103, the National Natural Science Foundation of China (No. 61603112) and the Self-Planned Task (No. SKLRS201721A) of State Key Laboratory of Robotics and System (HIT).

References

1. Ott, C., et al.: A humanoid two-arm system for dexterous manipulation. In: 2006 6th IEEE-RAS International Conference on Humanoid Robots. IEEE (2006)
2. Ogura, Y., et al.: Development of a new humanoid robot WABIAN-2. In: IEEE International Conference on Robotics and Automation, ICRA 2006, Orlando, Florida, USA, 15–19 May 2006 (2006)
3. Yamaguchi, J.-I., Takanishi, A., Kato, I.: Development of a biped walking robot compensating for three-axis moment by trunk motion. J. Robot. Soc. Jpn. **11**(4), 581–586 (1993)
4. Kaneko, K., et al.: Humanoid robot HRP-3. In: IEEE/RSJ International Conference on Intelligent Robots and Systems (2008)
5. Kaneko, K., et al.: Humanoid robot HRP-2. In: Proceedings of the IEEE International Conference on Robotics and Automation, ICRA 2004 (2004)
6. Kaneko, K., et al.: Cybernetic human HRP-4C. In: IEEE-RAS International Conference on Humanoid Robots 2009, Humanoids (2010)
7. Sakagami, Y., et al.: The intelligent ASIMO: system overview and integration. In: 2002 IEEE/RSJ International Conference on Intelligent Robots and Systems. IEEE (2002)
8. Chitta, S., Cohen, B., Likhachev, M.: Planning for autonomous door opening with a mobile manipulator. In: IEEE International Conference on Robotics and Automation (2010)
9. Ambrose, R.O., et al.: Robonaut: NASA's space humanoid. Intell. Syst. Appl. IEEE **15**(4), 57–63 (2000)
10. Diftler, M.A., et al.: Robonaut 2 - the first humanoid robot in space. In: 2011 IEEE International Conference on Robotics and Automation (ICRA) (2011)
11. Lemburg, J., et al.: AILA-design of an autonomous mobile dual-arm robot. In: 2011 IEEE International Conference on Robotics and Automation (ICRA). IEEE (2011)
12. Yang, Y.: Mechanism design of robonaut head and research on its object tracing control method. Master thesis, Harbin Institute of Technology (2011)

13. Zhu, J.: Research on humanoid robot head and its dynamic control. Master thesis, Harbin Institute of Technology (2010)
14. Huo, X., et al.: Design and development of a 7-DOF humanoid arm. In: 2012 IEEE International Conference on Robotics and Biomimetics (ROBIO). IEEE (2012)
15. Zhang, Z., et al.: A highly integrated joint controller for a humanoid robot arm. In: IEEE International Conference on Robotics and Biomimetics (2013)
16. Fan, S., et al.: The anthropomorphic design and experiments of HIT/DLR five-fingered dexterous hand. High Technol. Lett. **15**, 239–244 (2009)
17. Cao, B., et al.: Design and development of a two-DOF torso for humanoid robot. In: IEEE International Conference on Advanced Intelligent Mechatronics (2016)
18. Huo, X., Xia, Y., Liu, Y., Jiang, L., Liu, H.: Design and development of a rotary serial elastic actuator for humanoid arms. In: Zhang, X., Liu, H., Chen, Z., Wang, N. (eds.) ICIRA 2014. LNCS, vol. 8917, pp. 266–277. Springer, Cham (2014). doi:10.1007/978-3-319-13966-1_27
19. Huo, X., et al.: Humanoid arm with the integrated serial elastic actuator. In: 2014 IEEE International Conference on Robotics and Biomimetics (ROBIO). IEEE (2014)
20. Liu, H., et al.: The modular multisensory DLR-HIT-Hand: hardware and software architecture. IEEE/ASME Trans. Mechatron. **13**(4), 461–469 (2008)
21. Wu, K., et al.: Research of a novel miniature tactile sensor for five-finger dexterous robot hand. In: International Conference on Mechatronics and Automation (2010)
22. Zhou, Y., et al.: Rapid prototyping of real-time controllers for humanoid robotics: a case study. In: IEEE International Conference on Robotics and Biomimetics (2012)
23. Zhou, Y., et al.: A real-time controller development framework for high degrees of freedom systems. In: International Conference on Mechatronics and Automation (2012)

Design of a Robotic Laparoscopic Tool with Modular Actuation

Kai Xu[1]([⊠]), Huichao Zhang[2], Jiangran Zhao[2], and Zhengchen Dai[2]

[1] The State Key Laboratory of Mechanical System and Vibration,
Shanghai Jiao Tong University, Shanghai 200240, China
k.xu@sjtu.edu.cn
[2] The RII Lab (Lab of Robotics Innovation and Intervention),
UM-SJTU Joint Institute, Shanghai Jiao Tong University,
Shanghai 200240, China
{zhc_zju_sjtu, zjr318, zhengchen.dai}@sjtu.edu.cn

Abstract. This paper presents the development of a modular robotic laparoscopic tool for MIS (Minimally Invasive Surgery). A dual continuum mechanism is utilized in the tool design to ensure reliability as well as achieve enhanced distal dexterity, increased payload capability and actuation modularity under a simple construction. Via kinematics modeling, the laparoscopic tool could be maneuvered by a Denso manipulator to perform typical laparoscopic tasks and possesses the desired functionalities for MIS. Advantages of the implemented dual continuum mechanism lead to the performances of this attempt. Motivated by the commercial success of the da Vinci surgical system, this paper presents an alternative design to realize robotic laparoscopic surgeries, which could lead to possible future commercialization opportunities.

Keywords: Continuum mechanism · Dexterous wrist · Laparoscopic tools · Medical robotics · Surgical instruments

1 Introduction

Open surgery has been mostly replaced by MIS (Minimally Invasive Surgery) in treatments for various pathological conditions, due to the improved surgical outcomes, such as lower pain, reduced postoperative complications, and shorter hospital stay [1].

Although multi-port MIS is beneficial, manipulation of the manual tools could be challenging and exhausting, due to the lack of distal dexterity and the inversed tool-maneuvering motions. Numerous robotic systems were hence developed to assist surgeons in laparoscopic MIS for enhanced dexterity, higher motion precision, augmented tactile sensing, better ergonomics, etc. [2].

Among the existing surgical robotic systems, the da Vinci system has clinically enabled a wide spectrum of MIS procedures and dominates the market for laparoscopic surgical robots [3, 4]. Treating this system as a benchmark, the related researches primarily focus on improving (i) the tool distal dexterity [5, 6], (ii) the tactile sensing capability [7–9], and (iii) the system modularity and design compactness [10, 11]. The aforementioned systems and designs by no means exhausted all alternative design

© Springer International Publishing AG 2017
Y. Huang et al. (Eds.): ICIRA 2017, Part II, LNAI 10463, pp. 298–310, 2017.
DOI: 10.1007/978-3-319-65292-4_26

approaches. The SMARLT (Strengthened Modularly Actuated Robotic Laparoscopic Tool) was hence developed as shown in Fig. 1, aiming at realizing robotic multi-port laparoscopic surgeries with several performance enhancements. The SMARLT tool consists of an exchangeable effector and an actuation unit.

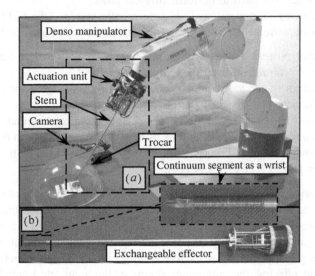

Fig. 1. The SMARLT mounted on a Denso manipulator: (a) the SMARLT tool that consists of the exchangeable effector and the actuation unit, and (b) the exchangeable effector

The SMARLT tool possesses two actuators for the wrist bending and one more actuator for the gripper. It can be attached to and maneuvered by a manipulator (e.g., the Denso manipulator in Fig. 1) for abdominal deployment through a trocar. The Denso manipulator acts as a programmable RCM (Remote Center of Motion) mechanism which positions and orients the SMARLT tool with respect to the trocar (namely, the skin incision point) in order to minimize possible tear to patient's abdominal wall. A comprehensive review of RCM mechanisms could be found in [12], while new designs were also proposed recently [13, 14].

Major contributions of this paper lie on (i) the SMARLT tool design using the concept of a dual continuum mechanism for actuation modularity and enhanced capabilities, and (ii) the analytical kinematics framework for the use of the SMARLT or similar tools with a generic manipulator under the motion constrains stemmed from the incision point. Existing results on the constrained-motion kinematics [15, 16] cannot be readily used since they don't directly include analytical formulation of the distal wrist motions. Minor contributions mainly include a compact actuation assembly design for driving the exchangeable effector.

This paper is organized as follows. With the design objectives and overview of the SMARLT tool summarized in Sect. 2, Sect. 3 describes the SMARLT design and the system components in detail. Section 4 presents a kinematics framework for the use of the SMARLT tool with a generic manipulator. Conclusions and future work are summarized in Sect. 5.

2 Design Objectives and Overview

The SMARLT tool was developed to facilitate multi-port robotic laparoscopic surgeries. It shall be attached to a manipulator so that the tool-manipulator system, shown in Fig. 1, could be tele-operated to perform surgical tasks.

Comparing to the aforementioned existing robotic systems, the SMARLT aims at realizing a few improvements. The design objectives are formulated as follows.

- The tool shall possess a wrist with small bending radius and at least two DoFs (Degrees of Freedom) for distal dexterity enhancement.
- The end effector of the tool could be changed during a surgery for different tasks and could be detached for sterilization.
- The tool could be deployed through a trocar with diameter less than 8 mm.
- Payload capability of the tool should be at least 300 g, according to the studies that the suture tension of a hand tie is less than 3 N [17, 18] and the tissue manipulation force ranges from 1.3 N to 3.5 N [19].

With the design objectives, the SMARLT tool is hence designed and constructed as in Fig. 1(a). It consists of an exchangeable effector and an actuation unit.

The exchangeable effector possesses a continuum segment as a distal wrist with a 2-DoF bending motion capability. It could be equipped with different distal surgical end effectors (e.g. gripper, scissors, cautery spatula, etc.). Different effectors (including the surgical end effector, the continuum segment, the stem, etc.) could be changed during a procedure. The effector is purely mechanical and it could be easily sterilized. The current design has a stem with a length of 400 mm and a diameter of 7 mm. The diameter is determined due to the availability of a critical component.

The actuation unit mainly includes three sets of servomotors and related actuation assemblies for driving the distal wrist and the gripper.

System descriptions of the SMARLT tool as well as the setup of the tool-manipulator system are detailed in Sect. 3. The derived kinematics is reported in Sect. 4. This kinematics framework could be applied to use the SMARLT or similar tools with a manipulator in tele-operated surgical tasks.

3 System Descriptions

The SMARLT is deployed and maneuvered by a Denso manipulator for surgical tasks in laparoscopic procedures. The SMARLT tool consists of an exchangeable effector and an actuation unit as shown in Fig. 1(a). Component descriptions of the exchangeable effector and the actuation unit are presented in Sect. 3.1 and Sect. 3.2 respectively. Controller architecture of the SMARLT tool, which allows its integrated control of Denso manipulator for teleoperation, is presented in Sect. 3.3.

3.1 Exchangeable Effector with a Continuum Wrist

The exchangeable effector shown in Fig. 1(b) is depicted in Fig. 2. It utilizes the concept of a dual continuum mechanism which is firstly proposed in [20]. The exchangeable effector consists of a gripper, a distal continuum segment, a stem, guiding cannulae, and a proximal continuum segment, as shown in Fig. 2. Both the distal and the proximal segments are structurally similar to the one shown in Fig. 5(b).

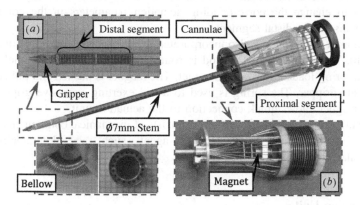

Fig. 2. The exchangeable effector of the SMARLT tool: (a) the distal segment as a wrist, and (b) the proximal segment with the gripper actuation

The segment in Fig. 5(b) consists of a base ring, several spacer rings, an end ring, and several backbones. The backbones are made from super-elastic nitinol rods. They are called backbones instead of tendons since they can be pushed and pulled while the tendons may only be pulled. Pushing and pulling these backbones bends the segment.

In the exchangeable effector shown in Fig. 2, both ends of a backbone are attached to the end rings of the distal and the proximal segments respectively, routing through the distal segment, the stem, the guiding cannulae and the proximal segment. The arrangement of the backbones in the distal segment is similar and scaled to that in the proximal segment. Hence, bending of the proximal segment always bends the distal segment in the opposite direction. This distal-proximal structure is referred to as a *dual continuum mechanism*.

Strength of the segment is affected by the diameter and the number of the backbones. In order to achieve a segment with higher payload capabilities and small bending radius, more and thinner backbones should be used. This design choice is also echoed by the experimental study in [21].

As explained by the kinematics in Sect. 4.2, bending of the distal segment is a 2-DoF motion. With the proximal segment, only two actuators are sufficient to drive the distal segment, no matter how many backbones are arranged.

Besides, the weight lifting experiments in [22] show that the payload capability of a continuum manipulator is also greatly affected by its torsional stability. Then two bellows, which can be easily bent but resist twisting as shown in Fig. 2, are used in the distal segment. The bellows' convolutions act as the spacer rings that prevent buckling of the backbones under compressive loads.

Actuation modularity is enabled by this dual continuum mechanism concept. The distal segment and the stem could be designed for different lengths, different diameters, and/or with different end effectors (e.g., grippers, scissors, cautery spatula, etc.). As far as the same proximal segment is used, the exchangeable effector could always be assembled into the actuation unit to bend the distal segment. The only modification required is to change the corresponding actuation parameters in the controller for different stems and/or distal segments.

Actuation of the gripper is also incorporated in the exchangeable effector as shown in Fig. 2(b). The gripper actuation rod is routed through a central channel and con-nected to a spring-loaded translating magnet. The magnet is pushed and pulled to close and open the gripper. The spring is used to avoid exerting excessive gripping force, while the magnet allows quick connection to the actuation unit.

The entire exchangeable effector only consists of mechanical components. It can be sterilized by emerging it in liquid agent such as glutaraldehyde and ortho-phthalaldehyde.

3.2   Actuation Unit

The actuation unit mainly consists of (i) one driving segment, (ii) two backbone driving assemblies, and (iii) a gripper driving assembly, as shown in Fig. 3. The actuation unit also includes casing and structural features that allow its attachment to a Denso manipulator.

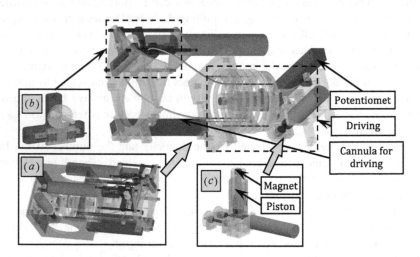

Fig. 3. Actuation unit of the SMARLT tool: (a) total assembly, (b) the backbone driving assembly, and (c) the gripper driving assembly

The driving segment is structurally similar to the one shown in Fig. 5(b), consisting of a base ring, several spacer rings, an end ring and four backbones made from ∅1 mm nitinol rods. It has matching geometries with the proximal segment in Fig. 2 so that the proximal segment can be securely assembled into the driving segment. Push-pull actuation of the driving backbones bends the driving and the proximal segments together so as to bend the distal segment. No matter how many backbones are arranged in the exchangeable effector, its actuation is always realized by the four driving backbones.

The backbone driving assembly pushes and pulls the driving backbones to bend the driving segment. According to the actuation kinematics in Eqs. (1) and (2), the two backbones that are 180° apart shall be pushed and pulled for the same amount at the same time.

As shown in Fig. 3(b), two driving backbones are connected to a rail-guided slider on which a rack is attached. The driving backbones are fixed to the end ring of the driving segment, routed through the guiding cannulae and the segment spacers respectively. A servomotor is connected to a pinion through a coupling to drive the rack to realize this push-pull actuation for the driving backbones.

A lead screw in the gripper driving assembly is driven by another servomotor through a gear train (including a pair of spur gears and a pair of bevel gears). A piston that is connected with the nut of the lead screw translates up and down. A magnet installed on the top of the piston allows quick connection to the translating magnet in the exchangeable effector. The attraction force between the magnets is big enough to pull the gripper open, while the piston pushes the magnets to close the gripper.

Potentiometers are installed in the gripper the backbone driving assemblies to sense the absolute positions of the lead screw and the driving backbones.

3.3 Control Infrastructure

The SMARLT's control infrastructure is set up so to allow teleoperation, which involves the control of the SMARLT tool and the Denso manipulator. A diagram of the control infrastructure is shown in Fig. 4.

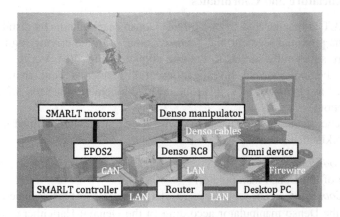

Fig. 4. Control infrastructure of the SMARLT and the Denso manipulator

A Phantom Omni device (Sensable Inc.) is connected to a desktop PC to acquire poses information. The SMARLT's central controller is an embedded system and it generates control reference signals for the SMARLT tool and the Denso manipulator according to the inputs from the Omni device and the inverse kinematics of the SMARLT and the Denso manipulator. The kinematics is detailed in Sect. 4.

The control references for the Denso manipulator are sent to the Denso RC8 controller (configured in a slave mode) via the LAN port using the UDP protocol so that the manipulator's joint references could be continuously updated.

The control signals for the SMARLT tool are sent to three Maxon EPOS2 24/2 digital controllers via the CAN bus. The EPOS2 controllers drive the servomotors in the actuation unit to drive the SMARLT tool.

The backbone driving assemblies use two Maxon A-max-22 motors. The gripper driving assembly uses one Maxon A-max-16 motor.

Three potentiometers in the actuation unit are read by the A/D ports of the EPOS2 controllers. The readings are sent back to the central controller via the CAN bus.

4   Kinematics Framework

As in Fig. 1, the continuum segment with a 2-DoF bending is incorporated in the SMARLT tool as a distal wrist. Kinematics of such a bending segment could be found in previous studies [7, 23, 24]. The segment's bending kinematics is summarized in Sect. 4.2 with the nomenclature and coordinates defined in Sect. 4.1.

The SMARLT tool is deployed and maneuvered by a manipulator (a 6-DoF Denso manipulator in this case) through a trocar. The Denso manipulator serves as a programmable RCM mechanism which positions and orients the SMARLT tool with respect to the trocar (the incision point, or the pivot point). The kinematics of the Denso manipulator system is presented in Sect. 4.3.

4.1   Nomenclature and Coordinates

The SMARLT is maneuvered by the 6-DoF Denso manipulator. Its distal segment is driven by the proximal segment that bends together with the driving segment. All the segments are structurally similar to the one shown in Fig. 5(b). Verified by the analytical and the experimental investigations, the segment's bent shapes could be approximated as circular arcs [7, 24]. The derived kinematics in Sect. 4.2 is based on this assumption.

Eleven coordinates are defined below with the nomenclature defined in Table 1 to describe the kinematics.

- *World Coordinate* $\{W\} \equiv \{\hat{\mathbf{x}}_W, \hat{\mathbf{y}}_W, \hat{\mathbf{z}}_W\}$ (or $\{D0\} \equiv \{\hat{\mathbf{x}}_{D0}, \hat{\mathbf{y}}_{D0}, \hat{\mathbf{z}}_{D0}\}$) is located at the base of the Denso manipulator.
- *Denso Coordinates* $\{Dj\} \equiv \{\hat{\mathbf{x}}_{Dj}, \hat{\mathbf{y}}_{Dj}, \hat{\mathbf{z}}_{Dj}\}$ $(j = 1, 2, \cdots, 6)$ are assigned to the joint axes of the Denso manipulator according to the Denavit-Hartenberg rules.

- *Segment Base Coordinate* $\{S1\} \equiv \{\hat{\mathbf{x}}_{S1}, \hat{\mathbf{y}}_{S1}, \hat{\mathbf{z}}_{S1}\}$ is attached to the segment's base ring. The XY plane is aligned with the base ring with its origin at the center. $\{S1\}$ is translated from $\{D6\}$ by a distance h in the $\hat{\mathbf{z}}_{D6}$ direction. $\hat{\mathbf{x}}_{S1}$ points from the center to the 1st backbone. The backbones are numbered according to the definition of δ_i.
- *Segment Base Bending Coordinate* $\{S2\} \equiv \{\hat{\mathbf{x}}_{S2}, \hat{\mathbf{y}}_{S2}, \hat{\mathbf{z}}_{S2}\}$ shares its origin with $\{S1\}$ and has the segment bending in its XY plane.
- *Segment Tip Bending Coordinate* $\{S3\} \equiv \{\hat{\mathbf{x}}_{S3}, \hat{\mathbf{y}}_{S3}, \hat{\mathbf{z}}_{S3}\}$ is obtained from $\{S2\}$ by a rotation about $\hat{\mathbf{z}}_{S2}$ such that $\hat{\mathbf{x}}_{S3}$ becomes the backbone tangent at the end ring. Origin of $\{S3\}$ is at the center of the end ring.
- *Segment Tip Coordinate* $\{S4\} \equiv \{\hat{\mathbf{x}}_{S4}, \hat{\mathbf{y}}_{S4}, \hat{\mathbf{z}}_{S4}\}$ is fixed to the end ring. $\hat{\mathbf{x}}_{S4}$ points from the end ring center to the first backbone and $\hat{\mathbf{z}}_{S4}$ is normal to the end ring.

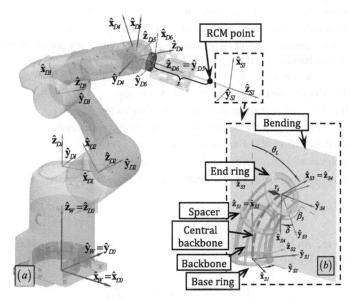

Fig. 5. Nomenclature and coordinates of the SMARLT-Denso system: (a) the Denso manipulator and (b) the bending segment

4.2 Kinematics of the Continuum Segment

As shown in Fig. 5(b), the backbones are pulled and pushed to bend the segment within the bending plane. The length and shape of the segment is indicated by a central backbone. According to previous investigations [7, 24], absence of the central backbone doesn't affect the bent shapes. In the case that the central backbone is removed to spare a lumen for passing through other components, a virtual central backbone still exists, governing the segment's length and shape.

The backbone shapes are approximated as circular arcs in planes parallel to the bending plane. The projection of one backbone on the bending plane has the same length as itself and is offset from the central backbone. The lengths of the central

Table 1. Nomenclature used in this paper

Symbol	Definition
j	Index of the Denso manipulator axes, $j = 1, 2, \cdots, 6$
φ_j	Joint variables of the Denso manipulator
$\boldsymbol{\Psi}_D$	$\boldsymbol{\Psi}_D \equiv [\varphi_1 \; \varphi_2 \; \cdots \; \varphi_6]^T$ is the manipulator's configuration vector
s	Distance along the SMARLT's stem from the $\{D6\}$ origin to the RCM point
h	Distance between the origins of $\{D6\}$ and $\{S1\}$ along the SMARLT's stem
i	Index of the segment backbones, $i = 1, 2, \cdots, m$
r_i	Distance from the virtual central backbone to the ith backbone
β_i	Division angle from the ith backbone to the 1st backbone; $\beta_1 = 0$ and β_i remain constant once the segment is built
L, L_i	Lengths of the central backbone and the ith backbone measured from the base ring to the end ring along the backbones
q_i	Push-pull actuation of the segment's ith backbone; $q_i \equiv L_i - L$
δ_i	A right-handed rotation angle about $\hat{\mathbf{z}}_{S1}$ from $\hat{\mathbf{y}}_{S2}$ to a ray passing through the central backbone and the ith backbone
δ	$\delta \equiv \delta_1$ and $\delta_i = \delta + \beta_i$
θ_L	The right-handed rotation angle from $\hat{\mathbf{x}}_{S2}$ to $\hat{\mathbf{x}}_{S3}$
$\boldsymbol{\Psi}_S$	$\boldsymbol{\Psi}_S \equiv [\theta_L \; \delta]^T$ is the segment's configuration vector
$\boldsymbol{\Psi}$	$\boldsymbol{\Psi} \equiv [\boldsymbol{\Psi}_D^T \; \boldsymbol{\Psi}_S^T]^T$ is the configuration vector of the entire system
$^{S1}\mathbf{p}_L$	Center position of the segment's end ring in $\{S1\}$
$^1\mathbf{R}_2$	Coordinate transformation matrix from frame 2 to frame 1
$^1\mathbf{T}_2$	Homogeneous transformation matrix from frame 2 to frame 1

backbone and the ith backbone are related as in (1), as well as the backbone actuation, according to the definition of q_i.

$$\begin{cases} L_i = L - r_i\theta_L\cos(\delta + \beta_i) \\ q_i = -r_i\theta_L\cos(\delta + \beta_i) \end{cases} \tag{1}$$

In order to bend the segment to a configuration specified by $\boldsymbol{\Psi}_s$, each backbone should be pushed or pulled according to (1). When many thin backbones are used in the distal segment of the SMARLT's exchangeable effector for (i) enhanced reliability and (ii) increased payload capability, it is more effective to use the proximal segment to actuate the distal segment since the bending possesses 2 DoFs.

The backbones arrangements (specified by r_i and β_i) could be arbitrary but should be similar and scaled in the distal and the proximal segments. This means $(r_i)_{\text{proximal}} = (\kappa r_i)_{\text{distal}}$ as well as $(\beta_i)_{\text{proximal}} = (\beta_i)_{\text{distal}}$. Then a bending of θ_L and δ on the distal segment requires a bending of θ_L/κ and $\delta + \pi$ on the proximal segment. The driving segment has the same configuration as the proximal segment. Then the four driving backbones should be pushed and pulled according to (2) with the corresponding bent configuration variables (θ_L/κ and $\delta + \pi$) and the structural parameters listed in Table 2.

$$\begin{cases} q_1 = -(r_i)_{driving} \frac{\theta_L}{\kappa} \cos(\delta + \pi) = -q_3 \\ q_2 = -(r_i)_{driving} \frac{\theta_L}{\kappa} \cos\left(\delta + \frac{3\pi}{2}\right) = -q_4 \end{cases}, \quad \kappa = \frac{(r_i)_{proximal}}{(r_i)_{distal}} \tag{2}$$

Table 2. Structural parameters of the SMARLT-Denso system

No.	α_{i-1}	a_{i-1}	d_i	φ_i	Distal segment	
	Denavit-Hartenberg parameters of the Denso manipulator				$r_i = 2.5$mm	$L = 40$mm
					Proximal segment	
1	0	0	473mm	φ_1	$r_i = 24$mm	$L = 35$mm
2	$-\pi/2$	180mm	0	$\varphi_2 - \pi/2$	Driving segment	
3	0	385mm	0	φ_3	$r_i = 30$mm	$L = 35$mm
4	$-\pi/2$	100mm	445mm	φ_4	$\beta_i = 0, \pi/2, \pi, 3\pi/2$	
5	$\pi/2$	0	0	φ_5	Translation	$h = 580$mm
6	$-\pi/2$	0	90mm	φ_6	Gripper tip	$g = 15$mm

The distal segment bends into circular arcs. Center position of the end ring is written as follows.

$$^{S1}\mathbf{p}_L = \frac{L}{\theta_L}\left[\cos \delta(1 - \cos \theta_L) \quad \sin \delta(\cos \theta_L - 1) \quad \sin \theta_L\right]^T \tag{3}$$

Where $^{S1}\mathbf{p}_L = [0 \quad 0 \quad L]^T$ when $\theta_L \to 0$.

Transformation matrix $^{S1}\mathbf{R}_{S4}$ relates $\{S4\}$ to $\{S1\}$

$$^{S1}\mathbf{R}_{S4} = {}^{S1}\mathbf{R}_{S2}{}^{S2}\mathbf{R}_{S3}{}^{S3}\mathbf{R}_{S4} \tag{4}$$

Where $^{S1}\mathbf{R}_{S2} = \begin{bmatrix} 0 & \cos \delta & \sin \delta \\ 0 & -\sin \delta & \cos \delta \\ 1 & 0 & 0 \end{bmatrix}$, $^{S2}\mathbf{R}_{S3} = \begin{bmatrix} \cos \theta_L & -\sin \theta_L & 0 \\ \sin \theta_L & \cos \theta_L & 0 \\ 0 & 0 & 1 \end{bmatrix}$, and

$^{S3}\mathbf{R}_{S4} = \begin{bmatrix} 0 & 0 & 1 \\ \cos \delta & -\sin \delta & 0 \\ \sin \delta & \cos \delta & 0 \end{bmatrix}$.

The segment's instantaneous kinematics (Jacobian) from the segment configuration space to the task space for the center of the end ring is as follows.

$$^{S1}\dot{\mathbf{x}} = \mathbf{J}_S \, \dot{\boldsymbol{\psi}}_S = \begin{bmatrix} \mathbf{J}_{vS} \\ \mathbf{J}_{\omega S} \end{bmatrix} \dot{\boldsymbol{\psi}}_S. \tag{5}$$

Where $^{S1}\mathbf{v} = \mathbf{J}_{vS}\dot{\boldsymbol{\psi}}_S$ and $^{S1}\boldsymbol{\omega} = \mathbf{J}_{\omega S}\dot{\boldsymbol{\psi}}_S$

$$\mathbf{J}_{vS} = L \begin{bmatrix} \cos \ \delta \left(\frac{\cos \theta_L - 1}{\theta_L^2} + \frac{\sin \theta_L}{\theta_L} \right) & \frac{\sin \delta}{\theta_L} \left(\cos \theta_L - 1 \right) \\ \sin \ \delta \left(\frac{1 - \cos \theta_L}{\theta_L^2} - \frac{\sin \theta_L}{\theta_L} \right) & \frac{\cos \delta}{\theta_L} \left(\cos \theta_L - 1 \right) \\ -\frac{\sin \theta_L}{\theta_L^2} + \frac{\cos \theta_L}{\theta_L} & 0 \end{bmatrix} \tag{6}$$

$$\mathbf{J}_{\omega S} = \begin{bmatrix} \sin \delta & \cos \delta \sin \theta_L \\ \cos \delta & -\sin \delta \sin \theta_L \\ 0 & \cos \theta_L - 1 \end{bmatrix} \tag{7}$$

4.3   Kinematics of the Denso Manipulator

Kinematics of the Denso manipulator could be easily described following the Denavit-Hartenberg parameters listed in Table 2. A general form of the homogeneous transformation matrix is as follows.

$$^{D(j-1)}\mathbf{T}_{Dj} = \begin{bmatrix} ^{D(j-1)}\mathbf{R}_{Dj} & ^{D(j-1)}\mathbf{p} \\ \mathbf{0}_{1\times3} & 1 \end{bmatrix}, \quad j = 1, 2, \cdots, 6 \tag{8}$$

Where $^{D(j-1)}\mathbf{R}_{Dj} = \begin{bmatrix} \cos \varphi_j & -\sin \varphi_j & 0 \\ \sin \varphi_j \cos \alpha_{i-1} & \cos \varphi_j \cos \alpha_{i-1} & -\sin \alpha_{i-1} \\ \sin \varphi_j \sin \alpha_{i-1} & \cos \varphi_j \sin \alpha_{i-1} & \cos \alpha_{i-1} \end{bmatrix}$,   and

$^{D(j-1)}\mathbf{p} = \begin{bmatrix} a_{j-1} & -d_j \sin \alpha_{i-1} & d_j \cos \alpha_{i-1} \end{bmatrix}^T$.

Jacobian matrix $\mathbf{J}_D$ of the Denso manipulator for the center of its distal flange could be derived in (9) to (11). The SMARLT tool is attached to the Denso manipulator through this flange.

$$^{D0}\dot{\mathbf{x}} = \mathbf{J}_D \dot{\boldsymbol{\psi}}_D = \begin{bmatrix} \mathbf{J}_{vD} \\ \mathbf{J}_{\omega D} \end{bmatrix} \dot{\boldsymbol{\psi}}_D \tag{9}$$

Where $^{D0}\mathbf{v} = \mathbf{J}_{vD} \dot{\boldsymbol{\psi}}_D$ and $^{D0}\boldsymbol{\omega} = \mathbf{J}_{\omega D} \dot{\boldsymbol{\psi}}_D$, $\mathbf{J}_{vD}, \mathbf{J}_{\omega D} \in \Re^{3\times6}$

$$\mathbf{J}_{vD} = \begin{bmatrix} ^{D0}\widehat{\mathbf{z}}_{D1} \times {}^{D0}\mathbf{p}_{D1D6} & ^{D0}\widehat{\mathbf{z}}_{D2} \times {}^{D0}\mathbf{p}_{D2D6} & ^{D0}\widehat{\mathbf{z}}_{D3} \times {}^{D0}\mathbf{p}_{D3D6} \\ \dot{\varphi}_4{}^{D0}\widehat{\mathbf{z}}_{D4} \times {}^{D0}\mathbf{p}_{D4D6} & \dot{\varphi}_5{}^{D0}\widehat{\mathbf{z}}_{D5} \times {}^{D0}\mathbf{p}_{D5D6} & \mathbf{0} \end{bmatrix} \tag{10}$$

$$\mathbf{J}_{\omega D} = \begin{bmatrix} ^{D0}\widehat{\mathbf{z}}_{D1} & ^{D0}\widehat{\mathbf{z}}_{D2} & ^{D0}\widehat{\mathbf{z}}_{D3} & ^{D0}\widehat{\mathbf{z}}_{D4} & ^{D0}\widehat{\mathbf{z}}_{D5} & ^{D0}\widehat{\mathbf{z}}_{D6} \end{bmatrix} \tag{11}$$

The 6th column of $\mathbf{J}_{vD}$ is zero because the rotation of the 6th joint does not induce additional linear velocity at the center of the distal flange.

5 Conclusions and Future Work

This paper presents the design and preliminary development of a modular robotic laparoscopic tool for MIS: the SMARLT tool. A dual continuum mechanism concept is utilized in the design to ensure reliability as well as achieve enhanced distal dexterity, increased payload capability and actuation modularity under a simple construction. With the kinematics derived, SMARLT tool would be able to be maneuvered by a Denso manipulator to perform typical laparoscopic tasks under teleoperation.

The SMARLT could provide an alternative option to realize robotic laparoscopic surgeries. The future efforts will primarily focus on (i) the derivation of kinematics with constrained motions and teleoperation, (ii) the compensation of continuum segment actuation, (iii) the stiffness characterization and representative surgical tasks demonstration.

Acknowledgments. This work was supported in part by the National Natural Science Foundation of China (Grant No. 51435010 and Grant No. 51375295), and in part by the Shanghai Jiao Tong University Interdisciplinary Research Funds (Grant No. YG2013MS26).

References

1. Cuschieri, A.: Laparoscopic surgery: current status, issues future developments. Surgeon 3(3), 125–138 (2005)
2. Taylor, R.H.: A perspective on medical robotics. Proc. IEEE **94**(9), 1652–1664 (2006)
3. Guthart, G.S., Salisbury, J.K.: The Intuitive^{TM} telesurgery system: overview and application. In: IEEE International Conference on Robotics and Automation (ICRA). San Francisco, CA (2000)
4. Guthart, G.: Annual report 2010, p. 108. Intuitive Surgical, Inc., Sunnyvale (2011)
5. Kanno, T., et al.: A forceps manipulator with flexible 4-DoF mechanism for laparoscopic surgery. IEEE-ASME Trans. Mechatron. **20**(3), 1170–1178 (2015)
6. Hong, M.B., In, Y.H.: Design of a novel 4-DoF wrist-type surgical instrument with enhanced rigidity and dexterity. IEEE-ASME Trans. Mechatron. **19**(2), 500–511 (2014)
7. Xu, K., Simaan, N.: An investigation of the intrinsic force sensing capabilities of continuum robots. IEEE Trans. Robot. **24**(3), 576–587 (2008)
8. Xu, K., Simaan, N.: Intrinsic wrench estimation and its performance index for multisegment continuum robots. IEEE Trans. Robot. **26**(3), 555–561 (2010)
9. He, C., et al.: Force sensing of multiple-DOF cable-driven instruments for minimally invasive robotic surgery. Int. J. Med. Robot. Comput. Assist. Surg. **10**(3), 314–324 (2014)
10. Leonard, S., et al.: Smart tissue anastomosis robot (STAR): a vision-guided robotics system for laparoscopic suturing. IEEE Trans. Biomed. Eng. **61**(4), 1305–1317 (2014)
11. Ma, R., et al.: Design and optimization of manipulator for laparoscopic minimally invasive surgical robotic system. In: 2012 International Conference on Mechatronics and Automation (ICMA). IEEE (2012)
12. Taylor, R.H., Stoianovici, D.: Medical robotics in computer-integrated surgery. IEEE Trans. Robot. Autom. **19**(5), 765–781 (2003)

13. Hadavand, M., et al.: A novel remote center of motion mechanism for the force-reflective master robot of haptic tele-surgery systems. Int. J. Med. Robot. Comput. Assist. Surg. **10**(2), 129–139 (2014)

14. Kuo, C.H., Dai, J.S.: Kinematics of a fully-decoupled remote center-of-motion parallel manipulator for minimally invasive surgery. J. Med. Devices-Trans. ASME **6**(2), 021008 (2012)

15. Nasseri, M.A., et al.: Virtual fixture control of a hybrid parallel-serial robot for assisting ophthalmic surgery: an experimental study. In: 2014 5th IEEE RAS & EMBS International Conference on Biomedical Robotics and Biomechatronics (BioRob), pp. 732–738 (2014)

16. Lopez, E., et al.: Implicit active constraints for robot-assisted arthroscopy. In: 2013 IEEE International Conference on Robotics and Automation (ICRA), pp. 5390–5395 (2013)

17. Dubrowski, A., et al.: Quantification of motion characteristics and forces applied to tissues during suturing. Am. J. Surg. **190**(1), 131–136 (2005)

18. Okamura, A.M.: Methods for haptic feedback in teleoperated robot-assisted surgery. Ind. Robot Int. J. **31**(6), 499–508 (2004)

19. Berg, D.R., et al.: Determination of surgical robot tool force requirements through tissue manipulation and suture force measurement. Trans. ASME-W-J. Med. Devices **5**(2), 027517 (2011)

20. Xu, K., Zhao, J., Fu, M.: Development of the SJTU unfoldable robotic system (SURS) for single port laparoscopy. IEEE/ASME Trans. Mechatron. **20**(5), 2133–2145 (2014)

21. Xu, K., Fu, M., Zhao, J.: An experimental kinestatic comparison between continuum manipulators with structural variations. In: IEEE International Conference on Robotics and Automation. HongKong (2014)

22. Zhao, J., et al.: An endoscopic continuum testbed for finalizing system characteristics of a surgical robot for NOTES procedures. In: IEEE/ASME International Conference on Advanced Intelligent Mechatronics (AIM), Wollongong, Australia, pp. 63–70 (2013)

23. Webster, R.J., Jones, B.A.: Design and kinematic modeling of constant curvature continuum robots: a review. Int. J. Robot. Res. **29**(13), 1661–1683 (2010)

24. Xu, K., Simaan, N.: Analytic formulation for the kinematics, statics and shape restoration of multibackbone continuum robots via elliptic integrals. J. Mech. Robot. **2**(1), 011006 (2010)

Preliminary Development of a Continuum Dual-Arm Surgical Robotic System for Transurethral Procedures

Kai Xu[1]([⊠]), Bo Liang[2], Zhengchen Dai[2], Jiangran Zhao[2], Bin Zhao[2],
Huan Liu[2], Liang Xiao[3], and Yinghao Sun[3]([⊠])

[1] State Key Laboratory of Mechanical System and Vibration,
School of Mechanical Engineering, Shanghai Jiao Tong University,
Shanghai 200240, China
k.xu@sjtu.edu.cn

[2] RII Lab (Lab of Robotics Innovation and Intervention),
UM-SJTU Joint Institute, Shanghai Jiao Tong University,
Shanghai 200240, China
{liangb_sjtu, zhengchen.dai, zjr318, zhaobin2014,
liuhuan_2013}@sjtu.edu.cn

[3] Changhai Hospital, The Second Military Medical University,
Shanghai 200433, China
shawn021@163.com, sunyhsmmu@126.com

Abstract. Bladder cancer, with the leading number of new cases in all urinary system cancers and a high recurrence rate, poses a substantial threat to human health. Even with the transurethral accessibility, current surgical tools have not fully allowed convenient resection of the bladder tumors. This paper presents the design and the preliminary development of a continuum dual-arm surgical robotic system for transurethral procedures. This development aims at improving the current surgical treatments by providing intravesicular imaging with enhanced distal dexterity. With the proposed system, new surgical techniques for bladder tumor resection could be explored. The clinical motivation, design overview, system descriptions and preliminary developments of this transurethral surgical robot are presented. With the system constructed in the near future, a series of ex-vivo and in-vivo experimentations would be carried out to verify the proposed functionalities.

Keywords: Continuum arm · Surgical robot · Transurethral procedures

1 Introduction

BLADDER cancer has the leading number of newly diagnosed cases in all urinary system cancers: about 74,000 new cases with 16,000 cancer-related deaths in US in 2015 [1]. About seventy percent of the bladder tumors are superficial when they are initially discovered. The primary treatment is transurethral resection (TUR) but the problem lies on the 3-month recurrence rate that could be as high as 75% [2]. Risk factors of the tumor recurrence include the number and size of the tumors: more and

© Springer International Publishing AG 2017
Y. Huang et al. (Eds.): ICIRA 2017, Part II, LNAI 10463, pp. 311–322, 2017.
DOI: 10.1007/978-3-319-65292-4_27

bigger tumors lead to a higher recurrence rate [3]. Chemotherapy, immunotherapy [2] or a repeat TUR (ReTUR) [4] could be carried out after the initial TUR to lower the tumor reoccurrence rate.

Even with the transurethral accessibility, current surgical tools have not fully allowed convenient resection of the bladder tumors so as to ensure a consistent treatment [2]. What's more, bigger bladder tumors are usually removed via multiple cuts. There is evidence that the floating tumor cells after electro-resection may also contribute to the tumor recurrence [2]. Clearly the clinical needs yearn for a surgical system with intravesicular dexterity and functionalities so that it could conveniently remove multiple bladder tumors unbrokenly even when the tumors are relatively big (e.g., bigger than 25 mm in diameter).

Contrast to the keen clinical needs, only incremental changes have been made to the surgical tools until recently Goldman *et al.* developed a continuum transurethral manipulator with integrated fiberscope, laser cautery and biopsy forceps for bladder tumor resection [5, 6]. On the related subjects, Hendrick *et al.* developed a dual-arm endoscopic robot for prostate surgery [7–9], while Russo *et al.* developed the ASTRO system for transurethral laser surgery of benign prostatic hyperplasia [10]. Several other systems were also developed for percutaneous brachytherapy [11, 12] and bladder urothelium examination [13]. As the main contribution, this paper proposes a dual-arm surgical system for transurethral procedures (DASSTUP) with enhanced intravesicular dexterity and imaging, as shown in Fig. 1, aiming at removing multiple superficial bladder tumors unbrokenly from their roots in the submucosa layer, even when the tumors are relatively large. The system consists of a customized multi-channel cystoscope, two exchangeable continuum arms for surgical interventions, and a lockable multi-joint system stand.

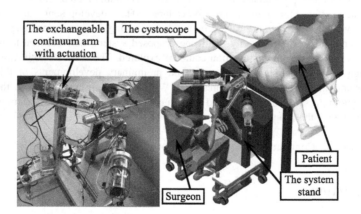

Fig. 1. Design and the partially constructed DASSTUP system

The paper is organized as follows. Section 2 presents the clinical motivations and the design overview. Section 3 presents the system descriptions while preliminary system fabrications and experimentations are reported in Sect. 4. The conclusions and the future work are summarized in Sect. 5.

2 Clinical Motivation and System Overview

A surgeon performing TUR should remove all visible tumors for the surgical treatment of bladder cancer, using a urologic resectoscope. The transurethral portion of a current resectoscope is a rigid tube with multiple telescoping components. A monopolar wire loop (or a laser fiber) is used to perform the tumor resection. Two critical hurdles are identified.

- Because of the rigidity of the resectoscope tube, it is quite difficult to pry the resectoscope to access the tumors on the side wall or near the entrance of the bladder. In some cases, the pubic bone prevents the resectoscope from being tilted to reach a bladder tumor.
- Large tumors have to be removed through multiple cuts and the floating cancer cells could potentially increase the tumor recurrence [2].

Despite the keen clinical needs, only incremental changes have been made to the resectoscope. For example, a flexible bending tip was integrated to enhance the intravesicular dexterity with laser resection [14], whereas the monopolar wire loop was made rotatable to facilitate tissue cutting [15].

Inspired by the recent advances in robotic systems for SPL (Single Port Laparoscopy) [16–20], The DASSTUP system with enhanced intravesicular dexterity and imaging is proposed in this paper, as shown in Figs. 1 and 2.

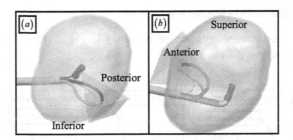

Fig. 2. Proposed uses of the DASSTUP system inside a bladder: (a) the forward-looking pose and (b) the backward-looking pose

A completely-redesigned cystoscope is firstly inserted into the bladder. With the vision unit extended and bent upwards, two miniature continuum arms could be inserted to form the forward-looking pose, as shown in Fig. 2(a). This working configuration primarily treats tumors in the posterior and inferior areas of the bladder.

The cystoscope could be flipped and inserted with the camera facing up. Then the vision unit could be extended and bent backwards. With the two continuum arms inserted, a backward-looking pose as shown in Fig. 2(b) is formed to treat tumors in the anterior and superior areas of the bladder. To be noted, the right arm in the DASSTUP system will appear as the left arm in the camera view under the backward-looking configuration. Proper mapping for the teleoperation should be implemented.

In either the forward-looking or the backward-looking poses, two manipulation arms could help to achieve complete resection of a large bladder tumor as a whole. For example, one arm could be used to push or grasp a large bladder tumor so as to expose its root. Then the other arm performs resection to separate the tumor from the (sub)-mucosa layer. The tumor could be placed into a pre-deployed bag for final extraction. In this way, possible floating tumors cells could be kept to a minimal level.

In order to lower the system complexity, the cystoscope is made manual, while the two continuum arms are motorized and controlled via a paradigm of teleoperation. As shown in Fig. 1, one surgeon will sit at the patient side to manipulate the cystoscope. Another surgeon will sit at a surgeon console (not shown) to tele-operate the continuum arms. The results presented in this paper focus on the patient-side development. Complete descriptions of the entire system would be introduced in a future publication.

3   System Description

The presented DASSTUP system consists of the following major components on the patient side: (i) one customized multi-channel cystoscope, (ii) two exchangeable continuum arms with actuation units, and (iii) a lockable multi-joint system stand. As mentioned above, the cystoscope is made manual, while the two continuum arms are motorized. These components are described in detail in this section.

3.1   Design of the Multi-channel Cystoscope

The customized multi-channel cystoscope is shown in Fig. 3. Its form was determined by carefully considering the intended use.

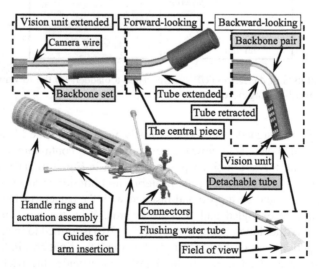

Fig. 3. (a) Cystoscope design of the DASSTUP system and (b) the cross section of the cystoscope

The primary function provided by a cystoscope is the illumination and visualization of the bladder. In the presented design in Fig. 3, the vision unit has a camera chip installed on the side wall (instead of on the end face like an ordinary cystoscope). The camera chip is surrounded by multiple LEDs (light-emitting diodes) for illumination. The vision unit is oriented by pushing and pulling three backbones that are made from super-elastic nitinol.

Since the DASSTUP system attempts to cover the entire bladder, ideally the camera chip could be oriented to look in any directions. If the camera chip is installed on the end face of the vision unit, the vision unit has to be rotated for nearly 180° to visualize the entrance area of the bladder. This could be hardly achieved by three backbones without any additional components to prevent the backbones from buckling.

With the camera chip installed on the side wall, the vision unit only needs to be re-oriented for about 90°. The actuation feasibility is demonstrated by the experiments that are presented in Sect. 4.1. Together with the rolling motion (rotation about the outer tube's axis) of the cystoscope, the entire bladder could be visualized.

This vision unit is oriented by pushing and pulling the backbones. These nitinol backbones and the vision unit essentially form a continuum parallel mechanism. Another continuum parallel robot could be seen in [21]. It is also similar to the DDU (Distal Dexterity Unit) firstly proposed in [22]. Since the vision unit is only subject to limited external loads, the use of a continuum structure is suitable.

Three adjustments could be realized for the vision unit, with respect to the cysto-scope, by telescoping the elements inside the detachable outer tube: (i) extend the central piece together with the vision unit; (ii) extend the backbones with respect to the central piece to change the distance between the vision unit and the central piece; and (iii) orient the vision unit upwards and downwards to form the forward-looking and the backward-looking poses, respectively.

Even though the vision unit only needs to be re-oriented within a plane, three backbones are used for better structural stability. Their arrangement is shown in the cross section in Fig. 4. Two backbones, forming the backbone pair, are arranged next to the camera wire and connected to the vision unit. In order to reduce possible tear to the camera wire, the backbone pair always has the same amount of translation as the camera wire. This feature is realized by the actuation assembly. The backbone set is arranged in the lower half of the cross section, formed by a nitinol rod inside a nitinol tube. The nitinol rod is attached to the vision unit, while the tube could slide with respect to the rod. To orient the vision unit upwards, the tube will be extended, sliding over the nitinol rod, to push the vision unit. To orient the vision unit downwards, the tube will be retracted first and the nitinol rod will then be pulled as shown in Fig. 3.

The reason for this particular design is to prevent backbone buckling while ori-enting the vision unit. While orienting the vision unit upwards, the backbone set would be strong enough to undertake the compressive load. Then the bending stiffness of the backbone set should be higher than the bending stiffness of the backbone pair plus the camera wire. While orienting the vision unit downwards, the bending stiffness of the backbone pair plus the camera wire should be higher than that of the backbone set. The retractable concentric tube-rod configuration could introduce such a desired change in the bending stiffness.

The cross section of the cystoscope is shown in Fig. 4. An inner tube is arranged inside the detachable ∅8 mm outer tube. This outer diameter is set equal to the French gauge number of 24 (instead of 26) to fit a greater group of patients.

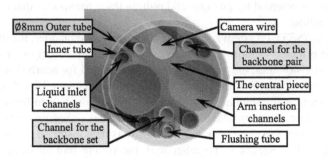

Fig. 4. Cross section of the DASSTUP's cystoscope

Besides the channels for the camera wire, the backbone pair and the backbone set, two channels are arranged next to the central piece for the two ∅3 mm exchangeable continuum arms. Four liquid inlet channels are used to circulate sterile liquid (e.g., water, saline, or glycine solution) during an operation. The gaps between the channels and between the outer and the inner tubes serve as the water outlets. There are four liquid circulation connectors in the cystoscope as shown in Fig. 3. The horizontal two are for inlets, while the vertical two are for liquid outlets.

One additional channel is reserved for the flushing tube. It will be used to flush the blood to clean the camera view at a surgical site.

The actuation assembly showed in Figs. 3 and 5 realizes the aforementioned three adjustments of the vision unit with respect to the cystoscope. There are three lead screws (A, B and C as shown in Fig. 5) in the actuation assembly. They are actuated by the three handle rings, via extensible shaft couplings.

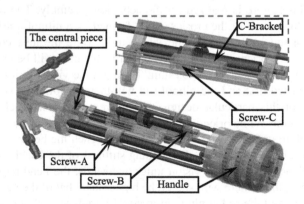

Fig. 5. Actuation scheme of the DASSTUP's cystoscope

For the deployment of the customized cystoscope, the detachable Ø8 mm outer tube would be firstly delivered into the bladder with the help of an obturator. Then the cystoscope is inserted into the outer tube with the vision unit at its initial pose. Under the initial pose, the backbones are completely retracted such that the vision unit axially pushes against the central piece.

After insertion, all the backbones are firstly extended from the central piece. Then the backbone set could be actuated to orient the vision unit, upwards or downwards, for a desired field of view.

3.2 Design of the Continuum Arm

Two Ø3 mm exchangeable continuum arms could be inserted through the channels in the cystoscope for surgical interventions.

The arm's DoF (Degree of Freedom) configuration is shown in Fig. 6(a). The arm consists of two continuum segments. Each segment possesses two bending DoFs. The arm is extended from the arm channel in the cystoscope. This is equivalent to actively change the length of the segment-I, giving the segment-I the third DoF. Moreover, the arm could be rotated about its axis to achieve a rolling motion. Thus in total, the arm possesses six DoFs plus one more actuation for the gripper. This DoF configuration is determined combining the advantages from previous developments [19, 23].

The arm has an outer diameter of 3 mm. Its structure is similar to the ones presented in [16, 19]. The arm cross section is shown in Fig. 6(b). Eight Ø0.5 mm holes are used for the arm backbones: four for each continuum segment.

The arm is connected to a transmission assembly as shown in Fig. 6(c). The arm backbones are routed by a set of guiding cannulae and connected to the nuts of the twin lead screws. The backbones arranged opposite to each other are connected to the nuts on a single twin lead screw so as to achieve the same amount of push-and-pull actuation to bend the corresponding segments.

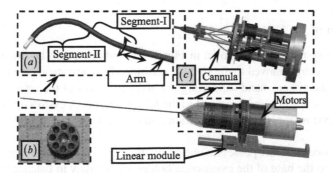

Fig. 6. Continuum arm of the DASSTUP system: (a) the DoF configuration, (b) the arm cross section, and (c) the transmission assembly

The arm rolling motion is actuated by a motorized gear ring in the actuation unit, whereas the translational motion of the Segment-I is actuated by a linear module that carries the motors and the arm.

The arm kinematics could refer to the kinematics presented in [23]. The kinematics is based on the modeling assumption that the segments bend into circular arcs. The details are not reported here for sake of brevity. Using the kinematics model, motions of the continuum arms could be simulated under teleoperation as shown in Fig. 7. It is shown that the arm tip could be controlled to trace the root area of a large bladder tumor in different locations of the bladder.

The root area assumes a diameter of 10 mm, while the bladder tumor assumes a diameter of 25 mm in the simulations in Fig. 7. The simulated tumors are located in the posterior and inferior areas of the bladder, while the DASSTUP system is assumed to be in its forward-looking pose.

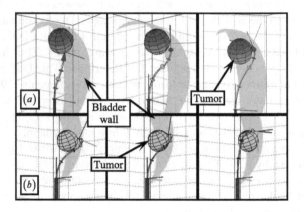

Fig. 7. Simulated movements of the continuum arm under teleoperation with the tumor in the (a) posterior and (b) inferior areas of the bladder.

3.3   Design of the System Stand

The lockable system stand is shown in Fig. 8. The wheels could be lifted by a lifting jack after the stand is moved to the suitable position. Two arm stands are located on the two sides of the system stand supporting the actuation units of the arms. A multi-joint lockable cystoscope holder is attached to the left arm stand. A cystoscope holder frame secures the cystoscope. One could position and orient the cystoscope by manipulating the handle.

The cystoscope holder possesses one prismatic joint and five rotary joints. A weight is connected to the base of the cystoscope holder via a pulley to balance the gravity of the cystoscope and the holder. Brakes are integrated to the pulley axis and to each rotary joint to lock the cystoscope holder, when the brakes are powered off. Once powered on, all the joints in the cystoscope holder could move freely. The brakes are switched on/off by a trigger in the handle.

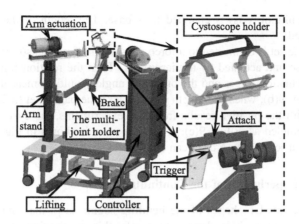

Fig. 8. Stand of the DASSTUP system

4   Preliminary Fabrications and Experimentations

With the DASSTUP system partially constructed, a series of experiments were carried out to verify the consistence between the actual system and the design expectation.

4.1   Verification of the Vision Unit

Before the cystoscope is fabricated, a series of experiments were firstly performed to verify the motion capability of the vision unit under the actuation by the backbones.

The experimental setup is shown in Fig. 9(a). The backbone arrangements are identical to the design as presented in Sect. 3.1. The camera wire and the backbone pair

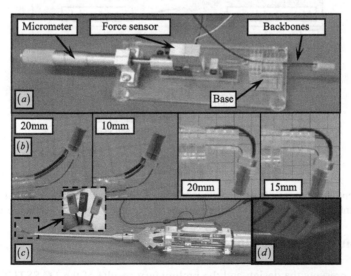

Fig. 9. Experimental verifications for the cystoscope: (a) the setup, (b) the forward-looking and backward-looking poses, (c) the fabricated cystoscope, and (d) a underwater view inside a coca can

(two ⌀0.5 mm nitinol rods) is clamped to a base. The length of the backbone pair could be adjusted.

The backbone set is connected to a force sensor. The force sensor could slide along a slot in the base and actuated by a micrometer. Then the relationship between actuation lengths, actuation forces and the orienting angles could be measured. The results are shown in Fig. 9(b), when the length of the backbone pair is set to 20 mm, 15 mm, 10 mm, etc. With the positive results from the experiments, the cystoscope was fabricated and assembled as in Fig. 9(c). A camera view is shown in Fig. 9(d) when the vision unit was inserted into a coca can filled with water.

4.2   Bending Experiments of the Continuum Arm

In the kinematics model, shapes of the bending segments in the continuum arm are assumed to be circular. It is desired to verify this assumption for this particular development since the arm size is quite small. With the assumption verified, the teleoperation motions as shown in Fig. 7 could be expected.

The continuum arm in a short version was firstly assembled as shown in Fig. 10(a). Then the 25 mm long segment was bent up to 90° as shown in Fig. 10(b). The bending of the two-segment continuum arm was also tested as shown in Fig. 10(c). The bent shapes are close to circular arcs.

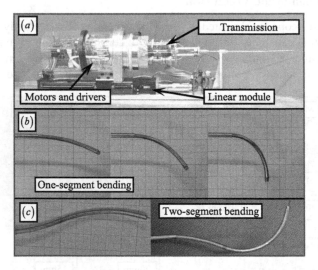

Fig. 10. Bending experiments of a continuum arm in a short version for modeling validation: (a) the assembly, (b) one-segment bending results and (c) two-segment bending results

5   Conclusions and Future Work

This paper presents the design and the preliminary results of the DASSTUP system, a continuum dual-arm surgical system for transurethral procedures. This development aims at improving the current surgical treatments of bladder cancer by providing intravesicular dexterity and imaging. With the proposed system, new surgical techniques for bladder tumor resection could be explored. The clinical motivation, design overview, component descriptions and system constructions are presented.

The immediate future work is to complete the system construction. Although most of the individual component have fabricated and tested, the integration could still be challenging. Fabrication of the long continuum arm could also become tricky. Arm actuation compensation would be expected. Eventually ex-vivo and in-vivo experimentations would be carried out to verify the proposed functionalities.

Acknowledgments. This work was supported in part by the National Natural Science Foundation of China (Grant No. 51435010, Grant No. 51375295 and Grant No. 91648103).

References

1. Siegel, R.L., Miller, K.D., Jemal, A.: Cancer statistics, 2015. CA: A Cancer J. Clin. **65**(1), 5–29 (2015)
2. Kurth, K.H., Bouffioux, C., Sylvester, R., Van der Meijden, A.P.M., Oosterlinck, W., Brausi, M.: Treatment of superficial bladder tumors: achievements and needs. Eur. Urol. **37**(Suppl. 3), 1–9 (2000)
3. Millán-Rodríguez, F., Chéchile-Toniolo, G., Salvador-Bayarri, J., Palou, J., Vicente-Rodríguez, J.: Multivariate analysis of the prognostic factors of primary superficial bladder cancer. J. Urol. **163**(1), 73–78 (2000)
4. Grimm, M.-O., Steinhoff, C., Simon, X., Spiegelhalder, P., Ackermann, R., Vögeli, T.A.: Effect of routine repeat transurethral resection for superficial bladder cancer: a long-term observational study. J. Urol. **170**(2), 433–437 (2003)
5. Goldman, R.E., Bajo, A., MacLachlan, L.S., Pickens, R., Herrell, S.D., Simaan, N.: Design and performance evaluation of a minimally invasive telerobotic platform for transurethral surveillance and intervention. IEEE Trans. Biomed. Eng. **60**(4), 918 (2013)
6. Pickens, R.B., Bajo, A., Simaan, N., Herrell, D.: A pilot ex vivo evaluation of a telerobotic system for transurethral intervention and surveillance. J. Endourol. **29**(2), 231–234 (2015)
7. Hendrick, R.J., Herrell, S.D., Webster III, R.J.: A multi-arm hand-held robotic system for transurethral laser prostate surgery. In: IEEE International Conference on Robotics and Automation (2014)
8. Hendrick, R.J., Mitchell, C.R., Herrell, S.D., Webster III, R.J.: Hand-held transendoscopic robotic manipulators: a transurethral laser prostate surgery case study. Int. J. Robot. Res. **34**(13), 1559–1572 (2015)
9. Mitchell, C.R., Hendrick, R.J., Webster III, R.J., Herrell, S.D.: Toward improving transurethral prostate surgery: development and initial experiments with a prototype concentric tube robotic platform. J. Endourol. **30**(6), 692–697 (2016)
10. Russo, S., Dario, P., Menciassi, A.: A novel robotic platform for laser-assisted transurethral surgery of the prostate. IEEE Trans. Biomed. Eng. **62**(2), 489–500 (2015)
11. Goldenberg, A.A., Trachtenberg, J., Kucharczyk, W., Yi, Y., Haider, M., Ma, L., Weersink, R., Raoufi, C.: Robotic system for closed-bore MRI-guided prostatic interventions. IEEE/ASME Trans. Mechatron. **13**(3), 374–379 (2008)
12. Mozer, P.C., Partin, A.W., Stoianovici, D.: Robotic image-guided needle interventions of the prostate. Rev. Urol. **11**(1), 7–15 (2009)

13. Yoon, W.J., Park, S., Reinhall, P.G., Seibel, E.J.: Development of an automated steering mechanism for bladder urothelium surveillance. J. Med. Devices **3**(1), 0110041 (2009)
14. Gao, X., Ren, S., Xu, C., Sun, Y.: Thulium laser resection via a flexible cystoscope for recurrent non-muscle-invasive bladder cancer: initial clinical experience. BJU Int. **102**(9), 1115–1118 (2008)
15. Wilby, D., Thomas, K., Ray, E., Chappell, B., O'Brien, T.: Bladder cancer: new TUR techniques. World J. Urol. **27**, 309–312 (2009)
16. Ding, J., Goldman, R.E., Xu, K., Allen, P.K., Fowler, D.L., Simaan, N.: Design and coordination kinematics of an insertable robotic effectors platform for single-port access surgery. IEEE/ASME Trans. Mech. **18**(5), 1612–1624 (2013)
17. Simi, M., Silvestri, M., Cavallotti, C., Vatteroni, M., Valdastri, P., Menciassi, A., Dario, P.: Magnetically activated stereoscopic vision system for laparoendoscopic single-site surgery. IEEE/ASME Trans. Mechatron. **18**(3), 1140–1151 (2013)
18. Kobayashi, Y., Sekiguchi, Y., Noguchi, T., Takahashi, Y., Liu, Q., Oguri, S., Toyoda, K., Uemura, M., Ieiri, S., Tomikawa, M., Ohdaira, T., Hashizume, M., Fujie, M.G.: Development of a robotic system with six-degrees-of-freedom robotic tool manipulators for single-port surgery. Int. J. Med. Robot. Comput. Assist. Surg. **11**(2), 235–246 (2015)
19. Xu, K., Zhao, J., Fu, M.: Development of the SJTU unfoldable robotic system (SURS) for single port laparoscopy. IEEE/ASME Trans. Mechatron. **20**(5), 2133–2145 (2015)
20. Zhao, J., Feng, B., Zheng, M.-H., Xu, K.: Surgical robots for SPL and NOTES: a Review. Minim. Invasive Ther. Allied Technol. **24**(1), 8–17 (2015)
21. Bryson, C.E., Rucke, D.C.: Toward parallel continuum manipulators. In: IEEE International Conference on Robotics and Automation (ICRA), Hong Kong, China, pp. 778–785 (2014)
22. Simaan, N., Taylor, R.H., Flint, P.: A dexterous system for laryngeal surgery. In: IEEE International Conference on Robotics and Automation (ICRA), New Orleans, LA, pp. 351–357 (2004)
23. Liu, S., Yang, Z., Zhu, Z., Han, L., Zhu, X., Xu, K.: Development of a dexterous continuum manipulator for exploration and inspection in confined spaces. Ind. Robot: Int. J. **43**(3), 284–295 (2016)

Optimal Design of a Cable-Driven Parallel Mechanism for Lunar Takeoff Simulation

Yu Zheng[1,2], Wangmin Yi[1,3(✉)], and Fanwei Meng[1,3]

[1] Beijing Institute of Space Environment Engineering, Beijing 100094, China
`yiwangmin79@hotmail.com`
[2] Science and Technology on Reliability and Environment Engineering Laboratory, Beijing Institute of Space Environment Engineering, Beijing 100094, China
[3] Beijing Engineering Research Center of the Intelligent Assembly Technology and Equipment for Aerospace Product, Beijing 100094, China

Abstract. The technology of taking off from the lunar is of great importance for the returning of spacecraft from lunar. To verify the stability of spacecraft when taking off lunar, the force state of the spacecraft needs to be simulated accurately. Traditional simulation mechanisms are unable to meet the simulation requirements of lunar takeoff. A 6-DOF lunar takeoff simulation device with nine cables is introduced and optimized in this paper. Firstly, the dynamic workspace of cable-driven parallel mechanism, which under the condition that the quality, the vertical acceleration and the horizontal acceleration are 15 kg, 2.4 m/s² and ±1.1 m/s² respectively, is analyzed when the acceleration requirements of simulation mechanisms is taken into consideration. Then the installation positions of cables are investigated in detail based on experimental design method and response surface method to achieve optimization results that could get the maximum dynamic workspace. Moreover, the virtual prototype experiment based on the multi-body dynamics model is utilized to verify the accuracy of the optimized results.

Keywords: Cable-driven parallel mechanism · Lunar takeoff simulation · Response surface method

1 Introduction

The technology of taking off from the lunar is the key for lunar exploration project, which is of great importance for the returning of spacecraft from lunar. To verify the stability of spacecraft when taking off lunar, the force state of the spacecraft needs to be simulated accurately, which is the key of the takeoff from lunar. Traditional simulation methods, such as computer simulation and rocket-propelled experiment, are too expensive or too complex for lunar takeoff simulation. In order to implement lunar takeoff simulation in an efficient way, a cable-driven parallel mechanism has been built [1]. Lunar takeoff simulation has high requirements on the acceleration and workspace of cable-driven parallel mechanism. In this paper, the installation positions of cables will be optimized based on the acceleration requirements and workspace requirements

© Springer International Publishing AG 2017
Y. Huang et al. (Eds.): ICIRA 2017, Part II, LNAI 10463, pp. 323–333, 2017.
DOI: 10.1007/978-3-319-65292-4_28

of simulation mechanism. It should be noted that in this paper, the layout of installation points on moving platform is "6–3".

Owing to the advantages of simple configuration, large workspace and high speed, cable-driven parallel mechanisms has been widely used [2–4]. However, the fact that cable can only bear tension but not compression, the installation positions of cables and the acceleration of moving platform all have effect on the dynamic workspace of cable-driven parallel mechanisms [5–7]. Lunar takeoff simulation requires the quality, the vertical and horizontal acceleration of moving platform are 15 kg, 2.4 m/s² and ±1.1 m/s² respectively. Therefore, the purpose of this paper is to optimize installation points of cables to maximize dynamic workspace of constraint condition in terms of cable tensions.

So far, great progress has been made in this field. Landsberger [8], studied the feasible workspace of cable-driven parallel mechanism. Barrette and Gosselin [9], carried out that the acceleration of moving platform will influence workspace and introduced the concept of dynamic workspace. Agrawal [10], introduced a cable driven upper arm exoskeleton and optimized the connection points of cables. Ouyang [11], maximum the wrench-feasible workspace of a parallel mechanism driven by 8 cables by optimizing the dimension of its frame. Rui [12], optimized a 3T cable-driven parallel mechanism with constraint condition in terms of cable force and stiffness.

The existing research results are lack of the study of the mathematic relationship between structural parameters and the volume of workspace. Response surface method (RSM) is a multi-parameter optimization method [13]. The main idea of RSM is establishing and analyzing the surrogate mathematical model between optimization variables and optimization index. This paper employs latin-hypercube sampling method to select experimental points and establishes surrogate mathematical model by RSM to achieve optimization results. The virtual prototype experiment is taken to verify the accuracy of the optimized results.

2 Kinematics and Force Equilibrium

The nine-cable parallel mechanism is shown in Fig. 1. The device is called cable-driven parallel simulator for lunar takeoff, CDPSL for short. CDPSL consists of a base frame equipped with motors and winches and a moving platform connected together with 9 cables, and the layout of cables is "6–3". The lower 3 cables are connected to the winches directly, while the upper 6 cables go through pulleys to be connected to the winches on the bottom. The length of cable and cable tension are controlled by winches driven by motors.

Figure 2 shows the mechanism model of CSPSL. In Fig. 2, two coordinates are defined as follow, the global frame K_b is fixed at the base frame, the moving frame K_p is connected to the moving platform. $B_i(i = 1,2,...,9)$ represent the cable installation positions on the base frame and $P_i(i = 1,2,...,9)$ denote the cable installation positions on moving platform. The platform-fixed vectors to the connecting points in K_p, the vectors to the fixed point on winches in K_b and the cable tension vectors are described by $p_i(i = 1,2,...,9)$, $b_i(i = 1,2,...,9)$ and f_i $(i = 1,2,...,9)$ respectively. f_p and τ_p represent extern forces and torques acting on the platform except for gravity. The position of origin of $\mathbf{K_p}$ in the fixed base frame $\mathbf{K_b}$ is defined as $\mathbf{op} = (x, y, z)$ and the posture of platform in $\mathbf{K_b}$ is defined as (Ψ, Φ, γ). Define $\mathbf{R}$ as the transformation matrix from $\mathbf{K_p}$ to $\mathbf{K_b}$.

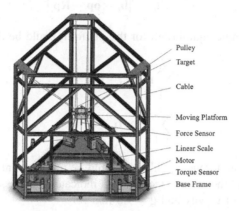

Fig. 1. CDPSL with nine cables.

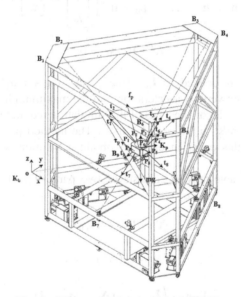

Fig. 2. The mechanism model of CDPSL

According to the vector loop method, the kinematics of each cable can be established in (1) with $i = 1, 2, 3, \ldots, 9$

$$\mathbf{l}_i = \mathbf{b}_i - \mathbf{op} - \mathbf{Rp}_i \qquad (1)$$

The unit vector of cable could be calculated by

$$\mathbf{u}_i = \frac{\mathbf{l}_i}{|\mathbf{l}_i|} = \frac{\mathbf{b}_i - \mathbf{op} - \mathbf{Rp}_i}{|\mathbf{b}_i - \mathbf{op} - \mathbf{Rp}_i|} \tag{2}$$

The force and torque equilibrium for the platform could be derived

$$\sum_{i=1}^{9} \mathbf{f}_i + \mathbf{f_p} + \mathbf{mg} = 0$$
$$\sum_{i=1}^{9} \mathbf{p}_i \times \mathbf{f_i} + \mathbf{\tau_p} = 0 \tag{3}$$

Then, the force and torque equilibrium could be rewritten into matrix form, where $\mathbf{A^T}$ is the transpose of the inverse Jacobian, m indicates the mass of moving platform, $\mathbf{g}$ is the acceleration of gravity and $\mathbf{g} = 9.8$ m/s^2.

$$\begin{bmatrix} \mathbf{u}_1 & \cdots & \mathbf{u}_9 \\ \mathbf{p}_1 \times \mathbf{u}_1 & \cdots & \mathbf{p}_9 \times \mathbf{u}_9 \end{bmatrix} \begin{bmatrix} t_1 \\ \vdots \\ t_9 \end{bmatrix} + \begin{bmatrix} \mathbf{f_p} \\ \mathbf{\tau_p} \end{bmatrix} = 0 \tag{4}$$

$$\mathbf{A^T} \cdot \mathbf{T} + \mathbf{w} = 0 \tag{5}$$

In this paper, $\mathbf{f_p}$ is defined as $\mathbf{f_p} = [ma_2, ma_2, mg + ma_1]^T$, where $\mathbf{a}_1$ and $\mathbf{a}_2$ represent the maximum vertical acceleration and the maximum horizontal acceleration respectively. In this article, on the basis of the simulation requirements of lunar takeoff, m = 15 kg, $\mathbf{a}_1 = 2.4$ m/s^2 and $\mathbf{a}_2 = \pm 1.1$ m/s^2. The method proposed by Lafourcade [14], is adopted in this paper in order to obtain a solution with continuous cable tension.

The solution of cable tension consists of two parts, $\mathbf{T}_{eff}$ shown in (7) and $\mathbf{T}_{null}$ shown in (8).

$$\mathbf{T} = \mathbf{T}_{eff} + \mathbf{T}_{null} \tag{6}$$

$$\mathbf{T}_{eff} = -\left(\mathbf{A^T}\right)^+ \cdot \mathbf{w} \tag{7}$$

$$\mathbf{T}_{null} = \left(\mathbf{I}_{m\times m} - \left(\mathbf{A^T}\right)^+ \mathbf{A^T}\right) \cdot \mathbf{t_d} \tag{8}$$

where, $\left(\mathbf{A^T}\right)^+$ is the M-P inverse of $\mathbf{A^T}$, $\mathbf{t_d} = [t_d\ t_d\ t_d\ t_d\ t_d\ t_d\ t_d\ t_d\ t_d]^T$ is a desired value of f_i. $\mathbf{t_d}$ is utilized to improve the performance of CDPSL as a control variable. The value of each element in $\mathbf{t_d}$ is identical because it is beneficial to simplify the control system and CDPSL is designed into modular part.

3   Workspace Determination and Single-Factor Test

3.1   Workspace Determination

For cable-driven parallel mechanism, the tension $t_i(i = 1, 2, ..., 9)$ (N) should be limited between the maximum tension t_{max}(N) and the minimum tension t_{min}(N), which is required to keep the cable tight. Thus, a condition shown in (9) should be added to (6).

$$t_{max} \geq t_i \geq t_{min} > 0 \tag{9}$$

The dimension of base frame is $3 \times 3 \times 4$ m. The space occupied by CDPSL is well-distributed into 562400 small cubes. Take the coordinate of the body-center of each cube as sample point, thus obtaining 562400 sample points.

The algorithm based on Monte-Carlo method can be used to analyze the workspace of CDPSL.

- Give a sample point and its position belonging to the convex set of fixed points;
- Calculate structure matrix $\mathbf{A}^T$
- $\mathbf{T}_{eff} = -\left(\mathbf{A}^T\right)^+ \cdot \mathbf{w}$
- Choose a desired tension $\mathbf{t}_d$
- $\mathbf{T}_{null} = \left(\mathbf{I}_{m\times m} - \left(\mathbf{A}^T\right)^+ \mathbf{A}^T\right) \cdot \mathbf{t}_d$
- Calculate $\mathbf{T} = \mathbf{T}_{eff} + \mathbf{T}_{null}$
- If $min(\mathbf{T}) \geq t_{min}$ and $max(\mathbf{T}) \leq t_{max}$, the sample point belongs to the workspace. Otherwise, the sample point does not belong to the workspace.
- Repeat the above process until all the sample points belonging to the convex set of fixed points are calculated.

The variable SUM is introduced to represent the number of sample points belonging to the workspace. Furthermore, SUM could be used to indicate the volume of workspace. Thus, we can optimize the structure of CDPSL by trying to find the maximum value of SUM.

3.2   Single-Factor Test

Lunar takeoff simulation requires the workspace of cable-driven parallel mechanism as large as possible. Hence, the target of structure optimization is set as following:

Maximize the value of SUM under the condition that the quality, the vertical and horizontal acceleration of moving platform are set as 15 kg, 2.4 m/s^2 and ± 1.1 m/s^2 respectively.

In this paper, $\mathbf{B}_i(i = 1, 2, ..., 9)$ are distributed around a circle with radius of 2 m, $\mathbf{P}_i(i = 1, 2, ..., 9)$ are distributed around a circle with radius of 0.25 m and the height of moving platform is 0.75 m. $\mathbf{B}_i$ are y axial symmetry in $\mathbf{K}_b$ and $\mathbf{P}_i$ are y axial symmetry in $\mathbf{K}_p$. The circle made up by $\mathbf{B}_i$ and the circle made up by $\mathbf{P}_i$ are in similarity relation. To optimize the installation positions of cables, $\theta_i(i = 1, 2, 3)$ are chosen as optimal variables. Optimal variables are shown in Fig. 3 and the range of optimal variables are shown in Table 1.

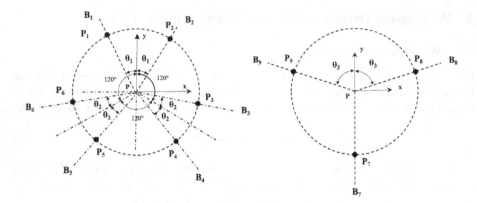

Fig. 3. Optimal variables

Table 1. Variable name and their lower and upper limits

Variable name	Lower limit	Upper limit
$\theta_1/^\circ$	0	60
$\theta_2/^\circ$	0	60
$\theta_3/^\circ$	0	90

Before optimizing structure, single-factor tests are taken to analyze the relationship between optimal variables and SUM qualitatively. Results of single-factor tests are shown in Fig. 4 and the meaning of each curve is illustrated in Table 2. It should be noted that only 8000 sample points are chosen to calculate SUM in this part. And in Table 2, "+" means taking a number of values in the range.

Figure 4 indicates that all optimal variables have influence on the workspace of CDPSL. With θ_1 increases, the values of SUM increases gradually, then decreases rapidly. With θ_2 increases, the values of SUM increases gradually. With θ_3 increases, the value of SUM increases gradually, then decreases gradually. Optimal variables are coupled to each other. Thus, it is necessary to use the optimal design method to optimize the value of optimal variables.

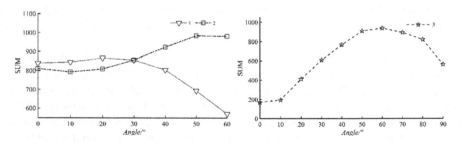

Fig. 4. Results of single-factor test

Table 2. The optimal variables corresponding to each curve in Fig. 4

	1	2	3
$\theta_1/°$	+	30	30
$\theta_2/°$	30	+	30
$\theta_3/°$	45	45	+

4   Optimal Design

4.1   Design of Experiment

In this paper, latin-hypercube method is taken to select experimental points. Response surface method is taken to optimize the installation positions of cables.

The latin-hypercube method is based on the principle of random probability orthogonal distribution principle. Thus, the response surface model with high precision is obtained by using fewer experimental points [15, 16]. The selected experimental points are shown in Fig. 5.

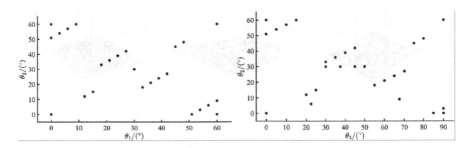

Fig. 5. Selected experimental points

4.2   Response Surface Model

Quadratic polynomial response surface model shown in (10) is taken to establish the surrogate mathematical model between optimization variables and optimization index.

$$y(x) = \beta_0 + \sum_{i=1}^{n} \beta_i x_i + \sum_{i=1}^{n} \beta_{ii} x_i^2 + \sum_{i=1}^{n-1} \sum_{j=i+1}^{n} \beta_{ij} x_i x_j \tag{10}$$

where, $y(x)$ is the predictive value of response surface model, x_i is the i-th component of the n-dimensional independent variable. β_0, β_i, β_{ii}, β_{ij} are the coefficients of the polynomial, which are calculated by least square method.

In this article, θ_1, θ_2 and θ_3 are set as independent variables, SUM is set as output response. Determinant coefficients R^2 and multiple fitting coefficient R_{adj}^2 are chosen to verify the surrogate mathematical model. The surrogate mathematical model is shown in (11) and calibration results are indicated in Table 2. It should be noted that all 562400 sample points are chosen to calculate SUM in this part.

$$SUM = 13800.76 + 1554.15\theta_1 + 149.15\theta_2 - 628.51\theta_3 - 73.96\theta_1\theta_2 - 8.79\theta_1\theta_3$$
$$+ 54.95\theta_2\theta_3 - 19.09\theta_1^2 - 0.12\theta_2^2 + 36.38\theta_3^2$$

$$(11)$$

From Table 3, the response surface model meets the accuracy requirements.

Table 3. Calibration results

Variable name	R^2	R^2_{adj}
SUM	0.9931	0.9865

The response surface is shown in Fig. 6.

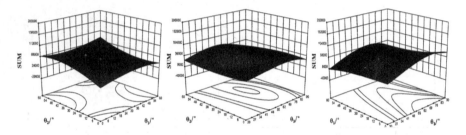

Fig. 6. Response surface

Figure 6 illustrates that the relationship between optimal variables and SUM can be approximated by established surrogate mathematical model. To maximize SUM, θ_1 and θ_2 should be close to 0 or 60 synchronously. θ_3 should be in the vicinity of 60.

Based on the above-mentioned results, multiple sets of optimal solutions could be obtained. Then, we obtain the optimized results while taking some restrictions on production, processing and installation into consideration. One of the solutions and the optimized results are indicated in Table 4.

Table 4. Optimal solution and optimized results

	$\theta_1/°$	$\theta_2/°$	$\theta_3/°$
Optimal solution	7	16	65
Optimized results	10	10	60

5   Virtual Prototype Experiment

Due to the disadvantages of physical prototype, such as high cost, long experimental cycle and the difficulty of structure adjustment, the virtual prototype experiment is utilized to verify the accuracy of the optimized results.

The virtual prototype model of CDPSL, shown in Fig. 7, is established using multi-body dynamics software ADAMS. The workspace of CDPSL could be determined through the analysis of each cable's tension condition when the moving platform are set at different positions under the condition that the vertical force and the horizontal force are 186 N and ±16.5 N respectively.

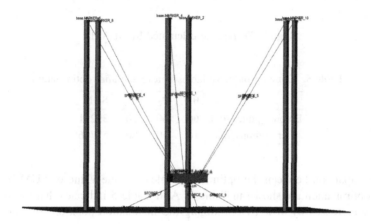

Fig. 7. Virtual prototype model

Figure 8 shows the workspace calculated by MATLAB according to Sect. 3.1 and Fig. 9 shows the workspace calculated by ADAMS.

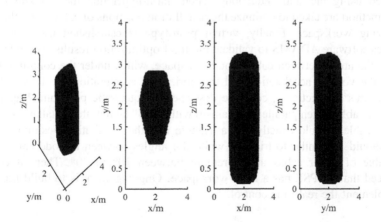

Fig. 8. Workspace calculated by MATLAB

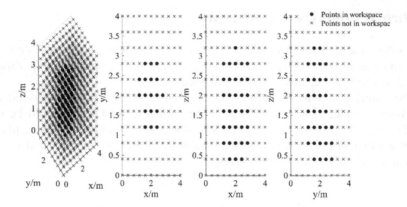

Fig. 9. Workspace calculated by ADAMS

Table 5. Comparison of variables before and after optimization

	$\theta_1/^\circ$	$\theta_2/^\circ$	$\theta_3/^\circ$	SUM
Before optimization	60	60	30	39281
After optimization	10	10	60	57786

The comparison between the optimal variables and the value of SUM before and after the optimization are shown in Table 5. And Table 5 indicates that the volume of workspace increased by 46% by optimizing the installation positions of cables.

6   Conclusion

In this paper, a cable-driven parallel mechanism with nine cables for lunar takeoff simulation (CDPSL) is presented and optimized. First, the workspace of CDPSL is calculated using the static equations. Then, latin-hypercube method and response surface method are taken to optimize the installation positions of cables with the goal of maximizing workspace. Finally, virtual prototype is established using multi-body dynamics software ADAMS to validate the final optimization results. The relationship between the angle between cables and workspace, which under the condition that the quality, the vertical acceleration and the horizontal acceleration are 15 kg, 2.4 m/s^2 and ±1.1 m/s^2 respectively, could be described by quadratic polynomial approximatively. For cable driven parallel mechanism with "6–3" layout, the installation positions of lower cables should exactly like a triangle and the installation positions of upper cables should be similar to triangle, while the angles between the odd cable and the even cable of upper cable should remain between 10° to 20°. Then, it could be guaranteed that CDPSL has a large workspace. Ongoing work is to build this device and implement the real-time control.

Acknowledgments. This research work is supported by the Natural Sciences Foundation Council of China (NSFC) 51405024.

References

1. Wang, W., Tang, X., Shao, Z., et al.: Design and analysis of a wire-driven parallel mechanism for low-gravity environment simulation. Adv. Mech. Eng. **2014**(1), 1–11 (2014)
2. Tang, X.: An Overview of the Development for Cable-Driven Parallel Manipulator. Adv. Mech. Eng. **2014**(1), 1–9 (2014)
3. Cao, L., Tang, X., Wang, W.: Tension optimization and experimental research of parallel mechanism driven by 8 cables for constant vector force output. Robot **37**(6), 641–647 (2015)
4. Bin, Z.: Dynamic modeling and numerical simulation of cable-driven parallel manipulator. Chin. J. Mech. Eng. **43**(11), 82–88 (2007)
5. Hiller, M., Fang, S., Mielczarek, S., et al.: Design, analysis and realization of tendon-based parallel manipulators. Mech. Mach. Theory **40**(4), 429–445 (2005)
6. Ming, A., Kajitani, M., Higuchi, T.: On the Design of Wire Parallel Mechanism. Int. J. Jpn Soc. Precis. Eng. **29**(4), 337–342 (1995)
7. Jamshidifar, H., Khajepour, A., Fidan, B., et al.: Kinematically-constrained redundant cable-driven parallel robots: modeling, redundancy analysis and stiffness optimization. IEEE/ASME Trans. Mechatron. **2016**(99), 1 (2016)
8. Landsberger, S.E.: Design and Construction of a Cable-Controlled Parallel Link Manipulator. Massachusetts Institute of Technology. vol. 2005, No. 22, p. 2317 (2005)
9. Barrette, G., Gosselin, C.M.: Determination of the dynamic workspace of cable-driven planar parallel mechanisms. J. Mech. Des. **127**(2), 242–248 (2005)
10. Agrawal, S.K., Dubey, V.N., et al.: Design and optimization of a cable driven upper arm exoskeleton. J. Med. Devices **3**(3), 437–447 (2009)
11. Ouyang, B., Shang, W.: Wrench-feasible workspace based optimization of the fixed and moving platforms for cable-driven parallel manipulators. Robot. Comput.-Integr. Manuf. **30**(6), 629–635 (2014)
12. Yao, R., Tang, X., Li, T., et al.: Analysis and design of 3T cable-driven parallel manipulator for the feedback's orientation of the large radio telescope. J. Mech. Eng **43**(11), 105–109 (2007)
13. Fu, D., Wang, F., Wang, X., et al.: Separation characteristics for front cover of ejection launch canister with low impact. J. Astronaut. **37**(4), 488–493 (2016)
14. Lafourcade, P., Libre, M., Reboulet, C.: Design of a parallel wire-driven manipulator for wind tunnels. In: Workshop on Fundamental Issues and Future Research Directions for Parallel Mechanisms and Manipulators, pp. 187–194 (2002)
15. Zhu, H., Liu, L., Long, T., et al.: Global optimization method using SLE and adaptive RBF based on fuzzy clustering. Chin. J. Mech. Eng. **25**(4), 768–775 (2012)
16. Xiong, F., Xiong, Y., Chen, W., et al.: Optimizing latin hypercube design for sequential sampling of computer experiments. Eng. Optim. **41**(8), 793–810 (2009)

Optimal Design of an Orthogonal Generalized Parallel Manipulator Based on Swarm Particle Optimization Algorithm

Lei Peng[⊠], Zhizhong Tong, Chongqing Li, Hongzhou Jiang, and Jingfeng He

Department of Mechatronics Engineering, Harbin Institute of Technology, Harbin 150001, China
pengleihit@163.com

Abstract. Study of methods to design optimal Gough–Stewart parallel manipulator geometries meeting orthogonality is of high interest. The paper will design the Orthogonal Generalized Gough-Stewart Parallel Manipulator (OGGSPM) based on the composite hyperboloids. Moreover, Particle Swarm Optimization (PSO) algorithm is introduced to perform the structure optimization as the Jacobian matrix condition number is used as the evaluation index. As an example, two different optimization cases are presented.

Keywords: Gough-Stewart parallel manipulator · Orthogonal design · A pair of composite hyperboloids · Optimization · PSO

1 Introduction

Gough-Stewart Parallel Manipulator (GSPM) is widely used in the field of motion simulation, micro operation, isolation vibration reduction, astronomical telescope, precise positioning and parallel machine tool since it possesses the advantages such as high rigidity, high bearing capacity, high precision etc. Above-mentioned applications require high performance of the parallel manipulator, thus the research of high performance GSPM is of tremendous interest.

Kozak et al. [1] proposed the lowest natural frequency of a GSPM as a performance index. Stocco et al. [2] studied the worst-case performance of a GSPM over the whole workspace by defining a "global isotropy index" (GII). Pittens and Podhorodeski [3] designed the GSPM numerically by minimizing the Jacobian matrix condition number at the design point. Zanganeh and Angeles [4] have found a 12-parameter family of isotropic GSPM designs.

The study of high performance GSPM usually leads to the research of the Orthogonal Gough-Stewart Parallel Manipulator (OGSPM). Because once the orthogonality is obtained at a single point, the GSPM can enjoy kinematic decoupling and the resulting higher performance throughout its workspace. In view of this, many scholars took the orthogonality as the design goal for high performance GSPM and proposed the corresponding evaluation index. McInroy and other scholars have carried on the thorough research on the OGSPM analysis and the optimized design [5–7]. McInroy used

© Springer International Publishing AG 2017
Y. Huang et al. (Eds.): ICIRA 2017, Part II, LNAI 10463, pp. 334–345, 2017.
DOI: 10.1007/978-3-319-65292-4_29

the elastic hinges to replace the rigid hinges. Based on the numerical method, three kinds of manipulator are presented [5]. Jafari et al. proposed the analytical description and design method of the orthogonal GSPM in a small workspace [6]. Yi et al. put forward a new design method of orthogonal Stewart parallel mechanism with even number struts [7]. However, the kinematic and static isotropy of degree of standard GSPM is 0.717 or 1.414, while that of the Generalized Gough-Stewart Parallel Manipulator (GGSPM) can reach 1 [8].

Generally speaking, multi-parameter and multi-objective optimization design of GSPM is still quite a challenging problem. Therefore, in this paper, a method combining the design of OGSPM with the Multi Objective Optimization Problem (MOOP) is proposed. Firstly, the description of the orthogonal generalized Gough-Stewart parallel manipulator based on the composite hyperboloids is introduced. Secondly, to improve search accuracy and speed, the Particle Swarm Optimization (PSO) algorithm is introduced to perform the structure optimization.

2 Description of GGSPM

Firstly, the definition of GGSPM is given:

Definition: A Generalized Gough-Stewart parallel Manipulator (GGSPM) is a spatial parallel manipulator consist of a movable platform and a fixed base and is connected by the 6 struts, which are categorized as two sets: the first set comprises strut 1, 3 and 5, and the second set comprises strut 2, 4 and 6. The attachment points for each strut are evenly distributed on the circumference of two circles on both the movable platform and the fixed base. The three struts in each group are rotated systematically every $2\pi/3$ (Fig. 1).

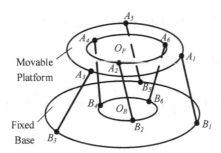

Fig. 1. Generalized Gough-Stewart parallel manipulator

Jiang et al. [9] proposed the method of describing GGSPM by using a pair of circular hyperboloids as follows:

(a) Given a pair of circular hyperboloids, surface equations of which are described as:

$$S_1 = \frac{(x^2 + y^2)}{r_1^2} - \frac{(z - a_{1z})^2}{c_1^2} = 1 \qquad (1)$$

$$S_2 = \frac{(x^2 + y^2)}{r_2^2} - \frac{(z - a_{2z})^2}{c_2^2} = 1 \qquad (2)$$

(b) Where: $r_1 > 0$, $r_2 > 0$, $a_{1z} \geq 0$, $a_{2z} \geq 0$, $c_1 > 0$, $c_2 > 0$. The throat radii of the hyperboloids are r_1 and r_2 respectively. The distance from the center of the hyperboloid to the coordinate origin are a_{1z} and a_{2z} respectively.

(c) Two straight-line generators $\Gamma 1$ and $\Gamma 2$, located on two hyperboloids, are selected. The projection of $\Gamma 1$ and $\Gamma 2$ on XOY plane remain mirror symmetry and the angle between $\Gamma 1$、$\Gamma 2$ and the x-axis is α.

(d) After rotating the line $\Gamma 1$ and $\Gamma 2 \pm 2\pi/3$ along the z-axis, two sets of straight lines are formulated. The first set comprises struts 1, 3 and 5, and the second comprises struts 2, 4 and 6 (Fig. 2).

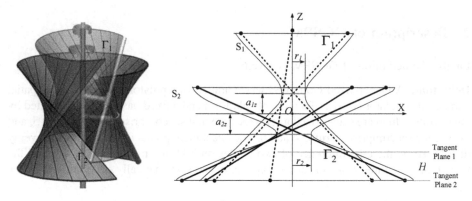

Fig. 2. GGSPM described by a pair of hyperboloids

Jacobian matrix of the GGSPM is given as:

$$\mathbf{J}_{l,x}(\alpha) = \left[\mathbf{p}_1(\alpha), \mathbf{p}_2\left(\alpha + \frac{2}{3}\pi\right), \mathbf{p}_1\left(\alpha + \frac{2}{3}\pi\right), \mathbf{p}_2\left(\alpha - \frac{2}{3}\pi\right), \mathbf{p}_1\left(\alpha - \frac{2}{3}\pi\right), \mathbf{p}_2(\alpha) \right]^T$$

$$(3)$$

where:

$$\mathbf{p}_1(\theta) = [-k_{a1}\sin\theta \quad k_{a1}\sin\theta \quad k_{c1} \quad r_1 k_{c1}\sin\theta \quad -r_1 k_{c1}\cos\theta \quad r_1 k_{a1}]^T$$
$$\mathbf{p}_2(\theta) = [-k_{a2}\sin\theta \quad -k_{a2}\sin\theta \quad k_{c2} \quad -r_2 k_{c2}\sin\theta \quad -r_2 k_{c2}\cos\theta \quad -r_2 k_{a2}]^T$$
$$k_{a1} = \frac{r_1}{\sqrt{r_1^2 + c_1^2}}, k_{c1} = \frac{c_1}{\sqrt{r_1^2 + c_1^2}}$$
$$k_{a2} = \frac{r_2}{\sqrt{r_2^2 + c_2^2}}, k_{c2} = \frac{c_2}{\sqrt{r_2^2 + c_2^2}}.$$

3 Evaluation Index

For GGSPM, when the input velocity vector is unity, the maximum and minimum output velocities of the manipulator could be described as:

$$\|v_{max}\| = \sqrt{\max(\|\lambda_{vi}\|)} \tag{4}$$

$$\|v_{min}\| = \sqrt{\min(\|\lambda_{vi}\|)} \tag{5}$$

Where $J_{l,x}$ refers to as the Jacobian matrix of GGSPM

λ_{vi} (i = 1, 2, 3, 4, 5, 6) are the eigenvalues of the matrix $J_{l,x}^T J_{l,x}$

Therefore, Velocity Transfer Index (VTI) is defined as:

$$\kappa_v = \frac{\|v_{max}\|}{\|v_{min}\|} = \text{cond}\left(J_{l,x}\right) = \frac{\lambda_{max}\left(J_{l,x}\right)}{\lambda_{min}\left(J_{l,x}\right)} \tag{6}$$

The Jacobian matrix condition number of GGSPM is used to describe the transfer characteristic of the input error to the terminal of GGSPM. The smaller the value calculated, the better the kinematic performance is supposed to be.

The study of orthogonality lies in the category of kinematic isotropy. Still, it provides the basis and reference for the study of complete isotropy. In view of the above-mentioned discourse, it is necessary to study the orthogonal Gough-Stewart parallel manipulator. When a GGSPM satisfies the orthogonal condition, each degree of freedom (DOF) is orthogonal at a single point.

If GGSPM is orthogonal, matrix G is supposed to be a diagonal matrix, namely:

$$G = J_{l,x}^T J_{l,x} = \begin{pmatrix} \lambda_1 & & 0 \\ & \ddots & \\ 0 & & \lambda_6 \end{pmatrix} \tag{7}$$

Taking into account of the hyperboloids parameter, then we can get the orthogonal condition as shown in Eqs. (8) and (9):

$$k_{a1}k_{c1}r_1 = k_{a2}k_{c2}r_2 \tag{8}$$

$$k_{a1}^2 a_{1z} = -k_{a2}^2 a_{2z} \tag{9}$$

4 Particle Swarm Optimization

4.1 Problem Description

The optimization problem is described as follows:

Objective: $\min y = f(x) = \text{cond}\left(J_{l,x}\right)$

Optimization variables: $[r_1 \ r_2 \ k_{c1} \ k_{c2} \ a_{1z} \ a_{2z} \ \alpha]$

$$\text{St.} \qquad k_{a1}k_{c1}r_1 = k_{a2}k_{c2}r_2$$
$$k_{a1}^2 a_{1z} = -k_{a2}^2 a_{2z}$$
$$k_{c1}^2 + k_{a1}^2 = 1$$
$$k_{c2}^2 + k_{a2}^2 = 1$$

$$\text{Where} \qquad r_{1min} < r_1 < r_{1max}$$
$$r_{2min} < r_2 < r_{2max}$$
$$a_{1zmin} < a_{1z} < a_{1zmax}$$
$$a_{2zmin} < a_{2z} < a_{2zmax}$$
$$\alpha_{min} < \alpha < \alpha_{max}$$

Note: *the value span of r_1, r_2 and a_{1z}, a_{2z} vary with different application cases. The boundaries of them are decided based on specific cases.*

k_{c1} and k_{c2} representing the projection of the unit direction of the strut in the Z axis are dimensionless quantity, range from 0 to 1.

4.2   Optimization Algorithm

The optimization problems can be optimized by direct search method, however, the efficiency of the direct search method is largely contingently on the selection of the initial value. The particle swarm optimization algorithm (PSO) is adopted in order to improve the search efficiency. As an evolutionary algorithm, PSO is remarkably useful since it has a faster convergence rate and higher computational efficiency.

Let the search space be D dimension, and the number of particles be N. The position of the ith particle is expressed as a D vector $X_i = (x_{i1}, x_{i2}, \cdots, x_{iD})$, and the velocity of the ith particle is expressed as a D dimensional vector $V_i = (v_{i1}, v_{i2}, \cdots, v_{iD})$. The optimal solution during the iteration process of the ith particle is marked as pBest and described as $P_i = (p_{i1}, p_{i2}, \cdots, p_{iD})$. And finally the iteration process is represented as follows [10]:

$$V_i^{j+1} = wV_i^j + c_1 r_1 [P_i^j - X_i^j] + c_2 r_2 [P_g^j - X_i^j] \qquad (10)$$

$$X_i^{j+1} = X_i^j + V_i^{j+1} \qquad (11)$$

Where
w -inertia factor;
c_1, c_2 -acceleration factor;
r_1, r_2 -random number in the interval of [0, 1].

The range of the dth dimension of the position and the velocity are $[-x_{d,min}, x_{d,max}]$ and $[-V_{d,min}, V_{d,max}]$ respectively. During the iteration process, the boundary value is obtained while the position or velocity of a particle in one certain dimension exceeds the boundary. The initial position and velocity of the particle swarm which are randomly generated are iterated according to Eqs. (10) and (11). And the iteration stops when the termination condition is met. The pseudo code of the PSO algorithm used for GGSPM optimization is as follows:

```
    Randomly initialized particle,7 variables per particle;
    Do
       For each particle
            Calculate the degree of fitness Xᵢ, described by
            The Objective function cond(J_{l,x});

            If  Xᵢ outperforms Pᵢ   Then update Pᵢ;
       End
       Select the particle of the optimal degree of fitness in
       the particle swarm (or neighborhood subgroup);
       If the optimal particle of the group outperforms P_g
       Then update P_g;
       For each particle
            Update the velocity of the particle Vi
            according to eqs.(10) and(11)
                If  V_{id} exceeds the boundary of [-V_{d, min}, V_{d, max} ]
                Then Assign Vᵢ to boundary value
                If  X_{id} exceeds the boundary of [-X_{d, min}, X_{d, max} ]
                Then Assign Xᵢ to boundary value
    End
    While  termination condition is satisfied
```

4.3 Configuration Selection

The configuration of the GGSPM can be classified according to the number of the circles of hinge points in the upper and lower platform. For instance, 2C-2C indicates that the upper and lower hinge points are distributed on 2 separate circles respectively. Therefore the GGSPM is categorized into the following three configurations:

As can be seen from Fig. 3, r_{a1} and r_{a2} represent the radii of the 2 upper circle respectively while r_{b1} and r_{b2} represent the radii of the 2 lower circle respectively.

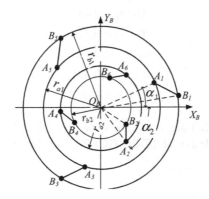

Fig. 3. Vertical view of generalized Gough-Stewart parallel manipulator

- Configuration 1: 1C-2C GGSPM

There are two kinds of configurations, namely 1C-2C GGSPM while $r_{a1} = r_{a2}$ and 1C-2C GGSPM while $r_{b1} = r_{b2}$. But for $r_{b1} = r_{b2}$, the size of the upper platform is larger than the lower platform, which may add difficulty to the manipulator feasibility. For 1C-2C GGSPM, the parameter of the two composite hyperboloids satisfies $k_{c1} = k_{c2}$, $r_1 = r_2$ and $a_{1z} = -a_{2z}(a_{1z} > 0)$.

- Configuration 2: 2C-2C GGSPM

The GGSPM which satisfies $r_{a1} \neq r_{a2}$ and $r_{b1} \neq r_{b2}$ is termed as 2C-2C GGSPM. This kind of structure belongs to the common form of generalized parallel manipulator.

- Configuration 3: 1C-1C GGSPM

The SGSPM belongs to this kind of configuration. while $r_{a1} = r_{a2}$ and $r_{b1} = r_{b2}$, the parameters of the two composite hyperboloid are same. The structural optimization design of configuration 3 belongs to the optimization problem of the standard parallel manipulator. and there has been a lot of research on it at home and abroad. To avoid repetition, it's not described in this article. As a result, configuration 1 and configuration 2 are selected to optimize the design.

5 Result

5.1 Parameters Initialization

The precision pointing equipment is a typical application of the parallel Manipulator. The following is an example of the design pattern of the precision pointing equipment, which is presented to verify the feasibility of the proposed design method. The geometric characteristic parameters of the load are as follows: the center of mass of the load is $\rho_c = \begin{bmatrix} 0 & 0 & 0.5 \end{bmatrix}^T$ m, the equivalent mass of the load is $m = 8.8$ kg, and the moment of inertia is $I_c = diag(0.821, 0.821, 1.76)$ kg·m^2. Maximum half cone angle, which usually measures the rotation capability, is no less than 10°.

The boundaries of PSO variables for this specific case are as follows:

$$0 < r_1 < 0.4\,m,$$
$$0 < r_2 < 0.4\,m,$$
$$0 < a_{1z} < 0.4\,m,$$
$$0 < a_{2z} < 0.4\,m,$$
$$0 \leq \alpha \leq \pi/2$$

The parameters of PSO algorithm are shown in Table 1, according to which the Structural optimization design of 1C-2C GGSPM and 2C-2C GGSPM are performed respectively.

Table 1. Parameters of PSO algorithm

Population number N	Maximum iteration	Acceleration factor **c1**	Acceleration factor **c2**	Initial inertia factor $w1$	Final inertia factor $w2$	Maximum residual error
30	50	1.2	1.2	0.8	0.2	0.0001

Note: the value of Population Number **N** and Acceleration Factor c1 and c2 are selected based on the past experience.

5.2 Structure Optimization of 1C-2C GGSPM

1C-2C GGSPM structure optimization result based on PSO is shown in Fig. 4. As it can be seen, the results of the evaluation function reach convergence as the times of iteration reaches over 25 times. At the same time, the evolution process of the structural parameters of the Manipulator is shown in Fig. 5.

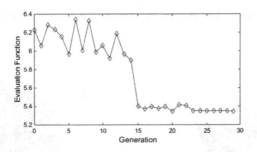

Fig. 4. Evolution process of evaluation function of 1C-2C GGSPM

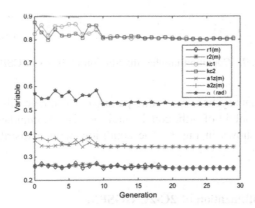

Fig. 5. Evolution process of evaluation variables of 1C-2C GGSPM

According to the optimization results, a set of composite circular hyperboloids parameters are chosen as shown in Table 2. The orthogonal center is located at the

Table 2. Structural parameter of 1C-2C GGSPM

Parameter	r_1 [m]	r_2 [m]	k_{c1}	k_{c2}	a_{1z} [M]	a_{2z} [m]	α [rad]
Final value	0.25	0.25	0.8	0.8	0.34	0.34	30

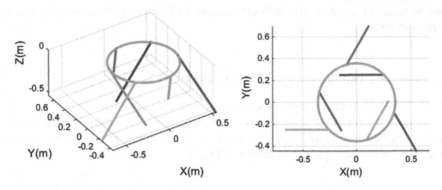

Fig. 6. Manipulator configuration of 1C-2C GGSPM

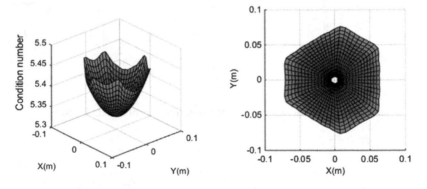

Fig. 7. Condition number distribution of 1C-2C GGSPM

origin of the coordinate system. And other parameters can be easily obtained based on the composite hyperboloid of orthogonal condition. The Manipulator configuration of 1C-2C GGSPM is shown in Fig. 6. The condition number distribution is shown in Fig. 7.

5.3 Structure Optimization of 2C-2C GGSPM

Similarly, based on the PSO algorithm for 2C-2C GGSPM structure optimization design, the evaluation function evolution curve is shown in Fig. 8, the final iteration results of each structural parameter are shown in Fig. 9.

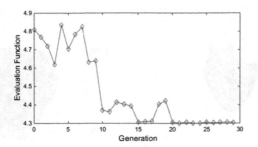

Fig. 8. Evolution process of evaluation function of 2C-2C GGSPM

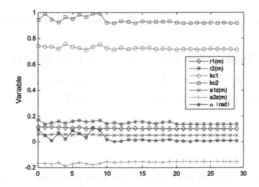

Fig. 9. Evolution process of evaluation variables of 2C-2C GGSPM

According to the results of the optimization design parameters, the generalized parallel Manipulator is shown in Fig. 10, the condition number distribution is shown in Fig. 11. According to the optimization results, a set of composite circular hyperboloids parameters are chosen as shown in Table 3.

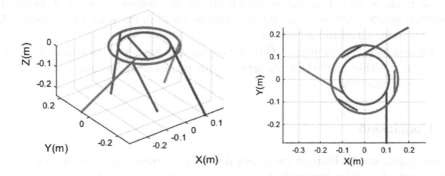

Fig. 10. Manipulator configuration of 2C-2C GGSPM

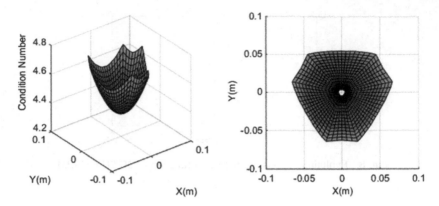

Fig. 11. Condition number distribution of 2C-2C GGSPM

Table 3. Structural parameter of 1C-2C GGSPM

Parameter	r_1 [m]	r_2 [m]	k_{c1}	k_{c2}	a_{1z} [m]	a_{2z} [m]	α [rad]
Final value	0.1	0.1364	0.7141	0.9165	0.05	-0.1531	0

5.4   Discussion

After simulation and check, in both cases, the half cone angle $>10°$, and there is no component interference, thus the design cases are proven to be valid.

As can be seen from the final optimized structural parameters, 1C-2C GGSPM configuration is relatively simple, and the strut length is equal, therefore it is easier to implement structural design and control system design. However, the Jacobian condition number of 2C-2C GGSPM is comparatively smaller, thus achieving better kinematic performance. In the practice, comprehensive consideration we should be taken to select the appropriate configuration.

The orthogonal design of 1C-2C GGSPM and 2C-2C GGSPM are decoupled in the neutral point. Moreover, as can be seen from the distribution of the Jacobian matrix condition number, the condition number remains small in the workspace which indicates favorable kinematic performance.

To conclude, by adopting GGSPM and on the basis of orthogonal designing approach, a viable method for designing high precision application such as flight simulator is presented.

6   Conclusion

In this paper, on the basis of the description of the generalized parallel manipulator with a pair of complex hyperboloids, we manage to design the orthogonal generalized parallel manipulator using the orthogonal condition. Furthermore, using the Jacobian matrix condition number as the evaluation index, we manage to perform 1C-2C

GGSPM and 2C-2C GGSPM optimization respectively based on the PSO algorithm. As can be seen, the final optimized Jacobian matrix condition numbers achieve minimum and the optimal performance of manipulator is obtained. Future study will focus the global evaluation index of GGSPM, such as the global working space condition index (GCI), based on which the structure optimization will be performed.

Acknowledgements. This research was supported by the Natural Science Foundation of China (Grant No. 51575121). The authors would like to thank the reviewers.

References

1. Kozak, K., Ebert-Uphoff, I., Voglewede, P., Singhose, W.: Concept paper: on the significance of the lowest linearized natural frequency of a parallel manipulator as a performance measure for concurrent design. In: Proceedings of the Workshop on Fundamental Issues and Future Research Directions for Parallel Mechanisms and Manipulators, Quebec City, QC, Canada, pp. 112–118, October 2002
2. Stocco, L., Salcudean, S., Sassani, F.: Fast constrained global minimax optimization of robot parameters. Robotica **16**, 595–605 (1998)
3. Pittens, K.H., Podhorodeski, R.P.: A family of Stewart platforms with optimal dexterity. J. Robot. Syst. **10**(4), 464–479 (1993)
4. Zanganeh, K., Angeles, J.: Kinematic isotropy and the optimum design of manipulator. Int. J. Robot. Res. **16**(2), 185–197 (1997)
5. McInroy, J.E., Hamann, J.: Design and control of flexure jointed hexapods. IEEE Trans. Rob. Automat. **16**(4), 372–381 (2000)
6. Jafari, F., McInroy, J.E.: Orthogonal Gough-Stewart platforms for micromanipulation. IEEE Trans. Rob. Automat. **19**(4), 595–603 (2003)
7. Yi, Y., McInroy, J.E., Jafari, F.: Optimum design of a class of fault-tolerant isotropic Gough–Stewart platforms. In: Proceeding of the IEEE International Conference on Robotics and Automation, New Orleans, LA, vol. 5, pp. 4963–4968, April 2004
8. Jiantao, Y., Yulei, H., Strait, M., Yongsheng, Z.: Isotropic analysis of structure analysis and optimization design of six axis force sensor. J. Mech. Eng. **45**(12), 22–28 (2009)
9. Jiang, H.Z., Tong, Z.Z., He, J.F.: Dynamic isotropy design of a class of Gough-Stewart parallel manipulator lying on a circular hyperboloid of one sheet. Manip. Mach. Theory **46**(3), 358–374 (2010)
10. Shi, Y., Eberhart, R.C.: A modified particle swarm optimizer. In: The 1998 Conference of Evolutionary Computation, IEEE Press, Piscataway, NJ, pp. 69–73 (1998)

Research on a 3-DOF Compliant Precision Positioning Stage Based on Piezoelectric Actuators

Guang Ren, Quan Zhang$^{(\boxtimes)}$, Chaodong Li, and Xu Zhang

School of Mechatronic Engineering and Automation, Shanghai University,
Shanghai 200072, China
lincolnquan@shu.edu.cn

Abstract. Precision positioning with multiple degree-of-freedoms (DOFs) is the core technology of nanometer manufacturing equipment. In this paper, a 3-DOFs monolithic parallel compliant manipulator is designed for relieving the conflicts among large workspace, high precision positioning and multi-DOFs. The 3-DOFs compliant micro-positioning manipulator has parallel structure and is composed of a static platform, a moving platform and three kinematic chains. Three kinematic chains are arranged symmetrically along the center of the moving platform with 120°, and each chain is actuated by a piezoelectric ceramic actuator through a variable cross-section symmetrical four-bar mechanism with low coupling and better sensitivity. Flexure hinges are utilized as the revolute joints to provide smooth and high accurate motion with nanometer level resolution. Based on the "pseudo-rigid-body model" method, the inverse kinematics model and dynamic model of the compliant manipulator are established. Finally, the natural frequency of the 3-DOFs compliant micro-positioning manipulator is obtained by the ANSYS software.

Keywords: 3-DOFs compliant manipulator · The Pseudo-rigid-body model · Piezoelectric actuator · Dynamic model

1 Introduction

Precision positioning technology is widely used in industrial production, weaponry, scientific research and other fields, and is also one of the most important technologies in supporting the equipment developed toward higher precision, higher efficiency, and intelligent. As a kind of new typed transmission mechanism, compliant manipulator mechanism is free assembly with flexure hinges instead of traditional joints. Compliant manipulator mechanism transforms motion and force with the elastic deformation of compliant component, and has the advantages of no mechanical friction, no clearance, no backlash, and higher sensitivity etc. [1, 2]. Hence it is suitable to use such transmission mechanism in precision positioning fields [3, 4].

At present, the research in compliant mechanism has been mainly concentrated in the design of the compliant manipulator based on pseudo-rigid-body model method [5] and topology optimization [6, 7]. In addition, 1-DOF and 2-DOFs micro-positioning platform have been investigated by many researchers. Zhu et al. [8] developed a

© Springer International Publishing AG 2017
Y. Huang et al. (Eds.): ICIRA 2017, Part II, LNAI 10463, pp. 346–358, 2017.
DOI: 10.1007/978-3-319-65292-4_30

compliant displacement inverter with topology optimization, and then analyzed the optimization problem of de facto hinges resulted by topology optimization in the created mechanisms. Li et al. [9] designed a new XY micro-motion stage with flexibility matrix method, and analyzed the static and dynamic performances of the stage via Finite Element Analysis (FEA). Besides, extensive experiments are conducted to test the performance of the micro stage, and the experimental results show that the input/output of the stage has a very good linearity and a good decoupling property [10]. Li et al. [11] designed a 2-DOFs Flexure-Based mechanism, then a closed-loop control method is developed to reduce the nonlinear behavior of the piezo-driven flexure-based mechanism. However, as to some more complex operations, the motions of the compliant manipulator is not limited in X and Y direction in the planar motion, but also need to adjust the rotation motion around the Z axis. Zhang et al. [12, 13] have an in-depth study in the essential theories and methods of the design of 3-DOFs compliant mechanism with topology optimization. Zhu et al. [14, 15] designed a 3-DOFs planar integrated fully compliant parallel mechanism with topology optimization, and investigated the stress distribution and first four vibration natural frequencies of the proposed mechanism via a contrast simulation study. Daniel et al. [16, 17] designed a 3-DOFs compliant parallel manipulator utilizing flexure hinges with pseudo-rigid-body model method, and the dynamic model of the mechanism with Lagrangian method was analyzed, at last the results show that the method can be simply applied to this more complex parallel mechanism.

In this study, a compliant manipulator which can achieve two translation motions (along X, Y-axis) and a rotation motion (Z-axis) in planar is designed and analyzed. This mechanism integrates the flexure hinges into a rigid material to form fully compliant manipulator. Based on the pseudo-rigid-body model method, the 3-DOFs compliant micro-positioning manipulator is designed, in which the semicircle shape is chosen as the notch profile of the flexure hinge pair and a symmetrical four-bar mechanism with variable cross-section is adopted as the compliant translational joints. The inverse kinematic models are established via the closed-loop vector constraint relationships of the 3-DOFs compliant micro-positioning manipulator, and then the dynamic model of the mechanism is analyzed based on the Lagrangian method. Finally, the 3-PRR compliant manipulator is simulated in ANSYS software to analyze the natural frequencies of the proposed manipulator.

2 Design of the 3-DOFs Compliant Manipulator

The 3-demensitional (3D) model of the 3-DOFs compliant micro-positioning manipulator is shown in Fig. 1. The 3-DOFs compliant micro-positioning manipulator transforms motion and force with the elastic deformation of compliant component instead of traditional joints. The 3-DOFs compliant micro-positioning manipulator is composed of a static platform, a moving platform and three kinematic chains. Three kinematic chains are arranged symmetrically along the center of moving platform with 120°. Each kinematic chain is actuated by a piezoelectric ceramic actuator and is composed of a variable cross-section symmetrical four-bar mechanism with low coupling and better sensitivity, a first-order displacement amplification mechanism, flexure hinges and linkages.

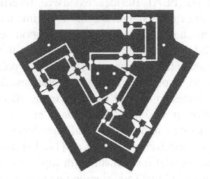

Fig. 1. The 3D model of the 3-DOFs compliant micro-positioning manipulator

3 Inverse Kinematics Modeling

When analyzing the kinematics and dynamics of the 3-DOFs compliant micro-positioning manipulator, the compliant manipulator model is assumed to be three first-order displacement amplification mechanisms and a 3-PRR compliant mechanism, as shown in Fig. 2.

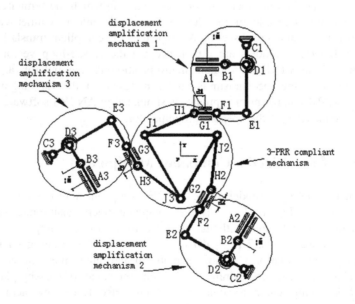

Fig. 2. The diagram of the 3-DOFs compliant micro-positioning manipulator

3.1 The Inverse Kinematics Model of the Displacement Amplification Mechanism

Figure 3 is the pseudo-rigid-body model of the first-order displacement amplification mechanism. Fixed coordinate frame $O'XY$ is established on the center of stationary frame point C_i ($i = 1, 2, 3$, the same below), and the X axis parallel to A_iB_i. θ_{i1} denote the angle between B_iD_i and Y axis, θ_{i2} denote the angle between D_iE_i and B_iD_i, θ_{i3} denote the angle between E_iF_i and Y axis. L_{GFi} denote the length between the center of prismatic pairs (point G_i) and flexure hinges F_i. ρ_i denote the input displacements of piezoelectric actuators, d_i denote the output displacements of point G_i through the first-order displacement amplification mechanism.

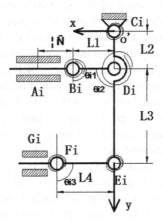

Fig. 3. The first displacement amplification mechanism based on lever principle

Based on the geometric and motion relationships under the coordinate frame of $O'XY$, the closed-loop constraint equation of the first-order displacement amplification mechanism can be written as follow:

$$l_{O'Gi} = l_{O'Bi} + l_{B_iD_i} + l_{D_iE_i} + l_{E_iF_i} + l_{F_iG_i} \tag{1}$$

$$l_{O'Ci} = l_{O'Bi} + l_{B_iD_i} + l_{D_iC_i} \tag{2}$$

Based on Eq. (1) and Fig. 3, the closed-loop constraint equation can be written as follow:

$$\begin{cases} L_4 + d_i = L_1 + \rho_i + L_1 cos\Delta\theta_{i1} + L_3 sin\Delta\theta_{i2} - L_4 cos\Delta\theta_{i3} + L_{FGi} \\ L_2 + L_3 = L_2 - L_1 sin\Delta\theta_{i1} + L_3 cos\Delta\theta_{i2} + L_4 sin\Delta\theta_{i3} \end{cases} \tag{3}$$

Based on Eq. (2) and Fig. 3, the closed-loop constraint equation can be written as follow:

$$\begin{cases} 0 = L_1 + \rho_i + L_1 cos\Delta\theta_{i1} - L_2 sin\Delta\theta_{i2} \\ 0 = L_2 - L_1 sin\Delta\theta_{i1} - L_2 cos\Delta\theta_{i2} \end{cases} \tag{4}$$

Since the angular displacement variations of the flexure hinges $\Delta\theta_{i1}$, $\Delta\theta_{i2}$, $\Delta\theta_{i3}$ are small, the following assumptions are given: $\sin\Delta\theta_i \sim \Delta\theta_i$, $\cos\Delta\theta_i \sim 1$. Taking the calculation formulas of the trigonometric function into Eqs. (3) and (4) can obtain the positive solution of the first-order displacement amplification mechanism as follow.

$$d_i = 2L_1 - 2L_4 + \frac{2L_1 \cdot L_3}{L_2} + L_{FGi} + \frac{L_2 + L_3}{L_2} \cdot \rho_i \tag{5}$$

3.2   Inverse Kinematics Model of the 3-PRR Compliant Mechanism

The end-effector of the 3-DOFs compliant micro-positioning manipulator has three degrees of freedom that translating along X, Y-axis and rotating about the Z-axis respectively. The input displacements of the PZT actuators ρ_i are magnified by the first-order displacement amplification mechanism, and then the output displacements of the first-order displacement amplification mechanism on the prismatic pairs (point G_i) can be regarded as the input displacements of the 3-PRR compliant mechanism. Vector $\boldsymbol{u} = (x_P, y_P, \varphi_P)$ denotes the output result of the center of the moving platform (point P).

The inverse kinematics model analysis of the 3-DOFs compliant micro-positioning manipulator is the basis of the positioning control. The input displacements of piezoelectric ceramic actuators can be obtained through the target location of the moving platform with the inverse kinematic analysis.

As shown in Fig. 4, the stationary frame XPY is established on the center of the regular triangle and its vertices are point G_1, G_2 and G_3. For the convenience of research, the initial position of each prismatic pair is fixed on the point G_i. A moving frame $X'PY'$ is also established on the moving platform, and the origin position of the moving frame lies on the center of the regular triangle which is consisted of three flexure hinges (point J_1, J_2 and J_3) on the moving platform. d_i denote the output displacements of point G_i through the displacement amplification mechanism, α_i denote the angles between output displacement of point G_i and x axis, β_i denote the angles

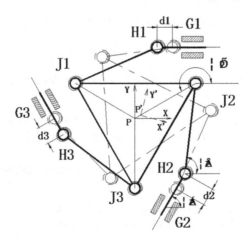

Fig. 4. The mathematical model of the 3-PRR compliant mechanism

between links H_iJ_i and X axis, φ_i denote the angles between links J_iP and X axis, point H_i and J_i denote the revolute joints respectively.

Based on the geometric and motion relationships of the 3-DOFs compliant micro-positioning manipulator, the closed-loop constraint equation can be written as

$$l_{pp'} + l_{p'J_i} = l_{pG_i} + l_{G_iH_i} + l_{H_iJ_i} \tag{6}$$

Based on Eq. (6) and Fig. 4, the closed-loop constraint equation can be written as follow:

$$x_{Gi} + (L_{GHi} - d_i)\cos\alpha_i + L_{HJi}\cos\beta_i = x_p + x'_{Ji}\cos\varphi_p - y'_{Ji}\sin\varphi_p \tag{7}$$

$$y_{Gi} + (L_{GHi} - d_i)\sin\alpha_i + L_{HJi}\sin\beta_i = y_p + x'_{Ji}\sin\varphi_p + y'_{Ji}\cos\varphi_p \tag{8}$$

where x'_{Ji} and y'_{Ji} are the coordinates of point J_i on the moving frame, x_{Gi} and y_{Gi} are the coordinates of point J_i on the stationary frame, L_{HJi} denote the length of links H_iJ_i. The initial position of stationary frame and moving frame are coincident. The output position and orientation of the end-effector (point P) is $\boldsymbol{u} = (x_p, y_p, \varphi_p)$ when the input displacements are given. Eliminating the intermediate variable β_i in Eqs. (7) and (8) yields the inverse kinematic solutions of the three piezoelectric actuators:

$$d_i = L_{GHi} \pm \sqrt{Q_{i1}^2 + Q_{i2}} - Q_{i1} \tag{9}$$

where:

$$\begin{cases} Q_{i1} = Q_{i3}\cos\alpha_i + Q_{i4}\sin\alpha_i \\ Q_{i2} = L_{HJ}^2 - Q_{i3}^2 - Q_{i4}^2 \\ Q_{i3} = x_p - x_{Gi} + x'_{Ji}\cos\varphi_p - y'_{Ji}\sin\varphi_p \\ Q_{i4} = y_p - y_{Gi} + x'_{Ji}\sin\varphi_p + y'_{Ji}\cos\varphi_p \end{cases} \tag{10}$$

As shown in Fig. 4, as to any pose of the moving platform, there are two possible solutions of input displacement about each chain. And hence there are eight solutions of input displacements totally. On the other hand, the solution should be confirmed based on the initial position and the continuity motion of piezoelectric ceramics actuator. Hence according to the structural characteristics of the compliant manipulator, the input displacements are selected as:

$$d_i = L_{GHi} - \sqrt{Q_{i1}^2 + Q_{i2}} - Q_{i1} \tag{11}$$

3.3   Jacobian Matrix

The Jacobian matrix reflects the linear transformation between input velocity and output velocity, it also be called the first-order kinematic influence coefficient matrix. According to the input displacement vector $\boldsymbol{\rho} = (\rho_1, \rho_2, \rho_3)$ of the piezoelectric ceramic actuators and the output displacement vector $u = (x_P, y_P, \varphi_P)$ of the moving platform, the position constraint equation of 3-PRR compliant mechanism can be obtained with differential transformation method [18] as follow.

$$F(\rho, u) = 0 \tag{12}$$

Generally, vector ρ and vector u have equal dimension, the derivative of Eq. (12) with respect to time can be written as follow:

$$J_\rho \dot{\rho} = J_u \dot{u} \tag{13}$$

where $J_\rho = \frac{\partial F}{\partial \rho}$, $J_\rho = \frac{\partial F}{\partial u}$, $\dot{\rho}$ and $\dot{u}$ denote the velocity vector of the piezoelectric ceramic actuators and moving platform respectively.

Equation (12) can be also written as follow:

$$\dot{\rho} = J_\rho^{-1} J_u \dot{u} = J \dot{u} \tag{14}$$

The Jacobian matrix J of the 3-DOFs compliant micro-positioning manipulator can be obtained based on Eqs. (5), (11) and (14) as follow:

$$J = J_\rho^{-1} J_u = \begin{bmatrix} V_1 & 0 & 0 \\ 0 & V_2 & 0 \\ 0 & 0 & V_3 \end{bmatrix}^{-1} \begin{bmatrix} M_{1x} & M_{1y} & M_{1\varphi} \\ M_{2x} & M_{2y} & M_{2\varphi} \\ M_{3x} & M_{3y} & M_{3\varphi} \end{bmatrix} \tag{15}$$

where:

$$M_{ix} = 2R(Q_{i1} - Q_{i3}) - cos\alpha_i \tag{16}$$

$$M_{iy} = 2R(Q_{i1} sin\alpha_i - Q_{i4}) + sin\alpha_i \tag{17}$$

$$M_{i\varphi} = [(2RQ_{i1} + 1)(Wsin\alpha_i - Ncos\alpha_i) + R(2NQ_{i3} - 2WQ_{i4})] \tag{18}$$

$$N = x'_{Ji} \cdot sin\varphi_P + y'_{Ji} \cdot cos\varphi_P \tag{19}$$

$$W = x'_{Ji} \cdot cos\varphi_P - y'_{Ji} \cdot sin\varphi_P \tag{20}$$

$$R = -\frac{1}{2} \left(Q_{i1}^2 + Q_{i2} \right)^{-\frac{1}{2}} \tag{21}$$

$$V_i = \frac{L_2 + L_3}{L_2} \tag{22}$$

4 Dynamic Modeling of the 3-PRR Compliant Mechanism

To further analyze the dynamic characteristics of the 3-PRR compliant mechanism, the dynamic model is developed based on the Lagrangian method. Figure 5 is the 3-PRR compliant mechanism with each flexure hinge rotating through a small angle. As shown

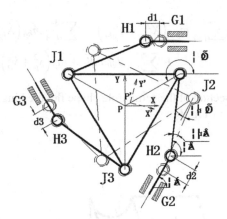

Fig. 5. The 3-PRR compliant mechanism with each flexure hinge rotating through a small angle

in Fig. 5, the mass of links G_iH_i and H_iJ_i are assumed as m_{ghi} and m_{hji} ($i = 1, 2, 3$, the same below) respectively, and the barycenter coordinates are assumed as (X_{ghi}, Y_{ghi}) and (X_{hji}, Y_{hji}) respectively, i denote the number of each chain.

Based on the geometric and motion relationships of compliant mechanism, the barycenter coordinate relationships of the compliant mechanism can be written as follow:

$$\begin{cases} X_{ghi} = X_{Gi} + \frac{(L_{GHi}-d_i)}{2}\cos\alpha_i \\ Y_{ghi} = Y_{Gi} + \frac{(L_{GHi}-d_i)}{2}\sin\alpha_i \end{cases} \tag{23}$$

$$\begin{cases} X_{hji} = X_{Gi} + (L_{GHi} - d_i)\cos\alpha_i + \frac{L_{HJi}}{2}\cos(\beta_i - \Delta\beta_i) \\ Y_{hji} = Y_{Gi} + (L_{GHi} - d_i)\sin\alpha_i + \frac{L_{HJi}}{2}\sin(\beta_i - \Delta\beta_i) \end{cases} \tag{24}$$

According to Eqs. (23) and (24), the barycenter velocity of links G_iH_i and H_iJ_i can be obtained as $(\dot{X}_{ghi}, \dot{Y}_{ghi})$ and $(\dot{X}_{hji}, \dot{Y}_{hji})$ respectively. The rotational inertia of links H_iJ_i around the axis of rotation as follow:

$$J_{ghi} = \frac{m_{gh}}{12} L_{ghi}^2 \tag{25}$$

The mass of moving platform is assumed as m_p, and the barycenter coordinate and barycenter velocity are assumed as (x_p, y_p) and $(\dot{x}_p, \dot{y}_p)$ respectively. Besides, The rotational inertia of the moving platform around the barycenter is J_p. Because the 3-DOFs compliant micro-positioning manipulator is in the plane, the gravitational potential energy of the 3-DOFs compliant micro-positioning manipulator can be ignored in the Lagrange equation. According to Eqs. (23)–(25), the total kinetic energy K of the micro positioning platform compliant mechanism can be obtained as follow:

$$
\begin{cases}
K = \dfrac{1}{2}[\sum_{i=1}^{3} m_{ghi}\left(\dot{X}_{ghi}^2 + \dot{Y}_{ghi}^2\right) + \sum_{i=1}^{3} m_{hji}\left(\dot{X}_{hji}^2 + \dot{Y}_{hji}^2\right) \\
\qquad + \sum_{i=1}^{3} J_{ghi}\Delta\dot{\beta}_i^2 + m_p\left(\dot{x}_p^2 + \dot{y}_p^2\right) + J_p\dot{\varphi}_p^2]
\end{cases}
\tag{26}
$$

The necked down section of the semicircle shape flexure hinge deflects to provide a small range, as shown in Fig. 6.

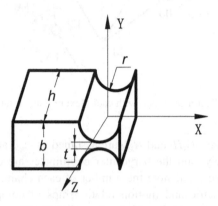

Fig. 6. Schematic of a flexure hinge

For the 3-DOFs compliant micro-positioning manipulator, any input of the piezoelectric ceramic actuators will lead the flexure hinges producing a small angular deformation. Thus, the potential energy stored in the flexure hinges of each kinematic chain. When the minimum thickness t of flexure hinge is much smaller than the flexure hinge radius r and the flexure hinge height h, the rotational stiffness of such flexure hinge can be approximately given as [19]:

$$
K_b = \frac{2Ebt^{5/2}}{9\pi r^{1/2}}
\tag{27}
$$

where t is the minimum thickness of flexure hinge, r is the radius of flexure hinge, h is the height of flexure hinge, b is the width of flexure hinge, E is the elastic modulus of the material.

Since each kinematic chain contains a prismatic pair(P) and two revolute pairs(R), the potential energy of each chain can be written as follow:

$$
P_i = \frac{1}{2}K_b\left(\Delta\beta_i^2 + \Delta\varphi_i^2\right)
\tag{28}
$$

where $\Delta\beta_i$ and $\Delta\varphi_i$ are the angular variation of the flexure hinges of each chain.

Based on the symmetric structure of the three kinematic chains of the compliant manipulator, the total potential energy is obtained as follow:

$$
P = \sum_{i=1}^{3} P_i = \frac{1}{2}K_b\left[\left(\Delta\beta_1^2 + \Delta\varphi_1^2\right) + \left(\Delta\beta_2^2 + \Delta\varphi_2^2\right) + \left(\Delta\beta_3^2 + \Delta\varphi_3^2\right)\right]
\tag{29}
$$

where:

$$\Delta\beta_i = \frac{x_p cot\varphi_i + y_p + 2d_i(sin\alpha_i + cot\varphi_i \cdot cos\alpha_i)}{L_{HJi}(sin\beta_i \cdot cot\varphi_i - cos\beta_i)} \tag{30}$$

$$\Delta\varphi_i = \frac{-x_p cos\beta_i - y_p sin\beta_i - 2d_i cos(\beta_i - \alpha_i)}{L_{Jip} sin\varphi_i(sin\beta_i \cdot cot\varphi_i - cos\beta_i)} \tag{31}$$

The input displacement vector and corresponding velocity vector are $\Delta d = (\Delta d_1, \Delta d_2, \Delta d_3)$ and $V_d = (\Delta\dot{d}_1, \Delta\dot{d}_2, \Delta\dot{d}_3)$ respectively. At the same time, the input force vector is assumed as $\Delta Q = (Q_1, Q_2, Q_3)$, then the dynamic model of 3-PRR compliant mechanism can be given by Lagrangian method as follow:

$$\frac{d}{dt}\frac{\partial k}{\partial \dot{d}_i} - \frac{\partial k}{\partial d_i} + \frac{\partial p}{\partial d_i} = \begin{bmatrix} Q_1 \\ Q_2 \\ Q_3 \end{bmatrix}, i = 1, 2, 3 \tag{32}$$

Based on Eq. (32), the final kinematic modeling of 3-PRR compliant mechanism can be given as follow:

$$\begin{bmatrix} \emptyset_1 & 0 & 0 \\ 0 & \emptyset_2 & 0 \\ 0 & 0 & \emptyset_3 \end{bmatrix}\begin{bmatrix} \ddot{d}_1 \\ \ddot{d}_2 \\ \ddot{d}_3 \end{bmatrix} + \begin{bmatrix} \delta_1 & 0 & 0 \\ 0 & \delta_2 & 0 \\ 0 & 0 & \delta_3 \end{bmatrix}\begin{bmatrix} d_1 \\ d_2 \\ d_3 \end{bmatrix} + \begin{bmatrix} \psi_1 \\ \psi_2 \\ \psi_3 \end{bmatrix} = \begin{bmatrix} Q_1 \\ Q_2 \\ Q_3 \end{bmatrix} \tag{33}$$

where:

$$\emptyset_i = \frac{1}{4}\sum\nolimits_{i=1}^{3} m_{ghi} + \sum\nolimits_{i=1}^{3} m_{hji} \tag{34}$$

$$\delta_i = \frac{1}{2}K_b \sum\nolimits_{i=1}^{3} \frac{8L_{Jip}^2 Q_{i5}^2 sin^2\varphi_i + 8L_{HJi}^2 Q_{i7}^2}{Q_{i6}^2 L_{HJi}^2 L_{Jip}^2 sin^2\varphi_i} \tag{35}$$

$$\begin{cases} \psi_i = \frac{1}{2}K_b \sum\nolimits_{i=1}^{3}[\frac{4(L_{Jip}^2 Q_{i5} sin^2\varphi_i cot\varphi_i + L_{HJi}^2 Q_{i7} cos\beta_i)}{Q_{i6}^2 L_{HJi}^2 L_{Jip}^2 sin^2\varphi_i}x_p \\ \qquad + \frac{4(L_{Jip}^2 Q_{i5} sin^2\varphi_i + L_{HJi}^2 Q_{i7} sin\beta_i)}{Q_{i6}^2 L_{HJi}^2 L_{Jip}^2 sin^2\varphi_i}y_p] \\ \qquad + \frac{1}{2}\sum\nolimits_{i=1}^{3} m_{hji}L_{hji}[\Delta\ddot{\beta}_i sin(\beta_i - \Delta\beta_i - \alpha_i) \\ \qquad + \Delta\dot{\beta}_i^2 cos(\beta_i - \Delta\beta_i - \alpha_i)] \end{cases} \tag{36}$$

$$Q_{i5} = sin\alpha_i + cot\varphi_i cos\alpha_i \tag{37}$$

$$Q_{i6} = sin\beta_i cot\varphi_i - cos\varphi_i \tag{38}$$

$$Q_{i7} = cos(\beta_i - \alpha_i) \tag{39}$$

5   Finite Element Analysis

To further analyze the dynamic characteristic of the proposed manipulator and validate the developed dynamic models, the finite element analysis is achieved in ANSYS software. Stainless steel + is chosen as the material of the 3-PRR compliant mechanism. The 3D model of the compliant manipulator is drawn by Solidworks software

Table 1. The main dimension parameters of compliant mechanism

Parameter	L_1	L_2	L_3	L_4	L_{GH}	L_{HJ}
Size/mm	16	6	40	16	7	35

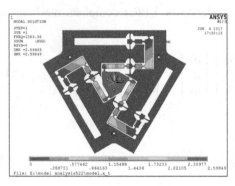

(a)modal 1 1383.36Hz

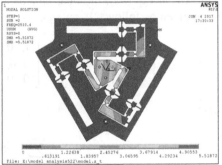

(b)modal 2 2510.4Hz

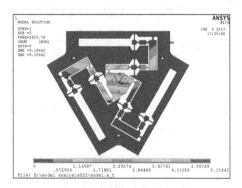

(c)modal 3 2520.76Hz

Fig. 7. Modal analysis

and then simulated in ANSYS software. Then the natural frequency of the compliant mechanism can be obtained. Table 1 denotes the main dimension parameters of the compliant mechanism. Figure 7(a), (b) and (c) denote the first three-orders natural frequency of the compliant mechanism respectively.

According to simulation results from ANSYS, the first three-orders natural frequencies of the compliant manipulator are obtained, as 1383.36 Hz, 2510.4 Hz, 2520.76 Hz respectively. In our future work, the experiments on the positioning control and vibration suppression based on the developed kinematic model and dynamic model will be conducted.

Acknowledgments. This work was supported by the National Natural Science Foundation of China (Grants nos. 51605271, 51577112, and 51575332).

References

1. Yue, Y., Gao, F., Zhao, X., et al.: Relationship among input-force, payload, stiffness and displacement of a 3-DOF perpendicular parallel micro-manipulator. Mech. Mach. Theory **45**(5), 756–771 (2010)
2. Li, Y.M., Huang, J.M., Tang, H.: A compliant parallel XY micromotion stage with complete kinematic decoupling. IEEE Trans. Autom. Sci. Eng. **9**(3), 538–553 (2012)
3. Xu, P., Yu, J.J., Zong, G.H.: Design of compliant straight-line mechanisms using flexural joints. Chin. J. Mech. Eng. **27**(1), 146–153 (2014)
4. Hu, J.F., Zhang, X.M.: Kinematical properties and optimal design of 3-DOF precision positioning stage. Opt. Precis. Eng. **20**(12), 2686–2695 (2012)
5. Howell, L.L.: Compliant Mechanisms. Wiley, Hoboken (2001)
6. Dorn, W.S., Gomory, R.E., Greenberg, H.J.: Automatic design of optimal structures. J. Mecanique **3**, 25–52 (1964)
7. Bendsoe, M.P., Sigmund, O.: Topology Optimization: Theory, Methods, and Applications. Springer Science & Business Media, Berlin (2013)
8. Zhu, B.L., Zhang, X.M.: A new level set method for topology optimization of distributed compliant mechanisms. Int. J. Numer. Meth. Eng. **91**(8), 843–871 (2012)
9. Li, Y.M., Xiao, S.L., Xi, L.Q.: Design, modeling, control and experiment for a 2-DOF compliant micro-motion stage. Int. J. Precis. Eng. Manufact. **15**(4), 735–744 (2014)
10. Li, Y.M., Xu, Q.S.: Design and analysis of a totally decoupled flexure-based XY parallel micromanipulator. IEEE Trans. Robot. **25**(3), 645–657 (2009)
11. Li, X., Tian, Y., Qin, Y., et al.: Design, identification and control of a 2-degree of freedom flexure-based mechanism for micro/nano manipulation. Nanosci. Nanotechnol. Lett. **5**(9), 960–967 (2013)
12. Jin, M., Zhang, X.: A new topology optimization method for planar compliant parallel mechanisms. Mech. Mach. Theory **95**, 42–58 (2016)
13. Wang, N.F., Liang, X.H., Zhang, X.M.: Pseudo-rigid-body model for corrugated cantilever beam used in compliant mechanisms. Chin. J. Mech. Eng. **27**(1), 122–129 (2014)
14. Zhu, D.C., Song, M.J.: Configuration design with topology optimization and vibration frequency analysis for 3-DOF planar integrated fully compliant parallel mechanism. J. Vib. Shock **3**, 006 (2016)
15. Zhu, D.C., Fang, W.J., An, Z.M.: Topology optimization integrated design of 3-DOF fully compliant planar parallel manipulator. J. Mech. Eng. **51**(5), 30 (2015)

16. Handley, D.C., Lu, T.F., Yong, Y.K., et al.: A simple and efficient dynamic modeling method for compliant micropositioning mechanisms using flexure hinges. In: Microelectronics, MEMS, and Nanotechnology. International Society for Optics and Photonics, pp. 67–76 (2004)

17. Lu, T.F., Handley, D.C., et al.: A three-DOF compliant micromotion stage with flexure hinges. Ind. Robot **31**(4), 355–361 (2004)

18. He, S.F., Tang, H., Qiu, Q., et al.: A novel flexure-based $XY\theta$ motion compensator: towards high-precision wafer-level chip detection. In: 2016 IEEE 18th Electronics Packaging Technology Conference (EPTC), pp. 381–387. IEEE (2016)

19. Paros, J.M., Weisbord, L.: How to design flexure hinges. Mach. Des. **37**, 151–156 (1965)

Analysis for Rotation Orthogonality of a Dynamically Adjusting Generalized Gough-Stewart Parallel Manipulator

ZhiZhong Tong$^{(\boxtimes)}$, Tao Chen, Lei Peng, Hongzhou Jiang, and Fengjing He

Department of Mechatronics Engineering, Harbin Institute of Technology, Harbin 150001, China
tongzhizhong@hit.edu.cn

Abstract. A Gough-Stewart parallel manipulator using point decoupled design method is orthogonal only at a single point, which means that it has small high-precision workspace. It is hard to break through its' restriction to be applied in the field of engineer applications with larger workspace. This paper formulates the problem to guarantee continuous orthogonality when the manipulator rotating along the z-axis, called rotation orthogonality. Compared to traditional optimum, the rotation orthogonality leads to better performances with enlarged workspace. A class of dynamically adjusting generalized Gough-Stewart parallel manipulators (DAGGSPM) is proposed. The dynamically adjusting mechanism for rotation orthogonality and the related algorithm are deduced analytically. Through analysis of three typical cases and numerical verifications, the results show that a DAGGSPM can pave the way for high-precision applications with large scale workspace including laser weapon pointing, scanning microscopes and integrated circuit fabrication.

Keywords: Dynamically adjusting generalized Gough-Stewart parallel manipulators · Rotation orthogonality · A pair of circular hyperboloids · Analytical algorithm · Optimum design

1 Introduction

A Gough-Stewart parallel manipulator (GSPM) can be designed, at a single configuration, to have better performances within a quite small workspace. It has been employed widely applied for micro systems. For the design of a GSPM, the geometry can be used to model the kinematics and to optimize desired performance through local design [1] or point design [2].

Isotropy or orthogonality is often considered a design objective [3–8]. Isotropic conditions are defined by Zanganch and Angeles [9]. A GSPM is isotropic if

$$\mathbf{J}_x^T \mathbf{J}_x = \begin{bmatrix} \alpha \mathbf{I} & 0 \\ 0 & \beta \mathbf{I} \end{bmatrix} \tag{1}$$

© Springer International Publishing AG 2017
Y. Huang et al. (Eds.): ICIRA 2017, Part II, LNAI 10463, pp. 359–370, 2017.
DOI: 10.1007/978-3-319-65292-4_31

at some point in the workspace ($\mathbf{J}_x$ is the inverse Jacobian matrix. α and β are scalars). If $\mathbf{J}_x^T \mathbf{J}_x$ is a diagonal matrix, a GSPM is orthogonal in general. An orthogonal Gough-Stewart parallel manipulator is easily to realize precision pointing, motion planning, control scheme, calibration and compensation. A numerical searching method is traditional to find unknown parameters and to achieve orthogonality under some structural constraints [3–5]. The second approach constructs the architecture after generating a desired Jacobian matrix. Jafari and McInroy explore various properties of the inverse Jacobian matrix to design orthogonal GSPMs among all the possible geometries [6]. Yi *et al.* presented a novel method for generating classes of orthogonal Gough-Stewart parallel manipulators [1, 7, 8]. However, these two approaches are complicated and the architectures generated may be hard to implement. Jiang and Tong *et al.* presented a geometrically intuitive and analytical method using circular hyperboloids [10–13].

The approaches mentioned above generate architectures only at one desire point. Coupling characteristics occur and decrease some performance indices when a GSPM moving away from the desire point. If the orthogonality would be reserved even if the motion has only one degree of freedom, it is more attractive, especially in laser weapon pointing and scanning microscopes. In this paper, we will propose the concept of dynamically adjusting generalized Gough-Stewart manipulators (DAGGSPMs). A DAGGSPM will exhibit the guaranteed orthogonality when it rotates along the z-axis. The analytical algorithm is derived and expressed explicitly. Finally, the presented algorithm will be verified.

2   DAGGSPMs Definition and Orthogonality Conditions

2.1   The Definition of DAGGSPMs

Recently a generalized Gough-Stewart parallel manipulator is considered a well-established option [9–12]. For a GGSPM, all gimbals are fixedly mounted on the movable platform and the fixed base respectively. If several gimbals can be driven to adjusted, the GGSPM can be reconstructed and it may be feasible in order to have some potential better performances.

Definition 1. For a GGSPM, three adjustable lower gimbals are driven actively and synchronously, which belong to even group. When the movable platform rotates along the z-axis, the vectors of even struts will vary smoothly and continuously. If the adjusting meets some orthogonal conditions, then this GGSPM leads to be orthogonal during the rotation. This kind of GGSPMs is defined as dynamically adjusting generalized Gough-Stewart parallel manipulators (DAGGSPM), as shown in Fig. 1.

For a GGSPM, all gimbals are attached to the movable platform and the fixed base, and each of them cannot be adjusted. Coupling occurs for a point orthogonal Gough-Stewart parallel manipulator when the movable platform rotates some angle along the z-axis. For a DAGGSPM, three adjustable lower gimbals are able to ascertain the configuration described by a pair of circular hyperboloids. During the rotational motion along the z-axis, the double circular hyperboloids can be guaranteed to meet orthogonal or decoupling condition by dynamically given adjusting mechanism.

A DAGGSPM introduces extra degrees of freedom, from decoupling points of view, while it exhibits potential abilities to meet some required applications and to enlarge high-precision workspace by guaranteeing orthogonality.

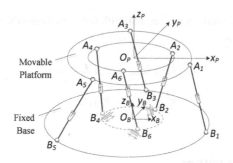

Fig. 1. The schematic view of a DAGGSPM

2.2   Jacobian Matrix and Orthogonal Conditions

With respect to line geometry, a DAGGSPM lies on dynamically changing double circular hyperboloids when it rotates along the z-axis at any time t_i. The odd number struts lie on the first hyperboloid Γ_{1,t_i}, and the even number struts lie on the second hyperboloid Γ_{2,t_i}. The surface equations of two hyperboloids can be given by

$$\Gamma_{1,t_i}: \quad \frac{x - r_{1,t_i}\cos\alpha_{t_i}}{-r_{1,t_i}\sin\alpha_{t_i}} = \frac{y - r_{1,t_i}\sin\alpha_{t_i}}{-r_{1,t_i}\cos\alpha_{t_i}} = \frac{z - a_{1z,t_i}}{c_{1,t_i}} \tag{2}$$

$$\Gamma_{2,t_i}: \quad \frac{x - r_{2,t_i}\cos\alpha_{t_i}}{-r_{2,t_i}\sin\alpha_{t_i}} = \frac{y - r_{2,t_i}\sin\alpha_{t_i}}{-r_{2,t_i}\cos\alpha_{t_i}} = \frac{z - a_{2z,t_i}}{c_{2,t_i}} \tag{3}$$

where r_{1,t_i} and r_{2,t_i} are the throat radii of the hyperboloids respectively. a_{1z,t_i} and a_{2z,t_i} are distances from the center of the hyperboloids to the frame origin. c_{1,t_i} and c_{2,t_i} can be called the shape parameters of the hyperboloids respectively and α_{t_i} as the sweep angle.

The Plücker coordinates of the lines, where the strut 1 and the strut 6 lie, are expressed at any time t_i in the form as

$$\begin{aligned}
\mathbf{p}_{1,t_i} = [&-k_{a1,t_i}\sin\alpha_{t_i} \quad k_{a1,t_i}\cos\alpha_{t_i} \quad k_{c1,t_i} \quad -a_{1z,t_i}k_{a1,t_i}\cos\alpha_{t_i} + r_{1,t_i}k_{c1,t_i}\sin\alpha_{t_i} \\
&-a_{1z,t_i}k_{a1,t_i}\sin\alpha_{t_i} - r_{1,t_i}k_{c1,t_i}\cos\alpha_{t_i} \quad r_{1,t_i}k_{a1,t_i}]^{\mathrm{T}}
\end{aligned} \tag{4}$$

$$\begin{aligned}
\mathbf{p}_{6,t_i} = [&-k_{a2,t_i}\sin\alpha_{t_i} \quad k_{a2,t_i}\cos\alpha_{t_i} \quad k_{c2,t_i} \quad -a_{2z,t_i}k_{a2,t_i}\cos\alpha_{t_i} + r_{2,t_i}k_{c2,t_i}\sin\alpha_{t_i} \\
&-a_{2z,t_i}k_{a2,t_i}\sin\alpha_{t_i} - r_{2,t_i}k_{c2,t_i}\cos\alpha_{t_i} \quad r_{2,t_i}k_{a2,t_i}]^{\mathrm{T}}
\end{aligned} \tag{5}$$

where $k_{a1,t_i} = \dfrac{r_{1,t_i}}{\sqrt{r_{1,t_i}^2 + c_{1,t_i}^2}}$, $k_{c1,t_i} = \dfrac{c_{1,t_i}}{\sqrt{r_{1,t_i}^2 + c_{1,t_i}^2}}$, $k_{a2,t_i} = \dfrac{r_{2,t_i}}{\sqrt{r_{2,t_i}^2 + c_{2,t_i}^2}}$ and $k_{c2,t_i} = \dfrac{c_{2,t_i}}{\sqrt{r_{2,t_i}^2 + c_{2,t_i}^2}}$.

Using screw property-alteration matrix, the Jacobian matrix at time t_i can be structured with six Plücker coordinates as

$$\mathbf{J}_{\mathrm{X},t_i} = \begin{bmatrix} \mathbf{P}_{1,t_i} & \mathbf{P}_{2,t_i} & \mathbf{P}_{3,t_i} & \mathbf{P}_{4,t_i} & \mathbf{P}_{5,t_i} & \mathbf{P}_{6,t_i} \end{bmatrix}^{\mathrm{T}} \tag{6}$$

If $\mathbf{J}_{\mathrm{X},t_i}^{\mathrm{T}} \mathbf{J}_{\mathrm{X},t_i}$ is a diagonal matrix, a manipulator is orthogonal at time t_i. The propositions can be given by

$$r_{1,t_i} k_{a1,t_i} k_{c1,t_i} = r_{2,t_i} k_{a2,t_i} k_{c2,t_i} \tag{7}$$

$$k_{a1,t_i}^2 a_{1z,t_i} = k_{a2,t_i} a_{2z,t_i} \tag{8}$$

3   Analytical Algorithm

During the movable platform rotating, it is ready to produce the other circular hyperboloids where the even struts lie according to the orthogonal conditions and a given circular hyperboloid yielded from the odd number struts. The rotational orthogonality will be approached in smooth and continuous time, instead of only one point. This calculation is listed step by step as follows.

Step 1:   A GGSPM with point orthogonality at a desired position is generated as an initial prototype, then choose α, two planes H_1 and H_2 ($H_2 < 0$), to cut the double hyperboloids of the DAGGSPM.

Step 2:   To meet the task requirements, for example, dexterity, singularity and workspace, it is unavoidable to verify its performance measures.

Step 3:   At time t_i two coordinates of the upper gimbals 1 and 6, denoted by $\mathbf{A}_{1,t_i} = \left(x_{A1,t_i}, y_{A1,t_i}, z_{A1,t_i} \right)^{\mathrm{T}}$ and $\mathbf{A}_{6,t_i} = \left(x_{A6,t_i}, y_{A6,t_i}, z_{A6,t_i} \right)^{\mathrm{T}}$ respectively, can be given by

$$\begin{cases} x_{A1,t_i} = -r_{1,t_i} \sin \alpha_{t_i} t_{A1,t_i} + r_{1,t_i} \cos \alpha_{t_i} \\ y_{A1,t_i} = r_{1,t_i} \cos \alpha_{t_i} t_{A1,t_i} + r_{1,t_i} \sin \alpha_{t_i} \\ z_{A1,t_i} = H_1 \end{cases} \tag{9}$$

$$\begin{cases} x_{A6,t_i} = -r_{2,t_i} \sin \alpha_{t_i} t_{A6,t_i} + r_{2,t_i} \cos \alpha_{t_i} \\ y_{A6,t_i} = -r_{2,t_i} \cos \alpha_{t_i} t_{A6,t_i} - r_{2,t_i} \sin \alpha_{t_i} \\ z_{A6,t_i} = H_2 \end{cases} \tag{10}$$

where $t_{A1,t_i} = \left(H_1 - a_{1z,t_i} \right) / c_{1,t_i}$ and $t_{A6,t_i} = \left(H_2 - a_{2z,t_i} \right) / c_{2,t_i}$.

Let $t_{B1,t_i} = \left(H_2 - a_{2z,t_i} \right) / c_{2,t_i}$ and $t_{B6,t_i} = \left(H_2 - a_{2z,t_i} \right) / c_{2,t_i}$, two coordinate of the upper gimbals 1 and 6 at time t_i, denoted by $\mathbf{B}_{1,t_i} = \left(x_{B1,t_i}, y_{B1,t_i}, z_{B1,t_i} \right)^{\mathrm{T}}$ and $\mathbf{B}_{6,t_i} = \left(x_{B6,t_i}, y_{B6,t_i}, z_{B6,t_i} \right)^{\mathrm{T}}$, can be expressed as forms

$$\begin{cases} x_{B1,t_i} = -r_{1,t_i}\sin\alpha_{t_i}t_{B1} + r_{1,t_i}\cos\alpha_{t_i} \\ y_{B1,t_i} = r_{1,t_i}\cos\alpha_{t_i}t_{B1} + r_{1,t_i}\sin\alpha_{t_i} \\ z_{B1,t_i} = H_2 \end{cases} \qquad (11)$$

$$\begin{cases} x_{B6,t_i} = -r_{2,t_i}\sin\alpha_{t_i}t_{B6,t_i} + r_{2,t_i}\cos\alpha_{t_i} \\ y_{B6,t_i} = -r_{2,t_i}\cos\alpha_{t_i}t_{B6,t_i} - r_{2,t_i}\sin\alpha_{t_i} \\ z_{B6,t_i} = H_2 \end{cases} \qquad (12)$$

Step 4: At next time t_{i+1}, the movable platform will rotate a very small angle θ along z-axis. The new coordinate of the upper gimbal 1 is obtained by

$$A_{1,t_{i+1}} = \begin{bmatrix} \cos\theta & -\sin\theta & 0 \\ \sin\theta & \cos\theta & 0 \\ 0 & 0 & 1 \end{bmatrix} A_{1,t_i} \qquad (13)$$

projecting $A_{1,t_{i+1}}$ into the plane H_2, then the new throat radius $r_{1,t_{i+1}}$ of the first hyperboloid $\Gamma_{1,t_{i+1}}$ can be represented as

$$r_{1,t_{i+1}} = \frac{x_{A1,t_{i+1}}y_{B1,t_{i+1}} - x_{B1,t_{i+1}}y_{A1,t_{i+1}}}{\sqrt{\left(x_{A,t_{i+1}} - x_{A1,t_{i+1}}\right)^2 + \left(y_{A,t_{i+1}} - y_{B1,t_{i+1}}\right)^2}} \qquad (14)$$

$$k_{c1,t_{i+1}} = \frac{-H_2}{\sqrt{\left(x_{A,t_{i+1}} - x_{B1,t_{i+1}}\right)^2 + \left(y_{A,t_{i+1}} - y_{B1,t_{i+1}}\right)^2 + \left(z_{A,t_{i+1}} - y_{B1,t_{i+1}}\right)^2}} \qquad (15)$$

Obviously, $k_{a1,t_i+1} = \sqrt{1-k_{c1,t_i+1}^2}$ and $c_{1,t_{i+1}} = r_{1,t_{i+1}}k_{c1,t_{i+1}}/k_{a1,t_{i+1}}$. Substituting $A_{1,t_{i+1}}$ into the surface equation of the hyperboloid $\Gamma_{1,t_{i+1}}$, then $a_{1z,t_{i+1}}$ is given by

$$a_{1z,t_{i+1}} = H_1 - \frac{c_{1,t_{i+1}}}{r_{1,t_{i+1}}}\sqrt{x_{A1,t_{i+1}}^2 + y_{A1,t_{i+1}}^2 - r_{1,t_{i+1}}^2} \qquad (16)$$

Step 5: Notice that $A_{6,t_{i+1}}$ remains on a new hyperboloid $\Gamma_{2,t_{i+1}}$, which satisfies

$$\frac{x_{A6,t_{i+1}}^2 + y_{A6,t_{i+1}}^2}{r_{2,t_{i+1}}^2} - \frac{\left(z_{A6,t_{i+1}} - a_{2z,t_{i+1}}\right)^2}{c_{2,t_{i+1}}^2} = 1 \qquad (17)$$

Let $f_1 = r_{1,t_{i+1}}k_{a1,t_{i+1}}k_{c1,t_{i+1}}$, $f_1 = k_{a1,t_{i+1}}^2 a_{1z,t_{i+1}}$ and $f_3 = -f_2/f_1$, the above equation can be rewritten as

$$l_a c_{2,t_{i+1}}^2 + l_b c_{2,t_{i+1}} + l_c = 0 \qquad (18)$$

where $l_a = f_3^2 + 1,$ $l_b = \left[\left(x_{A1,t_{i+1}}^2 + y_{A1,t_{i+1}}^2 \right) / f_1 + 2z_{A1,t_{i+1}} f_3 \right.$ and
$l_c = x_{A1,t_{i+1}}^2 + y_{A1,t_{i+1}}^2 + z_{A1,t_{i+1}}^2.$
Then we get:

$$c_{2,t_{i+1}} = \frac{-l_b + \sqrt{l_b^2 - 4l_a l_c}}{2l_a} \tag{19}$$

Next, the rest parameters of $\Gamma_{2,t_{i+1}}$ can be solved

$$a_{2z,t_{i+1}} = f_3 c_{2,t_{i+1}} \tag{20}$$

$$k_{a2,t_{i+1}} = \sqrt{f_1 / c_{2,t_{i+1}}} \tag{21}$$

$$k_{c2,t_{i+1}} = \sqrt{1 - k_{a2,t_{i+1}}^2} \tag{22}$$

$$r_{2,t_{i+1}} = k_{a2,t_{i+1}} c_{2,t_{i+1}} / k_{c1,t_{i+1}} \tag{23}$$

Step 6: After $\Gamma_{2,t_{i+1}}$ is calculated, the new joint coordinate of the upper gimbal 6 is given by

$$A_{6,t_{i+1}} = R(\theta) A_{6,t_i} \tag{24}$$

In Fig. 2, note $r_{1m} = r_{2,t_{i+1}}$ and $h = z_{A6,t_{i+1}} - a_{2z,t_{i+1}}$, then let l_{FG} denotes the projection length between F and G, l_{OG} denotes the projection distance between O and G. In the triangle ΔEFG, we have $l_{FG} = h\tan\varphi = hk_{a2,t_{i+1}} / k_{c2,t_{i+1}}$, $l_{OG} = \sqrt{x_{A1,t_{i+1}}^2 + y_{A1,t_{i+1}}^2}$.

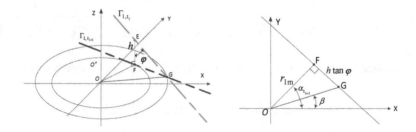

Fig. 2. The plane projection of leg 6th straight-line

Note that $\beta = \mathrm{asin} \frac{|y_{A1,t_{i+1}}|}{l_{OG}}$, then obtain

$$\alpha_{t_{i+1}} = \beta + \arcsin \frac{l_{FG}}{l_{OG}} \tag{25}$$

The coordinates of lower gimbal 6 can be given by

$$\begin{cases} x_{B6,t_{i+1}} = -r_{2,t_{i+1}} \sin \alpha_{t_{i+1}} t_{B6,t_{i+1}} + r_{2,t_{i+1}} \cos \alpha_{t_{i+1}} \\ y_{B6,t_{i+1}} = -r_{2,t_{i+1}} \cos \alpha_{t_{i+1}} t_{B6,t_{i+1}} - r_{2,t_{i+1}} \sin \alpha_{t_{i+1}} \\ z_{B6,t_{i+1}} = H_2 \end{cases} \qquad (26)$$

where $t_{B6,t_{i+1}} = \left(H_2 - a_{2z,t_{i+1}}\right)/c_{2,t_{i+1}}$.

With respect to rotational symmetry, the other two lower gimbal coordinates $B_{2,t_{i+1}}$ and $B_{4,t_{i+1}}$ can be obtained based on the rotation matrix.

The above procedure runs continuously, so that the orthogonality is guaranteed during the moving. The derived formulations are all analytical, so they can be executed efficiently in real-time.

4 Numerical Verifications

With the aid of numerical cases, we present how it is feasible to realize adjusting mechanism for a DAGSPM to guarantee the rotation orthogonality. Firstly, we have generated an optimal architecture with point orthogonality, and the surface equations of double hyperboloids have the forms

$$\begin{cases} \Gamma_{1,t_0} : \dfrac{x^2 + y^2}{0.04} - \dfrac{z - 0.1}{0.04} = 1 \\ \Gamma_{2,t_0} : \dfrac{x^2 + y^2}{0.0625} - \dfrac{z + 0.1}{0.25} = 1 \end{cases} \qquad (27)$$

The neutral architecture for the DAGGSPM is facilitated by setting $H_2 = -0.4$ m and $\alpha = 0°$.

Considering locations of the desired point may have three cases, so the relations of adjusting trajectories should be discussed. It is important to illustrate the differences among three cases and to propose some solution for mechanical design.

4.1 Case 1: $H_1 = 0$

In this case, $H_1 = 0$ means that the desired point is located at the geometric center of the movable platform. Assuming that the required rotation along z-axis is in the range from $-15°$ to $15°$, a numerical evaluation with fixed step $0.1°$ is executed.

The adjusting trajectories of even lower gimbals are calculated and drawn in Fig. 3. The circular hyperboloid, which the even number struts lie on, is constructed continuously during the rotational motion. The curves show that adjusting trajectories are nonlinear varying and satisfy rotational symmetry. The length of each trajectory is about 200 mm.

The relations of adjusting trajectories vs. rotation along the z-axis are illustrated in Fig. 4. X coordinate, y coordinate and z coordinate of each even lower gimbal are adjusted smoothly. It implies that the reconfiguration mechanism can be realized stably without singularity and cannot complicate the matters.

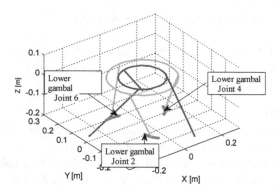

Fig. 3. Schematic of the reconfiguration mechanism for case 1

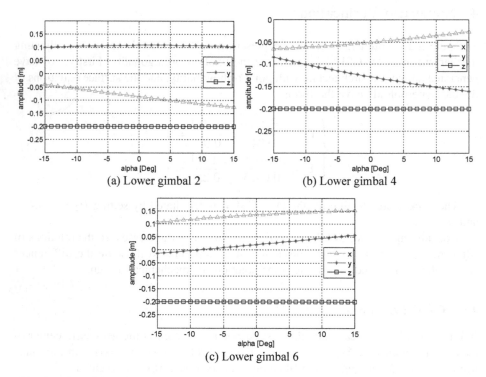

(a) Lower gimbal 2   (b) Lower gimbal 4

(c) Lower gimbal 6

Fig. 4. The adjusting trajectories VS. rotation along z-axis during reconfiguration of case 1

4.2   Case 2: $H_1 > 0$

If the desired point is over the movable platform, then $H_1 > 0$. This case is more general than Case 1, and occurs often in high-precision applications. Let $H_1 = 0.02$ m, the same rotation and step value as Case 1 are introduced. The results are shown in Figs. 5 and 6. Compare to Case 1, the adjusting trajectories of three lower gimbals have similar shapes but the length of each is shorter than the one of Case 1.

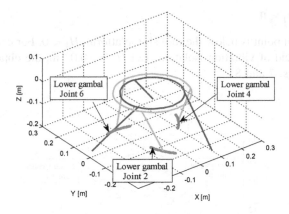

Fig. 5. Schematic diagram of reconfiguration mechanism for Case 2

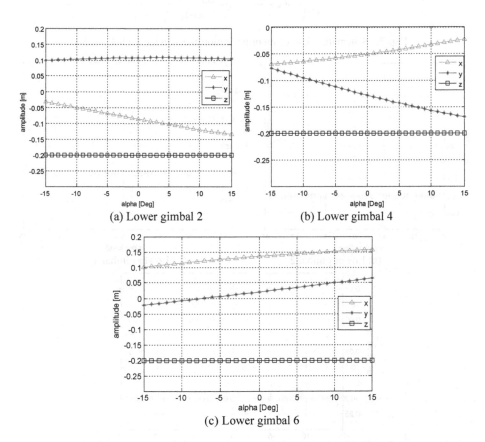

(a) Lower gimbal 2   (b) Lower gimbal 4

(c) Lower gimbal 6

Fig. 6. The trajectories VS. rotation along z-axis with Case 2

4.3   Case 3: $H_1 < 0$

When the desired point is below the movable platform, $H_1 < 0$. For example, it can be applied in the field of vibration isolation. Let $H_1 = -0.02$ m, the obtained results are illustrated in Figs. 7 and 8.

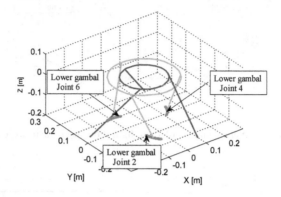

Fig. 7. Schematic view of reconfiguration mechanism with Case 3

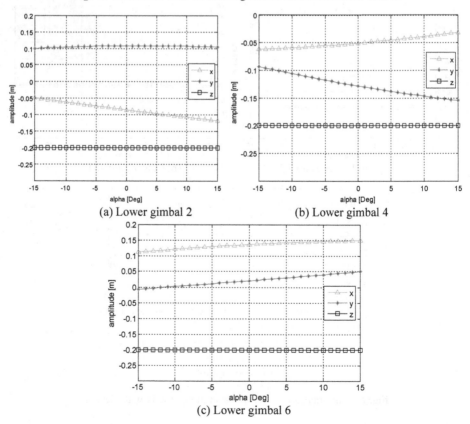

Fig. 8. The adjusting trajectories VS. rotation along z-axis with Case 3

In theoretical, it is the duality between Case 2 and Case 3. The former has better mechanical feasibility than the latter, because that the operator below the movable platform is prone to occurring collisions. The lengths of the adjusting trajectories during the reconfiguration are shorter compared with Case 1 and Case 2.

4.4   Discussion

The obtained results show that the rotation orthogonality is feasible, and a DAGGSPM has potential ability to enlarge the high precision workspace. To compare all the cases, the reconfigurations are illustrated in Fig. 9. Noticeably, it is observed that the length of adjusting trajectory decreases with H_1 increases. In three cases, Case 3 is more sensitive than the others. Furthermore, there are appropriate adjusting work areas, where the trajectories possess better linearity. It means that each adjustable gambal can be actuated by one degree of freedom actuator. Considering the rotation symmetry, three adjustable gambals can be underactuted, instead of introducing too many degrees of freedom and leading to a significant mechanical complication.

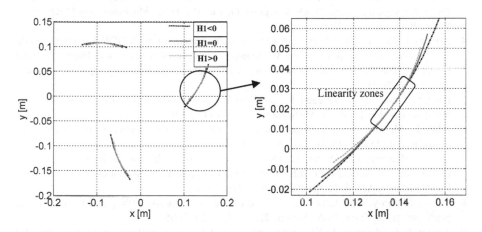

Fig. 9. Comparisons of the adjusting trajectories in the XOY plane of three cases

5   Conclusions

The design of GSPMs capable of meeting high-precision applications is considerable and significant. This paper formulates the problem to enlarge the workspace optimized by point or local optimization. A concept of DAGGSPM is proposed and the routine is discussed. The dynamically adjusting mechanism for rotation orthogonality is illustrated in depth. And also the related algorithm is deduced analytically, which executes in repetitive iteration smoothly and continuously. Thereby the rotational orthogonality along z-axis is approached, and the workspace with high-precision and low coupling is enlarged in full range of rotation. It may be a better solution to break down the restriction of point optimization. Finally, three cases provide numerical verifications in order that the reconfiguration mechanism is validated. A DAGGSPM gives more flexibility for the aspect of optimal design a parallel manipulator with high precision and larger workspace, especially in the engineering applications including laser weapon pointing, scanning microscopes and integrated circuit fabrication.

Acknowledgements. This research was supported by the Natural Science Foundation of China (Grant No. 51575121). The authors would like to thank the reviewers.

References

1. Yi, Y., McInroy, J.E., Jafari, F.: Generating classes of orthogonal Gough-Stewart platforms. In: IEEE International Conference on Robotics and Automation, New Orleans, pp. 4969–4974 (2004)
2. Legnani, G., Tosi, D., Fassi, I., Giberti, H., Cinquemani, S.: The "point of isotropy" and other properties of serial and parallel manipulators. Mech. Mach. Theory 45(10), 1407–1423 (2010)
3. Bernier, D., Castelain, J.M., Li, X.: A new parallel structure with 6 degrees of freedom. In: Proceedings of the Ninth World Congress on the Theory of Machines and Mechanisms, Milan, Italy, pp. 8–12 (2005)
4. Zabalza, I., Ros, J., Gil, J.J., et al.: A new kinematic structure for a 6-DOF decoupled parallel manipulator. In Proceedings of the Workshop on Fundamental Issues and Future Research Directions for Parallel Mechanisms and Manipulators, Quebec City, Quebec, Canada, pp. 12–15 (2002)
5. Briot, S., Arakelian, V., Guégan, S.: PAMINSA: a new family of partially decoupled parallel manipulators. Mech. Mach. Theory 44(2), 425–444 (2009)
6. Jafari, F., McInroy, J.E.: Orthogonal Gough-Stewart platforms for micromanipulation. IEEE Trans. Robot. Autom. 19(4), 595–603 (2003)
7. Yi, Y., McInroy, J.E., Jafari, F.: Optimum design of a class of fault-tolerant isotropic Gough-Stewart platforms. In: IEEE International Conference on Robotics and Automation, New Orleans, LA, vol. 5, pp. 4963–4968, April 2004
8. Yi, Y., McInroy, J.E., Jafari, F.: Generating classes of locally orthogonal Gough-Stewart platforms. IEEE Trans. Rob. Autom. 21(5), 812–820 (2005)
9. Zanganch, K., Angeles, J.: Kinematic isotropy and the optimum design of manipulators. Int. J. Robot. Res. 16(2), 185–197 (1997)
10. Jiang, H.Z., Tong, Z.Z., He, J.F.: Dynamic isotropy design of a class of Gough-Stewart parallel manipulators lying on a circular hyperboloid of one sheet. Mech. Mach. Theory 46(3), 358–374 (2010)
11. Jiang, H.Z., He, J.F., Tong, Z.Z.: Dynamic isotropic design for modified Gough-Stewart platforms lying on a pair of circular hyperboloids. Mech. Mach. Theory 46(9), 1301–1315 (2010)
12. Tong, Z., He, J., Jiang, H., et al.: Optimal design of a class of generalized symmetric Gough-Stewart parallel manipulators with dynamic isotropy and singularity-free workspace. Robotica 30(2), 305–314 (2012)
13. Tong, Z., He, J., Jiang, H., et al.: Locally dynamic isotropy of modified symmetric gough-stewart parallel micromanipulators. In: 13th World Congress in Mechanism and Machine Science, Guanajuato, México, 19–25 June 2011

A Delta-CU – Kinematic Analysis and Dimension Design

Jiayu Li, Huiping Shen$^{(\boxtimes)}$, Qinmei Meng, and Jiaming Deng

Changzhou University, Changzhou 213016, China
shp65@126.com

Abstract. According to the topological structure design method for parallel mechanism (PM) based on the position and orientation characteristics (POC) equation, a so-called Delta-CU, 2-$\underline{R}$//R//P(^{4R})//R⊕1-$\underline{R}$SS PM is first presented. It is proved that the PM can achieve three-translation output. Second, both the forward and inverse solutions for the position analysis are derived. Besides, the workspace analysis is then performed based on the inverse position equations. Furthermore, the matching relationship among $(R - r)$, l_1 and l_2 is obtained to ensure its maximum workspace. The PM has the advantages of simple structure, easy assembly and maintenance, etc., and can replace the Delta PM. This work lays a foundation for the dimension optimization, dynamic analysis and the prototype design of the Delta-CU PM.

Keywords: Parallel mechanism · Delta · Delta-CU · POC equation · Kinematics · Workspace

1 Introduction

Delta parallel manipulator or mechanism (PM) has been widely used because it adopts closed-loop branched chain [1, 2], which has the characteristics of high capacity, good stiffness and smaller errors. Many researchers investigate this PM [3–5]. Romdhane [6] first proposed that the positive position solutions of a class of 3 degrees of freedom (DOF) PM can be obtained using an analytical method. For example, the forward kinematics of linear actuated Delta mechanism was solved by using vector method [7]. Zhao and Zhu [8] simplified the forward kinematics solution of Delta with the solution of the three pyramid space geometry based on the geometric algorithm. Zhang [10] obtained the analytical expression of the kinematics equation by the space vector method based on the work in [8].

On the investigation of performance optimization of kinematics and dynamics, Alain [10, 11] established the dynamic model using the method that the 1/3 and 2/3 mass of the driven arm are respectively decomposed into the joint of moving platform and the actuated arms. Richard [12] proposed two optimization methods based on workspace: one method use the maximum workspace as the optimization goal. The second method uses the largest global condition index as the optimization objective.

According to the topological structure design method for PMs based on the position and orientation characteristics (POC) equations [13–15], a so-called Delta-CU,

© Springer International Publishing AG 2017
Y. Huang et al. (Eds.): ICIRA 2017, Part II, LNAI 10463, pp. 371–382, 2017.
DOI: 10.1007/978-3-319-65292-4_32

2-R//R//P(^{4R})//R⊕1-SPS PM with three-translation output is presented. Furthermore, kinematic analysis and dimension design based on the maximal workspace design are performed.

2   The Design of Delta-CU

2.1   Mechanism Design

The Delta PM consists of three identical mixed chains with complex architecture. While the Delta-CU PM proposed in this paper is shown in Fig. 1, which is composed of a base platform 0, a movable platform 1, only two same hybrid branched chains (HSOC) and an unconstrained chain (RSS). Each hybrid chain is composed of a driving arm and a parallelogram (4R), which can be recorded as P(^{4R}). The driving arm and the parallelogram are connected by a rotating pair. The two links of RSS chain are connected with each other through a spherical joint, and the RSS chain is connected with the base platform (0) through a rotating joint. The PM is denoted as 2-R//R//P(^{4R})// R⊕1-RSS. The length of the driving link of the three chains which is connected with the base platform 0 is equal to each other, and the length of the link in the unconstrained chain (RSS) which connects to the movable platform 1 is equal to the length of the longer link of the parallelogram in the hybrid chain.

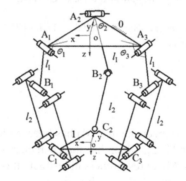

Fig. 1. Delta-CU PM

2.2   Topological Characteristics Analysis (POC, DOF, K)

The main topological characteristics such as POC, DOF and k will be analyzed below.

(1) The topology of the PM Unconstrained chain can be recorded as $SOC_3\{-R-S-S\}$ and two hybrid chains can be recorded as $HSOC_i\{-R//R(-P^{(4R)})//R-\}, i = 1, 2$.

(2) Select the geometric center o' of the moving platform 1 as basic point.

(3) Determine the POC of the end of each chain

$$M_{b_i} = \begin{bmatrix} t^1 \\ r^1 \end{bmatrix} \cup \begin{bmatrix} t^1 \\ r^0 \end{bmatrix} = \begin{bmatrix} t^1 \\ r^1 \end{bmatrix} = \begin{bmatrix} t^3 \\ r^1 \end{bmatrix} (i = 1, 2), \quad M_{b_3} = \begin{bmatrix} t^3 \\ r^3 \end{bmatrix}$$

(4) Determine the number of independent displacement equations of the first loop composed by $HSOC_1, HSOC_2$.

$$\xi_{L_1} = \dim.\left(\begin{bmatrix} t^3 \\ r^1 \end{bmatrix} \cup \begin{bmatrix} t^3 \\ r^1 \end{bmatrix} \right) = \dim.\left(\begin{bmatrix} t^3 \\ r^2 \end{bmatrix} \right) = 5$$

$$F_{(1 \sim 2)} = \sum_{i=1}^{m} f_i - \sum_{j=1}^{1} \xi_{L_j} = 8 - 5 = 3$$

$$M_{pa(1 \sim 2)} = \begin{bmatrix} t^3 \\ r^1 \end{bmatrix} \cap \begin{bmatrix} t^3 \\ r^1 \end{bmatrix} = \begin{bmatrix} t^3 \\ r^0 \end{bmatrix}$$

Obviously, the sub-structure composed by $HSOC_1, HSOC_2$ has realized three - translation output, so the third chains can be a simple unconstrained chain instead of adopting a complex hybrid chain. That is the proof of working principle of Delta-CU PM.

(5) Determine the number of independent displacement equations of the second loop composed by SOC_3.

$$\xi_{L_2} = \dim.(M_{pa(1 \sim 2)} \cup M_{b_3}) = \dim.\left(\begin{bmatrix} t^3 \\ r^0 \end{bmatrix} \cup \begin{bmatrix} t^3 \\ r^3 \end{bmatrix} \right) \dim.\left(\begin{bmatrix} t^3 \\ r^3 \end{bmatrix} \right) = 6$$

(6) Determine the degree of freedom of the mechanism

$$F = \sum_{i=1}^{m} f_i - \sum_{j=1}^{2} \xi_{L_j} = (4 + 4 + 6) - (5 + 6) = 3$$

The negative joint does not exist in this PM according to the criterion of the negative joint [13]. Therefore, when the three rotation joints ($A_1 \sim A_3$) on the base platform 0 are the driving one, the movable platform 1 can realize three-translation output.

3   Kinematic Analysis

3.1   Establishment of Coordinate System

The base and moving platforms are all equilateral triangles, their geometric center of which are O and O' points respectively. The relative position of the three driving arms in the base coordinate system can be represented by the angle $\alpha_1, \alpha_2, \alpha_3$ respectively, as shown in Fig. 2. For the convenience of calculation, the driven arm structure of the hybrid chain of the PM is equivalent to a single virtual link model, which is respectively connected with the driving arm and the movable platform 1 by a rotating pair R_B and R_C. A base Cartesian coordinate system $O - XYZ$ at the point O is established. Meanwhile a moving Cartesian coordinate system $O' - X'Y'Z'$ at the point O' is also established. Set R and r as the radius of the base platform 0 and the moving platform respectively, and set l_1 and l_2 as the length of the driving and the driven arm respectively, as shown in Fig. 3. Meanwhile, the input angle of the driving arm is $\theta_i (i = 1, 2, 3)$ and the output position of the movable platform 1 is $o'(x, y, z)$.

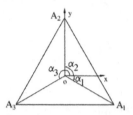

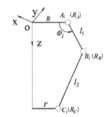

Fig. 2. Sketch of the base platform   **Fig. 3.** Geometric model of single chain

3.2   Solutions for Forward and Inverse Position

To analyze the position of each chain, we can obtain the vector equality from the Fig. 3 as follow:

$$\overrightarrow{B_iC_i} = (\overrightarrow{OO'} + \overrightarrow{O'C_i}) - (\overrightarrow{O'A_i} + \overrightarrow{A_iB_i}) \tag{1}$$

$$[x - (R \cos \alpha_i - r \cos \alpha_i + l_1 \cos \theta_i \cdot \cos \alpha_i)]^2 + [y - (R \sin \alpha_i - r \sin \alpha_i + l_1 \cos \theta_i \cdot \sin \alpha_i)]^2 \tag{2}$$
$$+ (z - l_1 \sin \theta_i)^2 = l_2^2 (i = 1 \sim 3)$$

When the central coordinate of the movable platform 1 is known, the inverse solution $\theta_i (i = 1, 2, 3)$ can be obtained by solving the equation of three variables $\theta_1, \theta_2, \theta_3$ according to Eq. (2).

Furthermore the following equations can be obtained from Eq. (2) as follow:

$$x^2 + y^2 + z^2 - 2b_{i1}x - 2b_{i2}y - 2b_{i3}z = l_2^2 - b_{i1}^2 - b_{i2}^2 - b_{i3}^2 \tag{3}$$

where

$$b_{i1} = (R - r) \cos \alpha_i + l_1 \cos \alpha_i \cdot \cos \theta_i$$
$$b_{i2} = (R - r) \sin \alpha_i + l_1 \sin \alpha_i \cdot \cos \theta_i$$
$$b_{i3} = l_1 \sin \theta_i$$

When $\theta_i (i = 1, 2, 3)$ are known, a set of the three-variable two-order equations can be obtained by Eq. (3), which can be solved to obtain the value of the positive solutions.

4 Workspace Analysis

4.1 Workspace Analysis of Single Chain

(a) Determine the spatial envelope equation of single chain interface

According to Eq. (2), when θ_i and $\alpha_i (i = 1, 2, 3)$ are given, the workspace of each chain is a sphere, whose radius is l_2 and its central coordinate is as follow:

$$B_i = [(R - r) \cos \alpha_i + l_1 \cos \theta_i \cdot \cos \alpha_i, (R - r) \sin \alpha_i + l_1 \cos \theta_i \cdot \sin \alpha_i, l_1 \sin \theta_i].$$

Theoretically, the swing range of the driving arm is from 0 to 360°, the workspace of each chain is a three dimensional circle which is accomplished by sphere whose radius is l_2 rotating around the point A_i. The boundary equation can be expressed as follow:

$$\begin{cases} x = [(l_1 + l_2 \cos t_i) \cos \theta_i + R - r] \cos \alpha_i, & t \in (0.2\pi), i = 1, 2, 3 \\ y = [(l_1 + l_2 \cos t_i) \cos \theta_i + R - r] \sin \alpha_i, \\ z = (l_1 + l_2 \cos t_i) \sin \theta_i \end{cases} \tag{4}$$

As a result, the workspace of the center of the moving platform is the intersection entity part formed by three envelopes.

(b) Workspace analysis of mechanism

Since the movable platform 1 and the base platform 0 is an equilateral triangle ($\alpha_1 = -30°$, $\alpha_2 = 90°$, $\alpha_3 = 210°$), the workspace of each chain in the base coordinate system around the Z axis is evenly distributed, and the workspace are exactly the same. Therefore, it is desirable to analyze any branch, such as the second chain (RSS). For the convenience of analysis, the three chains can be moved a distance of r to the center of the stable platform, thus $OA_2 = R - r$, as shown in the Fig. 4.

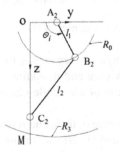

Fig. 4. Schematic diagram of simplified rod

The circle R_0 is obtained from the driving arm rotating around the point A_i from 0 to 360° with the radius of l_1.

R_3 represents the largest circle of the single chain workspace in the cross-section yoz and its radius is $(l_1 + l_2)$.

Since the workspace of each chain in the base coordinate system around the Z axis is evenly distributed, each branch maximum circumference and Z axis intersect at the point M, whose projection on the plane xoy coincide with circumcenter of the base platform 0. Only when $(l_1 + l_2) > R - r$, the workspace of each chain can intersect with the axis z.

According to Eq. (4), with the changing of the input angle, the size of each chain workspace and its position relative to the base platform 0 are determined by the size of $(R - r), l_1, l_2$. Furthermore, when the relative sizes of the three variables change, the intersection of each branch envelope will be formed. Further, the shape and volume of the workspace of the platform center also changed. At the same time, the author find that when $l_1 > l_2$, the effective workspace of the moving platform 1 is small, so the link length usually suggests $l_1 < l_2$ when considering practical applications.

Under the conditions such as $l_1 < l_2$, three different cases are analyzed and calculated as follows, so as to get the actual usable range of the workspace in the corresponding circumstances.

4.2   Cases I: Analysis of Workspace for $(R - r < l_1 < l_2)$

In Fig. 5, R_0 represents the circumference of the driving arm with the radius of l_1.

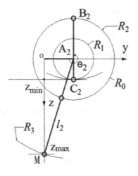

Fig. 5. Schematic diagram of the structure parameters in the yoz section

The radius of R_1 is $l_2 - l_1$, but the center of the moving platform cannot reach in the circle;

When the input angle θ is $3\pi/2$, R_2 represents the circle obtained from the driven arm rotating around B_2 of the driving and its radius is l_2.

R_3 represents the largest circle of the single chain workspace in the cross-section yoz and its radius is $(l_1 + l_2)$.

Thus the maximum circumference of three chain workspace and z axis intersect at the point M. At this time, the driving and the driven arm are collinear. The moving platform 1 reaches the maximum distance relative to the base platform 0, and it is easily get

$$z_{max} = \sqrt{(l_1 + l_2)^2 - (R - r)^2} \tag{5}$$

When $z < z_{min} = l_2 - l_1$, the center of the moving platform could not reach in the circumference R_1. Therefore, the workspace entity must have the hole in the cross section $x - y$, as shown in Fig. 6(b). Thus the range of the continuous workspace value should be

$$(l_2 - l_1) \leq z \leq \sqrt{(l_1 + l_2)^2 - (R - r)^2}$$

Define that the radius of the base and moving platform are $R = 200$ mm and $r = 100$ respectively. Try to take the numerical values such as $l_1 = 250$ (mm) and $l_2 = 400$ (mm), we can get the three-dimensional graph of the workspace in the software UG as show in Fig. 6(a). In Fig. 6(a), the small triangular plate at the center of the upper part represents the position and size of the base platform 0, form which the size relationship between the workspace and the overall shape of the PM can be judged roughly. Further, the cross-section of the workspace along the Z axis is obtained by the inverse solution Eq. (2) by using Matlab, as shown in Fig. 6(b)–(d).

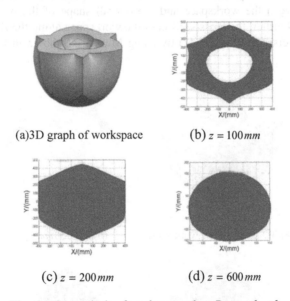

(a) 3D graph of workspace   (b) $z = 100$ mm

(c) $z = 200$ mm   (d) $z = 600$ mm

Fig. 6. The analysis of workspace when $R - r < l_1 < l_2$

4.3 Cases II: Analysis of Workspace for $(l_1 < R - r < l_2)$

The center of moving platform 1 could not reach in the circumference R_1 because of $z < z_{min} = l_2 - l_1$. Therefore, the workspace entity must have some holes in the cross-section $x - y$, as shown in Fig. 8(b). Thus the range of the continuous workspace value should be (Fig. 7):

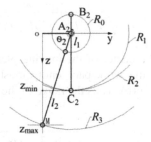

Fig. 7. Schematic diagram of the structure parameters in the *yoz* section

$$(l_2 - l_1) \leq z \leq \sqrt{(l_1 + l_2)^2 - (R - r)^2}$$

Define that the radius of base and movable platform are $R = 200\,(\text{mm})$ and $r = 500\,(\text{mm})$ respectively. Try to take the numerical values such as $l_1 = 100\,(\text{mm})$ and $l_2 = 400\,(\text{mm})$, we can get the three-dimensional graph of the workspace in the software UG as show in Fig. 8(a). In Fig. 8(a), the small triangular plate at the center of the upper part represents the position and size of the base platform 0, from which the size relationship between the workspace and the overall shape of the mechanism can be judged roughly. Further, the cross-section of the workspace along the Z axis is obtained based on the inverse solution Eq. (2) by using Matlab, as shown in Fig. 8(b)–(d).

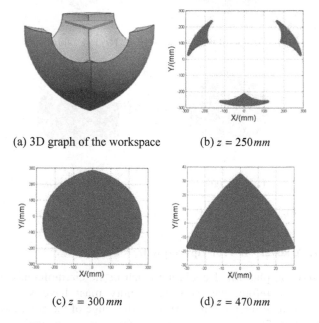

(a) 3D graph of the workspace   (b) $z = 250\,mm$

(c) $z = 300\,mm$ (d) $z = 470\,mm$

Fig. 8. Analysis of the workspace for $l_1 < R - r < l_2$

4.4   Case III: Workspace Analysis for $l_1 < l_2 < R - r$

When $l_1 < l_2 < R - r$, the structural parameters are as shown in Fig. 9.

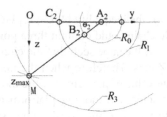

Fig. 9. Schematic diagram of the structure parameters in the yoz section

When $z = 0$, the section of the workspace for the single chain is as shown in Fig. 10.

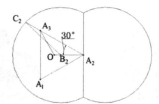

Fig. 10. The section of the workspace for the single chain ($z = 0$)

In Fig. 10, A_2B_2 and B_2C_2 stand for the driving and driven arm respectively, and that, $|A_2B_2| = l_1$, $|B_2C_2| = l_2$ and $|OA_2| = R - r$. The circumference represents the largest circle of the workspace. Only when $|A_2C_2| \geq |A_2A_3| = \sqrt{3}(R - r)$ is satisfied the workspace of each single chain could contain the three points A_1, A_2, A_3 belong to the base platform. At this time, the relation $z \geq z_{min} = l_2 - l_1$ can be satisfied and the workspace entity of the moving platform does not have some holes in the cross-section $x - y$. However, when $z < z_{min} = l_2 - l_1$, the workspace entity of the moving platform has some holes in the cross-section $x - y$.

In the triangle $\Delta A_2B_2C_2$, we can obtain the results as follows by cosine theorem:

$$|A_2C_2| = \frac{1}{2}(\sqrt{3}l_1 \pm \sqrt{4l_2^2 - l_1^2})$$

when $l_1 < l_2 < R - r$, then

$$|A_2C_2| = \frac{1}{2}(\sqrt{3}l_1 + \sqrt{4l_2^2 - l_1^2}) > 0.$$

Also

$$|A_2C_2| = \frac{1}{2}(\sqrt{3}l_1 + \sqrt{4l_2^2 - l_1^2}) < \frac{1}{2}(\sqrt{3}l_2 + \sqrt{3l_2^2 + (l_2^2 - l_1^2)}) < \sqrt{3}l_2 < \sqrt{3}(R - r)$$

That is, when $l_1 < l_2 < R - r$, the inequality $|A_2C_2| < |A_2C_3|$ is always satisfied. Thus, the workspace of every single chain could contain three points A_1, A_2, A_3 belong to the static platform, so $z_{min} \neq l_2 - l_1$. It can be seen that the circle R_1 whose radius is $l_2 - l_1$ has no intersection with the Z axis. Therefore when the input angle of the driving arm $\theta_2 = \pi$, the minimum ordinate value of the center of the moving platform 1 is obtained.

Define that the radius of static and moving platform are $R = 250$ (mm) and $r = 50$ (mm) respectively. Try to take the numerical values such as $l_1 = 100$ (mm) and $l_2 = 150$ (mm), we can get the three-dimensional graph of the workspace in the software UG as show in Fig. 11(a). In Fig. 11(a), the small triangular plate at the center of the upper part represents the position and size of the base platform 0, from which the size relationship between the workspace and the overall shape of the mechanism can be judged roughly. Further, the cross-section of the workspace along the Z axis is obtained by the inverse solution Eq. (2) by using Matlab, as shown in Fig. 11(b)–(d).

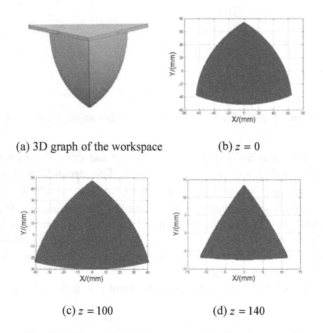

(a) 3D graph of the workspace   (b) $z = 0$

(c) $z = 100$ (d) $z = 140$

Fig. 11. The analysis of workspace for $(l_1 < l_2 < R - r)$

We can get from Fig. 11(a) that when $l_1 < l_2 < R - r$, the cross-sectional area of the workspace in plate $x - y$ is smaller than that of the base platform area and the maximum workspace is far less than the overall volume of the machine shape, this condition is not ideal.

4.5 Comprehensive Analysis of Workspace

From the analysis of the three cases above, we can find that:

(1) The ordinate value of the workspace should be larger as far as possible. The Eq. (5) shows that the smaller the value of $R - r$ is, the larger the value of z_{max} will be. Thus, when the value of $R - r$ is the minimal, the workspace is the maximal.

(2) In case III, when $R - r < l_1 < l_2$, the condition will appear that the installation space is far larger than the workspace of the moving platform, so the actual application should avoid this condition.

(3) For the case I $(R - r < l_1 < l_2)$ and case II $(l_1 < R - r < l_2)$, when $z < z_{min} = |l_2 - l_1|$, the workspace entity of the moving platform has some holes in the cross-section $x - y$. To maintain the continuity of the workspace, the value should be:

$$|l_2 - l_1| \leq z \leq \sqrt{(l_1 + l_2)^2 - (R - r)^2}.$$

And when $(R - r)_{min} = 0$, the workspace is the largest. Thus the actual application should be as far as possible to meet the requirement that the radius difference $(R - r)$ is less than the length of the rod including l_1 and l_2. Therefore, we should give priority to adopting the case I, i.e., $R - r < l_1 < l_2$.

5 Conclusions

Delta-CU PM is presented, which is proved that it can achieve three-translation output but it has the advantages of simple structure, easily assembly and repair. The Delta-CU PM can replace the Delta in a sense.

When the radius of the circumcircle of the base platform is equal to that of the moving platform and the driving arm is smaller than the driven arm, the distance of the workspace along the axial direction will be the max that can guarantee the maximum workspace.

Acknowledgments. This research is sponsored by the NSFC (Nos. 51475050, 51375062) and Jiangsu Key Development Project (No. BE2015043) and Jiangsu Scientific and Technology Transformation Fund Project (No. BA2015098).

References

1. Clavel, R.: Device for displacing and positioning an element in space. EP0250470 B1, Europe. 17 July 1991
2. Feng, L., Zhang, G., Gong, Z., et al.: Research and development of Delta series parallel robot. Robot (3), 375–384 (2014)

382   J. Li et al.

3. Gao Xiulan, L., Kaijiang, W.J.: Workspace resolution and dimensional synthesis on delta parallel mechanism. J. Agric. Mach. **39**(5), 146–149 (2008)
4. Zhang, L., Liu, X.: Geometry solving of maximum inscribed space on a parallel mechanism. Mach. Des. Res. (1), 20–22 (2002)
5. Yan, B., Liu, X., et al.: Study on workspace of 3-UPS parallel robot. Comb. Mach. Tool Autom. Manuf. Tech. (6), 3841 (2012)
6. Romdhane, L.: Design and analysis of a hybrid serial-parallel manipulator. Mech. Mach. Theory **34**(7), 1037–1055 (1999)
7. Ray, L., Blanpied, G.S., Coker, W.R.: Design and singularity analysis of a 3-translational-DOF in-parallel manipulator. J. Mech. Des. **124**(3), 419–426 (2002)
8. Zhao, J., Zhu, Y.: Geometric solution for direct kinematics of delta parallel robot. J. Harbin Inst. Technol. **35**, 25–27 (2003)
9. Zhang, L.: Integrated Optimal Design of Delta Robot using Dynamic Performance Indices. Doctoral Dissertation of Tianjin University, Tianjin (2011)
10. Codourey, A.: Dynamic modeling and mass matrix evaluation of the DELTA parallel robot for axes decoupling control. In: IEEE/RSJ International Conference on Intelligent Robots and Systems, pp. 1211–1218 (1996)
11. Alain, C.: Dynamic modeling of parallel robots for computed-torque control implementation. Int. J. Robot. Res. **17**(2), 1325–1336 (1998)
12. Stamper, E.R., Tsai, L.-W., Walsh, G.C.: Optimization of a three DOF translational platform for well-conditioned workspace. In: IEEE International Conference on Robotics and Automation, pp. 3250–3255 (1997)
13. Yang, T., Liu, A., Luo, Y., Shen, H.: Theory and Application of Robot Mechanism Topology. Science Press, Beijing

Research on the Synchronous Control
of the Pneumatic Parallel Robot with Two DOF

Shaoning Wang[✉], Tao Wang, Bo Wang, and Wei Fan

Beijing Institute of Technology, Beijing, China
2212325696@qq.com,
{wangtaobit, wangbo231, fanwei}@bit.edu.cn

Abstract. Basing on the structure of the pneumatic parallel robot with two degrees of freedom (DOF), this paper studies the control method of synchronous coordinated motion of two cylinders in the parallel mechanism to ensure the horizontal attitude of the end position of the parallel mechanism and the trajectory planning. Two active cylinders in the parallel mechanism cannot be controlled coordinately because of the compressibility of gas, the variety of the load force of parallel mechanism and the difference of cylinder pressure changing in the motion. The paper uses the fuzzy adaptive PID control algorithm with gravity compensation and the object-oriented coordinated method to control the motion of the active cylinder. Firstly, the pneumatic parallel mechanism in this paper is introduced. Then the paper studies on the control method when cylinders swing, which is the fuzzy adaptive PID control algorithm. In the end, two cylinder coordinated motion control method is researched based on the parallel mechanism. The synchronous motion control method lays a theoretical foundation for the study of pneumatic parallel mechanism and improves the pneumatic parallel robot technology.

Keywords: Two Degree Of Freedom · Pneumatic parallel robot · Synchronous coordinated control · Gravity compensation

1 Introduction

A parallel robot with two DOF is a two-dimensional space motion mechanism. It uses air as power source and cylinders as movement structure. There are a lot of researches about parallel mechanism from two to multi-degrees of freedom. Researches about kinematics and dynamics analysis of parallel robot are mature [1].

Based on the kinematic inverse model of Delta parallel robot, the static model of the mechanism is deduced by vector method, and the stiffness matrix of the mechanism is obtained [2]. In order to analyze the singular position of the parallel mechanism, the singularity of the workspace boundary is found by studying the Jacobi matrix of the parallel mechanism, which greatly reduces the complexity of the solution process [3].

Considering low weight, safe production process and high load capacity, parallel robots based on motor have limitations of application. With compressed air as power source, pneumatic parallel robot is safe and clean compared with parallel robot based

© Springer International Publishing AG 2017
Y. Huang et al. (Eds.): ICIRA 2017, Part II, LNAI 10463, pp. 383–393, 2017.
DOI: 10.1007/978-3-319-65292-4_33

on motor. Besides, pneumatic parallel robot has simple structure, strong load capacity, light weight, low cost.

It is important to establish synchronous coordination relationship in the parallel robot. A model-free cross-coupled controller is designed for the PD controller as the error compensated channel controller for multi-axis position synchronization problem [4]. Marvin applies linear quadratic optimal controller to multi-axis motion platform for high synchronization accuracy [5]. Takanori Miyosh proposed a real-time tuning scheme based on time-varying performance indicators and optimal control theory [6]. In the multi-axis motion platform, the fuzzy synchronization controller is designed by adaptive fuzzy control, and the effectiveness of the controller is verified by experiments [7].

There are two kinds of coordination control mode for coordinated motion control [8–10]. One is master-slave coordinated control mode and the other is oriented object coordinate control mode. The displacement coordinated motion control of this system is to ensure motion platform horizontal movement. And the two cylinders don't have relationship of master and slave. Oppositely, it needs to do coordinated control based on trajectory planning [11]. So this system needs the oriented object one.

This paper studies an algorithm suitable for parallel robots. The algorithm can make parallel robot achieve synchronous coordination movement, which paves the way for promotion of the pneumatic parallel robot technology.

2   Mechanism of Two DOF Parallel Robot

2.1   Structure of Two DOF Parallel Robot

The two DOF pneumatic parallel robots consist of two platforms, two parallel connection valve-controlled pneumatic cylinders and a freely adjustable telescopic support rod. Two cylinders can stretch independently. Cylinders and support rod connect with platforms up and down through U hinges. The motion platform can move freely in the direction of X axis and Y axis through the coordination control of two cylinders. The structure schematic diagram is shown in Fig. 1. And the parameters of the two DOF pneumatic parallel robots are shown in Table 1.

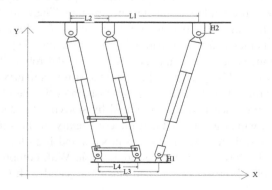

Fig. 1. The structure schematic diagram of parallel robot

Table 1. The parameters of the two DOF pneumatic parallel robot

Name	Value
Distance between cylinders on static platform L_1	322 mm
Distance between cylinder and telescopic auxiliary rod on static platform L_2	60 mm
Distance between cylinders on motion platform L_3	100 mm
Distance between cylinder and telescopic auxiliary rod on static platform L_4	60 mm
Height of T hinge on motion platform H_1	16 mm
Height of T hinge on static platform H_2	25 mm
The cylinders' stroke	150 mm
The shortest length of active branched chain L_{min}	294 mm
The longest length of active branched chain L_{max}	444 mm

2.2 Control System of Two DOF Parallel Robot

The control system of the two DOF parallel robot is shown in Fig. 2. The embedded controller compares the target signal received by the host computer with the sensor signal and outputs the corresponding control amount through the control algorithm. Then the control amount drives the electric/gas proportional servo valve by D/A conversion and signal amplification. The cylinder moves according to the target task. The motion control of the cylinder is realized by real-time acquisition of the movement state and real-time control of the cylinder.

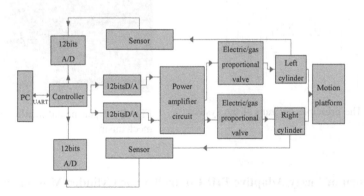

Fig. 2. The control system of two DOF parallel robot

3 Hardware Design and Control Method

3.1 Hardware Design of Parallel Mechanism

The principle diagram of the system is shown in Fig. 3. Cylinders' movement, displacement and speed are controlled by input and displacement of the cylinder. And the gas volume is related to the opening size of electric/gas proportional valve. Experimental facility of parallel mechanism is shown in Fig. 4. The two DOF pneumatic parallel robot developed by SMC-Beijing institute of technology pneumatics center uses low friction cylinders of SMC (SMC Corporation in Japan) for servo cylinders and LVDT displacement sensors for measuring cylinders' displacement. It also uses electric/gas proportional valves of FESTO (a German company playing an important role in the pneumatic industry) [12] for servo valves, whose type is MPYE-5-M5-010-B. Its output flow rate is 100 L/min. Two cylinders' type is CG1YB20-150Z. Its stroke is 150 mm with diameter 20 mm. STM32F103 is used as the controller, and PC is seemed as the host computer to send control instructions and display the relevant information. Data is transferred between controller and PC through UART. Two LVDT displacement sensors produced by SMWEI are used, whose type is SMW-WYDC-150L. The output voltage is 0–5 V, the range is 150 mm, the sensitivity is 100 mV/mm, the dynamic frequency is 0–200 Hz, the accuracy is 0.14% and using temperature is −20 °C to +70 °C. NI USB-6351 board is used to realize the acquisition of the displacement sensor output signal.

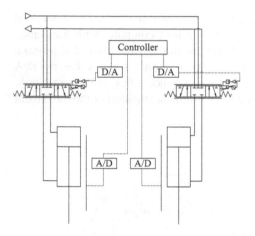

Fig. 3. The principle diagram of the system

Fig. 4. Experimental facility of parallel mechanism

3.2   Design of Fuzzy Adaptive PID Controller for Cylinder Movement

The mathematical model uncertainty is caused by the change of parallel robot motion posture and load due to the compressibility of gas and the influence of cylinder rod. Traditional PID method is simple but it is hard to find a suitable PID parameter making the parallel robot have a good control effect in the process of real-time motion [13]. To make the parallel robot move along the planning trajectory, we adopt an algorithm of fuzzy adaptive PID control. According to the actual situation of the parallel mechanism, the use of fuzzy reasoning achieves optimal PID parameters automatically.

Structure of Proposed Controller. Fuzzy controller is the core of fuzzy control. Fuzzy controller design process includes fuzzy interface, the establishment of fuzzy rules, fuzzy reasoning and anti-fuzzy processing. Designing the fuzzy controller structure is a key step in designing a fuzzy controller, which determines the input and output variables of the controller. The selection of the fuzzy controller should base on the controlled object. The structure of the fuzzy adaptive PID controller [14] is shown in Fig. 5. Displacement error e and error change ec are the inputs of the fuzzy controller. ΔK_p, ΔK_i, and ΔK_d are the outputs of the fuzzy controller. E and EC are the fuzzy variables of e and ec. Fuzzy sets of E, EC, ΔK_p, ΔK_i and ΔK_d are all {NB, NM, NS, ZO, PS, PM, PB}. NB denotes negative big. NM denotes negative medium. NS denotes negative small. ZO denotes zero. PS denotes positive small. PM denotes positive medium. PB denotes positive big.

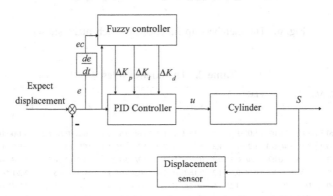

Fig. 5. The structure of fuzzy adaptive PID controller

Ranges of E, EC, ΔK_p, ΔK_i and ΔK_d are both {−3, −2, 1, 0, 1, 2, 3}. The corresponding conversion factor is shown as followed.

$$L_x = \frac{n}{X_x} \tag{1}$$

L_x is the conversion factor of corresponding variable, n is the number of variable domains, X_x is the range of corresponding variable.

Design of Membership Functions. Figure 6 shows the membership functions of E, EC, ΔK_p, ΔK_i and ΔK_d. Language functions of all fuzzy subsets are isosceles triangle.

Fuzzy Inference Rules. Establish the following Table 2 about ΔK_p, ΔK_i, ΔK_d for cylinder motion control in process of swinging, using the inference method of Mamdani.

Fuzzy Inference and Defuzzification. The range of e and ec are $[e_{min}, e_{max}]$ and $[ec_{min}, ec_{max}]$, the number of variable domains is n. The fuzzy calculation formula is as following.

$$\begin{cases} E = \frac{n \cdot e}{e_{max} - e_{min}} \\ EC = \frac{n \cdot ec}{ec_{max} - ec_{min}} \end{cases} \tag{2}$$

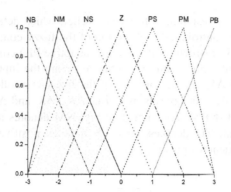

Fig. 6. The membership functions of each fuzzy subset

Table 2. The control rules

	$\Delta K_p/\Delta K_i/\Delta K_d$	EC						
		PM	PB	NB	NM	NS	Z	PS
E NB	6.6/−4.6/0.7	3.4/−3/−0.07	1.0/−1.3/−0.3	1.3/−1.2/−1.2	0.6/−0.6/−2.0	−0.2/−0.2/−1.2	−0.1/0/−0.6	
NM	5.2/−3.4/0.9	3.4/−3.4/−0.3	1.2/−1.2/−1.3	1.7/−1.7/−2.0	0.3/−0.3/−1.1	0/0/−0.9	0/−0.2/−0.2	
NS	3/−1.2/0.3	2/−1.5/−0.09	1.4/−0.3/−0.7	0.3/−0.2/−3	0.7/1.3/1.2	−0.7/0.2/−0.7	−1/0.8/−0.1	
Z	5/−3/−0.8	1.7/−1.3/−0.8	0.6/−0.6/−1.2	0.1/−0.1/−2.0	−0.5/0.6/−1	−1/0.7/−0.5	−2.2/1.5/−0.2	
PS	1.1/−1.2/0	1/−1/0.01	0.9/−0.2/−1.0	−0.3/0.3/−1.0	−0.6/0.7/0.1	−1.6/1.2/−1	−2/0.8/−0.01	
PM	1.6/−1.6/1.6	0.4/−0.4/0.2	−0.8/0.1/1	0.3/0.3/0.2	−1.1/1/1.1	−1.5/1.4/1.6	−2.8/1.1/1.6	
PB	0.3/0/0.1	0.1/0.2/1.4	−0.2/0.6/1.1	−1.2/1/1.2	−1.8/1.4/2	−2/2.1/2.3	−2.6/2.2/2.0	

E, EC are the values of *e* and *ec* after fuzzification. If *E* and *EC* are not integers, the values of ΔK_p, ΔK_i and ΔK_d can be ensured according to Table 2 after rounding approximately.

3.3   Gravity Compensation of Parallel Mechanism

The low friction cylinder is chosen as the drive cylinder, so the friction has little influence on the control of parallel mechanism motion and it even can be ignored. Parallel mechanism has different load capacity for the same load when it is under different positions and orientations. In order to ensure the force balance of parallel mechanism, it needs to carry out the gravity compensation of the load [15]. Based on dynamic model, this paper firstly calculates the gravity of each moment in the joint movement trajectory. Then it generates look-up tables, which can be used to compensate the gravity when controlling. The control law of the gravitational compensation is shown in Eq. (6).

4 Research of Double Cylinder Synchronous Coordinate Control

The two active cylinders constitute the main body of two DOF parallel robots. The two cylinders have constraint relations and form a physical coupling through the motion platform. It comes important to establish coordination relationship between the two cylinders as to the physical coupling and mechanical constraint. Only when the control variables of cylinders are coupled can two cylinders make coordinated movement during parallel mechanism motion and ensure the parallel mechanism a good effect.

4.1 Design of Object Oriented Double Cylinders Coordinate Controller

Figure 7 is the block diagram of double cylinders coordinated control system for two DOF parallel robots. Firstly, we ensure the object movement of the two DOF parallel robots and decide the motion path according to the trajectory planning. And we solve inverse kinematic problem of each cylinder based on the kinematics analysis of the parallel mechanism and the constraint relation of the parallel mechanism. Then an expected displacement of each cylinder on the corresponding planning trajectory is obtained. Besides, the system is controlled by using fuzzy adaptive PID controller with gravity compensation on the corresponding position of the parallel mechanism. Each cylinder controller is driven by the displacement error between desired displacement and actual displacement. The principle of the double cylinders coordination controller [16] is that the two cylinder controller each has an amount of error compensation according to a certain proportion through comparing the two cylinder displacement errors. Double cylinder controller each contains information of the other party, which achieves communication between each controller. Then it makes the deviation of the two cylinder displacement error become small and the two cylinders' motion tends to synchronization. The two cylinders can make coordinated movement. Finally, the

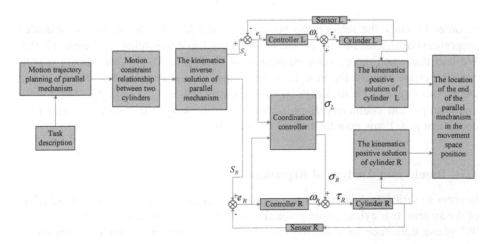

Fig. 7. Double cylinders object oriented coordinate control system of parallel mechanism

position and attitude of the parallel mechanism in Cartesian space are obtained by using the kinematics positive solution of the cylinder displacement [17].

The inputs of the fuzzy adaptive PID controller are e_L and e_R, and the outputs are ω_L and ω_R. The inputs of coordinate controller are e_L and e_R, and the outputs are σ_L and σ_R. The outputs of the control system are τ_L and τ_R.

The errors' relationships of the coordinate controller are as followed.

$$\sigma_L = \begin{cases} K_c\mu(e_L - e_R) & |e_L| \geq |e_R| \\ Sgn(e_L \cdot e_R)K_c(1 - \mu)(e_R - e_L) & |e_L| < |e_R| \end{cases} \quad (3)$$

$$\sigma_R = \begin{cases} K_c\mu(e_R - e_L) & |e_L| < |e_R| \\ Sgn(e_L \cdot e_R)K_c(1 - \mu)(e_L - e_R) & |e_L| \geq |e_R| \end{cases} \quad (4)$$

$Sgn(e_L \cdot e_R)$ is symbolic parameter in the above two formulas.

$$Sgn(e_L \cdot e_R) = \begin{cases} 1 & e_L e_R \geq 0 \\ -1 & e_L e_R < 0 \end{cases} \quad (5)$$

K_c is the coordinated proportion coefficient of cylinder, μ is the coordinate weighting coefficient of cylinder, whose range is [0,1].

The coordinated control law for the two cylinders of parallel mechanism:

$$\tau_X = \omega_X + \sigma_X = K_p e + K_i \int edt + K_d e' + G(s) + \sigma_X \quad (X = L, R) \quad (6)$$

$G(s)$ is the corresponding gravity compensation amount with parallel mechanism under a certain position corresponding, which can be obtained from look-up table [18, 19].

The synchronous coordination ability of the two cylinders is reflected by the difference of the actual tracking error of the two cylinders.

$$E_d = e_L - e_R \quad (7)$$

In order to study the relationship between E_d and K_c, we choose four coordinated proportion coefficients of 1, 6, 12 and 18. The coordinate weighting coefficient μ is 0.5. We find that when the coordinated proportion coefficient K_c is 6, the system coordination exercise effect is the best. In order to study the relationship between E_d and μ, we choose four coordinate weighting coefficient of 0.1, 0.3, 0.6 and 0.9. The coordinated proportion coefficient K_c is 6. We find that when the coordinate weighting coefficient μ is 1, the coordination effect is the best.

4.2   Results and Analysis of Experiments

In order to verify the fact that joining coordinated controller can improve control effect of the system, two cylinders are given sine desired displacement signal. And there is a 90° phase difference between the two sine signals. The two cylinder displacement errors difference changes reflect the effect of the control system. It can get the following curves compared with the system control effect without coordinated controller.

The sampling period is 1 ms. The pressure of gas source is 0.3 MPa. Inputs of two cylinders' displacement signal are shown in (8) and (9).

$$S_L = 70 \sin(\frac{\pi}{2}t) + 75 \tag{8}$$

$$S_R = 70 \sin(\frac{\pi}{2}t + \frac{\pi}{2}) + 75 \tag{9}$$

The greater the absolute value of E_d, the worse the synchronous coordination.

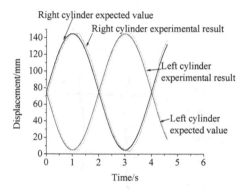

Fig. 8. The response curves to sine displacement inputs of the two cylinders

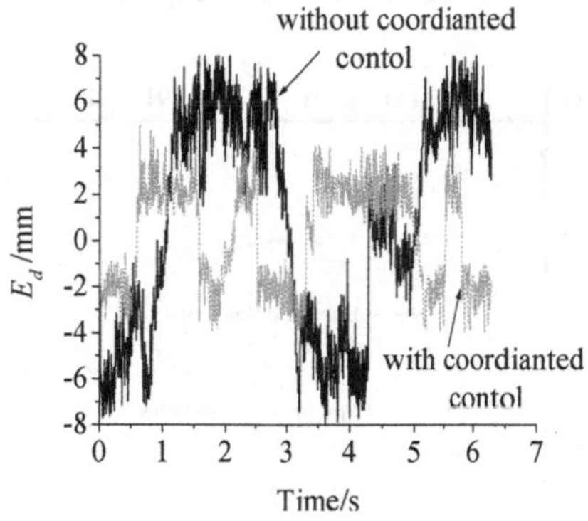

Fig. 9. The synchronization error response curves of with and without coordinated controller

Figure 8 is the two cylinders' sine displacement input tracking curve. The cylinder's experimental result can track the expected value effectively. Figure 9 is the curve of E_d with coordinated controller and without coordinated controller.

The E_d response curves of the above synchronous displacement errors indicate that the coordination controller can significantly reduce the displacement error of the two cylinder synchronous motion. The coordination control effect of the two cylinders can be enhanced obviously by adding the coordination controller and the motion control of the parallel mechanism can be improved.

4.3   Motion Experiment of a Predetermined Trajectory

The movement of the predetermined trajectory is moving platform motion trajectory of a fixed parallel robot designed in the motion space. Firstly the shape and location of the trajectory are determined. And then according to the motion trajectory and kinematic inverse solution, the cylinder motion control of this trajectory is realized in the control program. The fixed motion trajectory designed in the workspace is circular motion trajectory. The coordinates of the center of the circle is (161, −365), and the radius is 50 mm. Moving speed of end moving platform of parallel robot is 120 mm/s. Figure 10 shows the motion trajectory of the center point of the moving platform.

Through the analysis of circular motion trajectory of parallel robot, the maximum displacement deviation of the moving platform at the end of the robot is 4 mm. This motion error may be caused by the presence of small gaps in the mechanical joints.

4.4   Motion Experiment of a Teaching Trajectory

Teaching movement is to make the design of the trajectory in the host computer, and the parallel robot learns this trajectory and repeats the movement of the corresponding trajectory. In this experiment, a trajectory of "#" is designed in PC. The motion of the parallel robot is shown in Fig. 11.

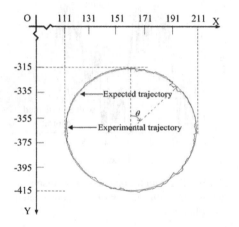

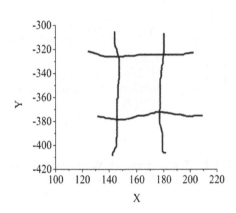

Fig. 10. Circular expected trajectory and experimental trajectory

Fig. 11. Results of teaching movement

5   Conclusion

In this paper, we study the design of two DOF pneumatic parallel robots and the synchronization control of the cylinder. Firstly, the structure and servo control system of the two DOF pneumatic parallel robots are introduced. Based on the parallel mechanism system, a fuzzy self-tuning PID control method is proposed, which is applied to the system, and the feasibility of this algorithm is verified by experiments. Finally, the parallel mechanism of double cylinder synchronous control is studied, and the double cylinder coordination controller is designed to achieve a good control effect. The research results lay the theoretical and experimental foundation for the research of the pneumatic parallel robot.

Acknowledgments. The research is supported by National Natural Science Foundation of China (Grant No. 51375045) and the State Key Laboratory Program (Grant No. GZKF-201214). The authors thank Dongge Zhang for helpful conversations, and producing several of the figures.

References

1. Craig, J.: Introduction to Robotics. Press of Mechanical Engineering, Beijing (2006)
2. Wahle, M., Corves, B.: Stiffness analysis of Clavel's DELTA robot. In: Jeschke, S., Liu, H., Schilberg, D. (eds.) ICIRA 2011. LNCS, vol. 7101, pp. 240–249. Springer, Heidelberg (2011). doi:10.1007/978-3-642-25486-4_25
3. Lopez, M., Castillo, E., Garcia, G., Bashir, A.: Delta robot: inverse, direct, and intermediate Jacobians. ARCHIVE Proc. Inst. Mech. Eng. Part C J. Mech. Eng. Sci. **220**(1), 103–109 (2006)
4. Sun, D., Shao, X.Y., Feng, G.: A model-free cross-coupled control for position synchronization of multi-axis motions: theory and experiments. IEEE Trans. Control Syst. Technol. **15**, 306–314 (2007)
5. Cheng, H.M., Mitra, A., Chen, C.Y.: Synchronization controller synthesis of multi-axis motion system. In: Fourth International Conference on Innovative Computing, Information and Control, pp. 918–921. IEEE (2009)
6. Miyoshi, T., Terashima, K., Maki, Y.: Optimum synchronous control for multiple-axis servo systems in terms of time-varying performance index. In: Conference of the IEEE Industrial Electronics Society, IECON 2000, vol. 2, pp. 1080–1086. IEEE (2002)
7. Yeh, Z.M., Tarng, Y.S., Lin, Y.S.: Cross-coupled fuzzy logic control for multiaxis machine tools. Mechatronics **7**(8), 663–675, 677–681 (1997)
8. Luh, J.Y.S., Zheng, Y.F.: Constrained relations between two coordinated industrial robots for motion control. Int. J. Robot. Res. **6**(3), 60–70 (1987)
9. Zheng, Y.F., Luh, J.Y.S., Jia, P.F.: A real-time distributed computer system for coordinate-motion control of two industrial robots. In: IEEE International Conference on Robotics and Automation, pp. 1236–1241. IEEE (1987)
10. Zheng, Y.F., Luh, J.Y.S.: Joint torque for control of two coordinated moving robots. In: IEEE International Conference on Robotics and Automation, pp. 1375–1380. IEEE (1986)
11. Zhou, J.P., Yan, J.P., Chen, W.J.: Research status and ponderation pertinent to dual arm robot. Robot **23**(2), 175–177 (2001)
12. Festo Group: Solutions for Maximized Productivity (2007). http://www.festo.com. Accessed 5 May 2016
13. Sharma, R., Gaur, P., Mittal, A.P.: Performance analysis of two-degree of freedom fractional order PID controllers for robotic manipulator with payload. ISA Trans. **58**, 279–291 (2015)
14. Mann, G.K.I., Hu, B.G., Gosine, R.G.: Analysis of direct action fuzzy PID controller structures. IEEE Trans. Syst. Man Cybern. **29**(3), 371–388 (1999)
15. Huo, W.: The Robot Dynamics and Control. Higher Education Press, Beijing (2005)
16. Nilakantan, A., Hayward, V.: The synchronization of multiple manipulators in Kali. Robot. Auton. Syst. **5**(4), 345–358 (1989)
17. Sun, Z.Y., Liu, G.F., Wang, M.: Applied research in cartesian impedance control for 6-UPS space docking mechanism. Appl. Mech. Mater. **540**, 363–367 (2014)
18. Matunaga, S., Onoda, J.: New gravity compensation method by dither for low-g simulation. J. Spacecr. Rocket. **32**(2), 364–369 (2015)
19. Mukherjee, K., Kar, I.N., Bhatt, R.K.P.: Adaptive gravity compensation and region tracking control of an AUV without velocity measurement. Int. J. Model. Identif. Control **25**(2), 154 (2016)

Analysis on Rigid-Elastic Coupling Characteristics of Planar 3-RRR Flexible Parallel Mechanisms

Qinghua Zhang$^{(\boxtimes)}$ and Qinghua Lu

Department of Mechatronics, Foshan University,
Foshan 528000, Guangdong, China
qinghuazhang411@163.com

Abstract. Based on the finite element method and Lagrange equation, high dimensions, time-invariant, nonlinear rigid-elastic coupling dynamic equations are established, in which the coupling influence of elastic deformation motion and the rigid body motion are considered. A comparative analysis of one-step method and two-step method for solving nonlinear rigid-elastic coupling dynamic equations is given. The simulation results show that the elastic vibrations of the flexible links have important effect on rigid body motion, particularly for angular velocity and angular acceleration. There are some errors in the amplitudes and phases of the elastic displacements and the elastic rotation angle for the two methods, rigid-elastic coupling of flexible links will critical influence elastic deformation displacements and elastic rotate angle of the moving platform, but change rule of the kinematic variables are basically consistent. So, one-pass method is used to solve dynamic equation of rigid-elastic coupling system that can reflect the dynamic characteristics well, it can provide guidance for controller design in future.

Keywords: Flexible parallel mechanical · Elastodynamics · Rigid-elastic coupling · One-pass method

1 Introductions

Recently, many researchers have paid attention to the light flexible parallel mechanisms with high speed, high acceleration, and high accuracy which are widely used in the assembly industry, the aerospace industry. Essentially, flexible parallel mechanism belong to flexible multibody systems, But dynamic modeling of flexible multibody system is a challenging task. A precise dynamic model in which not only the effect of the rigid body motions or nominal motions on elastic deformations must be studied, but also the influence of elastic deformations on rigid body motion and elastic deformations of other flexible links must be analyzed carefully. Amongst the published work which addresses dynamic modeling of flexible multibody systems, especially closed-loop flexible parallel multibody systems, the majority of the investigations assume that elastic deformations of the links of multibody systems are very small and do not have a significant effect on the rigid body motions, only considering the effect of rigid body motion on elastic motion. This approach called a two-pass method, usually requires that

Y. Huang et al. (Eds.): ICIRA 2017, Part II, LNAI 10463, pp. 394–404, 2017.
DOI: 10.1007/978-3-319-65292-4_34

the rigid body motions are prescribed or known a priori. Firstly, modeling rigid motion equation of the flexible multibody systems and solving variables of rigid body kinematics, then, these variables is substituted into the elastodynamics equation of flexible multibody systems for solving elastic deformation displacements of flexible links and the moving platform of the system.

Few research works [1–6] present one-pass methods to model closed-loop multibody systems with flexible links because of its complexity. The one-pass method takes into full account the dynamic coupling between the rigid body and elastic motions, and is a kind of more precise modeling and solving methods for flexible mutlibody systems, the governing rigid-elastic coupling nonlinear dynamic equation of the system is modeled through to combine dynamic equations of beam element.

2 Rigid-Elastic Coupling Dynamic Modeling

In this section, according to the finite element method (FEM) and Lagrange equation, the rigid-elastic coupling dynamic equation of planar 3-RRR flexible parallel mechanisms is modeled. Firstly, the flexible links are discretized into a series of beam elements by FEM. Dynamic equation of the beam element is modeled based on beam element theory. Then, the constraint equations of the flexible parallel system are investigated in detail, that include constraint equations of the rigid-body motion, constraint equations of the elastic deformation motion, dynamic constraints of the moving platform. Finally, considering constraint equations of the system and compatibility between beam elements, the governing rigid-elastic coupling nonlinear dynamic equation of the flexible multibody system is modeled.

2.1   The Sketch of Planar 3-RRR Parallel Mechanisms

The sketch of planar flexible 3-RRR parallel robots is constructed by the regular triangle moving platform $C_1C_2C_3$, the static platform $A_1A_2A_3$, and three symmetrical kinematic chains $A_1B_1C_1$, $A_2B_2C_2$, $A_3B_3C_3$, as shown in Fig. 1. Each kinematic chain has one active revolute (R) joint followed by two consecutive passive revolute (R) joints. The active revolute joints are installed at A_i (i = 1, 2, 3), A_1, A_2 and A_3 are the regular triangle's three vertices. $A_1B_1 = A_2B_2 = A_3B_3$, and $B_1C_1 = B_2C_2 = B_3C_3$. The vertexes O and P are centers of the regular triangles $A_1A_2A_3$, and $C_1C_2C_3$, respectively. The O-XY is the global fixed frame. Parameters $\alpha_i, \beta_i (i = 1, 2, 3)$ are the angles between the X-axis of the fixed frame and linkages A_iB_i, B_iC_i ($i = 1, 2, 3$), respectively. θ is the angle between the X-axis of the fixed frame and side C_1C_2 of the regular triangle $C_1C_2C_3$. L_1, L_2, L_3 , and L_4 are the length of the segments A_2B_2, B_2C_2, C_2P , and OA_2.

2.2   Dynamic Equation of the Beam Element

The flexible link can be modeled by connecting a series of beam elements. Figure 2 shows a beam element before and after deformation. The O-XY is the global fixed

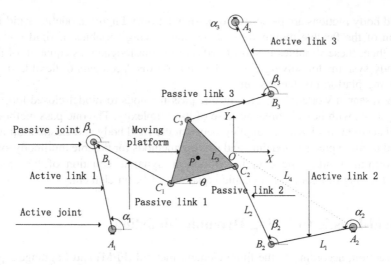

Fig. 1. The sketch of planar 3-RRR parallel robot.

frame and the A-xy is the local moving frame with Ax axis coincident with the neutral line of the beam element. Its original point A is located at one node of the beam element before deformation. B is another node of the beam element. The $O - x'y'$ system is an intermediate coordinate frame whose origin is rigidly attached to the origin of the O-XY and whose axes are parallel to the axes of the local moving frame A-xy. φ is the angle between the global fixed frame O-XY and the intermediate coordinate frame $O - x'y'$.

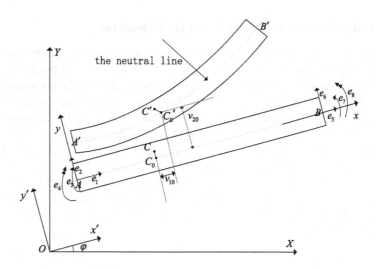

Fig. 2. Beam element deformation

Considering the general point C in the beam element, let point C_0 be the corresponding point on the neutral line. Points C' and C_0' are their positions after deformation, respectively. The elastic deformation of the point C_0 in the $A\text{-}xy$ is given by [7, 8].

$$v_0(x, t) = \begin{bmatrix} v_{10}(x, t) \\ v_{20}(x, t) \end{bmatrix} = \bar{N}(x)e_f \tag{1}$$

where $e_f = [e_1, e_2, e_3, e_4, e_5, e_6, e_7, e_8]^T$ is the nodal displacement vector, in which e_1 and e_5 are the axial displacements of two nodes A and B, respectively; e_2 and e_6 are the transverse displacements; e_3 and e_7 are elastic rotational angles; e_4 and e_8 are section curvatures; and the superscript 'T' indicates matrix transpose. $\bar{N}(x)$ is the shape function matrix.

The deformation displacements of C in $A\text{-}xy$ can be written as

$$v(x, t) = \begin{bmatrix} v_1(x, t) \\ v_2(x, t) \end{bmatrix} = \begin{bmatrix} v_{10}(x, t) - y\frac{\partial v_{20}(x,t)}{\partial x} \\ v_{20}(x, t) \end{bmatrix} \tag{2}$$

Setting $N = \begin{bmatrix} \bar{N}_1(x) - y\frac{\partial \bar{N}_2(x)}{\partial x} \\ \bar{N}_2(x) \end{bmatrix}$, then the displacement of C' can be expressed in $O - XY$ by

$$r_{c'} = r_A + R_\varphi(r_0 + Ne_f) \tag{3}$$

where r_A is the coordinate of the point A in $O - XY$, $R_\varphi = \begin{bmatrix} \cos\varphi & -\sin\varphi \\ \sin\varphi & \cos\varphi \end{bmatrix}$ is the direction cosine matrix, The vector $r_0 = [x \quad y]^T$ is the location coordinates of the point C in the $A - xy$ system, 'T' indicates the matrix transpose, $y\frac{\partial \bar{N}_2(x)}{\partial x}$ is the axial displacement caused by the transverse displacement of the beam.

Taking the first derivative of Eq. (3), yield:

$$\begin{aligned} \dot{r}_{C'} &= \dot{r}_A + R_\varphi(r_0 + Ne_f)\dot{\varphi} + RN\dot{e}_f \\ &= S\dot{q} \end{aligned} \tag{4}$$

where $R_\varphi = \frac{\partial R}{\partial \varphi}$, $S = \begin{bmatrix} I & R_\varphi(r_0 + Ne_f) & RN \end{bmatrix}$, $q = \begin{bmatrix} r_A^T & \varphi & e_f^T \end{bmatrix}^T$.

2.2.1 Kinetic Energy of the Beam Element

Kinetic energy of the beam element includes Translational kinetic energy and Rotational kinetic energy of beam element and Translational kinetic energy and Rotational kinetic energy of Lumped masses.

$$T = T_t + T_r + T_c = \frac{1}{2}\dot{q}^T m\dot{q} \tag{5}$$

2.2.2 Strain Energy of the Beam Element

Neglecting shear strain energy and buckling strain energy, according to Material mechanics, strain energy of the beam element can be expressed

$$
\begin{aligned}
U &= \frac{1}{2}\int_0^l EI(x)(v_{20}''(x,t))^2 dx + \frac{1}{2}\int_0^l ES(x)(v_{10}'(x,t))^2 dx \\
&= \frac{1}{2}q^T kq
\end{aligned}
\tag{6}
$$

where $v_{20}''(x,t) = \frac{\partial^2 v_{20}(x,t)}{\partial x^2}$, $v_{10}'(x,t) = \frac{\partial v_{10}(x,t)}{\partial x}$, E is elastic modulus of materials, $I(x)$, $S(x)$ are crosssectional moment of inertia and cross-sectional area of the beam element, respectively, and

$$
k = \begin{bmatrix} 0 & 0 & 0 \\ 0 & 0 & 0 \\ 0 & 0 & k_{ff} \end{bmatrix}, \; k_{ff} = \int_0^l EI(x){N_2''}^T(x)N_2''(x)dx + \int_0^l ES(x){N_1'}^T(x)N_1'(x)dx.
$$

2.2.3 Dynamic Equation of the Beam Element

According to Lagrange's equation, dynamic equation of the beam element can be written as

$$
m\ddot{q} + kq = p_e + p_v
\tag{7}
$$

where p_e are the generalized external forces and p_v is the quadratic velocity vector that contains the gyroscopic and the Coriolis force components, respectively.

The elemental dynamic Eq. (7) has been established in the A-xy system. Before forming the dynamic equation of the system, Eq. (7) must be expressed in the O-XY system. Define the coordinate transformation matrix B, then

$$
e_f = BE_f
\tag{8}
$$

where E_f is the elemental nodal coordinate vector in the O-XY system. Taking the first and the second derivative on (8) with respect to time yields

$$
\begin{aligned}
\dot{e}_f &= B_\varphi E_f \dot{\varphi} + B\dot{E}_f \\
\ddot{e}_f &= B_{\varphi\varphi} E_f \dot{\varphi}^2 + 2B_\varphi \dot{E}_f \dot{\varphi} + B_\varphi E_f \ddot{\varphi} + B\ddot{E}_f
\end{aligned}
\tag{9}
$$

Substitute (8) and (9) into (7) and premultiply by matrix $\begin{bmatrix} I & 0 & 0 \\ 0 & 1 & E_f^T B_\varphi^T \\ 0 & 0 & B^T \end{bmatrix}$, then

$$
\bar{m}(\bar{q})\ddot{\bar{q}} + \bar{k}\bar{q} = \bar{p}(\bar{q}, \dot{\bar{q}})
\tag{10}
$$

where $\bar{q} = \begin{bmatrix} r_A^T & \varphi & E_f^T \end{bmatrix}$.

According to generalized coordinate variables, Eq. (10) can be divided to rigid subsystem and elastic subsystem, namely

$$m_r \ddot{\varphi} = p_r (\tilde{q}, \dot{\tilde{q}}) + \tau_r \tag{11}$$

$$\tilde{m}_f \ddot{E}_f + c_f \dot{E}_f + k_f = p_f \tag{12}$$

2.3   Constraint Equations

2.3.1   Constraint Equations of the Rigid-Body Motion

Generalized coordinates are formed by rigid-body motion coordinates and elastic coordinates. As shown in Fig. 1, rigid body motion coordinates that include drive joint rotation angles $\boldsymbol{\alpha} = [\alpha_1 \quad \alpha_2 \quad \alpha_3]^T$, passive joint rotation angles $\boldsymbol{\beta} = [\beta_1 \quad \beta_2 \quad \beta_3]^T$, and translation displacement and rotation angle of the moving platform $X_P = [X_P \quad Y_P \quad \theta]^T$. The rigid body motion coordinate vectors $\boldsymbol{\alpha}, \boldsymbol{\beta}, X_P$ are not independent and exist some constraint, namely

$$OA_i + A_i B_i + B_i C_i + C_i P + PO = 0 \tag{13}$$

2.3.2   Constraint Equations of the Elastic Deformation Motion

Setting $\Delta X_{C_i}, \Delta Y_{C_i}, \Delta \eta_i$ are translational and rotational declinations of the point C_i, $\Delta X_P, \Delta Y_P, \varepsilon$ translational and rotational declinations of are moving platform because of elastic vibration of the flexible links, $(X_{C_i} \quad Y_{C_i})^T$ and $(X_{C_i'} \quad Y_{C_i'})^T$ are the coordinates of the points C_i, C_i' in the O-XY system, yields

$$\begin{pmatrix} \Delta X_{C_i} \\ \Delta Y_{C_i} \end{pmatrix} = \begin{pmatrix} X_{C_i'} \\ Y_{C_i'} \end{pmatrix} - \begin{pmatrix} X_{C_i} \\ Y_{C_i} \end{pmatrix} = \begin{bmatrix} I & T_1 \begin{pmatrix} -y_{C_i} \\ x_{C_i} \end{pmatrix}_P \end{bmatrix} \begin{pmatrix} \Delta X_P \\ \Delta Y_P \\ \varepsilon \end{pmatrix} \tag{14}$$

$$\varepsilon = \Delta \eta_1 + \Delta \eta_2 + \Delta \eta_3 \tag{15}$$

where I is 2×2 unit matrix, $T_1 = \begin{bmatrix} \cos \varphi & -\sin \varphi \\ \sin \varphi & \cos \varphi \end{bmatrix}$.

2.3.3   Dynamic Constraints of the Moving Platform

Assuming that F_i is the generalized joint constraint anti-force that the passive joint C_i are applied to the moving platform. M_v, J_v are the mass and moment of inertia of the moving platform. Then, dynamic constraint of the moving platform can be expressed by

$$\begin{bmatrix} M_v & & \\ & M_v & \\ & & J_v \end{bmatrix} \begin{bmatrix} \Delta \ddot{X}_P \\ \Delta \ddot{Y}_P \\ \ddot{\varepsilon} \end{bmatrix} = \sum_{i=1}^{3} F_i - \begin{bmatrix} M_v \ddot{X}_P \\ M_v \ddot{Y}_P \\ J_v \ddot{\phi} \end{bmatrix} \tag{16}$$

Setting U is the generalized coordinate vector, considering constraint equations and assembling all the element dynamic Eq. (12) with respect to the compatibility at the nodes, then, the elastic dynamic equation that describes the dynamic characteristic of system can be formed as

$$M(\alpha)\ddot{U} + C(\alpha,\dot{\alpha})\dot{U} + K(\alpha,\dot{\alpha},\ddot{\alpha})U = Q(\alpha,\dot{\alpha},\ddot{\alpha}) \qquad (17)$$

where $U = [U_{11}, U_{12}, \ldots, U_{1n}, \ldots, U_{3n}, \Delta X_P, \Delta Y_P, \Delta\varepsilon]^T$, $\alpha = [\alpha_1, \alpha_2, \alpha_3]^T$.

Assembling rigid dynamic Eq. (11), the governing rigid dynamic equation can be formed as

$$M_r(U)\ddot{\alpha} = Q_r(U,\dot{U},\ddot{U}) + \tau \qquad (18)$$

3   Simulation and Analysis

In order to quantitatively study for the rigid-elastic coupling characteristics of planar 3-RRR flexible parallel mechanism, the elastic dynamic Eq. (17) and the rigid dynamic Eq. (18) are simultaneously solved by iterative method, a comparative study is carried out to one-pass method and two-pass method.

It is assumed that the trajectory of the moving platform is described as

$$\begin{cases} X_P = 0.04 \cos(10\pi t) - 0.02 \\ Y_P = 0.04 \sin(10\pi t) \\ \theta = \pi/4 \end{cases} \qquad (19)$$

the trajectory of the moving platform is the circular trajectory with 0.2 s period. Assume that the material of the moving platform and links are an aluminum alloy, Young's Modulus (Pa) $E = 0.7 \times 10^5$ MPa, Poisson's Ratio $\mu = 0.3$, Density $\rho = 2712$ kg/m^3, the thickness of the moving platform c = 0.0171 m $L_1 = A_iB_i = 0.245$ m, $L_2 = B_iC_i = 0.242$ m, $L_3 = C_iP_i = 0.108$ m, $L_4 = A_iP = 0.4$ m, (i = 1, 2, 3), all the lumped mass of the joint is 0.15 kg, and all the lumped moment of inertia is 0.00005 kg·m^2, Width of link is 0.025 m, Thickness of link is 0.003 m, and all links is flexible.

Figures 3, 4, 5 and 6 are numerical simulation results. Figure 3 shows Rotation angle of active joint 1 with motion time, Fig. 4 shows Rotating angle velocity of active joint 1 with motion time, Fig. 5 shows rotation angle acceleration of active joint 1 with motion time, Fig. 6(a), (b), (c) are elastic displacement of moving platform along with X-axis and Y-axis, and elastic rotation angle in XY planar, respectively. In all figures, the results of one-pass method are indicated by real line with symbol '★', the results of two-pass method are indicated by real line.

From Figs. 3, 4 and 5, we can find that elastic vibration of flexible links have crucial influences on rigid motion of the system, especially for angular velocity and angular acceleration of rigid motion. So, one-pass method must be adopted to solve

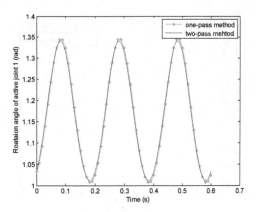

Fig. 3. Rotation angle of active joint 1

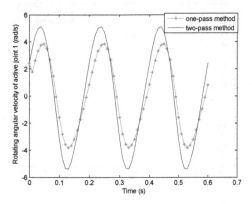

Fig. 4. Rotating angle velocity of active joint 1

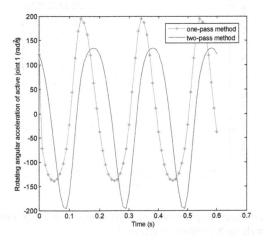

Fig. 5. Rotating angle acceleration of active joint 1

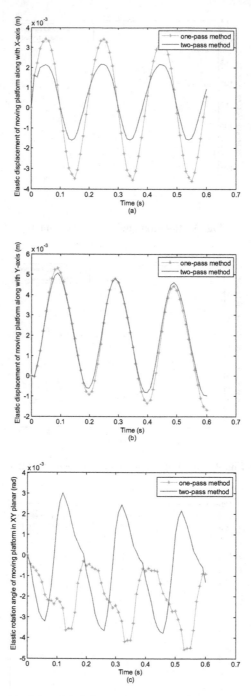

Fig. 6. (a), (b), (c) are elastic displacement of moving platform along with X-axis and Y-axis, and elastic rotation angle in XY planar, respectively

rigid-elastic coupling dynamic Eqs. (17) and (18) for getting an accurate dynamic characteristics of planar 3-RRR flexible parallel mechanisms.

From Fig. 6, we can find that elastic displacements and elastic angles of moving platform that are solve by two kinds of methods are different, and there are some offsets about their amplitudes, and their phases are also inconsistent, but their change rules are consistent. Rigid-elastic coupling is inherent characteristic for flexible multibody system, so one-pass method is used to solve rigid-elastic coupling dynamic equation can better reflect dynamic characteristic of the system.

4 Summary

(1) Rigid-elastic coupling dynamic equations of planar 3-RRR flexible parallel mechanisms are established be based on finite element method and Lagrange equation. The modeling method can be extended to the general flexible parallel mechanism system.

(2) Rigid-elastic coupling characteristics are studied, two different methods that includes one-pass method and two-pass method are used to solve rigid-elastic coupling dynamic equations. Numerical simulation results show that the calculation results of two kinds of method are different, so one-pass method is recommended to solve rigid-elastic coupling dynamic equations for getting accurate dynamic characteristics of the system. Meanwhile, it can provide guidance for controller design in future.

Acknowledgments. This work was supported by the National Natural Science Foundation of China (Grants nos. U1501247, 61603103, and 51505092), the of Natural Science Foundation Guangdong Province (Grants nos. 2014A030313616, 2015A030310181, and 2016A030310293), the Science and Technology Planning Project of Guangdong Province (2014A010104017 and 2015B010101015), and the Science and Technology Innovation Project of Foshan (2015AG10018). These supports are greatly acknowledged.

References

1. Gasparetto, A.: On the modeling of flexible-link planar mechanisms: experimental validation of an accurate dynamic model. ASME J. Dyn. Syst. Meas. Control **126**(2), 365–375 (2004)
2. Nagarajan, S., Turcic, D.A.: Lagrangian formulation of the equations of motion for elastic mechanisms with mutual dependence between rigid body and elastic motions. Part I: element level equations. ASME. J. Dyn. Syst. Meas. Control **112**(2), 203–214 (1990)
3. El-Absy, H., Shabana, A.A.: Coupling between rigid body and deformation modes. J. Sound Vibr. **198**(5), 617–637 (1996)
4. Yang, Z., Sadler, J.P.: A one-pass approach to dynamics of high-speed machinery through three-node lagrangian beam elements. Mech. Mach. Theory **34**(7), 995–1007 (1999)
5. Karkoub, K., Yigit, A.S.: Vibration control of a four-bar mechanism with a flexible coupler link. J. Sound Vibr. **222**(2), 171–189 (1999)

6. Xuping, Z., Mills, J.K., Cleghorn, W.L.: Coupling characteristics of rigid body motion and elastic deformation of a 3-PRR parallel manipulator with flexible links. Multibody Syst. Dyn. **21**, 167–192 (2009)
7. Zhang, Q.H., Zhang, X.M.: Dynamicmodeling and analysis of planar 3-RRR flexible parallel robots. J. Vibr. Eng. **26**(2), 239–245 (2013)
8. Zhang, Q., Fan, X., Zhang, X.: Dynamic analysis of planar 3-RRR flexible parallel robots with dynamic stiffening. Shock Vibr. **2014**, 1–13 (2014). doi:10.1155/2014/370145. Article ID: 370145

Advanced Parallel Robot with Extended RSUR Kinematic for a Circulating Working Principle

Stefan Tobias Albrecht, Hailin Huang, and Bing Li$^{(\boxtimes)}$

Harbin Institute of Technology, Shenzhen, China
libing.sgs@hit.edu.cn

Abstract. Parallel robots are manipulators who use closed kinematic loops to create a movement at the Tool Center Point (TCP). Due to this kind of kinematic loops, it causes the mechanism to have high stiffness, high ratio of load to self-weight and low inertia, even though it will have good repeatability and high dynamics. A very well-known example of a parallel manipulator is the Delta Robot which is commonly used in the pick and place industry. Delta Robots are known to normally use a pivoting working principle for the actuating upper arms, meaning, its design does not make it possible for the upper arm actuators to make a complete revolution. This paper introduces a new type of an advanced Delta Robot with extended RSUR kinematic which allows a complete revolution of the actuators what will lead to the advantages of storing rotational energy in cycles comparatively like a flywheel what is a machine element that stores rotational energy. The extended kinematic torque curve of actuators therefore runs more smoothly and thus the payload by same cycle time increases. The following work will introduce the kinematic model description of such an advanced Delta Robot that is working with a circulating working principle, and discuss several simulation results of such a kinematic to prove the efficiency of the circulating working principle. Furthermore, a comparison with a real Delta Robot hardware and the advanced Delta Robot simulation model is introduced.

Keywords: Delta Robot · RSUR kinematic · Circulating working principle

1 Introduction

In this paper, the Delta Parallel Robot which was developed and patented by R. Clavel in 1988 [1] is defined as Standard Delta Robot (SDR). The SDR consists of a baseplate, moving plate, lower legs and upper arms whereby each upper arm is directly driven by actuators. Thus, all motors have a fixed mounting position at the baseplate. Compared to serial manipulators this is a great advantage for parallel robots because of its significant reduction of the active moving mass. The special nature of the lower arms from the SDR are: they create a parallelogram position when the upper arms are moving which inhibit the rotational degrees of freedoms at the TCP and release only translational motion. These parallel rods also assure that the moving plate always remains parallel to the robot baseplate. Hence, the SDR has three translational degree-of-freedom (DOF).

© Springer International Publishing AG 2017
Y. Huang et al. (Eds.): ICIRA 2017, Part II, LNAI 10463, pp. 405–416, 2017.
DOI: 10.1007/978-3-319-65292-4_35

Due to the SDR design which does not allow a full rotation of the upper arms around its axle the output from the drive unit has a very limited range of angular movement. Therefore, the drive units must do rapid changes of direction which has the disadvantages of spending much kinetic energy into these direction changes and bring vibrations into the system, such a working principle is called *Pivoting Working Principle*.

Based on the knowledge about that disadvantage, we present an Advanced Delta Robot (ADR) which was first designed and patented by Revobotik GmbH, Freital (Germany) [2]. The main design idea for the ADR is to extend the SDR kinematic such that it is possible to realize a full rotation about its axis for the drive units. At best, the output of drive units does not change the direction and run continuously and its only controlled by different angular speed. This kind of working principle is defined as *Circulating Working Principle*. A possible technical solution to realize a Circulating Working Principle for the drive units is an extension with a Revolute- Sphere- Universal- Revolute (RSUR) four bar linkage. A detailed explanation about RSUR kinematic will be discussed in next chapter. Classification for the ADR is still a parallel manipulator because only when all the driven actuators move simultaneously the Tool Center Point moves as well, what is an indicator of closed kinematic loops structure.

2   RSUR Kinematic for the ADR

The revolute joints of the RSUR- Kinematic are placed at a fixed location and the two bars are connected by a spherical joint. Upper arms and the second bar from the RSUR-Kinematic are linked through a universal joint. We can use the following Kutzbach equation to prove the mobility of the RSUR- Kinematic [3–5].

$$M = 6(N - 1) - 5J_1 - 4J_2 - 3J_3 \tag{1}$$

Setting input parameters N = 4 (total links), $J_1 = 2$ (one- DOF joints), $J_2 = 1$ (two-DOF joints) and $J_3 = 1$ (three- DOF joints), the solution of mobility is 1DOF for the RSUR kinematic. Hence, the degree of freedom is sufficient to move the upper arms up and down. Since the motion pattern of the upper arms is like those of the SDR, resulting that the whole ADR manipulator system will still have three translational degree of freedoms at the TCP.

The origin of the ordinates is located at the center of the base frame whereby the Y-axis lies on the same line as one upper arm of the SDR and the angular displacement between each upper arm is 120° around the Z-axis. To develop the mathematically model for the RSUR, several design parameters and fixed design points are defined which are listed in the following table (Table 1):

In Fig. 1, a kinematic diagram of the RSUR mechanism is shown including the design parameters. At this diagram, the points A and B describes the fixed revolute joints positions, C is the spherical joint and D the universal joint. The bar between B and C has the length R_1 and the other the length R_2.

The constant tilting angle β is necessary to let the actuators which are mounted at point B slightly tilted. Therefore, the actuators are not parallel to the base frame. The

Table 1. Design variables for the RSUR kinematic

Variable	Definition
R_1	Link between actuator and point C
R_2	Link between point C and D
B	Actuator position
A	Revolute joint position from standard Delta Robot
β	Tilting angle from actuator
L_1	Length between point A and D
L_2	Length between point A and E

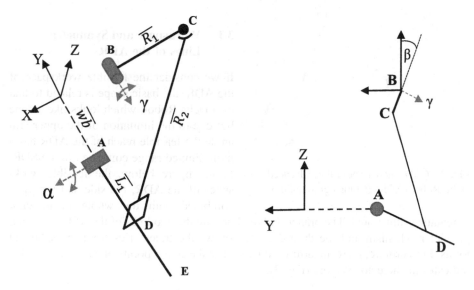

Fig. 1. Kinematic scheme RSUR parallel four bar linkage

advantage of this mounting positon is that the angle between the two links which impact the critical angle of spherical joints will decrease. This will be a significant benefit, if we want to increase the dynamic of the system.

3 Analysis of the Advanced Delta Robot

The Advanced Delta Robot manipulator is a combination of SDR and RSUR- Kinematic which allows the output of the actuator to move in a circular motion although the upper arms of the SDR is still moving in an up and down motion. As a result, the ADR fulfill the requirement to operate with a Circulating Working Principle for the drive units. Similar like the Delta Robot the ADR have a design symmetry around the Z- axis displaced by 120° and therefore three symmetric lines along the upper arms are existing.

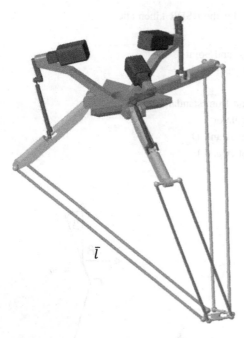

Figure 2 shows a CAD model of the ADR. It is clearly seen that the lower part of the ADR is based on the SDR mechanism. The black blocks represent the driven actuators mounted at the base frame of the ADR. Attached to the black blocks are two red links which are connected through a spherical joint and followed by a universal joint connection to the upper arm of the SDR.

Fig. 2. CAD- model about the Advanced Delta Robot (ADR) (Color figure online)

3.1   Workspace and Symmetric Lines of the ADR

If we consider the feasible workspace of the ADR, the basic shape is related to that of a Delta Robot, which looks like a cone but due to the limitation of the upper arm angle the feasible reach of the ADR has a more limited range compared to the SDR. In the figure below, the feasible workspace of the ADR, the side and top view combined with the whole manipulator kinematic is illustrated. The orange and red lines are the two links of the RSUR four bar parallel mechanism and the dashed circles shows the trajectories from the spherical joints. Furthermore, there are arrows illustrated in the center points of each circle which indicates the actuator vectors (Fig. 3).

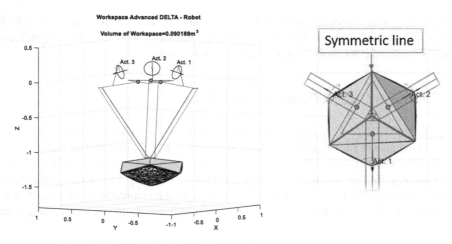

Fig. 3. View of feasible workspace and symmetric lines of the ADR

However, for the trajectory planning it must considered about the symmetric line of the advanced Delta Robot because if the end effector is travelling much far from the symmetric line a Circulating Working Principe is not possible. Since when the TCP is far away from the symmetric line and the actuator will do a full turning an unpredictable trajectory of the end effector will occur. Thereby, the area far from the symmetric line and the driven actuator near that area has a relationship. For example, if the end- effector is moving more far from the symmetric lines towards the area of actuator one, then it will be better if actuator one working in pivoting working principle to avoid a much unpredictable end- effector motion.

Therefore, a manipulator using Circulating Working Principle have always a *continuous* and *closed* resulted TCP Trajectory.

3.2   Inverse Position Kinematics Solution of the ADR

The Inverse Position Kinematic (IPK) solution [6] contributes in establishing the value of the articular coordinates corresponding to the end- effector (TCP) configuration. IPK solution is essential for the position control of parallel robots. For the inverse kinematic problem, the Cartesian position vector from the middle point of the moving platform $\vec{P}_{TCP} = \{xyz\}^T$ is known and therefore allows us to solve the calculation to find the required revolute joint angles $\vec{\gamma} = \{\gamma_1 \gamma_2 \gamma_3\}^T$.

The position from revolute joints $\vec{A}_k$, for $k = 0, 120, 240$ are constant and fixed in the base frame. Therefore, the following vectors hold:

$$\vec{A}_0 = \begin{pmatrix} 0 \\ -w_B \\ 0 \end{pmatrix};$$
(2)

$$\vec{A}_{120} = \begin{pmatrix} \frac{\sqrt{3}}{2} w_B \\ \frac{1}{2} w_B \\ 0 \end{pmatrix};$$
(3)

$$\vec{A}_{240} = \begin{pmatrix} -\frac{\sqrt{3}}{2} w_B \\ \frac{1}{2} w_B \\ 0 \end{pmatrix}$$
(4)

Constant positions for the actuators are:

$$\vec{B}_k = \boldsymbol{Rotz}(\delta_i) * \vec{B}_0$$
(5)

with:

$$\boldsymbol{Rotz}(\delta_i) = \begin{bmatrix} \cos(\delta_i) & -\sin(\delta_i) & 0 \\ \sin(\delta_i) & \cos(\delta_i) & 0 \\ 0 & 0 & 1 \end{bmatrix} \text{ for } \delta_i = \begin{cases} 0 \\ 120 \\ 240 \end{cases}$$
(6)

Points $\vec{D}_k$ describes the position of the universal joint connected to the upper arms with the second link of the RSUR mechanism. However, this position is dependent on revolute joint angle α_i, for $i = 1, 2, 3$ which can be calculated according to the inverse kinematic problem of a Delta Robot.

$$\vec{D}_0 = \begin{pmatrix} x_{D0} \\ y_{D0} \\ z_{D0} \end{pmatrix} = \begin{pmatrix} 0 \\ -w_B - L\cos\alpha_1 \\ -L\sin\alpha_1 \end{pmatrix} \tag{7}$$

$$\vec{D}_{120} = \begin{pmatrix} x_{D120} \\ y_{D120} \\ z_{D120} \end{pmatrix} = \begin{pmatrix} \frac{\sqrt{3}}{2}(w_B + L\cos\alpha_2) \\ \frac{1}{2}(w_B + L\cos\alpha_2) \\ -L\sin\alpha_2 \end{pmatrix} \tag{8}$$

$$\vec{D}_{240} = \begin{pmatrix} x_{D240} \\ y_{D240} \\ z_{D240} \end{pmatrix} = \begin{pmatrix} -\frac{\sqrt{3}}{2}(w_B + L\cos\alpha_3) \\ \frac{1}{2}(w_B + L\cos\alpha_3) \\ -L\sin\alpha_3 \end{pmatrix} \tag{9}$$

Equally the moving position from the spherical joint $\vec{C}_i$ can be mathematically defined as:

$$\vec{C}_i = \begin{pmatrix} R\cos\alpha_i\sin\gamma_i - \sin\alpha_i(y_B - R\cos\gamma_i\sin\beta_i) \\ \cos\alpha_i(y_B - R\cos\gamma_i\sin\beta_i) + R\sin\alpha_i\sin\gamma_i \\ z_B + R\cos\gamma_i\cos\beta_i \end{pmatrix} \quad \text{for} \quad i = 1, 2, 3 \tag{10}$$

Whereby, γ_i is the searched angle position which solves the inverse kinematic problem. The vector loop closure holds the condition:

$$\left| \vec{C}_i - \vec{D}_i \right| = R_2^2 \tag{11}$$

The approach which leads us to Eq. (11) is known as vector- loop closure method and is commonly used in research about parallel robots. Dissolving the Euclidian distance for γ_i it is possible to apply the Tangent Half-Angle Substitution method.

$$G_i\cos\gamma_i - H_i\sin\gamma_i - F = 0 \tag{12}$$

with:

$$G_i = \cos\beta(z_{B_k} - z_{D_i}) + \sin\beta(y_{D_i}\cos\alpha_i - x_{D_i}\sin\alpha_i - y_{B_k}) \tag{13}$$

$$H_i = x_{D_i}\cos\alpha_i + y_{D_i}\sin\alpha_i \tag{14}$$

$$F_i = \frac{a_i - R_1^2}{2R_1} \tag{15}$$

$$a_i = R_2^2 - z_{D_i}^2 - z_{B_k}^2 - x_{D_i}^2 - y_{D_i}^2 + 2z_{D_i}z_{B_i} - 2x_{D_i}y_{B_k}\sin\alpha_i + 2y_{D_i}y_{B_k}\cos\alpha_i - y_{B_k}^2 \tag{16}$$

Equation (13) substituted with Tangent Half- Angle results for:

$$\cos\gamma_i = \frac{1 - t_m^2}{1 + t_m^2} \tag{17}$$

$$\sin\gamma_i = \frac{2t_m}{1 + t_m^2} \tag{18}$$

Therefore:

$$-(F_i + G_i)t_m^2 - 2H_i t_m + (G_i - F_i) = 0 \tag{19}$$

Which can be solved using quadrat formula:

$$t_{m1/2} = \frac{2H_i + \sigma\sqrt{4H_i^2 - 4(-F_i - G_i)(G_i - F_i)}}{2(-F_i - G_i)}; \ \sigma = \left\{ \begin{array}{c} 1 \\ -1 \end{array} \right. \tag{20}$$

The solution for the IPK is:

$$\gamma_i = 2tan^{-1}(t_m) \tag{21}$$

If considered about the singularity of the ADR, and compare the singularity of the SDR, the singularity analysis looks very different. In fact, that the RSUR kinematic especially the universal joint connection between the upper arm of the SDR and the second link of the RSUR kinematic limited the moving area of α_i significant, only the right practical solution will occur. Hence, because of the design constraints, the RSUR kinematic will cancel the singularity problem for the lower SDR part because it holds: $\det(J_{SDR}) \neq 0$ for all α_i in the limited and constrained range of $[\alpha_{min}; \alpha_{max}]$.

Following from this benefit, we only must consider about the singularity positions for the RSUR kinematic. However, this will lead to a very simple problem because there existing only two possible configurations for γ_i: one positive and one negative solution with the controllable factor σ from Eq. (20).

3.3   Jacobian Matrix of the ADR

The Jacobian matrix of a robot in general describes the relationship between operation space velocities and joint velocities. In our case, the previous expressed vector- loop closure constraint equation was used and differentiated to obtain the velocity relationship for the ADR such as:

$$\dot{\vec{P}}_{TCP} = J_{ADR}\dot{\vec{\gamma}} \tag{22}$$

Procedure for developing the Jacobian matrix for the ADR (J_{ADR}), is to derive the vector-closure equation and as a result it is possible to find the relationship between the joint angles γ_i and α_i:

$$R^2 + 2Rcos(\gamma_i)G_i - 2Rsin(\gamma_i)(y_{B_i} - y_{D_i}) = \alpha_i \tag{23}$$

Derivation thereof holds as:

$$\frac{d\alpha}{dt}[x_{D_i}y_{B_i}\cos(\alpha_i) + y_{D_i}y_{B_i}\sin(\alpha_i) - Rcos(\gamma_i)\sin(\beta)[y_{D_i}\sin(\alpha_i)$$
$$+ x_{D_i}\cos(\alpha)]]R\frac{d\gamma}{dt}[\sin(\gamma_i) - G_i - \cos(\gamma_i)(x_{D_i}y_{D_i})] \tag{24}$$

In matrix form $i = 1, 2, 3$ multiplied by the diagonal matrix results in:

$$R\,diag(Q_i)\vec{\dot{\gamma}} = diag(T_i)\vec{\dot{\alpha}} \tag{25}$$

where:

$$Q_i = \sin(\gamma_i) - G_i - \cos(\gamma_i)(x_{D_i}y_{D_i}) \tag{26}$$

$$T_i = x_{D_i}y_{B_i}\cos(\alpha_i) + y_{D_i}y_{B_i}\sin(\alpha_i) - Rcos(\gamma_i)\sin(\beta)$$
$$\times [y_{D_i}\sin(\alpha_i) + x_{D_i}\cos(\alpha)] \tag{27}$$

$$\vec{\dot{\alpha}} = J_{SDR}\vec{\dot{P}}_{TCP} \tag{28}$$

As assumed, the Jacobian matrix is heavily dependent on the Jacobian matrix from the SDR (J_{SDR}). Several methods of how to solve inverse kinematic and develop the Jacobian matrix for the SDR are represented in [4, 5].

Consequently, the Jacobian matrix of the Advanced Delta Robot can be written as:

$$J_{ADR} = R \times diag(Q_i) \times diag(T_i)^{-1} \times J_{SDR}^{-1} \tag{29}$$

3.4 Dynamic Model of the ADR

Similar like for the kinematic model, the dynamic model of the Advanced Delta Robot consists of the two SDR and RSUR parts. For the RSUR kinematic we have two links that considers the dynamic development. Needed dynamic parameters thereby are the mass of the first arm and the mass of the second arm of the kinematic, m_{r1} and m_{r2}. The moment of inertia is computed by applying the parallel axis theorem for a thin rod and moment of a rigid body. The following torque equations holds for the ADR:

$$\vec{T}_\gamma = \vec{T}_{r1} + \vec{T}_{r2} + \vec{T}_\alpha \tag{30}$$

Whereas, $\vec{T}_{\alpha_i}$ are the input torque from the Standard Delta Robot kinematic and $\vec{T}_{r1_i}$ and $\vec{T}_{r2_i}$ are the computed torque from the two links of the RSUR kinematic. The torque at revolute joint α_i we can compute using:

$$\vec{T_\alpha} = I_{rf}\ddot{\vec{\alpha}} + J_{SDR}(\vec{\alpha})^T m_{et}\overrightarrow{\ddot{P}_{TCP}} - J_{SDR}(\vec{\alpha})^T m_{et}\begin{pmatrix} 0 \\ 0 \\ -g \end{pmatrix} - m_b r_{Ga}(-g)\begin{pmatrix} \cos(\alpha_1) \\ \cos(\alpha_2) \\ \cos(\alpha_3) \end{pmatrix} \quad (31)$$

whereas I_{rf} is the square diagonal mass matrix with $I_{rfi} = 1/3 m_{rf} rf^2 + (m_a + r * m_{re})rf^2$ matrix elements items, see [8]. Table 2 listed thereby the needed dynamic parameter.

Table 2. Dynamic parameters

Variable	Definition
m_{re}	Mass of lower arms
m_a	Mass of spherical joints
$m_{Payload}$	Mass of payload at TCP
m_{rf}	Mass of upper arms
m_e	Mass of moving platform
$m_{et} = m_e + m_{Payload} + 3(1 - r)m_{re}$	

4 Simulation and Discussion

The simulation model for the ADR is implemented in MATLAB for: Kinematic, Jacobian and inverse dynamic model. Furthermore, motion controller output data from a real hardware in the form of a Delta Robot were available. Following the results of the simulation and the comparison between the real hardware and simulated results are illustrated. That will underlie the advantage of the ADR over the SDR hardware.

Following table shows the used parameters used in simulation (Table 3).

Table 3. Parameters for simulation

Variable	Value	Variable	Value
m_{re}	60 g	L_1	205 mm
m_a	9.3 g	L_2	524 mm
$m_{Payload}$	5000 g	l	1244 mm
m_{rf}	320 g	β	15°
m_e	65.5 g	R_1	100 mm
m_{r1}	35 g	R_2	300 mm
m_{r2}	45 g		

First simulation shows the when the TCP is moving along Z- axis up and down, causing that all three actuators of the ADR mechanism turning in CWP. The following illustrations shows the simulation result for such TCP movement. On the left side, the measured actuator angle position and actuator torque data from the SDR hardware is illustrated. On right side, the simulated results for the upper arm angle position and actuator torque from the ADR is shown (Fig. 4).

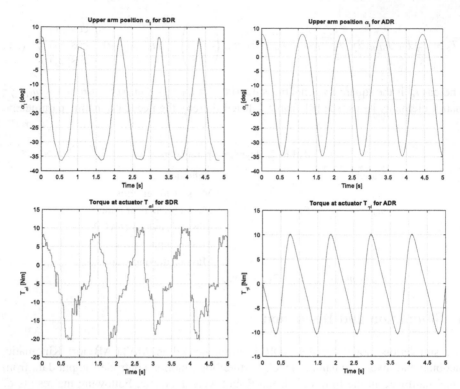

Fig. 4. Simulation comparison. Left top: α_i SDR, right top: α_i ADR. Bottom left: actuator torque T_{α_i} SDR real hardware, bottom right: actuator torque T_{γ_i} ADR

From the simulation, the upper arms approximately move in the same motion patterns for both manipulators and create therefore a good basis for comparison. It can significantly be seen that the torque curve for the ADR has a very harmonious progression between the plus and minus 10 Nm. This result is as expected because the actuators turn continuously in one direction and do not change in direction. On the other hand, the torque curve for the SDR show very quick direction changes and peaks up to −20 Nm.

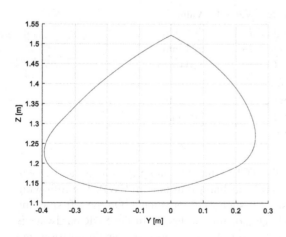

Fig. 5. Approximate pick and place trajectory

The second simulation example shown a trajectory approximately like a pick and place path with a length of 0.65 m and height of 0.3 m, see Fig. 5. As defined previously, the trajectory is closed and continuously and therefore applicable for CWP. Trough the simulation it can be seen how a trajectory, working for CWP is depend on a suitable application and obstacles in the workspace because the motion flexibility of the TCP is very limited. The following figure illustrates the simulated results for the ADR on left side and SDR on right side for a TCP trajectory which comply the trajectory in Fig. 5 (Fig. 6).

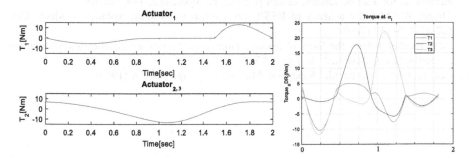

Fig. 6. Simulated torques comparison between ADR and SDR from a pick and place trajectory

Second simulation example match the same results like in the first simulation, the torque curve runs more smoothly and the provided input torque is lower. Therefore, a Circulating Working Principle is a workable solution for vibration and torque reduction.

However, it must be considered that the trajectory for CWP is very limited in flexibility and therefore not always a practical solution. In fact, the actuators of the ADR can also do turn the direction and working in Pivoting Working Principle which means similar like the Standard Delta Robot. If the actuators change thereby the direction, other results for the torque curve will occur.

5   Conclusion

This paper introduced and described a Delta kinematic where its design allowed a full rotation of the actuators which is a suitable solution for Circulating Working Principle. The kinematic and inverse dynamic model of the ADR are compute and the feasible workspace with resulting symmetric lines and their necessity for trajectory planning is described. Through a simulation which demonstrated the theory that if the drive unit does not change directions and it can continuously turn the torque curve which will run more smoothly and therefore need less driven torque is represented. Booth shown simulation examples represent thereby only if the actuators of the ADR kinematic does not change the directions.

Acknowledgement. This work was financially supported by the Joint Funds of the Natural Science Foundation of China (Grant No. U1613201), State Key Laboratory of Robotics and Systems (HIT) (SKLRS201701B), and in part by the Foundation for Innovative Research Groups of the National Natural Science Foundation of China (Grant No. 51521003), and in part by Shenzhen Research Funds (Grant Nos. JCYJ 20150529141408781 and JCYJ20160427183553203)

References

1. Revobotik Homepage. https://revobotik.de/de. Accessed 21 May 2017
2. Neugebauer, R.: Parallelkinematische Maschinen: Entwurf, Konstruktion, Anwendung. Springer, Berlin (2005)
3. Mazzotti, C., Troncossi, M., Parenti-Castelli, V.: Dimensional synthesis of the optimal RSSR mechanism for a set of variable design parameters. Meccanica, 1–9, (2016)
4. Chung, W.: Mobility analysis of RSSR mechanisms by working volume. J. Mech. Des. 127(1), 156–159 (2005)
5. Delta Parallel Robot - the Story of Success Homepage. http://www.parallemic.org/Reviews/Review002.html. Accessed 16 June 2017
6. Rey, L., Clavel, R.: Parallel Kinematic Machines. Springer, Berlin (1999)

Design and Analysis of a New Remote Center-of-Motion Parallel Robot for Minimally Invasive Surgery

Jingyuan Sun, Shuo Wang, Hongjian Yu[(⊠)], and Zhijiang Du

State Key Laboratory of Robotics and System, Harbin Institute of Technology,
Harbin 150001, China
yuhongjian99@126.com

Abstract. Surgical robot with a remote center of motion (RCM) plays an important role in minimally invasive surgery (MIS) field. To make the structure more compact and have higher rigidity and accuracy, a new type of parallel surgical robot with three degrees of freedom (DOFs) is developed. The detailed design of the mechanism is provided in the paper. To better study the characteristics of the mechanism, static analysis is proposed in the paper. Firstly, establish the equations of the force and torque through the analysis of the branch and the mobile platform. Next, obtain the expression of the active force and constraining force of the joints. One numerical example is given. In order to analyze the dynamic performance of the UPR-2UPRR system, the Kane method is used to build the dynamic model. The speed, acceleration, partial velocity and partial angular velocity are calculated. Then the dynamic equation is given after analyzing the generalized active force and inertial force. At last, the driving force of each telescopic rod is given.

Keywords: Parallel robot · Remote center-of-motion · Statics analysis · Dynamic model · Kane's dynamic equation

1 Introduction

In effective health care, basic science, engineering and medical collaborative effort provides physicians with improved tools and technologies. Minimally Invasive Surgery (MIS) has been thoroughly reform the ways of the surgical treatment, and in the progress of surgical robots, again to overhaul the MIS [1]. Bring in surgical robots to the operating room, reduces the trauma and short recovery time. During the process of a long-time operation, the doctor's fatigue value is small. Especially in the movement of the flexibility, surgical instruments are placed in the tip of the manipulator, which can provide a full range of mobility and allow the instrument rotate over 360° through tiny incisions.

Due to the limitation of the incision, surgical instruments need move around a fixed point at the end, which is called remote center-of-motion (RCM). RCM plays an important role in MIS by avoiding additional trauma except the incision. The request of the numbers of the DOFs is four, including two rotational motions around the incision shown in Fig. 1, one spin motion around the instrument axial direction, and one

© Springer International Publishing AG 2017
Y. Huang et al. (Eds.): ICIRA 2017, Part II, LNAI 10463, pp. 417–428, 2017.
DOI: 10.1007/978-3-319-65292-4_36

transition along the instrument. For safety, all rotation must be centered on incision, while the translational motion should always move along the line via the incision point.

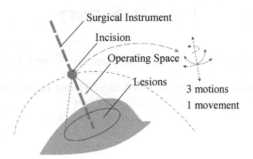

Fig. 1. Remote center of motion (RCM).

Typical structures of RCM contain parallelogram mechanism, multi-axis linkage, cambered mechanism, spherical, etc. The da Vinci uses double parallelogram mechanism in minimally invasive surgery, which is the most widely used [2]. Tianjin University developed a new type of parallelogram structure. The main characteristic of this structure is that the two axis of the rotation locate in two moving planes respectively, forming a unique intersection, which makes their cross curve always pass through the fixed point [3].

Static stiffness analytical model is to establish the mapping relations between the operating force and the deformation of the end-effector. In the early 1900s, Cho presented a complete analytical model of general parallel system by using the method of kinematic influence coefficient [4]. Lee deduced the formula of the static stiffness model of parallel robot through the geometrical relationship [5]. Depending on the principle of virtual work, Gosselin established the mapping relations between the operating force and the deformation of the end-effector [6].

The dynamic model could be used for the evaluation of the actual robot performance, the high precision motion control and movement simulation. Dynamics modeling is a fairly complex process, and the computational efficiency is different through different methods, including Newton-Eruler, Lagrangian, Gauss, Spinor and Kane equation methods, etc. Kane's method is a new dynamic modeling method for complex mechanical system [7]. Kane's method is very suitable for machinery multibody system modeling, because it could meet the requirements of real-time control. Nia et al. applied Kane's method in an under-actuated snake-like robot, improving the modeling efficiency [8].

In this paper, a 3DOF parallel robot arm is proposed for MIS. The roll motion is realized by the revolute joint and the universal joint, while the translation of the surgical tool is realized by the prismatic joint. The structure of the parallel robot is presented firstly. Then, static analysis based on the vector method is presented. And finally, one numerical example is implemented. To analyze the dynamic performance of the UPR-2UPRR system, the Kane's method is used to build the dynamic model. Firstly, the speed, acceleration, partial velocity and partial angular velocity are calculated. Then, after analyzing the generalized active force and inertial force, the dynamic equation is given. At last, a typical curve is used to represent the movement path of the upper platform, and the driving force of each telescopic rod is given.

2 Structure Characteristic of the Parallel Robot

2.1 Structure of the Parallel Robot

The parallel mechanism is shown in Fig. 2. The mechanism is composed of the base platform, telescopic link, swing link, the mobile platform and the surgical instrument. The base platform includes the fixed platform and the universal joints. Three universal joints uniformly distribute on the fixed platform, providing two directions of rotations along the axis. In the universal joints, the first rotation axes intersect in one point, which determines the location of the telocentric point. The second rotation axis is perpendicular to the axis of the first rotation, and is tangent to the cylindrical base. The DOF could be calculated by Eq. (1).

$$M = 6(n - g - 1) + \sum_{i=1}^{g} f_i = 6(11 - 12 - 1) + 15 = 3 \tag{1}$$

The lower end of the telescopic link is connected by universal joint to the base, while the upper end is connected with the swing link by revolute joint. The actuator is located on the telescopic link, and drives the platform by changing the length of the telescopic link, which can realize the movement of the instrument of the RCM. The mechanism can achieve two-dimensional rotation and one-dimensional transition along the axis of the surgical instrument. Besides, the mechanism has the isolated degree of freedom by rotating around its own axis. To solve this problem, fix one of the swing links with the surgical instrument. The configuration of the robot is transformed from three-universal-prismatic-revolute-revolute (3UPRR) into universal-prismatic-revolute-two-universal-prismatic-revolute-revolute (UPR-2UPRR). The parallel robot is 3-DOF mechanism without any isolated degree of freedom.

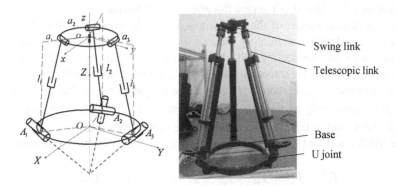

Fig. 2. A prototype of the parallel robot for MIS: (a) Free model sketch of UPR-2UPRR. (b) The fact picture of UPR-2UPRR parallel.

2.2   Workspace Analysis

In clinical application, the mechanism should cover the angle between $0 - \pi/4$ and the distance between 150–250 mm below the patient' incision. Parallel operation space is limited by many factors, such as the interference between the branch and the platform. In the UPR-2UPRR mechanism, the telescopic link limits the range of movement. Under the condition of the condition of the design requirements, the constraints of the working space are as follows: the swing angle of the surgical instruments is $\pm(\pi/4)$, and the length of the telescopic link is 330–610 mm. The method is based on the exhaustive method, so take over all angles between $0 - \pi/4$ around the horizontal axis or rotation. List all the heights of the distance between the center of the mobile platform and the telescopic center between 300–600 mm. The end of the instrument below the telescopic center consists of an approximate cone area in Fig. 3.

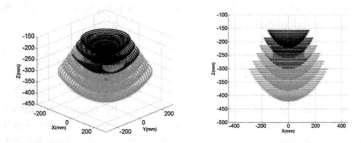

Fig. 3. The workspace: (a) The workspace of the UPR-2UPRR. (b) The workspace in x-z plane.

3   Static Analysis

3.1   Static Model of the Branch

The structure of the UPR-2UPRR can be simplified as symmetric structure, firstly develop the statics analysis of a single branch of UPR. Select the entire link branch as the research object, the connecting rod can be divided into the upper scale bar μ and lower fixed bar d two parts. The fixed platform connected with the connecting rod by the universal joints that don't accept the effect of the torque.

The mechanical model of the connecting rod is shown in Fig. 4(a). Establishing right Angle coordinate system along the axis direction, the direction of z_i-axis is along the direction of the connecting rod, the direction of x_i-axis is perpendicular to the second rotation axis of the universal joint. f_{uc}, T_u are the force and the torque effect on the upper linkage μ. f_{dc}, T_d are the force and the torque effect on the lower linkage d. $f_r^{(i)}$, $f_p^{(i)} f_u^{(i)}$ respectively express the reaction forces of constraints that revolute joint, prismatic joint, and universal joint inflict on the connecting rod.

The mechanical and torque equilibrium equation for connecting rod branch $A_i a_i$:

$$f_r^{(i)} + f_u^{(i)} + f_{uc}^{(i)} + f_{dc}^{(i)} + f_i^A + f_i^R = 0, i = 1, 2, 3 \tag{2}$$

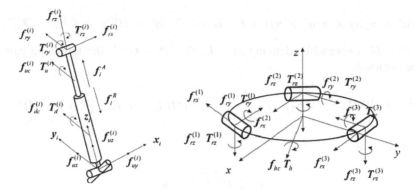

Fig. 4. Free body diagram: (a) The limb. (b) The upper platform.

$$l_i \times f_r^{(i)} + r_u^{(i)} \times f_{uc}^{(i)} + r_d^{(i)} \times f_{dc}^{(i)} + T_r^{(i)} + T_u^{(i)} + T_d^{(i)} = 0, \; i = 1, 2, 3 \qquad (3)$$

In the equation, f_i^A, f_i^R are the force that respectively effect on the upper linkage μ and the lower linkage d from the motor actuator. They are in same numerical value and the opposite direction. f_i^A is the active force of the connecting rod branch.

3.2 Static Model of the Mobile Platform

Assuming all the external force and torque effecting on the mobile platform, through the simplification to the center of mass can be expressed by a six-dimensional vector F_h, that is, $F_h = (f_h \, F_h)^T = (f_{hx} f_{hy} f_{hz} T_{hx} T_{hy} T_{hz})^T$, which is the external force of the mechanism. The force diagram of the mobile platform is shown in Fig. 4(b). Mobile platform receives constraints from three revolute joints and each revolute joint provides constrains of the force $f_r^{(i)}$ and the torque $T_r^{(i)}$ ($i = 1, 2, 3$). During the procession of the analysis of the mobile platform, the six mechanical and torque equilibrium equations can be obtained:

$$f_h - \sum_{i=1}^{3} f_r^{(i)} = f_h - (Cf^A + Df_{rx} + E) = 0 \qquad (4)$$

$$T_h - \sum_{i=1}^{3} a_i^c \times f_r^{(i)} - T_r^{(i)} = T_h - (Ff^A + Gf_{rx} + H) = 0 \qquad (5)$$

Equations (4) and (5) have six equations. The six unknown quantities are the reaction force $f_{rx}^{(1)} f_{rx}^{(2)} f_{rx}^{(3)}$ along the x_i-axis and the active force $f_1^A f_2^A f_3^A$.

In that equilibrium equations, $f^A = (f_1^A \, f_2^A \, f_3^A)^T$, $f_{rx} = (f_{rx}^{(1)} \, f_{rx}^{(2)} \, f_{rx}^{(3)})^T$, $C = -(z_1 \; z_2 \, z_3)$, $D = (u_1 \, u_2 \, u_3)$, $E = \sum_{i=1}^{3} (f_{ry}^{(i)} y_i + f_{rz}^{(i)} z_i)$, $F = (a_1^c \times z_1 \, a_2^c \times z_2 \, a_3^c \times z_3)$,

$$G = (a_1^c \times u_1 a_2^c \times u_2 a_3^c \times u_3) + I, \quad H = - \sum_{i=1}^{3} [a_1^c \times (f_{ry}^{(i)} y_i + f_{rz}^{(i)} z_i) - T_{rz}^{(i)}]. \quad \text{If}$$

$FC^{-1}D - G$ is reversible, defining $M = D(FC^{-1}D - G)^{-1}$, the following equation can be obtained:

$$f^A = C^{-1}[MFC^{-1} + E_{3\times3}M]\begin{bmatrix} f_h \\ T_h \end{bmatrix} - C^{-1}(MFC^{-1} + E_{3\times3})E + C^{-1}MH \quad (6)$$

Rewritten as:

$$f^A = J^T \begin{bmatrix} f_h \\ T_h \end{bmatrix} + K \quad (7)$$

The above equation describes the relationship between the motivation and the load and the load torque, which is the statics equation of the UPR-2UPRR parallel mechanism. J^T is defined as the Jacobi matrix, K is defined as the additional matrix generated by considering the external force and torque of the bar itself.

3.3   Example of the Static Analysis

The quality of related parts can be calculated through the software. The mass of the mobile platform is 0.223 kg, the quality of the fixed connecting rod is $m_1 = 0.353$ kg, the quality of the telescopic connecting rod is $m_2 = 0.451$ kg. The structure size is: the radius of moving platform is $r = 38$ mm, the radius of the fixed platform is $R = 120$ mm, the starting position for the mobile platform is parallel with the base platform, the coordinate of the mobile platform center compared with the center O of the base platform is $[0 \quad 0 \quad 400]^T$, the length of the rod is $l_u = 0.06$ m, $l_d = 0.28$ m (Table 1).

Table 1. The calculation result of the constraining force

Branch		f_r/N	$T_r/N \cdot m$	f_p/N	$T_P/N \cdot m$	f_u/N	$T_u/N \cdot m$
First	x	−43.44	0	43.44	0.03	43.44	0
	y	−0.48	14.77	−0.33	−17.38	−1.27	0
	z	−65.83	0	−	0	70.83	0
Second	x	61.82	0	−61.82	0.03	−61.82	0
	y	−0.48	−21.02	−0.33	24.73	−1.27	0
	z	14.25	0	−	0	−9.25	0
Third	x	−37.78	0	37.78	0.03	37.78	0
	y	−0.47	13.06	−0.33	15.33	−1.27	0
	z	54.38	0	−	0	−49.38	0

The calculation result of the active force is: $f_1^A = 65.83, f_2^A = 14.25, f_3^A = -54.38$.

4 Dynamic Analysis

4.1 Speed Analysis

The coordinate diagram of the limb is shown in Fig. 5. The fixed coordinate system O-xyz is located at the center of the lower platform, while the moving coordinate system M - $x_m y_m z_m$ is located at the center of the upper platform. The coordinate system U - $x_u y_u z_u$ and L - $x_1 y_1 z_1$ are at the center of the upper platform and lower platform respectively.

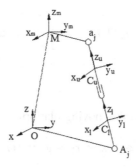

Fig. 5. The coordinate diagram of the limb.

The equation is as follows:

$$\mathbf{L_j} = \mathbf{R} \times \mathbf{r}_{a_j} + \mathbf{T} - \mathbf{r}_{A_j} , j = 1 , 2 , 3 \tag{8}$$

In the equation, $\mathbf{L_j}$ represents the vector $\mathbf{A_j a_j}$; $\mathbf{r}_{a_j}$ represents the position vector of joint a_j in coordinate system M - $x_m y_m z_m$; $\mathbf{r}_{A_j}$ represents the position vector of joint A_j in coordinate system O-xyz; $\mathbf{T}$ represents the position vector of the center of the moving platform in coordinate system O-xyz; $\mathbf{R}$ is the rotation matrix of the coordinate system M - $x_m y_m z_m$ relative to the coordinate system O-xyz.

The velocity of joint a_j in coordinate system O-xyz is as follows:

$$\mathbf{v_{a_j}} = \mathbf{v} + \boldsymbol{\omega} \times \mathbf{r_j} \tag{9}$$

In the equation, $\mathbf{v}$ represents the velocity of the moving platform and $\boldsymbol{\omega}$ represents the angular velocity of the moving platform, where $\mathbf{r_j} = \mathbf{R} \times \mathbf{r}_{a_j}$. The velocity of joint a_j in coordinate system O-xyz is shown as follows:

$$\mathbf{v_{a_j}} = \boldsymbol{\omega}_{l_j} \times \mathbf{L_j} + \dot{l}_j \mathbf{t_j} \tag{10}$$

Where $\mathbf{t_j} = \mathbf{L_j}/l_j$, $\boldsymbol{\omega}_{l_j}$ represents the angular velocity of the limb, l_j represents the length of the limb, $\dot{l}_j$ represents the relative velocity of the telescopic connecting rod and fixed connecting rod, and t_j represents the unit vector of the limb. A new equation can be derived with Eqs. (9) and (10):

$$\omega_{l_j} = \frac{t_j \times v_{a_j}}{l_j} = \frac{t_j \times (v + \omega \times r_j)}{l_j} \tag{11}$$

The velocity of the telescopic connecting rod's center, the fixed connecting rod's center and the lower and upper led could be presented as follows:

$$v = \begin{bmatrix} v_{mu} & v_{ml} & v_r \end{bmatrix}^T \tag{12}$$

Where $v_{mu} = \omega_{l_j} \times \left[(l_j - l_u/2)t_j + \dot{l}_j t_j \right]$, $v_{ml} = \omega_{l_j} \times (l_l/2 \cdot t_j)$, $v_r = v_{a_j} t_j$.

4.2 Acceleration Analysis

The acceleration of the joint on the moving platform is as follows:

$$a_{a_j} = a + \varepsilon \times r_j + \omega \times (\omega \times r_j) \tag{12}$$

$$a_{a_j} = \varepsilon_{l_j} \times L_j + \omega_{l_j} \times (\omega_{l_j} \times L_j) + \ddot{l}_j t_j + 2\omega_{l_j} \times \dot{L}_j t_j \tag{13}$$

In this equation, a is acceleration of the moving platform, ε is angular acceleration of the moving platform. Two equations can be derived with Eqs. (13) and (14):

$$\ddot{l}_j = a_{a_j} \cdot t_j + l_j \omega_{l_j} \cdot \omega_{l_j} \tag{14}$$

$$\varepsilon_{l_j} = (t_j \times a_{a_j} - 2\omega_{l_j} \times t_j)/l_j \tag{15}$$

The acceleration of the fixed and the telescopic connecting rod's center is:

$$a_{lj} = l_l(\varepsilon_{l_j} \times t_j)/2 + l_l[\omega_{l_j} \times (\omega_{l_j} \times t_j)]/2 \tag{16}$$

$$a_{uj} = \left\{ [\omega_{l_j} \times (\omega_{l_j} \times t_j)] + (\varepsilon_{l_j} \times t_j) \right\} \times (l_j - l_u/2) + \ddot{l}_j t_j + 2\omega_{l_j} \times \left(\dot{l}_j t_j \right) \tag{17}$$

4.3 The Analysis of Partial Velocity and Partial Angular Velocity

Normally, the motion of a mechanical system which has N degrees can be described by N generalized coordinates or their velocity. In order to simplify the analysis process, defining the generalized coordinates as $q = [x, y, z, \alpha, \beta, \theta]^T$. In this coordinate, $[x, y, z]$

is the position of the moving coordinate in fixed coordinate, $[\alpha, \beta, \theta]$ is the attitude of the moving coordinate. Thus the system's generalized velocity can be described as follows:

$$\dot{\mathbf{q}} = \left[\dot{x}, \dot{y}, \dot{z}, \dot{\alpha}, \dot{\beta}, \dot{\theta}\right] = \left[v_x, v_y, v_z, \omega_x, \omega_y, \omega_z\right] \tag{18}$$

The partial angular velocity of the limb can be derived from $\boldsymbol{\omega}_{l_j}$, while the partial velocity of the telescopic and the fixed connecting rod can be derived from $\mathbf{v}_{mu}, \mathbf{v}_{ml}$. And the partial velocity of the relative velocity of the telescopic connecting rod and fixed connecting rod can be derived from $\mathbf{v}_r$.

The partial velocity and the partial angular velocity of moving platform's center is:

$$\mathbf{v}'_c = [\mathbf{I}_3; \mathbf{O}_3] \tag{20}$$

$$\boldsymbol{\omega}'_c = [\mathbf{O}_3; \mathbf{I}_3] \tag{9}$$

4.4   Generalized Active Force and Generalized Inertial Force

The generalized velocity of this system includes 6 components, the generalized active force is $\mathbf{F}^a_j$ and generalized inertial force is $\mathbf{F}^b_j$; the mass of the moving platform, the fixed connecting rod and the telescopic connecting rod is m, m_l, m_u, respectively. The fixed connecting rod and telescopic connecting rod of the 3 limbs are same. The driving force of the 3 limbs is $\mathbf{F}_j (j = 1, 2, 3)$, the Gravitational acceleration is $\mathbf{g}$.

The generalized active force which corresponds to the generalized velocity v_x, v_y, v_z and the generalized angular velocity $\omega_x, \omega_y, \omega_z$ can be described as:

$$\mathbf{F}^a_i = m\mathbf{g} \cdot \mathbf{v}'_{c,i} + \sum_{j=1}^{3} \mathbf{F}_j \left(\frac{\dot{l}}{j} \mathbf{t}_j\right)'_j + \sum_{j=1}^{3} \left(m_l \mathbf{g} \cdot \mathbf{v}'_{lj,i} + m_u \mathbf{g} \cdot \mathbf{v}'_{uj,i}\right), \ i = 1, 2, 3 \tag{22}$$

$$\mathbf{F}^a_k = m\mathbf{g} \cdot \mathbf{v}'_{c,k} + \sum_{j=1}^{3} \mathbf{F}_j \left(\frac{\dot{l}}{j} \mathbf{t}_j\right)'_k + \sum_{j=1}^{3} \left(m_l \mathbf{g} \cdot \mathbf{v}'_{lj,k} + m_u \mathbf{g} \cdot \mathbf{v}'_{uj,k}\right), \ k = 4, 5, 6 \tag{10}$$

In the equation, $\mathbf{v}'_{c,i}$ represents the number i column of the vector of the partial velocity of moving platform's center, $\mathbf{v}'_{c,k}$ represents the number k column of the vector of the partial velocity of moving platform's center. The generalized inertial force which corresponds to the generalized velocity v_x, v_y, v_z can be described as:

$$\mathbf{F}^b_i = -m\mathbf{a} \cdot \mathbf{v}'_{c,i} - \sum_{j=1}^{3} \left(m_l \mathbf{a}_{lj} \cdot \mathbf{v}'_{lj,i} + m_u \mathbf{a}_{uj} \cdot \mathbf{v}'_{uj,i}\right) - \mathbf{A} \cdot \boldsymbol{\omega}'_{lj,i} - \mathbf{B} \cdot \boldsymbol{\omega}'_{lj,i} \tag{24}$$

In the equation, $A = \sum_{j=1}^{3} \left(I_{lj}\varepsilon_j + \omega_{l_j} \times I_{lj}\omega_{l_j} \right)$, $B = \sum_{j=1}^{3} \left(I_{uj}\varepsilon_j + \omega_{l_j} \times I_{uj}\omega_{l_j} \right)$,
I_{lj}, I_{uj} are the inertial tensor matrix of the fixed connecting rod and telescopic connecting rod. The generalized inertial force which corresponds to the generalized angular velocity $\omega_x, \omega_y, \omega_z$ can be described as:

$$F_k^b = -(I\varepsilon + \omega \times I\omega) \cdot \omega_k' - \sum_{j=1}^{3} \left(m_l a_{lj} \cdot v_{lj,k}' + m_u a_{uj} \cdot v_{uj,k}' \right) - A \cdot \omega_{lj,k}' - B$$
$$\cdot \, \omega_{lj,k}'$$

$$(11)$$

4.5   Dynamic Equation

The Kane equation can be described as: the sum of the Generalized Active Force and Generalized Inertial Force which act on a rigid body j equals 0.

$$F_j^a + F_j^b = 0 \qquad (26)$$

If the partial velocity and partial angular velocity of j are v_j' and ω_j', the active force and active torque which act on j are F_j and M_j, the inertial force and inertial torque that act on j are F_j' and M_j', then two equations can be derived:

$$\begin{cases} F_j^a = F_j v_j' + M_j \omega_j' \\ F_j^b = F_j' v_j' + M_j' \omega_j' \end{cases} \qquad (12)$$

Based on the Kane equation, we have:

$$JF = G \qquad (28)$$

In this equation, J is the Jacobian matrix. If f_j is the vector of joint a_j in coordinate O-xyz system, then we have:

$$f_j = R \times r_{a_j} + T \qquad (29)$$

If g_j represents the vector of joint A_j in coordinate system O - xyz, then we have:

$$J = \begin{bmatrix} \frac{g_1 - f_1}{|g_1 - f_1|} & \frac{g_2 - f_2}{|g_2 - f_2|} & \frac{g_3 - f_3}{|g_3 - f_3|} & e_1 & e_2 & e_3 \\ \frac{f_1 \times g_1}{|f_1 \times g_1|} & \frac{f_2 \times g_2}{|f_2 - g_2|} & \frac{f_3 \times g_3}{|f_3 - g_3|} & 0 & 0 & 0 \end{bmatrix} \qquad (30)$$

In the equation, $\mathbf{e}_1 = [1, 0, 0]^T$, $\mathbf{e}_2 = [0, 1, 0]^T$, $\mathbf{e}_3 = [0, 0, 1]^T$. $\mathbf{F}$ represents the force on the moving platform, which is provided by 3 limbs. $\mathbf{G}$ represents the sum of all the components in Eq. (28) except $\mathbf{F}$.

5 Dynamic Simulation

When the surgical instrument which is carried by the moving platform performs the remote center motion under certain resistance, the driving force of each telescopic connecting rod will change. The magnitude and change trend of this force is crucial to the design of the structure and control system. In order to carry the simulation, we exert a remote center movement on the surgical instrument, then we record the relative position change of each telescopic connecting rod. In the third step, we use the data to create 3 spline curves in Adams/PostProcessor, namely s1, s2, s3. After that, we remove the initial movement and apply movement on the motion joint between telescopic connecting rod and fixed connecting rod, making the displacement curves of 3 telescopic connecting rods the same as 3 spline curves. In this step, the functions we use are cubspl (time, 0, s1, 0), cubspl (time, 0, s2, 0), cubspl (time, 0, s3, 0). Then we exert a force $\mathbf{F} = (3, 3, 3)$ N at the end point of the surgical instrument. After the simulation, the driving forces of 3 telescopic connecting rods are shown in Fig. 6.

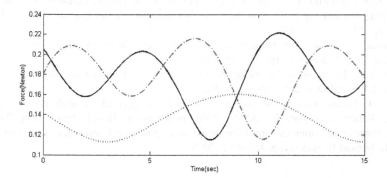

Fig. 6. The driving force of each telescopic rod.

6 Conclusion

In this paper, a 3-DOF parallel robot is proposed, and the design of the structure is available. A physical prototype of the 3-DOF robot is being developed, and the analysis of the statics model is given. Firstly, build the statics model of the branch, the upper linkage and the mobile platform. Then, obtain the equations of the active force and constraining force of the joints. Two numerical analyses are given in the paper. The dynamic model has been built in this paper, by using the Kane's method. The speed, acceleration, partial velocity and partial angular velocity are calculated. After analyzing the generalized active force and inertial force, the dynamic equation is given. A typical curve is given to represent the movement path, and the driving force of each telescopic rod is given.

Acknowledgements. This paper was supported by grants from National Natural Science Foundation of China (No. 61403108). And the work was carried out at State Key Laboratory of Robotics and System, Harbin Institute of Technology. The authors would like to thank all the supports and helps for this research.

References

1. Diana, C.W., Lum, M.J.H., et al.: The RAVEN: design and validation of a telesurgery system. Int. J. Robot. Res. **28**(9), 1183–1197 (2009)
2. Guthart, G.S., Salisbury, Jr., J.K.: The intuitiveTM telesurgery system: overview and application. In: IEEE International Conference on Robotics and Automation, pp. 618–621, San Francisco, CA (2000)
3. Li, J.M., et al.: Kinematic design of a novel spatial remote center-of-motion mechanism for minimally invasive surgical robot. J. Med. Devices **9**, 1–8 (2015)
4. Cho, W., Tesar, D., Freeman, R.A.: The dynamic and stiffness modeling of general robotic manipulator systems with antagonistic actuation. In: Proceedings of the IEEE International Conference on Robotics and Automation, pp. 1380–1387. Scottsdale, USA (1989)
5. Lee, K.M., Johnson, R.: Static characteristics of an in parallel actuated manipulator for clamping and bracing application. In: IEEE International Conference on Robotics Automation, pp. 1408–1413 (1989)
6. Gosselin, C.M.: Stiffness mapping for parallel manipulator. In: IEEE Transactions on Robotics and Automation, pp. 377–382 (1990)
7. Kane, T.R., Levinson, D.A.: Formulation of equations of motion for complex spacecraft. J. Guid. Control Dyn. **3**(2), 99–112 (1980)
8. Nia, H.T., Pishkenari, H.N., Meghdari, A.: A recursive approach for analysis of snake robots using Kane's equations. In: ASME 2005 International Design Engineering Technical Conferences and Computers and Information in Engineering Conference. American Society of Mechanical Engineers, pp. 855–861 (2005)

Accuracy Synthesis of a 3-R2H2S Parallel Robot Based on Rigid-Flexible Coupling Mode

Caidong Wang[1,2(✉)], Yihao Li[1,2], Yu Ning[1], Liangwen Wang[1,2], and Wenliao Du[1]

[1] College of Mechanical and Electrical Engineering,
Zhengzhou University of Light Industry, Zhengzhou 450002, China
vwangcaidong@163.com
[2] Henan Key Laboratory of Mechanical Equipment Intelligent Manufacturing,
Zhengzhou, China

Abstract. On the basis of accuracy analysis of rigid-flexible coupling model of 3-R2H2S parallel robot, the pose error analysis and synthesis is studied systematically in this paper. In terms of rigid analysis, the accuracy mathematical model of 3-R2H2S parallel robot is established by the differential theory based on the inverse solution. The independent and equal effect principle of error is applied to the accuracy synthesis of the robot. Through the computer simulation, the distribution of structure parameter errors is analyzed according to the allowable range of pose error. The simulation results show the active arm error has the most remarkable influence on the pose error of the parallel robot. As for flexible analysis, the rigid-flexible coupling model of 3-R2H2S is established by the software ADAMS and ANSYS. The simulation results of rigid-flexible coupling model show the elastic deformation of the fore arms plays a major role in the robot pose error. The methods proposed in this paper have provided theoretical basis for accuracy synthesis for parallel robot of this type.

Keywords: Parallel robot · Accuracy synthesis · Rigid-flexible coupling mode · Error analysis

1 Introduction

Parallel robots have many advantages, such as higher stiffness, strong payload capacity, small error, high precision, small ratio of weight load, good dynamic performance, easy to control and so on. Especially the symmetrical parallel mechanism has good isotropy, and it has a good performance in terms of movement and force transmission [1, 2]. According to these characteristics, the parallel robots have been applied in the fields of high rigidity, high accuracy or large load with small workspace. And they are becoming a niche product in several robotics research areas, such as in machine tools, assembly lines, and medical applications. Parallel robots have attracted the attention of many researchers in recent decades.

The accuracy is one of the important indexes to evaluate the working performance of robot. So the accuracy research for the parallel robots has attracted the attention of many researchers. The deviation value between the actual pose and the theory pose of

© Springer International Publishing AG 2017
Y. Huang et al. (Eds.): ICIRA 2017, Part II, LNAI 10463, pp. 429–440, 2017.
DOI: 10.1007/978-3-319-65292-4_37

robot is called the pose error. The main purposes of the analysis of robot pose error are: (a) to establish the pose error model of robot; (b) to design strategy for reducing the error; (c) to allocate the errors reasonably; (d) to improve the positioning accuracy of the end-effector. The accuracy of parallel robot mainly includes accuracy analysis and synthesis problems. The accuracy synthesis is more complex than accuracy analysis, because it is a multi-objective, multivariable and non-linear optimization problem.

At present, the accuracy synthesis of parallel robot has become a hot topic in this field, and has attracted more and more researchers. Kim has studied the analysis and synthesis of the Stewart parallel robot based on range of motion error [3]. Merlet and Daney [4] dealt with the dimensional synthesis of parallel robots with a guaranteed given accuracy over a specific workspace. Yao et al. [5] described a procedure for estimating the accuracy of a parallel robot. The method is based on the estimation of the pose accuracy from the synthesis. Yu et al. [6] established the pose error model based on the inverse kinematics solution model of parallel robot and analyzed the relationship between the error sources and pose errors of parallel robot through the solution of the error transmission matrix. Jauregui presented a methodology for determining the pose accuracy along the workspace of a six degrees-of-freedom parallel robot [7]. The method was applied to estimate the accuracy of a high precision Stewart platform and to verify it experimentally. Tsai introduced a generalized method for error analysis of multi-loop mechanisms with joint clearance [8]. Alexander [9] presented a simple geometric approach to computing the exact local maximum position and orientation error of 3- degrees of freedom (DOF) planar parallel robots. Zhao [10] proposed a method for the dimensional synthesis of the three translational DOF parallel robot while considering the kinematic anisotropic property.

Due to the inherent flexibility of lightweight linkages, inertial forces acting on the linkages due to high-acceleration motion, will lead to unwanted elastic deformation. The deformation will affect the accuracy of the parallel robot. But there are few studies focusing on the elastic deformation. Thus, this paper mainly studies the effect of structure parameters and the elastic deformation on the pose accuracy of the parallel robot's end-effector. On the basis of the accuracy analysis of the rigid robot, combining with the simulation of the flexible robot, a comprehensive analysis for the robot pose errors under a variety of factors was performed in this paper. First, the 3-R2H2S parallel robot is considered as an ideal rigid body, and the accuracy mathematical model is established by the differential theory based on the inverse solution. The distribution of structure parameter errors is analyzed according to the allowable range of pose error, and the correctness of the accuracy model is verified by the numerical simulation through matlab software. The algorithm is simple and viable, and it is of great significance for it changes parallel robot accuracy synthesis problem from a multi-object, multivariable and non-linear problem into a simple linear problem [11]. Then the rigid-flexible coupling model of 3-R2H2S robot is established based on the software ADAMS and ANSYS. The elastic deformation influence of the arms on the pose error is analyzed by the flexible dynamic simulation.

2 Inverse Solution of Position

The 3-D model of 3-R2H2S robot is shown in Fig. 1. Compared with structure of the DELTA robot, the connection joints of active and passive arm are replaced by two Hooke joints. The static platform and moving platform are connected through three same branches, and the branches are spread uniformly. Each branch concludes one rotational joint (R joint), two Hooke joints (H joint) and two spherical joints (S joint) connected in sequence. The simulation result shows that the new structure not only overcomes the traditional structure's problem that the connecting bar which is connected by two spherical joints rotates around its central axis [12], but also moves smoothly and has larger workspace and good kinematics performance.

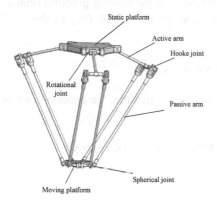

Fig. 1. 3-D model of 3-R2H2S robot.

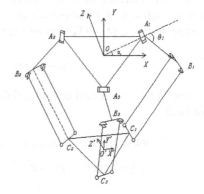

Fig. 2. Structure diagram of 3-R2H2S.

First, the 3-R2H2S parallel robot is considered as an ideal rigid body. The inversion solution of the robot is solved under the condition that there is no elastic deformation in the motion process.

As is shown in Fig. 2, setting the circumscribed circle radius of the base platform is R, the circumscribed circle radius of the moving platform is r, $A_iB_i = l_a$, $B_iC_i = l_b$. The fixed coordinate frame $O - XYZ$ and the moving coordinate frame $O' - X'Y'Z'$ is attached to the geometric center of the base and the moving platform, respectively.

According to the geometrical relationship, the position vector of the rotational joint's center A_i in fixed coordinate system $O - XYZ$ is:

$$a_{io} = R(\cos \alpha_i, \sin \alpha_i, 0)^T \quad (i = 1, 2, 3) \tag{1}$$

Where α_i represents the angle between OA_i and the positive direction of X axis. $\alpha_i = (4i - 3)\pi/6 \quad (i = 1, 2, 3)$

Similarly, the position vector of the equivalent center C_i on the moving platform in moving coordinate system can be obtained:

$$c_{io'} = r(\cos \alpha_i', \sin \alpha_i', 0)^T \quad (i = 1, 2, 3) \tag{2}$$

Where α_i' represents the angle between $O'C'$ and the positive direction of X' axis. $\alpha_i = (4i - 3)\pi/6$ $(i = 1, 2, 3)$

Due to the fact that the structure of base and moving platform are designed analogously, thus $\alpha_i = \alpha_i'$.

Assuming that the rotate angle form OA_i to A_iB_i is θ_i. Thus from the geometrical relationship, and then the position vector of B_i in fixed coordinate system can be obtained:

$$b_{io} = [(R + l_a \cos \theta_i) \cos \alpha_i, (R + l_a \cos \theta_i) \sin \alpha_i, -l_a \sin \theta_i]^T \tag{3}$$

Assuming that the coordinate of the moving platform center is (x, y, z) with respect to the fixed reference frame $O - XYZ$. The rotate angle of the moving platform relative to base platform in the direction of Z axis is zero, so the vector OC_i in the fixed coordinate frame $O - XYZ$ can be expressed as:

$$c_{io} = [r \cos \alpha_i + x, r \sin \alpha_i + y, z]^T \tag{4}$$

For $|B_iC_i| = l_b$, based on the definition of the vector module, the equation can be obtained:

$$[(R + l_a \cos \theta_i - r) \cos \alpha_i - x]^2 + [(R + l_a \cos \theta_i - r) \sin \alpha_i - y]^2 \\ + (-l_a \sin \theta_i - z)^2 = l_b^2 \tag{5}$$

3   Error Model of 3-R2H2S Robot

From the Eq. (5), the inverse solution of the robot can be expressed as:

$$\begin{cases} f_1(x, y, z, R, r, l_{1a}, l_{1b}, \theta_{11}) = 0 \\ f_2(x, y, z, R, r, l_{2a}, l_{2b}, \theta_{21}) = 0 \\ f_3(x, y, z, R, r, l_{3a}, l_{3b}, \theta_{31}) = 0 \end{cases} \tag{6}$$

Where l_{ia}, l_{ib} and θ_{i1} $(i = 1, 2, 3)$ represent the length of active arm, the length of driven arm, the pendulum angle of the active arm in the ith chain, respectively.

The position errors that caused by the structure parameters of the robot can be calculated by the differential theory [13]. And then the Eq. (6) can be written in matrix form by partial differentiation.

$$\begin{bmatrix} dx \\ dy \\ dz \end{bmatrix} = \begin{bmatrix} \frac{\partial f_1}{\partial x} & \frac{\partial f_1}{\partial y} & \frac{\partial f_1}{\partial z} \\ \frac{\partial f_2}{\partial x} & \frac{\partial f_2}{\partial y} & \frac{\partial f_2}{\partial z} \\ \frac{\partial f_3}{\partial x} & \frac{\partial f_3}{\partial y} & \frac{\partial f_3}{\partial z} \end{bmatrix}^{-1} \times \begin{bmatrix} -\frac{\partial f_1}{\partial R} dR - \frac{\partial f_1}{\partial r} dr - \frac{\partial f_1}{\partial l_{1a}} dl_{1a} - \frac{\partial f_1}{\partial l_{1b}} dl_{1b} - \frac{\partial f_1}{\partial \theta_{11}} d\theta_{11} \\ -\frac{\partial f_2}{\partial R} dR - \frac{\partial f_2}{\partial r} dr - \frac{\partial f_2}{\partial l_{2a}} dl_{2a} - \frac{\partial f_2}{\partial l_{2b}} dl_{2b} - \frac{\partial f_2}{\partial \theta_{21}} d\theta_{21} \\ -\frac{\partial f_3}{\partial R} dR - \frac{\partial f_3}{\partial r} dr - \frac{\partial f_3}{\partial l_{3a}} dl_{3a} - \frac{\partial f_3}{\partial l_{3b}} dl_{3b} - \frac{\partial f_3}{\partial \theta_{31}} d\theta_{31} \end{bmatrix} \tag{7}$$

Where

$$
\begin{bmatrix}
\frac{\partial f_1}{\partial x} & \frac{\partial f_1}{\partial y} & \frac{\partial f_1}{\partial z} \\
\frac{\partial f_2}{\partial x} & \frac{\partial f_2}{\partial y} & \frac{\partial f_2}{\partial z} \\
\frac{\partial f_3}{\partial x} & \frac{\partial f_3}{\partial y} & \frac{\partial f_3}{\partial z}
\end{bmatrix}
=
\begin{bmatrix}
x-(R+l_{1a}\cos\theta_{11}-r)\cos\alpha_1 & y-(R+l_{1a}\cos\theta_{11}-r)\cos\alpha_1 & z+l_{1a}\sin\theta_{11} \\
x-(R+l_{2a}\cos\theta_{21}-r)\cos\alpha_2 & y-(R+l_{2a}\cos\theta_{21}-r)\cos\alpha_2 & z+l_{2a}\sin\theta_{21} \\
x-(R+l_{3a}\cos\theta_{31}-r)\cos\alpha_3 & y-(R+l_{3a}\cos\theta_{31}-r)\cos\alpha_3 & z+l_{3a}\sin\theta_{31}
\end{bmatrix}
$$ is the

inverse Jacobi matrix J^{-1} which is related to the structure size and pose of 3-R2H2S parallel robot. When the parallel robot is in the singular pose, it means $|J^{-1}|=0$. The Eq. (7) is meaningless, because of the moving platform is in the singular pose. And the mechanism will lose more than one degree of freedom, and the driven motor of the parallel mechanism can't transmit any horizontal force on the moving platform. So in order to avoid singular pose, we must guarantee $|J^{-1}| \neq 0$ when designing this type of robot.

The Eq. (7) can be written as

$$
\begin{pmatrix} dx \\ dy \\ dz \end{pmatrix} = J\Delta\varepsilon
\tag{8}
$$

Assuming that,

$$
h = [R, r, l_{ia}, l_{ib}, \theta_i] \ i = 1,2,3
\tag{9}
$$

To facilitate the calculation, J can be written as $J = \begin{bmatrix} a_{11} & a_{12} & a_{13} \\ a_{21} & a_{22} & a_{23} \\ a_{31} & a_{32} & a_{33} \end{bmatrix}$

Calculating partial differential to arbitrary element of both sides of Eq. (9), such as to R, the result is $\frac{\partial f}{\partial h}dh = \frac{\partial f}{\partial R}dR$.

Then the Eq. (8) can be written as

$$
dx = \sum_{i=1}^{3} -a_{1i}\sum_{j=1}^{5}\frac{\partial f_i}{\partial h_{ij}}dh_{ij}, \ dy = \sum_{i=1}^{3} -a_{2i}\sum_{j=1}^{5}\frac{\partial f_i}{\partial h_{ij}}dh_{ij}, \ dz = \sum_{i=1}^{3} -a_{3i}\sum_{j=1}^{5}\frac{\partial f_i}{\partial h_{ij}}dh_{ij}
$$
$$
\tag{10}
$$

Where $i = 1,2,3$, $j = 1,2,3,4,5$, $\frac{\partial f_i}{\partial h_{ij}}dh_{ij}$ denotes the partial differential of j element in h_i.

4 Accuracy Synthesis and Simulation of 3-R2H2S Robot

The 3-R2H2S parallel robot has three same branches, and each branch has five original structure errors which are dR, dr, dl_{ia}, dl_{ib} and $d\theta_i$, respectively. Within the workspace, the pose of parallel robot has three component, the corresponding error are dx, dy and dz, respectively. The original structure errors can be assumed as mini quantity for they are much smaller compared with the robot structure parameters. Here assume that the pose errors are minuteness, namely the differential of the errors are tend to be infinitely small. The structure errors of the parallel robot are usually random variables that follow the normal distribution in the actual application. And for most cases, the random errors of the structures are independent with each other. If not, they may have weak correlation at most. In view of existence and rationality of this situation, thus the original structure errors can be assumed as mini quantity for they are smaller compared with the robot structure parameters.

In this paper, the correlation of structure errors is not considered, and the robot pose errors is analyzed based on the independent principle. The error range of the structure size is also distributed according to the principle that the original errors have equal effect on the pose accuracy.

From the Eq. (7), there are five original errors in each chain of the robot. Assuming that the original errors are only exist in the row i and column j. According to Eq. (10), the follow equations can be obtained:

$$dh_{ij}(x_{max}) = \frac{dx}{-a_{1i}\sum_{j=1}^{5}\frac{\partial f_i}{\partial h_{ij}}}, dh_{ij}(y_{max}) = \frac{dy}{-a_{2i}\sum_{j=1}^{5}\frac{\partial f_i}{\partial h_{ij}}}, dh_{ij}(z_{max}) = \frac{dz}{-a_{3i}\sum_{j=1}^{5}\frac{\partial f_i}{\partial h_{ij}}} \quad (11)$$

Where $i = 1, 2, 3, j = 1, 2, 3, 4, 5$

Actually, the pose error of the 3-R2H2S parallel robot is caused by the 15 original errors. According to the equal effect principle of the original errors, the follow equations can be obtained:

$$dh_{ij}(x_{max}) = \frac{dx}{-15a_{1i}\sum_{j=1}^{5}\frac{\partial f_i}{\partial h_{ij}}}, dh_{ij}(y_{max}) = \frac{dy}{-15a_{2i}\sum_{j=1}^{5}\frac{\partial f_i}{\partial h_{ij}}}, dh_{ij}(z_{max}) = \frac{dz}{-15a_{3i}\sum_{j=1}^{5}\frac{\partial f_i}{\partial h_{ij}}}$$

$$(12)$$

Where, $i = 1, 2, 3, j = 1, 2, 3, 4, 5$.

The above results are the accuracy synthesis from a component pose accuracy index of the parallel robot. The comprehensive accuracy indexes can be written as

$$|dh_{ij}| = \min\{|dh_{ij}(x)|, |dh_{ij}(y)|, |dh_{ij}(z)|\} \quad (13)$$

Where $dh_{ij} = (-|dh_{ij}|, |dh_{ij}|)$ is the accuracy synthesis result of the 3-R2H2S parallel robot.

According to the above algorithm, the parameters of the structure and the initial position accuracy are determined as follow: $R = 105\,mm$, $r = 50\,mm$, $l_{ia} = 220\,mm$, $l_{ib} = 495\,mm$, $dx = 0.3\,mm$, $dy = 0.3\,mm$, $dz = 0.5\,mm$. Accuracy synthesis was simulated by MATLAB software. Assuming that the end-effector's workspace is varying within the range of $z = -500$, $-200 \leq x \leq 200$, $-200 \leq y \leq 200$, we can get the allowable errors of the structure parameters. The simulation results are shown in Figs. 3, 4, 5, 6 and 7, in which the z axis denotes the absolute value of allowable errors.

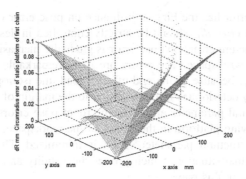

Fig. 3. Cricumradius error of static platform.

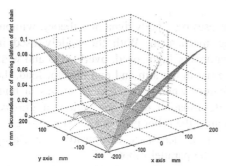

Fig. 4. Cricumradius error of moving platform.

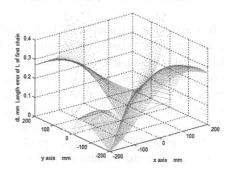

Fig. 5. Length error of l_{1a}.

According to the simulation results, the allowable manufacturing errors of dR, dr, dl_{1a}, dl_{1b} and $d\theta_1$ are 0.1 mm, 0.09 mm, 0.32 mm, 0.055 mm and 0.003 rad, respectively. Form the simulation results, the following conclusions can be achieved.

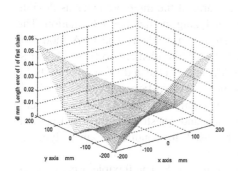

Fig. 6. Length error of l_{1b}.

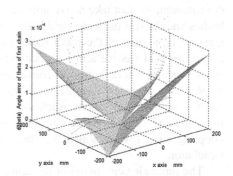

Fig. 7. Allowable control angle error of first chain.

The active arm error has the biggest influence on pose error of the parallel robot. Thus, the active arm error dl_{ia} should be strictly controlled in design process.

The effect of structure errors on pose error is gradually increasing along the axis of $(y = x)$ in the robot workspace. The errors are more remarkable at the edges of the workspace. In reality, because of the end effector load, the isotropic of parallel robot mechanism gradually deteriorates from the center to the margin of the workspace. And the impact of structural parameter errors on the robot pose errors is marked increasingly. When the moving platform in the center of the workspace, the isotropic is best and the impact of structural parameter errors is minimized. The simulation results coincide with the actual situation, which proves the validity and effectiveness of the analysis method used in this paper.

5 Elastic Deformation Analysis of 3R2H2S Robot

The conventional dynamics analysis of robot uses rigid body dynamics theory, without considering the elastic deformation of the mechanism. With the widely used in industrial production, the robot keeps the features of high speed, overloaded and wide range of movement. Because of the inertia exist in the arm and end effector load, the arms are generally has a certain flexible feature when parallel robot moves at high speed. The members will produce elastic deformation under the action of gravity, inertia force and workload, and which can cause the robot pose error. To improve the motion accuracy of the high speed robot, it is necessary to consider the elastic deformation of the members when designing the parallel robot.

Here the combination of finite element method software ANSYS and dynamics analysis software ADAMS is applied to build the rigid-flexible coupling model of 3-R2H2S. The robot motion process is simulated, and the influence of the elastic deformation on robot trajectory is studied. Through the flexible simulation, the impact of the members' elastic deformation on the pose error of end-effector is analyzed by comparing it with pure rigid analysis on the robot pose error.

Because the deformation error of end-effector is mainly caused by the arms' deformation, we just take active arm and passive arm of the three branches as flexible bodies and the rest parts are rigid bodies when analyzing the flexible deformation. The material properties of the robot members are shown in Table 1. The FEA model of active arm and passive arm has been established by finite element software ANSYS. After meshing and definition of contact points, the arm is changed as flexible body by its physical properties. Then the modal neutral file that ADAMS needs can be obtained. After a series of operations, such as, importing neutral file into ADAMS software, replacing the original rigid arm, and setting load and constraints, the rigid-flexible coupling model is established, which can be used to conduct flexible dynamic simulation.

The finite element model of the arm is shown in Fig. 8. The flexible arms files are imported into ADAMS, and the invalid rigid arms are replaced. Then, the rigid-flexible coupling model of the robot are established in the ADAMS, as is shown in Fig. 9. Considering the effect of the workload during the simulation, a vertical downward force

60 N is added in the center of the robot end-effector. Through the simulation, the deviations of the end-effector center can be obtained by compared with the rigid model simulation.

Table 1. The material properties of the robot members

Item	Material	Elasticity modulus	Density	Poisson ratio
Active arm	Aluminium alloy	7.1×10^{10} pa	2.77×10^{-6} kg/mm^3	0.33
Driven arm	Carbon fiber	1.16×10^{11} pa	1.7×10^{-6} kg/mm^3	0.31
Rest parts	Alloy steel	2×10^{11} pa	7.85×10^{-6} kg/mm^3	0.29

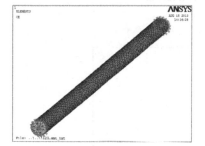

Fig. 8. FEA model of arm in ANSYS

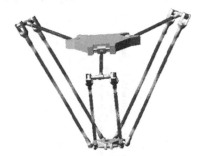

Fig. 9. The rigid-flexible coupling model

Due to the simulation of flexible body model is slow, the simulation time is set to be 20 s. The coordinate values of the end-effector center (x, y, z) are recorded in the process of the rigid model simulation and rigid-flexible coupling model simulation, respectively. Subtracting the pose coordinate values in the corresponding direction of the two simulations, then the pose errors that caused by the deformation of the arms in x, y, z direction can be obtained. The results are shown in Figs 10, 11 and 12. Due to the vibration of the robot at the beginning of movement, the deflection of the end-effector is big and it can't reflect the actually pose error correctly. In order to observe the exact pose errors after the robot moves stability, the error curves in x, y, z direction that in the whole 20 s are given. And the error curves in x direction after 8 s and the error curves in y, z direction after 6 s when the robot runs stability are also given.

According to the simulation results, the pose error is bigger in the first six seconds of the rigid-flexible coupling model simulation, which is due to the influence of vibration at the start step of the robot motion. When the robot moves smoothly in the rest time, the pose error ranges of the end-effector in the x, y, z direction are -0.006–0.003 mm, -0.11–0.14 mm and -0.04–0.25 mm, respectively. Due to the definition of coordinate system, and the vertical downward load added on the end-effector, the error in z direction is biggest, and the error in x direction is smaller. This simulation results are consonant with the theoretical analysis.

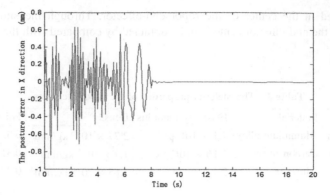

Fig. 10. The pose error in X direction

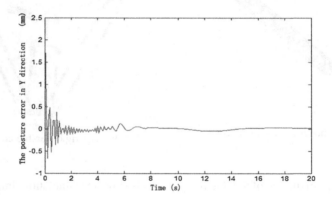

Fig. 11. The pose error in Y direction

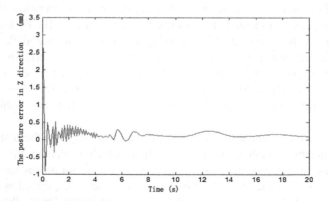

Fig. 12. The pose error in Z direction

From the analysis results of rigid and rigid-flexible coupling models, some conclusions can be obtained as follows,

(1) The active arm error dl_{ia} should be strictly controlled, for it has the biggest influence on pose error compared with other structural parameters.
(2) In order to get higher position control accuracy, the flexible deformation error in the z direction should be compensated in real-time.

6 Conclusions

The influence of the structural parameters and the arms' deformation on the end-effector's pose error of the parallel robot has been comprehensively studied in this paper. The rigid and rigid-flexible coupling models are established. As for rigid accuracy analysis, the error model is established by the differential theory and influence of the structure errors on pose error is analyzed. The algorithm is simple and viable, and it has practical use for it changes parallel robot accuracy synthesis problem from a multi-object, multivariable and non-linear problem into a simple linear problem. The rigid-flexible coupling model is built by the software ANSYS and ADAMS. The simulation results show that the active arm error dl_{ia} should be strictly controlled and the pose error in the z direction caused by the arm deformation should be compensated. The methods proposed in this paper has provided theoretical basis for error compensation and high accuracy control of this kind parallel robot. The further study of the methods is of great significance for designing and improving the transmission performance of parallel robot.

Acknowledgement. This work is supported by Scientific and Technological project of Henan Province (152102210353) and Program for IRTSTHN (2012IRTSTHN013).

References

1. Meng, X.D., Gao, F., Wu, S.F.: Type synthesis of parallel robotic mechanisms: Framework and brief review. Mech. Mach. Theory **78**(4), 177–186 (2014)
2. Feng, L.H., Zhang, W.G., Gong, Z.Y.: Developments of delta-like parallel manipulators – a review. Robot **36**(3), 375–384 (2014)
3. Kim, H.S., Choi, Y.J.: The kinematics error bound analysis of the Stewart platform. J. Rob. Syst. **17**(1), 63–73 (2000)
4. Merlet, J.P., Daney, D.: Dimensional synthesis of parallel robots with a guaranteed given accuracy over a specific workspace. In: Proceedings of the 2005 IEEE International Conference on Robotics and Automation, pp. 942–947. IEEE, Barcelona (2005)
5. Yao, T., Wang, J., Zhou, F., Zhou, X.I.: Accuracy synthesis of a 3-UCR parallel robot using genetic algorithms. J. Theoret. Appl. Inf. Technol. **53**(1), 24–29 (2013)
6. Yu, D.Y., Han, J.W., Li, H.R.: Study on pose errors analysis of parallel robot. J. Harbin Univ. Commer. (Nat. Sci. Ed.) **20**(3), 278–280 (2004)

7. Jauregui-Correa, J.C.: Validation process of pose accuracy estimation in parallel robots. J. Dyn. Syst. Meas. Contr. **137**(6), 484–503 (2015)
8. Tsai, M.J., Lai, T.H.: Accuracy analysis of a multi-loop linkage with joint clearances. Mech. Mach. Theory **43**(9), 1141–1157 (2008)
9. Yu, A., Ilian, A.B., Paul, Z.M.: Geometric approach to the accuracy analysis of a class of 3-DOF planar parallel robots. Mech. Mach. Theory **43**(3), 364–375 (2008)
10. Zhao, Y.J.: Dimensional synthesis of a three translational degrees of freedom parallel robot while considering kinematic anisotropic property. Rob. Comput.-Integr. Manuf. **29**(1), 169–179 (2013)
11. Zhao, Y.J., Zhao, X.H., Ge, W.M.: An algorithm for the accuracy synthesis of a 6-SPS parallel manipulator. Mech. Sci. Technol. **23**(4), 392–395 (2004)
12. Zhu, Y.H., Zhao, J., Cai, H.G.: Configuration design of a kind of new typed double parallel main manipulator. J. Mach. Des. **20**(8), 34–36 (2003)
13. Liu, D.J., Che, R.S., Ye, D.: Error modeling and simulation of 3-DOF parallel link coordinate measuring machine. China Mech. Eng. **12**(7), 752–755 (2001)

A Singularity Analysis Method for Stewart Parallel Mechanism with Planar Platforms

Shili Cheng[1](✉) 📵, Guihua Su[1], Xin Xiong[1], and Hongtao Wu[2]

[1] Yancheng Institute of Technology, Yancheng 224051, China
meecsl@126.com
[2] Nanjing University of Aeronautics and Astronautics, Nanjing 210016, China

Abstract. Singularity analysis is an important problem in the field of parallel mechanism, how to obtain a concise and analytical expression of the singularity locus has been the focus of the study for a long time. This paper presents a new method for the singularity analysis of the Stewart platform parallel mechanism. Firstly, the rotation matrix is described by quaternion; utilizing length constraint equations of the extensible limbs and properties of the quaternion, seven equivalent equations are obtained, and the variables of position and orientation are decoupled preliminarily. Secondly, by taking the derivative of seven the equivalent equations with respect to time, a new kind of Jacobian matrix is obtained, which reflects the mapping relationship between the change rate of the lengths of extensible limbs and the change rate of variables of position and orientation of the moving platform. Finally, the analytical expression of the singularity locus is derived from calculating the determinant of the new Jacobian matrix; when the quaternion are transformed into Rodriguez parameters, there are only 258 items in the fully expanded analytical expression of the singularity locus. Not only can this method be used to study the singularity problem of Stewart platform parallel mechanism, but it can also be used to study the singularity free workspace of the mechanism.

Keywords: Stewart platform · Parallel mechanism · Singularity analysis · Quaternion · Jacobian matrix

1 Introduction

It is well known that Stewart Platform parallel mechanism (also known as Gough platform), the typical representative of the six degrees of freedom spatial mechanism, is composed of a fixed base and a moving platform driven by six extensible limbs. Every limb is connected to the fixed base by spherical joint and to the moving platform by a universal joint. While working, the moving platform obtains three translational degrees of freedom and three rotational degrees of freedom by changing the lengths of the six extensible limbs, while the fixed base remains static [1]. Compared with the traditional serial mechanism, parallel mechanism possesses many advantages, such as higher precision of mobility, lower inertia, higher stiffness, larger payload capacity and better dynamic performance etc. It has been widely applied in the field of airplane simulators, parallel kinematic machines, parallel robots, micro displacement positioning devices, and medical and entertainment equipment, etc. [2].

© Springer International Publishing AG 2017
Y. Huang et al. (Eds.): ICIRA 2017, Part II, LNAI 10463, pp. 441–452, 2017.
DOI: 10.1007/978-3-319-65292-4_38

Since the mid-20th century, parallel mechanism has become a hot research topic in the field of mechanism. Many scholars at home and abroad have been studying the Stewart Platform parallel mechanism from different aspects, such as kinematics analysis, singularity analysis, workspace and dexterity, dynamics and control. In both theoretical research and engineering applications, great progress has been made in Stewart Platform parallel mechanism. However, there are still a lot of problems having not been solved to this day, especially forward kinematics, singularity and workspace, which have been named as the three basic problems of parallel mechanism by Merlet [1]. When singularity occurs, the moving platform would like to get or lose some extra degrees of freedom in a certain direction. As a result of the moving platform getting some extra degrees of freedom, the position and orientation of the moving platform would be out of control; the driving forces of joint, which is used to balance the load effects on the moving platform, will tend to infinity, parallel mechanism would not work normally, it can even be damaged seriously [3]. Therefore, the singularity analysis of Stewart parallel mechanism cannot be avoided, neither in theoretical research nor engineering application.

There are some methods can be used to analyze the singularity of parallel mechanism, such as Jacobian matrix analysis, Grassman geometry and screw theory, etc. Gosselin has divided the singularity of parallel mechanism into three categories by using Jacobian matrix [4]. In addition, Gosselin studied the singularity representation and the maximal singularity-free zones in the six-dimensional workspace of the general Gough–Stewart platform [5, 6]. Huang obtained the analytical expression of singularity locus, which can be used to analyze both position singularity locus and orientation singularity locus of the 6-SPS parallel mechanism [7, 8]. Coste studied rational parameterization of the singularity locus of Gough–Stewart platform which the fixed base and the moving platform are both general planar hexagons [9]. Karimi studied singularity-free workspace analysis of general 6-UPS parallel mechanisms via Jacobian matrix and convex optimization [10]. Cao studied the position singularity characterization of a special class of the Stewart parallel mechanisms based on the Jacobian matrix [11]. Hunt studied the singularity of parallel mechanism by using the screw theory [12]. Huang analyzed the singularity of parallel mechanism by the method of general linear bundles [13]. Merlet proposed the method of singularity analysis based on Grassmann geometry [14]. Caro analyzed the singularity of a six-dof parallel manipulator using grassmann-cayley algebra and Gröbner bases [15]. Doyon studied the Gough–Stewart Platform with constant-orientation and obtained the singularity locus by vector expression [16]. Based on screw theory, Liu studied two types of singularity of 6-UCU parallel manipulator which are caused by both the active joints and passive universal joints [17]. Shanker obtained singular manifold of the Stewart platform [18]. Kaloorazi combined the study of maximal singularity-free sphere to the workspace analysis [19]. Some scholars pointed out that singularity can be avoided by kinematic redundancy [20], or by means of suitable control scheme [21], or by means of reconfigurable mass parameters [22], or by means of trajectory optimization and multi-model control law [23]. Some other scholars studied the singularity of different kinds of lower mobility parallel mechanisms by using different methods [24–26].

The existing research results cannot express singular locus concisely or yield higher computational efficiency, especially they cannot distinguish the singularity of given configurations quickly and easily.

The remainder of this paper is organized as follows: in Sect. 2, coordinate parameters of the hinge points and rotation matrix used in this paper are introduced; in Sect. 3, seven equivalent equations are obtained by utilizing length constraint equations of the extensible limbs and properties of the quaternion; in Sect. 4, a new kind of Jacobian matrix is derived from the equivalent equations, and the analytical expression of the singularity locus is derived from calculating the determinant of the new Jacobian matrix; in Sect. 5, a numerical example is introduced to verify the method presented in this paper. Finally, Sect. 6 draws the conclusions.

2   Coordinate Parameters and Rotation Matrix

As shown in Fig. 1, a spatial 6-dof mechanism known as the Stewart parallel mechanism. The hinge points a_i for $i = 1$ to 6 are sketched on a circle symmetrically as the moving platform; while the hinge points b_i for $i = 1$ to 6 are sketched on a circle symmetrically as the fixed base. Due to the symmetry, coordinates of hinge points, neither on the fixed base nor on the moving platform can be described by four parameters, namely, r_1, r_2, θ_1, θ_2, as shown in Table 1.

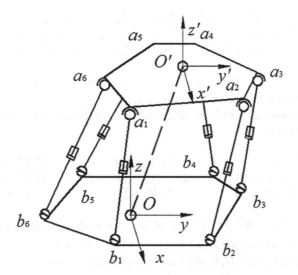

Fig. 1. Stewart Platform parallel mechanism. This mechanism consists of a moving platform, a fixed base and six extendable links. The moving platform is driven by six extendable links $a_ib_i(i = 1 \sim 6)$, and every one of them is connected to the moving platform by a spherical joint and connected to the base by a universal joint. Static coordinate system O-xyz and moving coordinate system O'-$x'y'z'$ are fixed to the base and the moving platform respectively.

Table 1. Coordinate parameters of hinge points.

	$a_{x,k}$	$a_{y,k}$	$b_{x,k}$	$b_{y,k}$
1	$r_2Cos[-\pi/6 - \theta_2]$	$r_2Sin[-\pi/6 - \theta_2]$	$r_1Cos[-\pi/6 - \theta_1]$	$r_1Sin[-\pi/6 - \theta_1]$
2	$r_2Cos[-\pi/6 + \theta_2]$	$r_2Sin[-\pi/6 + \theta_2]$	$r_1Cos[-\pi/6 + \theta]$	$r_1Sin[-\pi/6 + \theta_1]$
3	$r_2Cos[-\pi/2 - \theta_2]$	$r_2Sin[-\pi/2 - \theta_2]$	$r_1Cos[\pi/2 - \theta_1]$	$r_1Sin[\pi/2 - \theta_1]$
4	$r_2Cos[-\pi/2 + \theta_2]$	$r_2Sin[-\pi/2 + \theta_2]$	$r_1Cos[\pi/2 + \theta_1]$	$r_1Sin[\pi/2 + \theta_1]$
5	$r_2Cos[-7\pi/6 - \theta_2]$	$r_2Sin[-7\pi/6 - \theta_2]$	$r_1Cos[7\pi/6 - \theta_1]$	$r_1Sin[7\pi/6 - \theta_1]$
6	$r_2Cos[-7\pi/6 + \theta_2]$	$r_2Sin[-7\pi/6 + \theta_2]$	$r_1Cos[7\pi/6 + \theta_1]$	$r_1Sin[7\pi/6 + \theta_1]$

Therefore, coordinates of the hinge points on the moving platform can be expressed in the moving coordinate system $O'\text{-}x'y'z'$ as:

$$a_k = \begin{pmatrix} a_{x,k} & a_{y,k} & 0 \end{pmatrix}^T \tag{1}$$

Coordinates of the hinge point on the fixed base can be expressed in the static coordinate system $O\text{-}xyz$ as:

$$b_k = \begin{pmatrix} b_{x,k} & b_{y,k} & 0 \end{pmatrix}^T \tag{2}$$

The rotation matrix, which is described by quaternion [27], is shown as:

$$R = \begin{bmatrix} \varepsilon_0^2 + \varepsilon_1^2 - \varepsilon_2^2 - \varepsilon_3^2 & 2\varepsilon_1\varepsilon_2 - 2\varepsilon_0\varepsilon_3 & 2\varepsilon_1\varepsilon_3 + 2\varepsilon_0\varepsilon_2 \\ 2\varepsilon_1\varepsilon_2 + 2\varepsilon_0\varepsilon_3 & \varepsilon_0^2 - \varepsilon_1^2 + \varepsilon_2^2 - \varepsilon_3^2 & 2\varepsilon_2\varepsilon_3 - 2\varepsilon_0\varepsilon_1 \\ 2\varepsilon_1\varepsilon_3 - 2\varepsilon_0\varepsilon_2 & 2\varepsilon_2\varepsilon_3 + 2\varepsilon_0\varepsilon_1 & \varepsilon_0^2 - \varepsilon_1^2 - \varepsilon_2^2 + \varepsilon_3^2 \end{bmatrix} \tag{3}$$

where, ε_1, ε_2, ε_3, $\varepsilon_0 \in R$, $i^2 = j^2 = k^2 = -1$, and $ij = -ji = k$, $jk = -kj = i$, $ki = -ik = j$. Similar to unit vectors, $\varepsilon^T\varepsilon = 1$.

The rotation matrix, which is described by the Rodriguez parameters, can be expressed as follows:

$$R = \frac{1}{\Delta} \cdot \begin{bmatrix} U^2 - V^2 - W^2 + 1 & 2UV - 2W & 2UW + 2V \\ 2UV + 2W & -U^2 + V^2 - W^2 + 1 & 2VW - 2U \\ 2UW - 2V & 2VW + 2U & -U^2 - V^2 + W^2 + 1 \end{bmatrix} \tag{4}$$

where, $\Delta = U^2 + V^2 + W^2 + 1$, U, V, W are the Rodriguez parameters. It can be proved that the transformation relationship between quaternion $\varepsilon = (\varepsilon_1, \varepsilon_2, \varepsilon_3, \varepsilon_0)^T$ and the Rodriguez parameters (U, V, W) can be expressed as follows:

$$\varepsilon_0 = \frac{1}{\sqrt{\Delta}}; \varepsilon_1 = \frac{U}{\sqrt{\Delta}}; \varepsilon_2 = \frac{V}{\sqrt{\Delta}}; \varepsilon_3 = \frac{W}{\sqrt{\Delta}} \tag{5}$$

The position of the origin of the moving coordinate system respect to the static coordinate system can be described by a vector $P = \begin{pmatrix} P_x & P_y & P_z \end{pmatrix}^T$. Then linkage vector between a pair of hinge points is written as:

$$l_k e_k = P + R \cdot a_k - b_k \qquad (k = 1 \sim 6) \tag{6}$$

where, l_k is the length of kth extensible limb, e_k is the unit vector along the axis of kth extensible limb, a_k is the position vector of the hinge point of the moving platform in the moving coordinate system, b_k is the position vector of the hinge point of the fixed base in the static coordinate system, P is the position vector of the reference point of the moving platform in the static coordinate system, R is the rotation matrix.

3 Construction of Equivalent Equations

In order for the elements of the Jacobian matrix to have a simple form, the rotation matrix is described by the quaternion as shown in Eq. (3). Substituting the position vectors a_k, b_k and P, the length of kth extensible limb can be obtained by taking the dot product of $l_k e_k$ with itself. After reorganizing, because the z component of a_k and b_k is zero, square linkage length equation, it can be expressed as (for conciseness of expression, omit the subscripts k):

$$\begin{aligned} l^2 - r_1^2 - r_2^2 &= P_P - 2b_x P_x - 2b_y P_y + 2a_x W_x + 2a_y W_y - 2(\varepsilon_1^2 - \varepsilon_2^2)(a_x b_x - a_y b_y) \\ &\quad - 2(\varepsilon_0^2 - \varepsilon_3^2)(a_x b_x + a_y b_y) + 2(2\varepsilon_0 \varepsilon_3)(a_y b_x - a_x b_y) - 2(2\varepsilon_1 \varepsilon_2)(a_y b_x + a_x b_y) \end{aligned} \tag{7}$$

where, $\quad P_P = P_x^2 + P_y^2 + P_z^2, \quad W_x = P_x(\varepsilon_0^2 + \varepsilon_1^2 - \varepsilon_2^2 - \varepsilon_3^2) + P_y(2\varepsilon_1 \varepsilon_2 + 2\varepsilon_0 \varepsilon_3) + P_z$
$(2\varepsilon_1 \varepsilon_3 - 2\varepsilon_0 \varepsilon_2), \quad W_y = P_x(2\varepsilon_1 \varepsilon_2 - 2\varepsilon_0 \varepsilon_3) + P_y(\varepsilon_0^2 - \varepsilon_1^2 + \varepsilon_2^2 - \varepsilon_3^2) + P_z(2\varepsilon_2 \varepsilon_3 + 2\varepsilon_0 \varepsilon_1)$.
It can be seen from Eq. (7) that P_P, P_x, P_y, W_x, W_y, $\varepsilon_0^2 - \varepsilon_3^2$, $\varepsilon_1^2 - \varepsilon_2^2$, $2\varepsilon_0 \varepsilon_3$ and $2\varepsilon_1 \varepsilon_2$ can be regarded as nine new unknown variables. These nine unknown variables can be arranged into two groups, namely $\eta_1 = \begin{pmatrix} P_P & P_x & P_y & W_x & W_y & 2\varepsilon_0 \varepsilon_3 \end{pmatrix}^T$ and $\eta_2 = \begin{pmatrix} \varepsilon_0^2 - \varepsilon_3^2 & \varepsilon_1^2 - \varepsilon_2^2 & 2\varepsilon_1 \varepsilon_2 \end{pmatrix}^T$. Equation (7) is equivalent to the expressions below:

$$\begin{aligned} P_P &= P_{P0} + k_0 \left(\varepsilon_0^2 - \varepsilon_3^2 \right) \\ P_X &= P_{x0} + 2k_1 \varepsilon_1 \varepsilon_2 \\ P_Y &= P_{y0} + k_1 \left(\varepsilon_1^2 - \varepsilon_2^2 \right) \\ W_X &= W_{x0} + 2k_2 \varepsilon_1 \varepsilon_2 \\ W_Y &= W_{y0} + k_2 \left(\varepsilon_1^2 - \varepsilon_2^2 \right) \\ 2\varepsilon_0 \varepsilon_3 &= C_0 \end{aligned} \tag{8}$$

where parameters k_0, k_1, k_2 are all constants, which are determined by parameters $(r_1, r_2, \theta_1, \theta_2)$ of hinge points on both the fixed base and the moving platform; that is to say, once the structural parameters of the mechanism are given, parameters k_0, k_1, k_2 are all constants.

$$k_0 = 2r_1r_2Cos[\theta_1 - \theta_2]; k_1 = -r_2Csc[\theta_1 - \theta_2] Sin[\theta_1 + 2\theta_2];$$
$$k_2 = -r_1Csc[\theta_1 - \theta_2] Sin[2\theta_1 + \theta_2] \tag{9}$$

where the six parameters P_{P0}, P_{x0}, P_{y0}, W_{x0}, W_{y0}, C_0, are determined by parameters $(r_1, r_2, \theta_1, \theta_2)$ of hinge points and the lengths of extensible limbs l_i (for i = 1 to 6); their specific expressions are shown as follows:

$$P_{P0} = \frac{-6r_1^2 - 6r_2^2 + l_1^2 + l_2^2 + l_3^2 + l_4^2 + l_5^2 + l_6^2}{6}$$

$$P_{x0} = \frac{Csc[\theta_1 - \theta_2]((l_1^2 - l_2^2 - 2l_3^2 + 2l_4^2 + l_5^2 - l_6^2)Cos[\theta_2] + \sqrt{3}(l_1^2 + l_2^2 - l_5^2 - l_6^2)Sin[\theta_2])}{12r_1}$$

$$P_{y0} = \frac{Csc[\theta_1 - \theta_2](\sqrt{3}(l_1^2 - l_2^2 - l_5^2 + l_6^2)Cos[\theta_2] - (l_1^2 + l_2^2 - 2l_3^2 - 2l_4^2 + l_5^2 + l_6^2)Sin[\theta_2])}{12r_1}$$

$$W_{x0} = \frac{Csc[\theta_1 - \theta_2]((l_1^2 - l_2^2 - 2l_3^2 + 2l_4^2 + l_5^2 - l_6^2)Cos[\theta_1] + \sqrt{3}(l_1^2 + l_2^2 - l_5^2 - l_6^2)Sin[\theta_1])}{12r_2} \tag{10}$$

$$W_{y0} = \frac{Csc[\theta_1 - \theta_2](\sqrt{3}(l_1^2 - l_2^2 - l_5^2 + l_6^2)Cos[\theta_1] - (l_1^2 + l_2^2 - 2l_3^2 - 2l_4^2 + l_5^2 + l_6^2)Sin[\theta_1])}{12r_2}$$

$$C_0 = \frac{(l_1^2 - l_2^2 + l_3^2 - l_4^2 + l_5^2 - l_6^2)Csc[\theta_1 - \theta_2]}{12r_1r_2}$$

4 Establishment of Singular Locus Equation

When the Stewart platform parallel mechanism is working, the velocity and angular velocity of the moving platform are obtained by controlling the telescopic speeds of the six extensible limbs; the variables of input and output are connected by Jacobian matrix. The singularity of the Stewart platform parallel mechanism can be analyzed by studying the Jacobian matrix. The position and orientation of the moving platform is described by the position vector and quaternion respectively; so the output variables of the moving platform can be expressed with the derivative of the variables of position and orientation, it is $(\dot{P}_X \quad \dot{P}_Y \quad \dot{P}_Z \quad \dot{\varepsilon}_0 \quad \dot{\varepsilon}_1 \quad \dot{\varepsilon}_2 \quad \dot{\varepsilon}_3)$. That is to say, a mapping matrix can be established between the input variables $\dot{l}_i(i = 1 \sim 6)$ and the output variables $(\dot{P}_X \quad \dot{P}_Y \quad \dot{P}_Z \quad \dot{\varepsilon}_0 \quad \dot{\varepsilon}_1 \quad \dot{\varepsilon}_2 \quad \dot{\varepsilon}_3)$. Actually, this mapping matrix is a new kind of Jacobian matrix.

For a specific Stewart platform parallel mechanism, there are only six variables in $P_{P0}, P_{x0}, P_{y0}, W_{x0}, W_{y0}, C_0$, and the others are all constants; so Eq. (8) can be rewritten as:

$$P_{P0} = P_x^2 + P_y^2 + P_z^2 - k_0(\varepsilon_0^2 - \varepsilon_3^2)$$
$$P_{x0} = P_X - 2k_1\varepsilon_1\varepsilon_2$$
$$P_{y0} = P_Y - k_1(\varepsilon_1^2 - \varepsilon_2^2)$$
$$W_{x0} = P_x(\varepsilon_0^2 + \varepsilon_1^2 - \varepsilon_2^2 - \varepsilon_3^2) + P_y(2\varepsilon_1\varepsilon_2 + 2\varepsilon_0\varepsilon_3) + P_z(2\varepsilon_1\varepsilon_3 - 2\varepsilon_0\varepsilon_2) - 2k_2\varepsilon_1\varepsilon_2 \tag{11}$$
$$W_{y0} = P_x(2\varepsilon_1\varepsilon_2 - 2\varepsilon_0\varepsilon_3) + P_y(\varepsilon_0^2 - \varepsilon_1^2 + \varepsilon_2^2 - \varepsilon_3^2) + P_z(2\varepsilon_2\varepsilon_3 + 2\varepsilon_0\varepsilon_1) - k_2(\varepsilon_1^2 - \varepsilon_2^2)$$
$$C_0 = 2\varepsilon_0\varepsilon_3$$

And according to the nature of the quaternion, in order to maintain a unified form with Eq. (11), that is:

$$1 = \varepsilon_0^2 + \varepsilon_1^2 + \varepsilon_2^2 + \varepsilon_3^2 \tag{12}$$

In both Eqs. (11) and (12), there are a total of seven equations about seven variables, that is P_X, P_Y, P_Z, ε_0, ε_1, ε_2, ε_3, which are changing with time as well as the length of the extensible limbs $l_i (i = 1 \sim 6)$. Therefore, the mapping relationship between the variables of input and output can be obtained by taking the derivative of Eqs. (11) and (12) with respect to time yields:

$$\left(\frac{M_l}{\vec{0}} \right) \left(\dot{l}_1 \quad \dot{l}_2 \quad \dot{l}_3 \quad \dot{l}_4 \quad \dot{l}_5 \quad \dot{l}_6 \right)^T = M(P, \varepsilon) \cdot \left(\dot{P}_X \quad \dot{P}_Y \quad \dot{P}_Z \quad \dot{\varepsilon}_0 \quad \dot{\varepsilon}_1 \quad \dot{\varepsilon}_2 \quad \dot{\varepsilon}_3 \right)^T \tag{13}$$

where $\vec{0} \in R^{1 \times 6}$, $M_l \in R^{6 \times 6}$, $M_l(i,j) = f(l_j)$ is a constant when the length of the extensible limbs are given.

$$M(P, \varepsilon) = \begin{pmatrix} P_X & P_Y & P_Z & -k_0\varepsilon_0 & 0 & 0 & k_0\varepsilon_3 \\ 1 & 0 & 0 & 0 & -2k_1\varepsilon_2 & -2k_1\varepsilon_1 & 0 \\ 0 & 1 & 0 & 0 & -2k_1\varepsilon_1 & 2k_1\varepsilon_2 & 0 \\ m_{41} & m_{42} & m_{43} & m_{44} & m_{45} & m_{46} & m_{47} \\ m_{51} & m_{52} & m_{53} & m_{54} & m_{55} & m_{56} & m_{57} \\ 0 & 0 & 0 & \varepsilon_3 & 0 & 0 & \varepsilon_0 \\ 0 & 0 & 0 & \varepsilon_0 & \varepsilon_1 & \varepsilon_2 & \varepsilon_3 \end{pmatrix} \tag{14}$$

where,

$m_{41} = \varepsilon_0^2 + \varepsilon_1^2 - \varepsilon_2^2 - \varepsilon_3^2$

$m_{42} = 2\varepsilon_1\varepsilon_2 + 2\varepsilon_0\varepsilon_3$

$m_{43} = -2\varepsilon_0\varepsilon_2 + 2\varepsilon_1\varepsilon_3$

$m_{44} = 2P_X\varepsilon_0 - 2P_Z\varepsilon_2 + 2P_Y\varepsilon_3$

$m_{45} = 2P_X\varepsilon_1 + 2P_Y\varepsilon_2 - 2k_2\varepsilon_2 + 2P_Z\varepsilon_3$

$m_{46} = -2P_Z\varepsilon_0 + 2P_Y\varepsilon_1 - 2k_2\varepsilon_1 - 2P_X\varepsilon_2$

$m_{47} = 2P_Y\varepsilon_0 + 2P_Z\varepsilon_1 - 2P_X\varepsilon_3$

$m_{51} = 2\varepsilon_1\varepsilon_2 - 2\varepsilon_0\varepsilon_3$

$m_{52} = \varepsilon_0^2 - \varepsilon_1^2 + \varepsilon_2^2 - \varepsilon_3^2$

$m_{53} = 2\varepsilon_0\varepsilon_1 + 2\varepsilon_2\varepsilon_3$

$m_{54} = 2P_Y\varepsilon_0 + 2P_Z\varepsilon_1 - 2P_X\varepsilon_3$

$m_{55} = 2P_Z\varepsilon_0 - 2P_Y\varepsilon_1 - 2k_2\varepsilon_1 + 2P_X\varepsilon_2$

$m_{56} = 2P_X\varepsilon_1 + 2P_Y\varepsilon_2 + 2k_2\varepsilon_2 + 2P_Z\varepsilon_3$

$m_{57} = -2PX\varepsilon_0 + 2P_Z\varepsilon_2 - 2P_Y\varepsilon_3$

Equation (13) states that $M(P, \varepsilon)$ is the mapping transformation matrix between the driving speed and the change rate of position and orientation of the moving platform, namely the Jacobian matrix. When singularity occurs, the determinant of Jacobian matrix should be equal to zero, that is:

$$f = Det[M(P, \varepsilon)] = 0 \tag{15}$$

It can be seen from Eq. (13) that the elements of the Jacobian matrix are very simple monomials; Except for the elements of the 4th and 5th rows, in which rows every element is more complex. This feature of the new Jacobian matrix makes the calculation of the determinant faster, compared to the calculation of determinant of the traditional Jacobian matrix; the terms of polynomial is much less after fully expanded.

There are four components in quaternion, without loss of generality, define $\varepsilon_0 > 0$, as a result, it can be obtained:

$$\varepsilon_0 = \sqrt{1 - \varepsilon_1^2 - \varepsilon_2^2 - \varepsilon_3^2} \tag{16}$$

A new singular trajectory equation can be obtained by substituting Eq. (16) into Eq. (15), it is shown as follows:

$$f(\varepsilon_1, \varepsilon_2, \varepsilon_3, P_X, P_Y, P_Z) = 0 \tag{17}$$

Alternatively, Eq. (5) shows the transformation relationship between the quaternion and the Rodriguez parameters. Another singular trajectory equation, which is expressed of the Rodriguez parameter, can be obtained by substituting Eq. (5) into Eq. (15), it is shown as follows:

$$f(U, V, W, P_X, P_Y, P_Z) = 0 \tag{18}$$

In the case of any three variables of position and orientation are given, the variation of singular locus with respect to the remaining three variables can be studied, through neither Eq. (17) nor Eq. (18). When the variables of orientation are known, the singular locus of position is solved, and vice versa. As for the set of specific parameters of position and orientation, it can be determined whether the mechanism is in the singular pose or not. The two singular locus equations are equivalent in nature. In addition to the variables of position and orientation, there are only three parameters k_0, k_1, k_2 in them. The three parameters k_0, k_1, k_2 are constants and depend on the mechanism parameters. There are only 258 items in Eq. (18) when the symbolic expression is expanded completely. If the singularity is analyzed in the case of the symbolic expression factorized, the calculation speed will be further improved.

5   Numerical Example

As mentioned above, the structure parameters of the Stewart platform parallel mechanism can be described by $\theta_1, \theta_2, r_1, r_2$. The values of these four parameters are $\pi/5$, $\pi/9$, 1, 0.618. It can be obtained that $k_0 = 1.18812$, $k_1 = -2.17548$, $k_2 = -3.62575$.

(1) When the variables of orientation of the moving platform are determined, the change of the singular locus with respect to position variables can be studied by Eq. (17) or Eq. (18). The singular locus obtained by these two equations are identical in theory. For example: the Rodriguez parameters ($U = 0.7$, $V = 0.3$, $W = 0.4$)and the quaternion ($\varepsilon_0 = 0.758098$, $\varepsilon_1 = 0.530669$, $\varepsilon_2 = 0.227429$, $\varepsilon_3 = 0.303239$) can be used to describe the same given pose. The computer simulation shown that the singular locus equations are 3th polynomials with respect to the variables of position, and in these two cases, the singular locus can be expressed in in Fig. 2.

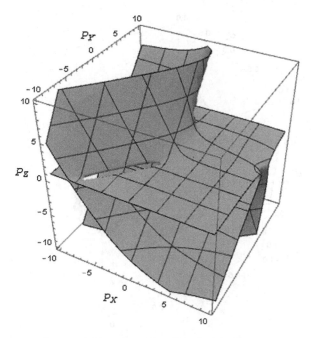

Fig. 2. Singular locus about variables of position. Whether Rodriguez parameters or quaternion can be used to describe the rotation matrix. The singular locus is identical.

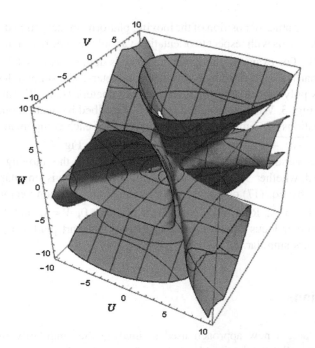

Fig. 3. Singular locus about variables of orientation at the given position. The variables of orientation are described by Rodriguez parameters.

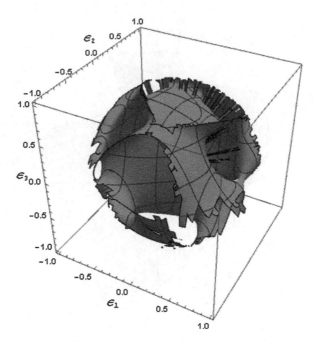

Fig. 4. Singular locus about variables of orientation at the given position. The variables of orientation are described by quaternions.

(2) When the variables of position of the moving platform are determined, the change of the singular locus with respect to orientation variables can be researched by Eq. (17) or Eq. (18). For example, the position vector is defined as $P = (2 \quad 2 \quad 4)^T$, the rotation matrix is described by Rodriguez parameters; The singular locus equations is 6 times polynomial about of Rodriguez parameters U, V, W, singular locus is shown in Fig. 3. When the rotation matrix is described by quaternion; The singular locus equations is 8 times polynomial about that of each component of quaternion $(\varepsilon_0 = \sqrt{1 - \varepsilon_1^2 - \varepsilon_2^2 - \varepsilon_3^2})$ singular locus is shown in Fig. 4.

(3) When the variables of position and orientation of the moving platform are determined, whether Stewart platform parallel mechanism is singular or not can be researched by Eq. (17) or Eq. (18). For example, the position vector is defined as $P = (0 \quad 0 \quad 5)^T$; Rodriguez parameters are $U = 0$, $V = 0$ and $W = 1$ respectively; singular locus equation is equal to zero, Stewart platform parallel mechanism occurs singular.

6 Conclusions

(1) In this paper, a new approach used to analyze the singularity of the Stewart Platform parallel mechanism is studied. There are only 258 items in analytical expression of the singular locus equation when it is expanded completely.

(2) The rotation matrix is described by quaternion, and the kinematics equation of the Stewart Platform parallel mechanism is transformed into a new form in this paper. In addition, normalization of quaternion is used and 7 equivalent equations are obtained, which can be used to study the singularity and the forward kinematics of the Stewart Platform parallel mechanism.

(3) Based on these equivalent equations, this paper presents a new Jacobian matrix, which states the mapping relationship between the driving speed and the change rate of position and orientation of the moving platform. Every component of the Jacobian matrix is a rather simple monomial.

(4) Evaluating the determinant of Jacobian matrix, the analytical expression of the singular locus equation can be derived. These singular trajectory equations can not only be used to study the distribution of singular locus in the workspace, but can also be used to judge whether a set of parameters of position and orientation corresponds to a singularity pose. The singularity analysis approach presented in this paper provides an important theoretical basis for studying the singularity free workspace.

Acknowledgment. The authors gratefully acknowledge the supported of the National Natural Science Foundation of China (Grant No. 51405417), the supported of Natural Science Foundation of Jiangsu Province, China (Grant No. BK20140470).

References

1. Merlet, J.P.: Parallel Robots. Springer, Netherlands (2006)
2. Tsai, L.W.: Robot Analysis: The Mechanics of Serial and Parallel Manipulators. Wiley, New York (1999)
3. Cheng, S.L., Wu, H.T., Wang, C.Q., et al.: A novel method for singularity analysis of the 6-SPS parallel mechanisms. Sci. China Technol. Sci. **54**(5), 1220–1227 (2011)
4. Gosselin, C.M., Angeles, J.: Singularity analysis of closed-loop kinematic chains. IEEE Trans. Robot. Autom. **6**(3), 281–290 (1990)
5. Mayer St-Onge, B., Gosselin, C.M.: Singularity analysis and representation of the general Gough-Stewart platform. Int. J. Robot. Res. **19**(3), 271–288 (2000)
6. Jiang, Q.M., Gosselin, C.M.: Determination of the maximal singularity-free orientation workspace for the Gough-Stewart platform. Mech. Mach. Theory **44**(6), 1281–1293 (2009)
7. Huang, Z., Cao, Y.: Property identification of the singularity loci of a class of the Gough-Stewart manipulators. Int. J. Robot. Res. **24**(8), 675–685 (2005)
8. Huang, Z., Zhao, Y.Z., Zhao, T.S.: Advanced spatial mechanism. Higher Education Press, Beijing (2014)
9. Coste, M., Moussa, S.: On the rationality of the singularity locus of a Gough-Stewart platform - Biplanar case. Mech. Mach. Theory **80**, 82–92 (2015)
10. Karimi, A., Masouleh, M.T., Cardou, P.: Singularity-free workspace analysis of general 6-UPS parallel mechanisms via convex optimization. Mech. Mach. Theory **80**, 17–34 (2014)
11. Cao, Y., Wu, M.P., Zhou, H.: Position-singularity characterization of a special class of the Stewart parallel mechanisms. Int. J. Robot. Autom. **28**(1), 57–64 (2013)

12. Hunt, K.H.: Kinematic Geometry of Mechanisms. The Clarendon Press, Oxford University Press, New York (1978)
13. Huang, Z., Zhao, Y., Wang, J., et al.: Kinematic principle and geometrical condition of general-linear-complex special configuration of parallel manipulators. Mech. Mach. Theory **34**(8), 1171–1186 (1999)
14. Merlet, J.P.: Singular configurations of parallel manipulators and Grassmann geometry. Int. J. Robot. Res. **8**(5), 45–56 (1989)
15. Caro, S., Moroz, G., Gayral, T., et al.: Singularity analysis of a six-dof parallel manipulator using grassmann-cayley algebra and gröbner bases. In: Angeles, J., Boulet, B., Clark, J.J., Kövecses, J., Siddiqi, K. (eds.) Proceedings of an International Symposium on the Occasion of the 25th Anniversary of the McGill University Centre for Intelligent Machines 2010, pp. 341–352. Springer, Berlin Heidelberg (2010)
16. Doyon, K., Gosselin, C., Cardou, P.: A vector expression of the constant-orientation singularity locus of the Gough-Stewart platform. ASME J. Mech. Robot. **5**(3), 1885–1886 (2013)
17. Liu, G.J., Qu, Z.Y., Liu, X.C., et al.: Singularity analysis and detection of 6-UCU parallel manipulator. Robot. Comput.-Integr. Manuf. **30**(2), 172–179 (2014)
18. Shanker, V., Bandyopadhyay, S.: Singular manifold of the general hexagonal stewart platform manipulator. In: Lenarcic, J., Husty, M. (eds.) Latest Advances in Robot Kinematics 2012, pp. 397–404. Springer, Dordrecht (2012)
19. Kaloorazi, M.H.F., Masouleh, M.T., Caro, S.: Interval-analysis-based determination of the singularity-free workspace of Gough-Stewart parallel robots. In: Iranian Conference on Electrical Engineering 2013, IEEE, pp. 1–6. IEEE, Mashhad (2013)
20. Gosselin, C.M., Schreiber, L.-T.: Kinematically Redundant Spatial Parallel Mechanisms for Singularity Avoidance and Large Orientational Workspace. IEEE Trans. Robot. **32**(2), 286–300 (2016)
21. Agarwal, A., Nasa, C., Bandyopadhyay, S.: Dynamic singularity avoidance for parallel manipulators using a task-priority based control scheme. Mech. Mach. Theory **96**, 107–126 (2016)
22. Parsa, S.S., Boudreau, R., Carretero, J.A.: Reconfigurable mass parameters to cross direct kinematic singularities in parallel manipulators. Mecha. Mach. Theory **85**, 53–63 (2015)
23. Pagis, G., Bouton, N., Briot, S., et al.: Enlarging parallel robot workspace through Type-2 singularity crossing. Control Eng. Pract. **39**, 1–11 (2015)
24. Aminea, S., Masoulehb, M.T., Caro, S., et al.: Singularity analysis of 3T2R parallel mechanisms using Grassmann-Cayley algebra and Grassmann geometry. Mech. Mach. Theory **52**, 326–340 (2012)
25. Kim, J.S., Jin, H.J., Park, J.H.: Inverse kinematics and geometric singularity analysis of a 3-SPS/S redundant motion mechanism using conformal geometric algebra. Mech. Mach. Theory **90**, 23–36 (2015)
26. Chai, X.X., Xiang, J.N., Li, Q.C.: Singularity analysis of a 2-UPR-RPU parallel mechanism. J. Mech. Eng. **51**(13), 144–151 (2015)
27. Kuipers, J.B.: Quaternions and Rotation Sequences: A Primer with Applications to Orbits. Aerospace and Virtual Reality. Princeton University Press, Princeton (1999)

Performance Research of Planar 5R Parallel Mechanism with Variable Drive Configurations

Weitao Yuan[1,2], Zhaokun Zhang[1], Zhufeng Shao[1(✉)],
Liping Wang[1,2], and Li Du[2]

[1] Beijing Key Lab of Precision/Ultra-Precision Manufacturing Equipments
and Control, State Key Laboratory of Tribology, Tsinghua University,
Beijing 100084, China
shaozf@mail.tsinghua.edu.cn
[2] School of Mechatronics Engineering, University of Electronic Science
and Technology of China, Chengdu 611731, China

Abstract. All joints of the serial mechanism are active ones, while only some joints are active ones for the parallel mechanism. Thus, there is a variety of drive configurations with different active joint selections. Drive configuration will affect the performance of parallel mechanism and the effect deserves further study. A planar 5R parallel manipulator with the capability of adjusting drive configurations in real time is proposed in this paper. Based on the deduced Jacobian matrices, performances of the 5R parallel manipulator are analyzed, considering three symmetrical drive configurations. Typical kinematic performance indices, such as local conditioning index and kinematic manipulability, are adopted to give a complete illustration of the performance variation. Then, the optimal adjustment of driven configurations is proposed and studied, considering local conditioning index. Simulation results reveal that the performance of the 5R parallel manipulator improves obviously through the reasonable combination of different drive configurations in the workspace. The drive configuration adjustment proposed in this paper provides a novel potential way to enhance the comprehensive performance of the parallel mechanism.

Keywords: Planar 5R parallel mechanism · Drive configuration · Performance analysis · Kinematics

1 Introduction

Mechanisms can be divided into two categories: serial and parallel mechanisms. Serial mechanisms consist of drive joints with single degrees of freedom (DoFs). Each joint is an independent active joint. Joints are connected serially, and only a kinematic chain links the framework (base) and the terminal (end effector). The parallel mechanism is recognized as a significant complement to the serial mechanism because of the closed-loop kinematic chains in its structure [1, 2]. Parallel mechanisms are usually composed of a base and an end effector, which are connected with at least two identical

© Springer International Publishing AG 2017
Y. Huang et al. (Eds.): ICIRA 2017, Part II, LNAI 10463, pp. 453–463, 2017.
DOI: 10.1007/978-3-319-65292-4_39

kinematic chains (branches) [3, 4]. Some joints are active ones, and passive joints that result in multiple available drive configurations exist. To reduce the mass and inertia of moving parts, the joints near the base are usually selected as active ones [5]. Studies on the performance analysis and optimal adjustment of different drive configurations are limited. However, drive configuration obviously affects the performance of parallel mechanism [6] and deserves further study.

Planar 5R parallel mechanism is preferred for the initial study of the parallel mechanism field because it possesses the advantages of simple structure, large workspace, good flexibility, absence of singularity in the reachable workspace [7], and easy control. For example, Zhang et al. [8] completed the theoretical and experimental research on the trajectory planning and kinematics simulation of the 5R mechanism. Ouyang et al. [9] performed dynamic analysis and studied the locomotive simulation problem of the 5R mechanism. Liu et al. [10–12] discussed the kinematics, singularity, and workspace of the 5R mechanism in detail. In general, the existing research on 5R mechanism covers position analysis, workspace, assembly pattern, singular research, dynamic analysis, and so on [13–16]. However, the research on the performance changes in different drive configurations is limited. By adding a facile transmission in the planar 5R mechanism, the real-time and flexible adjustment of drive configuration can be achieved without any change in the mass and inertia property of the mechanism. Variable drive configurations provide new possibilities for the performance improvement of the parallel mechanism and lead to new questions. How will the drive configuration affect the performance of parallel mechanism? How can the drive configuration be adjusted to achieve performance improvement? This study focuses on these questions.

In the process of analysis and optimal design, several indices, which are mainly kinematic indices, are inevitably adopted to illustrate the performance of the parallel manipulator [17]. The number of conditions in the kinematic Jacobian matrix is acknowledged as a comprehensive kinematic performance index and has been used by several researchers for optimum design. In addition, kinematic manipulability is also widely used as the supplement. The local conditioning index (LCI) was proposed by Gosselin and Angeles and has been used to evaluate the output isotropy of the parallel mechanism [18–20]. Yoshikawa defined the kinematic manipulability of a robot and analyzed the velocity transmission of the mechanism [21]. LCI and kinematic manipulator are adopted in this study to demonstrate the kinematic performance of the 5R parallel manipulator in different drive configurations.

The remainder of this paper is organized as follows. In the next section, the 5R parallel mechanism with variable drive configurations is introduced in detail, and the kinematic analysis is discussed. In the third section, the kinematic performances of the 5R parallel mechanism are illustrated and analyzed. LCI is taken as an example to verify the effect of performance improvement on the 5R parallel mechanism considering drive configuration adjustment. The conclusions drawn in this study are enumerated in the last section.

2 5R Mechanism with Variable Drive Configurations

2.1 Research Object

The virtual prototype of our object is shown in Fig. 1, which is actuated with two sets of AC servomotors and reducers. The basic structure is a planar 5R parallel mechanism, and the kinematic model is illustrated in Fig. 2. Two RRR branches exist (R represents revolute joint). In each branch, rod A_iB_i ($i = 1, 2$) and rod B_iP are connected directly by the revolute joint B_i, and rod A_iB_i is attached to the static framework with revolute joint A_i. The two branches connected by revolute joint P, which is also the end effector. To facilitate further analysis and description, the base coordinate system $O-XYZ$ is located at the center of A_1A_2 with the X-axis pointing toward point A_2, and the Y-axis is perpendicular to line A_1A_2. The geometric relations are defined as follows. $OA_1 = OA_2 = r_1$. The lengths of rods A_iB_i and B_iP are r_2 and r_3, respectively. $\boldsymbol{m}_i$ and $\boldsymbol{n}_i$ are the unit direction vectors of rods A_iB_i and B_iP, respectively. θ_i is the angle between vector $\boldsymbol{m}_i$ and the positive direction of the X-axis, while ψ_i is the angle between vectors $\boldsymbol{m}_i$ and $\boldsymbol{n}_i$. In this paper, vectors and matrices are in bold italics, while variables are in italics. The branch number is illustrated with the corresponding subscripts.

Fig. 1. Virtual prototype.

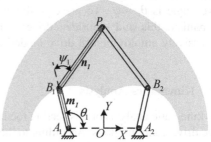

Fig. 2. Kinematic model.

Figure 3 illustrates the detailed structure of the branch. To adjust the drive configuration, the mechanism is equipped with the extra transmission system. The drive unit (servomotor and reducer) is attached to a cabin, which connects the static framework and the rod A_iB_i through electromagnetic clutches I and II, respectively. The power output shaft of the drive unit connects rod A_iB_i and pulley 1 with clutches III and IV separately. Pulley 2 is attached to rod B_iP as shown in the figure. The timing belt is adopted to transfer the rotation between two pulleys. The four clutches above are coaxial with joint A_i and pulley 2, while pulley 1 is coaxial with joint B_i. The clutches are controlled with the voltage signal, and can achieve real-time online control, which implements the power transmission path change and drives configuration variation.

When clutches I and III clasp (clutches II and IV release), the drive unit is equivalently fixed to the static framework, and the output shaft drives rod A_iB_i to rotate around joint A_i. Then, joint A_i becomes the active joint and the drive angle is θ_i, which

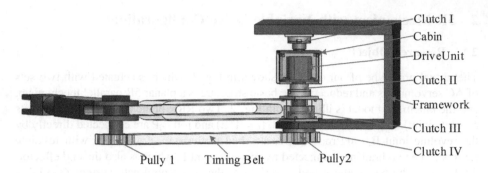

Fig. 3. Detailed drive structure of the branch.

is called branch drive mode 1. If clutches II and IV clasp (clutches I and III release), the drive unit is attached to rod A_iB_i and the output shaft connects rod B_iP through clutch IV and the belt system. In this situation, the drive unit is equivalently installed in joint B_i, the branch is driven by revolute joint B_i with drive angle ψ_i, which is called the branch drive mode 2. In addition, when clutches I and IV clasp (clutches II and III release), the drive unit is fixed to the static platform and the output shaft is connected to rod B_iP through clutch IV and the belt system. This is the branch drive mode 3, and the drive angle is $\theta_i - \psi_i$. To preserve the symmetrical characteristic of the 5R parallel mechanism, this study considers drive configurations I, II, and III when both branches successively employ branch drive modes 1, 2, and 3.

2.2   Kinematics Analysis

The kinematic analysis of this section focuses on deducing the Jacobian matrices under three drive configurations. According to the kinematic model in Fig. 2, the vector chain equation can be expressed as

$$p = \mu_i + r_2m_i + r_3n_i, \tag{1}$$

where μ_i is the position vector of point A_i.

The above equation can be simplified as

$$\tan\frac{\theta_i}{2} = \frac{-C_{i1} \pm \sqrt{C_{i1}^2 - (C_{i3}^2 - C_{i2}^2)}}{C_{i3} - C_{i2}}. \tag{2}$$

In addition, $C_{i1} = -2r_2(p - \mu_i) \cdot k_2$, $C_{i2} = 2(-1)^i r_2(p - \mu_i) \cdot k_1$, and $C_{i3} = (p - \mu_i)^2 + r_2^2 + r_3^2$ k_1 and k_2 are the unit direction vectors of the X-axis and the Y-axis, respectively. According to the assembly pattern of the 5R parallel mechanism, the positive and negative signs in Eq. (2) can be determined. Then, the inverse position solution of drive configuration I can be deduced as

$$\theta_i = 2 \arctan \frac{-C_{i1} + \sqrt{C_{i1}^2 - (C_{i3}^2 - C_{i2}^2)}}{C_{i3} - C_{i2}}. \tag{3}$$

When the position vector p of the moving platform is provided, vector n_i can be deduced with Eqs. (1) and (3). At the same time, vector n_i can be expressed as

$$n_i = \left[(-1)^{i+1}\cos(\theta_i - \psi_i) \quad \sin(\theta_i - \psi_i) \quad 0\right]^{\mathrm{T}}. \tag{4}$$

Input angle ψ_i, which is also the inverse position solution of drive configuration II, can be derived as

$$\psi_i = \theta_i - \arcsin n_i(2), \tag{5}$$

where $n_i(2)$ is the second element of the n_i vector.

By sorting Eq. (5), $\theta_i - \psi_i$ can be obtained, which is the inverse position solution of drive configuration III. Taking the derivative of Eq. (1) with respect to time yields

$$\dot{p} = r_2 \dot{\theta}_i (k_3 \times m_i) + r_3 \left(\dot{\theta}_i - \dot{\psi}_i\right)(k_3 \times n_i), \tag{6}$$

where $\dot{p}$ is the velocity vector of the moving platform. $\dot{\theta}_i$, $\dot{\psi}_i$ and $\dot{\theta}_i - \dot{\psi}_i$ denote the input angular velocities of drive configurations I, II, and III, respectively.

By taking the dot product of both sides of Eq. (6) with n_i, $r_2 m_i + r_3 n_i$, m_i, the following are respectively derived:

$$J_1 = \left[\frac{n_1^{\mathrm{T}}}{r_2(m_1 \times n_1) \cdot k_3}; \frac{n_2^{\mathrm{T}}}{r_2(m_2 \times n_2) \cdot k_3}\right], \tag{7}$$

$$J_2 = \left[\frac{(r_2 m_1 + r_3 n_1)^{\mathrm{T}}}{r_3[n_1 \times (r_2 m_1 + r_3 n_1)] \cdot k_3}; \frac{-(r_2 m_2 + r_3 n_2)^{\mathrm{T}}}{r_3[n_i \times (r_2 m_2 + r_3 n_2)] \cdot k_3}\right], \tag{8}$$

$$J_3 = \left[\frac{m_1^{\mathrm{T}}}{r_3(n_1 \times m_1) \cdot k_3}; \frac{m_2^{\mathrm{T}}}{r_3(n_2 \times m_2) \cdot k_3}\right], \tag{9}$$

Finally, we derive J_1, J_2, and J_3, which are the Jacobian matrices of drive configurations I, II, and III, respectively. In the following section, the following dimension parameters are adopted: $r_1 = 0.55$ m, $r_2 = 0.8$ m, and $r_3 = 1.6$ m.

3 Kinematic Performance Analysis

3.1 LCI

As deduced in the previous section, the Jacobian matrix is the mapping between the input and output velocities. The LCI of the parallel mechanism is defined based on the

Jacobi matrix, which is adopted to describe the isotropy degree of the output velocity under a certain position. The number of conditions of the parallel manipulator can be described as

$$\kappa = \frac{\sigma_{max}}{\sigma_{min}}, \tag{10}$$

where σ_{max} and σ_{min} are the maximum and the minimum singular values of the Jacobian matrix, respectively. Thus, $1 \leq \kappa \leq \infty$. The LCI is defined as the reciprocal of the number of conditions of the Jacobian matrix. Therefore, when LCI = 1, the mechanism is isotropic. A large LCI value ensures a good control accuracy.

The LCI value distributions of the 5R mechanism with different drive configurations are illustrated in the reachable workspace, as shown in Fig. 4. From these figures, we know that:

(1) The drive configuration has an obvious impact on the LCI value distribution of the 5R parallel mechanism. In general, large values are in the center area for symmetrical drive configurations. This area is slightly low for configuration III.
(2) The maximum LCI values for drive configurations I and III are similar and nearly 1. However, the maximum LCI value for drive configuration II is significantly less. The average value of the LCI for drive configuration I is the largest among the average values of the configurations.
(3) Noticeable differences can be observed between the LCI distributions of drive configuration I and II. The 5R mechanism can improve the LCI value distribution and the isotropic performance by properly transforming the drive configurations in the reachable workspace.

3.2   Kinematic Manipulability

The product of the singular values of the Jacobian matrix defines the kinematic manipulability index. For the planar 5R parallel mechanism, the kinematic manipulability index can be written as

$$\Gamma = \sqrt{\det\left(\boldsymbol{JJ}^{\mathrm{T}}\right)} = \sqrt{\lambda_1 \lambda_2} = \sigma_1 \sigma_2, \tag{11}$$

where σ_1 and σ_2 are the singular values of the Jacobian matrix. λ_1 and λ_2 are the matrix eigenvalues. Kinematic manipulability is a local performance index, which indicates the velocity transmission characteristics of the mechanism from the active joints to the terminal. A large kinematic manipulability index value ensures a good high-speed feature and an improved velocity amplification ability for the mechanism under the position. Therefore, a high output velocity can be obtained with the same input velocity.

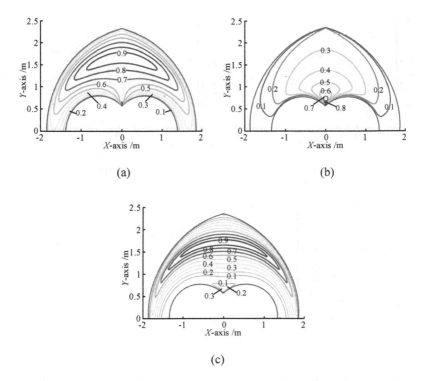

(a)   (b)

(c)

(a) Drive configuration I. (b) Drive configuration II. (c) Drive configuration III.

Fig. 4. The LCI value distribution.

The value distributions of the kinematic manipulability index for the planar 5R parallel mechanism are illustrated in Fig. 5. The comparative analysis of these curves indicates the following:

(1) Drive configuration significantly affects the value distribution of the kinematic manipulability index of the 5R parallel manipulator. Generally, the index values of drive configuration I are the smallest, while the index values of drive configuration III are the largest.

(2) In the value distribution of the index, the large values of drive configurations I and III are in the workspace center, while the large values of drive configuration II appear on both sides.

In all, the performance of the planar 5R parallel manipulator varies significantly with the drive configuration. In drive configuration I, the manipulator possesses the best isotropic property, while the best kinematic manipulability is acquired in drive configuration III. Drive configuration II is moderate in all. In addition, the performance of the parallel mechanism could be further improved by adjusting the drive configuration in the workspace appropriately.

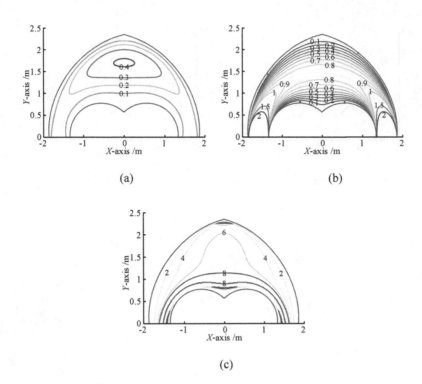

(a) Drive configuration I. (b) Drive configuration II. (c) Drive configuration III.

Fig. 5. The value distribution of the kinematic manipulability index.

4 Drive Configuration Adjustment

In this section, the kinematic performance improvement of the planar 5R mechanism is verified with the LCI, considering the drive configuration adjustment. The simulation parameters are given as $r_1 = 0.55$ m, $r_2 = 1.24$ m, and $r_3 = 2.34$ m. The LCI values in the reachable workspace are divided into five ranges. Figure 6 illustrates the area ratios of the LCI value ranges in four drive configurations, namely I, II, III, and the adjustment. Figure 7 shows the adopted drive configurations in the reachable workspace for the drive configuration adjustment. Drive configurations I and III are adopted in most of the reachable workspaces.

The comparative analysis in Fig. 6 shows that kinematic performance can be improved by adjusting the drive configuration. For example, the area ratios of the LCI value ≥ 0.8 for drive configurations I, II, and III are 19%, less than 1%, and 12%, respectively. The area ratios of the LCI values that are greater than or equal to 0.8 under the drive configuration adjustment achieves 24%, which is better than any single drive configuration. At the same time, the area ratio of the LCI value below 0.6 under the drive configuration adjustment is reduced to 39%. Drive configuration adjustment is a potential way to improve the performance of the parallel mechanism.

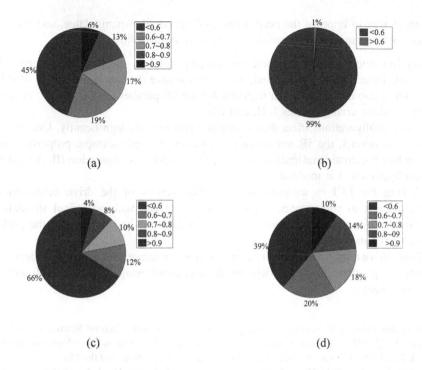

(a) Drive configuration I. (b) Drive configuration II. (c) Drive configuration III. (d) The adjustment of drive configuration

Fig. 6. Area ratios of the LCI value ranges.

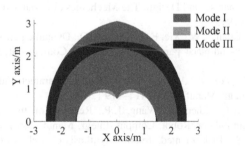

Fig. 7. Configuration distribution of the drive configuration adjustment.

5   Conclusions

Kinematic performances of the planar 5R parallel mechanism under three symmetrical drive configurations are studied in the reachable workspace by adopting the LCI and kinematic manipulability index. The influence of drive configurations on the kinematic performance is illustrated and analyzed. Drive configuration adjustment is proposed as a potential way to improve the performance of the parallel manipulator, and the simulation verification is given. Specific conclusions are as follows:

(1) By introducing the timing belt system and clutches, the planar symmetric 5R parallel mechanism is proposed, which can realize the real-time change of drive configurations. The Jacobian matrices for the 5R parallel mechanism are deduced under drive configurations I, II, and III.
(2) Drive configurations affect the mechanism performance significantly. Under drive configuration I, the 5R manipulator possesses the best isotropic property, while the best kinematic manipulability is acquired in drive configuration III. And, drive configuration II is moderate.
(3) Taking the LCI as an example, the effectiveness of the drive configuration adjustment to performance improvement is verified with numerical simulation. The drive configuration adjustment is a potential method to improve the performance of the parallel mechanism.
(4) More drive configurations will be analyzed in the future, and the optimal design of the 5R parallel mechanism will be studied considering the drive configuration adjustment.

Acknowledgments. This research is jointly sponsored by National Natural Science Foundation of China (No. 51575292), China's National Key Technology Research and Development Program (No. 2015BAF19B00), National Science and Technology Major Project of the Ministry of Science and Technology (No. 2015ZX04003004), and National Scholarship Fund (No. 201606215004).

References

1. Tsai, L.W.: Robot Analysis and Design: The Mechanics of Serial and Parallel Manipulators. Wiley, New York (1999)
2. Mo, J., Shao, Z.F., Guan, L.W., Xie, F.G., Tang, X.Q.: Dynamic performance analysis of the X4 high-speed pick-and-place parallel robot. Robot. Comput.-Integr. Manuf. **46**, 48–57 (2017)
3. Wu, J., Wang, J.S., You, Z.: An overview of dynamic parameter identification of robots. Robot. Comput. Integr. Manuf. **26**(5), 414–419 (2010)
4. Shao, Z.F., Tang, X.Q., Chen, X., Wang, L.P.: Research on the inertia matching of the Stewart parallel manipulator. Robot. Comput.-Integr. Manuf. **28**(6), 649–659 (2012)
5. Kuo, C.H., Dai, J.S.: Task-oriented structure synthesis of a class of parallel manipulators using motion constraint generator. Mech. Mach. Theory **70**(6), 394–406 (2013)
6. Chen, X.: Research on Transmission, Constraint and Output Performance Evaluation of Parallel Manipulator and Its Applications. Tsinghua University, Beijing (2015)
7. Liu, X.J., Wang, J.S.: Parallel Kinematics: Type, Kinematics and Optimal Design. Springer, Berlin (2014)
8. Zhang, L.J., Guo, Z.M., Li, Y.Q.: Trajectory planning, kinematic simulation and experiment research on planar 5R parallel manipulator. Mach. Des. Res. **29**(1), 10–13 (2013)
9. Quyang, P.R., Zhang, W.J., Wu, F.X.: Nonlinear PD control for trajectory tracking with consideration of the design for control methodology. In: Proceedings of the 2002 IEEE International Conference, vol. 4, pp. 4126–4131. Robotics and Automation, Washington DC, USA (2002)

10. Liu, X.J., Wang, J.S., Prischow, G.: Performance atlases and optimum design of planar 5R symmetrical parallel mechanisms. Mech. Mach. Theory **41**(2), 119–144 (2006)
11. Liu, X.J., Wang, J.S., Pritschow, G.: Kinematics singularity and workspace of planar 5R symmetrical parallel mechanisms. Mech. Mach. Theory **41**(2), 145–169 (2003)
12. Liu, X.J., Wang, J.S., Xie, F.G.: Analysis and kinematic optimization of planar 2-DoF 5R parallel mechanisms considering the force transmissibility. Optimization **41**(2), 453–461 (2001)
13. Alici, G.: An inverse position analysis of five-bar planar parallel manipulators. Robotica **20**(2), 195–201 (2002)
14. Cervantes-Sanchez, J.J., Hemandez-Rodriguez, J.C., Rendon-Sanchez, J.G.: On the workspace, assembly configurations and singularity curves of the RRRRR-type planar manipulator. Mech. Mach. Theory **35**(8), 1117–1139 (2000)
15. Gao, F., Zhang, X.Q., Zhao, Y.S.: A physical model of the solution space and the atlases of the reachable workspaces for 2-DoF parallel plane manipulator. Mech. Mach. Theory **31**(2), 173–184 (1996)
16. Park, F.C., Kim, J.W.: Singularity analysis of closed kinematic chains. J. Mech. Design **121**(1), 32–38 (1999)
17. Shao, Z.F., Tang, X.Q., Wang, L.P.: Atlas based kinematic optimum design of the Stewart parallel manipulator. Chin. J. Mech. Eng. **28**(1), 20–28 (2015)
18. Gosse, L.C., Angeles, J.: The optimum kinematic design of a spherical three-degree-of-freedom parallel manipulator. J. Mech. Design **111**(2), 202–207 (1989)
19. Khan, W.A., Angeles, J.: The kinetostatic optimization of robotic manipulators: the inverse and the direct problems. J. Mech. Design **128**(1), 168–178 (2006)
20. Shao, Z.F., Tang, X.Q., Wang, L.P.: Optimum design of 3-3 Stewart platform considering inertia property. Adv. Mech. Eng. **2013**, 249121 (2013)
21. Yoshikawa, T.: Manipulability of robotic mechanisms. Int. J. Robot. Res. **4**(2), 3–9 (1985)

Experimental Study on Load Characteristics of Macro-Micro Dual-Drive Precision Positioning Mechanism

Jing Yu, Ruizhou Wang, and Xianmin Zhang[✉]

Guangdong Provincial Key Laboratory of Precision Equipment and Manufacturing Technology, School of Mechanical and Automotive Engineering, South China University of Technology, Guangzhou 510640, China
tonyyu_scut@163.com, zhangxm@scut.edu.cn

Abstract. This paper presents a macro-micro two-stage parallel experimental platform. The 3RRR macro planar parallel mechanism driven by the YAS-WAKA Σ-V series motor is to ensure the system's large workspace. The 3RRR micro parallel mechanism based on compliant mechanism is driven by the pack-aged piezoelectric ceramics actuators (PPCA) which effectively guarantee the precision and accuracy of nanoscale positioning. In fact, the load characteristic is a basic index that affects the performance of the macro/micro combination mechanism. A displacement experiment is designed for the macro parallel mechanism to move in a straight line under different loads. The displacement is measured by the Renishaw laser interferometer. It is proved by the experiment that the precision and stability can be affected under different loads. The load characteristics research results and analysis in this paper are significant to the optimal design of a macro-micro dual-drive precision positioning planar parallel mechanisms for different applications.

Keywords: Precision positioning · Macro-micro combination · Parallel mechanism · Load characteristics

1 Introduction

With the development of science and technology, Macro-micro combination planar nanopositioning platforms play an important role in precise and accurate nanoscale positioning. The use of macro-micro dual-drive positioning mechanism can make the system have a higher positioning accuracy in large stroke. In 1984, 1988 and 1993, Andre Sharon used macro-micro combination of ways to control and improve the accuracy of the robot in the United States Massachusetts Institute of Technology [1]. In 2009, John-Ahn Kim developed a large travel scanning station for the atomic force microscope [2]. In 2012, Ronnie Fesperman of University of North Carolina at Charlotte in the United States developed a macro-micro combination system by using a multi-linear motor to drive the macro moving platform and micro platform driven by Piezoelectric [3]. Many domestic units have also developed a variety of forms of macro-micro positioning system. In 2006, Jie Degang of Harbin Institute of Technology

© Springer International Publishing AG 2017
Y. Huang et al. (Eds.): ICIRA 2017, Part II, LNAI 10463, pp. 464–471, 2017.
DOI: 10.1007/978-3-319-65292-4_40

developed a high-speed & high-precision positioning system of Marco/-Micro Driving, the macro table was driven by VCM and the micro table by PZT [4].

In recent years, planar nanopositioners have been widely employed in precise and accurate nanoscale positioning, which involves nanoimprint lithography [5], precision mechanical scanning in scanning probe microscopy (SPM) [6], micro/nano manipulation [7], precision machining and micro-nano surface metrology measurement [8]. An operating system that can achieve nanoscale precision in the context of a large stroke in the process of research and machining is needed.

The accuracy of the large stroke drive (precision screw drive, linear motor, voice coil motor) is generally limited to the micron level. The positioning accuracy of micro-positioning platform represented by piezoelectric ceramic actuator can reach nanometer positioning accuracy. So the macro/micro dual-drive operating system possess both large stroke by motors and response fast, sub-nanometer resolution characteristics by piezoelectric ceramics actuators.

The macro/micro dual-drive system will develop into a variety of integrated ways in the future, especially in two dimensions and three dimensions. The macro-micro parallel mechanism can improve the positioning accuracy, but certain coupling characteristics will have impact in marco-micro combination. One of the most basic coupling characteristic is the effect of the micro platform as a load on the macro platform. Different weight of loads can greatly affect the precision and accuracy of the end positioner.

The paper presents a 3-DOF macro-micro combination mechanism. The remainder of the paper is organized as following: A macro-micro planar parallel mechanism is proposed in Sect. 2. Based on kinematic equation and Lagrange equation, the mathematical dynamics model of macro-parallel positioning mechanism is established and the effect factor of the load is analyzed in Sect. 3. Load effect experiment on macro planar parallel mechanism is introduced in Sect. 4, followed by a brief conclusion and prospect in Sect. 5.

2 Design of a Macro/Micro Planar Parallel Mechanism

As is shown in the Fig. 1, the configuration of the macro-micro combination system is expected to be designed to include the following parts: A macro 3RRR mechanism and a micro 3RRR mechanism to ensure both large stroke positioning and nanoscale positioning, the machine vision makes the system a closed loop feedback.

The basic structure of both the 3RRR micro and macro parallel mechanism are shown in the Fig. 2, which contains the PRBM and RBM model of both platform, followed with a macro-micro mechanism model.

The controlled model of each part of the macro-micro combination system is shown in Fig. 3. The macro-parallel mechanism ensured for large stroke positioning is 3-RRR mechanism, which consists of a three-way chain, including the moving platform as an end positioner, the base platform and the connection. Each branch contains three rotary hinges and two connecting rods, where R is the active hinge and the rest are the passive hinge. The macro-parallel mechanism is driven by the YASWAKA servo motor, controlled by the DMC-1886PCI bus motion controller. The YASKAWA motor has a

24-bit incremental encoder, and each pulse in the position control mode rotates 5.23 μrad so that the end accuracy of the mechanism can reach 1 μm.

The micro-parallel mechanism has three symmetrical kinematic chains connecting three actuators and a moving plate. Three packaged piezoelectric ceramic actuators (PPCA) (P-841.30, from PI Ceramic GmbH) with a closed-loop stroke of 45 μm were selected to drive the platform [9].

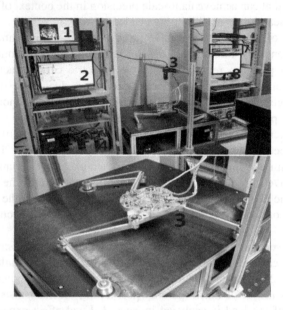

Fig. 1. Configuration of the macro-micro combination system. Number components are: 1, 2, 8 computer for machine vision, macro 3RRR mechanism and micro 3RRR mechanism; 3 machine vision; 4 YASWAKA motor; 5 macro/micro combination; 6 PPCA controller; 7 capacitive sensors

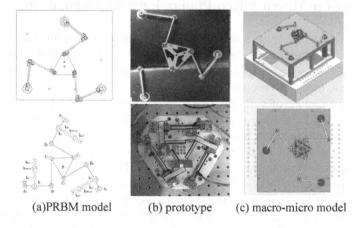

(a)PRBM model   (b) prototype   (c) macro-micro model

Fig. 2. (a) Rigid body model and pseudo rigid body model of 3RRR macro-micro mechanism (b) Prototype of 3RRR macro/micro mechanism (c) macro-micro combination model

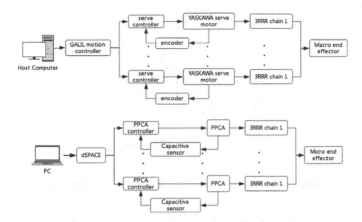

Fig. 3. Controlling model of the 3RRR macro/micro mechanism

The PPCAs controlled by dSPACE are respectively mounted on three modular fixtures axially configured and located symmetrically on the circumference of a circle. The flexible hinge guide mechanism transmits the movement and force of the micro actuator's output to the mobile platform. The selection and design guidance mechanism directly affects the final displacement output of the micro positioning platform. The micro-platform builds the mapping relationship between 16 size variables and seven main performance targets by establishing a multi-objective optimization model. The Pareto solution set is calculated by using the multi-objective particle swarm algorithm to determine the best individual under specific application requirements. The kinematics and dynamics model of the modeling compensation mechanism are obtained by the loss of the preload and the input displacement. In the drive SGS, the input and end of the establishment of three loops of the servo controller. Combining PI control algorithm, the accuracy and accuracy of end trajectory tracking has been improved.

3   Dynamic Modelling of the Macro-Parallel Mechanism

The Lagrange equation of the macro 3RRR mechanism can be written as [10]:

$$\frac{1}{3}m_1L_1^2\sum_{i=1}^{3}\ddot{q}_{ai}^2 + m_2L_1^2\sum_{i=1}^{3}\ddot{q}_{ai}^2 + \frac{1}{2}m_2L_1L_2\sum_{i=1}^{3}\left(\sin(q_{bi}-q_{ai})(\dot{q}_{ai}-\dot{q}_{bi})\dot{q}_{ci} + \cos(q_{ai}-q_{bi})\ddot{q}_{ci}\right) + \ldots$$

$$\ldots m_3\dot{X}_P\frac{\partial X_P}{\partial q_{ai}} + m_3\dot{Y}_P\frac{\partial Y_P}{\partial q_{ai}} + m_3L_3^2\dot{q}_{ci}\frac{\partial \theta_P}{\partial q_{ai}} = F_i, (i = 1, 2, 3)$$

where m_1 m_2 m_3 represent the mass of connecting rod A_iB_i B_iC_i and moving platform P, respectively. L_1 and L_2 stand for the length of connecting rod A_iB_i B_iC_i. $\frac{\partial X_P}{\partial q_{ai}}, \frac{\partial Y_P}{\partial q_{ai}}, \frac{\partial \theta_P}{\partial q_{ai}}, \frac{\partial \beta_i}{\partial q_{ai}}$ can be solved by kinematics equation (Fig. 4).

From the analysis of system dynamics equation, we can infer that the macro 3RRR mechanism in different motion trajectory under the joint drive torque and drive energy

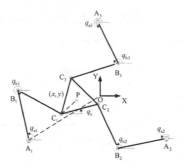

Fig. 4. Schematic diagram of macro 3RRR mechanism

consumption is very different. When the trajectory is straight and the load on the macro 3RRR mechanism increases, the driving torque and energy consumption of all active joints are significantly increased. Due to the PID parameters of YASKAWA motor which drives the 3RRR macro mechanism cannot be adjusted, the only way to improve positioning performance is to adjust the weight of the load to a suitable value via experiment. It is in Sect. 4 that we describes how to use experimental methods to optimize the design of the macro/micro combination mechanism.

4   Load Effect on Macro Planar Parallel Mechanism

We fixed the 1 kg, 2 kg, 3 kg weight as different quality loads on the macro planar parallel mechanism compared with the no load situation. First we use the software to control the host computer to adjust the Macro planar parallel mechanism until the end platform moving to the initial position and the corner angle of the platform is 0°. Then the host computer is used to guide the macro mechanism moved to [0, −13]. Then Start the preset run program to control the motion of the platform moving back and forth from [0, −13] to [0, 13], The macro planar parallel mechanism has a stop time of 1 s after each movement of 1 mm. The laser interferometer (XL-80, from Renishaw Inc.) recorded the entire process at 2500 Hz sampling frequency, recording 2500 dots per second (Figs. 5 and 6).

As the macro positioning mechanism stop for 1 s time's interval after each 1 mm displacement, we can obtain 27 stop points' data from [−13, 0] to [13, 0]. Firstly, the value filter of measured data by the laser interferometer was realized with MATLAB software system, which eliminate the coarse error. And then in the MATLAB software system, 400 consecutive points are selected from the 2500 sampling points, and the maximum difference of these continuous sampling points is less than 1 μm. The average value of 400 sampling points is taken as the absolute position of the end of macro 3RRR planar parallel mechanism at the displacement point.

Selecting 20 consecutive points in each of the 27 displacement points from [0, −13] to [0, 13], we can get 5 sets of data from −13 to 13 in the five round trip test. The five sets of data are put together to find the standard deviation of each displacement point. As the Fig. 7 shows, we compare the standard deviation to evaluate the stability of

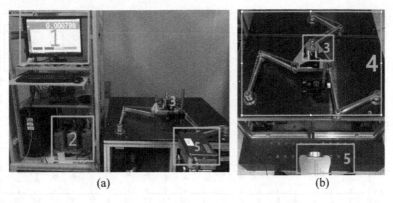

Fig. 5. Test setup of positioning system. Number components are: 1 host computer; 2 YASWAKA motor; 3 weight; 4 macro 3RRR mechanism; 5 Renishaw laser interferometer

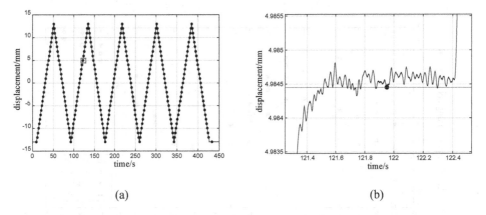

(a)   (b)

Fig. 6. (a) Experimental displacement diagram of macro positioning mechanism (b) Partial magnification diagram at the displacement of [0, 5]

macro 3RRR planar parallel mechanism under 4 different loads when it passes through each displacement point in the same way.

Besides, as the Fig. 8 show, we take the difference between the absolute position of the macro 3RRR planar parallel mechanism at each point and the point on the coordinate axis as the absolute positioning accuracy.

As shown above, we can get conclusions about the macro-micro combination mechanism under 4 different weight loads displacement and the same speed (2 mm/s), Within a certain range (0–3 kg) with the increase of the load quality of the simulated micro-motion platform, the repeatability of the macro-micro combination mechanism is also increased.

However, different from repeat positioning accuracy, the absolute positioning accuracy of the macro and micro combination mechanism decreases with the increase weight of load in the range of 0–3 kg.

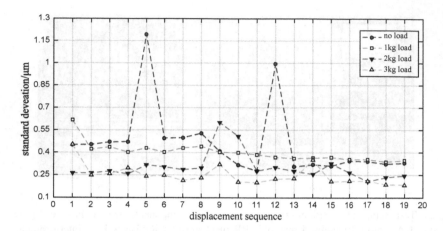

Fig. 7. Standard deviation of 19 point

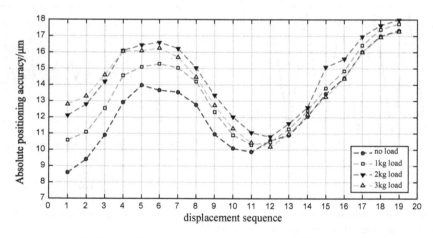

Fig. 8. Absolute positioning accuracy of 19 points

Therefore, the design of a load ranging from 1 kg to 2 kg in the micro 3RRR can make macro-micro combination mechanism to obtain a good repeatability positioning accuracy and absolute positioning accuracy.

5   Summary and Prospect

The paper presents the experimental investigation of a macro-micro combination mechanism. A macro-micro combination mechanism which can achieve plane 3DOF movement is proposed. The experimental results demonstrate that that the positioning precision and stability can be greatly affected by different loads. The main contribution of the study is the experimental methods to find the weight of the loads between certain range of weight. The method provided an effective solution. The experimental results are important for the implementation of mechanism design and further macro-micro-combination experiments.

Future research will focus on the coupling characteristics of the macro-micro combination system. In order to realize the closed loop feedback, one can add machine vision to the system. Moreover, applying different intelligent control strategies to improve the positioning precision, trajectory tracking accuracy and dynamic performance at a higher velocity is needed.

Acknowledgments. This research is supported by the National Natural Science Foundation of China (Grant Nos. U1501247, U1609206), the Natural Science Foundation of Guangdong Province (Grant No. S2013030013355), the Scientific and Technological Project of Guangzhou (Grant No. 2015090330001), and the Science and Technology Planning Project of Guangdong Province (Grant No. 2014B090917001). The authors gratefully acknowledge these support agencies.

References

1. Sharon, A., Hardt, D.: Enhancement of robot accuracy using endpoint feedback and a macro-micro manipulator system. In: American Control Conference, pp. 1836–1845. IEEE (1984)
2. Ahn, J., Kim, J.A., Kang, C.S., et al.: A passive method to compensate nonlinearity in a homodyne interferometer. Opt. Express **17**(25), 23299–23308 (2009)
3. Fesperman, R., Ozturk, O., Hocken, R., et al.: Multi-scale alignment and positioning system – MAPS. Precis. Eng. **36**(4), 517–537 (2012)
4. Degang, J.: Research of the high-speed/high-precision positioning system of marco/micro driving. Harbin Institute of Technology (2006)
5. De Jong, B.R., Brouwer, D.M., De Boer, M.J., et al.: Design and fabrication of a planar three-DOFs MEMS-based manipulator. J. Microelectromech. Syst. **19**(5), 1116–1130 (2010)
6. Eigler, D.M.: Positioning single atoms with a scanning tunnelling microscope. Nature **344** (6266), 524–526 (1990)
7. Howell, L.L.: Compliant mechanisms. In: 21st Century Kinematics, pp. 457–463. Springer, London (2013)
8. Fatikow, S., Wich, T., Hulsen, H., et al.: Microrobot system for automatic nanohandling inside a scanning electron microscope. IEEE/ASME Trans. Mechatron. **12**(3), 244–252 (2007)
9. Wang, R., Zhang, X.: Optimal design of a planar parallel 3-DOF nanopositioner with multi-objective. Mech. Mach. Theory **112**, 61–83 (2017)
10. Qinghua, Z.: Research on elastodynamics modeling and active vibration control of planar 3-RRR flexible parallel robots mechanism. South China University of technology (2013)

Machine and Robot Vision

Solving a New Constrained Minimization Problem for Image Deconvolution

Su Xiao[✉], Ying Zhou, and Linghua Wei

School of Computer Science and Technology, Huaibei Normal University,
Huaibei 235000, China
csxiaosu@163.com

Abstract. Though nature images generally consist of edges and homogeneous regions, most image deconvolution approaches will result in unnaturally sharp image estimations, since they tend to overemphasize sharpness while ignoring smoothness. To balance these two significant image properties, this paper presents a novel analysis-based image deconvolution approach based mainly on alternating minimization and variable splitting. The presented approach focuses on a new method of solving a constrained minimization problem which is derived from a universal model and is used to model the image deconvolution. By alternating minimization with variable splitting, the proposed image deconvolution model is firstly converted to an equivalent model, and then decoupled into a number of sub-problems. These sub-problems are alternately handled using a corresponding iterative method to obtain the solution for the proposed image deconvolution model. The presented approach has been demonstrated to be effective and superior to several state-of-the-art approaches for image deconvolution applications.

Keywords: Image deconvolution · Analysis-based approaches · Synthesis-based approaches · Variable splitting · Alternating minimization · Shrinkage operator · Proximal operator

1 Introduction

It is a universal phenomenon that digital images can be corrupted by blur and noise. This phenomenon generally can be explained by a degradation model of the form $y = Ax + n$, where $x \in R^N$ represents an unknown sharp image, or its transform coefficients generated by decomposing x using some transform basis, $y \in R^M$ represents a known measured image, $A \in R^{M \times N}$ represents a known linear operator and $n \in R^M$ represents white Gaussian noise. Estimating a sharp image from a given measured image is an ill-posed problem, since multiple feasible solutions are likely to be found. To acquire the optimal solution under particular evaluation criteria, the regularization technique, also known as Tikhonov regularization [1], has shown to be the most effective tool. The regularization methodology formulates the image deconvolution as a minimization

© Springer International Publishing AG 2017
Y. Huang et al. (Eds.): ICIRA 2017, Part II, LNAI 10463, pp. 475–485, 2017.
DOI: 10.1007/978-3-319-65292-4_41

problem $\{min_x J(x), \quad s.t. \quad x \in S\}$ (**P0**), where the regularizer $J(x) : R^N \to R$ plays an important role in edge-preservation and noise-suppression, and both $J(x)$ and S are convex. Initially, the quadratic forms of $J(x)$ are considered to be the most popular candidates [2–4], because this type of regularizer is involved, close-formed solutions of x can be obtained without using iterative solvers. However, the estimated images that are provided by image deconvolution approaches with quadratic regularizers are inevitably over-smoothed. Therefore, ultimately, most image deconvolution approaches turn to total variation (TV) regularizers [5–7], due to their outstanding detail-preservation performance. In TV models, significant staircasing effects are unavoidably observed in estimated images that should be piecewise continuous. With the rise of sparse representation [8], sparsity-inducing regularizers have now become the favorite choice [9–12] because of the advantages that they can offer. The corresponding image deconvolution approaches can be divided into two main types: synthesis-based approaches and analysis-based approaches. Since an image can be regarded as a linear combination of elements of a transform basis (such as a wavelet frame), when synthesis-based approaches are used, image deconvolution turns into estimation of the optimal combination coefficients (i.e., transform coefficients) of sharp images. In other words, synthesis-based approaches concentrate on handling a problem **P0** which has the form $\{min_x ||x||_1, \quad s.t. \quad ||Ax - y||_2 \le \epsilon\}$ (**P1**) or $\{min_x 0.5 \times ||Ax - y||_2^2 + \mu ||x||_1\}$ (**LP1**), where **LP1** is the equivalent Lagrangian form of problem **P1**, $|| \cdot ||_1$ and $|| \cdot ||_2$ denote the l_1-norm and l_2-norm, respectively, $||x||_1$ is a sparse-inducing regularizer, the linear operator A is the product of a blur matrix and a transform matrix, and x are the transform coefficients of an unknown sharp image. Figueiredo and Nowak [9] have presented an expectation-maximization based approach which adopts two alternating steps, namely E-steps and M-steps, to solve the problem **LP1**. To speed up the computation of this problem, Bioucas-Dias and Figueiredo [10] have presented a faster TwIST (two-step iterative shrinkage/thresholding) approach. TwIST uses the previous two iterative results to compute x in current iteration, which successfully accelerates the convergence rate. In the approach presented by Pan and Blu [11], the sharp image is treated as a linear combination of a series of known LET (linear expansion of thresholds) functions to reduce the scale of solving the problem **LP1**. Since the computing costs of the optimal LET coefficients are quite low, an image deconvolution problem denoted by **LP1** can be tackled very quickly. In order to code images more robustly and more sparsely, a NCSR (nonlocally centralized sparse representation) approach [12] has been used to learn a new PCA (principal component analysis) dictionary (i.e., a transform basis) which is applied to image deconvolution. As a representative approach of its type, NCSR has demonstrated that in comparison with analytically designed transform bases, learned dictionaries encourage image deconvolution approaches to obtain better results. Although synthesis-based approaches generally produce images with sharp edges, images that have been produced using these methods will inevitably contain undesirable artifacts, because this type of approach heavily penalizes the sparseness of the transform coefficients of unknown sharp images. Due to a difference in emphasis of the penalty, analysis-based approaches focus on addressing the problem **P0** which has the form $\{min_x||Wx||_q, \quad s.t. \quad ||Ax - y||_2 \le \epsilon\}$ (**P2**) or $\{min_x 0.5 \times ||Ax - y||_2^2 + \mu||Wx||_q\}$ (**LP2**), where **LP2** is the equivalent Lagrangian form of problem **P2**, W is an analysis operator (e.g., a wavelet frame), x is the unknown sharp image, $||Wx||_q$ is a sparse-inducing regularizer with $q = 0$ or $q = 1$, and the linear operator A denotes the convolution with a point spread function. To tackle the issue of the non-derivability of the l_q-norm with $q = 1$, Cai et al. [13] have employed a split Bregman method to divide and conquer problem **P2**. With only a single inner iteration and a "warm start", the approach of Cai et al. has been shown to be highly efficient for image deconvolution. Although the estimated images may be smoother than those generated by synthesis based approaches, since their method is an analysis-based approach, it has demonstrated superior performance to synthesis-based approaches in terms of noise suppression and artifact reduction. To balance the sharpness and smoothness of estimated images, in [14] the often-used l_1-norm has been replaced by the l_0-norm, which leads to a non-deterministic polynomial hard problem $\{min_x 0.5 \times ||Ax - y||_2^2 + \mu||Wx||_0\}$. To solve this NP-hard problem, Zhang et al. have introduced a penalty decomposition method and a block coordinate descent method, and have theoretically proved convergence of their presented approach. Experimental results obtained by their approach illustrate an inherent advantage of analysis-based approaches, and also show the balance that their work has achieved. More recently, Matakos et al. [15] and Almeida and Figueiredo [16] have both reported two independently studied but similar approaches which both employ an alternating direction method of multipliers. These two approaches both also consider unknown boundary conditions to eliminate artifacts, which makes them different from most existing approaches. As mentioned above, synthesis-based approaches and analysis-based approaches both have their own advantages and disadvantages and thus different approaches may be applicable to different types of images and different fields. Therefore, it cannot be concluded that either type of approach is superior [13].

Motivated by recent advances in approaches to the image deconvolution problem **P2**, this paper is devoted to solving the same problem using a novel analysis-based approach. To balance the smoothness and sharpness of image estimations and remove artifacts, the presented approach adopts a new equivalent formulation for the problem **P2**. By reformulation and decomposition, the new problem **P2** is converted into several tractable sub-problems. A good solution to the original problem is acquired by iteratively computing these sub-problems. The remainder of the paper is summarized by the following outline. In Sect. 2, image deconvolution is modeled as a new constrained minimization problem based on problem **P2**. In Sect. 3, the proposed image deconvolution model is efficiently tackled by the presented approach, and the computation cost of the presented approach is concisely analyzed. In next section, the validity and advantages of the presented approach are experimentally verified. Concluding remarks are drawn in the last section.

2   Deconvolution Problem

As mentioned above, this paper is devoted to addressing the image deconvolution problem **P2**. The constraint of this problem indicates that the feasible solutions of x lie in the ball

$$B(y, A, \epsilon) = \{x \in R^N : ||Ax - y||_2 \leq \epsilon\} \tag{1}$$

centered at point y, where the radius ϵ is proportional to the noise level. Thus, the problem **P2** becomes the following minimization problem

$$\min_x ||Wx||_1 + \chi_{B(0,I,\epsilon)}(Ax - y), \tag{2}$$

where the analysis operator $W \in R^{L \times N}$ is a tight frame which satisfies $W^t W = I$ [17] with t denoting the transposition, and the indicator function $\chi_C : R^M \to R$ is a proper convex function that is defined as:

$$\chi_C(c) = \begin{cases} 0 & if \quad c \in C \\ +\infty & if \quad c \notin C \end{cases} \tag{3}$$

with C denoting a closed nonempty convex set. Due to non-quadratic l_1-norm and inseparable structure, the solutions of problem (2) cannot be derived directly. To make problem (2) solvable, this paper suggests using variable splitting to tackle the issues of non-derivability and inseparability.

Consider the unconstrained minimization problem

$$\min_z f(z) + g(Pz), \tag{4}$$

where P is an arbitrary linear operator, $f : R^N \to R$ and $g : R^L \to R$ are closed and convex, and the function to be minimized is called a cost function which is generally non-quadratic. Variable splitting adopts a simple technique that introduces new splitting variables, i.e., $b1$ and $b2$, to replace the original arguments of problem (4), which results in the constrained minimization problem as follows:

$$\min_{b1,b2} f(b1) + g(b2) \quad s.t. \quad b1 = z, b2 = Pz \tag{5}$$

Problem (5) is equivalent to problem (4), but can be tackled more easily, which is really the rationale behind variable splitting. Besides image deconvolution, variable splitting has also been applied in several other fields, such as image inpainting, image reconstruction and image denoising. In [18], Zuo and Lin have jointly used variable splitting and a generalized accelerated proximal gradient to speed up image inpainting. Chen et al. [19] have employed variable splitting to transform the TV-based image reconstruction problem into a more easily handled problem. To efficiently remove image noise, Qin et al. [20] have presented an approach that involves an alternating direction augmented Lagrangian and variable splitting. According to the variable splitting methodology, problem (2) can be rewritten as

$$\min_{u,v} ||u||_1 + \chi_{B(0,I,\epsilon)}(v) \qquad s.t. \quad u = Wx, v = Ax - y \qquad (6)$$

with $u \in R^L$ and $v \in R^M$ denoting the splitting variables.

3 Presented Approach

To deal with problem (6), the constraints are relaxed by introducing data-fitting terms $||Wx - u||_2^2$ and $||Ax - y - v||_2^2$, yielding the unconstrained problem

$$\{\widehat{x}, \widehat{u}, \widehat{v}\} = \underset{x,u,v}{\operatorname{argmin}} ||u||_1 + \chi_{B(0,I,\epsilon)}(v) + \frac{\mu_1}{2}||Wx - u||_2^2 + \frac{\mu_2}{2}||Ax - y - v||_2^2 \quad (7)$$

which is also the Lagrangian of problem (6), where $\mu_1 > 0$ and $\mu_2 > 0$ are the Lagrangian multipliers. By using a continuation strategy to gradually increase the Lagrangian multipliers from very small values, the solutions of problem (7) will converge to that of problem (6) [21]. Since $||u||_1$ and $\chi_{B(0,I,\epsilon)}(v)$ are convex, and the data-fitting terms are convex and quadratic, the existence of a minimum value for problem (7) can be guaranteed. But the cost function is inseparable and non-quadratic and it has more than one argument, so the minimization problem (7) is not trivial in any case. This paper proposes adopting alternating minimization to address problem (7). Alternating minimization employs an iterative procedure that alternately minimizes the cost function jointly with respect to x, u and v, leading to

$$x^{k+1} = \underset{x}{\operatorname{argmin}} \mu_1||Wx - u^k||_2^2 + \mu_2||Ax - y - v^k||_2^2, \qquad (8)$$

$$u^{k+1} = \min_{u} ||u||_1 + \frac{\mu_1}{2}||Wx^{k+1} - u||_2^2, \qquad (9)$$

$$v^{k+1} = \arg\min_{v} \chi_{B(0,I,\epsilon)}(v) + \frac{\mu_2}{2}||Ax^{k+1} - y - v||_2^2 \qquad (10)$$

where the superscripts denote the iteration counters.

Obviously, when u and v are fixed, problem (8) becomes a least squares problem with a result given by

$$x^{k+1} = (\mu_1 I + \mu_2 A^t A)^{-1}[\mu_1 W^t u^k + \mu_2 A^t(y + v^k)], \qquad (11)$$

where A is a block-circulant matrix which can be diagonalized using the discrete Fourier transform. The FFT (fast Fourier transform) can be used as a tool for quick computation of x^{k+1}. With the help of this tool, the computation cost of the inversion of $\mu_1 I + \mu_2 A^t A$ can be reduced to $O(n\log n)$. Applying the FFT to the right side of Eq. (11) leads to

$$x^{k+1} = F^{-1}\left(\frac{\mu_1 F^*(W) \odot F(u^k) + \mu_2 F^*(A) \odot F(y + v^k)}{\mu_1 I + \mu_2 F^*(A) \odot F(A)}\right), \qquad (12)$$

where $F(\cdot)$ denotes the two-dimensional FFT, $F^*(\cdot)$ is the complex conjugate of $F(\cdot)$, $F^{-1}(\cdot)$ denotes the inverse FFT, and the symbol "$\odot$" denotes the dot product.

As both the l_1-norm and the l_2-norm have favorable separable structures, problem (9) can be equivalently written as

$$u^{k+1} = \underset{u}{\arg\min} \sum_t [|u_i| + \frac{\mu_1}{2}(W_i x^{k+1} - u_i)^2], \tag{13}$$

where $|\cdot|$ is the absolute value sign, u_i is the ith element of u, and W_i is the ith row of W. Therefore, for an arbitrary element u_i, problem (9) can be simplified to the equation given below

$$u_i^{k+1} = \underset{u_i}{\arg\min} |u_i| + \frac{\mu_1}{2}(W_i x^{k+1} - u_i)^2. \tag{14}$$

It is well-known that the close-formed solution of problem (14) follows a well-known shrinkage operator [22]

$$u_i = max(|W_i x^{k+1}| - \frac{1}{\mu_1}, 0) \times sgn(W_i x^{k+1}), \tag{15}$$

where sgn denotes the sign function. According to Eq. (15), the solution of the minimization problem (13) is given by

$$u^{k+1} = max(|W x^{k+1}| - \frac{1}{\mu_1}, 0) \odot sgn(W x^{k+1}), \tag{16}$$

In the above equation, a shrinkage operator is employed to compute the elements of u^{k+1} individually, thus there is an $O(n)$ cost for the computation of u^{k+1}.

Consider the minimization problem

$$\min_{\alpha}\{g(\alpha) = \frac{1}{2}||\beta - \alpha||_2^2 + \gamma\Phi(\alpha)\}, \tag{17}$$

where $\alpha \in R^Q$ and $\beta \in R^Q$ belong to some Hilbert space, $\Phi : R^Q \rightarrow R$ is a convex and proper function, and $g(\alpha)$ is also convex and proper. As given by Moreaus theory [23], the existence and uniqueness of the solution to the minimization problem (17) can be ensured, and this solution is a so-called proximal operator, that is

$$\Gamma_{\gamma\Phi}(\alpha) = \underset{\alpha}{\arg\min} \frac{1}{2}||\beta - \alpha||_2^2 + \gamma\Phi(\alpha). \tag{18}$$

Since the minimization problem (10) is projecting v onto an l_2-ball, its proximal operator has the following closed form [23]:

$$v^{k+1} = \Gamma_{\chi_B}(v) = \begin{cases} E & if \quad ||E||_2 \leq \epsilon \\ \epsilon \times E/||E||_2 & if \quad ||E||_2 > \epsilon \end{cases} \tag{19}$$

where $E = Ax^{k+1} - y$, and Eq. (19) can be computed with a cost of $O(n)$.

The presented approach is summarized below based on the solutions of problems (8), (9) and (10), which are given by Eqs. (12), (16) and (19), respectively.

Presented Approach

(1) **Input:** 1) μ_1 and μ_2; 2) u^0 and v^0; 3) W, A and y
(2) **Precompute:** $D = \mu_1 I + \mu_2 F^*(A) \odot F(A)$
(3) **Output:** x^{k+1}
(4) **for** $k = 0$ to K.
   a. $D_1 = \mu_1 F^*(W) \odot F(u^k), D_2 = \mu_2 F^*(A) \odot F(y + v^k)$
   b. $x^{k+1} = F^{-1}((D_1 + D_2)/D)$
   c. $d = W x^{k+1}$
   d. $u^{k+1} = max(|d| - 1/\mu_1, 0) \odot sgn(d)$
   e. Compute v^{k+1} according to equation (19)
(5) **end**

For each iteration of the presented approach, if FFT is used, x^{k+1} can be computed with $O(n)$ cost. Matrix-vector products and vector additions (or subtraction) require an $O(nlogn)$ cost and an $O(n)$ cost, respectively. Hence, in summary, the computation complexity of the presented approach is relatively low.

4   Numerical Experiments

To demonstrate the effectiveness and superiority of the presented approach, in this section we will conduct experiments on image deconvolution. Two famous gray test images, illustrated in Fig. 1, are adopted as the sharp images. All images have a resolution of 256×256. According to the degradation model, the measured images, shown in Fig. 2, are created by perturbing the sharp images with blurs and Gaussian noise. Both Gaussian blur and uniform blur are considered in the experiments, and the blur kernels of them are generated by using the MATLAB functions fspecial ('Gaussian', 9, 5) and fspecial ('average', 9), respectively, where the parameter "9" denotes that the size of the blur kernel is 9×9, and the parameter "5" is the value of the standard deviation. For comparison, several state-of-the-art approaches [11–14] are also introduced as competitors into the experiments, in addition to the presented approach. As stated in the first section, these competitors can be classified into two types: synthesis-based approaches [11,12] and analysis-based approaches [13,14]. All approaches are run using a notebook computer with the following configuration: MATLAB R2012b, Windows 7 Sp1 x64, and Intel Core i5-4258U CPU, and 4 GB RAM.

To maximize the performance of the presented approach, a redundant Haar frame with 4-level decompositions is selected as the analysis operator W, and the values of the arguments ϵ, μ_1 and μ_2 are manually adjusted to be 0.01, 0.1 and 1.5, respectively. For an equal comparison, the settings for the competitive approaches are kept the same as their corresponding references. As a

(a)   (b)

Fig. 1. Sharp images: (a) Boats, (b) Cameraman.

(a)   (b)   (c)   (d)

Fig. 2. Measured images: (a) to (b) Gaussian measured images, and (c) to (d) uniform measured images.

widely applicable objective criterion, the PSNR (peak signal-to-noise ratio) in dB defined as $10 \times log_{10}(N \times 255^2/||x^{k+1} - x||_2^2)$ serves as the evaluation tool for the quality of estimated images. The deconvolution results in terms of visual quality, PSNR and speed are reported in Figs. 3, 4, Tables 1 and 2. The "Time" rows in the tables reports the average time in seconds consumed by running the corresponding approach ten times. Although only two of the most popular types of blurs are considered, it is guaranteed that all of the image deconvolution approaches in experiments are independent the type of blurs. The experimental outputs comprehensively verify the validity of the presented approach, and also demonstrate the robustness of the presented approach. In comparison with its competitors, the presented approach provides better estimation results for sharp images, according to the PSNRs illustrated in Tables 1 and 2. Between all of the approaches, the presented approach shows the second best performance in terms of speed. Considering the difference in iterations between the presented app-roach and the Pan approach, the speed discrepancy demonstrated by these two approaches is reasonable. As illustrated in Figs. 3 and 4, compared with the pre-sented approach, the Pan approach and the Dong approach produce images with more artifacts, and the Cai approach and the Zhang approach produce smoother images. Of the five image deconvolution approaches used in experiments, only the presented approach balances both the smoothness and the sharpness of out-put images while significantly reducing the artifacts, which is a second advantage that the presented approach has over its competitors. Since it has been shown

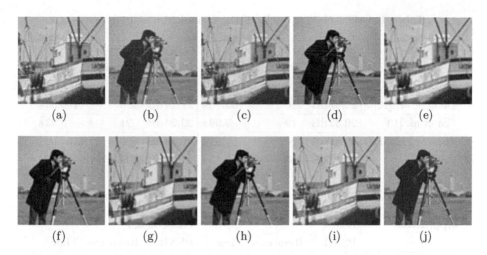

Fig. 3. Estimated sharp images by deconvolving Gaussian measured images: (a) to (b) outputs of the Pan approach [11], (c) to (d) outputs of the Dong approach [12], (e) to (f) outputs of the Cai approach [13], (g) to (h) outputs of the Zhang approach [14], and (i) to (j) outputs of the presented approach.

Fig. 4. Estimated sharp images by deconvolving uniform measured images: (a) to (b) outputs of the Pan approach [11], (c) to (d) outputs of the Dong approach [12], (e) to (f) outputs of the Cai approach [13], (g) to (h) outputs of the Zhang approach [14], and (i) to (j) outputs of the Presented approach.

from the PSNR values and visual quality that the competitive approaches are all data-dependent, the presented approach thus has a third advantage over its competitors since the output stability is data-independent.

Table 1. Deconvolution results of the Gaussian measured images.

Approaches	Boats			Cameraman		
	PSNR	Iterations	Time	PSNR	Iterations	Time
Pan and Blu [11]	29.07 dB	6	0.31 s	26.59 dB	6	0.29 s
Dong et al. [12]	29.57 dB	480	163.16 s	27.24 dB	480	153.33 s
Cai et al. [13]	29.32 dB	19	5.69 s	26.24 dB	21	5.93 s
Zhang et al. [14]	29.41 dB	500	67.82 s	26.46 dB	500	64.62 s
Presented	30.17 dB	300	3.12 s	27.84 dB	300	3.16 s

Table 2. Deconvolution results of the uniform measured images.

Approaches	Boats			Cameraman		
	PSNR	Iterations	Time	PSNR	Iterations	Time
Pan and Blu [11]	29.39 dB	6	0.33 s	26.98 dB	6	0.30 s
Dong et al. [12]	30.57 dB	480	150.71 s	28.26 dB	480	153.71 s
Cai et al. [13]	29.90	19	4.72 s	26.67 dB	21	5.84 s
Zhang et al. [14]	29.85 dB	500	59.08 s	27.13 dB	500	55.54 s
Presented	30.60 dB	300	3.07 s	28.43 dB	300	2.88 s

5   Concluding Remarks

To deal with a new image deconvolution problem modeled by Eq. (2), this paper presents a novel image deconvolution approach based on alternating minimization, variable splitting and other relevant methods. In the numerical experiments, the presented approach is used to estimate the sharp versions of two types of measured images, and is compared with four state-of-the-art approaches. The experimental results clearly show the effectiveness and the advantages of the presented approach in terms of suppressing artifacts and preserving both smoothness and sharpness of the estimated images. Although the presented approach is only applied to image deconvolution, it can be extended to other inverse imaging problems, such as image inpainting and image denoising.

Acknowledgments. This work is supported by Anhui Provincial Natural Science Foundation (No. 1608085QF150).

References

1. Hamarik, U., Palm, R., Raus, T.: A family of rules for parameter choice in Tikhonov regularization of ill-posed problems with inexact noise level. J. Comput. Appl. Math. **236**, 2146–2157 (2012)
2. Bertero, M., Boccacci, P., Talenti, G., Zanella, R., Zanni, L.: A discrepancy principle for poisson data. Inverse Prob. **26**, 105004 (2010)
3. Reichel, L., Sgallari, F., Ye, Q.: Tikhonov regularization based on generalized Krylov subspace methods. Appl. Numer. Math. **62**, 1215–1228 (2012)

4. Neuman, A., Reichel, L., Sadok, H.: Implementations of range restricted iterative methods for linear discrete ill-posed problems. Linear Algebra Appl. **436**, 3974–3990 (2012)
5. Paul, G., Cardinale, J., Sbalzarini, I.F.: Coupling image restoration and segmentation: a generalized linear model/Bregman perspective. Int. J. Comput. Vis. **104**, 69–93 (2013)
6. Chan, R.H., Tao, M., Yuan, X.: Constrained total variation deblurring models and fast algorithms based on alternating direction method of multipliers. SIAM J. Imaging Sci. **6**, 680–697 (2013)
7. Perrone, D., Favaro, P.: Total variation blind deconvolution: the devil is in the details. In: IEEE Conference on Computer Vision and Pattern Recognition, pp. 2909–2916. IEEE Press, New York (2014)
8. Wright, J., Ma, Y., Mairal, J., Sapiro, G., Huang, T.S., Yan, S.: Sparse representation for computer vision and pattern recognition. Proc. IEEE **98**, 1031–1044 (2010)
9. Figueiredo, M.A.T., Nowak, R.D.: A bound optimization approach to wavelet-based image deconvolution. In: IEEE International Conference on Image Processing, pp. 782–785. IEEE Press, New York (2005)
10. Bioucas-Dias, J.M., Figueiredo, M.A.T.: A new TwIST: two-step iterative shrinkage/thresholding algorithms for image restoration. IEEE Trans. Image Process. **16**, 2992–3004 (2007)
11. Pan, H., Blu, T.: An iterative linear expansion of thresholds for $\ell 1$-based image restoration. IEEE Trans. Image Process. **22**, 3715–3728 (2013)
12. Dong, W., Zhang, L., Shi, G., Li, X.: Nonlocally centralized sparse representation for image restoration. IEEE Trans. Image Process. **22**, 1620–1630 (2013)
13. Cai, J., Osher, S., Shen, Z.: Split Bregman methods and frame based image restoration. Multiscale Model. Simul. **8**, 337–369 (2010)
14. Zhang, Y., Dong, B., Lu, Z.: l0 minimization for wavelet frame based image restoration. Math. Comput. **82**, 995–1015 (2013)
15. Matakos, A., Ramani, S., Fessler, J.A.: Accelerated edge-preserving image restoration without boundary artifacts. IEEE Trans. Image Process. **22**, 2019–2029 (2013)
16. Almeida, M.S.C., Figueiredo, M.A.T.: Deconvolving images with unknown boundaries using the alternating direction method of multipliers. IEEE Trans. Image Process. **22**, 3074–3086 (2013)
17. Mallat, S.: A Wavelet Tour of Signal Processing: The Sparse Way. Academic Press, Cambridge (2009)
18. Zuo, W., Lin, Z.: A generalized accelerated proximal gradient approach for total-variation-based image restoration. IEEE Trans. Image Process. **20**, 2748–2759 (2011)
19. Chen, Y., Hager, W., Huang, F., Phan, D., Ye, X.: Fast algorithms for image reconstruction with application to partially parallel MR imaging. SIAM J. Imaging Sci. **5**, 90–118 (2012)
20. Qin, Z., Goldfarb, D., Ma, S.: An alternating direction method for total variation denoising. Optim. Method Softw. **30**, 594–615 (2015)
21. Wright, S.J., Nowak, R.D., Figueiredo, M.A.T.: Sparse reconstruction by separable approximation. IEEE Trans. Signal Process. **57**, 2479–2493 (2007)
22. Yang, J., Zhang, Y., Yin, W.: An efficient TVL1 algorithm for deblurring multichannel images corrupted by impulsive noise. SIAM J. Sci. Comput. **31**, 2842–2865 (2009)
23. Combettes, P.L., Wajs, V.R.: Signal recovery by proximal forward-backward splitting. Multiscale Model. Simul. **4**, 1168–1200 (2005)

An Object Reconstruction Method Based on Binocular Stereo Vision

Yu Liu[✉], Chao Li, and Jixiang Gong

School of Aerospace Engineering, Xiamen University, Xiamen 361005, China
mseliuyu@xmu.edu.cn

Abstract. This paper presents an object reconstruction method based on a binocular stereo vision system. First, the relative position and orientation between the two cameras of the system are obtained by a binocular calibration process. Then feature pixels are extracted from images of the two cameras of the same scene according to the SIFT (Scale Invariant Feature Transform) feature. Feature pixels on the image of one camera are matched with those of the other camera according to differences between their SIFT features. And the RANSAC (Random Sample Consensus) algorithm is used to eliminate incorrect matched pixels. Then a 3D coordinate point is obtained from each pair of matched pixels. Finally, 3D models are constructed from the 3D coordinate points through triangulation and texture mapping. In the above processes, a uniform method of calculating coordinates of 3D points from pixel pairs is introduced, which is fitted for arbitrarily orientated optical axes of the left and the right cameras. Experiment results show that the proposed method can obtain 3D points sampled from real objects and produce 3D models consistent with reality.

Keywords: Stereo vision · Binocular calibration · Stereo matching · Point cloud acquisition

1 Introduction

Object reconstruction based on binocular stereo vision is a process that builds spatial structures of a real object through recovering depth information from two images of the two cameras simulating human vision. Stereo matching is a vital step for recovering depth information [1, 2] in a binocular stereo vision system, which can be classified into local matching methods and global matching methods. The local matching methods can be further classified into the feature based method [3], the phase based method [4], and the region based method [5] according to different matching elements. The feature based method matches image features such as points and lines. The number of possible matching features can be reduced and the matching process can be simplified by using specified structures in different images [6]. However the sparsity of features results in fewer matching points and the matching performance depends on the accuracy of extracted features. The phase based method gets disparity maps by processing phase information of a bandpass filter. It can suppress high frequency noises of images and has problems of phase singularities and phase multi-values. The region based methods takes pixels in a certain sized window as a matching element. The

© Springer International Publishing AG 2017
Y. Huang et al. (Eds.): ICIRA 2017, Part II, LNAI 10463, pp. 486–495, 2017.
DOI: 10.1007/978-3-319-65292-4_42

region based method is used widely. But the proper size of the window is difficult to be determined. Global matching methods include the graph cut based method [7], the efficient belief propagation method [8], and the dynamic programming method [9]. Compared with local matching methods, global matching methods are easier to be influenced by image distortions and illumination changes. Our method extracts image feature points by the SIFT detection algorithm [10] and eliminates incorrect matched points by the RANSAC algorithm [11].

There are different methods of calculating coordinates of 3D points from matched image points. If optical axes of the left and right cameras are parallel or intersected with each other, then the 3D coordinates can be deduced from the similarity of triangles in the camera model [12]. In order to eliminate the vertical disparity in stereo matching, Epipolar correction methods [13, 14] are applied to align epipolar lines with horizontal scan lines of images. References planes in scenes are usually used in the epipolar correction methods. The quality of the epipolar correction depends much on how to choose the reference planes. In this paper, a uniform formula to calculate coordinates of 3D points from matched image points is introduced, which can be used when the optical axes of the left and right cameras are arbitrarily orientated and the inner parameters of the two cameras have the different values.

Our method consists of four main steps (shown by Fig. 1): binocular calibration, stereo matching, 3D point cloud acquisition, triangulation and texture mapping. First, the two cameras of our binocular stereo vision system are calibrated separately to get values of their inner and outer parameters. And the relative position and orientation of the two cameras are calculated according to their outer parameters. Secondly, two images of the same scene are collected by the two cameras. Features are extracted from the two images through the SIFT detection algorithm. Pixels of one image are matched with those of the other image according to differences between their SIFT features. And the RANSAC algorithm is used to eliminate bad matched pixels. Then a 3D coordinate point is calculated by our uniform formula from each pair of matched pixels. Finally, 3D models are constructed from the 3D coordinate points through triangulation and texture mapping. The remainder of this paper is organized as follows: Sects. 2, 3, 4 introduce the binocular calibration, the stereo matching, and the 3D point cloud acquisition respectively. Section 5 gives some experiment results. Conclusions are offered in Sect. 6.

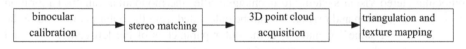

Fig. 1. Steps of our method

2 Binocular Calibration

Pinhole camera model is used to describe the imaging process of the left and the right cameras of our binocular stereo vision system. As shown in Fig. 2, there are four coordinate systems in the pinhole camera model: the image coordinate system O-uv, the image plane coordinate system O_1-xy, the camera coordinate system O_C-$X_C Y_C Z_C$, and the world coordinate system O_W-$X_W Y_W Z_W$.

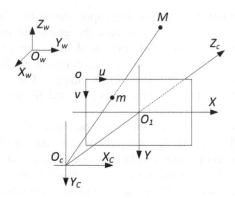

Fig. 2. The imaging model of a pinhole camera

Assume (x_w, y_w, z_w) and (x_c, y_c, z_c) are the coordinates of a 3D spatial point M under the world coordinate system and the camera coordinate system respectively. Assume (x, y) and (u, v) are the coordinates of the image point m of M under the image plane coordinate system and the image coordinate system respectively. Then $u = x/dx + u_0$ and $v = y/dy + v_0$, where dx and dy are distances between pixels on photosensitive components of cameras, (u_0, v_0) is the coordinates of the point O_1 under the image coordinate system. The projection relationship between point M and its image point m can be given by [15]

$$z_c \begin{bmatrix} u \\ v \\ 1 \end{bmatrix} = \begin{bmatrix} f/dx & 0 & u_0 \\ 0 & f/dy & v_0 \\ 0 & 0 & 1 \end{bmatrix} \begin{bmatrix} {}_w^c R & {}_w^c T \end{bmatrix} \begin{bmatrix} x_w \\ y_w \\ z_w \\ 1 \end{bmatrix} \tag{1}$$

where f is the focal distance of the camera, ${}_w^c R$ and ${}_w^c T$ denote the rotation matrix and the translation vector from the world coordinate system to the camera coordinate system respectively. The inner parameters $(f/dx, f/dy, u_0, v_0)$ and the outer parameters $\left({}_w^c R, {}_w^c T \right)$ can be obtained by a monocular camera calibration process.

Relative position and orientation between the left and the right cameras of our binocular stereo vision system are unchanged after the monocular camera calibration process. The position and orientation can be calculated from the outer parameters of the two cameras. Assume ${}_w^c R_l$ and ${}_w^c T_l$ is the outer parameters of the left camera C_l. And ${}_w^c R_r$ and ${}_w^c T_r$ is the outer parameters of the right camera C_r. Assume (x_{cl}, y_{cl}, z_{cl}) and (x_{cr}, y_{cr}, z_{cr}) are the coordinates of the 3D spatial point M under the left camera coordinate system and the right camera coordinate system respectively. Then

$$\begin{bmatrix} x_{cl} \\ y_{cl} \\ z_{cl} \\ 1 \end{bmatrix} = \begin{bmatrix} {}_w^c R_l & {}_w^c T_l \\ 0 & 1 \end{bmatrix} \begin{bmatrix} x_w \\ y_w \\ z_w \\ 1 \end{bmatrix}, \tag{2}$$

$$\begin{bmatrix} x_{cr} \\ y_{cr} \\ z_{cr} \\ 1 \end{bmatrix} = \begin{bmatrix} {}^c_w R_r & {}^c_w T_r \\ 0 & 1 \end{bmatrix} \begin{bmatrix} x_w \\ y_w \\ z_w \\ 1 \end{bmatrix}. \tag{3}$$

From Eqs. (2) and (3), we have

$$\begin{bmatrix} x_{cr} \\ y_{cr} \\ z_{cr} \\ 1 \end{bmatrix} = \begin{bmatrix} {}^c_w R_r \left({}^c_w R_l \right)^T & {}^c_w T_r - {}^c_w R_r \left({}^c_w R_l \right)^T {}^c_w T_l \\ 0 & 1 \end{bmatrix} \begin{bmatrix} x_{cl} \\ y_{cl} \\ z_{cl} \\ 1 \end{bmatrix} = \begin{bmatrix} {}^r_l R & {}^r_l T \\ 0 & 1 \end{bmatrix} \begin{bmatrix} x_{cl} \\ y_{cl} \\ z_{cl} \\ 1 \end{bmatrix}, \tag{4}$$

where the right superscript 'T' denotes matrix transpose. Equation (4) gives the relative position and orientation between the left camera C_r and the right camera C_l.

3   Stereo Matching

In order to calculate coordinates of 3D points from 2D images obtained from the left and the right cameras, the pixel correspondence between the left camera image and the right camera image needs to be established. The two corresponding pixels should be the image points of the same 3D spatial point. Our method gets corresponding pixels through matching feature points in the two images.

Feature points are extracted from images of the two cameras of the same scene according to the SIFT (Scale Invariant Feature Transform) feature [10]. Feature points on the image of one camera are matched with those of the other camera according to differences between their SIFT features. Then the RANSAC algorithm [11] is applied to eliminate bad matched pixel pairs. Assume (u_{li}, v_{li}) and (u_{ri}, v_{ri}) are the two pixels in a matched pixel pair, where $1 \leq i \leq m$, m is the number of matched pixel pairs. Then there exits the following transformation

$$\begin{bmatrix} u_{ri} \\ v_{ri} \\ 1 \end{bmatrix} = H \begin{bmatrix} u_{li} \\ v_{li} \\ 1 \end{bmatrix} = \begin{bmatrix} h_{11} & h_{12} & h_{13} \\ h_{21} & h_{22} & h_{23} \\ h_{31} & h_{32} & h_{33} \end{bmatrix} \begin{bmatrix} u_{li} \\ v_{li} \\ 1 \end{bmatrix}. \tag{5}$$

The transformation matrix H in the above equation can be obtained through at least four matched pixel pairs. The consistence of the matched pixel (u_{li}, v_{li}) and (u_{ri}, v_{ri}) can be evaluated by the deviation

$$e_H = \left\| \begin{bmatrix} u_{ri} \\ v_{ri} \\ 1 \end{bmatrix} - H \begin{bmatrix} u_{li} \\ v_{li} \\ 1 \end{bmatrix} \right\|, \tag{6}$$

where $\|\cdot\|$ denotes two norm of a vector. If $e_H < e_0$, (u_{li}, v_{li}) and (u_{ri}, v_{ri}) are considered as good matched pixels, otherwise they are bad matched pixels and need to be discarded, where e_0 is a predefined threshold.

The following is the steps of the RANSAC algorithm of eliminating bad matched pixel pairs. First, several (≥ 4) matched pixel pairs are selected from all the matched pixel pairs randomly and are used to calculate the transformation matrix H in Eq. (5). Then based on Eq. (6), all the matched pixel pairs are divided into two sets: one consists of good matched pixel pairs and the other consists of bad matched pixel pairs. Repeating the above process k times produces k different sets each of which consists of good matched pixel pairs. Finally, the set that has the maximum number of good matched pixel pairs is chosen as the results of eliminating bad matched pixel pairs.

4   3D Point Cloud Acquisition

Traditional methods [12–14] of calculating coordinates of 3D points from images assume that the inner parameters of the left camera are the same as the right camera or the optical axis of the left camera is parallel to or intersects with that of the right camera. However, in fact, the inner parameters of the left camera are often different from that of the right camera and their optical axes may be interlaced with each other. A uniform formula to calculate coordinates of 3D points from matched pixels will be introduced in the following, which can be used when the optical axes of the left and the right cameras are arbitrarily oriented and the inner parameters of the two cameras have different values.

As shown in Fig. 3, m_l and m_r are image points of the 3D spatial point M on the left and the right image planes respectively. Let (x_l, y_l) and (u_l, v_l) be the coordinates of m_l under the left image plane coordinate system and the left image coordinate system respectively. Let (x_r, y_r) and (u_r, v_r) be the coordinates of m_r under the right image plane coordinate system and the right image coordinate system respectively.

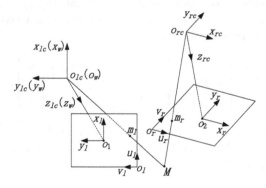

Fig. 3. The imaging model of our binocular stereo system

Assume the world coordinate system $(O_w - x_w y_w z_w)$ coincides with the left camera coordinate system $(O_{lc} - x_{lc} y_{lc} z_{lc})$. Then

$$z_w \begin{bmatrix} u_l \\ v_l \end{bmatrix} = \begin{bmatrix} f_l/dx_l & 0 & u_{10} \\ 0 & f_l/dy_l & v_{10} \end{bmatrix} \begin{bmatrix} x_w \\ y_w \\ z_w \end{bmatrix}, \tag{7}$$

$$z_{cr} \begin{bmatrix} u_r \\ v_r \\ 1 \end{bmatrix} = \begin{bmatrix} f_r/dx_r & 0 & u_{20} \\ 0 & f_r/dy_r & v_{20} \\ 0 & 0 & 1 \end{bmatrix} \begin{bmatrix} {}^r_l R & {}^r_l T \end{bmatrix} \begin{bmatrix} x_w \\ y_w \\ z_w \\ 1 \end{bmatrix}, \tag{8}$$

where (u_{10}, v_{10}) is the coordinates of the point O_1 under the left camera image coordinate system $(O_l - x_l y_l)$, (u_{20}, v_{20}) is the coordinates of the point O_2 under the right camera image coordinate system $(O_r - x_r y_r)$. From Eqs. (7) and (8), we have equations about (x_w, y_w, z_w):

$$\begin{bmatrix} H_r {}^r_l R \\ H_l \end{bmatrix} \begin{bmatrix} x_w \\ y_w \\ z_w \end{bmatrix} = \begin{bmatrix} -H_r {}^r_l T \\ 0 \end{bmatrix}, \tag{9}$$

where $H_r = \begin{bmatrix} f_r/dx_r & 0 & u_{20} - u_r \\ 0 & f_r/dy_r & v_{20} - v_r \end{bmatrix}$ and

$H_l = \begin{bmatrix} f_l/dx_l & 0 & u_{10} - u_l \\ 0 & f_l/dy_l & v_{10} - v_l \end{bmatrix}$. The solution of Eq. (9) is

$$\begin{bmatrix} x_w \\ y_w \\ z_w \end{bmatrix} = - \begin{bmatrix} {}^r_l R^T H_r^T H_r {}^r_l R + H_l^T H_l \end{bmatrix}^{-1} {}^r_l R^T H_r^T H_r {}^r_l T. \tag{10}$$

Equation (10) is the uniform formula to calculate coordinates of 3D points from matched pixels. The (u_l, v_l) and (u_r, v_r) in Eq. (10) are the coordinates of two matched pixels.

5 Experiments

Figure 4 shows our binocular stereo vision system which consists of two cameras, a tripod, a camera mount, a computer, and corresponding software system. The model of the cameras is Point Grey GRAS-20S4 M whose maximum resolution is 1624 × 1224. The computer has a 2.4 GHz CPU and 2 GB of RAM.

Fig. 4. Binocular stereo vision system

Fig. 5. Feature points extracted from images of a spoon and a cup.

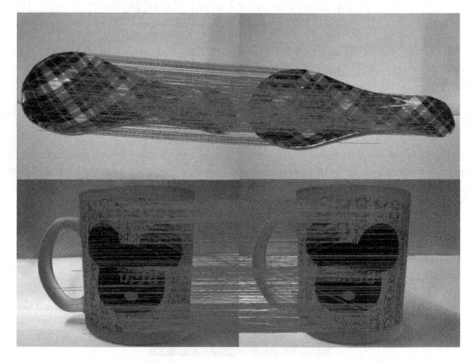

Fig. 6. Initial matched pixels between images of the left and the right camera.

Fig. 7. Pixel pairs after eliminating incorrect matched pixels.

Parameters of the left and the right cameras of the stereo vision system are obtained by the binocular calibration. Images of a chessboard calibration plate are used to get the parameters of the two cameras and relative positions between the two cameras. The results are $f_l/dx_l = 1884.62$, $f_l/dy_l = 1884.48$, and $(u_{10}, v_{10}) = (821.39, 585.95)$ for the left camera; $f_r/dx_r = 1871.52$, $f_r/dy_r = 1870.74$, and $(u_{20}, v_{20}) = (824.79, 608.45)$ for the right camera. The relative positions between the two cameras are

$$_l^r R = \begin{bmatrix} 0.999912 & -0.009713 & -0.009008 \\ 0.009650 & 0.999929 & -0.006985 \\ 0.009075 & 0.006897 & 0.999935 \end{bmatrix}, _l^r T = \begin{bmatrix} 69.24 \\ 0.32 \\ 4.42 \end{bmatrix}.$$

Then images of two real objects are collected and models of the two objects are reconstructed. Figure 5 shows feature pixels obtained by the SIFT detection algorithm, where '+' denotes feature pixels, the left and the right side by side images are collected from the left and the right cameras respectively. The number of feature pixels in the images of the spoon is 3274. The number of feature pixels in the images of the cup is 3274.

Feature pixels in the left image of each object are matched with those in the right image. Figure 6 shows the matched pixels, where the two corresponding pixels are connected by a line. The numbers of pixel pairs in Fig. 6 are 419 and 623 for the spoon and the cup respectively. The pixel pairs after eliminating incorrect ones are shown in Fig. 7. The numbers of pixel pairs in Fig. 7 are 284 and 498 for the spoon and the cup respectively.

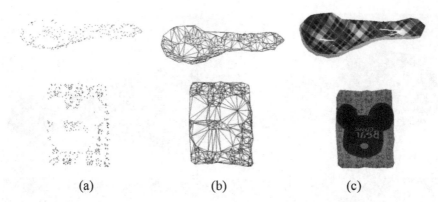

Fig. 8. Reconstructed models of the spoon and the cup: (a) point cloud; (b) triangle mesh; (c) triangle mesh after texture mapping.

Coordinates of 3D points corresponding to each pixel pair are calculated according to Eq. (10). These points are converted into triangle meshes. And texture mapping is applied to obtain a realistic display. Figure 8 shows results obtained by our method from images of the spoon and the cup.

6   Conclusion and Future Work

This paper presents an object reconstruction method based on a binocular stereo vision system. The method gets values of the inner parameters and relative position and orientation of the two cameras of the stereo vision system through a binocular camera calibration process. Then the method matches pixels of images of a real object by SIFT features and eliminates bad matched pixel pairs by the RANSAC algorithm. And coordinates of 3D points sampled from the object are calculated from matched pixel pairs. Finally, 3D model of the object is obtained from the 3D points through triangulation and texture mapping. There is no assumption about the orientation and the inner parameters of the two cameras. Then the method can be used when the optical axes of the left and the right cameras are arbitrarily orientated and the inner parameters of the two cameras have the same or the different values. Experiment results show that the proposed method can obtain 3D points sampled from real objects and produce 3D models consistent with reality. For the future work, we would like to analyze the measurement accuracy of 3D points obtained by the binocular stereo vision system and develop the method of producing dense 3D sampled points.

Acknowledgment. This work is supported by the National Natural Science Foundation of China (Grant No. 51205332), the SRF for the Returned Overseas Chinese Scholars, and Fujian Science and Technology Major Project (No. 2015HZ0002-1).

References

1. Luo, G.: Some issues of depth perception and three dimention reconstruction from binocular stereo vision. Central South University, China (2012)
2. Geng, Y.: Research on stereo matching algorithms. Jilin University, China (2014)
3. Han, H., Han, X., Fang, F.: A new stereo matching method based on edges and corners. J. Comput. Inf. Syst. **8**(14), 6041–6048 (2012)
4. El-etriby, S., Al-hamadi, A.K., Michaelis, B.: Dense depth map reconstruction by phase difference-based algorithm under influence of perspective distortion. Mach. Graph. Vis. **15**(3), 349–361 (2006)
5. Zhang, G., Hua, W., Qin, X., et al.: Stereoscopic video synthesis from a monocular video. IEEE Trans. Vis. Comput. Graph. **13**(4), 686–696 (2007)
6. Liu, K., Zhou, C., Wei, S., et al.: Optimized stereo matching in binocular 3D measurement system using structured light. Appl. Opt. **53**(26), 6083–6090 (2014)
7. Bleyer, M., Gelautz, M.: Graph-cut-based stereo matching using image segmentation with symmetrical treatment of occlusions. Sig. Process. Image Commun. **22**(2), 127–143 (2007)
8. Felzenszwalb, P., Huttenlocher, D.: Efficient belief propagation for early vision. Int. J. Comput. Vis. **70**(1), 41–54 (2006)
9. Gong, M., Yang, Y.H.: Fast unambiguous stereo matching using reliability-based dynamic programming. IEEE Trans. Pattern Anal. Mach. Intell. **27**(6), 998–1003 (2005)
10. Lowe, D.G.: Distinctive image features from scale invariant key points. Int. J. Comput. Vis. **60**(2), 91–110 (2004)
11. Martin, A.F., Robert, C.B.: Random sample consensus: a paradigm for model fitting with application to image analysis and automated cartography. Commun. ACM **24**(6), 381–395 (1981)
12. Shrivasthava, P., Vundavilli, P.R., Pratihar, D.K.: An approach for 3D reconstruction of environment using stereo-vision system. In: IEEE Region 10 and the Third International Conference on Industrial and Information Systems, pp. 1–7 (2008)
13. Hartley, R.I.: Theory and practice of projective rectification. Int. J. Comput. Vis. **35**(2), 115–127 (1999)
14. Al-Zahrani, A., Ipson, S.S., Haigh, J.G.B.: Applications of a direct algorithm for the rectification of uncalibrated images. Inf. Sci. **160**(1–4), 53–71 (2004)
15. Zhang, Z.: A flexible new technique for camera calibration. IEEE Trans. Pattern Anal. Mach. Intell. **22**(11), 1330–1334 (2000)

A New Method of Determining Relative Orientation in Photogrammetry with a Small Number of Coded Targets

Hao Wu[1], Xu Zhang[2(✉)], and Limin Zhu[1]

[1] Robotics Institute, Shanghai Jiao Tong University,
No. 800, Dongchuan Road, Minhang District, Shanghai 200240, China
[2] School of Mechatronic Engineering and Automation, Shanghai University,
No. 99, Shangda Road, BaoShan District, Shanghai 200444, China
xuzhang@shu.edu.cn

Abstract. In traditional close range photogrammetry, more than eight coded targets are adopted to determine the relative orientation. Generally, dozens or hundreds are needed for higher precision. It is a tedious task to place them in the scene. In this paper, a new method with less than eight coded targets is proposed. The key idea is the coded pattern not only provides identification information, but also contains substantial location information. A new detection method of coded targets is developed to recognize the coded targets and excavate the location information in the coded pattern. All this location information is adopted to determine the relative orientation. The experiment results show that two coded targets are enough to determine the relative orientation, and three are enough for high precision.

Keywords: Photogrammetry · Coded targets · Relative orientation · Corner information

1 Introduction

Close range photogrammetry is an important and popular method for 3D coordinates measurement in industry [1–6]. One fundamental task in these photogrammetric systems is determining relative orientation with coded targets, which have great advantage in the correspondence.

Some research institutes have proposed different systems of close range photogrammetry. In a method of photogrammetric measurement automation using TRITOP system and industrial robot introduced by Koutecky et al. [1], 26 coded markers are placed on the experimental object. Drap et al. [2] use 20 coded targets in the measurement of red coral by underwater photogrammetry. A relative orientation method for large-scale photogrammetry with local parameter optimization is proposed by Li et al. [3], in which 20 coded targets are distributed evenly in the measurement scene and the number of identical coded targets shared by two images is asked to be more than 8. There are also some optical 3D measurement systems on the market, such as the V-STARS N Series which is the latest innovation in affordable turnkey image-based

© Springer International Publishing AG 2017
Y. Huang et al. (Eds.): ICIRA 2017, Part II, LNAI 10463, pp. 496–507, 2017.
DOI: 10.1007/978-3-319-65292-4_43

industrial 3D coordinate measurement systems developed by the American company Geodetic System, Inc. (GSI). The V-STARS N Series offers more than three hundred coded targets.

Generally speaking, the number of coded targets used in the existing photogrammetric measurement systems varies from dozens to hundreds. One specific requirement is that at least 8 identical coded targets appear in each pair of images. This is because the eight-point algorithm is used for computing the fundamental matrix for determination of relative orientation [7].

The most common types of coded targets are shown in Fig. 1. Identification information is embedded in the form of linear, radial or angular bar codes. To use the coded target, the circle or cross mark in the center provides location information for determining the relative orientation, while the surrounding patterns provide identification information for recognizing the coded target.

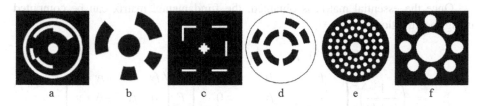

Fig. 1. Various types of coded targets.

In fact, what the surrounding patterns can provide is not only identification information, but also location information of their corners of bars or centers of small circles. From this point of view, the conventional way of using coded targets creates much waste of information.

Thanks to the development of computer performance and machine vision algorithm, the processing speed of measurement has been increased a lot. It costs only hundreds of milliseconds to process all of the images taken from one optical measurement project. By contrast, setting a large number of coded targets in the measurement scene can be a more time consuming work. Developing a relative orientation technique which requires less coded targets is thus very important.

In order to shorten the overall duration of measurement procedure, this paper analyzes the possibility of using only a small number of coded targets to determine the relative orientation between camera poses and proposes a practical method to implement it by utilizing the location information of surrounding patterns in coded targets. This method is validated experimentally afterwards.

2 Determination of Relative Orientation Using Coded Targets

By studying the morphological characteristics of coded targets, an innovative way of using coded targets for determination of relative orientation is put forward whose main idea is to utilize location information of surrounding patterns in coded targets.

2.1 Methods to Determine Relative Orientation

The fundamental matrix is a common tool in the analysis of scenes taken with two uncalibrated cameras. The eight-point algorithm is a frequently cited method for computing the fundamental matrix from a set of eight or more point matches. Because of its advantage of simplicity of implementation, the eight-point algorithm for computing the fundamental matrix is widely used in photogrammetric measurement systems for determination of relative orientation, which generates the requirement that eight or more identical coded targets should be shared by two views.

Under the circumstance where the cameras are calibrated, the essential matrix instead of the fundamental matrix can be used to describe relative motion between two views. One of the important properties of the essential matrix is that it encapsulates the epipolar geometry of the image configuration. One common way to compute the essential matrix is the five-point algorithm [8], in which at least five point correspondences are required.

Once the essential matrix is obtained, the fundamental matrix can be computed using the equation

$$
\mathbf{F} = K^{-T} \mathbf{E}\, K^{-1}
$$

$$
= \left(\frac{1}{fx * fy}\right)^2
\begin{pmatrix}
fy & 0 & 0 \\
0 & fx & 0 \\
-fy * u & -fx * v & fx * fy
\end{pmatrix}
E
\begin{pmatrix}
fy & 0 & -fy * u \\
0 & fx & -fx * v \\
0 & 0 & fx * fy
\end{pmatrix}
\tag{1}
$$

where F is the fundamental matrix, K is the camera intrinsic matrix, E is the essential matrix, fx and fy are the focal lengths, u and v are the coordinates of the center of projection.

It is obvious that less point matches are needed in the five-point algorithm than in the eight-point algorithm. In order to reduce the number of coded targets needed in photogrammetric measurement system, the five-point algorithm is adopted in this paper.

An additional advantage of using the five-point algorithm is that it can deal with the situation where the points are coplanar, while the eight-point algorithm cannot give correct results because of the planar degeneration.

2.2 Morphological Characteristics of the Coded Targets

Out of consideration for robust and reliable image processing, coded targets in photogrammetric measurement systems are geometrically constructed only of circular elements in most situations. The identification number of the target is coded down into the circular arrangement of the coding patterns around the point mark at the center of the target.

The coded targets used in this paper are shown in Fig. 2. The circle in the center with a radius of 5 mm allows to reach high accuracy of point position estimation due to subpixel target center coordinates determination. The circular belt around the central circle is equally divided into eight segments. Each white segment indicates the number "1",

while each black segment indicates the number "0". Eight segments altogether form a binary number, which can be converted to a decimal number as the unique identification number of the coded target. Table 1 shows the identification numbers of the coded targets in Fig. 2.

Table 1. Identification numbers of the coded targets used in this paper.

Coded target	Binary identification number	Binary identification number
Figure 2a	00000001	1
Figure 2b	00100111	39
Figure 2c	01010111	87
Figure 2d	01011111	95
Figure 2e	01111111	127

2.3   Corner Information in the Coded Targets

As the requirement of five-point algorithm for computing the essential matrix, each pair of images need to have at least five point matches. As far as the existing photogrammetric measurement systems are concerned, this requirement implies that at least five coded targets appear in the scene. Noticing that every segment of circular belt has four corners, one coded target can provide more than one point for computing the essential matrix. For example, the coded target in Fig. 2a has four corners, combining with the center of central circle, it can provide a total of five points. Similarly, the coded target in Fig. 2e also has five available points, the coded targets in Fig. 2b and d have nine available points respectively and the coded target in Fig. 2c has thirteen available points.

From this point of view, the circular belts in coded targets contain not only the coding information, but also the valuable location information of the corners. Theoretically one coded target can provide enough points for computing the essential matrix by the five-point algorithm.

2.4   Ordering Rules of the Corners

In order to use the corners in coded targets for computing the essential matrix, they have to be sorted so that one-to-one correspondence can be built between them in different images. The ordering rules of the corners are as follows:

1. The corners of smaller circular belt are prior to the corners of bigger circular belt. If there are circular belts with the same size, the circular belts are sorted in a way where the counterclockwise angle from center of the prior one to center of the posterior one is less than 180°.
2. As far as the corners in one circular belt are concerned, the corners on inner circle are prior to the corners on outer circle.
3. About the corners in one circular belt as well as on the same circle, the corners are sorted in a way where the counterclockwise angle from the prior corner to the posterior corner is less than 18°.

These rules are logically explicit and generic to all coded targets. The results of corner sorting are shown in Fig. 3.

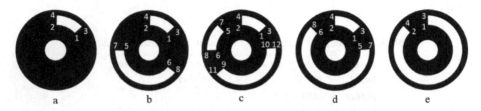

Fig. 3. Corners in order.

3   Corner Extraction and Ordering Algorithms

This chapter describes algorithms for corner extraction as well as corner ordering, which are verified by experiments in the following chapter.

3.1   Corner Extraction Algorithm

The corner extraction algorithm used in this paper consists of three steps.

1. In order to improve the speed of corner extraction process, regions containing the coded targets are trimmed from the whole image in advance as the regions of interest, using the method of pattern recognition.
2. A method of corner determination on the pixel level is applied among the regions of interest, which can find the most prominent corners in the specified image regions [9]. It calculates the corner quality degree at every source image pixel by calculating the minimal eigenvalue of gradient matrices [10]. A non-maximum suppression is performed to retain the local maximums in 3 × 3 neighborhood. The corners with the minimal eigenvalue less than a certain value are rejected. The remaining corners are sorted by the quality degree in the descending order. Each corner for which there is a stronger corner at a distance less than a predefined value is removed.
3. Using the corners detected in step 2 as centers, small square windows are built on the image, in which a method of corner determination on the subpixel level is applied in order to refine the corner locations. This algorithm iterates to find the subpixel accurate location of corners or radial saddle points.

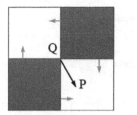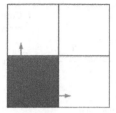

Fig. 4. Subpixel corner extraction algorithm. (Color figure online)

As shown in Fig. 4, red arrows indicate the gradient directions. The accurate corner locator at subpixel level is based on the observation that every vector from the center Q to a point P located within a neighborhood of Q is orthogonal to the image gradient at P subject to image and measurement noise.

By minimize the expression

$$D_{Pi}^T(Q - Pi),\tag{2}$$

where Pi is any point in the neighborhood of Q and D_{Pi} is an image gradient at Pi, the positon of Q can be found. More specifically, by summing the gradients within its neighborhood window, Q can be calculated using the equation

$$Q = \left[\sum_i \left(D_{Pi}D_{Pi}^T\right)\right]^{-1}\sum_i \left(D_{Pi}D_{Pi}^T Pi\right).\tag{3}$$

The center of the neighborhood window is set at this new center Q. The algorithm iterates until the center stays within a predefined threshold.

3.2 Corner Ordering Algorithm

After extracting corners from the coded target, they have to be sorted in order to build one-to-one point correspondence between different images. The corner ordering algorithm is implemented following the ordering rules of the corners described in Sect. 2.4. The flow chart shown in Fig. 5 displays the main processes of the corner ordering algorithm.

4 Experiments

Theoretically, only one coded target is needed to provide enough points for computing the essential matrix using the five-point algorithm. Experiments were conducted in order to find the minimum number of coded targets from which a satisfactory result of relative orientation between two camera poses can be drawn.

The coded targets used in the experiments are shown in Fig. 6a. They have LED light source, which improves recognition rate in complex environments. The experiment scene is shown in Fig. 6b, in which the camera is approximately three meters away from the coded targets and the objects to be measured.

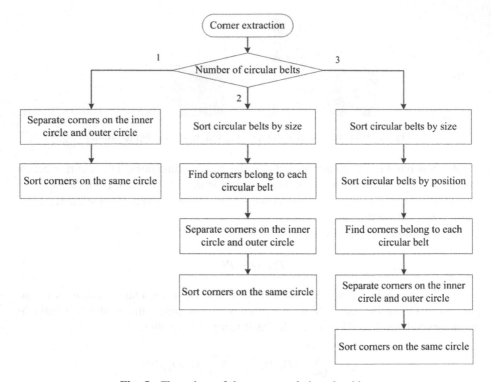

Fig. 5. Flow chart of the corner ordering algorithm.

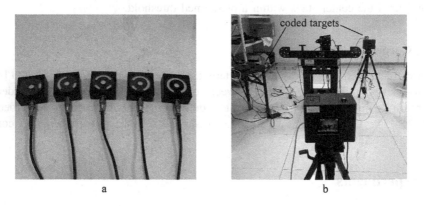

a   b

Fig. 6. The coded targets and experiment scene

4.1   Scenario with One Coded Target

In this experiment, only one coded target was used. One Canon 700D digital single lens reflex camera, by which two pictures were taken, was calibrated in advance.

As shown in Fig. 7a, uncoded circles in one picture were detected and marked in red. The essential matrix was computed using the corners and the center of central circle on the coded target from both pictures. After that, the fundamental matrix was derived from the essential matrix and the camera intrinsic matrix. Finally, epipolar lines of all the centers of uncoded circles in the Fig. 7a were drawn in the other picture in Fig. 7b.

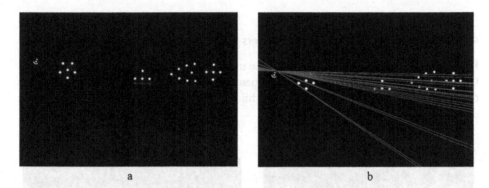

Fig. 7. Experiment with one coded target. (Color figure online)

If the essential matrix used to compute the fundamental matrix is accurate, each epipolar line of the center of an uncoded circle in one picture should go through the center of the same uncoded circle in the other picture. However the epipolar lines deviate from the center of the corresponding uncoded circles evidently, some of them even do not pass through the circle, which indicates that the essential matrix is not accurate. It is probably because the corners and the center of central circle of the coded target are too close to each other. Even though the number of point matches meets the requirement of five-point algorithm, it cannot produce a satisfactory result.

4.2 Scenario with Two Coded Targets

In this experiment, two coded targets were used. Other experimental factors were kept the same as that in the experiment with one coded target. Uncoded circles in one picture were detected and marked in red in Fig. 8a, while the epipolar lines of their centers were drawn in the other picture in Fig. 8b.

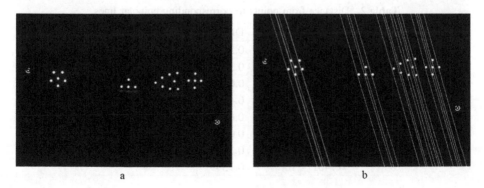

Fig. 8. Experiment with two coded targets. (Color figure online)

As shown in Fig. 8, almost every epipolar line passes throuth a circle, which is a better result than that shown in Fig. 7. Nevertheless, not an epipolar line passes through the center of a circle. Using a proper threshold, one-to-one correspondence between points from two images can be esablished, but the correctness is not guaranteed.

4.3   Scenario with Three Coded Targets

In this experiment, three coded targets were used. Other experimental factors were kept the same as that in the previous experiments. Uncoded circles in one picture were detected and marked in red in Fig. 9a, while the epipolar lines of their centers were drawn in the other picture in Fig. 9b.

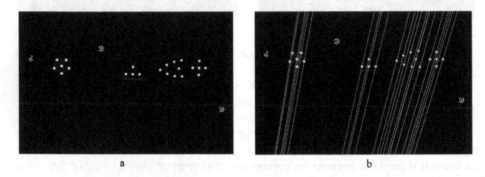

a   b

Fig. 9. Experiment with three coded targets. (Color figure online)

As shown in Fig. 9, every epipolar line passes through the center of a circle precisely, which ensures the one-to-one correspondence between points from two images. This result can be further investigated by computing the distance from centers of uncoded circles in Fig. 9b to their corresponding epipolar lines. The results are shown in Table 2. The unit for distance is the pixel.

Table 2. Distance from points to corresponding epipolar lines.

Point no.	Distance	Point no.	Distance	Point no.	Distance
1	0.2656	9	0.3391	17	0.4801
2	0.2739	10	0.3320	18	0.3372
3	0.1506	11	0.0005	19	0.3594
4	0.1680	12	0.4775	20	0.0142
5	0.1326	13	0.6251	21	0.3652
6	0.0398	14	0.0488	22	0.2343
7	0.2993	15	0.1018	23	0.2836
8	0.7401	16	0.1399		

Considering that the image size is 5184×3456 pixels and the average radius of the uncoded circles in the image is 18.0405 pixels, the average distance from one point to its corresponding epipolar line being 0.2699 pixel implies a high accuracy.

In order to compare the accuracy of proposed method and traditional method, further calculation is conducted by using only the centers of coded targets and eight uncoded circles to derive the essential matrix. The epipolar lines of other uncoded circles are computed again, and the average distance from one point to its corresponding epipolar line is 0.8290 pixel. In comparison, the proposed method has better accuracy.

To conclude, the minimum number of coded targets from which a satisfactory result of relative orientation between two camera poses can be drawn is three.

4.4   Structure from Motion Pipeline

After determining the one-to-one correspondence between points from each pair of two images, 3D coordinates of the points can be calculated by the method of Structure from Motion (SfM). SfM computes an external camera pose per image and a 3D point cloud representing the pictured scene, using matched feature points between image pairs and the camera intrinsic matrix as inputs.

The sequential SfM pipeline is used in this paper, which is a growing reconstruction process. It starts from an initial two-view reconstruction that is iteratively extended by adding new views and 3D points, using pose estimation and triangulation [11]. Non-linear refinements including bundle adjustment and Levenberg-Marquardt steps are performed to refine rotations, translations and structure.

In order to test the performance of this photogrammetric measurement system, an experiment was conducted. Three pictures were taken from different angles. One of the uncoded targets used in this experiment, shown in Fig. 10, was precisely machined using laser lithography method with a precision of one micrometer. This uncoded target was used to evaluate the result of 3D coordinate measurement.

The radius of each large circle in Fig. 10 is 50 mm. By setting the measured distance between point 1 and point 7 (P_1P_7) to 50 mm, the length of other line segments representing the radius is computed by PolyWorks. Three examples of length measurement are shown in Fig. 11. All of the measured length of line segments is shown in Table 3. The unit for length is the millimeter.

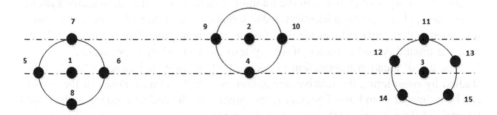

Fig. 10. Drawing of the uncoded target.

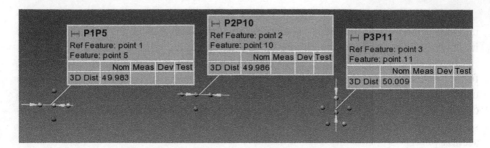

Fig. 11. Length measurement by PolyWorks.

Table 3. Measured length of line segments.

Segment	Length	Segment	Length	Segment	Length
P_1P_5	49.983	P_2P_9	49.961	P_3P_{13}	49.967
P_1P_6	49.980	P_2P_{10}	49.986	P_3P_{14}	49.996
P_1P_8	49.998	P_3P_{11}	50.009	P_3P_{15}	49.971
P_2P_4	50.069	P_3P_{12}	50.005		

Using the formula

$$S = \sqrt{\frac{\sum_{i=1}^{n}(s_i - \bar{s})^2}{n}}, \tag{4}$$

to calculate the standard deviation from the mean value, the result is 0.028 mm.

5   Conclusion

A close range industrial photogrammetric measurement system is proposed in this paper. One prominent feature of this system is that only a small number of coded targets are needed for the determination of relative orientation between camera poses by utilizing corner information in the coded targets. More specifically, each pair of images have to contain three identical coded targets for determining the relative orientation precisely. Compared with the traditional close range photogrammetry systems which demand at least eight identical coded targets for each pair of images, the system presented in this paper reduces the time for measurement procedure remarkably. In the end, the feasibility and accuracy of this system are verified by experiments.

The use of location information of coding patterns in coded targets can be further studied by considering the relative orientation as a PnP problem. Poses of the camera can be estimated from known geometric information of the coding patterns in the coded targets and their location information in the pictures. This is another possible way to reduce the number of coded targets needed for determining relative orientation in photogrammetry.

Acknowledgement. This work was partially supported by the National Natural Science Foundation of China under grant Nos. 51575332 and 61673252, and the National key research and development program under grant No. 2016YFC0302401.

References

1. Koutecký, T., Paloušek, D., Brandejs, J.: Method of photogrammetric measurement automation using tritop system and industrial robot. Optik – Int. J. Light Electron Opt. **124**(18), 3705–3709 (2013)
2. Drap, P., Merad, D., Mahiddine, A., Seinturier, J., Gerenton, P., Peloso, D., et al.: Automating the measurement of red coral in situ using underwater photogrammetry and coded targets. ISPRS – Int. Arch. Photogram. **XL-5/W2**, 231–236 (2013)
3. Li, W., Dong, M., Sun, P., Wang, J., Yan, B.: Relative orientation method for large-scale photogrammetry with local parameter optimization. Chin. J. Sci. Instrum. **35**(9), 2053–2060 (2014)
4. Ahn, S.J., Rauh, W., Recknagel, M.: Circular coded landmark for optical 3D-measurement and robot vision. In: Proceedings of the IEEE/RSJ International Conference on Intelligent Robots and Systems, IROS 1999, vol. 2, pp. 1128–1133. IEEE (1999)
5. Knyaz, V., Sibiryakov, A.: Non contact 3D model reconstruction using coded targets. In: International Conference of Graphicon, Moscow (1998)
6. Shortis, M.R., Seager, J.W.: A practical target recognition system for close range photogrammetry. Photogram. Rec. **29**(147), 337–355 (2014)
7. Hartley, R.I.: In defense of the eight-point algorithm. IEEE Trans. Pattern Anal. Mach. Intell. **19**(6), 580–593 (1997)
8. Nistér, D.: An efficient solution to the five-point relative pose problem. IEEE Trans. Pattern Anal. Mach. Intell. **26**(6), 756–770 (2004)
9. Shi, J., Tomasi, C.: Good features to track. In: IEEE Computer Society Conference on Computer Vision & Pattern Recognition, vol. 600, pp. 593–600. CiteSeer (2000)
10. Harris, C.: A combined corner and edge detector. In: Proceedings of the Alvey Vision Conference 1988, no. 3, pp. 147–151 (1988)
11. Moulon, P., Monasse, P., Marlet, R.: Adaptive structure from motion with a contrario model estimation. In: Lee, K.M., Matsushita, Y., Rehg, J.M., Hu, Z. (eds.) ACCV 2012. LNCS, vol. 7727, pp. 257–270. Springer, Heidelberg (2013). doi:10.1007/978-3-642-37447-0_20

A Camera Calibration Method Based on Differential GPS System for Large Field Measurement

Haijun Jiang[✉] and Xiangyi Sun[✉]

Hunan Key Laboratory of Videometrics and Vision Navigation,
College of Aerospace Science and Engineering,
National University of Defense Technology, Changsha, China
jianghaijun007@126.com, sunxiangyi_hn@163.com

Abstract. Aiming at the problem that the traditional space control point cannot be laid in the large field of view when calibrate the camera's parameters, a high precision calibration method of the camera is proposed. In this paper, the rover part of the differential GPS system is loaded on the unmanned aerial vehicle (UAV) and the base part is placed at the origin of the world coordinate system. The UAV is controlled to move within the field of view of the camera which to be calibrated and we can obtain the three-dimensional coordinates of the control point and the corresponding image coordinate. And then according to the "High Accuracy Calibration Method Based on Collinear Equation" to calculate the initial value of the camera's internal and external parameters, and finally we can solve the accurate calibration parameters after optimization. The measurement accuracy of differential GPS system can reach centimeter level, the control points should distributed evenly in the field of view to solve the exact results. Experiments show that the accuracy of the intersection of 300 m and 800 m can reach 0.6 m and 1 m respectively, so the method have the characteristics of high precision, practicality and simple operation.

Keywords: Camera calibration · Large field of view · Differential GPS system · On-field calibration

1 Introduction

Photogrammetry technology is used for the formation of image of objects using the camera and other equipment. It can measure many features of objects, such as structure, motion parameters and so on [1]. Its main advantage over other measurement techniques is its high accuracy, non-contact, motion and dynamic measurement, real-time measurement and other notable features [2, 3]. In the development of the camera calibration, the classical calibration methods include Tsai [4], Weng [5], and Zhang [6] and others. These methods are calibrate camera based on the references which its

---

This work is partially supported by Scientific Research Program of National University of Defense Technology (NO. ZK16-03-27).

Y. Huang et al. (Eds.): ICIRA 2017, Part II, LNAI 10463, pp. 508–519, 2017.
DOI: 10.1007/978-3-319-65292-4_44

structures is known, such as the spatial control points or control lines and so on. Furthermore, calibration results can be calculated by shoot a number of checkerboard images from different aspects. However, owing to the size of references are small, these calibration methods are only applicable for the laboratory and other small field of view, and it cannot be used in the large field condition, so their limitations are obvious. Since the camera in large field is fixed and it is difficult to obtain a Kruppa equation describing the constraint relationship between multiple imaging, these methods are not applicable for the large field condition, such as the camera self-calibration methods proposed by Faugeras et al., and active vision calibration methods based on camera-controlled motion [7–10]. In the author's engineering practice, the calibration problem in the large field is often encountered, and the range of camera field of view is about 3 km × 200 M. The traditional calibration method requires that the control points or control lines be distributed evenly in the field of view of the camera in order to obtain more stable and accurate calibration result. It is easily met under small field of view or laboratory conditions, such as the commonly used cross- Target, checkerboard plate and so on. However, in the large field of view, especially in the open and have no tall buildings or other objects, you cannot easily access control points that are distributed evenly in the camera field of view.

In the existing large-field camera calibration method, the method proposed by Ming et al. [11] is only applies to the range of field of view within 8000 mm × 8000 mm, so it is not applicable for the field of view about kilometers. The method proposed by Bowen et al. [12] can only apply to the small field of view. It needs a high-precision coordinate measuring machine, and the operation is complex, so this method is not suitable for large-field calibration. The method proposed by Banglei et al. [13] is similar to that of this paper, it uses the azimuth camera and wide beam laser distance measuring instrument to obtain the spatial control point coordinates, but its accuracy is only 0.25–1 m. The result has a certain impact to the accuracy of camera calibration. In order to solve the problems encountered in the actual project in large-field of view camera calibration, it is necessary to artificially set control points distributed evenly in the camera field of view, and then collect the image that have control point by the camera to be calibrated and extract the image points of the control points. Finally, calculate the camera's internal and external parameters according to the spatial coordinates of the control point and the corresponding image coordinates. In this paper, we proposed to use the UAV as a flight platform to load the rover station of the differential GPS system on the UAV, and put the base station on the coordinate origin of the world coordinate system. Thus, we can obtain accurately spatial coordinate values in the world coordinate system of rover station antenna on the UAV. Experiment shows that the distance between the base station and the rover station within a hundred kilometers, the differential accuracy can reach 2–5 cm, so that the spatial coordinate accuracy of the control point can reach centimeter level. The method is simple and practical in practical engineering. In theory, it only need four or more control points which are evenly distributed in the field of view that calibration parameters can be calculated. Moreover, due to the method is easy to obtain control points, it is easy to obtain dozens of spatial control point and it can get high precision calibration parameters.

2 The Differential GPS Positioning Measurement System

Over the past decades, with the rapid development of Global Navigation Satellite System (GNSS), the application of satellite positioning technology has been more and more extensive in all walks of life. At present, several major countries in the world are developing and improving their own navigation satellite system and they are playing a huge role both in the military and civilian. Although the accuracy of single point positioning can reach about 1 m, but single point positioning cannot meet the requirements in some applications require high precision.

2.1 The Basic Principles of the Differential GPS System

The accuracy of GPS single point positioning is not enough, the main factors are as follow: satellite ephemeris error, atmospheric delay (Ionospheric delay, Tropospheric delay) error, multi-path error and satellite clock error [14]. In order to eliminate these errors, if a fixed GPS receiver (called a base station) can be placed at a known point, GPS observations are taken with the user (called a rover station), and then sent the error corrections of the base station to the rover station. Consequently, the positioning accuracy can be greatly improved, and this is the basic principle of differential GPS. In the condition of dynamic differential positioning, the code pseudorange is $\tilde{\rho}_r^j$, which to be measured from the antenna of the base station to the satellite. The base station location is known, using the satellite ephemeris data, we can calculate the distance from the base station to the satellite ρ_r^j, ρ_r^j also contains the same satellite ephemeris error. The two distances are subtracted, that is:

$$\delta\rho_r^j = \tilde{\rho}_r^j - \rho_r^j \tag{1}$$

$\delta\rho_r^j$ contain the clock error, atmospheric refraction error. When the distance between the rover station and the base station is less than 100 km, $\delta\rho_r^j$ is sent as a distance correction number to the rover station receiver, and the rover station can effectively reduce or even eliminate the influence of some common errors. The distance relationship between the three-dimensional coordinates of the rover station receiver and the satellite is:

$$\tilde{\rho}_r^j - \delta\rho_r^j = \sqrt{(X^j - X_k)^2 + (Y^j - Y_k)^2 + (Z^j - Z_k)^2} + c(\delta t_k - \delta t_r) \tag{2}$$

In formula (2), there are four unknown parameters, that is, the three-dimensional coordinates X_k, Y_k, Z_k at the time t of the rover station, and the difference of clock error between the base station (denoted by the subscript r) and the rover station (denoted by the subscript k). Simultaneous observation of four satellites, we can find a unique solution to achieve dynamic positioning (Fig. 1).

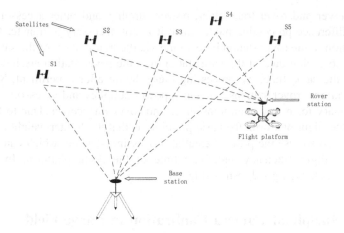

Fig. 1. Differential GPS schematic

2.2 Hardware Composition of the Differential GPS System

The differential GPS system is mainly used to obtain the spatial coordinates of the control point in the world coordinate system, which is difficult to get in large field conditions. As the differential GPS positioning measurement system has two measurement modes, so the hardware composition is different. A real-time carrier phase difference measurement, that is, RTK mode (Real Time Kinematic), which can provide real-time centimeter-level accuracy of the spatial coordinates in the measurement process. In RTK mode, the system consists of base stations, rover stations, antennas,

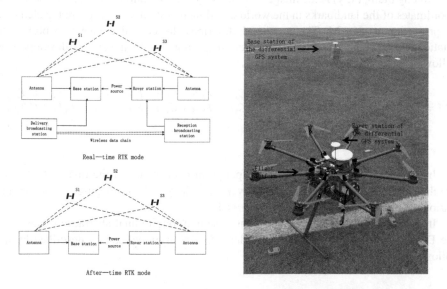

Fig. 2. Schematic diagram of two modes of the differential GPS systems and some physical maps

main transceiver and rover transceiver, power supplies and other accessories. In the latter post-difference processing mode, since it is not necessary to transfer the data of the base station to the rover station in real time via the transceiver. At the same time, as long as the observing data of the base station and the rover station are independently observed at the same time, we processing these datum after experiment. So we only need base stations, rover stations, antennas, power supplies and accessories and there are no necessary for heavy main transceiver and rover transceiver. Due to the limited payload of the flight platform, the basic principle is that the lighter weight of the rover part is better, so we use the post-differential processing scheme, which can effectively increases the flight efficiency and flight time of the flight platform. In the actual operation has achieved good results (Fig. 2).

3 The Principle of Camera Calibration in Large Field

3.1 The Solution of Camera's Internal and External Parameters

In this paper, a high-precision calibration method based on collinear equation is adopted. According to the imaging model that center perspective projection mix with nonlinear lens distortion, the collinear equation described by the elements of the projection matrix is:

$$
\begin{cases}
x = \frac{m_0 X + m_1 Y + m_2 Z + m_3}{m_8 X + m_9 Y + m_{10} Z + m_{11}} + \delta_x \\
y = \frac{m_4 X + m_5 Y + m_6 Z + m_7}{m_8 X + m_9 Y + m_{10} Z + m_{11}} + \delta_y
\end{cases}
\tag{3}
$$

Among them, (x, y) is the image coordinates with aberration. (X, Y, Z) is the spatial coordinates of the landmarks in the world coordinate system. $m_0 \sim m_{11}$ are the elements of projection matrix M. The matrix M describes the central perspective projection relationship of the spatial point to the image point, which can be developed as follows:

$$
M = \begin{bmatrix} m_0 & m_1 & m_2 & m_3 \\ m_4 & m_5 & m_6 & m_7 \\ m_8 & m_9 & m_{10} & m_{11} \end{bmatrix} = \begin{bmatrix} F_x r_0 + C_x r_6 & F_x r_1 + C_x r_7 & F_x r_2 + C_x r_8 & F_x T_X + C_x T_Z \\ F_y r_3 + C_y r_6 & F_y r_4 + C_y r_7 & F_y r_5 + C_y r_8 & F_y T_Y + C_y T_Z \\ r_6 & r_7 & r_8 & T_Z \end{bmatrix}
\tag{4}
$$

In formula (4), (C_x, C_y) is the principal point coordinates of the image. F_x and F_y are the equivalent focal lengths for transverse and longitudinal imaging. $r_0 \sim r_8$ are the various elements of camera rotation matrix R.

In formula (3), δ_x and δ_y are the transverse and longitudinal aberrations caused by the D.Crown lens distortion. D.C.rown lens distortion model can be described by the following formula:

$$\begin{cases} x_d = \left(1 + k_1 r^2 + k_2 r^4 + k_5 r^6\right) x_n + 2 k_3 x_n y_n + k_4 \left(r^2 + 2x^2\right) \\ y_d = \left(1 + k_1 r^2 + k_2 r^4 + k_5 r^6\right) y_n + 2 k_4 x_n y_n + k_3 \left(r^2 + 2y^2\right) \end{cases} \quad (5)$$

$$\begin{cases} x = F_x x_d + C_x \\ y = F_y y_d + C_y \end{cases} \quad (6)$$

In formula (5), $k_1 \sim k_5$ are the aberration coefficient, (x_n, y_n) is the normalized image coordinate of the ideal point:

$$\begin{cases} x_n = (\tilde{x} - C_x)/F_x \\ y_n = (\tilde{y} - C_y)/F_y \end{cases} \quad (7)$$

$$r^2 = x_n^2 + y_n^2 \quad (8)$$

In formula (7), $(\tilde{x}, \tilde{y})$ is the ideal image coordinate, (x, y) in formula (6) is the image coordinates with D.C.rown lens aberrations. The following formula can be obtained by the coordinate transformation relation:

$$\begin{bmatrix} 0 \\ 0 \\ 0 \end{bmatrix} = R \begin{bmatrix} X_0 \\ Y_0 \\ Z_0 \end{bmatrix} + T \quad (9)$$

In formula (9), (X_0, Y_0, Z_0) is the coordinates of the camera optical center in the world coordinate system.

Substituting formula (9) into formula (3) and remove aberrations, we can obtain:

$$\begin{cases} \tilde{x} = \frac{m_0(X-X_0) + m_1(Y-Y_0) + m_2(Z-Z_0)}{m_8(X-X_0) + m_9(Y-Y_0) + m_{10}(Z-Z_0)} \\ \tilde{y} = \frac{m_4(X-X_0) + m_5(Y-Y_0) + m_6(Z-Z_0)}{m_8(X-X_0) + m_9(Y-Y_0) + m_{10}(Z-Z_0)} \end{cases} \quad (10)$$

To transfer the formula (10), We can get linear homogeneous equations for the nine projection elements of $m_0 \sim m_2$, $m_4 \sim m_6$, $m_8 \sim m_{10}$, and by the nature of the rotation matrix:

$$m_8^2 + m_9^2 + m_{10}^2 = 1 \quad (11)$$

We can eliminate an unknown parameter due to formula (11), and for each spatial control point can list two equations. So when known the optical coordinates and four or more different control points, we can solution linearly to the nine projection elements: $m_0 \sim m_2$, $m_4 \sim m_6$, $m_8 \sim m_{10}$. Then the above-obtained elements are substituted into the formula (3) in which the aberration is removed, we can solve the remaining projection elements: m_3, m_7, m_{11}. And the initial value of the projection matrix is solved.

Yu and Yang [1] proposed a method that decomposing the linear parameters of the camera from the projection matrix and we can solve the internal parameters of the camera, such as: the main point, the equivalent focal length; camera external parameters, such as: rotation matrix and translation vector. The aberration coefficient is calculated from the camera distortion model described in formula (5) to (8). At this point, the initial value of the internal and external parameters of the camera is solved.

Then, the non-linear least squares method is used to optimize the internal and external parameters of the initial value, so that the deviation between the reprojection image point and the actual image point is the smallest. At this point, the cameras' accuracy solution of the inside and outside parameters are completed.

3.2 Calibration Procedure

In this paper, a typical procedure for camera parameter calibration under large field conditions is:

- The base station antenna is placed on the origin of the world coordinate system, and the rover station part is fixed on the flight platform.
- Control the UAV flight slowly in the field of view of the camera to be calibrated, as far as possible so that the UAV flight path to cover the entire field of view of camera to be calibrated, and have a certain depth changes.
- The image coordinates of the rover station antenna on the UAV are extracted by means of artificial method or target tracking. Select the image coordinate point which evenly distributed in the field of view as the calibration control point, so that the internal and external parameters of the camera can be calibrated (Fig. 3).

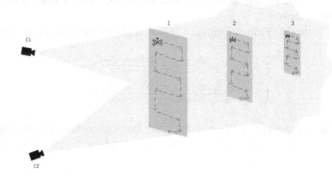

Fig. 3. The flight path of flight platform

4   Simulation Experiment and Error Analysis

- Simulation to generate two sets of camera internal and external parameters, and their focal lengths are 35 mm and 85 mm respectively. And then generate two sets of spatial coordinates, the distances between spatial coordinates of 35 mm camera and the corresponding camera are 100–300 m. At the same time, the distances between spatial coordinates of 85 mm camera and the corresponding camera are 500–800 m.

- The two sets of spatial coordinates are calculated by using the simulated camera parameters to obtain the ideal image coordinates, plus the D.C.Brown aberration to obtain the image coordinates with aberration.
- The error of the extracted image points in the actual situation is simulated by adding the random noise to the image coordinates with aberration. Set the random image error to 0.1–2 pixel, the step size is 0.1 pixel, and each error pixel is calculated 1000 times to get the image coordinates with random noise and aberration.
- The image coordinates in step 3 are removed the aberration to obtain the ideal point image coordinates with random noise.
- Two sets of image cameras which their focal length are 35 mm and 85 mm are used for binocular intersection, and the spatial coordinates of intersection are obtained.
- The spatial coordinates obtained by intersection are compared with the corresponding space coordinates in step 1, and the error values of the spatial coordinates of intersection are obtain.

From Fig. 4 we can see that the intersection error increases with the increase of the pixel error. For the 35 mm camera, the spatial point distance from the camera in the range of 100–300 m, pixel error is 2 pixel, the maximum distance error from the direction of the intersection is 0.6 m, while the other two directions of the error is less than 0.1 m. For the 85 mm camera, the spatial point distance from the camera in the range of 500–800 m, pixel error is also 2 pixel, the maximum distance error from the direction of the intersection is 2.2 m, while the other two directions of the error is less than 0.2 m. It can be drawn the following conclusions: The error of extracting the image points have greatly impact on the intersection error of the two cameras. So in the actual experiment, we are not only to complete the precise camera calibration, but also to improve the accuracy of the target point as much as possible. Such as, if we can ensure that the precision of pixel error is within 1 pixel, the intersection error of the two sets of cameras in the distance direction will be less than 1 m, and the results meet the experimental accuracy requirements.

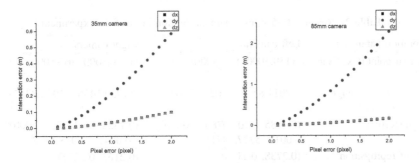

Fig. 4. Schematic image noise and intersection error of 35 mm and 85 mm camera

5 Out-Field Experiment

5.1 The Conditions of Out-Field Experiments and Implementation Process

In out-field experiment, we selected two pairs of cameras for calibration and the focal length of cameras were 35 mm and 85 mm. The distance between the two images is about 75 m, the frame rate is about 20 Hz, the height is about 0.8 m, and the angle between the optical axis of the two cameras is about 45°. As the cameras in out-field experiment have the GPS timing module, UAV itself also has a differential GPS time, so we can unified strictly the time of the control point coordinates and the image coordinates of the corresponding image points.

5.2 Results of Out-Field Experiment

We were calibrated two pairs of cameras and their focal length were 35 mm and 85 mm, calibration parameters as shown in Tables 1 and 2:

Table 1. Results of 35 mm-camera calibration in out-field experiment

Calibration parameters	Left camera	Right camera
Principal point (Cx, Cy)/ pixel	(1264.2035, 776.7317)	(1084.9563, 820.6420)
Equivalent focal length (Fx, Fy)/pixel/m	(5150.2914, 5153.3605)	(5162.9888, 5156.7727)
Lens distortion	(0.2865, −24.0956, 0.0031, 0.0066, 372.5262)	(−0.4769, 15.8208, 0.0032, −0.0039, −195.234)
RMS of reprojection errors (du, dv)/pixel	(0.3725, 0.2597)	(0.3285, 0.2924)

Table 2. Results of 85 mm-camera calibration in out-field experiment

Calibration parameters	Left camera	Right camera
Principal point (Cx, Cy)/ pixel	(1159.5000, 862.5000)	(1159.5000, 862.5000)
Equivalent focal length (Fx, Fy)/pixel/m	(12011.8521, 12020.1010)	(11930.1435, 11938.8317)
Lens distortion	(0.7585, −104.0780, 0.0039, 0.0018, 5227.1871)	(0.1428, 76.3326, −0.0008, −0.0041, −7078.1392)
RMS of reprojection errors (du, dv)/pixel	(0.2738, 0.2190)	(0.2163, 0.2275)

The calibration result indicates that the calibration re-projection error does not exceed 0.4 pixels. Compared intersection results with the GPS data, the results shown in Figs. 5 and 6:

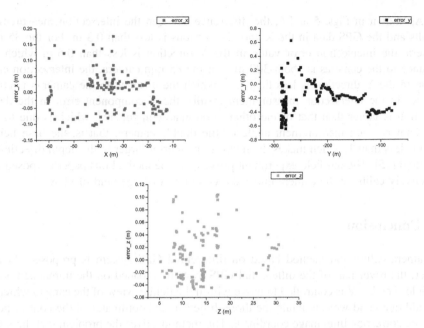

Fig. 5. The differences between intersection result and the GPS data in three directions in 35 mm cameras

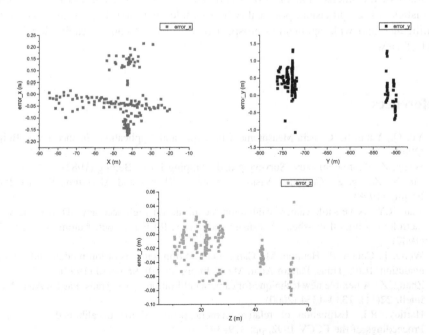

Fig. 6. The differences between intersection result and the GPS data in three directions in 85 mm cameras

As shown in Figs. 5 and 6, the difference between the intersection measurement results and the GPS data in the X and Z directions is less than 0.3 m. For the 35 mm camera, the intersection error value in the Y direction is less than 0.6 m which the distance to the cameras is about 300 m. For the 85 mm camera, the intersection error value in the Y direction is less than 1 m which the distance to the cameras is about 750 m. In the intersection measurement results, the Y component error in the depth direction is larger than that of the other coordinate components, which is due to the small intersection measurement angle of the double camera, that is, the base height ratio is less than 1, which makes the measurement error larger in the depth direction in out-field [15]. The out-field experiment proves that the method this paper proposed can effectively calibrate the camera under the condition of large field of view.

6   Conclusion

A camera calibration method based on differential GPS system is proposed. In this paper, the rover part of the differential GPS system is loaded on the unmanned aerial vehicle. The UAV is controlled to move within the field of view of the camera which to be calibrated and we can obtain the three-dimensional coordinates of the control point and the corresponding image coordinate. The method solves the problem that the space control point is difficult to obtain in the large field of view. The measurement accuracy of differential GPS system can reach centimeter level, the control points should distributed evenly in the field of view to solve the exact results. Experiments show that the accuracy of the intersection of 800 m can reach 1 m, so the method have the characteristics of high precision, practicality and simple operation. It is of great theoretical significance and wide application prospect in the field of camera calibration in large field of view.

References

1. Yu, Q., Yang, S.: Camera Measurement Principle and Application. Science Press, Beijing (2009)
2. Wang, Z.: Photogrammetry. Surveying and Mapping Press, Beijing (1982)
3. Ma, S., Zhang, Z.: Computer Vision Computing Theory and Algorithm. Science Press, Beijing (1998)
4. Tsai, R.Y.: A versatile camera calibration technique for high-accuracy 3D machine vision metrology using off-the-shelf TV camera and lenses. IEEE J. Robot. Autom. 3(4), 232–344 (1987)
5. Weng, J., Cohen, P., Hemiou, M.: Camera calibration with distortion models and accuracy evalution. IEEE Trans. Pattern Anal. Mach. Intell. 14(10), 965–980 (1992)
6. Zhang, Z.: A flexible new technique for camera calibration. IEEE Trans. Pattern Anal. Mach. Intell. 22(11), 1330–1334 (2000)
7. Hartley, R.I.: Estimation of relative camera positions for uncalibrated cameras. In: Proceedings of the ECCV 1992, pp. 379–397 (1992)

8. Maybanks, S., Faugeras, O.: A theory of self-calibration of a moving camera. Int. J. Comput. Vis. **8**(2), 123–151 (1992)
9. Ma, S.D.: A self-calibration technique for active vision system. IEEE Trans. Robot Autom. **12**(1), 114–120 (1996)
10. Hu, Z., Wu, F.: Camera calibration method based on active vision. Chin. J. Comput. Sci. **25** (11), 1149–1156 (2002)
11. Ming, Y., Qian, Z.X., Ju, H., et al.: High precision field calibration method for large field of view camera. Electron Opt. Control **18**, 56–61 (2011)
12. Yang, B., Zhang, L., Ye, N., et al.: Facing the camera with wide field of vision measurement calibration technology. Acta Optica Sin. **32**(9), 915001 (2012)
13. Banglei, G., Sun, X., Shang, Y., Yu, Q., et al.: Shaking on the platform of air shooting calibration method of camera with wide field of view. Acta Optica Sin. **35**(7), 712003 (2015)
14. Li, Z., Huang, J.: GPS Measurement and Data Processing. Wuhan University Press, Wuhan (2010)
15. Shang, Y., Sun, X., Yang, X., Qifeng, Yu.: A camera calibration method for large field optical measurement. Optik-Int. J. Light Electron Opt. **2013**(124), 6553–6558 (2013)

A New Pixel-Level Background Subtraction Algorithm in Machine Vision

Songsong Zhang, Tian Jiang, Yuanxi Peng$^{(\boxtimes)}$, and Xuefeng Peng

State Key Laboratory of High Performance Computing, College of Computer, National University of Defense Technology, Changsha 410073, China
13908473894@163.com, {tjiang,pyx}@nudt.edu.cn, pengxf@163.com

Abstract. In the machine vision, the algorithm of background subtraction is used in target detection widely. In this paper, we reproduce some algorithms such as ViBe, Local Binary Similar Pattern (LBSP), Local Ternary Pattern (LTP) and so on. In view of the problem of inaccurate edge in target detection, we propose a method of combining color and LBSP for background model. It can obtain information both in pixel and texture (marked as *improved*-LBSP in the paper). On this basis, we propose a new method marked as BFs-method in the paper, which have a new persistence consists of color, LBSP, and time (t). The key advantage of this method lie in its highly robust dictionary model as well as it's ability to automatically adjust pixel-level segmentation behavior, which improves the ability to remove the shadow of the target and the hole inside the target.

Keywords: Background modeling · Machine vision · Background foreground segmentation

1 Introduction

Background subtraction is an important research direction in the field of computer vision. It is widely used in multi-modal video registration, target detection, target tracking. Background modeling is very important for the processing of video, and even directly affects the success or failure of a system. And it can be considered as a classification problem in a certain degree, also known as background foreground segmentation. There are many ways to model the background in background subtraction, including: Single Gaussian (SG) used in real-time tracking of the human body, Mixture of Gaussian Model (MGS) [15], Running Gaussian Average(RGA) [3], CodeBook [8], SOBS-Self-organization background subtraction [17], SACON [19], ViBe algorithm [1], Color [5]. Temporal Median filter [10], W4 and so on. Barnich et al. proposed a visual background extraction (ViBe) algorithm, which just a frame image can established background model, and use the conservative update strategy, the algorithm is simple and real-time. The LBP algorithm uses the imaging principle of human eye to create

© Springer International Publishing AG 2017
Y. Huang et al. (Eds.): ICIRA 2017, Part II, LNAI 10463, pp. 520–531, 2017.
DOI: 10.1007/978-3-319-65292-4_45

a background. There is a trajectory-controlled watershed segmentation algorithm which effectively improves the edge-preserving performance and prevents the over-smooth problem [20].

Our method improved the original LBSP, our main contributions are: (1) combination color and LBSP, which can increase the accuracy of the foreground edge. (2) redefine persistence [12,14], which consists of color, LBSP, and t in time and space respectively, can remove the shadow of the target and the hole inside the target.

2   Relation Work

According to the needs of the project, our method is mainly dealing with static background video, in which the basic idea of the pixel-based model is to establish a background model for each pixel, comparing with the successive frames constantly. Detecting the target based on the difference between the two, and the pixels which is detected as background will be used to update the background model. As time goes on, using the frame to compare with background, updating the background model constantly, and after several frames, the establishment of the background model is stable gradually, and reflect the background more accurately and can adapt to the background scene changes, such as illumination variation.

The general method Includes four steps: Pre-processing, Background modeling, Foreground segmentation, and Data verification. At first, pre-processing consists of drying, enhancement, etc. And then background modeling, followed by foreground segmentation. In the process of updating the background model, the segmented data is re-applied to the background model in the form of feedback, complete the background (bg) and foreground (fg) segmentation at last. And there are some models we have learned and improved:

2.1   ViBe

The ViBe algorithm considers background modeling as a classification problem. When the selection is made on the color space, it is judged that the pixel is the foreground pixel by comparing the value of the pixel at the current position with the value of its neighborhood pixel, v_1, v_2, v_3, v_n represent the value of the pixel at the periphery of each pixel or adjacent frame, and each background pixel is modeled by a collection of background sample values, representation such as Eq. 1:

$$M(x) = \{v_1, v_2, \ldots, v_N\}. \tag{1}$$

According to its corresponding model $M(x)$, we can classify a pixel value $v(x)$. The specific approach is as follows:

Firstly, define a sphere $S_R(v(x))$ of radius R centered on $v(x)$. Second, Compare $v(x)$ to the closest values with the $S_R(v(x))$. At last, We use #min for the threshold, we use # for the number of the intersections between $M(x)$ and $S_R(v(x))$, see Eq. 2, if # is larger than or equal to the #min, the pixel value $v(x)$

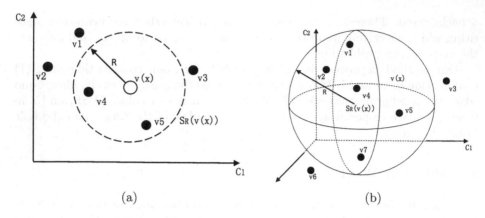

Fig. 1. (a), (b) respectively represent the algorithm in grayscale images and color images

is classified as background, else as foreground. And the algorithm representation respectively in grayscale images and color images as shown in Fig. 1:

$$\# \left\{ S_R\left(v\left(x\right)\right) \bigcap \{v_1, v_2, \ldots, v_N\} \right\}. \tag{2}$$

2.2   Local Binary Pattern: LBP

One of the significant algorithm of the background subtraction is the LBP [11]. LBP is an algorithm based on feature, which can create a descriptor for each pixel, the specific approach as follows: every pixel as the center pixel to create a 3*3 window, only consider the eight neighbor of a pixel in the basic version, and from the top left corner, clockwise, the eight pixels around the window will be a descriptor of the center pixel, where if the pixel is larger than or equal to center pixel, then marked as 1, else as 0, see Eq. 3:

$$s\left(x\right) = \begin{cases} 1 & I_x \geq I_c \\ 0 & I_x < I_c \end{cases} \tag{3}$$

The eight flag bits will produce an eight-bit binary number, which is written down from the upper left corner and clockwise to the binary number. This binary number is converted to a decimal number. The decimal number represents the LBP value of the center pixel. Each pixel will get an LBP according to the neighborhood information through the method. Experiments show that the LBP operator has strong robustness to illumination variation, but the disadvantage is the coverage area is too small to accurately describe the texture requirements.

2.3   Local Binary Similar Pattern: LBSP

The original LBSP is an improvement based on the LBP, and the difference is that it is not a constant that determines whether a pixel (recorded as *observe-pixel* in the paper) is a background pixel (recorded as feature *feature-pixel* in the paper), comparing the absolute value of the two instead, see Eqs. 4 and 5.

$$LBSP(x,y) = \sum_{p=0}^{p-1} d(v_p - v_c) \cdot 2^p \tag{4}$$

$$d(v_p - v_c) = \begin{cases} 1 & if \ |v_p - v_c| \leq T \\ 0 & otherwise \end{cases} \tag{5}$$

The advantage of the improvement is that there is a broader and more meaningful range in the background segmentation, which further enhances the robustness of the background modeling algorithm to illumination variation.

3   The Generation of New Methods

Through the study of three background modeling methods, we can find there are many ideas worth learning, the three of them in common is that establishing a background descriptor for each pixel, and comparing the corresponding background pixels with the next frame, except that ViBe is established under the assumption that the background model of a pixel can be described by randomly selecting pixels, and for this reason, the ViBe algorithm does not perform a good segmentation on the initialization if the target in the background model. Both LBP and LBSP use background pixels around a pixel to describe the background, and this description is more representative, the key points of this approach are the selection of the neighborhood and the determination of the threshold. Next, we will introduce our new method detailed.

3.1   Combine Color and LBSP Features to Establish Background Models

The Disadvantage of only Using LBSP Feature Modeling. In the segmentation process of using LBSP, there are some disadvantages of only using LBSP feature modeling: In the segmentation process, the *observe-pixel* segmentation using *feature-pixel* background model is to compare the pixels around *observe-pixel* (marked as *observe-pixel* LBSP in the paper) with the background model (marked as *feature-pixel* LBSP in the paper) [2]. Assuming that the intensity value of pixel i is 100 in background modeling and the neighborhood pixels are 40, the L1 distance of *feature-pixel* LBSP is 60, the *feature-pixel* LBSP of the pixel is 000000000000000; if the neighborhood pixels are 160, the L1 distance of the *observe-pixel* LBSP is still 60, and the *observe-pixel* LBSP is 0000000000000000. It is clearly that the segmentation result is wrong because losing the *observe-pixel* own pixel information. In this case, the experiment results get the noise (stain) shown in (e) of Fig. 2.

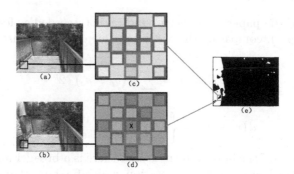

Fig. 2. The drawback of LBSP. (a) Nobody on the road. (b) A pedestrian on the road. (c) The *feature-pixel* LBSP. (d) The *observe-pixel* LBSP. (e) The noise in the results (marked with a green box). (Color figure online)

Combining Color and LBSP Information to Establish the Background Model. In the previous section, we have experimentally demonstrated that the segmentation result is wrong because losing the *observe-pixel* own pixel information. In this part, the LBSP method has been improved (marked as *improved-LBSP* in the paper), combining LBSP and color for background modeling is based on the pixel-level model [13]. We also refer to the color attribute in [9]. The specific process of modeling is taking each pixel as a center, making a 5*5 window, the description includes 25 neighbor pixels and the color value.

The pixel at the center position is marked as v_c. First, the center pixel is described by LBSP, as shown in Eq. 6:

$$LBSP(x, y) = \sum_{p=0}^{p-1} d(v_p - v_c) \cdot 2^p \tag{6}$$

Where v_p is the pixel intensity of the *p-th* neighbor of v_c on the predefined *P-length* LBSP pattern. Taking into account the distance between the surrounding pixels and the central pixel are different, in the LBSP character description is treated differently, the specific definition see Eqs. 7 and 8:

$$d(v_p - v_c) = \begin{cases} 1 & if \ |v_p - v_c| \leq T \cdot \alpha(p, c) \\ 0 & otherwise \end{cases} \tag{7}$$

$$\alpha(p, c) = \left(\frac{2 \cdot d(p, c)}{d(p, c) + 1} \right)^2 \tag{8}$$

Where T is the relative LBSP threshold, α(p,c) expresses the different weight due to the difference distance, which is used to determine the threshold, and $d(p, c)$ represents the Euclidean distance between pixels.

Combine the Color and LBSP Information for Pixel Segmentation. In this paper, we have the following conventions: When v_p and v_c belong to the same frame, use *feature-improved*-LBSP. When v_c is the pixel in the background model, v_p is the pixel around the frame waiting for segmentation, use *observe-improved*-LBSP. The specific process of segmentation is: If the color of the pixel to be segmented is far from the color of the background pixel, it is classified as the foreground directly. Otherwise, the distance between the *observe-improved*-LBSP and the *feature-improved*-LBSP is calculated, and the smaller the Hamming distance the more similar, see Eq. 9, and we will show the process in the next.

$$F(x) = \begin{cases} bg & if \ (H(a,b) < T_H) \\ fg & if \ (H(a,b) \geq T_H) \end{cases} \tag{9}$$

Algorithm 1. Establish the background model for fg/bg segmentation

---

1: *Establish the background model*
2: **procedure** ANY PIXEL:V(X,Y)
3:     **if** $|i_{(x,y)observe} - i_c| > T_I$ **then**
4:         **return** $v_{(x,y)}$ *is foreground*
5:     **else**
6:         *input* v_p, v_c
7:         $\alpha(p,c) \leftarrow (\frac{2 \cdot d(p,c)}{d(p,c)+1})^2$
8:         **if** $|v_p - v_c| \leq T \cdot \alpha(p,c)$ **then**
9:             **return** $d(v_p, v_c) \leftarrow 1$
10:        **else**
11:            **return** $d(v_p, v_c) \leftarrow 0$
12:        **end if**
13:        $improved\text{-}LBSP_{(x,y)} = \sum_{p=0}^{p-1} d(v_p, v_c) \cdot 2^p$
14:        **if** $H() > T_H$ **then**
15:            **return** $v(x,y)$ *is foreground*
16:        **else**
17:            **return** $v(x,y)$ *is background*
18:            $v(x,y)$ *will replace any pixel in the background model of* v_c *at a certain probability*
19:        **end if**
20:     **end if**
21: **end procedure**

Where $H()$ represents the Hamming distance, a, b respectively represent of *observe-improved*-LBSP of v_p and *feature-improved*-LBSP of v_c, and T_I represents the intensity threshold. V(X,Y) is segmentation pixel, whose color information is $i_{(x,y)observe}$, v_c, i_c respectively represent the central pixel, and its intensity, v_p is the pixel intensity of the *p-th* neighbor of the v_c. The update strategy draws on the reference [13] and will not repeat them.

3.2 Modeling the Background with Background Feature (BFs)

BFs Establishment and Foreground Segmentation. We propose a new background modeling method based on the modeling of Local Self Similary

(LSS) [16] and codebook model [4]. We refer to the expression method in reference [12], and changed the conditions for establishing the background model. For each background pixel, we simply store the data from its surroundings LBSP information derived from the pixel clearly not represent the texture information in the small area, and an LBSP feature is established for each pixel, according to the mapping relation of the position, the eight neighborhood LBSP signatures of each position are taken as the position feature of the pixel, this feature can reflect the central pixel of the local neighborhood of the texture features. In addition, in the segmentation, taking into account the time and space information, we combine color, LBSP, t information as a persistence index, the number of occurrences of a pixel at time t, when the pixel from the last time. In this paper, the background model of the pixel is presented with the background feature (BF), and the background of the pixel is described as the background feature of the pixel. The degree of good and bad is expressed by persistence. The establishment of BFs for each pixel is shown in Fig. 3:

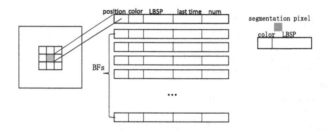

Fig. 3. The process of building BFs

In the pixel segmentation, we first need to calculate the resolution of the pixels to be segmented, that is, the pixels to be split with the corresponding position of the pixels of the similarity of BFs, the expression such as Eqs. 10, 11 and 12:

$$Q_t(x) = \sum_{\omega \in C_L(x)} q(\omega, t) \mid (d(I_t(x), \omega) < D(x)) \tag{10}$$

Where $C_{L(x)}$ is the current pixels BFs, and ω is a BF, $d(I_t(x), w) < D(x)$ indicating that the pixel to be segmented with the ω in the BFs meet the color/LBSP threshold and $Q_t(x)$ is persistence.

$$q(w, t) = \frac{n \cdot \left(p \cdot \alpha + \frac{1}{t - t_{last}} \cdot \beta\right)}{S_{num}} \tag{11}$$

Where n is the number of occurrences of w in BFs, p is the position information of w, α is the position weight, tlast is the time of the last occurrence of w, β is the time weight, S_{num} is the total number, who is a constant in the experiment.

$$S_t(x) = \begin{cases} 1 & if\ Q_t(x) < W(x) \\ 0 & otherwise \end{cases} \tag{12}$$

Where $W(x)$ represents the threshold of the background pixel persistence, and if the pixel is divided into a foreground, the result is 1; otherwise, the result is 0. $D(x)$ and $W(x)$ are dynamically adjusted, and the result of the segmentation is fed back to the BFs for the next segmentation [12].

Update Strategy. We compare the colors through L1 and compare the LBSP by the Hamming distance when calculates persistence. Combination the three information can improve the robustness to illumination variation. When the light of the observation scene has changed, we need to modify a small part of the BFs immediately. Only when the color/LBSP distortion is very small, we make random updates. If the change is only temporary, the small part that has not changed will restore the model to its original state quickly. We also draw on the sampling consistency model mentioned in [7,18], that is, if a pixel is divided into a background, the neighboring pixel models are replaced as their local expression. This update strategy is different from the improvement of the spatial background mentioned in codebook [11], which is more predictable because it can be carried out under the influence of external conditions such as camera shake [12].

4   Experiments and Conclusions

The experimental platform is Intel(R) Core(TM) i5-4570, CPU@3.20 GHz, RAM 8 GB, the development tool of the algorithm is OpenCV2.4.10 + VS2010.

4.1   Experiment

Experiment1: In the improved part of LBSP described in 3.1, we used the data set CDnet dataset [6]. This data set contains a rich set of video sequences. We selected one of these data for experimentation, and there are qualitative and quantitative analysis of the experimental results. We refer to the experimental parameters in [2], and the experimental parameters we use are: $T_H = 4, T = 30$, $T_I = 30$. The results are shown in Fig. 4:

In fact, we also change the size of the parameters in the experiment, $T_H = 1(\text{low})$, 28, 50(high) and with $T = 3(\text{low})$, 30, 90(high). And we will evaluate the model quantitatively from the following six index mentioned in [11].

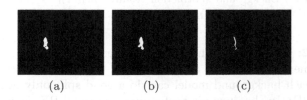

(a)   (b)   (c)

Fig. 4. (a) The results of the reference [1] (b) The results of this method in 3.1 (c) The difference between the two (edge information)

(1) Recall (Re): TP/(TP + FN)
(2) Specificity (Sp): TN/(TN + FP)
(3) False Positive Rate (FPR): FP/(FP + TN)
(4) False Negative Rate (FNR): FN/(TN + FP)
(5) Precision (Pr): TP/(TP + FP)
(6) F-measure: 2 *Pr * Re/(Pr + Re)

Where TP is the number of pixels at the correct location, TN is the number of true negatives, FN is number of false negatives, and FP is the number of pixels at the wrong location.

It can be seen T has a greater effect on the results from the experimental results. When T is small, the model is not robust for the noise. When T becomes larger, our model is more robust for the noise. The experimental results are shown in Table 1 and Fig. 5:

Table 1. The value of the index under different thresholds

Threshold	Index					
	Recall	Specifity	FPR	FNR	Precision	F-Measure
$T_H = 1, T = 3$	0.9841517331	0.9603088139	0.0396911861	0.0158482669	0.1975477899	0.3290464217
$T_H = 1, T = 30$	0.9250248188	0.9996272409	0.0003727591	0.0749751812	0.9609958018	0.9426672834
$T_H = 1, T = 90$	0.8704055379	0.9996680584	0.0003319416	0.1295944621	0.9630098738	0.9143690207
$T_H = 28, T = 3$	0.9837537377	0.9830047332	0.0169952668	0.0162462623	0.3649595923	0.5324042627
$T_H = 28, T = 30$	0.8557587104	0.9998242836	0.0001757164	0.1442412896	0.9797378360	0.9135611709
$T_H = 28, T = 90$	0.3790183376	0.9999999408	0.0000000592	0.6209816624	0.9999842689	0.5496905857
$T_H = 50, T = 3$	0.9837537377	0.9830047332	0.0169952668	0.0162462623	0.3649595923	0.5324042627
$T_H = 50, T = 30$	0.8557587104	0.9998242836	0.0001757164	0.1442412896	0.9797378360	0.9135611709
$T_H = 50, T = 90$	0.3790183376	0.9999999408	0.0000000592	0.6209816624	0.9999842689	0.5496905857

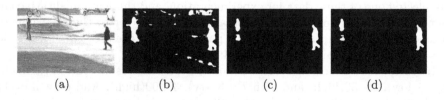

| (a) | (b) | (c) | (d) |

Fig. 5. (a) The original image in the dataset (b) Background subtraction result, when $T_H = 1, T = 3$ (c) Background subtraction result, when $T_H = 1, T = 30$ (d) Background subtraction result, when $T_H = 1, T = 90$

Experiment2: We also used the dataset CDnet dataset [6] to model the BFs. In the experiment, the number of BFs is fixed, the default is 50. In the static background, each background model can do a good split only need to activate two or three BFs. In the dynamic background, up to 20 BFs is activated. From the experimental results, our method can make the target complete which is blocked and camouflaged. Our improved method is denoted as BFs-method, the experimental results are shown in Fig. 6:

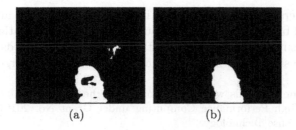

(a)   (b)

Fig. 6. (a) Experimental result of [12] (b) Experimental result of BFs-method

We also used the quantitative assessment in the previous section. The results are shown in Table 2. In this table, it show that BFs-method has a higher value in Recall, and has a good result in Precision (Pr). In addition, FPR has decreased comparing with the other six methods, and FNR is lower than PAWCS. This result shows that the persistence consists of color, LBSP, t can reflect the characteristics of the pixel background better, of which the number of occurrences can reflect the life of a pixel. In addition, the segmentation results clearly exceed the *improved*-LBSP which is described in Sect. 3.1.

Table 2. Quantitative assessment of several methods

Method	Index					
	Recall	Specifity	FPR	FNR	Precision	F-Measure
PBAS	0.78	0.990	0.010	0.009	0.82	0.75
ViBe+	0.69	0.993	0.007	0.017	0.83	0.72
SC-SOBS	0.80	0.983	0.017	0.009	0.73	0.73
LOBSTER	0.81	0.982	0.019	0.010	0.77	0.75
KDE	0.65	0.993	0.007	0.025	0.77	0.64
GMM	0.51	0.995	0.005	0.029	0.82	0.59
LSS	0.94	0.851	0.149	0.002	0.41	0.51
ED	0.71	0.969	0.031	0.017	0.62	0.61
PAWCS	0.85	0.991	0.050	0.140	0.87	0.85
Improved-LBSP	0.88	0.999	0.001	0.120	0.13	0.92
BFs-method	0.93	0.999	0.001	0.075	0.96	0.94

4.2 Conclusion

We change the condition of establishing LBSP by adding the weight of distance based on [1]. From the experimental results, we can see that our method can get more accurate edge of the targets and remove the noise in the background when we set suitable value of parameters. In addition, BFs-method combines

LBSP, color, persistence to build BFs, in which persistence is an index of the classification of background words, because it combines information of time and space, so it is superior to the classification of background words. BFs-method can remove the shadow of the target and the hole inside the target.

Acknowledgements. This work is supported by the Research Fund of State Key Laboratory of High Performance Computing under Grant No. 201612-01, National University of Defense Technology.

References

1. Barnich, O., Droogenbroeck, M.V.: Vibe: a universal background subtraction algorithm for video sequences. IEEE Trans. Image Process. **20**(6), 1709–24 (2011). A Publication of the IEEE Signal Processing Society
2. Bilodeau, G.A., Jodoin, J.P., Saunier, N.: Change detection in feature space using local binary similarity patterns. In: International Conference on Computer and Robot Vision, pp. 106–112 (2013)
3. Danelljan, M., Hager, G., Khan, F.S., Felsberg, M.: Adaptive decontamination of the training set: a unified formulation for discriminative visual tracking (2016)
4. Fan, Z., Chen, X., Chen, L.: Improvement strategy based on brightness ranges in codebook algorithms. J. China Univ. Metrol. (2013)
5. Fernndez, E., Besuievsky, G.: Efficient inverse lighting: a statistical approach. Autom. Constr. **37**(37), 48–57 (2014)
6. Goyette, N., Jodoin, P., Porikli, F., Konrad, J.: Changedetection.net: a new change detection benchmark dataset. In: Computer Vision and Pattern Recognition Workshops, pp. 1–8 (2012)
7. Hofmann, M., Tiefenbacher, P., Rigoll, G.: Background segmentation with feedback: the pixel-based adaptive segmenter. In: Computer Vision and Pattern Recognition Workshops, pp. 38–43 (2012)
8. Kim, K., Chalidabhongse, T.H., Harwood, D., Davis, L.: Real-time foreground-background segmentation using codebook model. Real-Time Imaging **11**(3), 172–185 (2005)
9. Liang, Z., Liu, X., Liu, H., Chen, W.: A refinement framework for background subtraction based on color and depth data. In: IEEE International Conference on Image Processing, pp. 271–275 (2016)
10. Lo, B.P.L., Velastin, S.A.: Automatic congestion detection system for underground platforms. In: International Symposium on Intelligent Multimedia, Video and Speech Processing, pp. 158–161 (2001)
11. Nonaka, Y., Shimada, A., Nagahara, H., Taniguchi, R.: Evaluation report of integrated background modeling based on spatio-temporal features. In: Computer Vision and Pattern Recognition Workshops, pp. 9–14 (2012)
12. St-Charles, P.L., Bilodeau, G.A., Bergevin, R.: A self-adjusting approach to change detection based on background word consensus. In: IEEE Winter Conference on Applications of Computer Vision, pp. 990–997 (2015)
13. Stcharles, P.L., Bilodeau, G.A.: Improving background subtraction using local binary similarity patterns. In: Applications of Computer Vision, pp. 509–515 (2014)
14. Stcharles, P.L., Bilodeau, G.A., Bergevin, R.: Online multimodal video registration based on shape matching. In: IEEE Conference on Computer Vision and Pattern Recognition Workshops, pp. 26–34 (2015)

15. Suo, P., Wang, Y.: An improved adaptive background modeling algorithm based on Gaussian mixture model. In: International Conference on Signal Processing, pp. 1436–1439 (2008)
16. Torabi, A., Bilodeau, G.A.: Local self-similarity-based registration of human rois in pairs of stereo thermal-visible videos. Pattern Recogn. **46**(2), 578–589 (2013)
17. Van Droogenbroeck, M., Barnich, O.: Visual background extractor (2011)
18. Van Droogenbroeck, M., Paquot, O.: Background subtraction: experiments and improvements for vibe. In: Computer Vision and Pattern Recognition Workshops, pp. 32–37 (2012)
19. Wang, H., Suter, D.: Background subtraction based on a robust consensus method. In: International Conference on Pattern Recognition, pp. 223–226 (2006)
20. Yin, X., Wang, B., Li, W., Liu, Y., Zhang, M.: Background subtraction for moving cameras based on trajectory-controlled segmentation and label inference. KSII Trans. Internet Inf. Syst. **9**(10), 4092–4107 (2015)

A New Chessboard Corner Detection Algorithm with Simple Thresholding

Qi Zhang[✉] and Caihua Xiong

State Key Lab of Digital Manufacturing Equipment and Technology,
Institute of Rehabilitation and Medical Robotics,
Huazhong University of Science and Technology,
Wuhan 430074, Hubei, People's Republic of China
{edisonzhangqi,chxiong}@hust.edu.cn

Abstract. Chessboard detection and corner extraction are imperative during the camera calibration process, which is a fundamental work in computer vision. This paper describes a new chessboard corner detection algorithm using the amplitude spectrum feature of circular sampling at each point as corner response function. A simple thresholding technique is that only the points near a chessboard corner will get positive corner response. However, image noise will bring a lot of false positives. The distribution of false positive response with respect to image noise is given, this distribution also has a property like the three-sigma rule in Gaussian distribution, that can be used as thresholding rule. The only parameter needed is the standard deviation of image noise, which can be measured using some patches of the image for once. Experiment is presented showing the efficiency of the proposed method against noise, compared with existing algorithm under simulated image.

Keywords: Chessboard detection · Discrete Fourier transform · Corner response · Difference of Rayleigh distribution

1 Introduction

Chessboard pattern is commonly used in camera calibration, which is a necessary step in many computer vision tasks. Since such pattern can provide simple and clear corners with known geometrical parameters, it is also suitable for 3D pose estimation and localization in robot vision. Thus extracting chessboard corners fast, accurately and automatically is still a significant challenge. The available methods for chessboard corner detection often require hand-tuning of parameters under different image conditions to filter out false corners. Although some methods give empirical parameters for corner filtering, they may still fail due to different noise level or light condition. Therefore, methods need manual intervention prohibit automated use.

In this paper, we will present a chessboard corner detection method which only need a priori knowledge of the image noise variance. This can be estimated

© Springer International Publishing AG 2017
Y. Huang et al. (Eds.): ICIRA 2017, Part II, LNAI 10463, pp. 532–542, 2017.
DOI: 10.1007/978-3-319-65292-4_46

using a bright and uniform image patch, e.g. an image of a white paper or some bright squares of the chessboard, taken from the same camera with fixed exposure gain. This prerequisite is reasonable, and easy to operate for only one time. Experiment with simulated images and real images shows the efficiency and feasibility of the proposed method.

Following a review of published chessboard corner detection methods in Sect. 2, in Sect. 3 we use circular sampling strategy proposed in [1] and rearrange the sampled points into a 1D vector. We discuss the time-frequency characteristic of the sampled vector at different location on the chessboard. Then we present a chessboard corner response function. In Sect. 4, we reveal that image noise will cause false positive response, and we show the distribution of undesired corner response caused by image noise. Surprisingly, this distribution is similar to Gaussian distribution and has a property like the three-sigma rule, which can be used to choose a threshold for false positive filtering. Experiment is conducted in Sect. 5, we present both simulated and real-world image to verify the proposed method's efficiency and feasibility. This paper ends with some conclusions and extensions.

2   Related Works

Several methods have been published for chessboard corner detection. These methods usually consist two main steps, locate all corners roughly and select the chessboard corners. General corner detector, such as the Harris [7], SUSAN [14], or FAST [13] features, can be used to locate corners roughly. Harris corner detector uses corner metric threshold to filter out most noise points, and it is especially suitable for L corner. However, the chessboard corner is an X corner, which will looks like two L corners when the image is blurred, thus applying Harris corner detector to a blurred chessboard image may get two corners for each chessboard vertex. FAST use an intensity difference threshold and the number of contiguous disparate pixels to classify a point as or not corner. If the intensity difference threshold is too small, there will be more than one FAST features near a chessboard corner, and if the value is too large, some chessboard corners may be missed due to nonuniform luminance. SUSAN detector only use non-maximal suppression to select corners, but it should be pointed out that SUSAN will treat a chessboard corner without any rotation or distortion as an edge point.

Although general corner detector always need proper threshold to get all chessboards corners, several methods are based on this detector to roughly locate all corners and use different techniques to select chessboard corners. Wang et al. [16] uses Harris detector to select all corners and apply two rotating orthogonal masks to filter out non X corners, then using line intersection to select chessboard internal points. This method will be affected by lens distortion. Kassir et al. [9] use multi-scale Harris detector and improved filtering method. Ha [6] use Lucas-Kanade feature detector to track corners and use a radial accumulator to select the chessboard corner. These two methods always need a lot of computing resource. Zhu et al. [19] propose an improved SUSAN detector with predefined threshold. He et al. [8] adopt Hessian feature to detect points, and propose an adaptive thresholding technique to extract X corners, then apply a circular template to eliminate blobs. However, This method also use some predefined threshold.

Besides general corner detector, several local feature based methods have been proposed. Yu et al. [17] insert 5 double-triangle patterns as reference into a chessboard and detect them using rotated templates, then recover chessboard corners using the double-triangle patterns. The performance of their method depends on the success of detecting the double-triangle patterns. Sun et al. [15] rearrange points in a square window centered at a point into several 1D vectors called layers, apply binarization and special morphology operation to select chessboard corners. This method will produce false corners from noise and is rather slow. Zhao and Wang [18] divide a window centered at a point into four quadrants, the summed intensity difference of adjacent quadrant is larger than the given threshold may contain a chessboard corner. This method needs hand-tuning threshold and may have a lot of false corners. Geiger et al. [5] use four pairs of convolution kernels to produce a corner response map, after non-maximal suppression for candidates and apply windowed gradient histogram to each candidate, they generate a new template for that chessboard. This method can detect multiple chessboard, but it needs several convolutions, which is quite slow. Donné et al. [3] propose a method based on convolution neural networks, this can be explained as using several weighted convolution kernels to detect chessboard. This may be a generalized method for detecting every single object, but it requires a lot of computing power. Bennett and Lasenby [1] advance a simple, fast and efficient corner response function using only circular points around sampling center. The main drawback is that the corner response function will trigger a lot of false corners, although thresholding will eliminate those false positives, it is nontrivial to give a threshold. Bok et al. [2] also use the features of circular sampled points to discard candidates. This method include fraction coordinates and needs a lot of interpolate intensities which will slow down the computation.

Accurate chessboard corner can also be obtained or selected by line intersection. de la Escalera and Maria [4] proposed a method which adopt Hough line transform twice to obtain edge lines on chessboard. This method also suffers from lens distortion. A complex technique, called ROCHADE proposed by Placht et al. [12], combines both point and edge features. This method is stable but very slow. This paper aims to offer a fast and minimal hand-tuning approach for chessboard corner detection.

3   Chessboard Corner Response Function

Although conventional corner detector like Harris and FAST can be used to extract chessboard corner roughly. It still needs further operation to exclude outliers. Most published filtering methods take advantage of the local characteristic of a chessboard corner, like a circular boundary centered at a corner [1,2,15]. The main benefit of circular sampling is rotationally invariant, also it can carry a lot of information to classify corners when samples enough points. Unlike most methods mentioned above, we use circular sampling to roughly extract points, which have similar local feature to chessboard corner, then filter out false corners. The characteristic of circular sampled data should be discussed first. Sampler in the following paragraphs refers to the circular sampler. Sampling center means the center of circular sampler.

Consider an ideal chessboard image with no noise, good light condition and only three gray levels which appears as a binary image. Apply a circular sampler with resolution one point per degree to the ideal image, three main cases will appear as listed below.

- When sampler moves to a chessboard corner, see Fig. 1a, the black circle indicates the sampling path. Suppose start sampling counter clockwise from the most right point, and put sampled intensities into a 1D vector in the sampling order. The 1D vector, also a 1D signal, will contain two cycles of square wave whose period is half the vector length, see Fig. 1b. If the sampling center is near a chessboard corner, the sampled vector will be slightly different from the one sampled at a corner, it will wrap two square waves with different period.
- As sampler goes to a block edge on the chessboard, see Fig. 1c, if apply the same sampling order mentioned above, we will obtain a wide, one cycle square wave whose period is the vector length, see Fig. 1d. When the sampler is near the block edge, the sampled vector is also square wave with a narrower or wider rectangle.
- While the sampler is on a flat region, e.g. a bright block on chessboard, the sampled vector will appear as straight line.

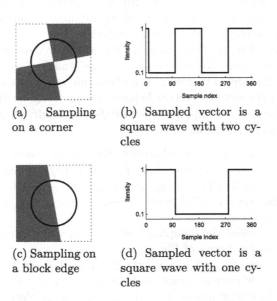

(a)   Sampling on a corner

(b) Sampled vector is a square wave with two cycles

(c) Sampling on a block edge

(d) Sampled vector is a square wave with one cycles

Fig. 1. Different sampling result

It is obvious that the sampled 1D vector, which behaves as a two cycle square wave, indicate the sampling center is on a chessboard corner. Thus, the chessboard corner detection can be implemented by finding the two cycle square wave in the sampled vectors, which can be solve by template matching. However, the template matching may fail due to the circular shifting of the sampled vector cause by chessboard rotation. So a circular shifting template matching methods is needed, but this method is always time-consuming. Nevertheless, we realize that the Fourier transform of the sampled vector is a reasonable solution to the matching problem. One of the basic properties of Fourier transform is time shifting does not change the magnitude of any frequency component, which means that the magnitude of a frequency component in the amplitude spectrum of the sampled vector will not change under circular shifting cause by chessboard rotation.

For computational efficiency, we adopt the discrete circular sampling strategy proposed by [1]. The sampled 1D vector is denoted as $\boldsymbol{x} = (x_1, x_2, \ldots, x_M)^T$, where T means transpose, x_i is sampled intensity. The sampling point (u_i, v_i) centered at (u_c, v_c), is defined as

$$u_i = \lfloor r \cos \left(2\pi \tfrac{i}{M} \right) + u_c \rfloor \tag{1}$$
$$v_i = \lfloor r \sin \left(2\pi \tfrac{i}{M} \right) + v_c \rfloor \tag{2}$$

where $i = 1, 2, \ldots, M$. M is the number of samples, r is the radius of the sampler, and $\lfloor x \rfloor$ return the rounded value of x, which ensures generating integer coordinates. For a blurred image, r may need to be larger. In our case, $r = 5$ pixels. The discrete Fourier transform of $\mathbf{x}$ is given as $\mathbf{f} = (f_0, f_1, \ldots, f_{M-1})^T$, where f_k encodes both amplitude and phase of a complex sinusoid which frequency is k cycles per M samples. $|f_k|/M$ is the amplitude of the sinusoid with frequency k, a large $|f_k|/M$ means the source signal mainly depend on the sinusoid with frequency k. This is the key idea of our method.

While dealing with an ideal image mentioned above, see Fig. 2a, apply the circular sampler to each pixel in the ideal image, and perform discrete Fourier transform for each sampled vector, it could be found out that the discrete amplitude spectrum of the sampled vector behaves distinctively at different position. Using only too frequency component, f_1 and f_2, we classify these behaviors into four main cases as listed below.

- When sampling at a chessboard corner, the sampled vector is a two cycles square wave, it is easy to verify that $|f_1| = 0$ and $|f_2|$ is relative big, see the center of Fig. 2b and c. The image center contains a chessboard corner.
- $|f_1| > |f_2|$ when the sampling center is near a block edge, see Fig. 2b bright vertical strip at the center.
- $|f_1| = |f_2| = 0$ when the sampling center is on a flat area, such as bright block, check the for corner of Fig. 2b and c.
- Relation between $|f_1|$ and $|f_2|$ is complicated. $0 < |f_1| < |f_2|$ while the sampling center is near a corner, see the points near the center of Fig. 2d. When sampling center moves toward four corners or along a block edge, it can be seen that $|f_2| < |f_1|$.

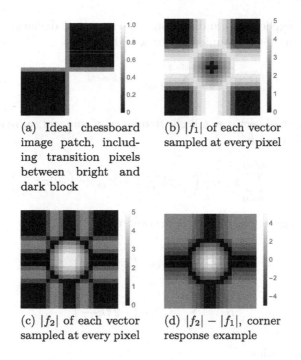

(a) Ideal chessboard image patch, including transition pixels between bright and dark block

(b) $|f_1|$ of each vector sampled at every pixel

(c) $|f_2|$ of each vector sampled at every pixel

(d) $|f_2| - |f_1|$, corner response example

Fig. 2. Amplitude spectrum analysis of ideal chessboard image

It is obvious that $|f_1| < |f_2|$ only if sampling center is near the chessboard. So we define a chessboard corner response function as

$$R = |f_2| - |f_1| \qquad (3)$$

One simple criteria for the point may be a chessboard corner is $R > 0$. And local maximal of R indicate that it is a chessboard corner. However, $R > 0$ is a necessary but not sufficient condition for a point to be a chessboard corner. Consider the real-world image, there always be noise, which can cause $R > 0$ even on a uniform region such as bright block on the chessboard. So it is necessary to eliminate the influence of noise.

4   Corner Response Filtering

A straightforward way to filter out false positives caused by noise is image filtering, such as Gaussian smoothing. But this can not eliminate all noise and will blur the corners, which will affect corner detection mention in Sect. 3. Although image noise may cause false positives, the corner response with respect to noise has a limited range, which can be used as a threshold to filter out false positives. Assume the images have additive Gaussian noise with standard deviation σ, denote a random variable x obeys Gaussian distribution as $x \sim \mathcal{N}(\mu, \sigma^2)$, where μ is the expectation and σ represents standard deviation. The discrete Fourier transform of the sampled vector $\mathbf{x}$ is defined as

$$f_k = \sum_{m=0}^{M-1} x_m e^{-i2\pi km/M} \tag{4}$$

where M is length of vector $\mathbf{x}$, k is the frequency of sinusoid, substituting Euler's formula in Eq. 4 we get

$$f_k = \sum_{m=0}^{M-1} x_m \cos\left(-2\pi k \frac{m}{M}\right) + i x_m \sin\left(-2\pi k \frac{m}{M}\right) \tag{5}$$

the real and imaginary part of f_k is denoted as

$$R_k = \sum_{m=0}^{M-1} x_m \cos\left(-2\pi k \frac{m}{M}\right) \tag{6}$$

$$I_k = \sum_{m=0}^{M-1} x_m \sin\left(-2\pi k \frac{m}{M}\right) \tag{7}$$

It is worth to note that

$$\sum_{m=0}^{M-1} \cos\left(-2\pi k \frac{m}{M}\right) = \sum_{m=0}^{M-1} \sin\left(-2\pi k \frac{m}{M}\right) = 0 \tag{8}$$

$$\sum_{m=0}^{M-1} \cos^2\left(-2\pi k \frac{m}{M}\right) = \sum_{m=0}^{M-1} \sin^2\left(-2\pi k \frac{m}{M}\right) = \frac{M}{2} \tag{9}$$

Suppose the sampling center is on a uniform region with noise mention above, then $\mathbf{x}$ is a random vector, its elements obey Gaussian distribution, denoted as $x_i \sim \mathcal{N}(\mu, \sigma^2)$, where μ is intensity of the uniform region, σ is the standard deviation of image noise. It can be derived that $R_k \sim \mathcal{N}(0, \sigma^2 M/2)$, $I_k \sim \mathcal{N}(0, \sigma^2 M/2)$. Both R_k and I_k obey Gaussian distribution. The magnitude of frequency k component is given by

$$|f_k| = \sqrt{R_k^2 + I_k^2} \tag{10}$$

Due to the fact that R_k and I_k both obey the same Gaussian distribution, $|f_k|$ obeys Rayleigh distribution, it's probability density function is given as

$$p(z) = \frac{z}{\tau^2} e^{-\frac{z^2}{2\tau^2}}, \quad x \geq 0 \tag{11}$$

where $z = |f_k|$ in our case, $\tau = \sigma^2 M/2$ is a parameter of Rayleigh distribution, it is also the standard deviation of R_k and I_k. Then the corner response function defined above, see Eq. 3, obeys distribution of difference of two variables

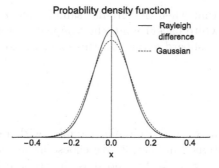

Fig. 3. Rayleigh difference distribution and Gaussian distribution, the standard deviation is 0.1

that obey Rayleigh distribution. We call this distribution as Rayleigh difference distribution, its probability density function is derived as

$$\gamma(y) = \int_{-\infty}^{+\infty} p(y)p(t-y)\,\mathrm{d}t \tag{12}$$

$$= \frac{e^{-\frac{y^2}{2\tau^2}}}{8\tau^3}\left(2|y|\tau + e^{\frac{y^2}{4\tau^2}}\sqrt{\pi}\left(2\tau^2 - y^2\right)\mathrm{Erfc}\left(\frac{|y|}{2\tau}\right)\right) \tag{13}$$

where $y = |f_2| - |f_1|$ in this case, while $|f_1|$ and $|f_2|$ are random variables obey a Rayleigh distribution with parameter τ as defined in Eq. 11, $\mathrm{Erfc}(y)$ is the complementary error function defined as

$$\mathrm{Erfc}(y) = \frac{2}{\sqrt{\pi}}\int_{y}^{\infty} e^{-t^2}\,\mathrm{d}t \tag{14}$$

This distribution is similar to Gaussian distribution, see Fig. 3, and it also has the three-sigma rule, however the sigma mentioned here is not the standard deviation as the one in Gaussian distribution, it is a parameter of the distribution, which is denoted as τ in our case mentioned above. Thus, we call this property as the three-tau rule. Define the cumulative distribution function of Rayleigh difference distribution as

$$P(y) = \int_{-\infty}^{y} \gamma(t)\,\mathrm{d}t \tag{15}$$

where $\gamma(t)$ is defined as Eq. 13.

Using numerical calculation, we can get that $P(3\tau) = 0.99919456$. This means that if we use 3τ as the threshold to filter out false positives, the ratio of filtered out point to all false positives is about 99.919456%. It need to note that for a VGA image, there are 640×480 points, if we choose the 3τ as threshold, there will be about 247 points that may not be filtered out. However, we can use 5τ as a threshold because of $P(5\tau) = 0.99999988$, which means for VGA resolution, all the false positives will be eliminated.

It is worth noting that when the circular sampler move to a strip like place, such as dark wire on white background, the sampled vector will also behave as the one sampled at chessboard corner. In this case, we adopt the ChESS in [1] to solve this problem.

5   Experiment

In our implementation, we use a circular sampler with radius 5 pixel and 16 sample points as mentioned in Sect. 3. We choose 5τ as the threshold for false positive filtering, where $\tau = 2\sqrt{2}\sigma$, and σ is the standard deviation of image noise. Our method is compared with the ChESS algorithm proposed in [1] under simulated image to demonstrate its feasibility and efficiency.

We generate a simulated grayscale image which is composed of 6 by 6 squares. Each square is 30 pixels wide. The intensities of the bright and dark block are 0.8 and 0.2, while the intensity range is $[0, 1]$. Between bright and dark block we add transition pixels, whose intensity is 0.5. To demonstrate the feasibility of eliminating false positives, we add Gaussian noise to the image while the noise standard deviation σ is 0.05. Then we apply our method and ChESS to this image.

Figure 4 shows the corner response of our method and ChESS. The image is cropped for recognizable, and it only contains 4 true chessboard corners. The image is not smoothed and the corner response maps are not filtered. Notice that the result of ChESS, see Fig. 4c, contains a lot of local maximum response caused by image noise. Although these false positives can be eliminated by setting a proper threshold, it will need hand-tuning, and may miss the true chessboard corner. However, the corner response of our method is quite clean, see Fig. 4b, only points near the corner have positive response as analyzed in Sect. 3. Small local maximum caused by noise are all filter out using the technique proposed in Sect. 4.

Figure 5 shows the detection result of our method under real-world, poor-lit image. After calculating corner response, we use non-maximum suppression to get the points with the highest local response to represent chessboard corners.

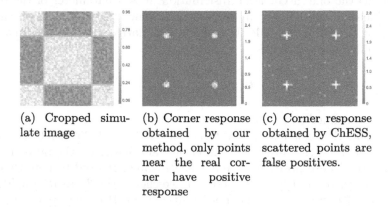

(a) Cropped simulate image

(b) Corner response obtained by our method, only points near the real corner have positive response

(c) Corner response obtained by ChESS, scattered points are false positives.

Fig. 4. Corner response comparison between ChESS and our method, the image is cropped for recognizable

Standard deviation σ of image noise is about 0.02, which is measured by choose some black block on the chessboard from this image. It is important to notice that some false positives still exist, see the red circle at the upper left corner of the image. These false positives are not caused by image noise. The main reason leads to this result is that we only use one circular sampler and the first simple threshold strategy proposed in Sect. 3 is a necessary but not sufficient rule. Thus, those points which have same circular sampling feature as the chessboard corner but are even not corners will be detected by our method.

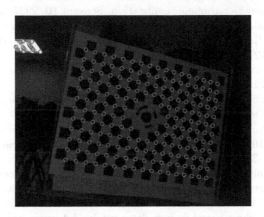

Fig. 5. Test our method using real-world image, red circles indicate chessboard corner, the standard deviation of image noise is about 0.02 which is estimated using dark blocks (Color figure online)

6   Conclusion

In this paper, we have proposed a chessboard corner detection algorithm based the magnitude spectrum characteristic of the circular sampled data. Given the standard deviation of image noise, we derived the distribution of the corner response with respect to image noise, which we name it as Rayleigh difference distribution. This distribution is similar to Gaussian distribution, and it has a property like the three-sigma rule in Gaussian distribution, which we call it the three-tau rule. However, we choose five taus to be the threshold due to fact that the probability covered in five taus is nearly 1. Experiment showed our method can efficiently eliminate false positives caused by noise, and is also robust against poor light condition.

Due to the strip image may have the same circular sampling characteristic as the chessboard corner. Our method can be used to extract strip center, such as the center line of a black wire on a bright background.

Acknowledgment. This work was partially supported by the National Natural Science Foundation of China (Grant No. 91648203 and No. 51335004), the International Science & Technology Cooperation Program of China (Grant No. 2016YFE0113600), and the Science Foundation for Innovative Group of Hubei Province (Grant No. 2015CFA004).

References

1. Bennett, S., Lasenby, J.: ChESS - quick and robust detection of chess-board features. Comput. Vis. Image Underst. **118**, 197–210 (2014)
2. Bok, Y., Ha, H., Kweon, I.S.: Automated checkerboard detection and indexing using circular boundaries. Pattern Recognit. Lett. **71**, 66–72 (2016)
3. Donné, S., et al.: MATE: Machine learning for adaptive calibration template detection. Sensors **16**, 1858 (2016)
4. de la Escalera, A., Armingol, J.M.: Automatic chessboard detection for intrinsic and extrinsic camera parameter calibration. Sensors **10**, 2027–2044 (2010)
5. Geiger, A., et al.: Automatic camera and range sensor calibration using a single shot. In: 2012 IEEE International Conference on Robotics and Automation, pp. 3936–3943 (2012)
6. Ha, J.-E.: Automatic detection of chessboard and its applications. Optical Eng. **48**, 067205–067205-8 (2009)
7. Harris, C., Stephens, M.: A combined corner and edge detector. In: Proceedings of Fourth Alvey Vision Conference, pp. 147–151 (1988)
8. He, J., et al.: Automatic corner detection and localization for camera calibration. In: IEEE 2011 10th International Conference on Electronic Measurement & Instruments, pp. 312–315 (2011)
9. Kassir, A., Peynot, T.: reliable automatic camera-laser calibration. In: Proceedings of the 2010 Australasian Conference on Robotics & Automation (2010)
10. Liu, Y., Chen, Y., Wang, G.: Robust and accurate calibration point extraction with multi-scale chess-board feature detector. In: Advances in Image and Graphics Technologies: Chinese Conference, IGTA 2014, Beijing, China, 19–20 June, Proceedings, pp. 153–164 (2014)
11. Liu, Y., et al.: A practical algorithm for automatic chessboard corner detection. In: 2014 IEEE International Conference on Image Processing (ICIP), pp. 3449–3453 (2014)
12. Placht, S., et al.: ROCHADE: Robust checkerboard advanced detection for camera calibration. In: Computer Vision - ECCV 2014: 13th European Conference, Zurich, Switzerland, 6–12 September, Proceedings, Part IV, pp. 766–779 (2014)
13. Rosten, E., Porter, R., Drummond, T.: Faster and better: a machine learning approach to corner detection. IEEE Trans. Patt. Anal. Mach. Intell. **32**, 105–119 (2010)
14. Smith, S.M., Brady, J.M.: SUSAN—a new approach to low level image processing. Int. J. Comput. Vis. **23**, 45–78 (1997)
15. Sun, W., et al.: Robust checkerboard recognition for efficient nonplanar geometry registration in projector-camera systems. In: Proceedings of the 5th ACM/IEEE International Workshop on Projector Camera Systems, pp. 1–7 (2008)
16. Wang, Z., et al.: Recognition and location of the internal corners of planar checkerboard calibration pattern image. Appl. Math. Comput. **185**, 894–906 (2007)
17. Yu, C., Peng, Q.: Robust recognition of checkerboard pattern for camera calibration. Optical Eng. **45**, 093201–093201-9 (2006)
18. Zhao, F., Wang, C.W.J.: An automated X-corner detection algorithm (AXDA). J. Soft. **6**, 791–797 (2011)
19. Zhu, W., et al.: A fast and accurate algorithm for chessboard corner detection. In: 2nd International Congress on Image and Signal Processing, CISP 2009, pp. 1–5 (2009)

FPGA-Based Connected Components Analysis Algorithm Without Equivalence-Tables

Luxiang Ling, Zhong Chen$^{(\boxtimes)}$, Shuai Li, and Xianmin Zhang

Guangdong Provincial Key Laboratory of Precision Equipment
and Manufacturing Technology, School of Mechanical and Automotive
Engineering, South China University of Technology, Guangzhou 510640, China
linglx@foxmail.com, {mezhchen, zhangxm}@scut.edu.cn

Abstract. Connected components analysis plays an important role in real-time target localization. In this paper, we propose a fast connected-components-analysis algorithm without equivalence-tables based on FPGA (Field Programmable Gate Array). The algorithm runs in a pixel-scanning mode and labels all run-length pixels. As a result, the geometric features of the connected components can be extracted by only one-times image-scanning. A mapping structure between a run-length and a connected components is designed to replace the equivalence table in order to reduce its resource consumption and promotes its running speed. The experiment results demonstrate that our algorithm occupies less resources than state-of-the-art algorithms and runs faster than personal-computer based algorithm. Besides, this FPGA-based algorithm can deal with any connected components in complicated shape or quantity.

Keywords: Connected component analysis · Target localization · FPGA · Run-length · Equivalence table · Data mapping

1 Introduction

Region-features extraction, such as the center of gravity, play a key role in a vision-based target localization. Connected components analysis (CCA) algorithm can extract the center of the target region quickly. Typically there are four operations of CCA algorithm: (1) scan the pixels and label them with the given rule [1–5]; (2) detect the pixels' label between adjacent lines and estimate whether they are belong to the same connected components or not; (3) merge pixels that are belong to the same connected component; (4) extract their geometric features, such as area, center of gravity. When the CCA algorithm is implemented on a personal computer, the above operations only can be executed in serial, which will reduces the execution speed of the algorithm. Because FPGA chips function as key characteristics of programmable runs in parallel, the four operations can be executed in parallel on a FPGA chip, which will improves the execution speed.

The basic principles of FPGA-based CCA algorithm [6–11] can be described as follows: (1) Scanning and labelling the pixels, the pixels in adjacent rows are evaluated according to the four or eight neighborhood principle. (2) The pixel labels which belong to a same connected component in adjacent lines are recorded. (3) The recorded

© Springer International Publishing AG 2017
Y. Huang et al. (Eds.): ICIRA 2017, Part II, LNAI 10463, pp. 543–553, 2017.
DOI: 10.1007/978-3-319-65292-4_47

pixel labels should be used to merge the labeled connected components. (4) Finally the geometric features of the connected components should be calculated. Both increasing the execution speed and reducing the resource occupation are two important goals of the algorithm, which mainly depend on the scanning mode and the data structure that are chosen in the CCA algorithm. The improved scanning mode can compress the pixels and reduce the space for storing pixels. The good data structure can manage the correspondence of the labels in adjacent rows, and can make it easy to extract various shapes of connected components (U shape, W shape, O shape), and increase the speed and reduce the resource occupations at the same time.

At present, there are two kinds of scanning modes, pixel-scanning and run-scanning. A pixel-scanning mode [6, 7] is based on pixels and set pixels as scanning and labeling objects, while run-scanning mode [8–11] adopts pixel-running method to scan and label the objects. The run is a management method of adjacent foreground pixels in one row, which can compress pixels and reduce its resource consumption. When the area and quantity of connected components are large, the scanning mode based on pixel will occupy more memory. So it is a better choice to use run-scanning mode in a FPGA-based algorithm in which the resources are limited. In this paper, we use run-scanning mode in a CCA algorithm.

On the other hand, the most-adopted data structures are equivalence tables [6, 7] and derivative structures based on equivalence tables [8–10] in the existing algorithms. As shown in Fig. 1, the equivalence table is abstracted as a list of data, and the correspondence of pixel labels in adjacent rows are written as an equivalence pair, which occupies one row in the table. The smallest label of pixels and label of connected components where the located pixels are regarded as a equivalence pair, and other remaining adjacent labels of pixels are written as equivalent pairs, so that algorithm can find the labels of the connected components sequentially where the pixels are located. However, there are two shortcomings for using equivalence table: Firstly, when the quantity of the connected components in an image is large, it needs to save many equivalence pairs and occupy much memory. Secondly, when the shape of a connected component in an image is complex, it is necessary to scan the equivalence table

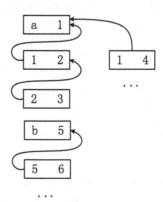

Fig. 1. Equivalence table, letters represent the label of connected components, the numbers represent labels of run-lengths, where each box indicates an equivalence pair

iteratively to obtain the minimum label of the connected components, which can reduces the execution speed and efficiency of the algorithm. A multi-layer-index structure [11] instead of equivalence table was proposed to manage the correspondence of labels in the adjacent rows, but it requires more complex logical judgments and the resource occupancy is higher.

In order to avoid those problems in the existing algorithm, we proposed a mapping structure between run-lengths and connected components (RL-CC) to replace the equivalence table in a CCA algorithm. As shown in Fig. 2, in the RL-CC mapping structure, the runs do not index with each other, which avoid the iterative traversal in the equivalence table. So we can obtain the label of the connected components where the run is located from the labels of run directly. By benefit of this mapping structure, all the run labels in the previous row are needed to be buffered and be taken as references. Compared with the existing algorithm with equivalence table, our CCA algorithm can effectively reduce resource consumptions.

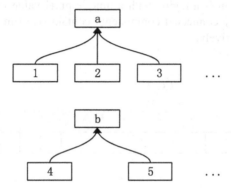

Fig. 2. RL-CC mapping structure, the letters represent the labels of the connected components, the numbers represent the labels of RL, and each box indicates an equivalence pair

2 Proposed Algorithm

2.1 Run Labelling

A run is a region with foreground pixels adjacent to each other in a same row. In this algorithm, the run label in each row is start from 1, and be independent with the run labels in previous row. As shown in Fig. 3, there are two runs in row n, labeled 1 and 2

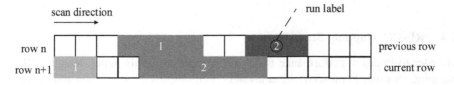

Fig. 3. The method of run labelling in this paper

respectively, as same as in row (n + 1). The traditional labelling method is shown in Fig. 4, the run labels in entire image is start from 1, but the first run in row (n + 1) is labeled as 3, and the second run is labeled as 1 because it is overlapped with the runs in previous row.

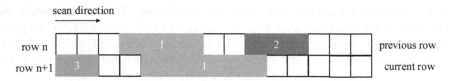

Fig. 4. The method of run labeling in traditional algorithm

2.2 Connected Components Labelling

A connected component is a region with a unique pixel value in an entire image. As shown in Fig. 5, every connected components contain one run individually, and are labeled from 1 successively.

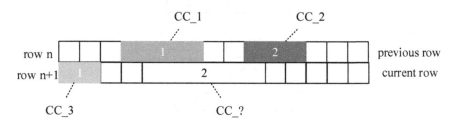

Fig. 5. Labelling method of connected components

2.3 Correspondence Detection of Runs Between Adjacent Rows

When a run is scanned at its end, the runs in the previous row are scanned to evaluate whether there is an overlap relationship between those runs. If a current run doesn't overlap any runs in the previous row, it should be assigned a unique label value as an independent connected component. If a current run overlaps with other runs in its previous line with 4-adjacency or 8-adjacency, these runs should be assigned a same connected component label. As shown in Fig. 6, connected components CC_1 and CC_2 in row *n* is merged into a same connected component with run_2 in row (*n* + 1).

2.4 RL-CC Mapping Structure

In order to improve performance of the CCA algorithm, this paper design a RL-CC mapping structure to preserve the equivalent relationships between runs and connected components. As shown in Fig. 7, three dual-port RAM are used as a medium to store

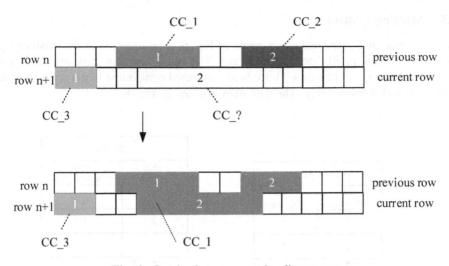

Fig. 6. Overlap between runs in adjacent row

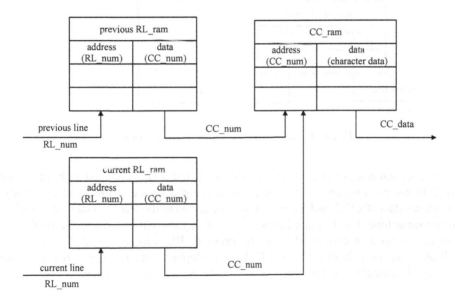

Fig. 7. Data mapping structure

the run labels in current and previous row, connected component labels in an entire image and their features respectively. In the RAM of run, the run label (RL_num) is used as an address and a connected component label (CC_num) is used as data. In the RAM of connected components, the connected component label is used as address and its features are used as data. In this way, we can obtain the connected component label directly through the run labels.

2.5   Merging Control

Take the runs and connected components in Fig. 6 as an example to demonstrate the evolution of data in RL_ram and CC_ram. The run_1 and run_2 in row n do not overlap with run_1 in row $(n + 1)$, so three connected component labels are created as label 1, 2 and 3 respectively. The data map is show as Fig. 8.

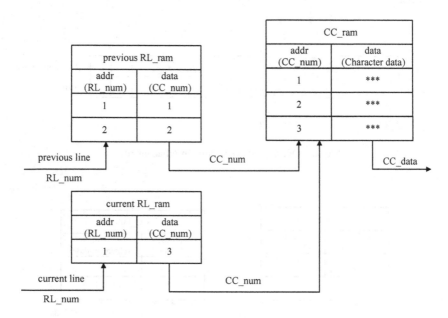

Fig. 8. Created three connected component labels

The second run in row $(n + 1)$ is labeled as 2, but it is overlapped with run_1 and run_2 in the previous row, so they need to be merged by the following steps. Firstly, the feature data of CC_2 and the current run are added to the feature data field of CC_1. At the same time, the data at address 2 in the CC_ram will be cleared. Secondly, the data at address 2 in current RL_ram and previous RL_ram are changed by 1, which indicates that they belong to the CC_1. The mapped data is shown in Fig. 9, the numbers in parentheses represent the previously connected component labels.

2.6   Algorithm Architecture

The architecture of hardware acceleration using FPGA is depicted as Fig. 10, which is divided into four operations: (1) Scan all pixels to evaluate whether they are foreground or background; (2) Detect the runs and label them; (3) Detect the run labels in the previous row to judge whether they are overlapped with the current run; (4) Merge them when an overlap happens between both runs. These four operations will be executed in parallel on FPGA.

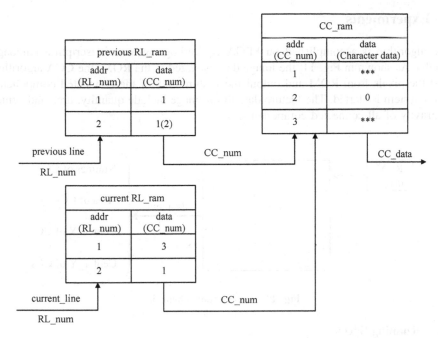

Fig. 9. Do an equivalent merger

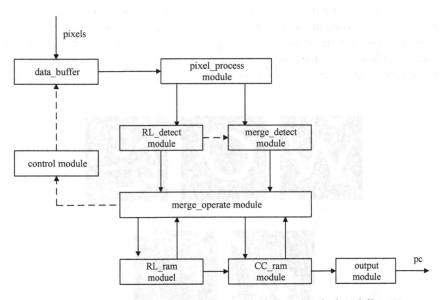

Fig. 10. Architecture of hardware, solid line represent data signal, dotted line represent the control signal

3   Experiments

The algorithm is accomplished on FPGA by verilog hardware description language (HDL). As shown in Fig. 11, the image data is stored in the ROM, the CCA algorithm read the pixels from ROM and output the feature data of the connected components when system is started. The feature data in an image include quantity, area and center of gravity of all connected components.

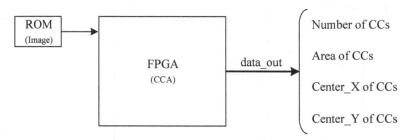

Fig. 11. Experiment schematic

3.1   Running Speed

In the experiments, the two platforms are FPGA (Cyclone III) and PC (3.4 GHz, 8 G memory, Windows 7 OS, C++). The images for performance evaluation are shown in Fig. 12, the processing time of the algorithm on two platforms are recorded in Table 1. The algorithm proposed in this paper is executed at a speed of one clock of one pixel, and the processing time of an image keep fixed, only dependent on system clock frequency. When the system clock frequency is 150 MHz, the execution time is as follows: $(256 * 256)/(150 * 10^6) = 0.44$ (ms).

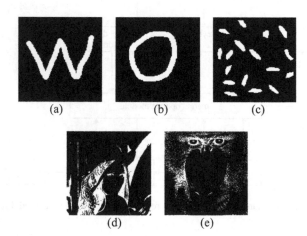

Fig. 12. Experiment images size of 256 * 256, (a) shape 'W', (b) shape 'O', (c) rice, (d) lena, (e) mandrill

Table 1. Experiment results

Image	Number of CCs	Time consumption of platform	
		Ours (ms)	PC (ms)
a	1	0.44	0.52
b	1	0.44	0.62
c	17	0.44	0.64
d	155	0.44	3.85
e	1168	0.44	4.39

When the size of the image is constant, as the number of CCs increases, the processing time of FPGA is unchanged, while the PC is increasing.

From the experiment, we can know that when the number of connected components is increased, the processing speed of PC platform will be lower because it needs more memory-access operations. So, FPGA has great advantages than PC for larger size and larger number of connected components of an image. At the same time, the proposed algorithm can deal with images with complex shape of connected components like 'W' and 'O'. Considering the domain center of CCA, the results of experiment are accurate compared with the PC, in five picels. Because the experiment time and resource consumption are carried out under this premise.

3.2 Resource Utilization

In our FPGA-based algorithm, the maximum quantity of runs and connected components are $M/2$ and $M*N/4$ respectively, which means that the length of the RL_ram and CC_ram is $M/2$ and $M*N/4$ respectively, the length of row buffer is same as the number of columns. The resources utilization in FPGA Cyclone III are shown in Table 2, and the resources used in [7 and 11] is presented, too.

Table 2. Experiment results

Size	Number of CCs	Time consumption of platform	
		FPGA (ms)	PC (ms)
256	1168	0.44	2.11
512	1168	1.75	5.12
1024	1168	6.99	16.51

When the size of the picture is changed, as the image size increases, the processing time of FPGA and PC are increasing, while the PC is longer than the FPGA.

Comparing with existing single-pass CCA algorithm proposed in [7 and 11], our algorithm only requires three RAMs to store the labels of runs, and the feature data of connected components respectively, which is lower than the [7 and 11]. The proposed RL-CC mapping structure simplifies the merge operation and logical judgments in the indexing, which reduce the utilization of LUT and Flip-flops. The resulted resource utilization indicate that our RL-CC mapping structure is better than multi-layer-index in [11] (Table 3).

Table 3. Resource utilization

Resource	Ours	Optimised single pass	Multi-layer-indexin
Block RAMs	3	4	13
Four-input LUTs	852	1757	4587
Flip-flops	436	600	3154

4   Conclusions

A connected components analysis algorithm without equivalence-tables based on FPGA was proposed in this paper. The algorithm runs in a pixel-scanning mode and labels all run-length pixels. Compared with the equivalence-table and multi-layer-index structure in the existing connected components analysis algorithms, the correspondence of labels in adjacent rows are managed by the RL-CC mapping structure which reduce the resource consumption. The proposed algorithm is simpler than the existing algorithm when operating connected components with complex shape, and runs faster than the connected components analysis algorithm on PC. This proposed algorithm can be applied in vision-based real-time target localization.

References

1. Suzuki, K., Horiba, I., Sugie, N.: Linear-time connected component labeling based on sequential local operations. Comput. Vis. Image Underst. **89**, 1–23 (2003)
2. Wu, K., Otoo, E., Shoshani, A.: Optimizing connected component labeling algorithms. In: Proceedings of the Medical Imaging 2005: Image Processing, pp. 1965–1976 (2005)
3. He, L., Chao, Y., Suzuki, K.: An efficient first-scan method for label-equivalence-based labelling algorithms. Pattern. Recognit. Lett. **31**, 28–35 (2010)
4. Kofi, A., Andrew, H., Patrick, D., Hongying, M.: Accelerated hardware video object segmentation: from foreground detection to connected components labelling. Comput. Vis. Image Underst. **114**, 1282–1291 (2010)
5. He, L., Chao, Y., Suzuki, K., et al.: Fast connected-component labeling. Pattern Recognit. **42**, 1977–1987 (2009)
6. Bailey, D.G., Johnston, C.T.: Single pass connected components analysis. In: Proceedings of the Image and Vision Computing New Zealand, pp. 282–287, Hamilton, New Zealand (2007)

7. Ni, M., Bailey, D.G., Johnston, C.T.: Optimised single pass connected components analysis. In: Proceedings of the International Conference on Field Programmable Technology, pp. 185–192, Taibei (2008)
8. Zhang, E.-H., Feng, J.: Run-list based connected components labeling for blob analysis. J. Appl. Sci. **5**, 019 (2008)
9. Kofi, A., Andrew, H., Patrick, D., Jonathan, O.: A run-length based connected component algorithm for FPGA implementation. In: Proceedings of the International Conference on Field-Programmable Technology, pp. 177–184, Taibei (2008)
10. Trein, J., Th, A., et al.: FPGA implementation of a single pass real-time blob analysis using run length encoding. In: Proceedings of MPC Workshop, February 2008
11. Zhao, F., et al.: Real-time single-pass connected components analysis algorithm. EURASIP J. Image Video **2013**, 21 (2013)

A Methodology to Determine the High Performance Area of 6R Industrial Robot

Nianfeng Wang, Zhifei Zhang[✉], and Xianmin Zhang

Guangdong Provincial Key Laboratory of Precision Equipment and Manufacturing Technology, School of Mechanical and Automotive Engineering, South China University of Technology, Guangzhou 510640, China
{menfwang,zhangxm}@scut.edu.cn, red13531@gmail.com

Abstract. To find an area where the robot can obtain a higher kinematic performance is a meaningful work for path planning and off-line programming. In this paper, a methodology was proposed to determine the high performance area (HPA) of a 6R industrial robot. Monte Carlo method was used to get a point cloud of the reference point of the end effector. The manipulability measure was selected as a performance index to filter the point cloud. A grid was defined to approach the filtered point cloud and its boundary was then extracted and smoothed. An example was presented to show that the proposed methodology is feasible to determine the HPA of a 6R industrial robot.

Keywords: High performance area · Performance index · Monte Carlo method

1 Introduction

The High Performance Area (HPA) was defined as, the cross-section of the high performance part of the workspace with a plane on which the first joint axis lies [10]. It was proposed as a measure of the high performance part of the workspace, and can provide designer with practical information about the performance of the robot structures. It can also provide guidance for path planning and workpiece placement.

To determine the high performance part of the workspace, a proper performance index must be selected first to tell this part of the workspace from the rest. Many performance indices were proposed in the previous works [5,7]. Yoshikawa proposed a measure of manipulability based on Jacobian of the robot [12]. The manipulability measure indicates the ability of the robot to move and apply forces in arbitrary directions. The condition index is another measure that was defined as the reciprocal of the condition number of the Jacobian and used to evaluate the control accuracy [8]. Other indices, like minimum singular value, isotropic index and so forth, were discussed in detail in this work [7].

The manipulability measure is one of the most commonly used performance indices [8]. Some researchers argued that the manipulability measure is a better indicator of dexterity than condition number or minimum singular value,

© Springer International Publishing AG 2017
Y. Huang et al. (Eds.): ICIRA 2017, Part II, LNAI 10463, pp. 554–563, 2017.
DOI: 10.1007/978-3-319-65292-4_48

because the manipulability measure considers the motion of the end effector in all directions instead of one or two [9]. Based on these ideas, we choose the manipulability measure as the performance index.

There are many methods to determine the workspace of a robot, which can be roughly classified into two types. Analytical methods determine closed form descriptions of workspace boundary [11], but these methods are usually complicated because of the nonlinear equations and inverse kinematic computation [2]. Numerical methods can get a approximate boundary of the workspace without inverse kinematic and are easier to operate [1,6]. Therefore, the methodology proposed by this paper will base on numerical method.

2 Determining the HPA

The overall processes of the proposed methodology to determine the HPA will be discussed in detail in this part.

Since the HPA was defined as the cross-section of the high performance part of the workspace with a plane on which the first joint axis lies, we can choose the xoz plane of the base frame as the plane where we compute the HPA. And we can know from previous work that the manipulability index does not depend on the first DOF [4]. Therefore, to simplify the computation, the first joint variable will always be set to 0.

2.1 The Manipulability Measure

The manipulability measure of a robot was defined as a scalar value w by [12]

$$w = \sqrt{det(J(\theta)J^T(\theta))} \tag{1}$$

where $J(\theta)$ and $J^T(\theta)$ are the Jacobian and transpose of the Jacobian respectively, and the det means determinant of a matrix.

If the robot has no redundant degree of freedom, which is like the case of this paper, the 6R industrial robot, the equation above becomes

$$w = |J(\theta)| \tag{2}$$

The manipulation Jacobian with respect to the base frame is given by [3]

$$^0J = \begin{bmatrix} ^0J_1(q) & ^0J_2(q) & \dots & ^0J_n(q) \end{bmatrix} \quad \text{with}$$

$$^0J_1(q) = \begin{bmatrix} z_{l-1} \times ^{l-1}p_n \\ z_{l-1} \end{bmatrix} \quad \text{for a revolute joint}$$

$$^0J_1(q) = \begin{bmatrix} z_{l-1} \\ 0 \end{bmatrix} \quad \text{for a prismatic joint}, \tag{3}$$

where z_{l-1} is the z-axis vector of the frame of link $l-1$ expressed in the base frame, $^{l-1}p_n$ is the position of the link n with respect to the frame of link $l-1$ and expressed in the base frame, $\times$ is the vector cross product.

Taking the end effector into account, $^{l-1}p_n$ in Eq. (3) will be replaced by $^{l-1}p_{end}$. The latter is the position of the reference point of the end effector with respect to the frame of link $l-1$ and expressed in the base frame.

2.2   Obtain and Filter the Point Cloud

Now, we will use the Monte Carlo method to get a point cloud from a set of random joint vector. Every component of these joint vectors was obtained from the following equation

$$q_l = \begin{cases} q_{l,low} + Rand \times (q_{l,up} - q_{l,low}), & l \neq 1 \\ 0, & l = 1 \end{cases} \tag{4}$$

where $q_{l,low}$ and $q_{l,up}$ is the lower and upper bound value of the l-th joint, $Rand$ stands for a random value between 0 and 1.

After that, we can compute the position and orientation of the end effector reference point by the forward kinematic equation

$$T_{end} = \left(\prod_{l=0}^{dof} {}^l T_{l+1}(q_l) \right) {}^{dof}T_{end} \tag{5}$$

where dof is the degree of freedom that the robot has, ${}^l T_{l+1}$ is the homogeneous matrix between the frames of the consecutive link l and link $l + 1$, and ${}^{dof}T_{end}$ is the homogeneous matrix of the end effector with respect to the frame of the last link.

For each point that was computed from Eq. (5), the Jacobian and manipulability measure of the robot in this point will be calculated. Then a threshold of manipulability measure will be introduced to handle those points. This step can simply be expressed by the following formula:

$$w = \begin{cases} w, & if\,(w > h) \\ 0, & otherwise \end{cases} \tag{6}$$

where h is the threshold, it is chosen in dependence on what kind of job the robot will do, which will not be discussed by this paper.

After the filtering, those points with a manipulability measure of 0 will be abandoned.

Although the first joint variable was set to 0, those points in the filtered point cloud may not locate at a single plane. This is because of the existence of the end effector, of which the reference point has a translation with respect to the frame of the last link.

Therefore, a transformation will be performed to transform these points into a single plane. The transformation is expressed by

$$p_s = \left[\sqrt{x_w^2 + y_w^2},\ 0,\ z_w \right]^T \tag{7}$$

where $p_w = [x_w, y_w, z_w]^T$ is the coordinates of a point from the filtered point cloud before the transformation, and p_s is the coordinates of this point after the transformation.

The nature of this transformation is a rotation around the z-axis of the base frame, or around the rotation axis of the first joint in terms of a 6R industrial robot. And as a result, all point will be transformed to xoz plane of the base frame. So actually this transformation will not affect the result we will get later because of the invariance of the manipulability measure and the symmetry characteristic of the workspace that we have discussed before.

2.3 Grid Handling

All those points in the point cloud are now at a single plane, i.e. xoz plane of the base reference frame. To define a grid in the plane that can overlap all points of the point cloud, the maximum dimensions of the point cloud along axis x and axis z have to be determined first. These two values can simply be calculated by the following formula:

$$\begin{cases} X_m = \max_{1 \leqslant k \leqslant N} (x_k) - \min_{1 \leqslant k \leqslant N} (x_k) \\ Z_m = \max_{1 \leqslant k \leqslant N} (z_k) - \min_{1 \leqslant k \leqslant N} (z_k) \end{cases} \tag{8}$$

where N is the number of points in the point cloud after filtering.

Now we define a grid with $m \times n$ cells, the size of each cell is Δx and Δz which can be considered as resolution dimensions. With lesser Δx and Δz we can obtain a more accurate result while the computation cost will be higher, so proper values should be chosen. In addition, each cell will have a flag value of either 0 or 1.

The number of cells along each axis can be derived by

$$\begin{cases} N_x = \lfloor \frac{X_m}{\Delta x} \rfloor + 1 \\ N_z = \lfloor \frac{Z_m}{\Delta z} \rfloor + 1 \end{cases} \tag{9}$$

where the operator $\lfloor \ \rfloor$ denotes the floor function that returns the nearest integer less than or equal to the real-valued argument inside the brackets.

After the grid was created, we decide which cell of the grid that a point p_k from the point cloud will locate in by

$$\begin{cases} i_k = \lfloor \frac{X_k}{\Delta x} \rfloor \\ j_k = \lfloor \frac{Z_k}{\Delta z} \rfloor \end{cases} \tag{10}$$

where i_k and j_k are the indices along x-axis and z-axis of the cell n_{i_k, j_k}, i.e. the cell where point p_k locates.

Then cell n_{i_k, j_k} will be colored and its flag value v_{i_k, j_k} will be set to 1. After all points from the point cloud have been handled like this, those cells with no point locates in will not be colored and their flag value remains 0.

2.4 Boundary Extraction

To extract the boundary of the cells that have been colored, two steps of sweeping operation will be applied.

In the first step, the sweeping operation is within a given row, each cells of this row will be swept to find the first and the last colored cells, which are called outer cells. The sweeping starts from cell $n_{0,j}$ to cell $n_{N_x,j}$. When we find a cell of which the index $i_{j,F}$ satisfies the following formula

$$\begin{cases} v_{i,j} = 0 & when \ \ i < i_{j,F} \\ v_{i,j} = 1 & when \ \ i = i_{j,F} \end{cases} \tag{11}$$

where $v_{i,j}$ is the flag value of cell $n_{i,j}$, then this cell $n_{i_{j,F},j}$ will be considered as the first colored cell of this row. Similarly, when we find a cell of which the index $i_{j,L}$ satisfies the following formula

$$\begin{cases} v_{i,j} = 0 & when \ \ i > i_{j,L} \\ v_{i,j} = 1 & when \ \ i = i_{j,L} \end{cases} \tag{12}$$

then this cell $n_{i_{j,L},j}$ will be considered as the last colored cell of the row. If all cells of a row have been swept and no proper $i_{j,F}$ and $i_{j,L}$ was found, then this row will be abandoned.

In the second step, the sweeping operation is within the whole grid, each row will be swept and its outer cells are considered to form the boundary. The indices of the outer cells of each row will be modified with the consideration of indices of the outer cells of the previous and following rows. This process expressed by

$$i_{j,F} = \begin{cases} i_{j,F}, & if \ \ j < 2 \ or \ j > N_z - 2 \\ \min(i_{j-1,F}, i_{j+1,F}), & if \ \ i_{j,F} < i_{j-1,F} \ and \ i_{j,F} < i_{j+1,F} \\ \max(i_{j-1,F}, i_{j+1,F}), & if \ \ i_{j,F} > i_{j-1,F} \ and \ i_{j,F} > i_{j+1,F} \\ i_{j,F}, & otherwise \end{cases} \tag{13}$$

and

$$i_{j,L} = \begin{cases} i_{j,L}, & if \ \ j < 2 \ or \ j > N_z - 2 \\ \min(i_{j-1,L}, i_{j+1,L}), & if \ \ i_{j,L} < i_{j-1,L} \ and \ i_{j,L} < i_{j+1,L} \\ \max(i_{j-1,L}, i_{j+1,L}), & if \ \ i_{j,L} > i_{j-1,L} \ and \ i_{j,L} > i_{j+1,L} \\ i_{j,L}, & otherwise \end{cases} \tag{14}$$

The purpose of this step is to ensure that there are no extreme sharps and hollows in the boundary we get.

Figure 1 illustrates the process of the two steps of operation.

After the two steps of operation, the center point of the outer cells of each row will be computed and added to an array, called boundary point array. Connecting points in this array in order with straight line, we will get a rough boundary of those colored cells.

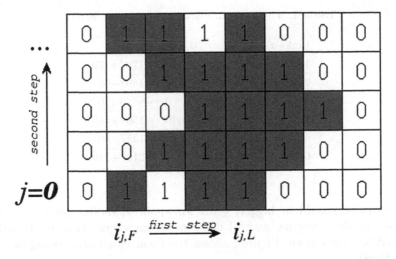

Fig. 1. The two steps of sweeping operation

2.5 Boundary Smoothing

The basic idea of smoothing the boundary is that processing the points in the boundary array with the least square method.

Suppose that the number of points in the boundary point array is N_B. Now we divide these points into N_l small point lists, each list will have $N_p = N_B/N_l$ points. For each list, the least square method is applied to determine the regression line.

$$Ls: \bar{z} = bx + a \quad \text{with} \quad \begin{cases} b = \dfrac{\sum\limits_{i=1}^{N_p}(x_i - \bar{x})(z_i - \bar{z})}{\sum\limits_{i=1}^{N_p}(x_i - \bar{x})^2} \\ a = \bar{z} - b\bar{x} \end{cases} \tag{15}$$

For two regression lines that their corresponding point lists are neighbor, their point of intersection will be computed. Then we will get N_l intersection points since we have N_l of regression lines. Connecting these intersection points in order with straight line and we will get the final approximate boundary of HPA. Moreover, rotating this boundary around the rotation axis of first joint, we will get the high performance part of the workspace.

3   An Example

In this part, an example will be presented to show that the proposed methodology is feasible to determine the HPA of a 6R industrial robot. The 6R industrial robot that this example will study on is Motoman MH12, of which the D-H parameters are given by Table 1.

Firstly, 5,000,000 random points were generated, and a threshold was applied to those points. For some applications that need more dexterity, like welding,

Table 1. The D-H parameters of Motoman MH12

Link	a/mm	$\alpha/\circ$	d/mm	$\theta/\circ$
1	0	0	0	$-170/+170$
2	155	-90	0	$-180/+65$
3	614	180	0	$-85/+150$
4	200	-90	-640	$-150/+150$
5	0	90	0	$-135/+90$
6	0	-90	0	$-210/+210$

the threshold h should be bigger, while for applications like carrying, a less h is proper. In this example, we will not aim at any application, so a moderate threshold 0.5 was chosen. Figure 2 shows the point cloud after being processed by the threshold.

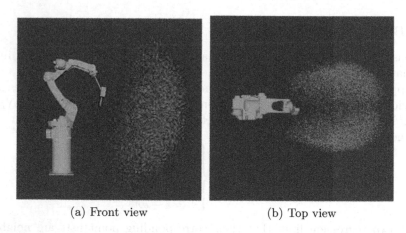

(a) Front view   (b) Top view

Fig. 2. The point cloud filtered from 5,000,000 points by a manipulability threshold of 0.5

Then a grid of which the cell size is $10\,mm \times 10\,mm$ was created. Figure 3 shows the colored cells of the grid we have got after the grid handling step.

After that, the two steps of sweeping operation was applied, a boundary array was got and its points were connected in order with straight line. The rough boundary was illustrated in Fig. 4.

Finally, the points in the boundary array were divided into 100 small lists and their regression lines were computed. 100 intersection points were obtained and connected in order with straight line to form the final boundary of HPA, which was illustrated in Fig. 5.

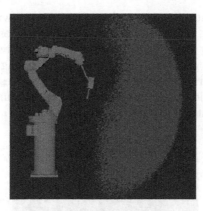

Fig. 3. The colored cells with the cell size of $10\,mm \times 10$ mm (Color figure online)

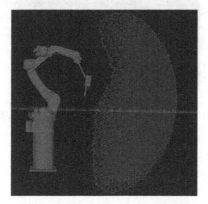

Fig. 4. Boundary before smoothing and the corresponding colored cells (Color figure online)

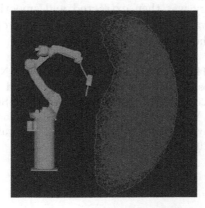

Fig. 5. The smoothed boundary and the corresponding colored cells (Color figure online)

The whole computation process takes about 10 min with a intel core i5 cpu, obtaining and filtering of the point cloud take the most of time, while grid handling and boundary extraction are much faster relatively.

Moreover, we can get several boundaries by using different thresholds. Figure 6 shows boundaries with their corresponding thresholds are 0.5, 0.6 and 0.7 from outside to inside respectively. We can get a rough idea about the distribution of the dexterity of the robot from this figure.

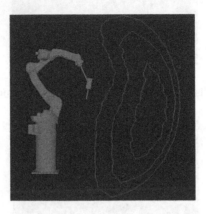

Fig. 6. Boundaries with the corresponding thresholds are 0.5, 0.6 and 0.7 from outside to inside respectively

4   Conclusion

In this paper, a methodology was proposed to determine the HPA of a 6R industrial robot. A point cloud was generated by Monte Carlo method and filtered by a manipulability threshold. A grid was defined to approach the filtered point cloud and the boundary was then extracted and smoothed with least square method. An example was presented to show that the proposed methodology is feasible to determine the HPA of a 6R industrial robot.

Acknowledgements. This work is supported by National Natural Science Foundation of China (Grant Nos. 51205134, 91223201), Science and Technology Program of Guangzhou (Grant No. 2014Y2-00217), Research Project of State Key Laboratory of Mechanical System and Vibration (MSV201405), the Fundamental Research Funds for the Central University (Fund No. 2015ZZ007) and Natural Science Foundation of Guangdong Province (S2013030013355).

References

1. Alciatore, D.G., Ng, C.C.D.: Determining manipulator workspace boundaries using the Monte Carlo method and least squares segmentation. ASME Robot.: Kinemat. Dyn. Control. **72**, 141–146 (1994)

2. Cao, Y., Qi, S., Lu, K., Zang, Y., Yang, G.: An integrated method for workspace computation of robot manipulator. In: International Joint Conference on Computational Sciences and Optimization, pp. 309–312 (2009)
3. Fu, K.S., Gonzalez, R.C., Lee, C.S.G.: Robotics: Control, Sensing, Vision, and Intelligence. Robotica (1987)
4. Gotlih, K., Troch, I.: Base invariance of the manipulability index. Robotica **22**(4), 455–462 (2004)
5. Kucuk, S., Bingul, Z.: Comparative study of performance indices for fundamental robot manipulators. Robot. Auton. Syst. **54**(7), 567–573 (2006)
6. Ottaviano, E.: A fairly general algorithm to evaluate workspace characteristics of serial and parallel manipulators. Mech. Based Des. Struct. Mach. **36**(1), 14–33 (2008)
7. Patel, S., Sobh, T.: Manipulator performance measures - a comprehensive literature survey. J. Intell. Robot. Syst. **77**(3), 1–24 (2015)
8. Salisbury, J.K., Craig, J.J.: Articulated handsforce control and kinematic issues. Int. J. Robot. Res. **1**(1), 4–17 (1981)
9. Tadokoro, S., Kimura, I., Takamori, T.: A dexterity measure for trajectory planning and kinematic design of redundant manipulators. In: Conference of IEEE Industrial Electronics Society, IECON 1989, vol. 2, pp. 415–420 (1989)
10. Valsamos, C., Moulianitis, V.C., Synodinos, A.I., Aspragathos, N.A.: Introduction of the high performance area measure for the evaluation of metamorphic manipulator anatomies. Mech. Mach. Theory **86**, 88–107 (2015)
11. Wang, Y., Chirikjian, G.S.: A diffusion-based algorithm for workspace generation of highly articulated manipulators. In: Proceedings of the IEEE International Conference on Robotics and Automation, ICRA, vol. 2, pp. 1525–1530 (2002)
12. Yoshikawa, T.: Manipulability of robotic mechanisms. Int. J. Robot. Res. **4**(4), 3–9 (1985)

Recognition of Initial Welding Position Based on Structured-Light for Arc Welding Robot

Nianfeng Wang$^{(\boxtimes)}$, Xiaodong Shi, and Xianmin Zhang

Guangdong Provincial Key Laboratory of Precision Equipment
and Manufacturing Technology, School of Mechanical and Automotive
Engineering, South China University of Technology, Guangzhou 510640, China
menfwang@scut.edu.cn, dongmyown@163.com

Abstract. This paper proposes a recognition of initial welding position for fillet weld. A structured-light vision system is presented. Using a laser scanning method, arc welding robot reciprocates along the end of weld seam with an incremental motion strategy, which makes the recognition of initial welding position fast and accurate. After given the endpoint types of fillet weld, an image processing is described. Firstly, the laser center stripe and the feature point extraction methods are given in detail, and then connected components in the image are extracted by using image segmentation algorithms. Finally, experiments are conducted with calibrated vision system to prove the effectiveness of the proposed method.

Keywords: Initial welding position · Structured-light vision · Image processing · Arc welding robot

1 Instruction

Industrial welding robots have been widely used for various automatic manufacturing process at present. However, most of welding robots work in the "teaching-and-playback" programming mode and off-line programming mode [1]. These robots must be taught in advance for the trajectory planning, work sequence and work condition etc. When the mounted position, orientation of the jigs and workpieces change, these types of welding robots must be taught and programmed again for lack of ability to adapt to circumstance changes in welding process. Therefore, to improve the automation level and intelligence of manufacture, many works have been done to realize intelligent functions in robot welding, such as seam tracking [2, 3] and weld quality control [4, 5]. However, few works pay attention to the initial welding point positioning. Recognizing the initial welding position intelligently and guiding the welding robot to move to the initial welding position automatically is not only the first step of automated welding process, but also the key technologies to achieve intelligent welding.

Some robot companies have adopt "touch method" to recognize the initial welding position. This method used the welding torch to touch the workpieces which must have obvious seam edges and special endpoint types of weld seam. And touch method is

© Springer International Publishing AG 2017
Y. Huang et al. (Eds.): ICIRA 2017, Part II, LNAI 10463, pp. 564–575, 2017.
DOI: 10.1007/978-3-319-65292-4_49

also time consuming and difficult to automatically search the initial welding position. Many researchers have used non-touch method to recognize the initial welding position [6–12]. In [6], an image-based visual servo control method was used to detect the initial welding position, but a special marker was needed for simplifying recognition and the error cannot meet the requirements of practical application. In [7, 8], an initial welding position recognition method based on the image pattern match was proposed, but the error was also too big for real production. In [9], a new method based on the geometric relationship between two seams at two different stages was presented, but the method was not suitable for 6-axis arc welding robot. In [10], a "single camera and double position" method was adopted to recognize the initial welding position, but the accuracy was significantly affected by the robot movement. In [11], a two-step method named "coarse-to-fine" was proposed. And in [12], an image processing method based on pulse-coupled neural network was adopted to recognize the initial welding position. All these researches mainly used passive vision technologies which are sensitive to the environment change and workpiece surface condition. And their big errors limit the application in real production. Meanwhile, passive vision is mainly used to detect the planar welds such as butt weld and not suitable for detection of some welds such as T-joint weld and lap weld. Therefore, active vision with high precision and adaptability is more acceptable for complex weld initial point detection and subsequent seam tracking.

In this paper, we aim at the recognition of initial welding position for fillet weld. A scanning method based on structured-light vision sensor is used to detect the initial welding position. The camera captures the continuous images with robot reciprocates along the end of weld seam once during the detection. In Sect. 2, the robot vision system is described. Section 3 gives the endpoint types of fillet weld and the detection process we used. In Sect. 4, an image processing method including laser center stripe extraction, feature point extraction and connected components extraction is introduced in detail. We have applied the proposed method to real experiments to recognize the initial welding position. The experimental results are showed in Sect. 5. And a conclusion is provided in Sect. 6.

2 Robot Structured-Light Vision System

2.1 System Configuration

The hardware of the robot structured-light vision system consists of a robot welding system and a vision system as shown in Fig. 1. The robot welding system mainly comprises arc welding robot, robot controller, welding equipment and so on. The vision system mainly comprises structured-light vision sensor, image capturing card, PC and so on. A 6-axis arc welding robot is utilized for trajectory planning during the detection. The structured-light vision sensor is designed which mainly consists of a CCD camera, a laser emitter and a filter. The vision sensor is installed on the welding torch by a fixed distance, and its viewing direction is kept along the weld seam during the detection process. A single line laser stripe is projected onto workpiece surface by the laser emitter and the CCD camera receives the reflection of the structured-light stripe. A PC is used for image processing and communication with robot controller.

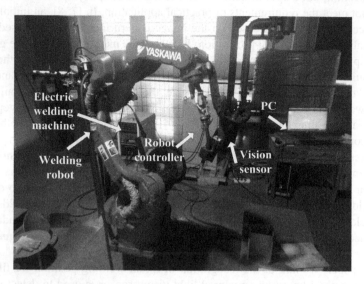

Fig. 1. Robot structured-light vision system

2.2 Vision System Model

The welding position of fillet weld in the image coordinate can be recognized through image processing. In order to obtain the initial welding position in robot base frame, the transform relation between camera and robot needs to be calibrated.

Suppose $p = (u, v)^{\mathrm{T}}$ is a point in the image coordinate and $P = (x_c, y_c, z_c)^{\mathrm{T}}$ is a corresponding point in the camera coordinate. According to the perspective projection model [13], the following equation between image coordinate and camera coordinate is set up:

$$s \begin{bmatrix} u \\ v \\ 1 \end{bmatrix} = K \begin{bmatrix} x_c \\ y_c \\ z_c \end{bmatrix}, K = \begin{bmatrix} a_x & \beta & u_0 \\ 0 & a_y & v_0 \\ 0 & 0 & 1 \end{bmatrix} \tag{1}$$

where s is an unknown scale, K is the camera intrinsic parameter matrix which can be obtained by camera calibration. (u_0, v_0) is the coordinate of camera principal point, a_x and a_y are effective focal length of the camera and β denotes the skewness of two image axes.

Meanwhile, configuration between camera and structured-light plane needs to be calibrated. The equation of structured-light plane in the camera coordinate is

$$ax_c + by_c + cz_c + 1 = 0 \tag{2}$$

where a, b and c are the coefficients of the light plane equation.

Suppose T_6 is the transformation matrix of robot tool frame to robot base frame, H is the robot tool frame to the camera frame, and R is the camera frame to robot base frame. Equation 3 denotes the relations of these three transformation matrixes.

$$R = T_6 H \tag{3}$$

T_6 can be obtained from the robot controller. H is a constant matrix which can be calculated through hand-eye calibration [14]. Therefore, points of laser line in the image can be transformed to space coordinate in robot base frame.

3 Endpoint Types and Detection Process

Fillet welds are mainly referred to as Tee joints which are two pieces of metal perpendicular to each other. The stripe in the image is formed by the laser as shown in Fig. 4. The endpoint types of fillet welds are shown in Fig. 2(a). The major difference between the two types is the different length of horizontal side of metal, which results in different shapes of laser stripe in the image (Fig. 2(b)) when the sensor is scanning the endpoint position of weld seam.

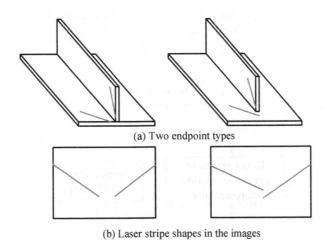

(a) Two endpoint types

(b) Laser stripe shapes in the images

Fig. 2. Two endpoint types of fillet weld and their laser stripe shapes

Since only one single line of structured-light stripe can be obtained in one image, a laser scanning method is used to capture continuous images. In order to recognize the initial position accurately and fleetly, welding robot reciprocates along the end of weld seam. Detection process is shown in Fig. 3. Firstly, welding robot is guided to the start detection position where a complete laser stripe can be captured by camera and intersection of two lines is located in the center area of the image. Then detection starts.

Weld seam point in the image can be recognized via image processing and its position in robot coordinates can be calculated with calibration parameters. At the same time, robot moves in the direction of endpoint of weld seam using an incremental motion strategy. In each interpolation cycle of robot motion, robot moves a specific distance along the weld seam automatically. In our experiments, each interpolation cycle is 4 ms, so robot performs about 250 interpolation motions in every second and initial welding position area can be found quickly with large interpolation motion. Then sensor detects the initial position of weld seam. When the laser stripe is broken in the image as shown in Fig. 2(b), the initial position area is found and robot starts moving in the opposite direction at low speed. Finally, the initial welding position can be recognized when the two laser stripes are integrated together again.

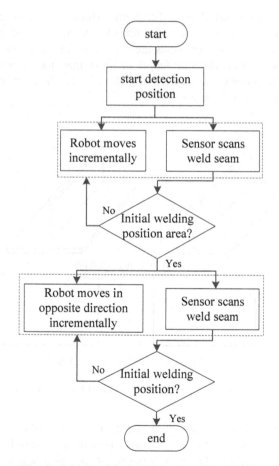

Fig. 3. Initial welding position detection process

4 Recognition of Initial Welding Position

When the laser scans the initial welding position, the laser stripe in the image changes suddenly. To recognize the initial welding position, an image processing method composed of three steps is proposed. Firstly, the laser center stripe is extracted. Then, the feature point of weld joint and dynamic region of interest (ROI) of the image are determined. Finally, connected components are extracted and their related information is computed to prompt the motion of robot.

4.1 Laser Center Stripe Extraction

The laser stripe has a nearly constant thickness and exhibits almost Gaussian distribution along the thickness direction in the image. A second derivate of Gaussian filter [15] which operates only in the direction of the columns of the image is applied. The following equation is a discrete approximation to the filter:

$$E(c,r) = \sum_{j=-1.5L_w}^{1.5L_w} w_j I(c,r+j)$$

$$w_j = \begin{cases} -0.5, j \in [-1.5L_w, -0.5L_w) \cup (0.5L_w, 1.5L_w] \\ 1, j \in [-0.5L_w, 0.5L_w] \end{cases}$$

(4)

where L_w denotes the thickness of the laser stripe in the image and $I(c, r)$ denotes the pixel value at the image coordinates (c, r), w_j are filter coefficients. The filter scans along the columns to convolve the intensity data and then the maximum response E_{max} and minimum response E_{min} are found. Then the following equation after the response is normalized can be acquired:

$$\psi = \frac{E_{max} - E(c,r)}{E_{max} - E_{min}} \times 100\%$$

(5)

if the value of ψ is closer to 1, the corresponding position (c, r_{max}) is more likely the center of the laser stripe in the column.

The accuracy of the above filter is only pixel accuracy. To obtain the higher measurement accuracy, the center of mass algorithm is used to extract the center line in each row with subpixel accuracy.

$$\hat{\delta} = \frac{\sum_{k=-[L_w/2]}^{[L_w/2]} kI(c,r+k)}{\sum_{k=-[L_w/2]}^{[L_w/2]} I(c,r+k)}, \hat{r} = r + \hat{\delta}$$

(6)

where $\hat{\delta}$ is the corrected value and $(c, \hat{r})$ is the new position of center stripe (Fig. 4).

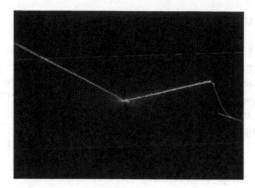

Fig. 4. Laser center stripe extraction

4.2 Weld Joint Feature and Dynamic ROI Determination

For fillet weld joints, the feature point denotes the intersection of two lines in the image. Random sample consensus (Ransac) algorithm is adopted to extract the two feature straight lines based on the center stripe earlier. After the two feature lines are obtained with Ransac, least-square line fitting technique is employed to improve the accuracy of the two lines. And the feature point of fillet weld joint can be recognized by computing the intersection of the two lines (Fig. 5(a)).

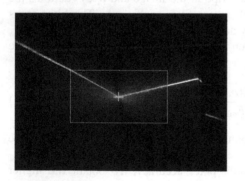
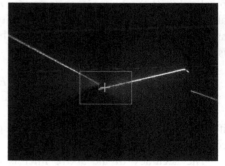

(a) ROI of first image and weld joint feature point

(b) New ROI of the ith image and weld joint feature point

Fig. 5. ROI of images and weld joint feature extraction

For the first frame image, a predefined ROI in the center of image is set up. This large ROI was chosen to ensure that the intersection of two lines is in the ROI (Fig. 5 (a)). Suppose the weld joint feature point of ith image $\boldsymbol{p}_i = (u_i, v_i)^T$ in the image coordinate is extracted by above algorithms. And a new ROI (Fig. 5(b)) is defined by

$$\begin{cases} [X_{\min}, X_{\max}] = [u_{i-1} - \Delta w, u_{i-1} + \Delta w] \\ [Y_{\min}, Y_{\max}] = [v_{i-1} - \Delta h, v_{i-1} + \Delta h] \end{cases} \tag{7}$$

Where $[X_{\min}, X_{\max}]$ and $[Y_{\min}, Y_{\max}]$ are respectively the x range and y range of the ROI, Δw and Δh are half width and height of ROI which are smaller than the ROI's in the first image. (u_{i-1}, v_{i-1}) denotes the weld joint feature of the previous frame image.

4.3 Connected Components Extraction

Connected components extraction is employed in the small region defined by the dynamic ROI. Firstly, segmentation operation is used to extract regions from the image that correspond to the objects we are interested in. As the background have significantly different gray values from the laser stripes, the threshold operation can be used to segment regions. And the image is made binary by setting a threshold of 180 (Fig. 6(a)). Since segmentation results often contain unwanted noisy parts because of reflections, a basic dilation morphology method with a 3×5 rectangle structuring element is chosen for the binary image. Laser stripe regions in the image are more uniform and less disconnected with the dilation operation. The laser stripes in the image we are interested in are characterized by forming a connected set of pixels. Hence, the connected components of the segmented regions must be computed to obtain the individual regions. The 8-connectivity definition and two-pass algorithm are adopted to label the connected components [16] in this paper. Finally, the individual connected regions can be extracted and theirs area and quantity can be computed (Fig. 6(b)).

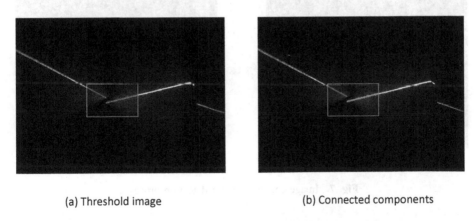

(a) Threshold image   (b) Connected components

Fig. 6. Connected components extraction

5 Experiments

Several experiments using the proposed methods were conducted based on robot vision system discussed above. Before experiments, calibration for robot vision system was carried out. Two endpoint types of fillet weld are scanned by the vision sensor. And

Fig. 7 shows the laser stripe changes in the image corresponding to the detection procedure (Fig. 3). When sensor scans the initial welding position area and two connected components are found (Fig. 7(b)), the computer prompts the robot to move in reverse. However, the welding robot cannot move in reverse immediately because of inertia. And as a result, the gap between two laser stripes become bigger in the image (Fig. 7(c)). Then, after the robot's speed is reduced to zero, it begins to move in reverse and the initial welding point can be searched (Fig. 7(d)). Figure 8 shows the robot's velocity during the detection process. The reverse movement speed of the robot is about 1/5 of the initial speed. Hence, the sensor can search the initial welding position area quickly with robot original movement and can extract the initial welding point precisely with robot reverse movement.

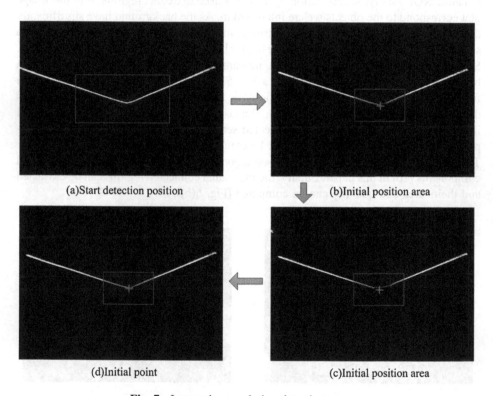

(a)Start detection position   (b)Initial position area

(d)Initial point   (c)Initial position area

Fig. 7. Image changes during detection process

After the initial welding point was recognized, it can be transformed to space coordinate in robot base frame with calibration parameters. The PC sends the initial welding position to the robot, then the robot is automatically guided to the initial welding position (Fig. 9). The experiments are repeated 20 times for each endpoint type. The exact position of initial welding position is measured by guiding the robot manually. For both two endpoint types of fillet weld, absolute errors in X and Y

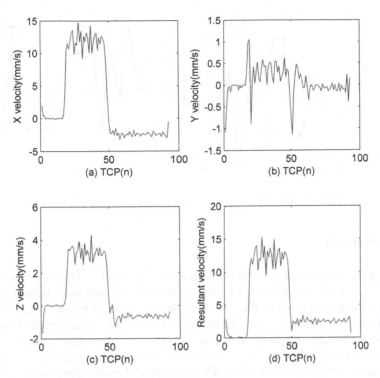

Fig. 8. The robot velocity during detection process

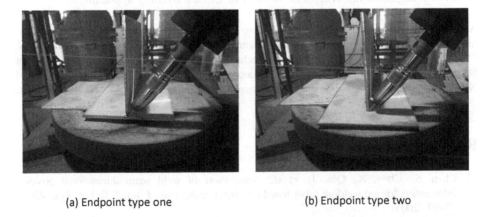

(a) Endpoint type one   (b) Endpoint type two

Fig. 9. The Robot is guided to the initial welding position automatically

directions are almost less than 0.8 mm. And absolute errors in Z direction are almost less than 0.3 mm. Figure 10 shows the absolute errors in three directions. The errors are acceptable in most of the arc welding applications.

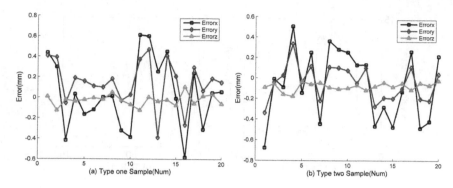

Fig. 10. Absolute error of recognition for two endpoint types

6 Conclusion

In this paper, we propose a method for recognizing initial welding position of fillet weld for arc welding robot. A robot vision system based on structured-light vision is set up and two endpoint types of fillet weld are defined. Due to the characteristic of the vision system, a scanning method for recognition of initial welding position based on incremental robot movement strategy is presented. And image processing is described in detail containing laser center stripe extraction, welding seam feature point extraction and connected components extraction. The experimental results show that the proposed method can realize the recognition of fillet welding initial position and that the welding robot can be guided to the welding initial position with the proposed method. And the errors of recognition are acceptable in most of the arc welding applications.

Acknowledgments. The authors would like to gratefully acknowledge the reviewers' comments. This work is supported by National Natural Science Foundation of China (Grant Nos. 51575187, 91223201), Science and Technology Program of Guangzhou (Grant No. 2014Y2-00217), the Fundamental Research Funds for the Central University (Fund No. 2015ZZ007) and Natural Science Foundation of Guangdong Province (S2013030013355).

References

1. Chen, S., Chen, X., Qiu, T., et al.: Acquisition of weld seam dimensional position information for arc welding robot based on vision computing. J. Intell. Robot. Syst. **43**(1), 77–97 (2005)
2. Gu, W., Xiong, Z., Wan, W.: Autonomous seam acquisition and tracking system for multi-pass welding based on vision sensor. Int. J. Adv. Manuf. Technol. **69**(1), 451–460 (2013)
3. Xu, Y., Fang, G., Lv, N., et al.: Computer vision technology for seam tracking in robotic GTAW and GMAW. Robot. Comput.-Integr. Manuf. **32**, 25–36 (2015)
4. Chen, X., Luo, H., Lin, W.: Integrated Weld Quality Control System Based on Laser Strobe Vision. Springer, Heidelberg (2007)

5. Huang, W., Kovacevic, R.: A laser-based vision system for weld quality inspection. Sensors **11**(1), 506–521 (2011)
6. Guo, Z., Chen, S., Qiu, T., et al.: A method of initial welding position guiding for arc welding robot based on visual servo control. China Weld. **12**(1), 29–33 (2003)
7. Zhu, Z., Liu, T., Piao, Y., et al.: Recognition of the initial position of weld based on the image pattern match technology for welding robot. Int. J. Adv. Manuf. Technol. **26**(7), 784–788 (2005)
8. Chen, X., Chen, S., Lin, T., et al.: Practical method to locate the initial weld position using visual technology. Int. J. Adv. Manuf. Technol. **30**(7), 663–668 (2006)
9. Fang, Z., Xu, D., Tan, M.: Vision-based initial weld point positioning using the geometric relationship between two seams. Int. J. Adv. Manuf. Technol. **66**(9), 1535–1543 (2013)
10. Chen, S., Lin, T., Wei, S., et al.: Autonomous guidance of initial welding position with "single camera and double positions" method. Sensor Rev. **30**(1), 62–68 (2010)
11. Chen, X., Chen, S.: The autonomous detection and guiding of start welding position for arc welding robot. Ind. Robot **37**(1), 70–78 (2010)
12. Yang, L., Lou, P., Qian, X.: Recognition of initial welding position for large diameter pipeline based on pulse coupled neural network. Ind. Robot. **42**(4), 339–346 (2015)
13. Wei, Z., Li, C., Ding, B.: Line structured light vision sensor calibration using parallel straight lines features. Optik – Int. J. Light Electr. Opt. **125**(17), 4990–4997 (2014)
14. Ulrich, M., Heider, A., Steger, C.: Hand-eye calibration of SCARA robots. In: Open German-Russian Worokshop on Pattern Recognition and Image Understanding (2014)
15. Kim, J., Koh, K., Cho, H.: Adaptive tracking of weld joints using active contour model in arc-welding processes. In: Proceedings of SPIE - The International Society for Optical Engineering, vol. 4190, pp. 29–40 (2001)
16. Steger, C., Ulrich, M., Wiedemann, C.: Machine Vision Algorithms and Applications. Wiley-VCH, Weinheim (2007)

A Dual-Camera Assisted Method
of the SCARA Robot for Online Assembly
# of Cellphone Batteries

Kai Feng[1,2], Xianmin Zhang[1,2(✉)], Hai Li[1,2], and Yanjiang Huang[1,2,3]

[1] Guangdong Provincial Key Laboratory of Precision Equipment
and Manufacturing Technology, Guangzhou 510640, China
[2] School of Mechanical and Automotive Engineering,
South China University of Technology, Guangzhou 510640, China
zhangxm@scut.edu.cn
[3] Research into Artifacts, Center for Engineering,
The University of Tokyo, Chiba, Japan

Abstract. A dual-camera assisted method of the SCARA robot for online assembly of cellphone batteries is proposed, to solve the problem of low success rate of SCARA robot used for assembling cellphone battery. In this method, both cellphone battery and cellphone base can be precisely located with the help of the dual-camera, thus the success rate of assembly can be significantly improved. Two steps are mainly included in this method: First, calibrating hand-eye relationships between the SCARA robot and each of the two cameras; second, extracting features of the assembled targets and controlling the robot to rectify the error. The experimental results show that the proposed method can well meet the requirements of cellphone battery assembly and improve the success rate.

Keywords: Dual-camera assistance · SCARA robot · Cellphone battery · Assembly

1 Introduction

Due to the needs for mass production, robot technology is widely used in modern automated production assembly work [1]. SCARA (Selective Compliance Robot Arm) robot, a cylindrical coordinate industrial robot invented by Makino in 1978 [2], which has four degrees of freedom, including translation along the X-, Y-, Z-axis, and rotation with Z-axis, is particularly suitable for assembly and handling work. Currently, it is widely used in the areas of automotive industry, electronics industry, pharmaceutical industry and food industry, etc. [3–8].

During the process of assembly using SCARA robot, a key step is to identify and locate the workpieces. In the early applications, this step was normally accomplished with the help of a dedicated carrier or fixture on the assembly line, meanwhile, the SCARA robot is in accordance with the pre-set parameters to finish the required procedure. These kind of methods are suitable for applications with low accuracy requirements, such as simple handling works and sorting works, etc. However, for

© Springer International Publishing AG 2017
Y. Huang et al. (Eds.): ICIRA 2017, Part II, LNAI 10463, pp. 576–587, 2017.
DOI: 10.1007/978-3-319-65292-4_50

applications where high precision are needed, such as cellphone battery assembly, these traditional locating methods are not applicable. Whereas, the rising of machine vision technology provides an alternative for SCARA robot's high precision assembly applications [9].

Robot hand-eye system (HES) is a kind of robot vision system that can provide visual feedback for robot. According to the position between the camera and the robot, the HES can be roughly divided into Eye-to-Hand (ETH) system and Eye-in-Hand (EIH) system. For the former, camera is mounted in a fixed position. As for the latter, the camera is fixed with the robot's end effector, and moves with the robot [10]. Nowadays, the single HES (SHES) is widely used in the field of robot automatic assembly. However, an obvious drawback of SHES is that only one pose of the assembly targets can be precisely obtained with the help of camera. Whereas the pose of the other assembly part is still not accurate enough. As a result, for the assembling tasks where quite high accuracy is needed, such as the assembly of cellphone batteries, robot combined with SHES still may be failed. In order to realize the assembly of cellphone batteries, this paper presents a dual-camera assisted method of the SCARA robot for online assembly of cellphone batteries. Both EIH and ETH HES are used, and each of them provides the location and gesture information of the cellphone base and the cellphone battery, respectively. Once the pose information is obtained, the robot is then manipulated to finish the high accurate assembly work. The experimental results show that the assembly system with dual-camera assist works better than the one with only single camera assist.

2   System Description

In this paper, a high precision automatic assembly system is constructed and employed to realize the assembly task of cellphone battery, as shown in Fig. 1. It consists of three modules: hardware module, control module and sensor module. The hardware module is composed of a conveyor belt, a vacuum suction device (VSD), and a SCARA robot; the control module is composed of industrial computer (IPC), a PLC controller, and servo controller; and the sensor module consists of a photoelectric switch and two industrial CCD cameras.

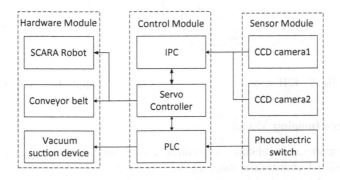

Fig. 1. The structure of the cellphone battery assembly system

The workflow of the system includes the following steps: (1) First, the PLC commands the servo controller to control the conveyor belt to transport the battery and the base to a specific location; (2) After that, the photoelectric switch is triggered, and a signal for starting assembly is sent to the PLC; (3) Once the starting signal is received, the robot moves to the top of the cellphone base and an image of the base is captured by camera 1 and sent to the IPC; (4) Straight after, the robot is controlled to suck the battery with the help of the VSD and then moves the cellphone battery to the top of the camera 2; (5) In succession, an image of the battery is acquired by camera 2 and sent to the IPC; (6) Finally, after the compensation values of position component and gesture component are calculated by processing the images acquired in the step 3 and step 5, they are sent to the servo controller to control the SCARA robot to accurately complete the cellphone battery assembly works.

3   Coordinate Transformation and Hand-Eye Calibration

3.1   Coordinate Transformation

The realization principle of this system is based on the coordinate transformation, and the position compensation and gesture compensation of the two assemblies are obtained by the two HESs to complete the high precision assembly work. Therefore, the transformation between different coordinate systems will be the focus of this paper. Assuming that O_w is the world coordinate system, O_t is the SCARA robot tool coordinate system, O_{c1} is the CCD camera 1 coordinate system, O_{b1} is the cellphone base coordinate system, O_{c2} is the CCD camera 2 coordinate system, O_{b2} is the cellphone battery coordinate system. And the relationships of each coordinate system are shown in Figs. 2 and 3, where Fig. 2 is the EIH system and Fig. 3 is the ETH system.

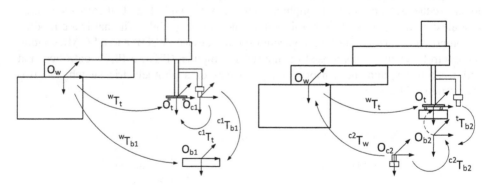

Fig. 2. EIH system   **Fig. 3.** ETH system

3.2   Camera Imaging Model

The use of cameras to obtain the position and gesture of objects is a process of three-dimensional coordinates mapping to two-dimensional coordinates, each single brightness on the image reflects the intensity of the emitted light at a point on the surface of the space object, and the position of the point on the image is related to the geometrical position corresponding to the surface of the space object. The relationship between these positions is determined by the camera imaging geometry model, we call the parameters of the geometric model as camera parameters, which must be determined by experiments and calculations, and the process of obtaining camera parameters called camera calibration.

The camera imaging model is built by introducing the pinhole model, as is shown in Fig. 4, O_w is the world coordinate system, O_c is the camera coordinate system, Z_c is the camera optical axis, which is perpendicular to the image plane, O_f is the image coordinate system.

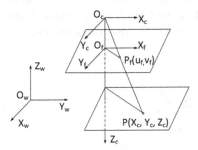

Fig. 4. Camera imaging model

There is a point $P_w(x_w, y_w, z_w)$ in the world coordinate system, it's coordinates in the camera coordinate system is $P_c(x_c, y_c, z_c)$, $P_f(u_f, v_f)$ is the corresponding image point of P_w. The image on the imaging plane is magnified to get the digital image. Accordingly, the imaging point on the plane is converted into an image point (u, v). The image coordinates of the intersection between the optical axis and the imaging plane are recorded (u_0, v_0), the homogeneous coordinates relationship of the image pixels and the points in the world coordinate system:

$$
z_c \begin{bmatrix} u_f \\ v_f \\ 1 \end{bmatrix} = \begin{bmatrix} k_x & 0 & u_0 & 0 \\ 0 & k_y & v_0 & 0 \\ 0 & 0 & 1 & 0 \end{bmatrix} \begin{bmatrix} R & T \\ 0^T & 1 \end{bmatrix} \begin{bmatrix} x_w \\ y_w \\ z_w \\ 1 \end{bmatrix} = M_1 M_2 \bar{x}_w \tag{1}
$$

In formula (1), k_x is the magnification of the X-axis direction, k_y is the magnification of the Y-axis direction. And R is the 3×3 rotation matrix and T is the translation matrix, they represent the positional relationship between the camera coordinate system and the world coordinate system. M_1 called camera intrinsic parameters and M_2 called camera extrinsic parameters. There are many methods for camera calibration, Zhang's classical calibration method is adopted in this paper, and the internal and external parameter matrix can be obtained by obtaining more than 4 coplanar corners on the plane calibration template [11].

3.3   EIH System Calibration

As discussed in the previous section, the position relationship between the camera coordinate system and the workpiece coordinate system can be obtained by camera calibration. In order to obtain the positional relationship between the SCARA robot tool coordinate system and the workpiece coordinate system, it is necessary to know the relative position relationship between the camera coordinate system and the SCARA robot tool coordinate system, which we use a rotation matrix R and a translation matrix T to represent, the process of solving R and T we called the hand-eye calibration. Since the camera is fixed at the end of the SCARA robot actuator, the calibration is also known as the EIH system calibration.

The idea of the EIH system calibration is to control the CCD camera, which is mounted on the SCARA robot end effector, to observe a known calibration reference in different locations, so as to deduce the R and T. Figure 5 shows the relative position of each coordinate system when the SCARA robot end effector moving from position P_a to position P_b.

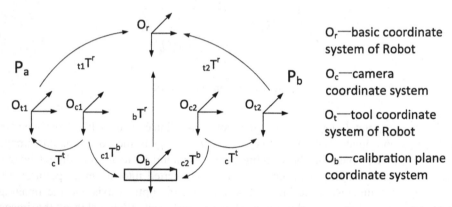

Fig. 5. The relative position relationship of each coordinate system when SCARA robot moving from P_a to P_b in EIH system

It is easy to know that $_{c1}T^b$ and $_{c2}T^b$ can be obtained when camera at the positions P_a and P_b by camera calibration, the position of the SCARA robot tool coordinate system O_{t1} and O_{t2} can be read by the robot motion controller, so $_{t1}T^r$ and $_{t2}T^r$ can also be obtained. $_cT^t$ is the positional relationship between the SCARA robot tool coordinate system and the camera coordinate system. The following relations can be obtained through the above coordinate diagram:

$$_{t1}T^r \cdot _cT^t = _bT^r \cdot _{c1}T^b \tag{2}$$

$$_{t2}T^r \cdot _cT^t = _bT^r \cdot _{c2}T^b \tag{3}$$

From formulas (2) and (3) we can get:

$$(_{t2}T^r)^{-1} \cdot {}_{t1}T^r \cdot {}_cT^t = {}_cT^t \cdot (_{c2}T^b)^{-1} \cdot {}_{c1}T^b \tag{4}$$

Let $A = (_{t2}T^r)^{-1} \cdot {}_{t1}T^r$, $B = (_{c2}T^b)^{-1} \cdot {}_{c1}T^b$, $X = {}_cT^t$:

$$AX = XB \tag{5}$$

Formula (5) is the basic equation of EIH system calibration. A, B can be derived from the known conditions, X is what we want. The principle is based on that the relative position of camera and SCARA robot end effector remains unchanged before and after the robot end effector moves. Let the SCARA robot end effect take multiple points and camera take photos of the same calibration plane, so as to solve the X.

3.4   ETH System Calibration

In the ETH system, the camera is fixed on a workbench in a certain position, the idea of hand eye calibration is to move the known objects to different positions, and to get the results of the calibration, so as to obtain R and T. Figure 6 shows the relative position of each coordinate system when the SCARA robot end effector moving from position P_a to position P_b.

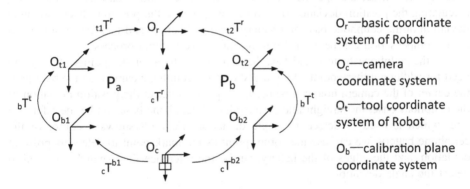

Fig. 6. The relative position relationship of each coordinate system when SCARA robot moving from P_a to P_b in ETH system

It is easy to know that $_{c1}T^b$ and $_{c2}T^b$ can be obtained when camera at the positions P_a and P_b by camera calibration, the position of the SCARA robot tool coordinate system O_{t1} and O_{t2} can be read by the robot motion controller, so $_{t1}T^r$ and $_{t2}T^r$ can also be obtained. $_cT^r$ is the positional relationship between the SCARA robot base coordinate system and the camera coordinate system. The following relations can be obtained through the above coordinate diagram:

$$_cT^r = {}_{t1}T^r \cdot {}_bT^t \cdot {}_cT^{b1} \tag{6}$$

$$_cT^r = {}_{t2}T^r \cdot {}_bT^t \cdot {}_cT^{b2} \tag{7}$$

From formula (6) and formula (7) we can get:

$$_{t2}T^r \cdot ({}_{t1}T^r)^{-1} \cdot {}_cT^r = {}_cT^r \cdot ({}_cT^{b2})^{-1} \cdot {}_cT^{b1} \tag{8}$$

Let $C = {}_{t2}T^r \cdot ({}_{t1}T^r)^{-1}$, $D = ({}_cT^{b2})^{-1} \cdot {}_cT^{b1}$, $X = {}_cT^r$:

$$CX = XD \tag{9}$$

Formula (9) is the basic equation of ETH system calibration. C, D can be derived from the known conditions, X is what we want. The principle is based on that the relative position of calibration object and SCARA robot end effector remains unchanged before and after the robot end effector moves. Fix a calibrator on the SCARA robot end effector to take a number of points and shoot the calibration plane with a camera fixed to the workbench, so as to solve the X.

4 Image Processing and Posture Rectification

In order to obtain the position and gesture of the cellphone base and battery, it is necessary to extract the specific mark points on the cellphone base and battery. By calculating the position deviation of several mark points, the position offset and angle deviation of the cellphone base and battery can be obtained, and then the position and angle compensation can be carried out during the assembling process.

As the cellphone battery and base have a lot of characteristic points, there is no need for additional mark points. Figure 7 shows the cellphone camera on the base part, the center of the camera hole can be the mark point to detect the position deviation of the cellphone base, the straight line on the edge of cellphone base can be the reference line, which is used to detect the angle deviation. Figure 8 shows a corner of the cellphone battery, we can use the corner point as the mark point to detect its position deviation, and the edges of the battery can be the reference lines, which are used to detect the angle deviation.

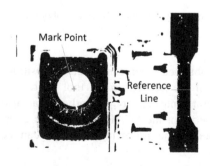

Fig. 7. Cellphone base

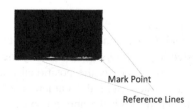

Fig. 8. Cellphone battery

4.1   Line and Circle Detection

The key of the assembly system to complete the high-precision assembly work of the cellphone battery is whether it can accurately find the mark points and reference lines. Hough transform is a classical algorithm for line detection and circle detection, which is one of the basic methods of image processing [12]. The core principle of Hough transform is the mapping relation between image space and image parameter space. In the standard parameterization, the line l expression in the image space:

$$\rho = x \cos \theta + y \sin \theta, \rho \geq 0, 0 \leq \theta < \pi \tag{10}$$

As shown in Fig. 9, ρ represents the vertical distance from the origin to the line, θ represents the angle between the vertical line and the X axis. According to the formula (10), it is easy to prove that the different points (x, y) on the line l are transformed into a set of sinusoidal curves which intersect at the same point $p(\rho, \theta)$ in the parameter space. Obviously, if we can determine the point p in the parameter space, we can realize the line detection in the image space. The method to derive the point p is to find the peak value of the accumulator by the cumulative voting in the parameter space, the corresponding point of the peak is the line that needs to be detected in the image space. Due to there are many linear features on the cellphone base, the Hough transform operation may take a lot of time, for which you can set a region of interest and perform a Hough linear detection in the region of interest. The main steps include: (1) Image banalization; (2) Set region of interest; (3) Extract the edge of the region of interest by Canny algorithm, set P as the set of edge points; (4) for $\theta \in (0, 180°)$, using the formula (10) to calculate corresponding polar diameter of each edge point; (5) Accumulate statistics on voting units $(\rho, \theta) : H(\rho, \theta) = H(\rho, \theta) + 1$; (6) For a unit with the highest number of cumulative votes is the line we need to find.

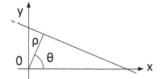

Fig. 9. Parametric form of line

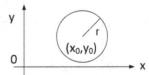

Fig. 10. Parametric form of circle

In the standard parameterization, the circle expression in the image space:

$$\begin{cases} x = x_0 + r \cos \theta \\ y = y_0 + r \sin \theta \end{cases} \tag{11}$$

As shown in Fig. 10, a circle needs to be described by three parameters, (x_0, y_0) is the center coordinates, r is the circle radius. It is easy to know that a point (x, y) in the image space is transformed into a three-dimensional cone in the parameter space, the points on the same circle are transformed into a set of three-dimensional cones which intersecting at a point in the parameter space. The computational complexity is increased due to the three-dimensional problem. For the center of the circle (x_0, y_0) we can limit it to a certain extent. As the cellphone camera hole has a certain degree of accuracy to ensure that r can be set to a fixed value, so that a three-dimensional problem can be reduced to a two-dimensional problem. The main steps of Hough circle detection include: (1) Image banalization; (2) Set region of interest; (3) Extract the edge of the region of interest by Canny algorithm; (4) Do regional segmentation according to the edge, establish the regional edge point list; (5) Find the location of the centroid of the region, and a possible range D of the center of the circle is obtained; (6) Calculate the distance r' from the edge points to all points in the D; (7) Accumulate for possible center coordinates which satisfy the condition: $r' = r$; (8) The unit with the highest votes is the circle to be detected.

4.2   Position Compensation and Gesture Compensation

After we get the mark points and reference lines, we can use these features to calculate the position offset and angle offset of the cellphone battery and the cellphone base, and then make the appropriate compensation to complete the high precision assembly of cellphone batteries. Taking the position offset and angle offset of the cellphone base as an example, the principle is shown in Fig. 11.

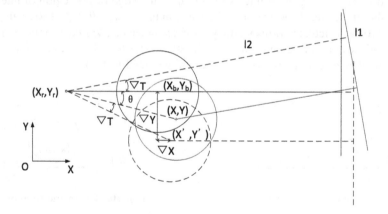

Fig. 11. Schematic diagram of position offset and angle offset of cellphone base

Where (X_r, Y_r) is the rotation center, which is fixed and set, manually. In the standard case, the connecting line between mark point and rotation center is perpendicular to the reference line. (X, Y) is the mark point position in the actual case, and l_1 is the reference line in the actual case, l_2 is an auxiliary line which pass through the center of rotation and perpendicular to the line l_1. θ is the angle between the connect line of the mark point and the center of rotation and the horizontal line. The position offset and angle offset compensation strategy of the cellphone base is to rotate around the rotation center of ΔT degree first, and then move the distance ΔX in the X-axis direction, lastly move the distance ΔY in the Y-axis direction. Set (X', Y') as the coordinates of the mark point, which has been rotated. k_1 is the slope of the line l_1, k_2 is the slope of the line l_2. According to the geometric relationship in the Fig. 11:

$$k_1 \cdot k_2 = -1 \tag{12}$$

$$\Delta T = \arctan k_2 \tag{13}$$

$$\theta = \arctan\left(\frac{Y_r - Y}{X_r - X}\right) \tag{14}$$

$$\Delta X = X_b - \sqrt{(X_r - X)^2 + (Y_r - Y)^2} \cdot \cos(\theta + \Delta T) \tag{15}$$

$$\Delta Y = Y_b - \sqrt{(X_r - X)^2 + (Y_r - Y)^2} \cdot \sin(\theta + \Delta T) \tag{16}$$

Since (X_r, Y_r) and (X_b, Y_b) are known, (X, Y) and k_1 can be derived from the image processing, so the position offset $\Delta X, \Delta Y$ and angle offset ΔT of cellphone base can be obtained according to the above formulas. Similarly, the position offset and angle offset of cellphone battery can also be obtained. The offset is fed back to the SCARA robot controller to make the corresponding compensation, so as to complete the high precision assembly of the cellphone battery.

5 Experiment and Results

The experimental device of SCARA robot high precision cellphone battery assembly system assisted by dual-camera is shown in Fig. 12. In the experiment, the SCARA robot is the YK400XG type multi-joint robot provided by Yamaha, and the repeat positioning accuracy is ± 0.01 mm [13]. CCD Camera 1 and CCD Camera2 are used SCI-CM500-GL-01 model 500 W pixel black and white camera with a resolution of 2592×1944 pixels, both lenses are 500 W pixel telecentric lens, model OPT-5M03-110, provided by OPT Company [14]. The cellphone battery is placed in a table, the position accuracy in the direction X and Y are ± 1 mm. The cellphone base is placed on

Fig. 12. The experimental system

a conveyor belt with a mobile carrier, its location accuracy is determined by the precision of the conveyor belt and the installation accuracy of the carrier, the accuracy in the X direction is ±0.05 mm, and ±0.15 mm in the Y direction.

The dimensional tolerance of cellphone battery and battery compartment is ±0.08 mm the unilateral reserved gap is 0.10 mm. In order to ensure that the cellphone battery can be installed in any case, the assembly accuracy should be at least 0.04 mm. To verify the effect of dual-camera assistant on SCARA robot assembly, repeat 100 times online assembly of the cellphone battery in the case of the system with dual-camera assisted, and do the same experiment in the case of the system with only CCD camera 2 assisted. The experimental results are shown in Table 1, and Table 2 is ten position and angle offset which are randomly selected in the experiments.

Table 1. Experimental results

Situation	Assembly successful	Assembly failure	Success rate
Dual-camera	100	0	100%
Only CCD camera 2	13	87	13%

Table 2. The random selection of the ten position and angle offset in the experiments

Offset results	Cellphone battery			Cellphone base		
	ΔX(mm)	ΔY(mm)	ΔT(°)	ΔX(mm)	ΔY(mm)	ΔT(°)
1	0.6640	0.6305	0.2811	−0.0031	0.1146	−0.1337
2	0.4777	0.2435	0.8222	−0.0128	0.0659	−0.1247
3	0.1268	−0.3113	−0.4751	0.0145	−0.0205	0.0765
4	0.5292	0.2944	0.8724	0.0109	0.0992	0.4227
5	−0.4013	0.2631	−0.5642	0.0321	−0,0895	0.2015
6	0.5269	0.2360	0.6155	0.0424	0.1458	0.5024
7	−0.9187	−0.3446	−0.6576	0.0212	0.1007	−0.1615
8	−0.3167	0.1582	0.3849	0.0446	0.1132	0.1495
9	0.4512	0.3022	0.2098	−0.0387	0.1392	0.1581
10	−0.2642	0.8743	0.9976	−0.0422	0.0916	0.2552

As can be seen from the Table 1, the cellphone battery assembly success rate can reach 100% when the assembly system is added with dual-camera assistance, its assembly accuracy can reach 0.04 mm. While the assembly success rate is only 13% when the assembly system is assisted by CCD camera 2 only. Known from the analysis, it can be found that the position offset compensation and angle offset compensation for both cellphone base and battery can be realized in the process of using the dual-camera assisted assembly, it can achieve a higher precision assembly work. For the single camera assisted assembly system, only the cellphone battery can be compensated for position and angle offset. However, the positioning accuracy of the cellphone base is limited by the accuracy of the carrier and the conveyor belt, which is difficult to guarantee the quality of assembly, so the assembly success rate is bad.

6 Summary

In this paper, we propose a dual-camera assisted method of the SCARA robot for online assembly of cellphone batteries. Both the hardware configuration of the assembly system and the mainly procedures for realizing the online assembly task are introduced. The experimental results verify that the proposed two-camera assisted method are effective and practical for the improvement of the success rate. Hence, this method can be an alternative for the real industrial application. Future works will focus on applying this scheme to accomplish more complex assembly tasks.

Acknowledgement. This work was supported by the Scientific and Technological Research Project of Guangdong Province (2014B090922001).

References

1. Mikkel, R.P., Lazaros, N., Rasmus, S.A., Casper, S., Volker, K., Ole, M.: Robot skills for manufacturing: from concept to industrial deployment. Robot. Comput.-Integr. Manuf. **37**, 282–291 (2016)
2. Furuya, N., Soma, K., Chin, E., Makino, H.: Research and development of selective compliance assembly robot arm (2nd report): hardware and software of SCARA controller. J. Japan Soc. Precis. Eng. **49**(7), 835–841 (1983)
3. Subhashini, P.V.S., Raju, N.V.S., Venkata, R.: Study on robotic deburring of machined components using a SCARA robot. In: International Conference on Robotics, pp. 91–103 (2015)
4. Nkomo, M., Collier, M.: A color-sorting SCARA robotic arm. In: International Conference on Consumer Electronic, pp. 763–768 (2012)
5. Kitahara, Y.: Development of compact assembly robot and its application. Robot Tokyo **110**, 9–14 (1996)
6. Li, W.B., Cao, G.Z., Guo, X.Q., Huang, S.D.: Development of a 4-DOF SCARA robot with 3R1P for pick-and-place tasks. In: International Conference on Power Electronics Systems and Applications, pp. 110–121 (2015)
7. Yang, X.: Robotic assembly of automotive wire harnesses: New research suggests that six-axis robots can be used to install automotive wiring harnesses. Assembly **57**(7), 7–13 (2014)
8. Gojin, M., Yanting, L., Zhong, L., Mingyu, G.: A machine vision based sealing rings automatic grabbing and putting system. In: IEEE International Conference on Industrial Informatics, pp. 202–206 (2016)
9. Lu, R., Lidai, W., Mils, J.K., Dong S.: 3-D automatic micro assembly by vision-based control. In: IEEE/RSJ International Conference on Intelligent Robots and System, 2017, IROS 2007, pp. 297–302 (2007)
10. Flandin, G., Chaumette, F., Marchand, E.: Eye-in-hand/eye-to-hand cooperation for visual servoing. In: IEEE International Conference on Robotics and Automation, vol. 3, pp. 2741–2746 (2000)
11. Zhang, Z.: A flexible new technique for camera calibration. IEEE Trans. Pattern Anal. Mach. Intell. **22**(11), 1330–1334 (2000)
12. Illingworth, J., Kittler, J.: A survey of the Hough transform. Comput. Vis. Graph. Image Process. **43**(1), 87–116 (1988)
13. The YAMAHA SCARA robot information. http://www.yamaha-motor.com.cn/robot/lineup/ykxg/small/yk400xr/index.html. Accessed 25 Apr 2017
14. The telecentric lens information. http://www.optmv.com/pro_listjt.aspx?ProductsCateId=175&CateId=175&page=2. Accessed 25 Apr 2017

A Fast 3D Object Recognition Pipeline in Cluttered and Occluded Scenes

Liupo Zheng, Hesheng Wang$^{(\boxtimes)}$, and Weidong Chen

Shanghai Jiao Tong University, Shanghai 200240, China
{wanghesheng,wdchen}@sjtu.edu.cn

Abstract. In this paper we propose a framework for instance recognition and object localization in cluttered and occluded household environment for robot grasping task. The whole system bases on a coarse to fine pipeline in combination with the state-of-the-art methods of RGBD-based object detection. We build a sparse feature model by extracting structure key points incorporating texture cues in the train procedure. After that, the paper demonstrates how the algorithm decreases the time complexity and simultaneously guarantees the accuracy of the recognition and pose estimation. Quantitative experimental evaluations are presented using both acknowledged ground truth dataset and real-world robot perception system.

Keywords: Object recognition · RGBD-image · Clutter · Pose estimation

1 Introduction

One of the essential challenges in robot perception field is the object recognition and localization in unstructured scenes. The robot in such environment encounters many restrictions such as occlusions, clutters, illumination changes, multiple objects,real-time limits and etc. The conservative recognition methods using 2D image features like SIFT, SURF, ORB, HOG [1–3] are more and more unable to satisfy the requirements with the improvement of system accuracy and speed. Recently, the work based on RGBD images has been fostered thanks to the availability of low-price and high-performance 3D sensors such as Intel RealSence and Microsoft Kinect.

Different from general computer vision, robotic vision system has the particular characteristics [4]. We pay more attention to instance recognition rather than at category levels [5] and we should identify the object and simultaneously estimate 6-DOF pose for the robot grasping. Because the number of object instance is usually huge, the real-time performance should also be considered mainly in practical application. One way of solving these is to introduce hierarchy which executes the recognition at different levels [6]. The other direction of the research is trying to find more effective feature descriptors in local [7–9] or global [10,11] aspect. The local approaches construct the correspondence model in a scene by

© Springer International Publishing AG 2017
Y. Huang et al. (Eds.): ICIRA 2017, Part II, LNAI 10463, pp. 588–598, 2017.
DOI: 10.1007/978-3-319-65292-4_51

extracting and matching feature points and afterwards these correspondences are clustered under certain rules to generate a object hypothesis [12]. The global algorithms on the other hand extract a single descriptor to describe the object in sight [13]. There is also some work dealing with performance improvements through combining multiple algorithms [14,15].

In this paper, we propose a coarse to fine object recognition and localization framework for robot perception in the complex environment. The first step is the off-line object modelling and training. We define a unique generalized eigenvector consisting of spatial and texture features to build a sparse model of the object. In order to search more effectively, we organize the training data in a specific structure and label each surface of the object. We conduct a novel search strategy to deliver the candidates in coarse pipeline after training. Then, several approaches are used to refine the initial result and estimate the 6-DOF pose of each object hypotheses. The framework ends with the output of object identity and transformation matrix with respect to camera frame.

The paper is structured as follows. The next section presents the steps of sparse feature model building and object training. Sect. 3 introduces more detailed components of the coarse to fine object recognition and pose estimation framework, followed by experiment and result in Sect. 4. Finally, the result is analysed and concluded in Sect. 5.

2   Object Modelling and Training

As the number of each point cloud data is huge, it is hard to guarantee the system real-time performance if we take all of the cues into consideration. For each object training process, we build a sparse model M_s including shape, size and texture information of an object which mainly characterizes its local feature. We define a novel generalized eigenvector f_s : $f = [cloud\_corrdinates, normals, feature\_descriptors]$ to describe these unique feature points. The item of $feature\_descriptors$ are extracted from two aspects: 3D points generated from the cloud data and 2D points which are back-projected from 3D space. More concretely, we use SHOT (Signature of Histograms of Orientations) [16] to describe the original 3D feature points and SIFT (Scale-invariant feature transform) to represent the 2D once.

The first step of our recognition pipeline is to build models M_s for each object. To do this, we use an off-the-shelf *Simultaneous Localisation and Mapping* approach [17,18] to merge the image data gathered by moving RGB-D camera around table-top object. In each frame, a color and a depth image are taken and the point cloud is generated through fusing both of information. The correspondences between two frames are estimated in color images, and the projected 3D points corresponding key points in 2D space are used to compute the transformation between two frames. The infinite points caused by the camera are first filtered, followed by the points which are too far away from the camera. To estimate the transformation, we use RANSAC (Random Sample Consensus) to eliminate outliers and ICP (Iterative Closest Point) to get more accurate results.

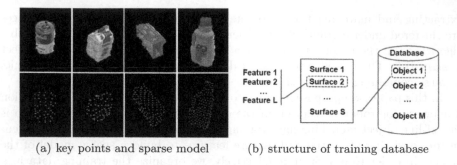

(a) key points and sparse model   (b) structure of training database

Fig. 1. The sparse model building and data training

They are further optimized by the ParallaxBA (Parallax Bundle Adjustment) proposed by Zhao [19]. Finally all the frames are transformed into one object coordinate system. For each surface $S^i = \{S_1^i, S_2^i, S_1^i, S_3^i, \ldots, S_s^i\}$ of one object M_i, we unify the formation of feature vector f_i to describe them. These points are stored in the training database $\mathbf{M}_d = \{M_1^d, M_2^d, \ldots, M_m^d\}$. In Fig. 1, the sparse model building and data training are shown in detail.

3   Object Recognition and Pose Estimation

In order to fully exploit the occluded objects and simultaneously compute 6-DOF pose for robot grasping, we propose a coarse to fine framework which consists of three major parts, namely $Off-line\ training$, $Coarse\ pipeline$ and $Fine\ pipeline$. The structure of the proposed method is outlined in Fig. 2. The output of the system is a cluster with recognized object label and its homogeneous transformation matrix with respect to the camera frame.

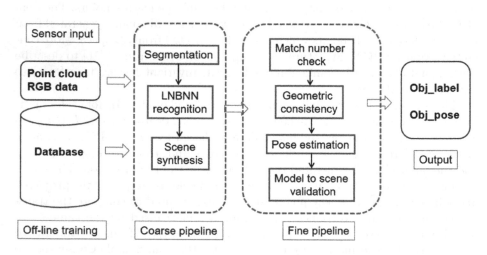

Fig. 2. Coarse to fine object recognition and pose estimation framework

3.1 Coarse Recognition Pipeline

Segmentation. In the testing case, the input point cloud P_j is firstly segmented into multiple object hypotheses $\mathbf{O}_j = \{O_1^j, O_2^j, \ldots, O_k^j\}$. We try to reduce the computational cost by focusing on the point cloud within a certain range on the supporting plane. Hence we use RANSAC to estimate the plane of the supporting table and cut the rest data below the plane.

For the objects on the plane, we conduct a bottom-up segmentation method proposed by Richtsfeld [20]. We over-segment the point cloud into supervoxels and build up a supervoxel adjacency graph using Voxel Cloud Connectivity Segmentation [21], followed by a pre-merging process, in which all the adjacent supervoxels are merged into patches based on their normal similarity. The clustered patches are fitted to a object hypotheses according to the local convexity and sanity criterion and using noise filtering procedure to merge the small noisy patches into the neighboring segment with the greatest size.

Local Naive Bayes Nearest Neighbor. The next step for coarse pipeline is to search object hypothesis candidates in the train set $\mathbf{M}_d = \{M_1^d, M_2^d, \ldots, M_m^d\}$ for each segmented point clusters $\mathbf{M}_s = \{M_1^s, M_2^s, \ldots, M_k^s\}$ which are the feature models extracted from $\mathbf{O}_j = \{O_1^j, O_2^j, \ldots, O_k^j\}$ using the same sparse expression as training.

The naive Bayes Nearest Neighbor algorithm is widely used in the image searching and classification [22]. The goal of the approach is to determine the most probable class $\hat{C}$ of a query image Q according to

$$\hat{C} = arg\ max\ P(C|Q) \tag{1}$$

Refer to this thought, we define our problem as flows. Each train model M^d is essentially a set of generalized eigenvectors $M^d = \{f_i : f_i = [cloud\_cor, normals, feature\_descriptors], i = 1, 2, \ldots, L\}$. In the on-line object recognition pipeline, we extract feature descriptors from segmented object hypotheses and obtain the sparse model M^s which contains N feature vectors $\{f_j, j = 1, 2, \ldots, N\}$. Then the recognition issue is transformed into the maximization of posterior probability shown in Eq. 2.

$$\hat{M}^d = arg\ max\ P(M^d|M^s) \tag{2}$$

Assuming a uniform prior probability over objects and independence of the descriptors f_j extracted from cluster M^s, applying Bayes' rules:

$$P(M^d|M^s) \propto P(M^d)P(M^s|M^d) \propto \prod_{j=1}^{N} P(f_j|M^d) \tag{3}$$

According to kernel density estimation and the descriptors f_j are highly dimensional, therefore distribute sparsely, the $P(f_j|M^d)$ can be rewritten approximately

$$P(f_j|M^d) = \frac{1}{L}\sum_{i=1}^{L} K\ (f_i - f_j) \approx \frac{1}{L}e^{-\|f_j - f_{NN}(f_j)\|^2} \tag{4}$$

where the $K(\cdot)$ is the *Gaussian − Parzen kernel* and $f_{NN}(f_j)$ is the nearest neighbor to f_j in the train database $M^d = \{f_i : i = 1, 2, \ldots, L\}$. Substituting Formula 4 into 3 and taking logarithm both sides of 3, we can get

$$\hat{M}^d = arg\,min \sum_{j=1}^{N} \|f_j - f_{NN}(f_j)\|^2 \tag{5}$$

We notice that the feature cluster M_m^s is actually one of the object surfaces, in order to match more accurately and provided more refined cues later for the pose estimation, we label each surface S^d in the train database, rewrite the generalized eigenvector $f_s : f = [surface\_label, cloud\_corrdinates, normals, feature\_descriptors]$ and search target in the category of S^d. Suppose the number of object is M, each object trains S surfaces and each surface has L features averagely. We conduct KD (K-Demensional) search strategy in a single loop and the complexity of the NBNN is $O(M \cdot S \cdot N \cdot log(L))$ and the time consumption increases linearly with the number of train database elements. The real-time will be influenced heavily with the object number increasing.

Considering the *Gaussian − Parzen kernel* $K(f_i - f_j) = e^{-\frac{\|f_i - f_j\|^2}{2\sigma^2}}$, $P(f_j|M^d)$ decreases exponentially with respect to $\|f_i - f_j\|^2$ and there is no need to search every surface S^d in the database, we just care several nearest neighbors of the f_j. Under this point, we merge all of the trained surfaces S^d into a new structure $\{f_k^{DB}\} = \{f_k^{S_1^d}\}\bigcup\{f_k^{S_2^d}\}\bigcup\cdots\bigcup\{f_k^{S_{M\cdot S}^d}\}$ and conduct the KD search in the merged database $\{f_k^{DB}\}$. Compare with the previous NBNN, our method just search the nearest neighbors in one structure instead of every object. We call this method *Local Naive Bayes Nearest Neighbor* (summarized in Algorithm 1). The complexity of LNBNN is $O(N \cdot log(M \cdot S \cdot L))$ and the time consumption increases logarithmically with the number of objects.

Scene Synthesis. After LNBNN search, we get candidates $\mathbf{C}_{candi}^{M^s} = \{C_1^{M^s}, C_2^{M^s}, \ldots, C_t^{M^s}\}$ of certain object cluster M^s. Since we have marked the category and surface label for each candidate, it is easy to distinguish which candidate belongs to the same object. We conduct a voting scheme to synthesize the coarse recognition result and select the final candidate set with the uniform classification label.

3.2   Fine Recognition Pipeline

Match Number Check and Geometric Consistency. The coarse pipeline output a set of most likely candidates of the object hypotheses, we next determine which is the best recognition result in the fine pipeline. Given the descriptors of object hypotheses and an appropriate candidate, we first use FLANN (Fast Library for Approximate Nearest Neighbors) strategy to discover the correspondent points and define a matching threshold to discarded correspondences with large distance. We then execute *match number check* stage to decide whether it is meet the requirement or not. Afterwards *geometric consistency*

Algorithm 1. Local Naive Bayes Nearest Neighbor

$\quad$ **Input**: $Train\ dataset\ \{f_k^{DB}\}\ and\ object\ hypotheses\ feature\ cluster\ M^s$

$\quad$ **Output**: $Candidates\ of\ object\ hypotheses\ \{C_1^{M^s},\ C_2^{M^s}, ..., C_t^{M^s}\} \in \{f_k^{DB}\}$

1 $\quad Initialize\ dist_m = 0\ (m = 1, 2, ..., M \cdot S)\ where\ dist_m\ represents\ the\ distance$
$\quad between\ object\ hypotheses\ M^s\ and\ trained\ surface\ S^d$

2 $\quad$ **for** $f_j \in M^s$ **do**

3 $\qquad search\ in\ \{f_k^{DB}\}\ and\ get\ r + 1\ nearest\ neighbors\ \{f_{NN1}, f_{NN2}, ..., f_{NNr+1}\}$
$\qquad merge\ eigenvectors\ belong\ to\ one\ surface\ and\ obtain\ set\ \{C_1^{M^s},\ C_2^{M^s}, ...C_s^{M^s}\}$
$\qquad set\ dist_0\ equal\ \|f_j - f_{NNr+1}\|^2$

4 $\qquad$ **for** $C_i \in \{C_1^{M^s},\ C_2^{M^s}, ...C_s^{M^s}\}$ **do**

5 $\qquad\quad | \quad dist_m = dist_m + \|f_j - f_{NN_i}\|^2 - dist_0$

6 $\qquad$ **end**

7 **end**

8 $\quad sort\ dist_m\ in\ ascending\ order\ and\ take\ the\ first\ t\ \{C_i^{M^s}\}\ (i = 1, 2, ..., t)$
$\quad as\ candidates$

9 **return** $\{C_i^{M^s}\}$

---

clustering algorithm will be conducted to enforce geometric constraints between pairs of correspondences and remove the mismatching points.

Pose Estimation. Since the correspondences of the key points have been extracted, we conduct SVD (singular value decomposition) to get the initial transformation T_m^c between the candidate surface C_i and object M^s. Because we use nearest neighbors of the descriptors to generate point-to-point relationships in 3D space, some of our correspondences are likely to be incorrect. We account for this using RANSAC algorithm and through dozens of iteration to get a better result which will be used in ICP stage as initial value to increase the accuracy of the transformation by $T_m^c = T_{ICP} \cdot T_m^c$. Therefore, the 6-DOF pose of the object hypotheses M^s with respect to camera is $T_s = T_{cam}^m \cdot T_m^c$, where the T_{cam}^m is the transformation matrix between object frame and camera frame.

Model to Scene Validation. For each object hypotheses M^s conducts these steps above, all the scene objects have been recognized and we will get a cluster of most likely candidates $\mathbf{C} : \{C_1, C_2, \ldots, C_k\}$ correspond to the segmented point cluster $\mathbf{M}_s : \{M_1^s, M_2^s, \ldots, M_k^s\}$. If more than one candidates C_i contain the same label, two conditions probably lead to this. One is the over segmentation and the other is false recognition. We distinguish this two situations by means of checking whether object hypotheses bounding volumes with same label are overlap. If so, we merge those hypotheses and estimate the pose again, otherwise we discard this false result and restart the recognition pipeline.

$\quad$ After checking scene consistency, we project each candidate into the scene using estimated pose matrix and conduct overlap ratio verification. We construct a two-dimensional histogram of $20 \cdot 20$ bins to represent the distribution and orientation of surface normals. Since the surface normals are normalized, the term n_z is determined by n_x and n_y, there is no need to contain n_z in the histogram. A direct method to evaluate the similarity of two histograms is

$$D(A, B) = \sum_{i,j} |a_{ij} - b_{ij}| \tag{6}$$

Normalizing the histogram and utilizing the following equation

$$min(a, b) = \frac{1}{2}(a + b) - \frac{1}{2}|a - b| \tag{7}$$

we can get a more effective way to compute the similarity of the histogram

$$S(A, B) = 1 - \frac{1}{2}D(A, B) = \sum_{i,j} min(a_{ij} - b_{ij}) \tag{8}$$

where $S(A, B)$ is the metric which has a positive correlation between the value and the similarity of histograms. Further more, it is also a indirect indication of the overlap ratio between the candidate model and the scene object. The bigger of the metric value, the more accurate of the recognition and pose estimation.

4   Experiment and Result

4.1   Recognition Experiment

We evaluated the efficiency of recognition framework on a famous household dataset *Willow Challenges* from ICRA *Perception Challenge* 2011. This dataset contained 35 rigid object instances and 39 scenes including both simple and complex case, each object consisted of 37 frames from different views. Our training pipeline built the sparse feature model using these point cloud instance while recognition pipeline was implemented with the given ground truth scenes using OpenCV and PCL.

After training all of the objects in dataset, we selected 14 scenes randomly for testing and each scene contained 4 different view ports. The whole process steps are shown in Fig. 3. After LNBNN searching and fine pipeline optimizing, the final recognition precision is summarized in Table 1. We also analysed the pose recovery of detected object with the benchmark, the statistics indicate the average errors of translation and rotation are under 4 cm and 8°. Figure 4 shows the line chart of translation and rotation errors around X-axis, Y-axis and Z-axis.

4.2   Grasping Experiment

The proposed method was then tested on the real-world environment under the ABB industrial manipulator with a calibrated Kinect sensor (shown in Fig. 5(a)). The objects were placed on a plane table in clutter and the goal was to recognize all instances in the scene and estimate the pose at the same time. For the robot-grasp planning, we followed the work of *Dogar* [23], by means of NGR (negative goal region) and relevant methods to deliver the motion trajectory and grasp gesture.

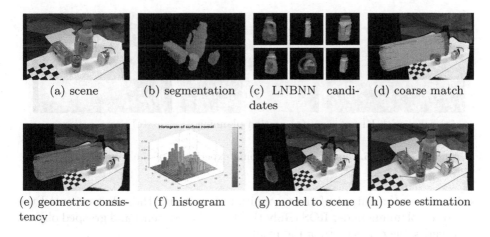

(a) scene (b) segmentation (c) LNBNN candi- (d) coarse match
dates

(e) geometric consis- (f) histogram (g) model to scene (h) pose estimation
tency

Fig. 3. Overview of object recognition and pose estimation pipeline

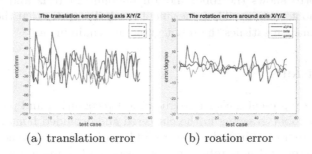

(a) translation error   (b) roation error

Fig. 4. The errors of translation and rotation around X-axis, Y-axis and Z-axis

Table 1. Quantitative recognition precision

Item	Value
True positive	96.76%
False positive	3.24%
False negative	2.18%
Recall	97.80%
Precision	96.76%

The experiment was designed as follows:

- Training the objects which are commonly used in household environment.
- Captured the original scene in clutter and conducted the segmentation stage to deliver the object hypotheses $\{M_1^s, M_2^s, \ldots, M_k^s\}$.
- Applied LNBNN and *scene synthesis* to obtained candidates in coarse pipeline.
- Conducted fine pipeline and iterated until all the hypotheses are recognized.

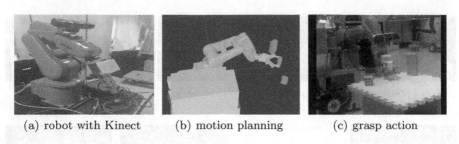

(a) robot with Kinect   (b) motion planning   (c) grasp action

Fig. 5. Snapshots for grasping experiment of the robot

- Combined the output of the recognition framework, the robot executed the motion planning under ROS (Robotic Operation System) and grasped objects to verity the correctness of the result.

The Fig. 5(b) shows the robot motion planning in Rviz and Fig. 5(c) is a snapshot of robot grasping. The experiment result illustrates that the recognition and pose estimation satisfies the robot grasping requirement.

5   Conclusions

In this paper, we present a coarse to fine object recognition and pose estimation framework for robot grasping in the cluttered and occluded household environment. Our system combines the texture and spatial cues constructing a generalized eigenvector and it is robust for occlusion and illumination changes. We exploit the segmented clusters by means of LNBNN to ensure the real time performance and through fine recognition pipeline to improve the accuracy of the pose estimation. The experiment illustrates the efficiency of the proposed method. In the future, we aim to extend our algorithm to non-rigid objects and a more effective segmentation approach is also need to explore deeply.

Acknowledgement. This work was supported in part by the Natural Science Foundation of China under Grant U1613218, 61473191 and 61503245, in part by the Science and Technology Commission of Shanghai Municipality under Grant 15111104802, in part by Shanghai Sailing Program under Grant 15YF1406300, in part by State Key Laboratory of Robotics and System (HIT).

References

1. Collet, A., Berenson, D., Srinivasa, S.S., Ferguson, D.: Object recognition and full pose registration from a single image for robotic manipulation. In: 2009 IEEE International Conference on Robotics and Automation, ICRA 2009, pp. 48–55. IEEE (2009)
2. Martinez, M., Collet, A., Srinivasa, S.S.: Moped: a scalable and low latency object recognition and pose estimation system. In: 2010 IEEE International Conference on Robotics and Automation (ICRA), pp. 2043–2049. IEEE (2010)

3. Rublee, E., Rabaud, V., Konolige, K., Bradski, G.: ORB: an efficient alternative to SIFT or SURF. In: 2011 IEEE International Conference on Computer Vision (ICCV), pp. 2564–2571. IEEE (2011)
4. Tang, J., Miller, S., Singh, A., Abbeel, P.: A textured object recognition pipeline for color and depth image data. In: 2012 IEEE International Conference on Robotics and Automation (ICRA), pp. 3467–3474. IEEE (2012)
5. Janoch, A., Karayev, S., Jia, Y., Barron, J.T., Fritz, M., Saenko, K., Darrell, T.: A category-level 3D object dataset: putting the kinect to work. In: Fossati, A., Gall, J., Grabner, H., Ren, X., Konolige, K. (eds.) Consumer Depth Cameras for Computer Vision. Advances in Computer Vision and Pattern Recognition, pp. 141–165. Springer, London (2013). doi:10.1007/978-1-4471-4640-7_8
6. Lai, K., Bo, L., Ren, X., Fox, D.: A large-scale hierarchical multi-view RGB-D object dataset. In: 2011 IEEE International Conference on Robotics and Automation (ICRA), pp. 1817–1824. IEEE (2011)
7. Mian, A., Bennamoun, M., Owens, R.: On the repeatability and quality of keypoints for local feature-based 3D object retrieval from cluttered scenes. Int. J. Comput. Vis. **89**(2–3), 348–361 (2010)
8. Papazov, C., Burschka, D.: An efficient RANSAC for 3D object recognition in noisy and occluded scenes. In: Kimmel, R., Klette, R., Sugimoto, A. (eds.) ACCV 2010. LNCS, vol. 6492, pp. 135–148. Springer, Heidelberg (2011). doi:10.1007/978-3-642-19315-6_11
9. Petrelli, A., Di Stefano, L.: On the repeatability of the local reference frame for partial shape matching. In: 2011 International Conference on Computer Vision, pp. 2244–2251. IEEE (2011)
10. Aldoma, A., Tombari, F., Rusu, R.B., Vincze, M.: OUR-CVFH – oriented, unique and repeatable clustered viewpoint feature histogram for object recognition and 6DOF pose estimation. In: Pinz, A., Pock, T., Bischof, H., Leberl, F. (eds.) DAGM/OAGM 2012. LNCS, vol. 7476, pp. 113–122. Springer, Heidelberg (2012). doi:10.1007/978-3-642-32717-9_12
11. Wohlkinger, W., Vincze, M.: Ensemble of shape functions for 3D object classification. In: 2011 IEEE International Conference on Robotics and Biomimetics (ROBIO), pp. 2987–2992. IEEE (2011)
12. Jiang, D., Wang, H., Chen, W., Wu, R.: A novel occlusion-free active recognition algorithm for objects in clutter. In: 2016 IEEE International Conference on Robotics and Biomimetics (ROBIO), pp. 1389–1394. IEEE (2016)
13. Aldoma, A., Tombari, F., Stefano, L., Vincze, M.: A global hypotheses verification method for 3D object recognition. In: Fitzgibbon, A., Lazebnik, S., Perona, P., Sato, Y., Schmid, C. (eds.) ECCV 2012. LNCS, vol. 7574, pp. 511–524. Springer, Heidelberg (2012). doi:10.1007/978-3-642-33712-3_37
14. Aldoma, A., Tombari, F., Prankl, J., Richtsfeld, A., Di Stefano, L., Vincze, M.: Multimodal cue integration through hypotheses verification for RGB-D object recognition and 6DoF pose estimation. In: 2013 IEEE International Conference on Robotics and Automation (ICRA), pp. 2104–2111. IEEE (2013)
15. Lutz, M., Stampfer, D., Schlegel, C.: Probabilistic object recognition and pose estimation by fusing multiple algorithms. In: 2013 IEEE International Conference on Robotics and Automation (ICRA), pp. 4244–4249. IEEE (2013)
16. Tombari, F., Salti, S., Stefano, L.: Unique signatures of histograms for local surface description. In: Daniilidis, K., Maragos, P., Paragios, N. (eds.) ECCV 2010. LNCS, vol. 6313, pp. 356–369. Springer, Heidelberg (2010). doi:10.1007/978-3-642-15558-1_26

17. Henry, P., Krainin, M., Herbst, E., Ren, X., Fox, D.: RGB-D mapping: using kinect-style depth cameras for dense 3D modeling of indoor environments. Int. J. Robot. Res. **31**(5), 647–663 (2012)
18. Herbst, E., Henry, P., Fox, D.: Toward online 3-D object segmentation and mapping. In: 2014 IEEE International Conference on Robotics and Automation (ICRA), pp. 3193–3200. IEEE (2014)
19. Zhao, L., Huang, S., Sun, Y., Yan, L., Dissanayake, G.: Parallaxba: bundle adjustment using parallax angle feature parametrization. Int. J. Robot. Res. **34**(4–5), 493–516 (2015)
20. Richtsfeld, A., Mörwald, T., Prankl, J., Zillich, M., Vincze, M.: Segmentation of unknown objects in indoor environments. In: 2012 IEEE/RSJ International Conference on Intelligent Robots and Systems, pp. 4791–4796. IEEE (2012)
21. Papon, J., Abramov, A., Schoeler, M., Worgotter, F.: Voxel cloud connectivity segmentation-supervoxels for point clouds. In: Proceedings of the IEEE Conference on Computer Vision and Pattern Recognition, pp. 2027–2034. IEEE (2013)
22. Tuytelaars, T., Fritz, M., Saenko, K., Darrell, T.: The NBNN kernel. In: 2011 IEEE International Conference on Computer Vision (ICCV), pp. 1824–1831. IEEE (2011)
23. Dogar, M., Srinivasa, S.: A framework for push-grasping in clutter. Robot.: Sci. Syst. VII **1** (2011)

Implementation of Multiple View Approach for Pose Estimation with an Eye-In-Hand Robotic System

Kai Li, Chungang Zhuang, Jianhua Wu, and Zhenhua Xiong[✉]

State Key Laboratory of Mechanical System and Vibration,
School of Mechanical Engineering, Shanghai Jiao Tong University,
Shanghai 200240, China
{leica8244,cgzhuang,wujh,mexiong}@sjtu.edu.cn

Abstract. This paper compares implementation of multiple view approach of pose estimation on an eye-in-hand robotic system. By combining RGB-D frames from multiple view with the eye-in-hand system, geometry information of target objects can be best recovered thus pose estimation performance can get improved. Two primary approaches for pose estimation, namely 3D point cloud registration and 2D image matching are implemented and compared. For the 3D method, we reconstruct target objects by taking advantage of the eye-in-hand system to get an accurate representation of target objects. For the 2D method, we discuss distance metrics and regression for 6DOF pose and apply RANSAC with it to fuse multiple estimation results. State-of-the-art pose estimation algorithms which cover both the 3D and 2D approaches are implemented and compared. Experiments show that the multiple view approach can provide more accurate and reliable pose estimation results when compared with conventional single view approach.

Keywords: Multiple view · Eye-in-hand system · Pose estimation · Object reconstruction · RANSAC

1 Introduction

Accurate localization and pose estimation of 3D objects play an essential role in many robotic applications such as object grasping and random bin picking. When an eye-in-hand system is set up by mounting the camera on the end effector of a robot manipulator, observations can be made from multiple view, which provides additional advantages for pose estimation, as information from different views can be used together to form a more accurate result. Although recent work has started to apply the multiple view approach [1,2], detailed investigations and discussions are still needed on how to perform pose estimation effectively with multiple view approach and the eye-in-hand system.

In computer vision, 3D point cloud registration and 2D image matching are two primary approaches for pose estimation. Reference [1,2] applies the multiple view approach to 3D point cloud based method by reconstructing the scene

© Springer International Publishing AG 2017
Y. Huang et al. (Eds.): ICIRA 2017, Part II, LNAI 10463, pp. 599–610, 2017.
DOI: 10.1007/978-3-319-65292-4_52

with depth data taken from different views. Reference [1] uses 3 mature point cloud registration algorithms to reconstruct the scene and applies superellipsoid-based pose estimation [3] on sweet pepper. Reference [2] takes advantage of robot encoder data to align each frame of depth data and implements principle component analysis (PCA) and iterative closest point (ICP) [4] for pose estimation. However, reference [2] also mentions that due to poor accuracy of eye-hand calibration, the reconstruction process may result in gross errors sometimes. As many implementation details are not given in [2], discussions are still needed to solve potential problems in practice. Except for superellipsoid and PCA, point pair feature (PPF) [5] is another state-of-the-art method for pose estimation on 3D point cloud. Reference [6, 7] extend PPF by adding boundary or color information. Although PPF has been proved to be a robust way for pose estimation, more accessible RGB-D camera can't provide high quality depth data like the 3D scanner used in [6]. Therefore when it comes to RGB-D depth data with no color extension, scene object reconstruction with multiple view approach can be a possible solution.

2D image matching is another primary approach for pose estimation. For objects with rich color variation, local image features have shown great success in recovering the 6DOF pose [8], while for objects with little texture, template matching is a frequently used method. Reference [9] applies chamfer matching to pose estimation by using a multi-flash camera to extract depth edges of textureless objects. The state-of-the-art LINEMOD/LINE-2D/LINE-3D [10], is another well-cited template matching method for pose estimation. However, all the above work takes only one frame into account and no attempt has been made to use the multiple view approach.

In this paper we investigate and compare the multiple view approach for pose estimation with a RGB-D camera and the eye-in-hand robotic system. The multiple view approach is applied on both 3D and 2D pose estimation methods. For the 3D method, we revisit the encoder-data based scene object reconstruction and choose to apply PCA and PPF on the reconstructed target objects for pose estimation. Implementation details and problems encountered in practice for both scene reconstruction and pose estimation are discussed. For the 2D method, we perform pose estimation from multiple view and use RANSAC [11] for robust regression on the multiple estimation results. Metrics for 6DOF pose on $SE(3)$ are also discussed for its importance in RANSAC. State-of-the-art 2D method LINE-2D is chosen to apply for its high matching speed. Experiments and comparisons show that our multiple view approach can overcome problems of single view approach and yield more reliable pose estimation results.

The outline of this paper is as follows, Sect. 2 introduces the overall workflow of the multiple view approach for both 3D and 2D methods. Section 3 discusses specific pose estimation algorithms and details for their implementations. Section 4 shows the experiment results for both 3D and 2D methods. Finally, Sect. 5 gives conclusion for this paper.

2   Multiple View Approach for Pose Estimation

2.1   3D Point Cloud Method

3D point cloud registration is a primary approach for pose estimation. 6DOF pose of the target object can be recovered by aligning a prior known object model and the target object in the scene. Geometry features of the target objects may not be well-recovered when viewed from one single angle as only partial surfaces can be seen. Moreover, depth data generated from a low-cost RGB-D camera contains large amount of noises and loses many detailed features. This obviously brings negative effects for point cloud registration. The eye-in-hand system can provide a solution to this problem by observing the scene from multiple views and perform scene-reconstruction before pose estimation.

Typical way of 3D reconstruction such as RGB-D SLAM [12] estimates camera poses through image features in multiple frames. When it comes to an eye-in-hand robotic system, camera pose can be easily extracted as the joint encoder data and forward kinematics provide the end effector pose, and eye-hand calibration is done before the pose estimation task. Reference [2] notices this point, yet they also mention that poor accuracy of eye-hand-calibration may result in huge errors for registration. In order to overcome this problem, they teach the robot to observe the scene from several fixed angles and calibrate each camera pose following the standard RGB-D SLAM [12] routine. If the object container is moved or the experiment environment is changed, re-calibration for camera pose is needed, which involves a lot of extra work.

In our approach, we reconstruct the scene object by taking advantage of the robot data and eye-hand calibration result. We teach the robot to perform a pre-scanning motion along a fixed trajectory, during which depth data is taken and aligned. In order to overcome misalignment due to vibrations of the end effector during scanning motion, communication delay between the PC and robot controller, or poor accuracy for eye-hand calibration, we add a registration refinement by ICP.

Although ICP has been extensively used for registration refinement, different implementation procedure may have huge effects on the final registration result. In our approach, target point cloud for ICP is chosen to be the reconstructed scene object and source point cloud the newly-captured depth data. Instead of setting a fixed distance threshold for point pair matches, we set an inlier ratio to make the registration robust to false matches. Statistical outlier removal is applied after each frame is aligned to remove point cloud noises. Voxel grid filtering is also applied to the reconstructed point cloud to make it have an even distribution of points, which avoids ICP local minimum in high density area. All the above procedure overcomes adverse effects from noises, outliers and local minimum and makes best use of ICP for scene reconstruction.

Another important part of the pre-processing is object segmentation. For simple registration method such as PCA, object segmentation is a must as PCA is very vulnerable to outliers and missing data. Although robust registration methods like PPF have been proposed, segmentation still plays an important role for improving performance in terms of both accuracy and run-time speed. While reference [1] uses a reconstruction-then-segmentation approach, we segment the object on each frame and only take object points into account to speed up the registration process. Reference [2] uses a pre-trained fully convolutional network to segment target objects from the scene, which requires large amount of training data and GPU-support. In our approach, we segment target objects from the scene through pixel intensity directly, as the objects and backgrounds have obvious color contrast. Figure 1 shows the overall workflow of the 3D multiple view approach and the experimental setup.

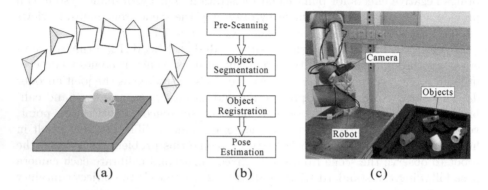

Fig. 1. (a) Multiple view approach. (b) Overall workflow for the 3D method. (c) Experiment setup.

2.2 2D Image Matching Method

2D image matching is another primary approach for pose estimation. As mentioned in Sect. 1, previous work on 2D pose estimation focuses on how to extract object pose from one single frame. In our approach, we take advantage of the eye-in-hand system and perform 2D pose estimation from multiple view. RANSAC is applied to the multiple estimation results for outlier removal and pose regression.

RANSAC has been well-adopted in computer vision for robust regression. Typical applications of RANSAC such as finding homography between two images uses l_2 norm to measure pixel distances. However, when extended for robust regression on 6DOF poses, l_2 norm would not work as the pose no longer belongs to the vector space $\mathbb{R}^6$. Here we give a discussion about metrics and regression for 6DOF pose on the $SE(3)$ group.

A 6DOF pose on $SE(3)$ contains a rotation part $\mathbf{R} \in SO(3)$ and a translation part $\mathbf{t} \in \mathbb{R}^3$. Distance metric on the l_2 norm sense for two translations $\mathbf{t}_1, \mathbf{t}_2 \in \mathbb{R}^3$ can be represented by

$$d_{\mathbf{t}}(\mathbf{t}_1, \mathbf{t}_2) = \|\mathbf{t}_1 - \mathbf{t}_2\|_2 \tag{1}$$

Accordingly, weighted regression on the l_2 norm least square sense for m given translations $\mathbf{t}_1, \ldots, \mathbf{t_m}$ can be given by

$$\mathbf{t_a} = \frac{1}{m} \sum_{i=1}^{m} w_i \mathbf{t_i} \qquad (2)$$

As for the $SO(3)$ rotation part, many different metrics exist and we adopt the Riemannian distance, namely the length of the geodesics connecting $\mathbf{R}_1$ and $\mathbf{R}_2$.

$$d_{\mathbf{R}}(\mathbf{R}_1, \mathbf{R}_2) = \frac{1}{\sqrt{2}} \|\log(\mathbf{R}_1^{\mathrm{T}} \mathbf{R}_2)\|_{\mathrm{F}} \qquad (3)$$

where

$$\log(\mathbf{R}) = \begin{cases} \mathbf{0} & if \ \theta(\mathbf{R}) = 0 \\ \frac{1}{2} \frac{\theta(\mathbf{R})}{\sin\theta(\mathbf{R})} (\mathbf{R} - \mathbf{R}^{\mathrm{T}}) & otherwise \end{cases} \qquad (4)$$

$$\theta(\mathbf{R}) = \cos^{-1}\left(\frac{\mathrm{trace}(\mathbf{R}) - 1}{2}\right) \qquad (5)$$

Accordingly, weighted regression on the Riemannian sense for m given rotations $\mathbf{R}_1, \ldots, \mathbf{R_m}$ can be given by

$$\mathbf{R_a} = \underset{\mathbf{R} \in SO(3)}{\arg\min} \sum_{i=1}^{m} w_i \|\log(\mathbf{R}_i^{\mathrm{T}} \mathbf{R})\|_{\mathrm{F}}^2 \qquad (6)$$

No close form solution for (6) exists and we solve it with an iterative method. First we initialize $\mathbf{R_a}$ as

$$\mathbf{R_a} = \sum_{i=1}^{m} w_i \log(\mathbf{R}_i) \qquad (7)$$

Then we compute the average on the tangent space of $\mathbf{R_a}$

$$\mathbf{r} = \sum_{i=1}^{m} w_i \log(\mathbf{R_a}^{\mathrm{T}} \mathbf{R}_i) \qquad (8)$$

Finally we update $\mathbf{R_a}$ by moving it towards $\mathbf{r}$.

$$\mathbf{R_a} \longleftarrow \mathbf{R_a} e^{\mathbf{r}} \qquad (9)$$

By iteratively solving (8) and (9), $\mathbf{R_a}$ gets converged and result of (6) can be extracted. It is noteworthy that in order to make it easy to implement, approximation for (6) can be represented on the Euclidean sense by orthogonal projection of $\overline{\mathbf{R_t}} = \frac{1}{m} \sum_{i=1}^{m} w_i \mathbf{R}_i^{\mathrm{T}} = \mathbf{U}\Sigma\mathbf{V}^{\mathrm{T}}$ onto $SO(3)$

$$\mathbf{R_a} = \begin{cases} \mathbf{V}\mathbf{U}^{\mathrm{T}} & if \ \det(\overline{\mathbf{R_t}}) > 0 \\ \mathbf{V}\mathbf{H}\mathbf{U}^{\mathrm{T}} & otherwise \end{cases} \qquad (10)$$

$$where \ \mathbf{H} = diag[1, 1, -1] \qquad (11)$$

Detailed description about the mathematical background of motion groups can be found in [13].

3   Implementations of Pose Estimation

The previous section introduces overall workflow for the multiple view pose estimation approach for 3D and 2D methods. In this section we discuss implementations for specific pose estimation algorithms which covers both the 3D and 2D methods (Fig. 2).

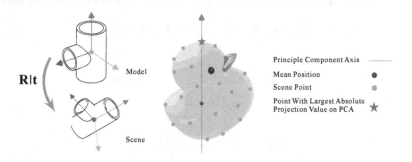

Fig. 2. Coarse registration can be achieved by aligning each corresponding principle component and mean position. Principle component sign uncertainty can be avoided by using a projection selection method.

For the 3D method, we first choose to implement PCA on the segmented scene object for coarse registration and add ICP as refinement. Despite being vulnerable to outliers, PCA is an effective and straightforward way for coarse registration. Although similar idea has been used in [2], a hidden problem which is not discussed is that PCA can't determine the sign of each principle component, therefore the result orientation might rotate 180° along certain axis. In [2]'s experiment environment, most target objects are symmetric along certain axis thus the sign ambiguity is not a big issue. However a general solution is needed to make sure that PCA can be applied to objects of all kinds of shapes. Here we provide a solution as follows. After PCA is applied for a given point cloud set, we traverse each mean-normalized point and calculate its projection on each principle component axis. For all projections on one axis, we find the point with the largest projection absolute value and make the axis direction pointing towards it. Workflow of this solution is shown in Algorithm 1. Note that when Algorithm 1 returns −1, it means the point set has equal projections on both directions of the input principle component. Result of Algorithm 1 and the corresponding rotation part of PCA pose estimation are shown in Table 1.

Except for PCA, PPF is another state-of-the-art pose estimation method focusing on 3D point cloud. Although reference [5,6] have demonstrated PPF's capability of dealing with heavily clustered environment, in practice we find that PPF can't give reliable estimation results using one single frame of RGB-D depth image. One obvious reason is that single depth image captured from a RGB-D camera can't provide enough geometry information for target objects, which can be improved by the multiple view scene-object-reconstruction approach. Another

Algorithm 1. Determine Sign of PCA

Input: $\mathcal{P} = \{p_1, p_2, p_3, ...p_m\}, r$
Output: r
1: **function** GETSIGNPCA(r,$\mathcal{P}$)
2:    $\mathcal{H} \leftarrow \{\phi\}$
3:    **for all** $p_i \in \mathcal{P}$ **do**
4:       $\mathcal{H} \leftarrow p_i \cdot r$
5:    **end for**
6:    $projMax \leftarrow$ FindMax($\mathcal{H}$)
7:    $projMin \leftarrow$ FindMin($\mathcal{H}$)
8:    **if** $projMax < (-1) * projMin$ **then**
9:       $r \leftarrow r * (-1)$
10:   **else**
11:      **if** $projMax = (-1) * projMin$ **then**
12:         **return** -1
13:      **end if**
14:   **end if**
15:   **return** r
16: **end function**

Table 1. Rotation matrix result of PCA

GetSignPCA($r_x, \mathcal{P}$)	GetSignPCA($r_y, \mathcal{P}$)	GetSignPCA($r_z, \mathcal{P}$)	R
else	else	else	$(r_x, r_y, r_x \times r_y)$
-1	else	else	$(r_y \times r_z, r_y, r_z)$
else	-1	else	$(r_x, r_z \times r_x, r_z)$
else	else	-1	$(r_x, r_y, r_x \times r_y)$
else	-1	-1	$(r_x, r_y, r_x \times r_y)$
-1	else	-1	$(r_x, r_y, r_x \times r_y)$
-1	-1	else	$(r_y \times r_z, r_y, r_z)$
-1	-1	-1	$(r_x, r_y, r_x \times r_y)$

reason is due to the uncertainty of surface normal sign. Surface normal of the point cloud serves as one dimension of PPF. The typical way of determining the sign of the normal vector for a given point p_i is by setting a vector $r = p_i - p_r$ pointing from a given reference point p_r to p_i. Then the sign of the normal vector can be determined in the way that its projection on r is positive. This can't ensure that each corresponding point on the object model and the scene have the same normal sign, as the reference point is randomly selected. In our approach, the reference point is set at the mean position for both the prior known model and the segmented scene object, which helps to overcome the problem of normal sign uncertainty. Also, pre-segmentation reduces the number of points needed to be matched thus speeds up the pose estimation process (Fig. 3).

For 2D image matching method, we choose to implement LINE-2D, which is the 2D version of LINEMOD [10] and has been extensively used in single frame pose estimation. As the matching process of LINE-2D is smartly designed, high

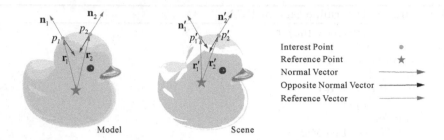

Fig. 3. Reference points for both the scene object and the model object are set at the mean position. Although positions of the two reference points are slightly different due to scene point cloud corruptions, normal sign remains consistent, both pointing from the inner to the outer.

processing speed can be achieved, making LINE-2D particularly suitable for real-time multiple view pose estimation. LINE-2D uses color gradient as features for matching. When viewed from certain angles, the object may contain very few features, resulting in unreliable estimation result. Multiple view approach for LINE-2D is as follows. Like the previous process, the robot first takes a scanning motion along a predefined route. Meanwhile, matching by LINE-2D is constantly running and a series of estimation results is extracted from frames with high matching score. Finally, all results with high confidence are fused with RANSAC to form an accurate final result (Fig. 4).

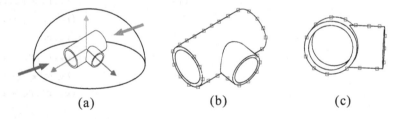

Fig. 4. (a) View angles of the target object. (b) View angle with rich selected features. Red rectangle represents color gradient features. (c) View angle with few selected features. (Color figure online)

4   Experiments

In this section, we test the above approach on four kinds of objects and evaluate pose estimation performance for the multiple view approach. Point cloud models of all objects are generated from the open source software Blender (Fig. 5).

We first test the overall performance of the proposed multiple view approach for PCA, PPF and LINE-2D. For PCA and PPF, 15 frames of color and depth images are recorded during the scanning motion to reconstruct scene objects. For LINE-2D, matching similarity threshold is set to 80% and 140 frames of estimation results are fused together for RANSAC.

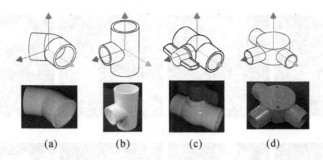

(a)   (b)   (c)   (d)

Fig. 5. Four types of objects used in our experiment. (a) 2-hole PVC pipe. (b) 3-hole PVC pipe. (c) PVC valve. (d) 3-hole PVC box

Figure 6 shows a typical experiment result on the 4 objects. As the ground truth pose of the target object is unknown, we follow the error evaluation method in [6] by repeating pose estimation process and choosing absolute deviation from the results median as the error metric. We first run our multiple view approach for 100 times for each tested object and get 100 estimation results. Next we calculate each result's deviation from the results median. Pose deviation metric is defined as discussed in Sect. 2. Histograms of pose estimation deviation are shown in Fig. 7. Note that the position metric is in meter and the rotation metric is in arbitary unit. Average deviation values are shown in Table 2.

From the histograms we can see that deviations from the estimation result median is generally reduced by the multiple view approach. For PCA, single view estimation result shows large deviation and the multiple view approach greatly helps to improve estimation accuracy. For PPF, which is robust to noise and missing data, performance on rotation is generally the same for both approaches, and the multiple view approach help to improve position estimation accuracy to some extent. For LINE-2D, the multiple view approach helps to narrow the deviation from result median for both rotation and position. Numeral comparison among the 3 algorithms shows that repeated experimental results of LINE-2D have the least variance among the 3 tested algorithms, which means that the LINE-2D's performance is the most stable one in terms of repeated accuracy. However, graphical display for the estimation results in Fig. 6 shows that LINE-2D also has the most failure cases. This is due to the fact that template library for LINE-2D can hardly cover all possible view angles and objects with extreme poses may be out of the matching scope. On the other hand, 3D methods are free of this problem and able to handle any object poses. For PCA which is vulnerable to outliers, misalignment case with opposite orientation still exists, as noise points may disturb the projection value on each principle axis and affect the final estimation result as a whole. PPF shows more stable and robust performance with no failure cases when compared with PCA. In terms of robotic picking task, PPF is the best choice for its reliability and robustness.

The experiment is implemented on a PC with Intel Core i5-4570 CPU and 12 GB RAM. Open source library OpenCV and PCL are used for image and

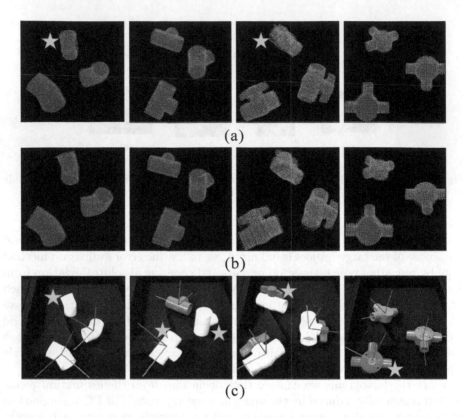

Fig. 6. Experiment result of the propsed approach. (a) PCA. (b) PPF. (c) LINE-2D. Failure cases with gross error or invalid matches are marked with yellow stars. (Color figure online)

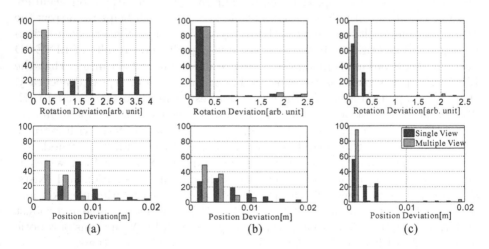

Fig. 7. Histograms of deviation from estimation results median. (a) PCA. (b) PPF. (c) LINE-2D.

point cloud processing and visualizing. We use open source code of PPF and LINE-2D provided by OpenCV. The experiment setup consists of a 6-DOF robot manipulator from Universal Robots and an ASUS Xtion Pro RGB-D camera. The runtime speeds for single frame pose estimation are as follows: 50 ms for PCA+ICP, 1500 ms for PPF+ICP and 100 ms for LINE-2D. Considering the pre-scanning motion, it takes 15 to 20 s for the whole process.

Table 2. Average deviation from the result median

DOF type	PCA	PPF	LINE-2D
Position[a](Multiple/Single)	**0.0042**/0.0100	**0.0044**/0.0069	**0.0017**/0.0024
Rotation[b](Multiple/Single)	**0.4192**/2.5421	**0.3305**/0.3106	**0.1735**/0.1857

[a]Metric follows (1), in meter.
[b]Metric follows (3), arbitrary unit.

5 Conclusions

In this paper, we discuss implementations of multiple view approach for pose estimation with an eye-in-hand robotic system. We apply the multiple view approach on both the 3D and 2D pose estimation methods. For the 3D method, we revisit the reconstruction-then-estimation scheme and take advantage of the eye-in-hand robotic system for camera pose extraction. For the 2D method, we adopt RANSAC for outlier removal and results regression. Metrics and regression are discussed in detail for the 6DOF pose. We choose to implement PCA, PPF and LINE-2D for pose estimation, which covers both the 3D and 2D approach. Experiment results show that the multiple view approach can improve pose estimation performance for both the 3D and 2D methods. Comparison among the 3 implemented methods also show that PPF has the most reliable performance.

Acknowledgements. This research was supported in part by National Natural Science Foundation of China under Grant 51675325 and National Science and Technology Support Program under Grant 2015BAF01B02.

References

1. Lehnert, C., Sa, I., McCool, C., Upcroft, B., Perez, T.: Sweet pepper pose detection and grasping for automated crop harvesting. In: 2016 IEEE International Conference on Robotics and Automation (ICRA), pp. 2428–2434. IEEE (2016)
2. Zeng, A., Yu, K.T., Song, S., Suo, D., Walker Jr., E., Rodriguez, A., Xiao, J.: Multi-view self-supervised deep learning for 6D pose estimation in the Amazon picking challenge. arXiv preprint arXiv:1609.09475 (2016)
3. Duncan, K., Sarkar, S., Alqasemi, R., Dubey, R.: Multi-scale superquadric fitting for efficient shape and pose recovery of unknown objects. In: 2013 IEEE International Conference on Robotics and Automation (ICRA), pp. 4238–4243. IEEE (2013)

4. Besl, P.J., McKay, N.D.: Method for registration of 3-D shapes. In: Robotics-DL Tentative, pp. 586–606. International Society for Optics and Photonics (1992)
5. Drost, B., Ulrich, M., Navab, N., Ilic, S.: Model globally, match locally: efficient and robust 3D object recognition. In: 2010 IEEE Conference on Computer Vision and Pattern Recognition (CVPR), pp. 998–1005. IEEE (2010)
6. Choi, C., Taguchi, Y., Tuzel, O., Liu, M.Y., Ramalingam, S.: Voting-based pose estimation for robotic assembly using a 3D sensor. In: 2012 IEEE International Conference on Robotics and Automation (ICRA), pp. 1724–1731. IEEE (2012)
7. Choi, C., Christensen, H.I.: RGB-D object pose estimation in unstructured environments. Robot. Autonom. Syst. **75**, 595–613 (2016)
8. Collet, A., Berenson, D., Srinivasa, S.S., Ferguson, D.: Object recognition and full pose registration from a single image for robotic manipulation. In: 2009 IEEE International Conference on Robotics and Automation (ICRA), pp. 48–55. IEEE (2009)
9. Liu, M.Y., Tuzel, O., Veeraraghavan, A., Taguchi, Y., Marks, T.K., Chellappa, R.: Fast object localization and pose estimation in heavy clutter for robotic bin picking. Int. J. Robot. Res. **31**(8), 951–973 (2012)
10. Hinterstoisser, S., Cagniart, C., Ilic, S., Sturm, P., Navab, N., Fua, P., Lepetit, V.: Gradient response maps for real-time detection of textureless objects. IEEE Trans. Pattern Anal. Mach. Intell. **34**(5), 876–888 (2012)
11. Fischler, M.A., Bolles, R.C.: Random sample consensus: a paradigm for model fitting with applications to image analysis and automated cartography. Commun. ACM **24**(6), 381–395 (1981)
12. Endres, F., Hess, J., Engelhard, N., Sturm, J., Cremers, D., Burgard, W.: An evaluation of the RGB-D slam system. In: 2012 IEEE International Conference on Robotics and Automation (ICRA), pp. 1691–1696. IEEE (2012)
13. Wang, Y., Chirikjian, G.S.: Nonparametric second-order theory of error propagation on motion groups. Int. J. Robot. Res. **27**(11–12), 1258–1273 (2008)

Research on Extracting Feature Points of Electronic-Component Pins

Yongcong Kuang[1(⊠)], Jiayu Li[1], Jinglun Liang[2], and Gaofei Ouyang[1]

[1] Guangdong Provincial Key Laboratory of Precision Equipment and
Manufacturing Technology, School of Mechanical and Automotive Engineering,
South China University of Technology, Guangzhou 510640, China
yckuang@scut.edu.cn, 753110373@qq.com
[2] School of Mechanical Engineering, Dongguan University of Technology,
Dongguan 523808, Guangdong, China

Abstract. Most of the plug-in machine adopts the visual positioning system
that is used by Surface Mount Technology (SMT), and obtains the bottom image
of component. For some electronic components, as the color of pin is similar to
the body, the pin is difficult to be distinguished. In order to improve the positioning accuracy of the rotational stereo vision positioning method for plug-in
machine, an electronic component pin edge tracking algorithm based on iterative
optimization and a sub-pixel edge extracted method based on the cubic curve
fitting are proposed. Furthermore, a center point calculation method is proposed
for component pin with different shape to improve the accuracy of feature point
extraction. Experimental results show that the proposed center point calculation
method performs higher accuracy than the method of taking average of borders
as center point directly. In addition, the proposed subpixel border extracting
method performs better accuracy than the Canny algorithm.

Keywords: Plug-in machine · Subpixel border · Center point · Feature point

1 Introduction

With the development of industrial technology, the automation of electronic products is
getting higher and higher. More and more electronic component was plugged into PCB
by machine [1]. Plug-in machine is the precision equipment that automatically plugs odd
electronic components into the PCB (printed circuit board) using machine vision positioning [2]. The non-contact visual positioning method is the key technology of the
plug-in machine, and its accuracy and stability are very important to plug-in machine [3].

Professor Liu et al. proposed a method for accurately locating the position and tilt
angle of electronic components, whose position accuracy is up to 0.027 mm for side
length of 15 mm square element [4]. Huang proposed a method of detecting the center
position of a positioning column that determining the position of the component pins
according to the central location [5]. The image pyramid model is used to accelerate the
search for targets to speed up computing [1].

As shown in Fig. 1(a), most of the plug-in machine adopts the visual positioning
system that is used by SMT(Surface Mount Technology), and obtains the bottom image

© Springer International Publishing AG 2017
Y. Huang et al. (Eds.): ICIRA 2017, Part II, LNAI 10463, pp. 611–622, 2017.
DOI: 10.1007/978-3-319-65292-4_53

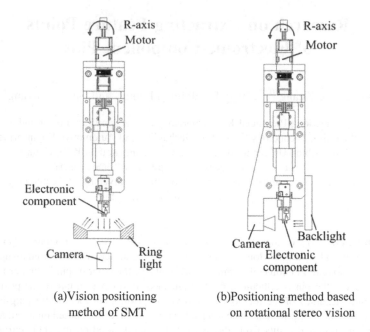

(a)Vision positioning
method of SMT

(b)Positioning method based
on rotational stereo vision

Fig. 1. The schematic diagram of component vision positioning system

of the component and is illuminated with ring light. However, Some electronic components cannot be detected because colors between pins and body are similar, as well as irregular shape of pins, as shown in Table 1.

Table 1. Comparison of pin image between two methods

Physical map	Method based on rotational stereo vision	Method of SMT

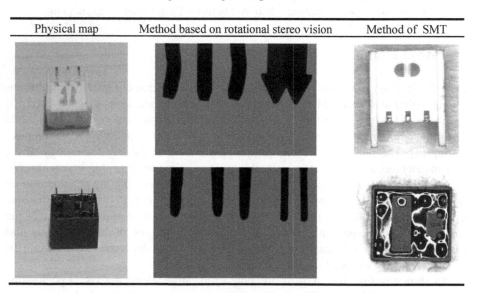

A fast and accurate image processing technique is particularly important for the component positioning method of Fig. 1(b). In image processing, the center point in the end of component pin which is considered as the feature point of component pin needs to be extracted in image [6]. Traditional methods such as Canny, Sobel algorithm only have pixel-level accuracy, which cannot meet the requirements of electronic-component pin positioning [7, 8]. Moreover, due to the pinhole imaging model of the common lens, there is a theoretical error in the method of directly taking average of both borders as the center of the pin.

In order to improve the positioning accuracy of the rotational stereo vision positioning method in Fig. 1(b), a pin sub-pixel edge extracted method based on the cubic curve fitting is proposed. Furthermore, a center point calculation method is proposed for component pin with different shape. Experimental results show that the proposed method of calculating the center point has higher accuracy than taking average of both borders, and has higher precision and comparable calculation speed than the Canny algorithm.

2   Edge Coarse Positioning

2.1   Determine the Initial Searching Interval

The initial interval of pin edge is determined by the threshold like the formula (1), and the initial search interval of the pin edge as shown in Fig. 2 is obtained, where Δj is length of edge search interval.

$$|I(i_0, j + \Delta j) - I(i_0, j)| > thresh \tag{1}$$

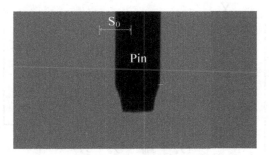

Fig. 2. Initial searching interval of pin edge

2.2   Edge Tracking

In order to speed up pin edge searching, jumping search was used for pin edge tracking, that upper search interval S_k was used as the initial interval of the next search interval S_{k+1} in steps Δi. But the upper and lower edge of the interval is not in the same column, as result from pin slight tilt, or chamfer and fillet in the end of pin, as shown in Fig. 3. Due to the accuracy of the sub-pixel edge is affected by the range of the edge interval, an iterative optimization method shown in Fig. 4 was proposed to obtain the

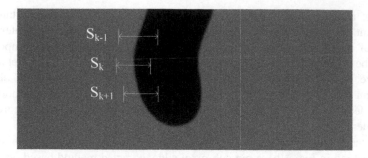

Fig. 3. The schematic diagram of pin edge tracking

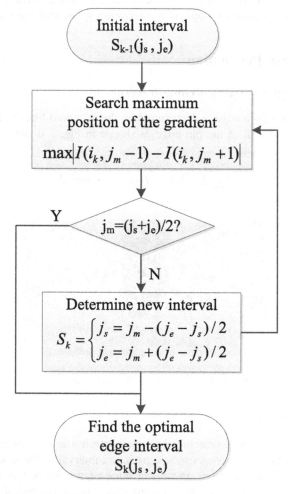

Fig. 4. The schematic diagram of pin edge interval iterative optimization

optimal position of the next edge interval, with upper edge interval as the initial position of the next edge interval, which makes the grayest gradient which max $|I(i_k, j_m + 1) - I(i_k, j_m - 1)|$ of edge pixel located in the middle of the edge interval, as result, pin edge was tracked.

3   Edge Fine Positioning

Obviously, gray of pin edge can be fitted with cubic curve, as shown in Fig. 5. Assume curve to be fitted as:

$$f = ax^3 + bx^2 + cx + d \tag{2}$$

where v is the abscissa of pixel, f is the gray of pixel, a, b, c are polynomial coefficients respectively. For fitting the cubic curve by optimization, the pixels in the coarse positioning interval obtained by the edge tracking are substituted into the formula (2) a, b and c are calculated with the algebraic distance of the pixel gray scatters to the curve as optimization target as formula (3) shown.

$$\min \sum_{j=j_s}^{j_e} [f_i - (ax_j^3 + bx_j^2 + cx_j + d)]^2 \tag{3}$$

Calculating the derivative of the obtained formula (2), the sub-pixel edge is the position of its maximum derivative as formula (4) shown.

$$\max |f'(x)| \tag{4}$$

Subpixel edge of pin is obtained as formula (5) shown:

$$X_k = -\frac{b}{3a} + j_s \tag{5}$$

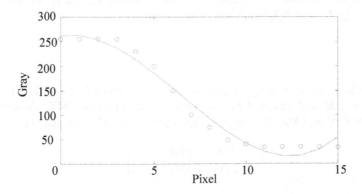

Fig. 5. The schematic diagram of pin edge model

4   Extract Center Point of Pin

Due to the pinhole imaging model of common lens as shown in Fig. 6, there is a theoretical error by the method of taking average of both borders as the center of the pin directly as formula (6) show.

$$error = OM - 1/2(OX_1 + OX_2) \tag{6}$$

where O is the main point of imaging plane, f is the focal length of lens, P is the geometric center of pin, whose projection point is M, X_1 and X_2 are the sub pixel points of pin calculated by formula (5). For improving positioning accuracy, a center point calculation method was proposed for component pin with different shape, including circle pin and flat pin.

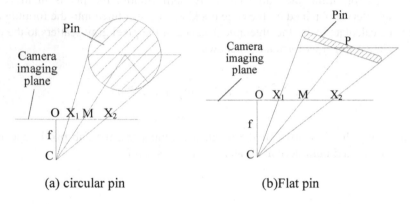

(a) circular pin   (b)Flat pin

Fig. 6. The schematic diagram of pin center projection

4.1   Calculate Center Point of Circular Pin

The circular pin center point calculation formula is derived as shown in the Fig. 7, l_1 and l_2 can be calculated as formula (7) shown, where l_1 and l_2 are the bevels of triangular PQ_1E_1 and PQ_2E_2,r is the radius of pin.

$$l_1 = \frac{r}{\sin \alpha_1}, \quad l_2 = \frac{r}{\sin \alpha_2} \tag{7}$$

According to the imaging projection model, the triangle CE_1P is similar to the triangular CX_1M, and triangle CPE_2 is similar to the triangular CMX_2. Assume that k is the ratio of X_1M and MX_2 on camera imaging plane, it can be calculated as formula (8).

$$k = \frac{X_1M}{MX_2} = \frac{OM - OX_1}{OX_2 - OM} = \frac{l_1}{l_2} \tag{8}$$

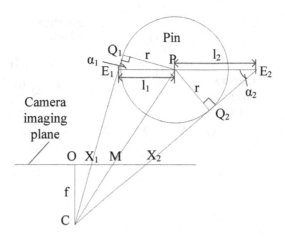

Fig. 7. The schematic diagram of calculation of circular pin center point

Center point of pin M can be calculated by formulas (7) and (8) as:

$$OM = \frac{k \cdot OX_2 + OX_1}{1 + k}$$

$$where, \ k = \frac{\sqrt{f^2 + OX_1^2}}{\sqrt{f^2 + OX_2^2}}$$

(9)

As result there is a theoretical error between formula (9) and the method of taking average of both borders directly as:

$$error == OM - 1/2(OX_1 + OX_2) = \frac{(k-1)}{2(k+1)} \cdot (OX_2 - OX_1) \qquad (10)$$

4.2 Calculate Center Point of Flat Pin

The flat pin center point calculation formula is derived as shown in the Fig. 8, l_1 and l_2 can be calculated as formula (11) shown, where l_1 and l_2 are the bevels of triangular PQ_1E_1 and PQ_2E_2,r is average of the pin width.

$$\frac{\sin \alpha_1}{\sin(\pi - \alpha_1 - t)} = \frac{r}{l_1}$$

$$\frac{\sin \alpha_2}{\sin(\pi - \alpha_2 - t)} = \frac{r}{l_2}$$

(11)

According to the imaging projection model, the triangle CE_1P is similar to the triangular CX_1M, and triangle CPE_2 is similar to the triangular CMX_2. Assume that k is the ratio of X1 M and MX2 on camera imaging plane, it can be calculated as formula (12).

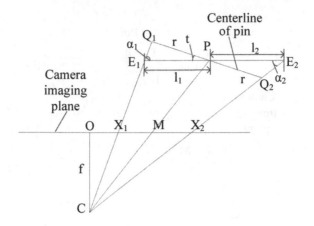

Fig. 8. The schematic diagram of calculation of flat pin center point

$$k = \frac{X_1 M}{M X_2} = \frac{OM - OX_1}{OX_2 - OM} = \frac{l_1}{l_2} \tag{12}$$

Center point of pin M can be calculated by formulas (7) and (8) as:

$$OM = \frac{OX_1 + (OX_2 - OX_1) \cdot k}{k+1}$$

$$where, k = \frac{\sin(\alpha_1 + t)/\sin(\alpha_1)}{\sin(\alpha_2 + t)/\sin(\alpha_2)} \tag{13}$$

$$\alpha_1 = \arctan(f/OX_1)$$

$$\alpha_2 = \arctan(f/OX_2)$$

There is an additional parameters t in formula (13) for Calculating center point M of flat pin, which is the angle between center line of the pin and camera imaging plane. Since the similarity of the grabbing posture in each time, t can be estimated by the posture of electronic component.

As result there is a theoretical error between formula (13) and the method of taking average of both borders directly as:

$$error == OM - 1/2(OX_1 + OX_2) = \frac{(1 - 3k)OX_1 + (K - 1)OX_2}{2(k+1)} \tag{14}$$

5   Extract Feature Point of Pin

In order to obtain the feature point of pin, which is the center point in the end of the pin. Firstly, a number of sub pixel edge points and center points were calculated with the method discussed above as shown in Fig. 9. Then fitting the center points using straight line. Finally, the feature point of pin can be calculated using cubic curve fitting as formula (5) in the fitted straight line.

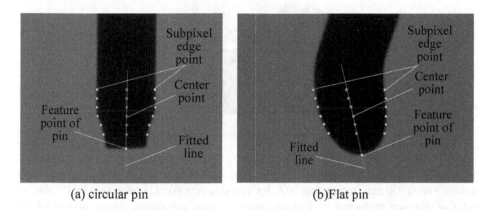

(a) circular pin   (b)Flat pin

Fig. 9. The schematic diagram of feature point extraction

6   Experiments

The plug-in machine visual experiment platform as Fig. 10 shown, the side camera is used to obtain the side image of pin for reconstruction based on rotational stereo vision [9–11], with the Image-DFK23G445 1290 × 960 camera and Computar-35 mm lens, and is lighted with backlight, the bottom camera is used to measure the relative distance of the pins as the standard value, with the Image-DFK23G445 camera, and the DKM125-90-AL telecentric lens, lighted with ring light, R-axis is driven with Panasonic-MMMA3ACN1A servo motor, the computer is equipped with CPU i5-5200U 2.20 GHz, 8 GB of memory.

6.1   Accuracy Experiment

The accuracy of the proposed feature point extraction is verified by pin reconstruction experiment based on rotational stereo vision [9–11] in the plug-in machine visual experiment platform as Fig. 10 shown. The accuracy of the proposed method was verified using relative distance error:

$$e = l - k \tag{15}$$

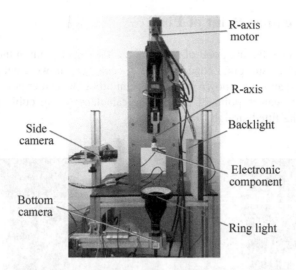

Fig. 10. The schematic diagram of experiment platform

where l is the relative distance between pins that reconstructed based on rotational stereo vision (R-axis rotate four 90° for reconstruction); k is the standard value of relative distance measured by bottom camera. Canny algorithm, the method of taking average of both borders as the center of the pin and the proposed method are used to reconstruct the circular pin and flat pin of the component as Table 2 shown for 100 groups of experiments, the reconstruction accuracy results are shown in Table 3.

Table 2. Electronic components for experiment (/mm)

	Circular pin	Flat pin
Standard distance k	5.10722	11.07950
Pin diameter(width)	0.57	2.4

Table 3. Accuracy analysis (/mm)

	Algorithm	Average of e	Standard deviation of e
Circular pin	Canny	0.01403	0.03278
	Method of taking average of borders	0.01268	0.00314
	Proposed method	0.01267	0.00314
Flat pin	Canny	0.05445	0.2901904
	Method of taking average of borders	0.01130	0.0038024
	Proposed method	0.00550	0.0037811

Table 4. Efficiency analysis (/ms)

	Algorithm	Average of calculate time
Circular pin	Canny	15.87174
	Method of taking average of borders	18.66956
	Proposed method	18.31794
Flat pin	Canny	19.92103
	Method of taking average of borders	15.52941
	Proposed method	18.88675

According to the experiments results as shown in Table 3, for flat pin, the proposed method performed obvious advantages of accuracy compared with the method of taking average of edge point, but not for the circular pin. Moreover, the proposed subpixel border extracted method performs better accuracy than Canny algorithm.

6.2 Efficiency Experiment

Efficiency of Canny algorithm, the method of taking average of both edge as the center of the pin, and the proposed method were verified using the images of the circular pin and flat pin of the components as Table 2 shown for 100 groups of experiments in C ++(not include picture reading time). According to the experiments results as shown in Table 4, the proposed method has almost the same efficiency as the method of taking average of borders and Canny algorithm, and all less than 20 ms.

7 Conclusion

An electronic component pin edge tracking algorithm based on iterative optimization and a sub-pixel edge extracted method based on the cubic curve fitting were proposed. Furthermore, a center point calculation method was proposed for component pin with different shape to improve accuracy of feature point extraction. Experimental results show that the proposed center point calculation method possesses higher accuracy than the method of taking average of edges directly. On the other hand, the proposed subpixel border extracted method performs better accuracy than the Canny algorithm.

Acknowledgments. This research was supported by the Natural Science Foundation of Guangdong Province (Grant No. 2015A030310415) and the Industrial Technologies Research and Development Program (1st NC) of Dongguan (Grant No. 2015222119).

References

1. Zhuang, D.: Examination of control systems of the automatic plug-ins machine. Qingdao University of Science and Technology (2011)
2. Song, J.: Research and application of visual localization technology based on machine vision. Dalian University of Technology (2014)
3. Choon, T.K.: The origin and prevention of post reflow defects in surface mount assembly. J. Electron. Manuf. **6**(01), 1–12 (1996)
4. Xie, Y., Liu, Q.: Research on locating algorithm of vision alignment system in automatic high precision chip mounter. Opt. Tech. **34**(3), 449–454 (2008)
5. Huang, Z.: Research of connector inspection and location technology based on machine vision. China University of Mining and Technology (2015)
6. Chen, Z., Zhou, D., Liao, H., Zhang, X.: Precision alignment of optical fibers based on telecentric stereo microvision. IEEE/ASME Trans. Mechatron. **21**(4), 1924–1934 (2016)
7. Qing, L., Tu-sheng, L.: Image edge detection algorithm based on mathematical morphology. J. South China Univ. Technol. Nat. Sci. Ed. **36**(9), 113–116 (2008)
8. Maini, R., Aggarwal, H.: Study and comparison of various image edge detection techniques. Int. J. Image Process. (IJIP) **3**(1), 1–11 (2009)
9. Binbin, Z.: Research on technology of 3D reconstruction of rotational image sequences based on microscopic vision. Zhejiang University of Technology (2013)
10. Zhang, Z.: A flexible new technique for camera calibration. IEEE Trans. Pattern Anal. Mach. Intell. **22**(11), 1330–1334 (2000)
11. Cai, J.-R., Zhao, J.-W.: Stereo vision system calibration based on dual-camera. J. Jiangsu Univ. (Nat. Sci. Ed.) **27**(1), 6–9 (2006)

Efficient Combinations of Rejection Strategies for Dense Point Clouds Registration

Shaoan Zhao, Lin Zuo, Chang-Hua Zhang, and Yu Liu$^{(\boxtimes)}$

University of Electronic Science and Technology of China,
Chengdu, Sichuan, China
yuliu@uestc.edu.cn

Abstract. The Iterative Closest Point (ICP) algorithm has been viewed as a standard approach to registering two point clouds. In the process of point clouds registration, the eliminating incorrect point pairs has important effect on the accuracy and stability of registration. In the past two decades, numerous strategies of excluding point pairs have been developed and various combinations of them have been applied to the variants of ICP algorithm. In this paper, an efficient combination of rejection strategies is proposed. It also is compared with other heuristic combinations. As shown in our case studies, the proposed combination can realize more accurate registration without sacrificing computational efficiency.

Keywords: Point cloud registration · Iterative Closest Point · Rejection strategies

1 Introduction

Point clouds registration algorithm is a core technique of computer graphics. Registering two point clouds consists in finding the rotation and the translation occurred in the overlapping regions of two point clouds, and it has widespread applications, such as simultaneous localization and mapping (SLAM). Therefore, the point clouds registration algorithms have received considerable attentions. The classic solution of registering two point clouds is the Iterative Closest Point (ICP) algorithm proposed by Besl and McKey [1]. The original ICP is not suitable for the situation where the point clouds to be registered overlap partially. However, partial overlap is a very common phenomenon in engineering practices, many variants of the ICP algorithm have been developed to fulfill the practical needs.

As reported in [2], the existing methods for registering two point clouds can be categorized in two groups, namely the sparse approaches and the dense approaches. The sparse approaches focus on a few meaningful points that have obvious geometry features or visual features and match point pairs by comparing the values of these features. Although matching feature is more robust than the dense approaches, it requires additional information and computational efforts. Most dense approaches only need the Cartesian coordinate of points which can be provided by tremendous sensors, such as 3D laser scanners, stereo cameras or depth cameras, etc. The dense approaches

© Springer International Publishing AG 2017
Y. Huang et al. (Eds.): ICIRA 2017, Part II, LNAI 10463, pp. 623–633, 2017.
DOI: 10.1007/978-3-319-65292-4_54

are, therefore, simple and easy for use. The dense approaches have to assume that a reasonable initial estimate of transform is available, while the sparse approaches don't need such information. Compared with the sparse approaches, the dense approaches are stronger in will but weaker in power with respect to the data association. The original ICP algorithm is a typical dense approach, and it establishes the point correspondences by finding the closest point in other meshes starting from the current transform estimate. It usually refers to the point-to-point alignment. The point-to-point alignment is inevitable to produce numerous spurious point pairs, resulting in the incorrect alignment of two point clouds. Several association methods [3, 4] were developed in the past years to partially remedy the limitation of the point-to-point alignment. An alternative way to cope with the limitation of the point-to-point alignment is to find suitable rejection strategies of the false point pairs.

In the last two decades, numerous strategies of excluding pairs have been developed and their various combinations have been applied to the variants of the ICP algorithm. Nevertheless, it is difficult to find a versatile combination as argued in [5]. This paper focuses on the heuristic rejection strategies used in dense approaches and compares the performances of their combinations. A simple but also efficient combination of the strategies of rejecting point pairs is proposed for accurate point cloud registration.

The remainder of this paper is organized as follow. Section 2 introduces the details of the existing strategies of rejecting pairs. Section 3 proposes a robust point clouds registration algorithm that combines strategies of excluding pairs used in the vanilla ICP and ICRP. In Sect. 4, the proposed algorithm is compared with several heuristic combinations of rejection strategies by real-world data, and it is applied to the consistent mapping. Section 5 draws a conclusion.

2   Previous Work

Besl and McKay [1] proposed the classic ICP algorithm which is a fundamental tool to point clouds registration. The algorithm was examined by experimental data, but it didn't consider the partial overlap between two point clouds. Various variants of the ICP algorithm have been proposed during the last two decades. Developing a suitable strategy of rejecting pairs is an important research direction of these studies. The simplest solution of rejecting incorrect point pairs is to preset a distance threshold. If the distance of the point pairs exceeds a pre-set threshold, they will be identified as noise point pairs and discarded. The original formulation of the ICP approach including a distance threshold is called the vanilla ICP approach as reported in [6]. It is reasonable that vanilla ICP algorithm decreases gradually distance threshold with the increase of the iterations. Because the distances of the correct point pairs trend to be zero with the increase of the iterations. Such strategy was used in [7], and later on, Pomerleau *et al.* [8] introduced the Relative Motion Threshold (RMT) for rejection, and a simulated annealing ratio was used to automatically change the rejection threshold.

Pajdla and Van Gool [9] developed the Iterative Closest Reciprocal Point (ICRP) algorithm. Their idea is that the distances of point pairs are theoretically zero and the closest point relation is symmetrical when the exact registration has been realized. Although the motion transform breaks this relation, the distance used to represent the relation should be less than a threshold ε. Point g in source point cloud has been associated with point m in target point cloud with nearest neighbor search, and then point m is associated with its closest point g' in source point cloud. If the distance between points g and g' exceeds the threshold ε, the point pair (g, m) will be excluded.

Rusinkiewicz and Levoy [10] compared several strategies of excluding pairs, and they concluded that rejecting pairs containing points on mesh boundary is especially useful for avoiding erroneous pairing when the overlapping between two point clouds is not complete.

Chetverikov et al. [5] developed the Trimmed Iterative Closest Point (TrICP) algorithm. It used the degree of overlapping to determine whether the point pairs should be retained. Assume that the degree of overlapping is known and represented by a variable n_o. By the TrICP, point correspondences are firstly found with point-to-point alignment method. The distances of point pairs are arranged in ascending order. The n_o pairs at the front of the queue are treated as the exact point pairs. Comparing it to the ICRP algorithm, the authors claimed that the TrICP outperformed the ICRP when the degree of overlapping is less than 50%.

Zinßer et al. [11] proposed the Picky ICP algorithm. The algorithm attempted to reject the point pairs except the point pair with smallest distance when a point appeared in different point pairs. May et al. [6] proposed the frustum ICP. The concept of *frustum* came from the field of computer vision, and applied to the ICP algorithm. It created a vision cone to describe degree of overlapping, and it showed that the algorithm was more robust than the vanilla ICP with respect to rejecting pairs. However, the cone is not suitable for some sensors, such as laser scanner. In other words, its versatility is weak.

Serafin and Grisetti [2] described a full approach, namely the Normal Iterative Closest Point (NICP), that can directly operate on raw 3D sensor data, either depth images or 3D laser scans. The NICP rejected false pairs using the combination of three rejection strategies: distance threshold, compatibility of normal and compatibility of curvatures.

All the aforementioned algorithms have important roles in point clouds registration. Many attempts have been made to compare performances of these algorithms to find out the best algorithm. There is no one attested that it is able to reject completely noise pairs. The existing algorithms actually are complementary or superfluous. May et al. [6] noted that the frustum ICP can be used together with other algorithms. However, they did neither give any strategy for combination nor analyze the difference of the algorithm with others except the vanilla ICP. Pomerleau et al. [8] combined several rejection strategies with the picky ICP to test their proposed algorithm. Various combinations of rejection strategies were used. Nevertheless, there is no reported work that was devoted to analyzing the performances of different combinations between the existing rejection strategies. This paper aims at comparing the performances of different combinations of rejection strategies, and then, proposes a robust point clouds registration algorithm which combines strategies of rejecting pairs used in the vanilla ICP and the ICRP. In the ensuing sections, our proposed algorithm will be firstly presented and compared to other heuristic combinations. Because the new algorithm combines the strategies of excluding pairs used in the vanilla ICP and the ICRP, it is called the vanilla Iterative Closest Reciprocal Point (vanilla-R ICP) algorithm in this paper. As Rusinkiewicz and Levoy [10] classified the variants of the ICP algorithm into six stages, i.e., selection of points, matching points, weighting the corresponding pairs appropriately, rejecting certain pairs, assigning an error metric and minimizing it. We, therefore, follow this framework to introduce the new algorithm.

3   The Vanilla-R ICP Algorithm

3.1   Selection of Points

The computational overhead of the ICP algorithm depends mainly on the number of points. The exhaustive pairing often is computationally unaffordable. The existing methods for data reduction include the random sampling, uniform sampling, normal-space sampling, covariance sampling, and other sampling methods using extra information such as color and intensity of points. Although the normal space sampling proposed by Rusinkiewicz and Levoy [10] may work better on accuracy, we choose the random sampling to reduce data in this study. One of the reasons for choosing the random sampling is that computing surface normals of points in certain meshes is not necessary, and it will reduce the computational effort. Another reason is that there are often enough points attached to different surfaces such that the distribution of normals among selected points is not too small. Sampling from the normal space of points might be more suitable in the case where the existing poor normals occurs. In our experiment, we randomly sample points in source point cloud without replacement until the number of eligible point pairs is up to a pre-set threshold n_e, and this sampling method is used to replace the sampling methods in other algorithms.

3.2   Matching Points

Rusinkiewicz and Levoy [10] summarized most correspondence finding methods. The commonly used methods include finding the closest point in the other mesh [1] and searching the point projected from source point cloud to destination mesh in the destination range image. The later might use a metric based on point to segment distance, point to plane distance, point to facet distance, the related mathematical tools were presented in [12]. Finding the correspondences through projection has an obvious shortcoming of efficiency due to the large amount of data to be manipulated. Serafin and Grisetti [2] introduced the concept of index image to reduce the memory movements. In this paper, we choose the point-to-point approaches due to two reasons: (1) There are enough numbers of points so that the influence of the discrete is small for data association, and (2) finding the closest point in the other meshes is much faster than finding the correspondences through projection [7]. In our algorithm, point correspondences are found by:

$$\min_{y_j \in T} f(y_j) = \|y_j - (\mathbf{R}x_i + \mathbf{t})\|_2 \tag{1}$$

where point x_i is a selected point from source points, whereas point y_j belongs to target point cloud. If the criterion described in Eq. (1) is reached, the pair (x_i, y_j) will be treated as the candidate pair. In our algorithm, the k-d tree was used to accelerate this computation. There are available variants of k-d tree to improve computational efficiency, such as the cached k-d tree [13].

3.3   Rejecting Pairs

The rejection principles of pairs used in the vanilla ICP and the ICRP are combined to realize accurate and robust registration. This combination is called the vanilla-R criterion for simplification in this paper. As to be demonstrated in Sect. 4, it works better than other algorithms introduced in Sect. 2. The method excludes the point pairs which have lager distance than the pre-set distance threshold that would degrade with iteration. It is noted that the distance of point pair refers to the distance between the point transformed by current transform estimate and its mate. Transform between two point clouds consists of a rotation $\mathbf{R}$ and a translation $\mathbf{t}$. $\mathbf{R}_k$ and $\mathbf{t}_k$ are obtained in the kth iteration of the algorithm. If the distance of point pair is lower than the distance threshold, the algorithm will decide whether the pair should be retained using the rejection strategy of the ICRP [9]. A candidate pair (x_i, y_j) is discarded if one of the following criterions holds.

The distance of pair is larger than the threshold d_k which gradually decrease with iteration according to value of *step*:

$$\|y_j - (\mathbf{R}_k x_i + \mathbf{t}_k)\|_2 > d_k \tag{2}$$

$$d_k = d_{k-1} - step \tag{3}$$

The distance between x_i and reciprocal point x_i' in the source point cloud S exceeds the pre-set threshold ε:

$$\min_{x_i' \in S} f(x_i') = \|y_j - (\mathbf{R}_k x_i' + \mathbf{t}_k)\|_2 \tag{4}$$

$$\|x_i - x_i'\|_2 > \varepsilon \tag{5}$$

The parameter settings are presented in Table 1.

Table 1. Parameters were set as optimal values derived by sampling the parameter space. The Vanilla-R ICP obviously outperforms others on accuracy and its computational efficiency is slightly lower than that of the vanilla-picky ICP.

Algorithms	Error	Run time	Related parameters
Vanilla ICP	37627.6	1.9035	$d_0 = 1010$, $step = 25$, $n_e = 5000$
ICRP	29698.1	2.76656	$\varepsilon = 10$, $n_e = 1000$
TrICP	24094	8.02731	$n_o = 20000$
Picky ICP	39966.9	2.88134	$n_e = 500$
Vanilla-R ICP	**13126.1**	1.86003	$d_0 = 1010$, $step = 25$ $\varepsilon = 10, n_e = 500$
Vanilla-Picky ICP	25038.1	**1.67218**	$n_e = 100$
Trimmed-R ICP	25599	9.14937	$n_o = 8000$, $\varepsilon = 10$
Trimmed-Picky ICP	31787	4.8373	$n_o = 8000$

3.4 Error Metric and Minimization

This step uses all the pairs filtered by rejection strategies to calculate the misalignment error and find the transform between two point clouds. Our algorithm uses the point-to-point error metric, and finds the transform using a method based on singular value decomposition [14].

3.5 The Design of Convergence Rules

The error usually fluctuated slightly with iterations because of the random sampling, and deceptive convergence might oftentimes occur. The convergence rule is designed to assure that our algorithm can reach an accurate result and prevent the algorithm from deceptive convergence and infinite loop. The mean registration error of eligible pairs should below an error threshold and change slightly with iteration. If the number of times the criterion has been satisfied is up to a pre-set value, the algorithm terminates. At the same time, the maximum number of iterations is designed to avoid the situation where the pre-set error threshold is too small for the algorithm to terminate, resulting in infinite loop.

4 Algorithm Evaluation and Application

4.1 Criterion of Evaluation

In this paper, we don't analyze our algorithm with experimental data. Rather, it is applied to registration of real-world data. There is no ground truth of transform between two point clouds for evaluation. The criterion of evaluation proposed here was inspired by the TrICP [5]. The sum of the registration errors from overlapping section theoretically trend to be zero when the accurate registration has been realized. Resulting registration errors of pairs are arranged in ascending order, and the result of summing registration errors of n_{ev} pairs at the front of the queue is used as the criterion of evaluation. The value of n_{ev} was set to 5000. In general, the smaller the summing results, the better the registration. It is noted that incorrect registration theoretically might produce a small value, although the situation didn't occur in our evaluation. Visual effect of registration is an assistant means to remedy this defect of the criterion.

4.2   The Algorithms for Comparison

The incorrect correspondences can be divided to two categories: (1) the points outside of the overlapping section took part in pairing, and (2) the inner points from over-lapping section matched wrongly each other. The pair containing an outside point usually has a huge distance. The Vanilla ICP uses a distance threshold to reject noise pairs. The TrICP only chooses the n_o pairs with the least value of distance as correct pairs. Therefore, the vanilla ICP and the TrICP can reject this class of pairs easily. It is noted that the results of the vanilla ICP and the TrICP will be completely equivalent if the distance threshold used in the vanilla ICP is equal to the distance of n_oth pair with the least value of distance. The pairs containing outside points generally occur in boundary mesh. Thus, rejecting pairs containing points in boundary mesh also is efficient to exclude this class pairs. In fact, the strategies of excluding pairs used in the ICRP and the Picky ICP are also useful to reject the point pairs containing an outside point, but they are obviously time-consuming due to searching nearest neighbor.

Figure 1 shows a general situation where a mobile robot moved in indoor plane, and one needs to register the source point cloud into target point cloud. We assume that in Fig. 1 all the points in source space belong to inner points, the outside points have been rejected with a distance threshold or other methods above. The green section denotes the favorable pairs for registration. The yellow area describes the bad pairs which usually lead to terrible result. The brown area presents the neutral point pairs which protect the algorithm from divergence but slow down the speed of convergence. We continue to evaluate the capability of available strategies of excluding pairs in yellow and brown areas. It is obvious that the vanilla ICP and the TrICP are both incapable of rejecting the pairs in yellow and brown areas. Compared to the picky ICP, the ICRP is more competent to exclude the point pairs in the yellow area because the ICRP can find the latent pairs by searching backwards. The method based on normals can also reject the pairs in yellow area, but computing the normals of points costs at least twice amount of the time which the ICRP costs. It is therefore not considered. All the methods above are inefficient for point pairs in brown area.

According to the capability of excluding pairs, the strategies of rejecting pairs are divided into two categories: (1) more competent to reject wrong point pairs containing outside points, such as the ones used in the vanilla ICP and the TrICP, and (2) The ones which work well in terms of reducing incorrect point pairs of overlapping section, such as the ones used in the ICRP and the picky ICP. Nevertheless, the combinations that consist of the rejection strategies of one category don't improve performance of the algorithm efficiently. Hence, we only combine the rejection strategies that play different role in the process of excluding point pairs. All the algorithms to be compared are presented in Table 1. The vanilla-R ICP is the proposed algorithm that combine the rejection strategies used in the vanilla ICP and the ICRP. The Trimmed-R ICP means

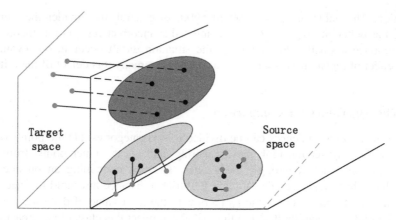

Fig. 1. The green section depicts the favorable pairs for registration. The yellow area describes the bad pairs which usually lead to terrible result. The brown area presents the neutral point pairs which protect the algorithm from divergence but slow down the speed of convergence. (Color figure online)

that we first use the rejection strategies of the TrICP and continue to exclude the pairs with rejection strategy used in the ICRP. The Vanilla-Picky ICP and the Trimmed-Picky ICP use the same principle of combination as the Trimmed-R ICP. At the same time, we compare these combinations with the algorithms that only use an individual excluding pairs strategy.

4.3   Results

All the programs are executed on an Intel Core i5-4460 at 3.2 GHz. The dataset used here comes from the website: http://kos.informatik.uni-osnabrueck.de/3Dscans/, and the number of dataset is 5. The results presented in Table 1 are analyzed from two aspects: computational effort and accuracy.

In terms of computational effort, the vanilla-Picky ICP took the least amount of time. It benefits from the rejection strategies used in the Picky ICP, which don't need to execute time-consuming search of closest neighbor. The Vanilla-R ICP is slightly less efficient than the Vanilla-Picky ICP, which benefits from random sampling, it just needs the number of eligible pairs to be up to a pre-set threshold. The TrICP expends plenty of time in the search of closest neighbor, therefore it is very time-consuming. Comparing to the Trimmed-Picky ICP and the TrICP, we found that the combination dramatically reduces the executing time. In fact, the rejection strategy used in the Picky ICP retains a fairly small portion of the pairs filtered by the trimmed ICP. In our experiment, there are about 300 pairs left after using the rejection strategy used in the Picky ICP whereas the TrICP kept the 8000 pairs as correct pairs. According these results, we can verify that the computational effort is influenced mainly by method of selecting point. Due to the need of ensuring the integrity of the method, some low efficient sampling methods have to be remained, such as the TrICP.

In terms of accuracy, the vanilla-R ICP is superior to other algorithms. As noted in Sect. 4.2 that the rejection strategy used in the ICRP is more competent to exclude incorrect pairs of inner points than used in the Picky ICP. It is expected that the vanilla-R ICP is more accurate than the vanilla-Picky ICP. This expectation is also confirmed by the data presented in Table 1. In process of the Trimmed-Picky ICP, small number of point pairs accelerated the solution of transform relation but resulted in poor accuracy compared to the TrICP. The combinations improved the performance of individual rejection strategy except combinations with the TrICP. Figure 2 shows the visual effects of registration of the vanilla ICP and the vanilla-R ICP, the results are consistent with the indicated error data. Others have more bad or similar visual effect to the vanilla ICP, and thus, they aren't exhibited anymore.

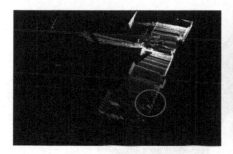

Fig. 2. The left was the result of the vanilla ICP, and the right is the result of the vanilla-R ICP. The red circle highlights part where is easy to find the registration error. It can be found that the vanilla-R ICP is better than the vanilla ICP. (Color figure online)

According to the above observations, we can conclude that the vanilla-R criterion is a more efficient combination of rejection strategies than the others.

4.4 Application

In this section, our algorithm was applied to realize the 3D consistent mapping. We use *incremental matching* method introduced by Chen [3]. The new scan is registered against a so-called *metascan*, which is the union of the previously acquired and registered scans. In general, using this method leads the accumulation of mapping error. To verify the consistent mapping ability of the vanilla-R ICP algorithm, we don't use any global optimization algorithm which usually improve the consistency of mapping. Figure 3 presents the visual effect of mapping the five data. No obvious error was found in the mapping, which illustrates the proposed algorithm is helpful to improve accuracy of consistent mapping. The octomap [15] was used to build the occupancy grid map which can be used to mobile robot navigation, the result is presented in Fig. 4.

Fig. 3. The vanilla-R ICP to 3D consistent mapping were used, and no obvious geometrical inconsistencies were found in the map. Additionally, no global optimization algorithm was used in the process of consistent mapping.

Fig. 4. Occupancy grid map. Blue represents the occupied area. (Color figure online)

5   Conclusion

In this paper, we summarize the available heuristic strategies of rejecting point pairs and compare the performances of various combinations of these strategies. We propose the vanilla-R ICP algorithm which combines exclusion strategies used in the vanilla ICP and the ICRP. The convergence rules are also designed to assure that the algorithm can converge to an accurate result. A new criterion for evaluating point clouds registration is designed for situation where the ground truth of transform between two point clouds to be registered is absent. Compared to other heuristic combinations of rejection strategies, the proposed algorithm can realize the more accurate registration without asking for additional computational effort.

Acknowledgement. The authors greatly acknowledge the grant support from the Fundamental Research Funds for the Central Universities under contract number ZYGX2015J082.

References

1. Besl, P.J., McKay, N.D.: Method for registration of 3-D shapes. IEEE Trans. Pattern Anal. Mach. Intell. **14**(2), 239–256 (1992)

2. Serafin, J., Grisetti, G.: NICP: dense normal based point cloud registration. In: 2015 IEEE/RSJ International Conference on Intelligent Robots and Systems, pp. 742–749. IEEE Press, Hamburg (2015)

3. Chen, Y., Medioni, G.: Object modelling by registration of multiple range images. Image Vis. Comput. **10**(3), 145–155 (1992)

4. Censi, A.: An ICP variant using a point-to-line metric. In: 2008 IEEE International Conference on Robotics and Automation, pp. 19–25. IEEE Press (2008)

5. Chetverikov, D., Svirko, D., Stepanov, D., Krsek, P.: The trimmed iterative closest point algorithm. In: 16th International Conference on Pattern Recognition, pp. 545–548. IEEE Press (2002)

6. May, S., Droeschel, D., Holz, D., Fuchs, S., Malis, E., Nüchter, A., Hertzberg, J.: Three-dimensional mapping with time-of-flight cameras. JFR **26**(11–12), 934–965 (2009)

7. Pulli, K.: Multiview registration for large data sets. In: The Second International Conference on 3-D Digital Imaging and Modeling, pp. 160–168. IEEE Press (1999)

8. Pomerleau, F., Colas, F., Ferland, F., Michaud, F.: FSR, pp. 229–238. Springer, Heidelberg (2010). doi:10.1007/978-3-642-13408-1_21

9. Pajdla, T., Van Gool, L.: Matching of 3-D curves using semi-differential invariants. In: The Fifth International Conference on Computer Vision, pp. 390–395. IEEE Press (1995). doi:10.1109/iccv.1995.466913

10. Rusinkiewicz, S., Levoy, M.: Efficient variants of the ICP algorithm. In: The Third International Conference on 3-D Digital Imaging and Modeling, pp. 145–152. IEEE Press (2001). doi:10.1109/im.2001.924423

11. Zinßer, T., Schmidt, J., Niemann, H.: A refined ICP algorithm for robust 3-D correspondence estimation. In: 2003 International Conference on Image Processing, vol. 2, pp. II-695. IEEE Press (2003). doi:10.1109/icip.2003.1246775

12. Armesto, L., Minguez, J., Montesano, L.: A generalization of the metric-based iterative closest point technique for 3D scan matching. In: 2010 IEEE International Conference on Robotics and Automation, pp. 1367–1372. IEEE Press (2010). doi:10.1109/robot.2010.5509371

13. Nuchter, A., Lingemann, K., Hertzberg, J.: Cached KD tree search for ICP algorithms. In: Sixth International Conference on 3-D Digital Imaging and Modeling, pp. 419–426. IEEE Press (2007). doi:10.1109/3dim.2007.15

14. Arun, K.S., Huang, T.S., Blostein, S.D.: Least-squares fitting of two 3-D point sets. IEEE Trans. Pattern Anal. Mach. Intell. **5**, 698–700 (1987). doi:10.1109/cdc.1997.649861

15. Hornung, A., Wurm, K.M., Bennewitz, M., Stachniss, C., Burgard, W.: OctoMap: an efficient probabilistic 3D mapping framework based on octrees. Auton. Robots **34**(3), 189–206 (2013). doi:10.1007/s10514-012-9321-0

Reconstructing Dynamic Objects via LiDAR Odometry Oriented to Depth Fusion

Hui Cheng, Yongheng Hu, Haoguang Huang, Chuangrong Chen,
and Chongyu Chen[✉]

School of Data and Computer Science, Sun Yat-Sen University,
Guangzhou 510006, Guangdong, China
chenchy47@mail.sysu.edu.cn
http://hcp.sysu.edu.cn

Abstract. LiDAR odometry is the key component of LiDAR-based simultaneous localization and mapping (SLAM). However, the low vertical resolution of LiDAR makes it difficult to produce pleasant mapping results. It is even more challenging to reconstruct the surface of dynamic objects from the raw LiDAR input. To address this problem, existing approaches typically divide it into several subproblems like object detection and tracking and then solve them individually, which greatly increases the complexity of LiDAR odometry as well as the SLAM framework. In this work, we propose to address this problem by improving LiDAR odometry with appropriate modifications to the depth fusion process and several additional lightweight components. Extensive evaluations on KITTI dataset and Velodyne HDL-16E laser scanner demonstrate the effectiveness of the proposed method. The results of the improved LiDAR odometry include abundant information about the dynamic objects, which can be used for many high-level tasks such as object recognition and scene understanding.

Keywords: LiDAR odometry · Depth fusion · 3D reconstruction · Simultaneous localization and mapping

1 Introduction

Simultaneous localization and mapping (SLAM) has been widely studied for years because of its irreplaceable importance to robot perception. A general assumption of LiDAR-based SLAM approaches [6,10] is that the static scene is scanned by a moving LiDAR, which is inaccurate when there are dynamic objects in the scene. Moreover, the vertical resolution of LiDAR is very low (e.g. 32 lines), making it difficult to perform surface reconstruction in mapping tasks.

To handle dynamic objects in the scene, researchers have made great efforts to improve the SLAM principles from different aspects, such as object detection and tracking. For example, LiDAR measurements on objects can be considered as outliers of the scene measurements. Therefore, the random sample consensus (RANSAC) [5] paradigm, which is originally designed for outlier detection, is

© Springer International Publishing AG 2017
Y. Huang et al. (Eds.): ICIRA 2017, Part II, LNAI 10463, pp. 634–646, 2017.
DOI: 10.1007/978-3-319-65292-4_55

widely used for object detection and removal [4]. Specific cues for LiDAR data, such as occupancy grid [14] and ray tracing [11], are also adopted for object detection in LiDAR scans. To handle dynamic objects continuously, tracking methods are also employed. Moosmann and Stiller [8] propose to jointly address tracking and self-localization, which typically requires dense depth input (e.g. 64-line LiDAR data) and an additional component of track management. Yang and Wang [16] propose a specific RANSAC oriented to spatiotemporal consistency to build links among moving object tracking, multi-scale segmentation, and LiDAR odometry. Although all the components work in a stable manner, the problem of reconstructing dynamic objects remains unsolved due to the sparsity of the raw LiDAR input. It can be observed that existing solutions tend to add more components to handle dynamic scenes, which will increase the computation amounts. This is not preferred in intelligent systems with limited computational resources.

In this paper, we study on the challenging problem of reconstructing dynamic objects from sparse depth data. The proposed solution, which is called *L iDAR odometry oriented to fusion* (LOOF), is built upon standard LiDAR odometry with the orientation to depth fusion for dynamic objects. Several lightweight algorithms are employed to fully exploit necessary information from the LiDAR odometry pipeline. In particular, spatial clustering and temporal association are employed to exploit spatiotemporal consistency of dynamic objects. Parameters for rendering and fusing depth maps are carefully chosen to be adaptive to the sensor characteristics. To the best of the authors' knowledge, the proposed method is the first solution to reasonable surface reconstruction of dynamic objects from sparse LiDAR data. Evaluations on public datasets verify the effectiveness of the proposed method from the aspects of reconstruction quality, odometry accuracy, and computational efficiency. Abundant information about object segmentations and trajectories can be easily extracted from the results of the proposed solution, which can be used for many high-level tasks such as object recognition, object tracking, and scene understanding.

2   Proposed Framework

In this section, we present our solution of employing lightweight algorithms to LiDAR odometry with orientation of depth fusion. As shown in Fig. 1, The employed algorithms are mainly for foreground-background separation and depth processing.

2.1   LiDAR Odometry

Starting with standard LiDAR odometry, the proposed method takes as input a sequence of depth measurements. The set of depth measurements obtained at the same time is defined as a data frame. By exploiting the coordinate information from LiDAR data, we build a multi-dimension map D_i for each data frame, where i is the frame number. In this section, we denote a data point in D_i as

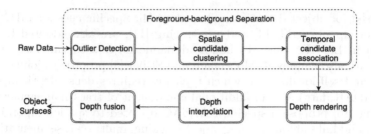

Fig. 1. The pipeline of the proposed LiDAR odometry oriented to fusion (LOOF).

$\mathbf{p}(u, v) = \{x, y, z, d\}$, where v is the measurement number related to the scanning azimuth, $u \in \{1, \ldots, u_{\max}\}$ is the number of scanning line, d is the depth value, and $\{x, y, z\}$ represents the 3D coordinates of the point in the global coordinate system. For the data provided by mainstream LiDAR, typical values of $u_{\max}$ are 16, 32, and 64. Small value of $u_{\max}$ brings in difficulties to both accurate frame registration and reasonable surface reconstruction.

For the registration between two sequential frames D_i and D_{i+1}, classical point-to-point iterative closest point (ICP) method [1] may produce incorrect registration results. The main reason is that the same LiDAR scanning line is probably on different positions of the object surface in sequential frames, while in computational point of view, the points on the same scanning line will probably be considered as correspondences between two frames. Therefore, we connect the 3D points in D_i as a mesh with normals and adopt standard point-to-plane ICP method [3] for frame registration.

2.2   Foreground-Background Separation for LiDAR Data

To reconstruct dynamic objects, we separate the object measurements from the scene measurements, which is done by an improved RANSAC paradigm with spatial clustering and temporal association.

Adaptive Spatial Clustering of Outliers. In the registration between frames of LiDAR data, the depth measurements from the ground usually affects the registration accuracy. Therefore, in this work, we adopt the RANSAC paradigm for two purposes, i.e., ground measurement removal as described in [2] and dynamic object detection. In particular, we divide the data frame into n point sets $\{P_1, \ldots, P_n\}$. For every point set P_k, we apply ICP individually to obtain a rigid transforms T_k. After applying T_k to all 3D points, the dynamic points are then detected by exploiting their distances with the corresponding points. In this work, the distance threshold is set to 0.05 m for the frame rate of 10 fps, aiming at detecting dynamic points with a speed no less than 0.5 m/s. The RANSAC iterations stop within 10 iterations. During the iterations, if the average error is smaller than the distance threshold, the iterations will stop immediately.

After the detection of outliers, we cluster the outlier points in the spatial domain, resulting in a preliminary estimation of foreground objects. The

Fig. 2. The relationship between scanning range and object-LiDAR distance with a fixed scanning area.

density-based spatial clustering of applications with noise (DBSCAN) [15] is employed because of its nice ability in preserving object continuity. There are two key parameters of DBSCAN, i.e., radius of search window r_{win} and the minimal number of points within the searching radius which is denoted as n_{min}. Since the density of laser-scanned measurements decreases as the depth increases, both r_{win} and n_{min} have to be adaptive to depth. In this work, we propose to compute r_{win} for the current point as:

$$r_{win}(d_2) = s * r_{win}(d_1) \tag{1}$$

where the scale

$$s = \frac{\arctan^2(1/d_1)}{\arctan^2(1/d_2)},$$

$r_{win}(d_1) = 0.5$ m is the searching radius for the depth of $d_1 = 10$ m, d_2 is the depth of the current 3D point. The reason for using Eq. (1) that the number of LiDAR measurements for the same scanning area become smaller as the distance increase. As illustrated in Fig. 2, for the same scanning area of πr^2, there are θ_1/θ_{int} measurements for the distance of d_1, while there are θ_2/θ_{int} measurements for the distance of d_2, where the scanning interval θ_{int} is a fixed value for a given LiDAR device.

It should be noticed that these separation results are not perfect that some foreground points will be considered as background and vice versa. An example of incorrect labelling is illustrated in Fig. 3(a) and (b).

Temporal Association of Object Candidates. The reconstruction of object surface usually requires continuous scanning from different views. Therefore, we need to build the links for the 3D points from the same object across different frames. Although the spatial clustering results are not perfect, they still provide nice object candidates when being applied to individual frames. Given the imperfect object candidates, we propose to exploit temporal consistency for linking object candidates across frames. In particular, we first divide the sequence into several segments of identical length. Then, close cluster centers across sequential frames are linked together, so that each cluster corresponds to a trajectory. The trajectories that lasts for more than 0.5 s (5 frames) with a breakup smaller than 0.5 s are preserved and marked as foreground. Other trajectories will be removed and the related clusters are marked as background. Then, we compass the points

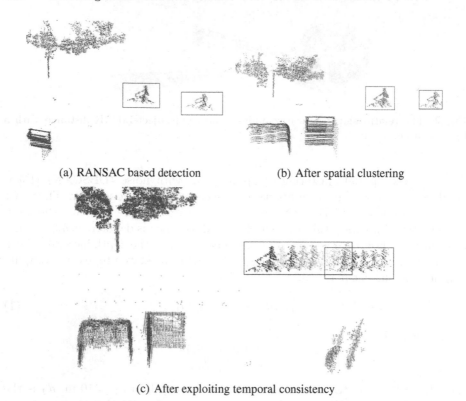

(a) RANSAC based detection　　　　　(b) After spatial clustering

(c) After exploiting temporal consistency

Fig. 3. Dynamic object detection after different steps, in which blue points represent the scene points and dots in other colors represent dynamic points. There are two moving object in the scene, which are marked by black rectangles. All frames of point clouds are shown in (c), with different colors indicating dynamic points from different frames. It can be observed that the proposed method with temporal considerations can well detect the moving objects while sorting out most points with incorrect labels. (Color figure online)

in cubes respectively determined by the maximum and minimum coordinate of the points with foreground label, by doing which the foreground labels are propagated to the surrounding 3D points, resulting in more labeled foreground points. An example of the objects extracted by applying spatial clustering and temporal association is shown in Fig. 3(c). It is shown that considering the temporal consistency can lead to more accurate foreground separation.

2.3　Depth Interpolation and Fusion

Inspired by the success of KinectFusion [9], we resort to depth fusion in the form of truncated signed distance function (TSDF) for reconstructing the surface of dynamic objects. All the foreground points are used to produce depth maps in an object-wise manner.

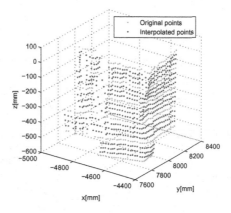

Fig. 4. Depth interpolation for the object surface.

Depth Interpolation. Considering that the depth measurements are sparse in the vertical direction, we propose to interpolate depth values in the index domain. That is, as for points $\mathbf{p}(u, v) = \{x, y, z, d\}$ in a data frame D_i, in every vertical line, we firstly interpolate the vertical scanning line index u uniformly and acquire the expected vertical index u^*, then we use the initial index u, expected index u^* and the depth d corresponding to index u as the input of the cubic spline interpolation promising the first and second derivative of the curve and get a denser data frame. As we can see in Fig. 4, the raw object points are very sparse in vertical direction. By applying depth interpolation, we can acquire denser point cloud about the object.

After interpolation, we compute the 3D coordinates for every interpolated point, resulting in a denser version of data frame. These denser point clouds bridge the gap between the depth measurements from LiDAR scan and depth camera, paving the way to utilize existing techniques for depth fusion.

Depth Rendering. Given a sequence of interpolated 3D points for the dynamic objects, we have to convert them to depth maps before surface reconstruction. In this work, we propose to generate depth maps by placing a virtual camera pointing at every object. The coordinate system of the virtual camera is shown in Fig. 5(a). Note that we only have to rotate the virtual camera around the y-axis to make z-axis point at the object because the LiDAR is usually placed on a horizontal plane and thus its x-axis is horizontal. The rotating angle is determined by exploiting the LiDAR coordinate system. For example, the Velodyne Laser scanner ihas 64 lines in the vertical direction and 4,000 points for every scanning line. It can be computed that the angle interval between neighboring scanning points (i.e. the horizontal precision) is $\theta_{int} = 360°/4000 = 0.09°$. Therefore, the rotating angle can be computed by $\theta = 0.09\bar{u}$, where $\bar{u}$ is the median index of the LiDAR measurements on the object.

With a virtual camera pointing at the object, we can generate a depth map by projecting the 3D points to 2D image plane according to the perspective

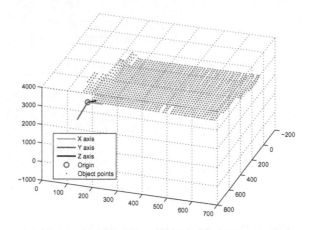

(a) The coordinate system of the virtual camera.

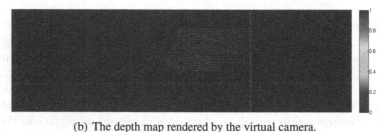

(b) The depth map rendered by the virtual camera.

Fig. 5. An example of generating depth maps from 3D points of the object.

imaging model. There are several considerations for choosing the focal length f_x and f_y and the resolution of the depth map $H \times W$:

1. The focal lengths f_x and f_y can be neither too small or too large because the generated depth maps will be either too dense or too sparse;
2. The width W and the height H of the depth map should be large enough so that the object can be completely shown;
3. The resolution $H \times W$ cannot be too large due to the limitations of computation and storage resources.

Let θ_l denote the vertical precision of LiDAR measurement. We propose to set f_x, f_y, W, and H with respect to the constraints described by the following approximations:

$$\frac{W}{H} \approx \frac{f_x}{f_y} \approx \frac{\theta_{int} * k}{\theta_l} \tag{2}$$

where k is an adaptive coefficient which is in inverse proportional to the LiDAR-object distance while in proportional to the object size. Figure 5(b) shows an example of the depth map rendered from a 16-line LiDAR input. In this example, we choose $f_x = 500$, $f_y = 100$, $W = 396$, $H = 96$ for the object distance of 4.0

m, horizontal precision of 0.2°, and vertical precision of 2°. It should be noted that for W and H that can be exactly divided by 32, one can get a higher computational efficiency in GPU implementation.

Depth Fusion. Given the rendered depth maps for every dynamic object, we propose to fuse them in the form of TSDF as described in KinectFusion [9]. Since the depth maps from LiDAR are different from that from Kinect, adaptions for the fusion process are necessary. Key issues include camera intrinsic parameters, volume size, and voxel size. Similar to depth rendering, TSDF fusion also requires a virtual camera. To maintain the rendering consistency, in depth fusion, we use a virtual camera with intrinsic parameters identical to that of the camera for depth rendering. Different from KinectFusion that uses a fixed volume with pre-defined voxel size, in this work, we use adaptive volume size and voxel size for each object. That is, the volume should contain the whole object, whose size is determined by the minimum and maximum coordinates of the object point cloud. More importantly, the voxel size is set according to the precision along each axis direction. For example, the voxel length along z-direction should be smaller than that along the y-direction because LiDAR has a low precision along the y-axis. Considering the data density is dependent on the distance, we suggest that the

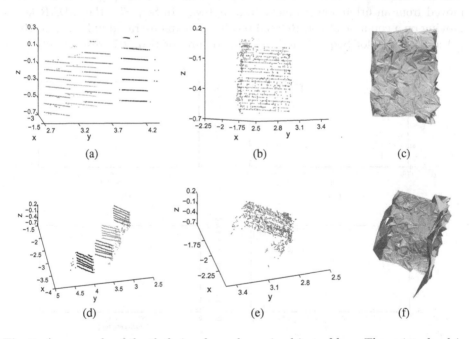

(a)   (b)   (c)

(d)   (e)   (f)

Fig. 6. An example of depth fusion for a dynamic object of box. The point cloud is obtained by Velodyne HDL-16E, which only has 16 scanning lines along the vertical direction. (a) shows 4 frames of the 10 frames of extracted LiDAR data. (b) is the fusion result of 10 frames. (c) is the mesh constructed by triangulating the point clouds extracted from the TSDF. (d), (e), and (f) are showing the same objects of (a), (b), and (c) from another view, respectively.

voxel size should increase as the distance between object and LiDAR increases. After determining these key parameters, we enable surface reconstruction with given LiDAR data by the TSDF module used in KinectFusion.

Taking the LiDAR data from Velodyne HDL-16E as an example, we perform the fusion process for an object, i.e. a box. The extracted data and fusion results are shown in Fig. 6, which demonstrate that the object surface is successfully reconstructed. Note that the triangulation algorithm [7] adopted for converting TSDF to mesh is also used in KinectFusion.

3 Experiments

In this section, the proposed method is evaluated from the aspects of odometry accuracy, reconstruction quality, and computational efficiency.

3.1 The Odometry Accuracy and Computational Efficiency

The odometry accuracy is evaluated by conducting quantitative experiments on the KITTI odometry datasets. First, we conduct experiments using the adopted LiDAR odometry on two sequences, Seq. #3 and #7. In Seq. #3, the LiDAR is moved from an urban environment to a highway. In Seq. #7, the LiDAR keeps moving on a common street. Related results are demonstrated in Fig. 7, in which (a) and (b) are for Seq. #3 and (c) and (d) are for Seq. #7, respectively. It is

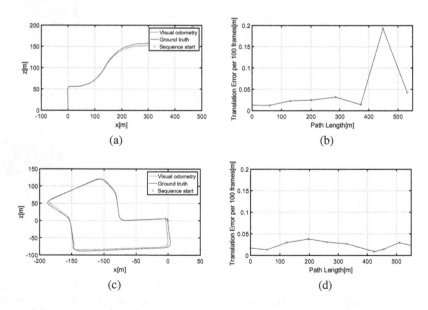

Fig. 7. Odometry results of the adopted LiDAR odometry on two data sets of the KITTI odometry data. (a) Comparison of odometry trajectories on Seq. #3; (b) The translation errors on Seq. #3; (c) Comparison of odometry trajectories on Seq. #7; (d) The translation errors on Seq. #7.

Table 1. Quantitative evaluation of the odometry accuracy and computational efficiency.

Data set	Frame number	GICP [12]	CLS [13]	CLS-M [13]	Proposed
Seq. #0	4540	**0.0315**	0.0622	0.0529	0.0931
Seq. #3	800	**0.0218**	0.0275	0.2390	0.0442
Seq. #4	270	0.0497	0.0316	0.0394	**0.2320**
Seq. #5	2760	**0.0228**	0.0726	0.0413	0.0884
Seq. #6	1100	0.0362	**0.0327**	0.0383	0.0384
Seq. #7	1100	0.0132	0.0222	**0.0117**	0.0208
Seq. #8	4070	0.0626	0.1001	0.0643	**0.0451**
Avg. time/frame	/	25.68 s	2.36 s	28.56 s	**1.58 s**

shown that the adopted LiDAR odometry achieves nice accuracy on these two data sets. The relatively big error occurs in the estimation of translation vector for Seq. #3 is caused by the scene change from urban environment to a highway. In open environments such highway, the LiDAR measurements will concentrate on far distances and thus ICP based odometry will be inaccurate.

For further evaluations, we compare the adopted LiDAR odometry with two state-of-the-art methods are compared, generalized iterative closest point (GICP) [12] and collar line segments (CLS) [13]. Note that the CLS method can be improved by processing multiple scans. This improved version, named CLS-M, is also included in our evaluations. Table 1 illustrate the overall odometry results and computational efficiencies of the compared methods on 7 data sets, where bold fonts indicate the state-of-the-art results. It is shown that the proposed method can achieve comparable odometry accuracy while keeping the lowest computational complexity. Note the failure cases, i.e., Seq. #1 and Seq. #2, of the adopted LiDAR odometry are not reported because these cases only include open environments. Considering other data sets are obtained from common urban environments, the reported results are believed to be sufficiently representative.

3.2 Computational Efficiency

Besides the LiDAR odometry, our solution also employ several lightweight algorithms. The employed algorithms are implemented in C/C++ without any code optimization. By analyzing the execution time of these algorithms on a common PC with a 3.6 GHz CPU using a small subset of the KITTI dataset, we obtain the following efficiency results.

1. The spatial clustering consumes 42 ms on average;
2. The temporal association consumes 0.35 ms on average;
3. The partial GPU implementation of depth fusion consumes 131 ms for fusing every depth map with around 7×10^3 non-zero pixels.

Note that all these employed algorithms are only applied on the short sequence of dynamic objects. They can be done in a separated thread and thus do not affect the real-time performance of the LiDAR odometry.

3.3   Surface Reconstruction for Objects

To verify the effectiveness of the proposed depth fusion, we take as an example a dynamic object extracted by the employed algorithms from Seq. #3. As shown in Fig. 8, the extracted object is a car at the distances ranging from 15 m to 18 m. According to the principles described in Eq. (2), we use a voxel size of $0.035 \times 0.07 \times 0.035$ m^3. The depth maps are rendered with a virtual camera whose focal lengths are $fx = 500$ mm and $fy = 180$ mm and a depth resolution of 352×96. Note that both 352 and 96 can be exactly divided by 32. The visual comparison between Fig. 8(a) and (b) indicates that, with proper parameter setup, the adopted TSDF depth fusion successfully merge the depth maps constructed from the sparse LiDAR data. The point cloud extracted from the TSDF representation is denser compared to the raw LiDAR input. Although the reconstructed surfaces are with artifacts, they still illustrate the major shape of the whole object. It is believed that when the object is closer or there are more LiDAR measurements, the reconstruction quality can be further improved.

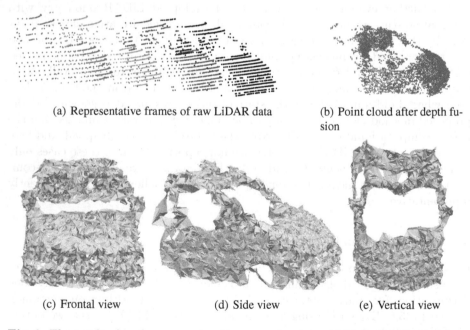

(a) Representative frames of raw LiDAR data

(b) Point cloud after depth fusion

(c) Frontal view   (d) Side view   (e) Vertical view

Fig. 8. The result of surface reconstruction on a dynamic object (car) extracted from Seq. #3. (a) The 1st, 3rd, 6th, and 9th frames of LiDAR data on the object; (b) The point cloud of the object constructed by fusing 10 frames of LiDAR data via depth fusion; (c), (d), and (e) illustrate different views of the reconstructed object surface.

4   Conclusions

In this paper, we have studied the challenging problem of reconstructing dynamic objects from sparse LiDAR data and proposed a pioneer solution named LiDAR odometry oriented to fusion (LOOF). Several lightweight algorithms are employed to exploit the information about dynamic objects from classical ICP-based odometry framework. It is shown that the extracted information is sufficient to reconstruct reasonable object surfaces without adding complex components. It is believed that LOOF can achieve better reconstruction when the depth input is denser. Future works may include faster implementation of the employed algorithms and integrating LOOF and LiDAR-SLAM in a multi-thread framework.

Acknowledgment. This work is partially supported by the National Natural Science Foundation of China (NSFC) under Grant 61602533, NSFC-Shenzhen Robotics Projects (U1613211), The Fundamental Research Funds for the Central Universities, and Science and Technology Program of Guangzhou, China (201510010126).

References

1. Besl, P.J., Mckay, N.D.: A method for registration of 3-D shapes. IEEE Trans. Pattern Anal. Mach. Intell. **14**(2), 239–256 (1992)
2. Cesic, J., Markovic, I., Juric-Kavelj, S., Petrovic, I.: Detection and tracking of dynamic objects using 3D laser range sensor on a mobile platform. In: 2014 11th International Conference on Informatics in Control, Automation and Robotics (ICINCO), pp. 110–119 (2014)
3. Chen, Y., Medioni, G.: Object modelling by registration of multiple range images. Image Vis. Comput. **10**(3), 145–155 (1992)
4. Dewan, A., Caselitz, T., Tipaldi, G.D., Burgard, W.: Motion-based detection and tracking in 3D LiDAR scans. In: 2016 IEEE International Conference on Robotics and Automation (ICRA), pp. 4508–4513. IEEE (2016)
5. Fischler, M.A., Bolles, R.C.: Random sample consensus: a paradigm for model fitting with applications to image analysis and automated cartography. ACM (1981)
6. Kohlbrecher, S., von Stryk, O., Meyer, J., Klingauf, U.: A flexible and scalable SLAM system with full 3D motion estimation. In: 2011 IEEE International Symposium on Safety, Security, and Rescue Robotics, pp. 155–160, November 2011
7. Marton, Z.C., Rusu, R.B., Beetz, M.: On fast surface reconstruction methods for large and noisy point clouds. In: IEEE International Conference on Robotics and Automation, ICRA 2009, pp. 3218–3223. IEEE (2009)
8. Moosmann, F., Stiller, C.: Joint self-localization and tracking of generic objects in 3D range data. In: 2013 IEEE International Conference on Robotics and Automation (ICRA), pp. 1146–1152. IEEE (2013)
9. Newcombe, R.A., Davison, A.J., Izadi, S., Kohli, P., Hilliges, O., Shotton, J., Molyneaux, D., Hodges, S., Kim, D., Fitzgibbon, A.: KinectFusion: real-time dense surface mapping and tracking. In: International Symposium on Mixed Augmented Reality (ISMAR), pp. 127–136 (2011)

10. Opromolla, R., Fasano, G., Rufino, G., Grassi, M., Savvaris, A.: LIDAR-inertial integration for UAV localization and mapping in complex environments. In: 2016 International Conference on Unmanned Aircraft Systems (ICUAS), pp. 649–656, June 2016

11. Pomerleau, F., Krüsi, P., Colas, F., Furgale, P., Siegwart, R.: Long-term 3D map maintenance in dynamic environments. In: 2014 IEEE International Conference on Robotics and Automation (ICRA), pp. 3712–3719. IEEE (2014)

12. Segal, A., Haehnel, D., Thrun, S.: Generalized-ICP. In: Robotics: Science and Systems (RSS), vol. 2 (2009)

13. Velas, M., Spanel, M., Herout, A.: Collar line segments for fast odometry estimation from velodyne point clouds. In: IEEE International Conference on Robotics and Automation (ICRA), pp. 4486–4495. IEEE (2016)

14. Vu, T.D., Burlet, J., Aycard, O.: Grid-based localization and online mapping with moving objects detection and tracking: new results. In: 2008 IEEE Intelligent Vehicles Symposium, pp. 684–689, June 2008

15. Wang, W.T., Wu, Y.L., Tang, C.Y., Hor, M.K.: Adaptive density-based spatial clustering of applications with noise (DBSCAN) according to data. In: International Conference on Machine Learning and Cybernetics, pp. 445–451 (2015)

16. Yang, S.W., Wang, C.C.: Simultaneous egomotion estimation, segmentation, and moving object detection. J. Field Robot. **28**(4), 565–588 (2011)

A Robot Teaching Method Based on Motion Tracking with Particle Filter

Yanjiang Huang[1,2,3], Jie Xie[1,2], Haopeng Zhou[1,2],
Yanglong Zheng[1,2], and Xianmin Zhang[1,2(✉)]

[1] Guangdong Provincial Key Laboratory of Precision Equipment
and Manufacturing Technology, Guangzhou, China
zhangxm@scut.edu.cn
[2] School of Mechanical and Automative Engineering,
South China University of Technology, Guangzhou, China
[3] Research into Artifacts, Center for Engineering,
The University of Tokyo, Chiba, Japan

Abstract. Particle filters are being widely used in tracking single object for its unique advantages. According to former studies, particle filters can solve the tracking problems in the circumstances of nonlinear and non-Gaussian observation. In this paper, we propose a robot teaching method based on multi-objective particle filter for motion tracking, in which targets can be recognized by the value of H in HSV color space. After the targets are recognized, particle filtering is applied to achieve motion tracking. PnP algorithm is another essential part of this method, which can obtain translation and rotation matrix of the moving objects. All these pose information is sent to robot to reproduce the targets' motions. The experiment in this paper is divided into three parts, firstly, a single robot arm achieve the linear motion along three-axis respectively, and then move along a particular trajectory. Finally, to achieve both arms tracking hands movements at the same time. When the single target is tracked, the experimental results are better compared with track hands simultaneously.

Keywords: Robot teaching · Computer vision · Particle filter · PnP pose estimation

1 Introduction

Robotic arms are widely used in industrial manufacturing [1]. However, manufacturing has gradually evolved from large-scale mass production to personalized customization, this makes the robot teaching complex and cumbersome. Although robot manufacturers supply teaching programs for planning, the trajectories use teaching panels [2] or joysticks [3]. The Omni Device sensor is a small arm model that records the angular variation of each joint when its end position changes. In [4], this sensor is used to control the movement of the real robot arm by changing the end position to complete the task of gripping and placing the workpiece, this teaching method is simple and easy to operate, but the accuracy is relatively poor. Off-line programming refers to the use of CAD systems to establish three-dimensional scenes of robots and workpieces after

© Springer International Publishing AG 2017
Y. Huang et al. (Eds.): ICIRA 2017, Part II, LNAI 10463, pp. 647–658, 2017.
DOI: 10.1007/978-3-319-65292-4_56

knowing the relative positional relationship between them. In the CAD system, extract the posture information of teaching point, and through the program, automatically generate the teaching procedures that robot can identify [5]. These traditional methods can't fulfill the demands. If combine the teaching methods with computer vision, the position and posture of targets can be estimated by computer vision, and the teaching of the robot can be achieved with coordinate transformation. It will be much more convenient to achieve robot teaching. To the robotic target grabbing and placing tasks, the literature [6] uses the camera to capture the process of human hand grabbing and placing the target, and repeat the grabbing and placing operations. This method combines computer vision, and has a certain degree of intelligence, but only limited to the completion of specific tasks of teaching operations.

We proposed such a robot teaching method based on particle filter tracking. The idea of a particle filter to apply a recursive Bayesian filter based on sample sets was independently proposed by several research groups [7–10]. According to the principle of particle filtering, color histogram is introduced into the target tracking. When the operator's hand doing different motions, it can be recognized by the value of H in HSV color space, the gesture action is then recorded by the Kinect sensor. Hand is tracked with a particle filter by comparing its histogram with the histograms of the sample positions using the Bhattacharyya distance. And the internal parameters of Kinect depth and color cameras are calibrated. Then we obtain the translation and rotation matrix by using solvePnP in OpenCv library which is on the basis of LM (Levenberg-Marquard). Lastly, all these pose information is transmitted to the robot, according to these data, robot arm achieves motion reproduction. Combined all of what's mentioned above, a vision-based robot teaching system software is designed. Our teaching method has the advantages of making the teaching simple, general and easy, reducing the requirements for operators. What we only need to do is implementing the action we want the robot to reproduce.

The outline of this article is as follows. In Sect. 2, we indicate how to recognize the target and describe particle filtering tracking. The calibration and PnP algorithm are explained in Sect. 3. In Sect. 4, we present some experimental results and finally, in Sect. 5, we summarize.

2   Target Tracking

2.1   Color Recognition of Targets Based on HSV Model

HSV is the most common cylindrical-coordinate representations of points in an RGB color model, which stands for hue, saturation, and value. And it is the simple transformation of device-dependent RGB models, the physical colors they define depend on the colors of the red, green, and blue primaries [11]. Compared with other color spaces, HSV has obvious advantages in distinguishing color. So in this paper, we designed the color judgment method based on the HSV model.

In order to be able to show all the colors, almost all the modern displays use RGB three primary colors mixed principle. Each frame obtained from the video is represented in RGB color mode. So first of all, we need to translate the RGB values of the points in video images to HSV values [12].

In HSV, hue represents color, and the value of H can basically determine a certain color. So we just need to get the H value of target area, and then use the threshold segmentation method, the target area can be distinguished from the rest.

2.2 Particle Filtering

The particle filter uses a weighted particle set to estimate the system state and uses a sequential importance sampling method to update the particle set [13].

The basic concept of particle filter [14] can be summarized as follows: At first, through the center of initial object region, a series of random sample particles and their particle weights is collected. And then, the position of the particles is adjusted and stratified resampling is proceeded to revise the weights and positions of particle. Finally, the object position can be estimated by the minimum mean square estimation. When the sample number of particles gets larger, the posterior probability density of particles obtained by particle filter will approximate to true posterior density of state variables.

2.3 Adaptive Tracking Algorithm Based on Particle Filtering

According to the principle of particle filter, combined with the adaptive color-based particle filter mentioned in [15], this paper improve and adapt it to the hand target tracking algorithm.

Firstly, the histogram of the image captured from video in HSV color space is used as a description of the target model and the candidate model. In our experiments, the histograms are typically calculated in the HSV space using 8 * 8 * 8 bins, and there are 45 particles in total. The proposed tracker employs the Bhattacharyya distance to update the a priori distribution calculated by the particle filter.

Each particle represents a state of the target, including the center point location, size, movement speed and other information of target. And the model can be given as

$$s = \left\{ x, y, \dot{x}, \dot{y}, H_x, H_y, a \right\} \tag{1}$$

where x, y specify the location of the target, $\dot{x}, \dot{y}$ are the movement speed of target in the direction of x, y, H_x, H_y represent target's width and height, and a is the corresponding scale change.

The sample set is propagated through the application of a dynamic model.

$$s_k = As_{k-1} + e_{k-1} \tag{2}$$

where A is the state transition matrix and e_{k-1} is the noise of the system.

Through the state transition Eq. (2) and the particle set of the moment $k - 1$, we can get the particle set of the moment k. The probability density function of the region represented by each particle can be described as

$$p_{s_k^i}^u = C \sum_{i=1}^n K\left(\left\| z_i^* \right\|^2 \right) \delta[b(z_i) - u] \tag{3}$$

where C is normalized coefficient and can be expressed by Eq. (4). K is Epannechnikov Kernel function. z_i^* denotes the normalized distance of the pixel point $z_i(x_i, y_i)$ when the target's center pixel $z_0(x_0, y_0)$ is considered as reference point, and z_i^* can be expressed by Eq. (5). $\delta[b(z_i) - u]$ is defined as Eq. (6), and u is the color index of histogram.

$$C = 1/\sum\nolimits_{i=1}^{n} K\left(\|z_i^*\|^2\right) \tag{4}$$

$$z_i^* = \sqrt{\frac{(x_i - x_0)^2 + (y_i - y_0)^2}{(h_x/2)^2 + (h_y/2)^2}} \tag{5}$$

$$\delta[b(z_i) - u] = \begin{cases} 1 & \text{The pixel value at } z_i \text{ is the component u of the histogram} \\ 0 & \text{other} \end{cases} \tag{6}$$

Similarity estimates need to be made between target area and the latest state observations, accordingly we can get the weight of each particle at moment k.

$$w_k^i = \frac{1}{\sqrt{2\pi}\sigma} e^{-\frac{\left(1-\rho\left[p_{s_k}^i, q\right]\right)}{2\sigma^2}} \tag{7}$$

where $\rho\left[p_{s_k}^i, q\right]$ represents Bhattacharyya coefficient, and it defines as follows

$$\rho(p, q) = \sum\nolimits_{u=1}^{m} \sqrt{p_u(f)q_u} \tag{8}$$

The average of the particle set $E(s_k)$ can be estimated from the weight of the particle, it is exactly the output of target location.

$$E(s_k) = \sum\nolimits_{i=1}^{n} w_k^i s_k^i \tag{9}$$

The last step is Re-sampling for particle degradation may happen in iterative updating.

In our experiment, tracking only one single object can't satisfies our need. According to the above-mentioned single-target tracking principle, the algorithm is directly extended to track multiple targets simultaneously.

3   Pose Estimation

3.1   Robot and Camera Calibration

After using the particle filter algorithm to track the position of hand in the image coordinates, it is necessary to transfer the two-dimensional image coordinate into three-dimensional coordinates of the space, so we need to establish the camera's mathematical model and calibrate camera internal parameters.

Zhang proposed a practical method of camera calibration in [16], the recommended calibration procedure is as follows: (1) print a pattern and attach it to a planar surface; (2) take a few images of the model plane under different orientations by moving either the plane; (3) detect the feature points in the images; (4) estimate all the intrinsic parameters and extrinsic parameters; (5) estimate the coefficients of the radial distortion by solving the linear least-squares; (6) refine all parameters by minimizing.

Kinect 2.0 sensor used in this paper is made up of a color camera, an infrared camera and an infrared generator. GML C++ Camera Calibration Toolbox [17] is a simple and efficient camera calibration tool, which uses Zhang and the form of the calibration plate has been improved. By using the toolbox, we can get the internal matrix as follows:

$$K_D = \begin{bmatrix} 365.370 & 0 & 257.070 \\ 0 & 264.727 & 210.577 \\ 0 & 0 & 1.000 \end{bmatrix} \tag{10}$$

$$K_C = \begin{bmatrix} 1035.749 & 0 & 948.879 \\ 0 & 1035.009 & 552.505 \\ 0 & 0 & 1.000 \end{bmatrix} \tag{11}$$

where K_D is the internal matrix of the infrared camera, and K_C is the internal matrix of the color camera.

So far, we have taken the coordinates of the target in the camera coordinate system. In order to achieve the robot teaching, what we need to obtain is the position of target in the robot coordinate system. Thus, it's necessary to calculate the conversion relationship between the robot coordinate system and the camera coordinate system. Arbitrarily select a point as the test point, respectively measure the pose of test point in these two coordinate system, and then we can get the second transformation matrix between camera coordinate system and robot coordinate system. The relationship can be described as follows:

$$P^R = T_C^R P^C \tag{12}$$

where P^R represents the point's pose in the robot coordinate system, P^C represents the point's pose in the camera coordinate system, and T_C^R is the second transformation matrix between camera coordinate system and robot coordinate system.

3.2 Optimized P4P Algorithm

As early as 1981, Fishchler et al. proposed a perspective n-point positioning problem (PnP) [18], which can accurately identify the pose parameters of the target. In this paper, four feature points that in the same plane are used to estimate the hand position parameters [19], and then using Dog-Leg optimized algorithm to solve the P4P problem. There are two key problems in solving the P4P problem: the corresponding relation between the target feature points and the P4P pose estimation algorithm.

Four coplanar Mark points with known positional relationships are needed to solve the P4P problem, and the Mark point of this paper is made up of four different color dots, the diameter of the dots and the relative positional relationship between the centers are shown in Fig. 1.

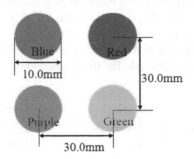

Fig. 1. Mark point (Color figure online)

Where four Mark points are defined as red (254, 0, 11), green (0, 255, 31), purple (206, 2, 237), blue (0, 162, 255) in the RGB color space. Place the Mark point on the back of the palm and define the target object coordinate system $O_o - xyz$, in which the origin O is the center of red Mark, the X-axis is the connection between the centers of red and blue Mark, the Y-axis is the connection between the centers of red and green Mark. The pose estimation is exactly to solve the description of the coordinate system $O_o - xyz$ under the camera coordinate system $O_c - xyz$.

Accurately identifying the coordinate position of the different Mark points in the image is important for establishing the correspondence of feature points on the target object in the object coordinate system and the image coordinate system. The image processing procedure [20] is shown in Fig. 2.

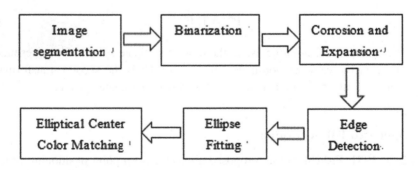

Fig. 2. Processing procedure

Nonlinear least squares methods include Gauss Newton, LM (Levenberg-Marquard) and DL (Dogleg). There is a dedicated solution solvePnP() for PNP problem in OpenCV (Open Source Computer Vision Library), this function uses the LM optimization method for iterative solution. In this paper, the position and posture parameters of the target object are regarded as a nonlinear least squares problem, and we use the function solvePnP() to get the pose information.

4 Experiment and Discussion

4.1 Experiment Platform

As shown in Fig. 3, the robot teaching platform mentioned in this paper includes: PCs, Kinect sensor, Mark marking plate, Baxter robot and RS232-USB adapter cable.

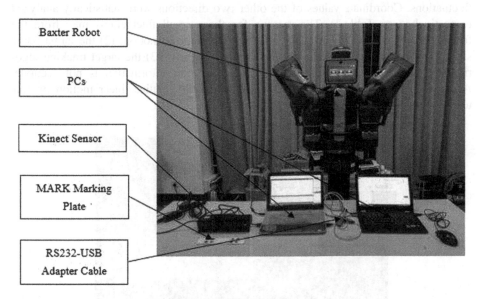

Fig. 3. Experiment platform

There are two PCs used in the experiment, the one whose operating system is Windows 10 is connected with the Kinect sensor, get real-time video and process through the designed program to obtain the target pose information. The other is equipped with Ubuntu system and controls the robot through the ROS robot operating system. The data transmission between them is realized through RS232-USB adapter cable. Baxter robot has two arms, each has seven rotary joints with a torque sensor, and the torque measured by the torque sensor for each joint under external forces, so that the joint can change angle. The Kinect sensor and Mark marking plate have been introduced above.

4.2 Single Arm Linear Motion Experiment

First, we analyze the motion performance of the robot by simple linear motion. When the target moves along the X-axis, Y-axis and Z-axis respectively, the robot arm will move linearly with the same trajectory. The Mark plate was stuck on a hard cardboard and moves with the straight slot (see Fig. 4). We described the relationship between the coordinate value of robot and the time when the target was linearly moving along the three coordinate axes, and took the movement along the X-axis as an example (see Fig. 5). When the target moved in a certain direction, the corresponding coordinate value increased or decreased with time, but because of the time lag caused by the solution of the transmitted data when the robot moved, the coordinate value didn't change continuously, but there was a short time stay. And the scatters on the coordinate axis represented the coordinate value of the target movement, indicating that the robot was moving in the same straight line as the target. The value of coordinates of the other two directions was almost a horizontal line, but there were still a small range of fluctuations. Coordinate values of the other two directions were statistically analyzed respectively (see Table 1). The reasons for these small fluctuations may from the following reasons: (1) the accuracy error of the robot in motion; (2) the target has a slight shaking in other directions when moving linearly; (3) the target tracking algorithm is not stable enough, resulting in the location information is not accurate. According to the analysis, the robot can successfully realize linear motion by this teaching method.

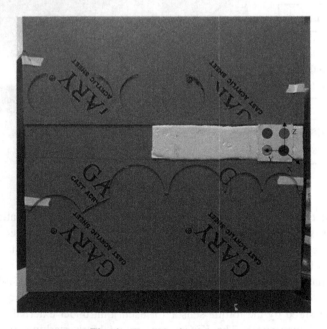

Fig. 4. The experiment target

Table 1. Variance in the other two directions

Moving direction	X		Y		Z	
Variance (mm)	Y	0.103	X	0.148	X	0.278
	Z	0.012	Z	0.004	Y	0.325

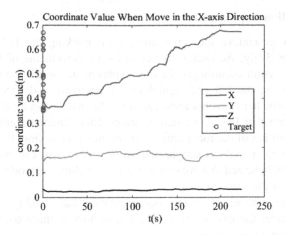

Fig. 5. Coordinate value when move in the X-axis direction

4.3 Single Arm Trajectory Motion Experiment

In this section, the experiment will be more complicated. The target moved along a certain trajectory rather than linear motion along only one direction, including the movement along X-axis, Y-axis and Z-axis. The position information was sent to the robot, and the robot arm reproduced the motion. We put the motion trajectories of target and robot in a same coordinate system (see Fig. 6) in order to compare. From the figure, we found that the movement trajectories of arm and target were almost coincide completely, which means the arm is able to basically realize simple motion tracking. It can be seen that the teaching method has the ability to achieve a single arm of the movement trajectory reproduction, and the performance is ideal.

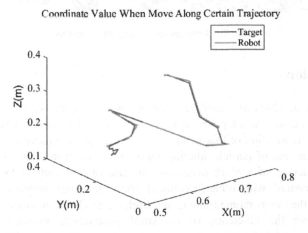

Fig. 6. Coordinate value when move along certain trajectory

4.4   Arms Motion Experiment

Lastly, we need to achieve the robot arms were tracking the left and right hand movements respectively. According to the different movements of hands, the robot should make the action accordingly. In this experiment, we put the two Mark plates mentioned above on the left and right hand respectively and tried to do different positions to test whether the robot arms can track the hands and do the same motions. The results is shown in Fig. 7, we demonstrated four different simple actions, compared with the position of the robot arms after the motion tracking, from which we can see when the operator make a simple action, the robot arm can basically copy the action. The results indicated that we could use this teaching method to control left and right robot arms at the same time.

But when the number of targets becomes two, the stability of the algorithm decreases, and under the circumstance, if the trajectory is more complex, the probability of loss of the target increases.

Fig. 7. Arms motion experiment results

5   Conclusion

In this paper, a novel teaching method combined with computer vision is proposed. The proposed method adds an adaptive particle filtering and the target can be automatically identified. The innovation of this method lies in the use of a more recognizable HSV color space, the use of particle filtering based on color histogram to achieve target tracking, the design of a Mark plate, and the use of LM method to solve the P4P problem. Compared with other traditional robot teaching systems, the improved method avoid the cumbersome training and high demands on the operators, which can greatly improve the efficiency of industrial production, saving manpower and resources.

The method in this paper can be applied in personalized production, and teach the industrial robot to reproduce different easy action according to different production demands.

Future works will be focused on improving the stability of the algorithm, and implementing robot teaching with more complicated action.

Acknowledgments. This work was partially supported by the Scientific and technological project of Guangzhou (2015090330001, 201707010318), partially supported by the Natural Science Foundation of China (51505151), partially supported by the Research Project of State Key Laboratory of Mechanical System and Vibration (MSV201605), and partially supported by the Natural Science Foundation of Guangdong Province (2015A030310239).

References

1. Brogårdh, T.: Present and future robot control development—an industrial perspective. Ann. Rev. Control **31**(1), 69–79 (2007)
2. Ishii, M.: A robot teaching method using Hyper Card system. In: Proceedings of the IEEE International Workshop on Robot and Human Communication, pp. 410–412. IEEE (1992)
3. Ikeura, R., Inooka, H.: Manual control approach to the teaching of a robot task. IEEE Trans. Syst. Man Cybern. **24**(9), 1339–1346 (1994)
4. Ju, Z., Yang, C., Li, Z., et al.: Teleoperation of humanoid baxter robot using haptic feedback. In: International Conference on Multisensor Fusion and Information Integration for Intelligent Systems, pp. 1–6. IEEE (2014)
5. Neto, P., Pires, J.N., Moreira, A.P.: CAD-based off-line robot programming. In: Robotics Automation and Mechatronics, pp. 516–521. IEEE (2010)
6. Lin, H.I., Chiang, Y.P.: Understanding human hand gestures for learning robot pick-and-place tasks. Int. J. Adv. Rob. Syst. **12**, 1 (2015)
7. Gordon, N., Salmond, D., Ewing, C.: Bayesian state estimation for tracking and guidance using the bootstrap filter. J. Guidance Control Dyn. **18**(6), 1434–1443 (2012)
8. Isard, M., Blake, A.: Contour tracking by stochastic propagation of conditional density. In: Buxton, B., Cipolla, R. (eds.) ECCV 1996. LNCS, vol. 1064, pp. 343–356. Springer, Heidelberg (1996). doi:10.1007/BFb0015549
9. Isard, M., Blake, A.: CONDENSATION—conditional density propagation for visual tracking. Int. J. Comput. Vis. **29**(1), 5–28 (1998)
10. Kitagawa, G.: Monte Carlo filter and smoother for non-gaussian nonlinear state space models. J. Comput. Graph. Stat. **5**(1), 1–25 (1996)
11. Schwarz, M.W., Cowan, W.B., Beatty, J.C.: An experimental comparison of RGB, YIQ, LAB, HSV, and opponent color models. ACM Trans. Graph. **6**(2), 123–158 (1987)
12. Cai, Z.Q., Luo, W., Ren, Z.N., et al.: Color recognition of video object based on HSV model. Appl. Mech. Mater. **143–144**, 721–725 (2011)
13. Zhou, Z., Wu, D., Zhu, Z.: Object tracking based on Kalman particle filter with LSSVR. Optik – Int. Journal Light Electron Opt. **127**(2), 613–619 (2015)
14. Chang, C., Ansari, R.: Kernel particle filter for visual tracking. Sig. Process. Lett. IEEE **12**(3), 242–245 (2005)
15. Nummiaro, K., Koller-Meier, E., Gool, L.V.: An adaptive color-based particle filter. Image Vis. Comput. **21**(1), 99–110 (2003)

16. Zhang, Z.: A flexible new technique for camera calibration. IEEE Trans. Pattern Anal. Mach. Intell. **22**(11), 1330–1334 (2000)
17. GML C++ Camera Calibration Toolbox [EB/OL], 4 December 2016. http://graphics.cs.msu. ru/en/node/909
18. Fischler, M.A., Bolles, R.C.: Random sample consensus: a paradigm for model fitting with applications to image analysis and automated cartography. Commun. ACM **24**(6), 381–395 (1981)
19. Horaud, R., Conio, B., Leboulleux, O., et al.: An analytic solution for the perspective 4-point problem. Comput. Vis. Graph. Image Process. **47**(1), 33–44 (1989)
20. Hough, B.P.: Methods and means for recognizing complex patterns. U.S. Patent No. 3069654 (2010)

Self Calibration of Binocular Vision Based on Bundle Adjustment Algorithm

Duo Xu[✉], Yunfeng Gao, and Zhenghua Hou

State Key Laboratory of Robotics and System, Harbin Institute of Technology,
Harbin 150001, Heilongjiang Province, China
xd2345tc@163.com

Abstract. At present, binocular vision system is widely used in unmanned aerial vehicle (UAV). However, there is a large vibration in the process of UAV's flight. It will lead to the change of the position relationship in the binocular vision system. To solve this problem, this paper proposes a method based on Bundle Adjustment optimization algorithm. It is based on the camera calibration data calibrated before out of factory. The rigid body transformation matrix between the two cameras is calibrated and optimized by the position of the feature points and the image information around the feature points. A series of experiments are conducted to test the algorithm. The experiment shows that the distance between calibrated 3D points and the ground truth is less than 5 mm when the length between target and binocular vision system is about to 2 m. It has fully satisfied the needs of the subsequent computation of disparity map. The algorithm has been applied to a UAV binocular vision system.

Keywords: Calibration of binocular vision · Bundle Adjustment · UAV

1 Introduction

At present, the UAV of consumption level has come from the laboratory to the vast number of consumers, it has been widely used in aerial photography. In UAV, binocular vision system has become an indispensable part [1]. It plays an irreplaceable role in navigation, positioning and attitude keeping. In addition, point cloud can be achieved base on the calculation of binocular disparity map, UAV will avoid obstacle in terms of the point cloud. However, in the actual application, it is found that due to the current design of the airfoil, there will be a lot of vibration and noise in the process of flight [2]. The practice shows that there is little influence on the intrinsic parameter of cameras, but a relative large influence happened on the extrinsic parameters of binocular vision system.

Previous researchers have done a lot to solve the problem. But they are all based on specific targets, such as checkerboard [3], circular and so on. However, in the process of UAV flight, the objects are various, it is almost impossible to see a specific target. The calibration algorithm for the non-specific environment, the traditional methods are five point algorithm [4], eight point algorithm [5], Direct Linear Transform [6], Bundle Adjustment [7], etc. These algorithms have been widely used in the field of robot SLAM (Simultaneous Location And Mapping). In recent years, a lot of calibration

© Springer International Publishing AG 2017
Y. Huang et al. (Eds.): ICIRA 2017, Part II, LNAI 10463, pp. 659–670, 2017.
DOI: 10.1007/978-3-319-65292-4_57

algorithms are proposed for special scene [8, 9]. However, the first three of these algorithms are based on no initial transformation matrix between the two cameras. These algorithms extract features and match them to get corresponding feature points. Some mathematical calculations are then performed to obtain results. However, Bundle Adjustment algorithm is applied to get further optimization using the initial results. This paper aims at the problem that the insufficient accuracy of Bundle Adjustment algorithm because of make no use of the pixel value information. We use a algorithm similar to the Bundle Adjustment but with higher accuracy. Both the position of the feature points and the pixel value information of the image are used to calibrate the camera. In the process of the experiment, a stereo calibration is made before the flight of UAV. After a few hours of flight, the proposed algorithm is used for real-time calibration, then the traditional calibration method is used as the ground truth. The answer shows that spatial 3D point re-positioning error is less than 2 mm when the length between target and binocular vision system is about to 2 m. The accuracy has fully satisfied the needs of the subsequent computation of disparity map.

2 Related Works

In order to solve the problem of calibrating extrinsic parameters in an unknown environment with known intrinsic parameters, the researchers have made more work in precision and speed. In the five point algorithm, the researchers first extract five matched feature points, then make use of four formula of multiple view of geometry. $m^T F m = 0$, $\det(F) = 0$, $K^{-T} E K^{-1} = F$, $2 E E^T E - tr(E E^T) E = 0$. The five points are put into the equation and solved to obtain the fundamental matrix, and then the essential matrix is obtained. The advantage of this method is that no matter the geometry relation among the five point, the answer will be obtained. But the disadvantage is that only five points are calculated, there is inevitable mismatch between these five points, so RANSAC [10] algorithm must be used to cull errors. The sampling times of RANSAC algorithm is generally at least 100. And each time we need to solve a large set of linear equations, so the method has no significant advantage in speed. Eight point algorithm is most commonly used, it make use of the property of fundamental matrix $m^T F m = 0$ and eight unknowns of the fundamental matrix. Using 8 equations to solve F,after that $K^{-T} E K^{-1} = F$ is used to calculate essential matrix, and $E = [t]_\times R$, so R and T can be solved. The advantage of the eight point algorithm is that the speed is faster than the five point algorithm, and the number of linear equations is less. Real-time computation is still maintained when adding RANSAC algorithm. The disadvantage is that there must be no 4 points in the same plane among the 8 points, which will lead to degenerate cases. The Direct Linear (DLT) algorithm is usually used when the space points are in the same plane. The algorithm only needs four points in space using homography matrix $x' = Hx$. The relative pose of two cameras can be calculated directly using homography matrix. The problem of the algorithm is similar to the eight point algorithm mentioned above, and the application conditions are too limited. The 4 points in the DLT algorithm must be on the same plane. This greatly limits the usage scenario of the algorithm. The Bundle Adjustment algorithm is a better way to solve this problem. This algorithm is based on nonlinear optimization, The usual cost function is the sum of the 3D projection errors on left and right camera. The optimization variables are the coordinates of the 3D point and transformation matrix between the left and right cameras. Compared with three methods above, the accuracy of this method is greatly improved, usually the re-projection error is within the range of 0. 4 pixels. The algorithm use Levenberg-Marquardt (LM) iteration algorithm until a certain number of iterations, or the decline of the cost function is less than a certain number. At this time, the rotation matrix and translation matrix are the final results, which is the optimal solution under the given initial value. However, the biggest problem is that the above four algorithms do not make full use of pixel value information, just use the location information of the feature points, which will cause the precision can not reach a higher level, and the matching error will cause a certain impact on the results. In this paper, in order to further improve the calibration accuracy, the cost function is not only considering the position of the feature points, but also taking into account the pixel information around the feature points. Joined the sub-pixel level optimization, make up for the Bundle Adjustment algorithm does not make full use of pixel value information.

3   Binocular Vision Model

3.1   Camera Model

The classic camera model uses the principle of pinhole. 3D point $P(X_w, Y_w, Z_w)$ is projected into image point $p(u, v)$. The relation between P and p is:

$$s \cdot p = KK[R \quad T]P \tag{1}$$

In the formula, KK is the intrinsic camera parameter, $p = (u, v, 1)^T$ Is the homogeneous coordinates of the image coordinate and $P = (X_w, Y_w, Z_w, 1)^T$ is the homogeneous coordinates in the world coordinate system. s is a proportional coefficient. R and T are respectively called rotation matrix and translation matrix.

3.2   Binocular Camera Model

As shown in the Fig. 1a 3D point coordinate can be calculated intrinsic parameters,the transformation matrix between left and right camera and the projected pixel coordinates in both cameras. $P(x, y, z)$ is the 3D coordinate and $p_l(u_l, v_l, 1)^T, p_r(u_r, v_r, 1)^T$ projected left and right pixel homogeneous coordinates, so the 3D point coordinate can be calculated as follows:

$$\left(\begin{bmatrix} p_1^\wedge \end{bmatrix} M_l \\ \begin{bmatrix} p_2^\wedge \end{bmatrix} M_r \right) P = 0 \tag{2}$$

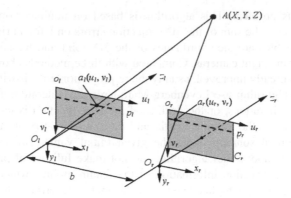

Fig. 1. Binocular camera model

$p_1 = KK_l^{-1}p_l$, $p_2 = KK_r^{-1}p_r$, $M = [R \quad T]$, $p^\wedge$ corresponds to its skew symmetric matrix

$$p^\wedge = \begin{pmatrix} 0 & -p_3 & p_2 \\ p_3 & 0 & -p_1 \\ -p_2 & p_1 & 0 \end{pmatrix} \tag{3}$$

4   Bundle Adjustment Algorithm

Bundle Adjustment originally originated in geodetic survey [7], and later was quickly applied to computer graphics. The Bundle Adjustment can be used to map the observed image position and the predicted image location point to the minimum error. Therefore, the minimum error is achieved by nonlinear least square method, and the best iterative algorithm is Levenberg -Marquardt (LM) algorithm [11]. Due to its simple implementation and damping strategy, it can quickly converge to the steady state from the initial hypothesis, thus becoming the best method.

The cost function of the usual Bundle Adjustment algorithm is the sum of error of predicted pixel coordinate and observed pixel coordinate.

$$\arg\min \sum_{i=1}^{m} \sum_{j=1}^{n} \left\| h(C_i, X_j) - x_{ij} \right\| \tag{4}$$

C_i is intrinsic parameter, X_j is 3D point coordinate and h() is project function, and x_{ij} is the observed feature coordinate. But the biggest problem of this traditional method is only used of feature point coordinate information. It abstract pixel information into the coordinates of feature points. So it will lose all the information of pixels. Once the feature point extraction is not accurate, the iteration will not converge. To solve the problems above, this paper proposes a new optimization method.

5 Improved Bundle Adjustment Algorithm

The method is mainly divided into two steps, the first step of the algorithm is still the most common Bundle Adjustment algorithm. The cost function of the first step is as follows:

$$T_{LR} = \arg\min \frac{1}{2} \sum_{i=1}^{m} \sum_{j=1}^{n} \left\| \pi(KK, T_{LR}, P_j) - x_{ij} \right\|^2 \tag{5}$$

KK is the intrinsic camera parameter, T_{LR} is the transformation matrix between the left and right camera. P_j is the coordinate of the 3D point. the first step is also use re-projection error as the cost function The second step is to fuse the pixel information into the cost function. The results of the first step is used as the initial value into the second step optimization. The cost function of the second step is pixel difference of the pixel on the left and right images with the same 3D point. The cost function of second step is as follows:

$$T_{LR} = \arg\min \frac{1}{2} \sum_{i=1}^{m} \left\| \delta I(\xi, P_i) \right\|_2^2 \tag{6}$$

The formula is a nonlinear optimization of T_{LR}, Gauss-Newton method will be applied here. The T_{LR} obtained by previous optimization is the initial value of the second iteration, and then update by twist matrix [12]. Step ξ and residual e are calculated using Inverse Compositional Image Alignment Algorithm [13].

$$\delta I(\xi, P_i) = I_R\left(\pi\left(\overset{\vee}{T_{LR}}, KK_R, P_i \right) \right) - I_L(\pi(T(\xi), KK_L, P_i)) \tag{7}$$

$I(u, v)$ is an small image slice which a feature is centered and the area is 3×3. $\pi(\cdot)$ represents projection function below.

$$z_c \begin{pmatrix} u \\ v \\ 1 \end{pmatrix} = KK[R \quad T] \begin{pmatrix} x \\ y \\ z \\ 1 \end{pmatrix} \tag{8}$$

That is, the 3D point project to image coordinates u, v. The update method of transformation matrix is:

$$\overset{\vee}{T_{LR}} = \overset{\vee}{T_{LR}} \cdot T(\xi)^{-1} \tag{9}$$

To find the best update step $T(\xi)^{-1}$, calculate the partial derivative of formula 6 and set it to zero. The result is below

$$\sum_{i=1}^{n} \frac{\partial \delta I(\xi, P_i)}{\partial \xi}^T \delta I(\xi, P_i) = 0 \tag{10}$$

To solve formula 10, we make Taylor expansion at zero:

$$\delta I(\xi, P_i) = \delta I(0, P_i) + \frac{\partial \delta I(0, P_i)}{\partial \xi} \xi \tag{11}$$

The Jacobian matrix is 9×6, because the transformation matrix have six dimensions of freedom. Apply the chain rule:

$$\frac{\partial \delta I(\xi, P_i)}{\partial \xi} = \frac{\partial I(a)}{\partial a}\bigg|_{a=u_i} \cdot \frac{\partial \pi(b)}{\partial b}\bigg|_{b=p_i} \cdot \frac{\partial T(\xi)}{\partial \xi}\bigg|_{\xi=0} \cdot P_i \tag{12}$$

$\frac{\partial I(a)}{\partial a}$ is pixel depth derivation of pixel position. It is like optical flow. In this paper, bi-linear interpolation is used to obtain the accurate sub-pixel values:

$$I(u, v) = \frac{1}{4} \sum_{i=0}^{3} I(u_i, v_i) |(u - u_{3-i})(v - v_{3-i})| \tag{13}$$

$\frac{\partial \pi(b)}{\partial b}$ is pixel position derivation of camera coordinates, $\pi(\cdot)$ is that

$$u = \frac{1}{z_c} (f_x x_c + z_c u_0)$$
$$v = \frac{1}{z_c} (f_y y_c + z_c v_0) \tag{14}$$

So u and v derivation of xc, yc, zc:

$$\frac{\partial \pi(b)}{\partial b} = \begin{pmatrix} \frac{f_x}{z_c} & 0 & -\frac{f_x x_c}{z_c^2} \\ 0 & \frac{f_y}{z_c} & -\frac{f_y y_c}{z_c^2} \end{pmatrix} \tag{15}$$

$\frac{\partial T(\xi)}{\partial \xi}$ is camera coordinate derivation of camera position which is 3×6. Commonly used transformation matrix (formula 16) is special European Group SE(n) is a member of Lie Group. Lie Group is a group of continuous properties.

$$SE(3) = \left\{ T = \begin{bmatrix} R & t \\ 0^T & 1 \end{bmatrix} \in \mathfrak{R}^{4 \times 4} | R \in SO(3), t \in \mathfrak{R}^3 \right\} \tag{16}$$

Each transformation matrix has six degrees of freedom. The rotation matrix belongs to another kind of Li Group called special orthogonal group SO(3).

$$SO(n) = \{R \in \mathfrak{R}^{n \times n} | RR^T = I, \det(R) = 1\} \qquad (17)$$

The corresponding Lie algebra is in $\mathfrak{R}^6$

$$se(3) = \{\Xi = \xi^\wedge \in R^{4 \times 4} | \xi \in R^6\} \qquad (18)$$

The relationship between Li Group and Lie Algebra is an exponential mapping relationship, namely applying exponential to Lie Algebra is Lie Group, applying logarithm base e to Lie Group is Lie Algebra. Lie Group is not closed in plus and minus but Lie Algebra is closed [14]. Here $\xi^\wedge$ is not skew-symmetric matrix above.

$$\xi^\wedge = \begin{bmatrix} \rho \\ \phi \end{bmatrix}^\wedge = \begin{bmatrix} \phi^\wedge & \rho \\ 0^T & 0 \end{bmatrix} \qquad (19)$$

$\phi^\wedge$ is the skew-symmetric matrix of $[\phi_1 \; \phi_2 \; \phi_3]^T$. We set $q = \exp(\xi^\wedge)P$, that

$$\frac{\partial q}{\partial \xi} = \frac{\exp(\delta \xi^\wedge) \exp(\xi^\wedge)P - \exp(\xi^\wedge)P}{\delta \xi} = \begin{bmatrix} I_{3 \times 3} & -q_3^\wedge \end{bmatrix} \qquad (20)$$

Put the formula 11 into the formula 10, and the Jacobi matrix into formula 10 we can get

$$\left(\sum_{i=1}^n J^T J \right) \xi = - \sum_{i=1}^n J^T \delta I(0, u_i) \qquad (21)$$

The twist matrix can be obtained by solving the above equation. After that the iteration can be calculated according to formula 9.

6 Experiments and Results

6.1 Binocular Calibration Steps

The binocular calibration process is shown in Fig. 2:

(1) Collect left and right images of cameras and extract features.
(2) Features matching and calculate 3D points.
(3) Culling unusual 3D points which is mismatched.
(4) Calibrate Binocular Cameras.

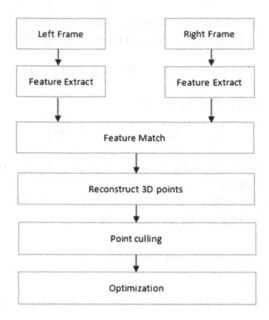

Fig. 2. Procedure of calibration

6.2 Algorithm Accuracy Experiment

This experiment uses USB camera OV2710 to compose binocular vision system (Fig. 3). UAV is M100 of DJI (Fig. 4). In the experiment, the binocular vision system of UAV is calibrated before the UAV flies for a long time. When UAV flies for a long time, using proposed algorithm to calibrate binocular cameras, Then let the UAV return. Using Matlab Calibration Toolbox to calibrate the binocular vision system as ground truth. Compare the two parameters and determinant the accuracy of the algorithm. Method of measure the accuracy of proposed algorithm is as follows (Fig. 5):

(1) The calibration method of Zhang is used to calibrate the initial value of the pose of the camera.
(2) After a long time flight (about 1 h), calibrate binocular camera using algorithm proposed in this paper.
(3) After UAV fly back, using Zhang's calibration method calibrate again as ground truth.
(4) Extract the corner of checkerboard in step (3) and calculate 3D points using the calibrated result in step (2)
(5) Compare the distance of the same point between result in step (3) and step (4) to determinate the accuracy of the algorithm.

Figure 6 is the image after correction about optical axis parallel, the upper two image is using the initial calibration result and the under two image is using the calibrated result. It is obvious that after a long time flight of UAV, the camera rotate around the X axis and the pitch angle is obvious. The same point is not in the same line of left and right images. This will cause matching error in disparity image. But after the calibration this situation has been significantly eased, the same point is in the same line of pixels (Table 1).

Fig. 3. OV2710 camera

Fig. 4. M100 UAV of DJI

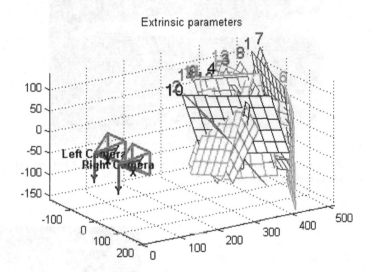

Fig. 5. Calibration result using Zhang's method

Table 1. Matrix of binocular vision

Original	Calibrated by algorithm	Ground truth
$R = \begin{bmatrix} 0.999994 & 0.000084 & 0.003417 \\ -0.000030 & 0.999877 & -0.015708 \\ -0.003418 & 0.015708 & 0.000871 \end{bmatrix}$	$R = \begin{bmatrix} 0.999994 & 0.000661 & 0.003367 \\ -0.000621 & 0.999929 & -0.011890 \\ -0.003374 & 0.011888 & 0.999923 \end{bmatrix}$	$R = \begin{bmatrix} 0.999994 & 0.000639 & 0.003395 \\ -0.000599 & 0.999928 & -0.011969 \\ -0.003403 & 0.011967 & 0.999922 \end{bmatrix}$
$T = \begin{bmatrix} -0.120207 & -0.000132 & -0.000434 \end{bmatrix}$	$T = \begin{bmatrix} -0.120207 & -0.000064 & -0.000438 \end{bmatrix}$	$T = \begin{bmatrix} -0.121305 & -0.000143 & -0.000521 \end{bmatrix}$

Table 2. Original coordinates and relocation coordinates mm

No		1	2	3	...	52	53	54
Original coordinates	X	0	0	0		320	320	320
	Y	0	40	80		120	160	200
	Z	0	0	0		0	0	0
Relocation coordinates	X	0	−0.02	−0.04		319.93	319.91	319.89
	Y	0.80	40.8	80.8		120.96	160.96	200.96
	Z	4.93	4.91	4.89		3.26	3.23	3.21

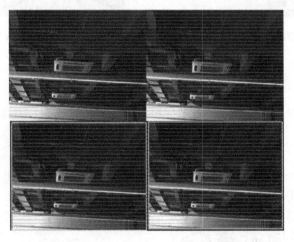

Fig. 6. Photo of left and right cameras calibrated or not after rectify 1, 2 left and right photos not calibrated 3, 4 left and right photos calibrated

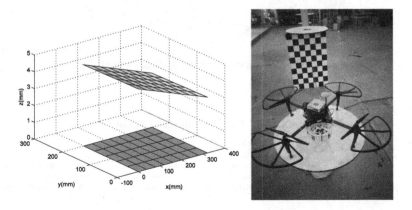

Fig. 7. Checkerboard relocation (the blue is ground true, red one is calibration result) (Color figure online)

When calculating the re-projection error of 3D point in the checkerboard, it is placed at 2 m away from the binocular camera. Using Zhang's algorithm as ground truth and put it as the origin of the world coordinate system (Fig. 7 blue one). After that extract the grid corners and in the checkerboard and calculate 3D points using transformation matrix calculated by proposed algorithm result when UAV was in the air (Fig. 7 red one). As shown in Fig. 7, it can be found that the target re-location error is less than 5 mm (Table 2).

6.3   Comparative Experiments of Different Algorithms

In order to differences on the performance between algorithms, we design a set of experiments to measure the accuracy and speed of 5 point algorithm, 8 point algorithm, DLT algorithm and Bundle Adjustment algorithm. The result are shown in Fig. 8. The algorithm proposed in this paper has a longer running time (CPU i5-5200U), but the precision is much higher than other algorithms (Fig. 8).

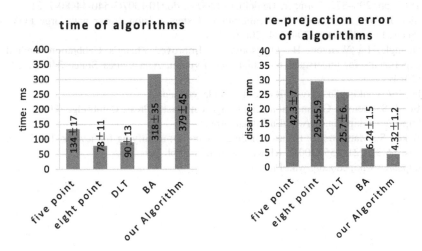

Fig. 8. Comparison between algorithms in accuracy and speed

7   Conclusions

Calibration algorithm is widely used in binocular camera. This calibration algorithm can calibrate extrinsic parameters which does not limited to specific scenes and specific targets. When the distance between target and camera is about to 2 m, the 3D point re-localization error is less than 5 mm. The results show that the algorithm improves the accuracy of calibration.

Acknowledgments. This work was financially supported by the National Natural Science Foundation of Heilongjiang province under Grant QC2014C072, and Postdoctoral Science Foundation of Heilongjiang under Grant LBH-Z14108.

References

1. Zhou, G., Fang, L., Tang, K., Zhang, H., Wang, K.: Guidance: a visual sensing platform for robotic applications. In: IEEE Conference on Computer Vision & Pattern Recognition Workshops (2015)
2. Liang, J., Zhang, W.G., Wang, X.N., Du, J.H.: Development for restraining oscillation device of the UAV model in 4 m × 3 m the wind tunnel. J. Exp. Fluid Mech. 21(4), 65–70 (2007)
3. Zhang, Z.: A flexible new technique for camera calibration. IEEE Trans. Pattern Anal. Mach. Intell. 22(11), 1330–1334 (2000)
4. Li, H., Hartley, R.: Five-point motion estimation made easy. In: International Conference on Pattern Recognition, vol. 1, pp. 630–633. IEEE (2006)
5. Longuet-Higgins, H.C.: A computer algorithm for reconstructing a scene from two projections. Nature 293, 133–135 (1981)
6. Hartley, R., Zisserman, A.: Multiple View Geometry in Computer Vision. Cambridge University Press, Cambridge (2004)
7. Triggs, B., McLauchlan, P.F., Hartley, R.I., Fitzgibbon, A.W.: Bundle adjustment — a modern synthesis. In: Triggs, B., Zisserman, A., Szeliski, R. (eds.) IWVA 1999. LNCS, vol. 1883, pp. 298–372. Springer, Heidelberg (2000). doi:10.1007/3-540-44480-7_21
8. Sun, J., Wu, Z., Liu, Q.: Field calibration of stereo vision sensor with large FOV. Opt. Precis. Eng. 22(11), 1330–1334 (2009)
9. Guangle, L., Wenyou, H., Qingsong, L.: Improved Zhang's Calibration method and experiments for underwater binocular stereo-vision. Acta Optica Sinica 34(12), 1215006 (2014)
10. https://en.wikipedia.org/wiki/Random_sample_consensus
11. Boyd, S.: Convex Optimization. Cambridge University Press, Cambridge (2013)
12. https://en.wikipedia.org/wiki/Euclidean_group
13. Baker, S., Matthews, I.: Lucas-Kanade 20 years on: a unifying framework. Int. J. Comput. Vis. 56(3), 221–255 (2004). Kluwer Academic Publishers
14. https://en.wikipedia.org/wiki/Lie_alGebra

Statistical Abnormal Crowd Behavior Detection and Simulation for Real-Time Applications

Wilbert G. Aguilar[1,4,5(✉)], Marco A. Luna[2], Hugo Ruiz[1,8],
Julio F. Moya[2], Marco P. Luna[3,7], Vanessa Abad[6],
and Humberto Parra[1,7]

[1] Dep. DECEM, Universidad de las Fuerzas Armadas ESPE,
Sangolquí, Ecuador
wgaguilar@espe.edu.ec
[2] Dep. DEEE, Universidad de las Fuerzas Armadas ESPE,
Sangolquí, Ecuador
[3] Dep. Tierra y Construcción, Universidad de las Fuerzas Armadas ESPE,
Sangolquí, Ecuador
[4] CICTE Research Center, Universidad de las Fuerzas Armadas ESPE,
Sangolquí, Ecuador
[5] GREC Research Group, Universitat Politècnica de Catalunya,
Barcelona, Spain
[6] Universitat de Barcelona, Barcelona, Spain
[7] PLM Research Center, Purdue University, Indiana, USA
[8] Universidad Politécnica de Madrid, Madrid, Spain

Abstract. This paper proposes a low computational cost method for abnormal crowd behavior detection with surveillance applications in fixed cameras. Our proposal is based on statistical modelling of moved pixels density. For modelling we take as reference datasets available in the literature focused in crowd behavior. During anomalous events we capture data to replicate abnormal crowd behavior for computer graphics and virtual reality applications. Our algorithm performance is compared with other proposals in the literature applied in two datasets. In addition, we test the execution time to validate its usage in real-time. In the results we obtain fast execution time of the algorithm and robustness in its performance.

Keywords: Abnormal crowd · Surveillance · Image analysis · Real-time applications

1 Introduction

According to [1] abnormal or unusual behavior are somehow interesting that catch the attention of human observers, and often quite easy to identify. In recent years, surveillance systems have been improved, the main reasons are the technological

© Springer International Publishing AG 2017
Y. Huang et al. (Eds.): ICIRA 2017, Part II, LNAI 10463, pp. 671–682, 2017.
DOI: 10.1007/978-3-319-65292-4_58

advances and the increasing availability of monitoring cameras in different environments [2–4]. The challenge of abnormal or unusual behavior detection for surveillance camera operators is that it requires a lot of attention without stop and is an exhausting process because the abnormal events occur with a low probability, making that the major part of the effort be wasted watching regular videos [1, 5]. Abnormal crowd behavior detection is a topic of interest in multiple fields as: computer vision, real-time applications [6–8] crowd behavior analysis, and others; capturing the attention of several research groups [9].

In the field of computer graphics [10], computer simulations [11, 12], and virtual reality [13]; models of people behavior have been used to simulate crowds in different scenarios [14, 15]. According to [16], the behavior of people in crowds could be different depending on situation; it makes more difficult to represent the dynamic models of a crowd.

Our approach is focused on fast abnormal events detection based on analysis of inter-frame pixel motion. We obtain a model of pixel density behavior in human crowds using statistical concepts, and replicate the abnormal crowd behavior in computer simulations.

This paper is organized as follows: Sect. 2 describes the related work on abnormal crowd behavior detection. Next, the statistical modelling and our algorithm are presented in the Sect. 3. In Sect. 4, we present the experimental results followed by the summary. Finally conclusions and future works are presented in Sect. 5.

2 Related Works

In the literature, several research groups have proposed methods to deal with the problem of abnormal human behavior detection in different contexts [1] like suspicious events [17], irregular behavior [18], abnormal behavior [19, 20], and others. ed to. Multiple authors have divided all these methods in two main approaches: model based detection and particle advection based detection. The choose of the model will depend of the density of the crowd to be analyzed [21].

Model based methods analyze individuals behavior using detection and segmentation, works like [22, 23] analyze the motion patterns using people tracking. In [24, 25] the extracted trajectories were used to represent normal patterns, atypical values were considered abnormal.

Particle advection based models [26–28] represents a holistic vision of a crowd. Common approaches in this field include: optical flow [14, 27, 28], gradients [29], spatio-temporal features [30], and others.

Our method for anomaly detection uses a particle advection based algorithm; we analyze pixel densities with statistical modelling to identify abrupt movements in the image and detect abnormal behavior.

3 Our Approach

For abnormal crowd behavior detection, we divide our method in three stages: Motion detection and motion analysis, statistical modelling of normal behavior, and abnormal crowd behavior detection and simulation.

3.1 Motion Detection and Motion Analysis

For motion detection, we use algorithm based on image differences in consecutive frames. First, we three consecutive frames are read: previous frame (or first frame), current frame and next frame (the algorithm could have one frame of delay). Then, we applicate RGB to Gray transformation to each frame for color channels simplification. We operate between images as color matrices obtaining the absolute difference between consecutive frames using the Eqs. 1 and 2:

$$absdiff_1 = |Current\ Frame - Previous\ Frame| \tag{1}$$

$$absdiff_2 = |Next\ Frame - Current\ Frame| \tag{2}$$

We apply the logical operation AND to these differences, finding pixels that have moved in current frame. This algorithm is graphically presented in Fig. 1.

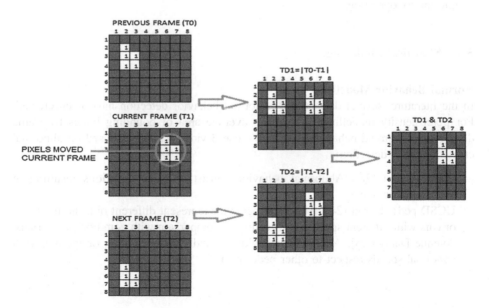

Fig. 1. Motion detection algorithm

Fig. 2. Motion pixel detection

With moved pixels, adjacent pixels are grouped to get little contours. Background subtraction approaches like [31] proposes that with the increase of the speed in the motion, increases the density of these grouped pixels. We group this moved pixels inside bounding boxes as show the Fig. 2.

The Fig. 2 shows that the algorithm generates bounding boxes around moved pixels; the computational advantage of this approach is to reduce the image analysis to simple matrix operations.

3.2   Statistical Modelling

Normal Behavior Modelling

In the literature, several datasets for abnormal behavior detection have been created. For pixel density modelling, we get the average area of bounding boxes by frame during normal crowd behavior scenes. We use 3 video datasets focused on abnormal crowd behavior detection:

- UMN dataset [28]: Abnormal behavior consist in panic situations resulting in evacuation or scape.
- UCSD ped1 dataset [24]: Abnormal behavior represent different objects like bikes or cars with different speed and direction in comparison with common pedestrians.
- Avenue Dataset [5]: Abnormal behavior are individuals running or moving with abnormal speeds respect to other pedestrians.

Pixel density histograms

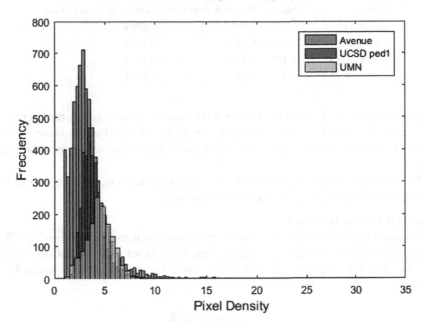

Fig. 3. Histograms of pixel density

In order to have small variation in data, we set all videos to the same size (320 × 240 pixels). We tabulate the data for each frame. In total, 12154 frames were analyzed. The results of histograms are presented in Fig. 3.

Additionally we compute the mean, standard deviation and coefficient of variation given by the Eqs. 3, 4 and 5 respectively:

$$\mu = \frac{1}{N} \sum_{i=1}^{N} x_i \tag{3}$$

$$\sigma = \sqrt{\frac{1}{N} \sum_{i=1}^{N} (x_i - \mu)^2} \tag{4}$$

$$CV = \frac{\sigma}{\mu} \tag{5}$$

Table 1. Statistical values of pixel density

Dataset	Mean	Standard deviation	Coefficient of variation
Avenue [5]	3.025	**2.087**	**0.6899**
UCSD ped1 [24]	3.6130	1.1341	0.3139
UMN [28]	**4.474**	1.9378	0.4331
All	3.401	1.888	0.5554

In the Eqs. 3 and 4, N represents the total number of data, and x_i represents each single data ($i = 1, 2, 3, \ldots N$). As we can see in Fig. 3, all datasets present similar trends. Data of mean, standard deviation and coefficient of variability are tabulated in Table 1.

According to standard deviation and coefficient of variation, Avenue dataset [5] is the most unstable for the application of this method.

Abnormal Crowd Detection

For abnormal crowd behavior we compute the atypical values in pixel densities. When an atypical value is detected, we will assume that crowd experiments an abnormal behavior. For this, we calculate the upper limit for atypical values given by the equation:

$$Ul = Q_3 + 3IQR \tag{6}$$

Where Q_3 represent the third quartile and IQR is the interquartile range. We only take values over the upper limit because there are not atypical values under lower limit. The interquartile range (IQR) is defined as:

$$IQR = Q_3 - Q_1 \tag{7}$$

Where Q_1 is the first quartile.

The value of upper limit for extreme atypical values in each dataset is presented in Table 2.

The Table 2 shows that UMN dataset have the maximum upper limit, so we will take this value to implement the algorithm.

Table 2. Upper limits for different datasets

Dataset	Upper limit
Avenue [5]	9.76
UCSD ped1 [24]	8.26
UMN [28]	**9.93**

3.3 Abnormal Behavior Detection and Simulation Algorithm

With the atypical values calculated, we determinate that abnormal crowd behavior occurs when different pixels exceed upper limit density.

Simulation

We obtain the center points of bounding boxes during abnormal crowd in function of time that they appear. In video screens the position $y = 0$ begins at up-left corner, for this reason, we must subtract 240 minus "y" positions to match it with Cartesian coordinates. We present graphical results in Fig. 4.

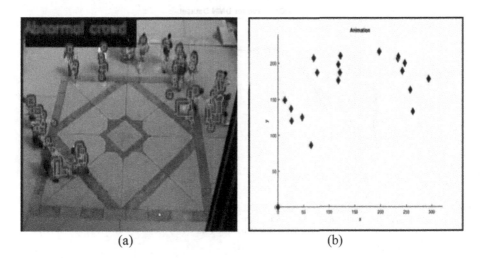

Fig. 4. Simulation of abnormal crowd. (a) Video detection (b) Simulation points

4 Results and Discussion

4.1 Execution Time

To evaluate the execution time we run the program in two computers with the following characteristics:

- PC1: Processor Intel (R) Core (TM) i3, 2.4 GHz, 4.0 GB of RAM memory
- PC2: Processor Intel (R) Core (TM) i7, 2.8 GHz, 7.7 GB of RAM memory

The results of evaluation are presented in Table 3.

Table 3 shows that in computers with low and high features the algorithm works fast, thus the algorithm can be used in real time applications.

Table 3. Execution time of the algorithm in datasets

Dataset	Execution time (fps)	
	PC1	PC2
Avenue [5]	**72.6**	**91.3**
UCSD ped1 [24]	71.02	88.6
UMN [28]	71.39	89.2

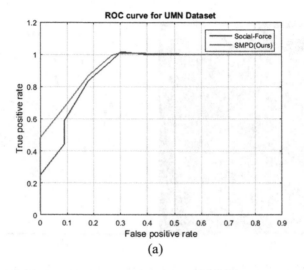

(a)

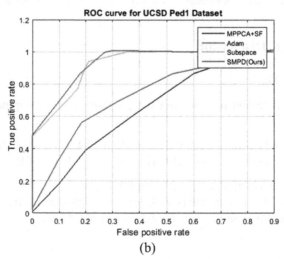

(b)

Fig. 5. ROC curves for abnormal crowd behavior datasets. (a) UMN dataset. (b) UCSD ped1 dataset

4.2   Algorithm Performance

For algorithm evaluation, we count the number of bounding boxes that present atypical values, defining a threshold of counted atypical values to determinate an abnormal behavior. In each threshold we tabulate the true positive and false positive rates. We compare our algorithm based on statistical pixel density model (SPDM) in two data-sets: UMN [28] and UCSD ped1 [24] according to their respective ground–truth annotations. In the first dataset we evaluate our approach and social-force model presented in [28]. In the second dataset we compare our method with: MPPCA + SF [24], Adam [32], y subspace [33]. Results are presented in Fig. 5.

In Fig. 5, our algorithm presents better performance that methods presented in the state of art for frame level analysis in two datasets. In UCSD ped1 [24], many approaches determinate the performance at pixel level, however this paper not deal with this task.

5   Conclusions and Future Work

Despite the videos for statistical modeling are different, they present similar features during density pixel estimation; it allows generating a global model that works in different applications.

Due to the low computational cost of image processing, the proposed method for motion detection works for real time applications as show the Table 3.

In our method, we use abrupt changes in pixel density to detect anomalies. This proposal for abnormal crowd behavior detection is robust compared with approaches in the literature.

In the future, we will improve the detection including pixel level analysis; however it is necessary to include more variables in the process. This will allow improving simulation and avoidance systems like [34–36].

Acknowledgement. This work is part of the project Perception and localization system for autonomous navigation of rotor micro aerial vehicle in gps-denied environments, VisualNav-Drone, 2016-PIC-024, from the Universidad de las Fuerzas Armadas ESPE, directed by Dr. Wilbert G. Aguilar.

References

1. Popoola, O.P., Wang, K.: Video-based abnormal human behavior recognition—a review. IEEE Trans. Syst. Man Cybern. Part C Appl. Rev. **42**, 865–878 (2012)
2. Aguilar, W.G., Luna, M.A., Moya, J.F., Abad, V., Ruiz, H., Parra, H., Angulo, C.: Pedestrian detection for UAVs using cascade classifiers and saliency maps. In: Rojas, I., Joya, G., Catala, A. (eds.) IWANN 2017. LNCS, vol. 10306, pp. 563–574. Springer, Cham (2017). doi:10.1007/978-3-319-59147-6_48
3. Aguilar, W.G., Luna, M.A., Moya, J.F., et al.: Pedestrian detection for UAVs using cascade classifiers with meanshift. In: 2017 IEEE 11th International Conference on Semantic Computing, pp. 509–514. IEEE (2017)
4. Rahmalan, H., Nixon, M.S., Carter, J.N.: On crowd density estimation for surveillance. In: IET Conference on Crime Security, pp. 540–545. IEEE (2006)
5. Lu, C., Shi, J., Jia, J.: Abnormal event detection at 150 fps in matlab. In: Proceedings of IEEE International Conference on Computer Vision, pp. 2720–2727 (2013)
6. Aguilar, W.G., Angulo, C.: Real-time model-based video stabilization for microaerial vehicles. Neural Process. Lett. **43**, 459–477 (2016)
7. Aguilar, W.G., Angulo, C.: Real-time video stabilization without phantom movements for micro aerial vehicles. EURASIP J. Image Video Process. **2014**, 46 (2014)
8. Aguilar, W.G., Angulo, C.: Robust video stabilization based on motion intention for low-cost micro aerial vehicles. In: 2014 11th International Multi-Conference Systems, Signals & Devices (SSD), pp. 1–6 (2014)
9. Silveira Jr., J., Musse, S.R., Jung, C.R.: Crowd analysis using computer vision techniques. IEEE Sig. Process. Mag. **27**, 66–77 (2010)
10. Cabras, P., Rosell, J., Pérez, A., et al.: Haptic-based navigation for the virtual bronchoscopy. IFAC Proc. **18**, 9638–9643 (2011)
11. Aguilar, W., Morales, S.: 3D environment mapping using the kinect V2 and path planning based on RRT algorithms. Electronics **5**, 70 (2016)
12. Aguilar, W.G., Morales, S., Ruiz, H., Abad, V.: RRT* GL based optimal path planning for real-time navigation of UAVs. In: Rojas, I., Joya, G., Catala, A. (eds.) IWANN 2017. LNCS, vol. 10306, pp. 585–595. Springer, Cham (2017). doi:10.1007/978-3-319-59147-6_50
13. Shendarkar, A., Vasudevan, K., Lee, S., Son, Y.-J.: Crowd simulation for emergency response using BDI agent based on virtual reality. In: Proceedings of the 38th Conference Winter Simulation, pp. 545–533 (2006)
14. Andrade, E.L., Blunsden, S., Fisher, R.B.: Modelling crowd scenes for event detection. In: Pattern Recognition, ICPR 2006, vol. 1, pp. 175–178 (2006)
15. Bellomo, N., Dogbé, C.: On the modelling crowd dynamics from scaling to hyperbolic macroscopic models. Math. Model Methods Appl. Sci. **18**, 1317–1345 (2008)
16. Lemercier, S., Jelic, A., Kulpa, R., et al.: Realistic following behaviors for crowd simulation. In: EUROGRAPHICS (2012)
17. Lavee, G., Khan, L., Thuraisingham, B.: A framework for a video analysis tool for suspicious event detection. Multimed. Tools Appl. **35**, 109–123 (2007)

18. Zhang, Y., Liu, Z.-J.: Irregular behavior recognition based on treading track. In: 2007 International Conference on Wavelet Analysis and Pattern Recognition, pp. 1322–1326. IEEE (2007)
19. Benezeth, Y., Jodoin, P.-M., Saligrama, V., Rosenberger, C.: Abnormal events detection based on spatio-temporal co-occurences. In: 2009 IEEE Conference Computer Vision and Pattern Recognition, pp. 2458–2465. IEEE (2009)
20. Park, K., Lin, Y., Metsis, V., et al.: Abnormal human behavioral pattern detection in assisted living environments. In: Proceedings of 3rd International Conference on PErvasive Technologies Related to Assistive Environments – PETRA 2010, p. 1. ACM Press, New York, USA, (2010)
21. Raghavendra, R., Cristani, M., Del Bue, A., Sangineto, E., Murino, V.: Anomaly detection in crowded scenes: a novel framework based on swarm optimization and social force modeling. In: Ali, S., Nishino, K., Manocha, D., Shah, M. (eds.) Modeling, Simulation and Visual Analysis of Crowds. TISVC, vol. 11, pp. 383–411. Springer, New York (2013). doi:10.1007/978-1-4614-8483-7_15
22. Basharat, A., Gritai, A., Shah, M.: Learning object motion patterns for anomaly detection and improved object detection. In: 2008 IEEE Conference on Computer Vision and Pattern Recognition, pp. 1–8. IEEE (2008)
23. Stauffer, C., Grimson, W.E.L.: Learning patterns of activity using real-time tracking. IEEE Trans. Pattern Anal. Mach. Intell. 22, 747–757 (2000)
24. Mahadevan, V., Li, W., Bhalodia, V., Vasconcelos, N.: Anomaly detection in crowded scenes. In: 2010 IEEE Computer Society Conference on Computer Vision and Pattern Recognition, pp. 1975–1981. IEEE (2010)
25. Wu, S., Moore, B.E., Shah, M.: Chaotic invariants of lagrangian particle trajectories for anomaly detection in crowded scenes. In: Computer Vision and Pattern Recognition (2010)
26. Mehran, R., Moore, B.E., Shah, M.: A streakline representation of flow in crowded scenes. In: Daniilidis, K., Maragos, P., Paragios, N. (eds.) ECCV 2010. LNCS, vol. 6313, pp. 439–452. Springer, Heidelberg (2010). doi:10.1007/978-3-642-15558-1_32
27. Ali, S., Shah, M.: A Lagrangian particle dynamics approach for crowd flow segmentation and stability analysis. In: Computer Vision and Pattern Recognition (2007)
28. Mehran, R., Oyama, A., Shah, M.: Abnormal crowd behavior detection using social force model. In: Computer Vision and Pattern Recognition (2009)
29. Ke, Y., Sukthankar, R., Hebert, M.: Event Detection in crowded videos. In: Comput Vision, 2007 ICCV 2007 pp. 1–8 (2007)
30. Kratz, L., Nishino, K.: Anomaly detection in extremely crowded scenes using spatio-temporal motion pattern models. Computer Vision and Pattern Recognition, pp. 1446–1453 (2009)
31. Zivkovic, Z., Van Der Heijden, F.: Efficient adaptive density estimation per image pixel for the task of background subtraction. Pattern Recognit. Lett. 27, 773–780 (2006)
32. Adam, A., Rivlin, E., Shimshoni, I., Reinitz, D.: Robust real-time unusual event detection using multiple fixed-location monitors. IEEE Trans. Pattern Anal. Mach. Intell. 30, 555–560 (2008)
33. Elhamifar, E., Vidal, R.: Sparse Subspace Clustering : In: 2009 IEEE Conference on Computer Vision and Pattern Recognition, pp. 2790–2797. IEEE (2009)
34. Aguilar, W.G., Casaliglla, V.P., Pólit, J.L.: Obstacle avoidance based-visual navigation for micro aerial vehicles. Electronics 6, 10 (2017)

35. Aguilar, W.G., Casaliglla, V.P., Pólit, J.L., Abad, V., Ruiz, H.: Obstacle avoidance for flight safety on unmanned aerial vehicles. In: Rojas, I., Joya, G., Catala, A. (eds.) IWANN 2017. LNCS, vol. 10306, pp. 575–584. Springer, Cham (2017). doi:10.1007/978-3-319-59147-6_49
36. Aguilar, W.G., Casaliglla, V.P., Polit, J.L.: Obstacle avoidance for low-cost UAVs. In: Proceedings of - IEEE 11th Int Conference on Semantic Computing, ICSC 2017 (2017)

Driver Fatigue Detection Based on Real-Time Eye Gaze Pattern Analysis

Wilbert G. Aguilar[1,3,4(✉)], Jorge I. Estrella[2], William López[3], and Vanessa Abad[5]

[1] Dep. DECEM, Universidad de las Fuerzas Armadas ESPE, Sangolquí, Ecuador
wgaguilar@espe.edu.ec

[2] Dep. Eléctrica y Electrónica, Universidad de las Fuerzas Armadas ESPE, Sangolquí, Ecuador

[3] CICTE Research Center, Universidad de las Fuerzas Armadas ESPE, Sangolquí, Ecuador

[4] GREC Research Group, Universitat Politècnica de Catalunya, Barcelona, Spain

[5] Universitat de Barcelona, Barcelona, Spain

Abstract. This paper introduces a real time non-intrusive method to determine driver fatigue by analyzing eye gaze patterns. Using a standard webcam and a personal computer, the proposed method combines different techniques in order to keep a low computational cost without a loss of performance. Facial features are identified in a reference image to extract a region of interest, around the eyes of the user, and tracked by an optical flow algorithm in subsequent frames. Color segmentation on the resulting images allow the system to extract data needed to determine ocular following, blink detection, frequency, and percentage of eye lid closure over time (PERCLOS). This approach, while simple, proves to be very efficient and accurate for the hardware restricted setup, allowing faster information processing on modest specifications systems. For safety reasons, our experiments are limited to different subjects, simulating fatigue in laboratory conditions as well as a real time test on a moving vehicle to analyze the blinking patterns.

Keywords: Fatigue detection · Real-time · Driver assistance · Eye-gaze patterns · PERCLOS

1 Introduction

Over the last decade, there is a growing concern in the increased rate of road accidents. The world health organization statistics shows road injury as one of the top ten causes of death worldwide. Last year, the registered number of deaths on the road was estimated at 1.25 million, reaching the same number of deaths caused by diseases like malaria or AIDS. The amount of road accidents has led to the creation of the Decade of Action for Road Safety in 2011 and is a case of study for the Global Status Reports on Road Safety (GSRRS). The Sustainable Development Goals include a target of 50% reduction in road traffic deaths by 2020 through campaigns on law enforcement,

© Springer International Publishing AG 2017
Y. Huang et al. (Eds.): ICIRA 2017, Part II, LNAI 10463, pp. 683–694, 2017.
DOI: 10.1007/978-3-319-65292-4_59

education and technology. The technological field of road safety has seen a continuous development on vehicles by the research of reactive methods, like the ABS breaking systems, and proactive methods, like road, driver monitoring, and autonomous navigation systems [1–5]. It is of particular interest to further develop the latter, considering almost 50% of the accidents involve car occupants and nearly 20% are related to driver fatigue [6, 7]. Driver fatigue is characterized by reduced alertness associated with diminished cognitive and motor performance of an individual [8] and can be monitored to an extent by various methods. Driver physical and physiological measures have been the subject of several studies, and vary according to their implementation [9]. The most relevant are, intrusive methods, which rely on biometrics like electroencephalogram (EEG) and electrocardiograph (ECG) signals [10, 11]; semi intrusive methods like eye tracking systems embedded on glasses, as seen in commercial applications such as Optalert or SmoothEyes; and non-intrusive methods, based on remote sensing and analysis of data. While intrusive and semi-intrusive methods present a highly reliable solution to the fatigue detection issue, their invasive nature and high price range turns away the population most affected by road accidents (low- and middle-income countries [12]). The rapid increase in computational power on compact electronic devices is starting to give remote sensing systems an edge over intrusive ones by reducing implementation costs and the ability to retrofit in any vehicle. In this scenario, vision based driver fatigue systems should continue to be researched to make them affordable and reliable worldwide.

At present, computer vision based systems [13, 14] can track several facial cues to determine the state of alertness and drowsiness of the driver. Some approaches focus on head position and face orientation by tracking feature points on subsequent frames [15] or use color segmentation techniques to identify facial expressions [16]. Other cues like nodding and yawning have also been analyzed in driver fatigue [17]. Finally, most recent studies focus on eye gaze features like eyelid movement, blinking frequency and percentage of eye closure over time (PERCLOS) [18], addressing more accurately the micro-sleep issue [19]. An updated survey of current and ongoing investigations on this subject can be found at [20]. While most of these methods use near-IR cameras to track the eye gaze features and patterns, few specify the limitations it has on daylight use [9, 18]. Considering various researches show that there are two daily peaks when fatigue accidents occur due to cognitive and motor performance, one early in the morning (2:00 AM to 6:00 AM), and the other between 2:00 PM and 4:00 PM [21, 22], the use of conventional cameras and methods should not be discarded.

In this paper, we propose a non-invasive fatigue monitoring system based on eye gaze activity and estimation patterns using a standard single webcam connected to a mobile computer. Our approach focuses on mobility and performance of common hardware to determine the accuracy and response of the methods in a real environment. The system starts by acquiring a single reference image of the driver where face and eyes are identified by the well-known Viola-Jones method [23]. This image is used to find key points in the eye region of the driver using the ORB algorithm [24]. After classifying the reference key points, they are introduced in a Kanade Lucas Tomassi (KLT) algorithm to track the eyes of the driver on subsequent frames. The optical flow of the points in the eye region allows the system to determine ocular following while a color map segmentation creates a binary mask to extract the pixel area of the eyes as well as the blinking patterns of the driver. Experiments involving a group of 5 persons aged 25–65, in laboratory conditions and on the road, show a high accuracy of blinking patterns in both scenarios, which allow us to calculate PERCLOS. It is to be noted that this method requires a steady frame rate and a low execution time to be reliable, which is the main reason we decided to focus on computational cost and performance, metrics rarely reported in current literature [20].

2 Our Proposal

The general architecture of our system can be divided in three stages. (1) Feature extraction, (2) Eye region tracking and (3) Eye gaze pattern analysis.

2.1 System Initialization (Feature Extraction)

The system starts by acquiring a reference image of the test subject inside the vehicle while his level of attention is high under stable lighting conditions, to quickly identify features by conventional methods.

The captured image goes through a facial feature detection algorithm and will be labeled as reference image once it contains all the necessary parameters to start the tracking algorithm. If it does not meet this criteria, another image will have to be captured and processed.

Over the last decade the Viola-Jones method has been continuously researched and well documented on the facial features detection field. By using a series of boosted cascades of classifiers to identify simple features this method proved to be very efficient on static images [23]. However, with a video input, the rate at which images are analyzed drops considerably, more so when trying to identify specific features like eyes, lips or nose. In a real time facial feature detection system, where lighting conditions and background content varies frame to frame, this results in a considerable decrease in performance over time. For this reason, we do not use the Viola-Jones method to track the facial features. Instead we only use it to identify the face and the eyes of the driver, creating a mask image to find key points which can be later tracked by a more efficient method. This image is obtained either on system start up or after a conditioned reset to re-acquire the reference parameters. Figure 1 shows the different stages to obtain the reference image using the Viola-Jones algorithm.

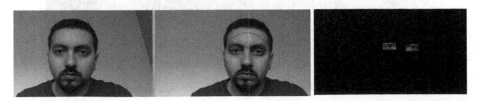

Fig. 1. Facial feature extraction using the Viola Jones algorithm. From left to right: input image is converted to greyscale to identify the face and the eyes of the subject. Image on the right shows the "eyes mask image" obtained.

After extracting the eye mask, points of interest are extracted to finally determine if the image acquired can be classified as a reference image. Feature extraction methods focus on specific regions of an image to determine whether it can be classified as a feature or not. They differ in aspects such as attribute invariance, compute efficiency, robustness among others. A survey on the various methods to extract features and descriptors can be found in [25]. For our application, the most convenient method was ORB due to its open source nature, low processing cost and fast execution time. As stated in [25], "ORB is a highly optimized and very well engineered descriptor, since the ORB authors were keenly interested in compute speed, memory footprint and accuracy." ORB can be used in real-time practical applications [26–29]. However, like most feature extractors and descriptors, ORB by itself is not designed to match or track specific facial features frame to frame. In order to do so, some pre-processing of the images is needed to help the detector find the interest points, increasing the computational cost and reducing the frame rate. Considering the latter, we used ORB only to extract the feature points in a reference image candidate.

We set up the ORB algorithm to extract 300 key points within the eye region. The Features From Accelerated Segment Test (FAST) extractor inside ORB, filters them and we select 50 key points with the highest response rating. The response rating specifies how strong a key point is, according to:

$$R = Det(H) - k(Trace(H))^2$$

Where,

R is the response rating

H is the matrix of the sums of the products of derivatives at each pixel

$$H(x,y) = \begin{pmatrix} S_{x2}(x,y) & S_{xy}(x,y) \\ S_{xy}(x,y) & S_{y2}(x,y) \end{pmatrix}$$

The higher the response value, the more likely the feature will be recognized among several instances of an object. In our system, the key points stack around the eye contour, easing the identification of outliers to set up reset conditions later on (Fig. 2).

The feature and key point extraction flowchart can be seen in Fig. 3.

Fig. 2. Key point extraction in the eye mask image using the ORB algorithm. 50 key points are selected from a set of 300 points.

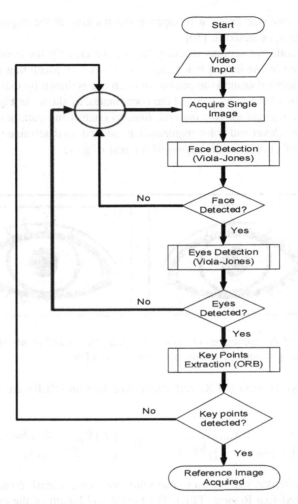

Fig. 3. Feature extraction algorithm flow chart. Several conditions have to be met before declaring an image as Reference Image.

2.2 Eye Region Tracking

Once the key point extraction is successful, the resulting points are inserted into an optical flow loop to track them over time in consecutive frames. We chose the KLT method for our purpose due to the nature of the experiment. The tracking of the facial features of a driver consists in slow head displacements and small eye movements, a temporal persistence. The brightness of the analyzed frames is rather constant in time, and the points we extracted before in the k instant ($S_k[j]$, $j = 50$) are classified according to the spatial coherence of the reference image.

By definition, the KLT algorithm relies on the local information of the surroundings of a key point to determine the relative position of the same point in subsequent frames. The surroundings of every key point in the k instant will be analyzed in a 30×30

pixels window, any size larger will improve the tracking at the expense of increased computational cost as stated in [30].

To extract both eye regions in every frame, we classify the coordinates of each tracked key point in the instant k + 1 ($S_{k+1}[j]$, $j = 50$) in ascending order. A rectangular region of interest around the contour of each eye is drawn by using the maximum and the minimum of the (x, y) value of the coordinates, as shown in Fig. 3. We need to separate the eye regions for two reasons; first, to control the parameters of each eye, such as brightness level and color segmentation; second, to determine positive tracking by the amount of outliers in the region of interest (Fig. 4).

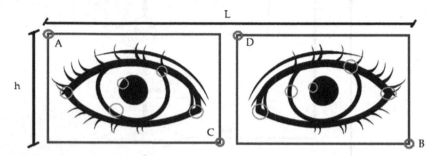

Fig. 4. Diagram of the eye regions of interest. In green, the tracked points, in red, the minima and maxima of the key points coordinates. (Color figure online)

The Left Eye Region (LER) and Right Eye Region (RER) are defined by two corners:

$$LER \begin{Bmatrix} A(X_{min}, Y_{min}) \\ C(X_{min} + \frac{X_{max}-X_{min}}{2}, Y_{max}) \end{Bmatrix}; RER \begin{Bmatrix} D(X_{max} - \frac{X_{max}-X_{min}}{2}, Y_{min}) \\ B(X_{max}, Y_{max}) \end{Bmatrix}$$

After extracting the eye region in real time, we create a mask containing LER and RER, the Tracked Eye Region (TER). The height and length of the eye region in the reference image are stored as reference dimension values, and allow us to estimate ocular following of the driver by comparing the dimensions of the TER with the reference eye region. If the values in the x or y axis vary over a threshold value, it means the distance between the reference key points has changed over time too, according to the optical flow equation:

$$\varphi_{j,k+1} = \frac{(P_{j,k+1} - P_{j,k})}{T_s}$$

Since the tracked key points remain stacked around the eye region, then the variation of the distance, over the threshold, between two points $P_{j,k+1} - P_{j,k}$ can be interpreted as a movement of ocular following, this is, if a person moves the head towards a point in the borders of his view range, the eyes will follow the same direction in order to maintain awareness. Using the same reasoning, we defined the limits of the distance variation to maintain the eye pattern analysis relevant only when the TER is within them.

2.3   Eye Gaze Pattern Analysis

The state of alert of the driver can be estimated by several factors such as distraction, drowsiness and fatigue. Our method can identify distraction through ocular following, but the most accurate way to determine state of alert, is through monitoring drowsiness and fatigue levels. We propose to use two parameters to determine driver fatigue: blinking patterns and PERCLOS.

To determine the blinking frequency and duration, the eye Region of Interest (ROI) of the reference image is transformed from the Blue Green Red (BGR) to Hue Luminosity Saturation (HLS) color space. By modifying the value of the L channel, we successfully threshold the eye ROI to obtain a binary image where the eye contour is represented by white pixels, as seen in Fig. 5.

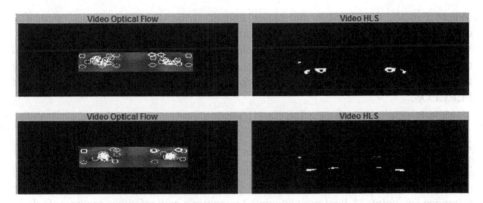

Fig. 5. Top: TER with open eyes converted to the HLS color space. Bottom: TER with closed eyes converted to the HLS color space

A pixel count in the area formed by the eye mask sets the reference value of the non-zero pixels when the driver eyes are open (NZ_{Ref}). The same procedure takes place in successfully tracked eyes ROI $(NZ_{Tracked})$. Lower white pixel values indicate driver eyelid closure, while values well above to the reference set point indicate the tracking conditions have been altered and a reset is required. The NZ values over time form a curve as (Fig. 6):

Studies on measuring the eye closure have determined the average time of normal blinks to take less than 300 ms, while slow blinks take over 300–500 ms [31] Since we are working at 20FPS, each frame is analyzed every 50 ms, enough time to distinguish both types of blinks.

"PERCLOS is the percentage of eyelid closure over the pupil over time" and several studies demonstrate the high correlation between this measure and the level of fatigue of an individual [32], according to:

$$PERCLOS = \frac{t_c}{t_c + t_o}$$

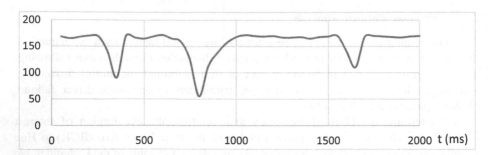

Fig. 6. White pixels over time in the TER of the driver. Valleys represent blinks, plateaus represent open eyes.

Where t_c is the time the eyes are closed and t_o. is the time the eyes are open. The common time frame to compute this measure is between 30 s and 1 min.

By calculating the area covered by each eye in every incoming binary image, and the variation of it in relation to the reference image, the system determines blinking frequency and duration accurately enough to calculate blinking duration, frequency and PERCLOS.

2.4   Reset Conditions

To keep the information relevant throughout time, the algorithm resets on several conditions. A soft reset replaces the key points in the tracking algorithm with the reference key points. A hard reset attempts to acquire a new reference image, calculates new key points and new HLS values for the new tracking parameters.

- Maximum number of outliers allowed reached inside the KLT tracking algorithm: occurs on partial occlusion, on driver distraction and on large lighting variations. A hard reset is required.
- Number of frames limit reached (refreshing): soft reset.
- The area of the eye region greatly exceeds the reference value: the Luminosity parameter on the HLS color map is recalculated.

3   Results

3.1   System Performance

The main focus of our research is to validate an algorithm able to perform on low cost commercial devices without losing accuracy. Although it can still be optimized, our algorithm worked using 6% of a single thread in the CPU and 125 MB of RAM memory. Without needing to resize the input images, our approach works at 16.1 FPS on a live video feed using the webcam and at 20.9 FPS working on a pre recorded video. Table 1 compares this results with the Viola Jones implementation, extracting facial features on every frame instead of tracking it.

Table 1. Average Frames Per Second using he Viola Jones method and using the KLT tracker + color segmentation method.

	Viola Jones	Our approach
Webcam IN	9.8	16.1
Recorded video IN	13.1	20.9

The execution time of the different parts of our algorithm is detailed in Table 2.

Table 2. Execution time of the various methods used in our system in comparison to the state of the art.

Execution time [ms]				
Feature extraction		Eye detection/tracking		Eye gaze analysis
Shi - Tomassi	ORB	Template matching	KLT tracker	
27	6	18	8	5

By adapting modern methods like ORB to the tracking optical flow algorithm, we managed to reduce the total execution time of our method to an average 42.5 ms/frame, guaranteeing real-time performance.

3.2 Detection and Tracking

Table 3 compares detection rate and tracking of facial features over time using our approach and using the Viola–Jones method on every frame. The recordings were segmented in 5 min videos.

Table 3. Identification and tracking of facial features over time.

Identification/tracking average rate [%]		
Subject	Viola Jones	ORB + Optical flow
1	35.7	92.1
2	38.5	90
3	47.8	96.4
4	42.1	93.8
5	44.9	96.1

Using the KLT tracking approach, the detection rate of facial features in subsequent frames is highly reliable due to the partial tracking of features that takes place during periods of occlusion. Soft reset conditions keep the eye region relevant throughout time during long periods of occlusion and allow rotation of the face to a certain degree before losing the tracking points. This allows our method to constantly compare the area of the eye region in subsequent frames for the PERCLOS calculation. If the eye region is not tracked on a high rate, the blinking pattern algorithm is rendered inaccurate.

3.3 PERCLOS and Blinking Parameters Calculations

The average accuracy rating of blinking and PERCLOS of our system is shown in Table 4 for every subject. It is to be noted that, while blinks can be counted to evaluate the detection rating, the PERCLOS measure cannot be manually measured. Instead we decided to calculate the number of false positives due to false tracking.

Table 4. Average accuracy rating (%).

Subject	Blinking	PERCLOS
1	97.2	98.1
2	93.4	92.7
3	96.7	95.3
4	96.1	94.2
5	95.4	95.8

4 Conclusions

Our proposal for driver fatigue detection proved to be very cost effective, reducing the execution times compared to state of the art methods on standard commercial hardware. Real-time performance at over 15 FPS was achieved without reducing the quality or the size of the input image. The experiment on vehicle showed that false positives encountered during the calculation of PERCLOS are mainly due to false tracking. This can occur during fast and strong lighting changes, when crossing below a bridge for instance, or during partial occlusion moments. To prevent this, we only calculate eye gaze parameters while the dimensions of tracked eye region are similar to the reference ones. A better tracking of the features would improve the PERCLOS calculation by having a more constant influx of data.

References

1. Aguilar, W.G., Verónica, C., José, P.: Obstacle avoidance based-visual navigation for micro aerial vehicles. Electronics **6**(1), 10 (2017)
2. Aguilar, W.G., Verónica, C., José, P.: Obstacle avoidance for low-cost UAVs. In: IEEE 11th International Conference On Semantic Computing (ICSC), San Diego (2017)
3. Aguilar, W.G., Morales, S., Ruiz, H., Abad, V.: RRT* GL based optimal path planning for real-time navigation of UAVs. In: Rojas, I., Joya, G., Catala, A. (eds.) IWANN 2017. LNCS, vol. 10306, pp. 585–595. Springer, Cham (2017). doi:10.1007/978-3-319-59147-6_50
4. Aguilar, W.G., Morales, S.: 3D environment mapping using the Kinect V2 and path planning based on RRT Algorithms. Electronics **5**(4), 70 (2016)
5. Cabras, P., Rosell, J., Pérez, A., Aguilar, W.G., Rosell, A.: Haptic-based navigation for the virtual bronchoscopy. In: 18th IFAC World Congress, Milano, Italy (2011)
6. The Royal Society for the Prevention of Accidents: The Royal Society for the Prevention of Accidents. http://www.rospa.com/road-safety/advice/drivers/fatigue/road-accidents/. Accessed 17 Aug 2016

7. Road Safey Observatory: Road Safety Observatory. http://www.roadsafetyobservatory.com/ Summary/drivers/fatigue. Accessed 17 Aug 2016
8. Williamson, A., Friswell, R.: Exploratory study of fatigue in light and short haul transport drivers in NSW Australia. Accid. Anal. Prev. **40**(1), 410–417 (2008)
9. Dawson, D., Searle, A.K., Paterson, J.L.: Evaluating the use of fatigue detection technologies within a fatigue risk management system for the road transport industry. Sleep Med. Rev. **18**, 1–12 (2013)
10. Zhao, C., Zhao, M., Liu, J., Zheng, C.: Electroencephalogram and electrocardiograph assessment of mental fatigue in a driving simulator. Accid. Anal. Prev. **45**, 83–90 (2012)
11. Saroj, A.C., Lal, K.L.: A critical review of the psychophysiology of driver fatigue. Biol. Psychol. **55**, 173–194 (2001)
12. World Health Organization: Global Status Report on Road Safety (2015). http://www.who.int/violence_injury_prevention/road_safety_status/2015/en/. Accessed 17 Aug 2016
13. Aguilar, W.G., Luna, M.A., Moya, J.F., Abad, V., Ruiz, H., Parra, H., Angulo, C.: Pedestrian detection for UAVs using cascade classifiers and saliency maps. In: Rojas, I., Joya, G., Catala, A. (eds.) IWANN 2017. LNCS, vol. 10306, pp. 563–574. Springer, Cham (2017). doi:10.1007/978-3-319-59147-6_48
14. Aguilar, W.G., Luna, M., Moya, J., Abad, V., Parra, H., Ruiz, H.: Pedestrian detection for UAVs using cascade classifiers with meanshift. In: IEEE 11th International Conference on Semantic Computing (ICSC), San Diego (2017)
15. Liu, K., Luo, Y., Gyomei, T.E.I., Yang, S.: Attention recognition of drivers based on head pose estimation. In: Vehicle Power and Propulsion Conference (2008)
16. Sanchez-Cuevas, M.C., Aguilar-Ponce, R.M., Tecpanecatl-Xihuitl, J.L.: A comparison of color models for color face segmentation. Procedia Technol. **7**, 131–141 (2013)
17. Li, L., Chen, Y., Li, Z.: Yawning detection for monitoring driver fatigue based on two cameras. In: 12th International IEEE Conference on Intelligent Transportation Systems (2009)
18. Bergasa, L.M., et al.: Visual monitoring of driver inattention. IEEE Trans. Intell. Transp. Syst. **1**(1), 63–77 (2006)
19. SleepDex: "Sleepdex."http://www.sleepdex.org/microsleep.htm. Accessed 18 Aug 2016
20. Al-Rahayfeh, A.M.E.R., Faezipour, M.I.A.D.: Eye tracking and head movement detection: a state-of-art survey. IEEE J. Transl. Eng. Health Med. **1**, 2100212 (2013)
21. Eskandarian, A., et al.: Advanced driver fatigue reasearch. US Department of Transportation (2007)
22. Horne, J., Reyner, L.: Vehicle accidents related to sleep: a review. Occup. Environ. Med. **56**, 289–294 (1999)
23. Viola, P., Jones, M.: Rapid object detection using a boosted cascade of simple features. In: Conference on Computer Vision and Pattern Recognition (2001)
24. Rublee, E., Rabaud, V., Konolige, K., Bradski, G.: ORB: An efficient alternative to SIFT or SURF. In: ICCV (2011)
25. Krig, S.: Interest point detector and feature descriptor survey. In: Computer Vision Metrics: Survey, Taxonomy, and Analysis. Apress (2014)
26. Aguilar, W.G., Angulo, C.: Real-time model-based video stabilization for microaerial vehicles. Neural Process. Lett. **43**(2), 459–477 (2016)
27. Aguilar, W.G., Angulo, C.: Real-time video stabilization without phantom movements for micro aerial vehicles. EURASIP J. Image Video Process. **1**, 1–13 (2014)
28. Aguilar, W.G., Angulo, C.: Robust video stabilization based on motion intention for low-cost micro aerial vehicles. In: 11th International Multi-conference on Systems, Signals & Devices (SSD), Barcelona, Spain (2014)

29. Aguilar, W.G., Casaliglla, V.P., Pólit, J.L., Abad, V., Ruiz, H.: Obstacle avoidance for flight safety on unmanned aerial vehicles. In: Rojas, I., Joya, G., Catala, A. (eds.) IWANN 2017. LNCS, vol. 10306, pp. 575–584. Springer, Cham (2017). doi:10.1007/978-3-319-59147-6_49
30. Tomasi, C., Kanade, T.: Detection and Tracking of Point Features. Carnegie Mellon University, Pittsburgh (1991)
31. Hammoud, R.I.: Passive Eye Monitoring, Algorithms. Applications and Experiments. Springer, Heidelberg (2008). doi:10.1007/978-3-540-75412-1
32. Dinges, D.F., Grace, R.: PERCLOS: a valid psychophysiological measure of alertness as assessed by psychomotor vigilance. US Department of Transportation (1998)

Onboard Video Stabilization for Rotorcrafts

Wilbert G. Aguilar[1,3,4(✉)], David Loza[2], Luis Segura[2],
Alexander Ibarra[2], Thomas Abaroa[2], and Ronnie Fuertes[2]

[1] Dep. DECEM, Universidad de las Fuerzas Armadas ESPE,
Sangolquí, Ecuador
wgaguilar@espe.edu.ec
[2] Dep. Energía y Mecánica, Universidad de las Fuerzas Armadas ESPE,
Sangolquí, Ecuador
[3] CICTE Research Center, Universidad de las Fuerzas Armadas ESPE,
Sangolquí, Ecuador
[4] GREC Research Group, Universitat Politècnica de Catalunya,
Barcelona, Spain

Abstract. The main goal of this article is a stabilization system for cameras onboard mini Rotor UAVs that minimize the effect of the undesired movements on captured videos through the selection of, an autopilot and a ground control station for data extraction, an inertial stabilized platform (gimbal), an onboard computer for image processing, the communication interface, the software involved, the implementation of the stabilization system algorithm and results.

Keywords: Multirotor UAV · Gimbal · Motion estimation · Embedded system · Kalman Filter

1 Introduction

For navigation [1–10], autonomous or teleoperated, of unmanned aerial vehicles (UAV) in applications [10–21] as rescue, transport or surveillance, a problem during the acquisition of images in on-board systems are undesired movements between consecutive photograms, as a result of complex aerodynamic characteristics of a UAV in flight [22–24].

The effect of the rolling shutter makes each row of an illustration to be exposed to a slight displacement in different time intervals, resulting in an unstable output video. This problem can be corrected by digital [12, 13, 25–27] and mechanical techniques based on inertial measurement units (IMU) to measure the rotation of the camera at high frequencies and with high precision. Two types of movements are distinguished with the camera as seen in [11, 14, 25]: Weak Shaking, app. $\pm$ 10° variation in the horizontal and vertical directions and/or some small fractions of Hertz; and Strong Shaking, more than $\pm$ 10° variation in the horizontal and vertical directions and/or tens of Hertz.

There also exists phantom movements that are mainly, false displacements in parameters of scale and/or translation as a result of compensation and smoothing of high frequency movements, where the smoothing process removes real movements and/or introduces a delay in them. Both cases are defined as phantom movements [25].

© Springer International Publishing AG 2017
Y. Huang et al. (Eds.): ICIRA 2017, Part II, LNAI 10463, pp. 695–702, 2017.
DOI: 10.1007/978-3-319-65292-4_60

Weak Shaking can be solved using software image stabilization (digital stabilization). Strong Shaking can be managed with an additional hardware, such as a servo drive or a pneumatic/hydraulic unit, capable of compensating the movement of the camera in the opposite direction [11]. There are several post processing algorithms designed for compensate undesired movements generated from Weak Shaking, such as L1 Optimal [13], Subspace Video Stabilization [12], Model based Video Stabilization [24] and Parrot's Director Mode. Our proposal combine mechanical and computational solution for video stabilization, i.e., we combine [24] with our designed gimbal.

2 Device Design

RUAS are categorized as seen in [21], based on attributes as size and payload. The STORM drone V3 was used in this project due to belongs to the category IV, this drone supports approximately 1.325 kg of payload, but there is not recommended to reach this limit because the time of flight is considerably reduced by the amount of energy required to lift this weight, and the flight is not stable, that could result in having unwanted collisions. Despite the UAV category, RUAS include two different integrated systems. One is the Autopilot system, that is a group of actuators that control the movement and control circuits make these actuators move. The second is the Flight Director (FD), that is the Autopilot system brain.

Single-Board Computer are complete computers built in a single board, that primary advantages are the dimensions, reduced weight and price. The single-board computer selected for the analysis of image, is the Odroid XU4, because of performance compared to other single-boards as seen in Fig. 1. The Odroid XU4 OS currently installed is the Ubuntu 15.04 robotics edition from the HardKernel Odroid Forum, because ROS Jade (bare bones) is previously installed, and contains a compatible OpenCV for image processing.

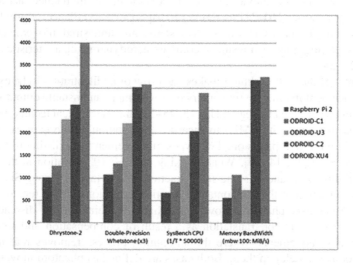

Fig. 1. Performance of the CPU/RAM of many SBC (HardKernel co, 2017)

3   Video Stabilization System

Kalman Filter is a smoothing method for the motion intention estimation. There is a combination of a low-pass filter of second order, using the last frame and the input of the control to estimate a reliable movement intention estimation [25, 28]. The transfer function is calculated using the control signal [1, 2, 25, 28–32], sent from the RC and the drone position. This information is obtained from the channel that controls the yaw rotation.

In order to obtain the visual motor model, there is necessary to compute the motion parameter from the warping generated by control actions, and relate calculated parameter and control data captured. The transfer function is estimated with input data of the RC control and the movement translation estimation previously obtained (Fig. 2).

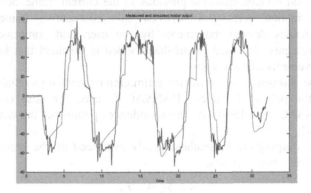

Fig. 2. The estimated transfer function Is based on the Yaw action control and the movement translation on the X axis. Blue Curve: Modeled simulated output. Black Curve: Model measured.

Once the estimated function is generated, the system is modeled as state space, acquiring the Kp, Tp1, Tp2, and Tz parameters:

$$G(s) = Kp * \frac{1 + Tz * s}{\left(1 + 2 * Zeta * Tp1 * s + (Tp1 * s)^2\right) * (1 + Tp2 * s)} \tag{1}$$

The state space model is applied to the Kalman Filter for estimating pilot motion intention (Fig. 3).

For interest point detection and description SURF is selected as detector, and BRIEF as descriptor. SURF has more computational cost than FAST, but is used due to the reliable points. This is evidenced during the matching process of the algorithm using both methods. For descriptors, BRIEF is selected instead of SURF because of computational cost and robustness are comparable. Even though a mechanical stabilizer (gimbal) is implemented for the onboard camera, a video stabilization algorithm is developed based on the real time stabilization algorithm [24].

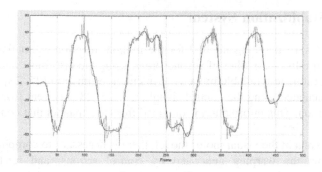

Fig. 3. Kalman Filter based estimated function.

This algorithm is based on the geometric transformation obtained, that represents the translation and rotation from the previous to the current frame, accumulating the variations on motion parameters between two consecutive frames. Trajectory smoothing estimates desired movements by the user, and compensates only the undesired movements. As result, a stabilized video is obtained that keeps the most significant movements of the camera.

Because the geometric transformation estimation depends on a reliable and correct matching of the points of interest, RANSAC is used for false correspondences rejection. RANSAC considers false correspondences points that are not in the model [18, 19, 33, 34].

Inter frame warping can be mathematically expressed by the geometric transformation that relates interest points:

$$It = Ht^{-1} * Isp \tag{2}$$

Where I_{sp} is a set of points of interest of the reference image and the non-compensated image, respectively. And H_t is the geometric transformation matrix.

The translation model is used, that refers to the movement of the image when the capturing device movement is of translation, parallel to the plane of the image. The model is as follows:

$$Ht = \begin{bmatrix} 1 & 0 & tx \\ 0 & 1 & ty \\ 0 & 0 & 1 \end{bmatrix} \tag{3}$$

$$Ht = \begin{bmatrix} s * \cos(\varnothing) & -s * \sin(\varnothing) & tx \\ s * \sin(\varnothing) & s * \cos(\varnothing) & ty \\ 0 & 0 & 1 \end{bmatrix} \tag{4}$$

Where t_x y t_y are translations, and $\varnothing$ is the rotation angle of "Roll".

This model is used, because from matrix the values of the variables t_x, t_y, and $\varnothing$ are extracted, so they can be processed using the Kalman Filter previously implemented, were the actual values are compared with the previous, estimating a confident movement intention. The affine model estimates four parameters, two displacements in the parallel plane of the image, like the translation model, "Roll" rotation, and scale s, that are proportional to the movement in the "Roll" axis orientation [26].

Finally, the original image and the processed on are stored so they can be managed to verify the results with the ITF method.

4   Results and Discussion

The following chapter present some of the frames and resultant videos of the tests of the stabilization system (Fig. 4).

Fig. 4. Offline Video Up: Original Video. Down: Stabilized Video.

To determine the performance of the algorithm applied, the ITF metric evaluation is considered as a reliable method to evidence that the system is working. To obtain the ITF, the RMSE evaluation is required, followed by the PSNR.

The RMSE evaluates the reliability of the estimated and observed movement. A low RMSE means that the estimated movement is similar to the real movement intention (Table 1).

$$RMSE(k) = \left(\frac{1}{M * N} \sum_{i=0}^{M-1} \sum_{j=0}^{N-1} ||Frame_k(i,j) - Frame_{k-1}(i,j)||^2 \right)^{\frac{1}{2}} \quad (5)$$

Table 1. Total RMSE and improvement percentage

	RMSE without stabilization	RMSE with stabilization	Improvement percentage
Static camera	723.66	871.79	−16.99
Camera with Gimbal	1002.36	1290.37	−22.31
Offline test	159.65	154.29	3.35

PSNR is an engineering term for the proportion between maximum energy possible of a signal and the noise power affecting the fidelity of representation (Table 2).

$$PSNR(k) = 10log_{10}\frac{Ip_{MAX}}{MSE(k)} \tag{6}$$

Table 2. PSNR by frame

Sample name	Original	Processed
Static camera with no video stabilization	12.72	16.64
Camera with gimbal and no video stabilization	13.43	16.55
Offline video frames 1222 and 1223	22.86	24.5

ITF is a method commonly used to measure the effectiveness and performance of a video stabilization, that formula is as follows:

$$ITF = \frac{1}{Nf - 1} * \sum_{k=1}^{Nf-1} PSNR(k) \tag{7}$$

And from the same, the following values are obtained (Table 3):

Table 3. ITF and improvement percentage

	ITF without stabilization	ITF with stabilization	Improvement percentage
Static camera	17.29	16.68	−4.82
Camera with Gimbal	16.34	15.41	−6.04
Offline test	23.65	24.22	2.3

5 Conclusions

The dimensions of the different electronic and mechanical components were considered to obtain the forces distribution to generate a minor inertial momentum in the reduced space we were provided. The base of the Yaw motor is the principal component that permits the union between the gimbal and the total base through shock absorbers, avoiding the Rolling Shutter effect. The system was evaluated comparing the original and the processed video, obtaining as a result, image improvement, such a 9.718% on RMSE, 16.706 & on PSNR and 20.618% on ITF, that proves that the video stabilization system.

Acknowledgement. This work is part of the projects VisualNavDrone 2016-PIC-024 and MultiNavCar 2016-PIC-025, from the Universidad de las Fuerzas Armadas ESPE, directed by Dr. Wilbert G. Aguilar.

References

1. Engel, J., Sturm, J., Cremers, D.: Scale-aware navigation of a low-cost quadrocopter with a monocular camera. Robot. Auton. Syst. **62**(11), 1646–1656 (2014)
2. Engel, J., Sturm, J., Cremers, D.: Camera-based navigation of a low-cost quadrocopter. In: 2012 IEEE/RSJ International Conference on Intelligent Robots and Systems, pp. 2815–2821. IEEE (2012)
3. Pierre-Jean, B.: The navigation and control technology inside the AR. drone micro UAV, pp. 1477–1484 (2011)
4. François, P.B., David, C., Jemmapes, D.: The navigation and control technology inside the AR. Drone micro UAV. In: 18th IFAC World Congress, pp. 1477–1484 (2011)
5. Fernandez, A., Diez, J., de Castro, D., Silva, P.F., Colomina, I., Dovis, F., Friess, P., Wis, M., Lindenberger, J., Fernandez, I.: ATENEA: advanced techniques for deeply integrated GNSS/INS/LiDAR navigation. In: 2010 5th ESA Workshop on Satellite Navigation Technologies and European Workshop on GNSS Signals and Signal Processing (NAVITEC), pp. 1–8 (2010)
6. Aguilar, W.G., Morales, S.G.: 3D environment mapping using the Kinect V2 and path planning based on RRT algorithms. Electronics **5**(4), 70 (2016)
7. Aguilar, W.G., Morales, S., Ruiz, H., Abad, V.: RRT* GL based optimal path planning for real-time navigation of UAVs. In: Rojas, I., Joya, G., Catala, A. (eds.) IWANN 2017. LNCS, vol. 10306, pp. 585–595. Springer, Cham (2017). doi:10.1007/978-3-319-59147-6_50
8. Cabras, P., Rosell, J., Pérez, A., Aguilar, W.G., Rosell, A.: Haptic-based navigation for the virtual bronchoscopy. In: IFAC Proceedings Volumes (IFAC-PapersOnline), vol. 18(PART 1) (2011)
9. Dong, Z., Zhang, G., Bao, H.: Robust monocular SLAM in dynamic environments. In: 2013 IEEE International Symposium on Mixed and Augmented Reality (ISMAR), pp. 209–218 (2013)
10. Geiger, A., Lenz, P., Urtasun, R.: Are we ready for autonomous driving? The KITTI vision benchmark suite. In: Proceedings of the IEEE Computer Society Conference on Computer Vision and Pattern Recognition, pp. 3354–3361 (2012)
11. Liu, S., Yuan, L., Tan, P., Sun, J.: SteadyFlow: spatially smooth optical flow for video stabilization. In: 2014 IEEE Conference on Computer Vision and Pattern Recognition, pp. 4209–4216 (2014)
12. Liu, F., Niu, Y., Jin, H.: Joint subspace stabilization for stereoscopic video. In: 2013 IEEE International Conference on Computer Vision, pp. 73–80 (2013)
13. Grundmann, M.: Computational video: post-processing methods for stabilization, retargeting and segmentation. Georgia Institute of Technology (2013)
14. Cho, S., Wang, J., Lee, S.: Video deblurring for hand-held cameras using patch-based synthesis. ACM Trans. Graph. **31**(4), 1–9 (2012)
15. Benenson, R., Hosang, J.: Ten years of pedestrian detection, What have we learned? Eccv, pp. 613–627 (2014)
16. Miksik, O., Mikolajczyk, K.: Evaluation of local detectors and descriptors for fast feature matching. In: 2012 21st International conference on Pattern Recognition (ICPR), pp. 2681–2684 (2012)

17. Dollár, P., Wojek, C., Schiele, B., Perona, P.: Pedestrian detection: an evaluation of the state of the art. IEEE Trans. Pattern Anal. Mach. Intell. **34**(4), 743–761 (2012)
18. Leutenegger, S., Chli, M., Siegwart, R.Y.: BRISK: binary robust invariant scalable keypoints. In: Proceedings of the IEEE International Conference on Computer Vision, pp. 2548–2555 (2011)
19. Rublee, E., Rabaud, V., Konolige, K., Bradski, G.: ORB: an efficient alternative to SIFT or SURF. In: Proceedings of the IEEE International Conference on Computer Vision, pp. 2564–2571 (2011)
20. Aguilar, W.G., Luna, M.A., Moya, J.F., Abad, V., Ruiz, H., Parra, H., Angulo, C.: Pedestrian detection for UAVs using cascade classifiers and saliency maps. In: Rojas, I., Joya, G., Catala, A. (eds.) IWANN 2017. LNCS, vol. 10306, pp. 563–574. Springer, Cham (2017). doi:10.1007/978-3-319-59147-6_48
21. Aguilar, W.G., Luna, M.A., Moya, J.F., Abad, V., Parra, H., Ruiz, H.: Pedestrian detection for UAVs using cascade classifiers with meanshift. In: Proceedings - IEEE 11th International Conference on Semantic Computing, ICSC 2017 (2017)
22. Kendoul, F.: Survey of advances in guidance, navigation, and control of unmanned rotorcraft systems. J. Field Robot. **29**, 315–378 (2012)
23. Blasco, X., García-Nieto, S., Reynoso-Meza, G.: Control autónomo del seguimiento de trayectorias de un vehículo cuatrirrotor. Simulación y evaluación de propuestas. Rev. Iberoam. Automática e Informática Ind. RIAI, **9**(2), 194–199 (2012)
24. Autonomous flight in GPS-denied environments using monocular vision and inertial sensors. J. Aerosp. Inf. Syst. **10**, (4), 172–186 (2013)
25. Aguilar, W.G., Angulo, C.: Real-time model-based video stabilization for microaerial vehicles. Neural Process. Lett. **43**(2), 459–477 (2016)
26. Aguilar, W.G., Angulo, C.: Real-time video stabilization without phantom movements for micro aerial vehicles. EURASIP J. Image Video Process. **2014**(1), 46 (2014)
27. Aguilar, W.G., Angulo, C.: Robust video stabilization based on motion intention for low-cost micro aerial vehicles. In: 2014 11th International Multi-Conference on Systems, Signals Devices (SSD), pp. 1–6 (2014)
28. Mourikis, A.I., Roumeliotis, S.I.: A multi-state constraint Kalman Filter for vision-aided inertial navigation. In: Proceedings of 2007 IEEE International Conference on Robotics and Automation, pp. 3565–3572 (2007)
29. Aguilar, W.G., Casaliglla, V.P., Pólit, J.L.: Obstacle avoidance based-visual navigation for micro aerial vehicles. Electronics **6**(1), 10 (2017)
30. Aguilar, W.G., Casaliglla, V.P., Pólit, J.L., Abad, V., Ruiz, H.: Obstacle avoidance for flight safety on unmanned aerial vehicles. In: Rojas, I., Joya, G., Catala, A. (eds.) IWANN 2017. LNCS, vol. 10306, pp. 575–584. Springer, Cham (2017). doi:10.1007/978-3-319-59147-6_49
31. Aguilar, W.G., Casaliglla, V.P., Polit, J.L.: Obstacle avoidance for low-cost UAVs. In: Proceedings - IEEE 11th International Conference on Semantic Computing, ICSC 2017 (2017)
32. Engel, J., Cremers, D.: Accurate figure flying with a quadrocopter using onboard visual and inertial sensing. In: IMU (2012)
33. Alahi, A., Ortiz, R., Vandergheynst, P.: FREAK: fast retina keypoint. In: Proceedings of the IEEE Computer Society Conference on Computer Vision and Pattern Recognition, pp. 510–517 (2012)
34. Derpanis, K.G.: Overview of the RANSAC algorithm. Image Rochester NY **4**, 2–3 (2010)

Evolutionary People Tracking for Robot Partner of Information Service in Public Areas

Wei Quan[✉] and Naoyuki Kubota

School of System Design, Tokyo Metropolitan University,
Asahigaoka 2-6-6, Hino, Tokyo, Japan
quan-wei1@ed.tmu.ac.jp, kubota@tmu.ac.jp

Abstract. The future would be full of artificial intelligence definitely. Since Olympic game 2020 would be held in Tokyo, it is overwhelming important to give a navigating service to the tourist from the entire world. Even there would be a large number of volunteers then, there would be a lack of position absent. This paper described a vision system for robot system for airport navigation that set in the airport or other places. This visual system contained the detecting part and human counting part that combined with evolutionary and clustering algorithms and the experiment shows an efficient result in some cases.

Keywords: Robot vision · Evolutionary algorithm · People tracking

1 Introduction

Airport is a public place and thousands and hundreds of tourists arrived from all of the world and ready to go the world everyday. Despite the flight service, the service at the airport is also significant to the tourists. Thus a good navigation service would lead to a good public order and relieve the pressure for the staff at the airport.

For example, since the Olympic game 2020 would be held in Tokyo, it is overwhelming important to give a navigation to the travelers from all of the world. It will also bring a challenge to the airport navigating management. It is not a easy task because that we have to consider so many problems. Language problems would be the first task that need to be solved. Since the travelers come from different countries, the communication would be the problems if they can only speak his(her) native language. The multi-language capability would be too strict to staffs of the airport.

And because of the Olympic, the number of travelers at the airport would have an instantaneous growth at that time. And the number of travelers that have troubles also increase. Because of the limitation of human resource, we can not expect that there are sufficient staffs in the airport that is ready for giving the assistance. Thus robot navigator would be a optional solution for solving this kind of shortness of human resource.

© Springer International Publishing AG 2017
Y. Huang et al. (Eds.): ICIRA 2017, Part II, LNAI 10463, pp. 703–714, 2017.
DOI: 10.1007/978-3-319-65292-4_61

For a certain case, if a traveler cannot find his/her destination, then he might be stopped in front of the robot, this will give a chance to the robot for detecting this wondering person. Then according to the situation, robot will give him the help such as showing the way to his destination or providing the detail information about his/her flight.

Based on these situations, we proposed an evolutionary people tracking system in a robot system that is used for giving the information service in not only airport but also public areas. And the robot has the appearance as Fig. 1 shows. The left one is the prototype of the robot and the right is well-designed type. This robot system has the ability to recognize the people in front of it and counting the number of them. For the people who might have some troubles, it would moves towards this people and then provides a information service automatically.

The rest of this paper is constructed as follows: Sect. 2 will give the referent research background such as evolutionary robot vision, and a short description for the system construction can be seen in Sect. 3, whereas Sect. 4 shows the detail of the evolutionary tracking system. Section 5 describes the detail of the experiment, and in the last section, the conclusion is summarized.

Fig. 1. Appearance of the airport robot. Left one is the prototype whereas right one is advanced designed shape.

2 Research Background

Image processing is a method to perform some operations on an image, in order to get an enhanced image or to extract some useful information from it [5]. It contains several of categories, and computer vision and visual system might be one of the most popular area.

Visual system can be discussed from the dynamic point of view since the visual image is changing over time [6,7,9]. The visual systems including image processing can be divided into passive vision and active vision. In general, the passive vision is used for the focused information extraction toward a specific direction, while the active vision is used for the information extraction by updating the sensing direction and range in the vast area where the range of visual scene is restricted.

Comes from computer vision but different from it, robot vision considers more about the trade-off between optimality and adaptability. Robot vision is usually applied in the dynamic environment, thus it is significant to fulfill the real-time requirement.

Evolutionary computation [8] such as genetic algorithm is a filed of simulating evolution on a computer. Evolutionary optimization methods are fundamentally iterative generation and alternation processes of candidate solutions. [3] has proposed evolutionary robot vision for people tracking which shows efficient performance of multiple human face detection [4]. In this paper, we improved it and applied it in the multiple human detection and tracking and also made a good performance.

3   System Description

The whole system is constructed by four parts: surveillance sensor; computation core; moving robot and communication interface, which can be seen on Fig. 2.

Surveillance sensor performs like the "eye" of the system and depth camera is usually used. We use ASUS Xtion live pro in this case. And it is deserved to be mentioned that we just rotate the Xtion to make it more suitable in the robot. Therefore the open source for referent applications such as toolkit for human skeleton extraction is not available in our case. We need to find a suitable solution that is less depends on other sources.

Computation core is the "brain" of the robot. It receives the data from the Xtion sensor, and recognizes the human in front of it, and also keeps calculating the coordinate of the target during time the robot moving towards the target person. And it is also the server of the TCP protocol that sends the referent command to the clients according to the situation.

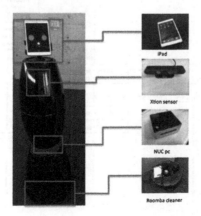

Fig. 2. Detail for the robot system.

And for robot's base, we used robot cleaner iRobot for robot moving component. It has lots of advantages such as that it provides its software development kit as the open source and we can control its moving at will.

Last but not least, good interface would leave a good impression to users. Therefore we developed an application under iOS system and ran it on iPad for the communicating interface between robot and users.

In this paper, we focus on the part of robot vision, and our main target is to detect the person in front of the robot correctly. For more detail, that detect the number of the person and make a simple judgment that whether these person need help or not. After then, it will send command to the iPad interface for making different operations. The detail of this part will be given in the following section.

4 Human Detection and Recognition

Human detection and recognition is the first and also the most important part in this robot system. Accurate detecting is the premise for the following computations. In this case, we applied depth camera for the detecting sensor and all of the human detection and recognition will based on the RGB-D images gotten from it (Fig. 3).

4.1 Coordinate Transform

The images captured by depth camera have extra information of the distance comparing with other single cameras. Thus it is possible to transform the pixels in this image into 3 dimensional voxels if we got the parameters of the depth

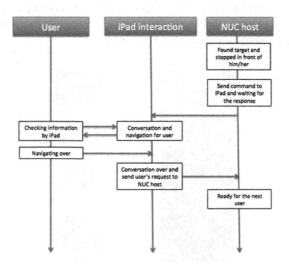

Fig. 3. Procedure flow for the whole system.

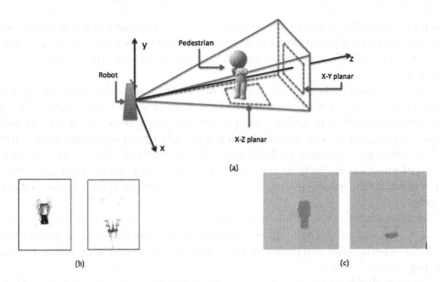

Fig. 4. RGB-D images and the projections on x-y and x-z planar. (a) Robot coordinate system; (b) The projection images on x-y and x-z planar respectively; (c) The discrete projection space of voxels on x-y and x-z planar respectively.

sensors once we got the RGB-D images. For a pixel $p(u, v, d)$ on the image, where u and v are the position of the pixel on camera screen coordinate and d is the depth of this pixel. We make the transform it into the 3 dimensional voxel that $p(u, v, d) \rightarrow v(x, y, z)$ under the camera coordinate system. And the coordinate system can be shown in Fig. 4(a).

4.2 Foreground Detection

It is important to detect the foreground target. In our system, we only concern about the target that within a certain area and we call it region of interest. The foreground voxel sets V can be detected as

$$V = \{v_i(x_i, y_i, z_i, b_i)\} \tag{1}$$

Where x_i, y_i, z_i are the coordinate that calculated as previous subsection, and b_i is a Boolean value and will be 1 if v_i satisfy $x_i \in [0, t_x], y_i \in [0, t_y]$ and $z_i \in [0, t_z]$ (t_x, t_y and t_z are the thresholds for region of interest at each direction respectively), and 0 otherwise.

4.3 Human Candidate Detection and Nearest Neighbor Clustering

In this robot system, we applied genetic algorithm for human detection and nearest neighbor is applied for the calculation of human number.

It is a little slow to calculate in the coordinate space described above. For the purpose of real-time approach, we shrink down this continuous 3 dimensional space into a smaller discrete 3 dimensional space. And the shrinking size should depend on different situations. In our case, we shrieked all the region of interest into $M \times M \times M$ discrete cubes, and each cube represents the smallest unit in this area.

First proposed by Holland [1] Genetic algorithm is a random-based algorithm inspired by the process of natural selection that belongs to the larger class of evolutionary algorithms, and it is commonly used to generate high-quality solutions to optimization and search problems by relying on bio-inspired operators such as mutation, crossover and selection. We attribute all the genetic units into these cubs and let all the units reproduce, crossover and mutate for a given times of iteration, and choose the units with the highest fitness values as the detected targets.

Suppose there are a series of genetic units $G = \{g_j(x_{gj}, y_{gj}, z_{gj})\}$, where $j = 1, 2 \ldots N$, and N is the number of genetic units. For initial part, we attribute all of the genetic units into region of interested averagely and with a little random shifting.

For each divide cube $c_i(l_{c_i}, m_{c_i}, n_{c_i})$ that satisfy $l_{c_i}, n_{c_i}, n_{c_i} = 1, 2 \ldots M$, its projection on x-y and x-z planar as Fig. 4(c) shows.

For conventional, we project all these cubes into x-y planar and x-z planar and calculate the density on these two planar respectively.

And we choose the density that foreground voxel project into two planar as the fitness function of genetic units as Fig. 2(c) shows, the x-y planar and x-z planar have been compressed into a series of grids, and each grid represents for a position on its corresponding planar. Let G_{x-y} and G_{x-z} represent the grids of grids at x-y and x-z planar respectively, and the size of grids are $m \times m$, the value of each grid represents the number of voxels that project into this grid:

$$G_{x-y} = \{g_{x-y}(j, k) | 0 < j < m, 0 < k < m\} \tag{2}$$

And for the fitness value of each grid is

$$f_1(j, k) = \sum v_i \tag{3}$$

where the coordinate of v_i satisfies $\frac{j*t_x}{M} < x_i < \frac{(j+1)*t_x}{M}, \frac{k*t_y}{M} < y_i < \frac{(k+1)*t_y}{M}$ and $b_i = 1$.

Similarity, we can also make the fitness value on x-z planar

$$G_{x-z} = \{g_{x-z}(j, l) | 0 < j < m, 0 < l < m\} \tag{4}$$

And the value for each grid on x-z planar calculated from

$$f_2(j, l) = \sum v_i \tag{5}$$

where the coordinate of v_i satisfies $\frac{j*t_x}{M} < x_i < \frac{(j+1)*t_x}{M}, \frac{l*t_z}{M} < z_i < \frac{(l+1)*t_z}{M}$, and $b_i = 1$.

According to the formulas above, the fitness function for a genetic unit could described as:

$$f(g_i) = \lambda_1 f_1(j, k) + \lambda_2 f_2(j, l) \tag{6}$$

Algorithm 1. Combined evolutionary multiple detection algorithm

INPUT
$G = \{g_1, g_2 ... g_N\}$
θ as the threshold filter
OUTPUT
$G_t \subset G$ for the set of candidate targets
K is the number of detected people
START
for $i \leftarrow 2$ to t for the maximum iteration time **do**
   for $j \leftarrow 1$ to N the number of genetic units **do**
      Calculating the fitness value $f(g_j)$ for each genetic g_j
      Reproduction the offspring according to the fitness value
      Generate offspring from the crossover of two parents
      Mutate some generated offspring
   end for
end for
for $j \leftarrow 1$ to N the number of genetic units **do**
   $G_c \leftarrow \{g_j\} \cup G_t$ if $f(g_j) \geq \theta$
end for
$K_1 = \{t_1\}$ add K_1 to K for initialization
$k = 1$
for $i \leftarrow 2$ to n **do**
   find the t_i in some cluster K_m in K such that $dis(t_i, k_m)$ is the smallest.
   if $dis(t_i, t_m) < \theta$ **then**
      $K_m \leftarrow K_m \cup \{t_i\}$
   else
      $k \leftarrow k + 1; K_k = \{t_i\}$; add K_k to K
   end if
end for

where the position $c(l, m, n)$ for a genetic unit represents the position within the discrete space, and λ_1, λ_2 are fixed controlling parameters.

The set of candidate targets contains all the positions that might be the targets. Nevertheless, there is a situation that some of these positions belong to one human because these position probably point out part of these human. Thus it is necessary to use clustering algorithm to cluster these candidate into one cluster.

In some cases, different genetic units would have found the same targets, thus it is important to cluster these units into one cluster. Clustering algorithms can be categories into two kinds: sequence-based and distance-based. Typical algorithms for sequence-based like k-means clustering algorithm, and nearest neighbor can be a good sample for distance-based algorithm. Both of these algorithms have advantages and defects. For instance, we have to get the exact number of the clusters before calculation if we want to applied k-means algorithm, and in most cases it is hard to get such information beforehand. Thus this is one of the reasons that we chose nearest neighbor instead.

As mentioned before, it is necessary to cluster all of the genetic units that have the high fitness values. And we tried the Nearest Neighbor clustering algorithm for this issue.

Combined with genetic algorithm and nearest neighbor algorithm, the pseudo code can be described as Algorithm 1.

The radius would be a significant value in this system. It largely affects the result of the tracking. Thus a proper value would be the ideal. The experiment section will show the affection of the radius value.

4.4   Targets Tracking

We have already recognized the number and positions of the targets by the process that described above, but it is still need to track the targets from different frames for a series of time. We also use the nearest neighbor for tracking in this part.

Suppose all the tracked targets at time $p-1$ are in the set $T^{p-1} = \{t_i^{p-1}\}, i = 1, 2m$, where t_i^{p-1} is the tracked target at time $p-1$ and has the feature $t_i^{p-1} = (j_i, k_i, l_i, d_i)$, where j_i, k_i, l_i are the position of the target at 3 different direction in discrete space respectively and d_i is time decay value.

For the next time p, we have detected candidate targets $C^p = \{c_o^p\}, j = 1, 2n$, and $c_o^p = \{j_o, k_o, l_o\}$ is the corresponding position. For each targets t_i^{p-1}, we try to find its current position among the candidate targets set C^p.

We try to find the best matching between current targets and candidate targets by bidirectional matching like

$$\theta_{t_i^{p-1}} = \arg\min_{c_o^p \in C^p} nn(t_i^{p-1}, c_o^p) \tag{7}$$

$$\delta_{c_o^p} = \arg\min_{t_k^{p-1} \in T} nn(c_o^p, t_k^{p-1}) \tag{8}$$

where the function $nn(\cdot)$ satisfies:

$$nn(a, b) = \begin{cases} \|a - b\| & if \ \|a - b\| \le r \\ \infty & otherwise \end{cases} \tag{9}$$

For r is the threshold for the radius.

The target t_i^{p-1} will update to t_i^p and put into set T^p only if it has found its corresponding candidate targets c_o^p, i.e., $\theta_{t_i^{p-1}} = c_o^p$ and $\delta_{c_o^p} = t_i^{p-1}$. And the time decay value d_i will be refreshed to 0.

And if t_i^{p-1} cannot find its corresponding candidate target or $\theta_{t_i^{p-1}} = c_o^p$ but $\delta_{c_o^p} \neq t_i^{p-1}$, then it means that this target is occluded or disappear. d_i would refreshed as $d_i = d_i + 1$. And if d_i has risen to ti that is the threshold, and then this target will be deleted.

For the candidate target c_o^p, if it can not find it corresponding target, or $\delta_{c_o^p} = t_i^{p-1}$ but $\theta_{t_i^{p-1}} \neq o_c^p$, then it will be regarded as the potential target. And if it has become the potential target continuously for a long time, it will be regarded as a new target and put into T^p.

5 Robot Moving Control and Human-Robot Communicating

The system should control the robot moving towards the target once it is been detected for a while. The power of movement for the robot is provided by the robot cleaner iRobot. Once it received the command from the system that a person that is interested in the robot appeared, it get the position of this target person, and keep moving towards to him/her until the distance to the target is less than a threshold.

It is also important to have a good user interface for communicating between users and robot. Travelers in the airport should get the necessary information as quick as possible once they consult to the robot.

Figure 5 shows an example for the navigation service. When the iPad showing the facial expression, press the button and then talk to it. Currently the system can recognize English, Japanese and Chinese languages. After talked by using one of these three languages, the menu for the same language will be shown to the user. And user start to chose the service. For security, the users have to scan their boarding pass before get service. And after then the service will be shown to the user, and the robot will be ready to provide the next service.

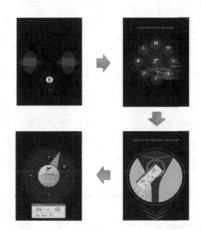

Fig. 5. Example of a navigation service.

6 Experiment

6.1 Target Detection

In order to make a comparison between different cases, we categorized the whole scene into case of single and multiple persons. For single case, the experiment shows a good performance. We tested the experiment with several different radius for nearest neighbor.

It is obvious to get the conclusion from the Table 1 that for single case, the accuracy rate would be higher if the size of the radius becoming larger. This is because in single case, we do not need to consider the influence from other targets during clustering. And the error of under estimated might caused in the part of foreground abstraction.

To make a comparison with single case, we also made the experiment with the case that have two people in the scene. According to the distance between these two people, we divide it into 3 sub cases: near, middle and far. And the result shows in Tables 2, 3 and 4.

It is much more complicating when there are two or more people appears in the scene at the same time. Radius size is not the only fact that affects the result anymore, and the distance between each target also has a strong for detecting. Different from the single case that larger radius has the more accuracy rate, for different distance between targets, and different radius would be the proper size. Thus the ideal radius size is no longer a fixed value any more.

Table 1. Accuracy rate for single person with different radius.

Radius size	50	100	200	400
Accuracy rate	86%	98%	99%	100%
Over estimated rate	14%	0	0	0
Under estimated rate	0	2%	1%	0

Table 2. Accuracy rate for double person with the large distances between them.

Radius size	50	100	200	400
Accuracy rate	83%	98%	99%	100%
Over estimated rate	17%	1%	0	0
Under estimated rate	0	1%	1%	0

Table 3. Accuracy rate for double person with the middle distance between them.

Radius size	50	100	200	400
Accuracy rate	86%	100%	98%	89%
Over estimated rate	14%	0	0	0
Under estimated rate	0	0	2%	11%

Table 4. Accuracy rate for double person with the short distance between them.

Radius size	50	100	200	400
Accuracy rate	89%	95%	61%	0%
Over estimated rate	11%	0	0	0
Under estimated rate	0	5%	39%	100%

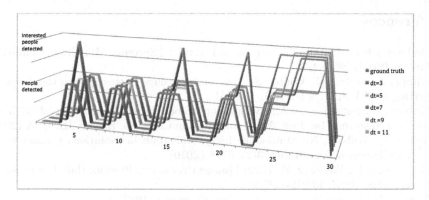

Fig. 6. Comparison for different value of threshold d_t.

6.2   People Tracking

Currently we introduce position as the only feature for people tracking. Therefore the radius is also significant in this stage.

Usually the travels pass through the screen by 5 to 10 frames according to the frames per second. And if there is a traveler that spends more time in the screen, he/she will be regarded as interested in this robot or having some problems. We made the experiment result of different time trigger.

In order to show the effect of the threshold value d_t, we made the experiment and the result can be seen in Fig. 6. In this case, the whole flow is divided into 3 status: none; people detected and interested people detected. We can see from the figure that smaller values for d_t will make the trigger for defining the human of interest earlier than usual whereas the larger values give the opposite affect. Therefore in this case, $d_t = 7$ is the suitable selection.

7   Conclusion

We constructed a robot navigating system for airport and other public areas and detecting interested pedestrian by heuristic and clustering algorithms. It shows a ideal performance on several cases, and the computation cost keeps in a low level so that this system can be applied into the platform which does not have high quality of the hardware device.

Currently the result is largely depending on the parameters setting such as the radius in nearest neighbor, and we set these parameters only by the experience. In the future we would focus on the generation of the self-adaptive parameters to reduce the affect from parameters setting as low as possible.

Nevertheless, we only considered the position of the target as the feature until now. It is not sufficient in some special cases. Later we will introduce other features to improve its robustness.

References

1. Holland, J.H.: Adaptation in Natural and Artificial Systems. MIT Press, Cambridge (1992). ISBN:0262082136
2. Klippe, A., Freksa, C., Winter, S.: You-are-here maps in emergencies the danger of getting lost. J. Spat. Sci. **51**(1), 117–131
3. Kubota, N., Sulistijono, I.A.: Evolutionary robot vision for people tracking based on local clustering. In: 2008 World Automation Congress, Hawaii, HI, USA (2008)
4. Yorita, A., Kubota, N.: Multi-stage fuzzy evaluation in evolutionary robot vision for face detection. Evol. Intell. **3**(2), 6778 (2010)
5. Gonzalez, R.C., Woods, R.E.: Digital Image Processing. Prentice Hall, Upper Saddle River (2008). ISBN 978013168728-8
6. Marr, V.D.: Vision. W. H. Freeman, San Francisco (1982)
7. Chen, C.H., Pau, L.F., Wang, P.S.P.: Handbook of Pattern Recognition and Computer Vision. World Scientific Publication Co. Pte Ltd., Singapore (1993)
8. Fogel, D.: Evolutionary Computation. IEEE Press, New York (1995)
9. Bülthoff, H.H., Wallraven, C., Lee, S.-W., Poggio, T.A. (eds.): BMCV 2002. LNCS, vol. 2525. Springer, Heidelberg (2002). doi:10.1007/3-540-36181-2

Robot Grasping and Control

Study on the Static Gait of a Quadruped Robot Based on the Body Lateral Adjustment

Qingsheng Luo[1(✉)], Bo Gao[1(✉)], and Rui Zhao[2(✉)]

[1] School of Mechatronical Engineering, Beijing Institute of Technology,
Beijing 100081, China
106788978@qq.com
[2] School of Mechanical Engineering, Beijing Institute of Technology,
Beijing 100081, China

Abstract. In order to improve the static gait performance of a quadruped robot, the method based on the body lateral adjustment has been proposed. In this paper, two kinds of optimization models were designed, which were based on the body intermittent lateral adjustment, and based on the body continuous lateral adjustment, designed a evaluation function with the aim of stable motion and fast walking speed; Besides, the simulations of the virtual prototype model based on those two kinds of optimization models have been finished in MATLAB and ADAMS, and the simulation results were analyzed by the de-signed function; Finally, we can find that the better optimization model to improve the static gait performance of a quadruped robot is based on the body continuous lateral adjustment. Those simulation results revealed that it greatly enhances the quadruped robot's static gait performance by using the designed optimization model adjusting its body lateral position continuously.

Keywords: Quadruped robot · Fitness function · The body lateral adjustment · Static gait

1 Introduction

Static gait is the gait that quadruped robot has at least three legs to the ground at any time, such as walk gait and so on. The static gait is a common gait of the quadruped robot movement. It is usually possible to obtain a high stability by using a static gait when the quadruped robot travels at a low speed or in the rugged road conditions.

At present, scholars at home and abroad carried out a lot of related re-searches on the static gait of quadruped robot. Stability margin is set as the evaluation criteria for the static gait planning research, they carry out static gait planning based on the support of the triangle theory, such as Xu and Wan, Lin, Yin, Tan, et al. [1–4]. This method only serves as a criterion for the static stability of the quadruped robot moving on the horizontal plane, which has certain limitations [5].

Further researches, such as Pan, Yang, Hirose, et al., they use the stability margin as the evaluation criteria in the research process, taking into account the impact of the body center of gravity for motion performance while the quadruped robot in the process of walking, the body lateral position adjustment in the process of walking is

© Springer International Publishing AG 2017
Y. Huang et al. (Eds.): ICIRA 2017, Part II, LNAI 10463, pp. 717–726, 2017.
DOI: 10.1007/978-3-319-65292-4_62

introduced, improved its motion performance to a great degree. However, these studies based on body lateral adjustment, usually used intermittent lateral adjustment. From the perspective of bionics analysis, the quadruped reptile is traveling in a continuously adjusted gait in the low speed static gait.

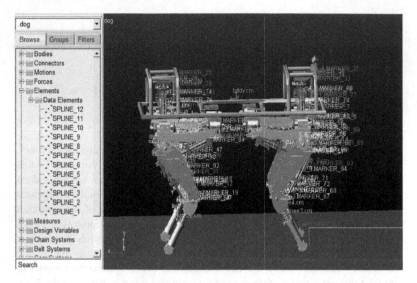

Fig. 1. Virtual prototype of the quadruped robot

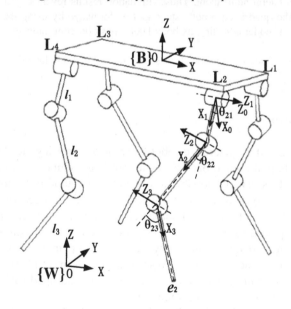

Fig. 2. Kinematic model of the quadruped robot

Therefore, we used MATLAB and ADAMS carrying out static gait simulation test on the quadruped robot virtual prototype, respectively based on the body intermittent lateral adjustment and body continuous lateral adjustment. And we compared and analyzed the simulation data. It is can be explained that the static gait based on body continuous adjustment can get better motion performance. The virtual model and kinematic model in ADAMS simulation experiment are shown in Figs. 1 and 2.

2 Static Gait Planning Based on Body Lateral Adjustment

Static gait based on the body lateral adjustment can be roughly divided into two categories, the body intermittent lateral adjustment and the body continuous lateral adjustment. The static gait based on the body intermittent adjustment is the gait that quadruped robot adjusts the body sideways to move the center of gravity only when the four legs touch the ground at the same time. Meanwhile the static gait based on the body continuous adjustment is the gait that quadruped robot is accompanied by a body lateral adjustment while completing the single leg lift.

2.1 Static Gait Planning Based on Body Intermittent Lateral Adjustment

In the continuous crawling static gait, the robot's static gait of motion performance is improved by adding the body intermittent adjustment. According to the order of 4-2-3-1 leg swing [5], the optimized gait diagram is shown in Fig. 3.

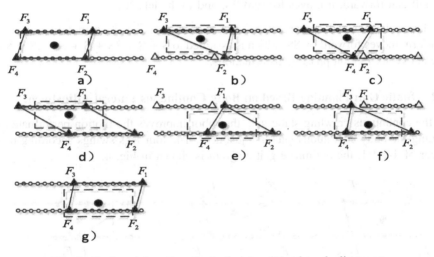

Fig. 3. Static gait based on the body intermittent lateral adjustment

The right direction is the forward direction of the robot in Fig. 3. The rectangular box represents the quadruped robot's four hip rotation joints in the horizontal plane projection, fixed in the body coordinate system, can be regarded as the quadruped robot body, long b, width a. In the direction of the robot's foot movement, the radius of the hollow circle is S, which marks the movement distance between the foot end and the body relative to the ground.

In a crawling gait based on the intermittent adjustment of the body, the robot moves the body center of gravity when the four legs provide support, as shown in (d) and (g) in Fig. 3. In these two phases, the body center of gravity forward distance is two times of the distance when the single leg swings, and only in these two phases, the body center of gravity has a lateral movement of 2L.

Quadruped robot platform forward movement is reduced to the body center of gravity forward, the specific implementation process is as follows:

(a) →(b): Four legs of the quadruped platform moves to the initial position to ensure stability requirements when the quadruped platform moves in a crawling gait. The center of gravity moves to the left and forward, it moves forward S, and to the left L;

(b) →(c): Leg F4 complete a lift action, span E, while the center of gravity moves forward S;

(c) →(d): Leg F2 complete a lift action, span E, while the center of gravity moves forward S;

(d) →(e): The robot's four legs touch the ground, the center of gravity moves to the right and forward, it moves forward 2S, and to the right 2L;

(e) →(f): Leg F3 complete a lift action, span E, while the center of gravity moves forward S;

(f) →(g): Leg F1 complete a lift action, span E, while the center of gravity moves forward S;

(g) →(b): The robot's four legs touch the ground, the center of gravity moves to the left and forward, it moves forward 2S, and to the left 2L.

From the planned movement process, we can see that in a motion period, the platform moves forward by 8S, which is assigned to S + S + 2S + S + S + 2S = 8S in the above six motion phase transformations.

2.2 Static Gait Planning Based on Body Continuous Lateral Adjustment

In the continuous crawling static gait, the robot improves the motion performance by adding the body continuous lateral adjustment. The four legs swings according to the order of 4-2-3-1, the optimized gait diagram is shown in Fig. 4.

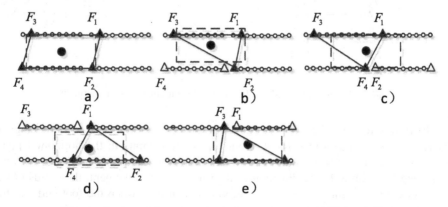

Fig. 4. Static gait based on the body continuous lateral adjustment

It can be seen from Fig. 4 that the robot moves the phase of the body center of gravity while completing the single leg lift action in the crawling gait based on the body continuous lateral adjustment, as shown in (b), (c), (d) and (e). In these four phases, the body center of gravity moves forward 2S, and lateral 2L.

Quadruple platform forward movement is simplified as the fuselage center of gravity forward, the specific implementation process is as follows:

(a) →(b): Four legs of the quadruped platform moves to the initial position to ensure stability requirements when the quadruped platform moves in a crawling gait. The center of gravity moves to the left and forward, it moves forward S, and to the left L;
(b) →(c): Leg F4 complete a lift action, span E, while the center of gravity moves forward 2S, and to the right 2L;
(c) →(d): Leg F2 complete a lift action, span E, while the center of gravity moves forward 2S, and to the right 2L;
(d) →(e): Leg F3 complete a lift action, span E, while the center of gravity moves forward 2S, and to the right 2L;
(e) →(b): Leg F1 complete a lift action, span E, while the center of gravity moves forward 2S, and to the right 2L.

From the planned movement process, we can see that in a motion period, the platform moves forward by 8S, which is assigned to 2S + 2S + 2S + 2S = 8S in the above six motion phase transformations.

3   The Design of Evaluation Function

The smooth, linear and other aspects of the robot movement should be balanced considerations, in order to make the quadruped robot movement with the best motion performance. Through the observation and analysis of the simulation experiment and the physical experiment for a long time, three evaluation criteria for the performance of the movement have been finally determined:

L_{max}:   Maximum forward distance (Movement speed index)
σ_y: Standard deviation of lateral fluctuation in horizontal plane (Lateral stability index)
σ_z: Fluctuation standard deviation in vertical direction (Longitudinal stability index)

The construction of the evaluation function should be paid attention to the following two points: (1) Set the weight coefficient according to the effects degree of the target volume on the final optimization. (2) Set the weight coefficient according to the values order of magnitude between the aim parameters. Setting the unreasonable weight coefficient will get a bias on the direction of the optimization target, and it is difficult to achieve the desired goal [9].

After several experiments and test data analysis, the evaluation function is:

$$f = 20 * \sigma_y + 2 * \sigma_z + 500/L_{\max}$$

The evaluation function is divided into two kinds of functions, global maximum and global minimum. According to the analysis of above three robot motion performance indexes, The smaller the σ_y (Horizontal fluctuation in the horizontal plane Standard deviation) and σ_z (the vertical direction of the volatility of the standard deviation) is, the more stable the quadruped robot will be, and the bigger the $L_{\max}$ (maximum forward distance) is, the faster the quadruped robot movement will be. Therefore, in order to make the quadruped robot moving faster. We can ensure that the evaluation function is the global minimum. That is, the smaller the evaluation function value is, and the better the quadratic robot's motion performance will be.

The online version of the volume will be available in LNCS Online. Members of institutes subscribing to the Lecture Notes in Computer Science series have access to all the pdfs of all the online publications. Nonsubscribers can only read as far as the abstracts. If they try to go beyond this point, they are automatically asked, whether they would like to order the pdf, and are given instructions as to how to do so.

Please note that, if your email address is given in your paper, it will also be included in the meta data of the online version.

4 Simulation Test

4.1 Simulation Description

In this paper, the experimental simulation model is established based on the laboratory hydraulic quadruped robot entity 1:1 ratio model. The main parameters are:

$$l_1 = 0.43\,\text{m}, l_2 = 0.43\,\text{m}, l_3 = 0.14\,\text{m}$$
$$a = 1.0854\,\text{m}, b = 0.6\,\text{m}, H_c = 0.8\,\text{m}$$

Where l_1, l_2, l_3 is the length of the legs joints (see Fig. 2), a, b, H are expressed as: robot body length, width, and the altitude of the robot centroid from the ground.

In this paper, the composite cycloid is used as the quadruped robot foot trajectory, and the cycloid can make the speed and acceleration of the foot end are 0 at the phase change point. In the oxz coordinate system (Fig. 5), plan the swing phase of the foot trajectory. Since there is a fixed phase difference and distance difference between the each foot trajectory, the foot trajectory [9] is described by taking number 1 leg as an example:

$$\begin{cases} x_1 = S_0 \cdot \left[\frac{t-kT}{T_S} - \frac{1}{2\pi} \sin\left(\frac{2\pi}{T_S}(t-kT) \right) \right] + kS_0, \, kT \leq t < T_S + kT \\ x_1 = (k+1)S_0, \, T_S + kT \leq t \leq T + kT \end{cases} \tag{1}$$

$$\begin{cases} z_1 = \frac{H_0}{4+\pi}\left[\frac{4\pi}{T_S}(t-kT) - \sin\left(\frac{4\pi}{T_S}(t-kT)\right)\right], kT \le t < \frac{T_S}{4} + kT \\ z_1 = \frac{4H_0}{4+\pi}\left[\sin\left(\frac{2\pi}{T_S}(t-kT) - \frac{\pi}{2}\right) + \frac{\pi}{4}\right], \frac{T_S}{4} + kT \le t < \frac{3T_S}{4} + kT \\ z_1 = \frac{4\pi H_0}{4+\pi}\left\{\left(1 - \frac{t-kT}{T_S}\right) - \frac{1}{4\pi} - \sin\left[4\pi\left(1 - \frac{t-kT}{T_S}\right)\right]\right\}, \frac{3T_S}{4} + kT \le t < T_S + kT \\ z_1 = 0, T_S + kT \le t \le T + kT \end{cases} \quad (2)$$

Where H_0 is the foot end maximum ground height, T is the gait cycle, and T_S is the duration of a swing phase.

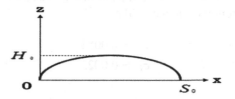

Fig. 5. The foot trajectory in oxz

4.2 Simulation Results

Using MATLAB and ADAMS for simulation analysis, we get the fluctuating pictures of quadruped robot virtual prototype centroid in X, Y, Z direction, which is based on the body intermittent lateral adjustment and the body continuous lateral adjustment static gait.

4.2.1 The Centroid Wave Pictures Based on the Body Lateral Adjustment

In Fig. 6, bx represents the fluctuating pictures of the centroid in the X-axis direction based on the body intermittent lateral adjustment. lx represents the fluctuating pictures of the centroid in the X-axis direction based on the body continuous lateral adjustment.

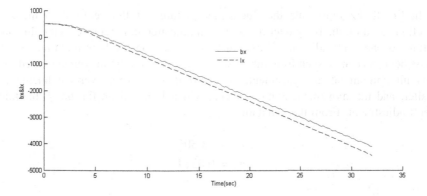

Fig. 6. The centroid motion based on the body lateral adjustment in the X direction

As we can know from the figure, the centroid moves forward longer distance and in a faster speed based on the body continuous lateral adjustment than the body intermittent lateral adjustment. From the simulation data:

$$L(bx)_{\max} = 4.012e + 03; L(lx)_{\max} = 4.925e + 03$$

In Fig. 7, by represents the fluctuating pictures of the centroid in the Y-axis direction based on the body intermittent lateral adjustment. ly represents the fluctuating pictures of the centroid in the Y-axis direction based on the body continuous lateral adjustment. As we can see from the figure, compared with the movement based on the body intermittent lateral adjustment, the centroid of the robot horizontal fluctuation is smaller, and the movement is more stable, which is based on the body intermittent lateral adjustment. From the simulation data:

$$\sigma_{by} = 0.0053$$
$$\sigma_{ly} = 0.0023$$

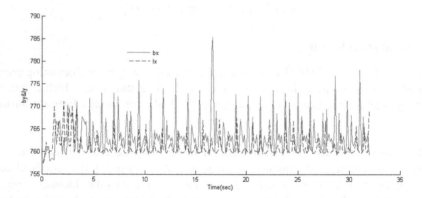

Fig. 7. The centroid motion based on the body lateral adjustment in the Y direction

In Fig. 8, bz represents the fluctuating pictures of the centroid in the Z-axis direction based on the body intermittent lateral adjustment. lz represents the fluctuating pictures of the centroid in the Z-axis direction based on the body continuous lateral adjustment. As can be seen from the figure, compared with the movement based on the body intermittent lateral adjustment, the centroid of the robot vertical fluctuation is smaller, and the movement is more stable, which is based on the body intermittent lateral adjustment. From the simulation data:

$$\sigma_{bz} = 0.0806$$
$$\sigma_{lz} = 0.0314$$

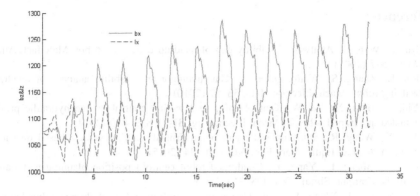

Fig. 8. The centroid motion based on the body lateral adjustment in the Z direction

4.3   Analysis of Experimental Data

Using the above designed evaluation function,

$$f_b = 20 * \sigma_y + 2 * \sigma_z + 500/L_{\max} \approx 0.3918$$
$$f_1 = 20 * \sigma_y + 2 * \sigma_z + 500/L_{\max} \approx 0.2103$$

Therefore we can conclude that,

$$f_1 < f_b$$

So the quadruped robot movement based on the body continuous adjustment, the center of mass performance is better.

5   Conclusion

In this paper, we carry out research and studies on the static gait based on the body lateral adjustment. The quadruped robot virtual prototype static gait simulation test based on the body intermittent adjustment and the body continuous adjustment are respectively conducted by the union of MATLAB and ADAMS. Analyze the experimental data graph, and use the designed evaluation function to evaluate the static gait motion performance. The conclusion is that the body continuous lateral adjustment is more conducive to improve the quadruped robot static gait motion performance. From the perspective of bionics, the body swings continuously and horizontally to ensure the stability of the movement while the quadruped animals is walking, so continuous body adjustment is more in line with the principle of bionics.

References

1. Xu, Y., Wan, L.: Analysis of stability gait of quadruped walking robot. Manufact. Autom. **23**(8), 5–7 (2001)
2. Lin, L., Zhang, S., Wang, Q., et al.: Gait planning and stability analysis of quadruped walking robot. Mach. Electron. **1**(1), 53–57 (2010)
3. Yin, Y., Bian, X., Lu, W.: Gait planning and motion control on the hydraulic pressure actuated quadruped robot. J. Mech. Electr. Eng. **31**(7), 839–843 (2014)
4. Tan, Y., Wang, R.: Study on static gait of quadruped walking vehicle based on support polygon. Agricu. Equip. Veh. Eng. **6**(06), 20–24 (2015)
5. Han, X., Shang, L., Yang, Y.: Quadruped robot centroid position planning and stability analysis. Comput. Simul. (06), 308–313 (2015)
6. Pan, S., Shi, J., Wang, J., et al.: Crawl gait optimization for quadruped robot based on gravity center lateral movement. Mech. Sci. Technol. Aerosp. Eng. **34**(06) (2015)
7. Yang, C.: Research on gait planning for a hydraulic quadruped robot. Beijing Institute of Technology (2015)
8. Hirose, S., Fukuda, Y., Yoneda, K.: Quadruped walking robots at Tokyo Institute of Technology. Robot. Autom. Mag. IEEE **16**(2), 104–114 (2009)
9. Hu, Y.: Practical Multiobjective Optimization. Shanghai Scientific & Technical Publishers, Shanghai (1990)
10. Han, B., Jia, Y., Li, H., et al.: Posture adjustment for quadruped robot troting on a slope. Trans. Beijing Inst. Technol. (03), 242–246 (2016)

Analyses of a Novel Under-Actuated Double Fingered Dexterous Hand

Rui Feng[1](✉) ⓘ and Yifan Wei[2] ⓘ

[1] Beijing University of Aeronautics and Astronautics,
Beijing 100191, People's Republic of China
fengrui886@126.com
[2] Farragut High School, Koxville, TN 37934, USA

Abstract. In the field of service robot, implementation of accurate and effective grab is usually the key step in operating process. Common grab of traditional robot is completed by special gripper or holder. Usually these special grippers are composed of mechanisms which are relatively simple, and only used for grab of specific objects, lacking of force feedback and strong universality. Nevertheless, dexterous hand could overcome these shortcomings, which can complete operation with more extensive adaptability. This manuscript designs a kind of double slider-crank mechanism based under-actuated double fingered dexterous hand, describes double-joint finger mechanism and analyzes movements of two grasp modes. According to the grasp of objects in small size, mechanism's output force performance is analyzed, and movement characteristics of finger mechanism in two grasp modes are described from perspective of joint angles. Degree of adaption is proposed and analyzed, and grasp experiments are designed and performed. At the end of this manuscript, experiments results are showed and analyzed. The results show that the mechanism designed by this manuscript has adaption ability to different sizes and shapes.

Keywords: Under-actuated · Dexterous hand · Self-adaption ability · Degree of adaption · Grasp mode

1 Introduction

Under-actuated mechanism is a kind of mechanism in which the number of independent actuations is less than the number of mechanism degrees of freedom (DOFs) [1]. Study and application on dexterous hand have important implications for reducing complexity of drive mechanism and control system, promoting compaction and lightness of mechanism as well as reducing energy consumption. Under-actuated mechanisms with more DOFs can realize coordination between joint angles by elastic components in general, which make it achievable to adjust grasp configuration of finger mechanism according to difference of objects in size and geometry.

To realize integration of actuators, tendon transmission is widely adopted by current studies. Such as IOWA hand [2], UB-III hand [3]. In past decade, DLR hand has complex tendon system to provide high dynamic performance [4]. ACT hand mimics human musculoskeletal [5]. X-Hand can replicate human grasping functions [6].

© Springer International Publishing AG 2017
Y. Huang et al. (Eds.): ICIRA 2017, Part II, LNAI 10463, pp. 727–738, 2017.
DOI: 10.1007/978-3-319-65292-4_63

Tendons imitate human physiological structure better but perform not well in motion stability and precision. Link transmission is more suitable to provide greater output force. Such as TBM hand, which uses four-bar linkages as its finger mechanism, bringing problem of joint motion coupling [7]. Some compliant mechanisms such as continuum differential mechanisms are proposed [8]. To keep the DOFs and reduce number of actuators, as well as to solve the problem of motion coupling, a kind of under-actuated mechanism is proposed in this manuscript.

This manuscript presents the design, description and dynamic analyses of a novel double slider-crank mechanism based under-actuated double-fingered dexterous hand, according to different grasping attitudes. The feasibility and self-adaptation ability are proved by experiments of grasp.

2   Design of Mechanism Configuration

With this chapter, design theory of the dexterous hand is shown and performance of the mechanism is discussed by dynamics and kinematics analysis.

2.1   Two Main Grasping Mode of Human Hand

This section shows two main modes extracted from the moving observation of human hand during grasp (Fig. 1).

Fig. 1. Two attitudes of finger grasp

Mode 1: If the grasped object is small in size, human hand takes the mode of contacting object by distal phalanxes. In this case, it is considered that the object is contacted by 2 points with fingers. Contact pressure is transmitted via these two points, which is combined with the friction employed to realize force closure. The movement can be described as a single phase that two fingers rotate about root joints in opposite directions with the fingers straight until two fingertips contact the object.

Mode 2: If the object is large in size, the fingers flex inward to wrap around the object in order to grasp it firmly. This mode adds the contact points between fingers and object up to 3 or 4. And the contact points is closer to the root of fingers making arm longer, so as to provide greater force on the premise that the motor output torque is constant, thus realize more grasp weight. The movement can be divided into 2 phases:

Phase 1 is just same to the single phase of Mode 1. Phase 2 starts at the moment that proximal phalanxes contact with object and stop to rotate. Then the distal phalanxes continue to rotate about middle joints until they touch the object, with fingers wrapping around the object.

2.2 Finger Mechanism of Dexterous Hand

Analysis of DOFs. To realize the most basic grasp function, the dexterous hand retains thumb and index finger as the fingers used to grasp. The two fingers are designed to be distributed symmetrically relative to palm with the same structure and geometrical dimensions. Figure 2(a) shows the kinematic structure of the dexterous hand. In which, each finger consists of 2 joints and 2 phalanxes. 4 DOFs are configured in mechanism, actuated by 2 motors. Therefore, 2 redundant DOFs are retained to realize hand's self-adaptation ability and to meet the demands of various size or geometry.

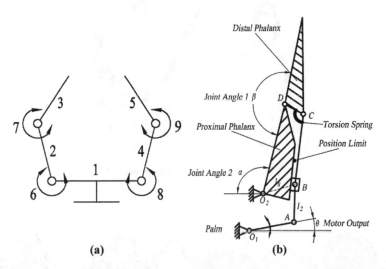

(a)   (b)

Fig. 2. (a) Diagram of mechanism DOFs. (1-palm; 2-thumb proximal phalanx; 3-thumb distal phalanx; 4-index finger proximal phalanx; 5-index finger distal phalanx; 6-thumb joint 1; 7-thumb joint 2; 8-index finger joint 1; 9-index finger joint 2) (b) Mechanism schematic of finger

Components of Finger Mechanism. As shown in Fig. 2(b), the proposed finger mechanism adopts double slider-crank mechanism as power transmission mechanism. It is a planar mechanism, mainly composed of 2 phalanxes: distal phalanx and proximal phalanx. The bottom of proximal phalanx is hinged at palm base. Distal phalanx is hinged at the top of proximal phalanx. In this mechanism, motor arm, linkage l_2, slider and proximal phalanx combine to make up one slider-crank mechanism, while proximal phalanx, distal phalanx, linkage BC and slider combine to make up another slider-crank mechanism. Link O_1A is fixed to motor output arm. With the actuation of the motor, different motions are generated for different grasp modes.

In order to coordinate the relation of joint angle 1 and joint angle 2, meanwhile to make the mechanism move according to the pre-planned sequence, a torsion spring is needed to keep the stretch torque, which keeps the finger stretching without external load. When grasping small objects in mode 1, joint 2 keeps stretching and the joint angle 2 maintains extreme position all the time. When grasping objects in large size, joint 2 keeps stretching in phase 1. When the movement of proximal phalanx is obstructed, joint 2 tends to bend because of the thrust of linkage *AB*. When the push of linkage is great enough, torsion spring occurs deformation and finger bends. Distal phalanxes begin to wrap around the object.

Analysis of Mechanism Movement. Grasp mode 1 is shown as Fig. 3: When grasping small object, two fingers keep joint 2 locked at the maximum angle within constraint of torsion spring. Two phalanxes of per finger can be seen as a single rigid body driven by actuations, rotating until the distal phalanxes contact with object.

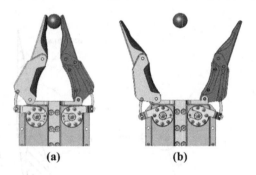

(a)   (b)

Fig. 3. Movement of grasp mode 1

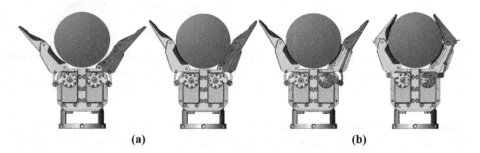

(a)   (b)

Fig. 4. Phase 1 (a) and Phase 2 (b) of grasp mode 2

Grasp mode 2: The process of grasping large object can be divided into 2 phases. Phase 1 is shown as Fig. 4(a). Before touching the object, no external load is applied to mechanism. Fingers keep stretching so that two phalanxes rotate together under the driving of motor, until proximal phalanxes stop moving for touching the object. Meanwhile, the distal phalanxes have not touched the surface of object, therefore, they continue moving without constraint. The push of slider and linkage is against the joint stretch torque applied by torsion spring, making joint 2 tend to bend. If the linkage generates enough moment which is greater than the preload of torsion spring, joint 2 bends and phase 2 begins.

Phase 2 is shown as Fig. 4(b), dark color part in figure represnts partial double slider-crank mechanism. With the push of linkages, distal phalanxes continue rotating about joint 2 until touching object. When the load detected by sensors reaches the set value, phalanxes have finished wrapping around the object.

2.3 Dynamic Analyses of Finger Mechanism

To facilitate analyses of dynamic characteristic and description of mechanics principle, dynamics diagram is shown as Fig. 5(a). θ is the output angle of motor $(-14° \leqslant \theta \leqslant 36.5°)$. Length and angle parameters are listed in Table 1.

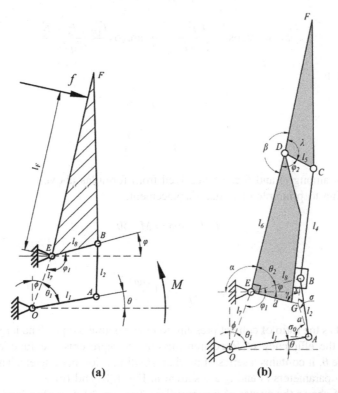

(a)   (b)

Fig. 5. Mechanism dynamics schematic (a) and Finger kinematics schematic (b)

Table 1. Geometry parameters of finger mechanism

Parameters	l_1	l_2	l_7	l_8	l_F	ϕ
Value (mm or °)	22	18	19	15	67	20.7°

In grasp mode 1, contact pressure is considered to be generated at fingertip (distal phalanx). In this case, joint 2 is locked by joint limit structure. Thus the whole finger can be seen as a single body. The mechanism degenerates into four-bar linkage. Figure 5(a) shows the force analysis. OA is the arm of motor, with a length of l_1. O is the output shaft. AB is the linkage linked with motor arm directly, with a length of l_2. BEF is the simplified geometry shape of the whole finger. And BE is the distance between two hinges, whose length is l_8. E is the hinge at joint 1, distance between O and E is l_7. θ is defined as the angle between the motor arm and the horizontal line. φ is defined as the angle between BE and the horizontal line. M is the output torque of motor. And f is the contact pressure between fingertip and object, length of f arm is l_F.

The formula (1) can be obtained by Cosine Law:

$$l_{AE} = \sqrt{l_1^2 + l_7^2 - 2l_1 l_7 \cos \theta_1} \tag{1}$$

$$\varphi_1 = \arccos \frac{l_{AE}^2 + l_7^2 - l_1^2}{2 l_{AE} l_7} + \arccos \frac{l_{AE}^2 + l_8^2 - l_2^2}{2 l_{AE} l_8} \tag{2}$$

In the formula,

$$\theta_1 = \frac{\pi}{2} - \theta - \phi \tag{3}$$

$$\varphi = \varphi_1 - \phi - \frac{\pi}{2} \tag{4}$$

The relation among φ and θ can be derived from formulas above.

According to principle of virtual displacement,

$$f \cdot l_F \cdot \delta\varphi = M \cdot \delta\theta \tag{5}$$

Therefore,

$$\mu = \frac{f}{M} = \frac{1}{l_f} \left(\frac{\partial \varphi}{\partial \theta} \right)^{-1} \tag{6}$$

μ is defined as the ratio of contact pressure to motor output torque. The larger this ratio, the higher the efficiency of force transmission. It represents the function of motor output angle θ. It contains a series of mechanism dimension parameters. The influences of the main parameters l_2 and l_8 are shown in Fig. 6(b) and (c).

Figure 6 shows the curves of μ versus θ. In diagram (b), l_2 is chosen as the variable parameter, varying from 13 mm to 19 mm with the interval of 1 mm. It is concluded that value change of l_2 has small contribution to the maximum of μ. But has a significant influence on the motor output angle θ when μ reaches a maximum.

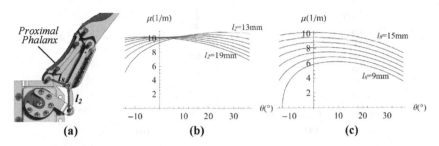

Fig. 6. Influence to μ by change of mechanism geometry parameters

In diagram (c), l_8 is chosen as the variable parameter, varying from 9 mm to 15 mm with the interval of 2 mm. It is concluded that value change of l_8 has greater contribution to the maximum of μ. l_8 represents the parameter of slideway. Therefore, choosing better slideway parameters can help to improve the output force performance of finger mechanism.

2.4 Analysis of Motion Stability

Kinematics diagram is established as Fig. 5(b), and dimension parameters are shown in Tables 2 and 3.

Table 2. Length parameters of finger mechanism

Parameters	l_1	l_2	l_4	l_5	l_6	l_7	d
Value (mm)	22	18	34	10	44	19	15

Table 3. Angle parameters of finger mechanism

Parameters	λ	γ	ϕ
Value (°)	115	82	20.7

The motions in first phase of grasp mode 2 and grasp mode 1 are identical, which can be described as a rotation motion about joint 1 hinge actuated by motor arm. The mechanism degenerates into four-bar linkage. In Fig. 5(b), α is angle of joint 1, defined as the angle between horizontal reference line and line AB (connecting line between hinges of joint 1 and joint 2). The mathematical relationship between mechanism motion and motor output angle can be characterized by graph of α–θ curve, shown as Fig. 7(a).

Notice from Fig. 7(a) that curve α–θ is smooth, so is the motion represented by the curve. In order to reflect variation of angular velocity intuitively, define ω_1 as the angular velocity of the finger rotating about joint 1, when the motor output speed is constant ω_0. ω_1 can be express as:

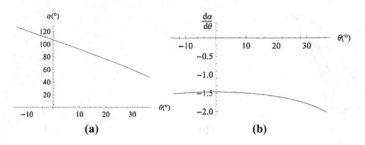

Fig. 7. Curves of joint angle 1 (a) and its rate of change (b) with motor output θ

$$\omega_1 = \frac{\mathrm{d}\alpha}{\mathrm{d}t} = \frac{\mathrm{d}\alpha}{\mathrm{d}t} \cdot \frac{\mathrm{d}\theta}{\mathrm{d}t} = \frac{\mathrm{d}\alpha}{\mathrm{d}\theta} \cdot \omega_0 \tag{7}$$

Therefore, when ω_0 is constant, stability of $\frac{\mathrm{d}\alpha}{\mathrm{d}\theta}$ value reflect the stability of output motion. Take the derivative of $\alpha(\theta)$ and plot the graph as Fig. 7(b). It can be seen that $\frac{\mathrm{d}\alpha}{\mathrm{d}\theta}$ keeps the value between -1.5 to -2.0, which illustrates the fluctuation of speed is small and the shock of motion is small.

In phase 2 of grasp mode 2, motion of proximal phalanx is obstructed so that cannot to continue. From this moment on, motor arm pushes the slider by linkage AB, then causes the rotation of distal phalanx through linkage BC. The motion can be characterized by angle β of joint angle 2. In the process, joint 1 angle α keeps constant. However, for objects in different size, α is also different. So it will be handled as a variable parameter in the latter analyses.

Joint angle 2 β is defined as the angle between line DF (connecting line between fingertip and joint hinge 2) and line DE (connecting line between hinges of joint 1 and joint 2) in Fig. 5(b). For different value of α, curve β–θ is also different.

As shown in Fig. 5(b), double slider-crank mechanism forms two quadrilaterals $ABEO$ and $BCDE$ geometrically with a shared side BE. At point O,

$$\theta_1 = \frac{\pi}{2} - \theta - \phi \tag{8}$$

According to Cosine Law in $\triangle OAE$,

$$l_{AE} = \sqrt{l_1^2 + l_7^2 - 2l_1 l_7 \cos \theta_1} \tag{9}$$

Slope of line a is:

$$\sigma_0 = \arctan \frac{l_7 \cos \phi + d \, \cos \alpha - l_1 \sin \theta}{l_1 \cos \theta - d \, \sin \alpha - l_7 \sin \phi} \tag{10}$$

Therefore,

$$\sigma = \frac{3}{2}\pi - \gamma - \alpha + \sigma_0 \tag{11}$$

Length of line a can be expressed as:

$$a = \sqrt{(l_7 \cos \phi + d \cos \alpha - l_1 \sin \theta)^2 + (l_1 \cos \theta - d \sin \alpha - l_7 \sin \phi)^2} \tag{12}$$

According to Cosine Law in $\triangle ABG$, a quadratic equation can be obtained and one efficient solution is:

$$x = a \cos \sigma + \sqrt{a^2 \cos^2 \theta - a^2 + l_2^2} \tag{13}$$

According to Cosine Law in $\triangle BEG$,

$$l_8 = \sqrt{x^2 + d^2 - 2xd \cos \gamma} \tag{14}$$

According to Sine Law in $\triangle BEG$, inclination angle of l_8 is:

$$\varphi = \arcsin \frac{x \sin \gamma}{l_8} + \frac{\pi}{2} - \alpha \tag{15}$$

At point E,

$$\theta_2 = \pi - \alpha - \varphi \tag{16}$$

According to Cosine Law in $\triangle BDE$,

$$l_{BD} = \sqrt{l_8^2 + l_6^2 - 2l_8l_6 \cos \theta_2} \tag{17}$$

According to Cosine Law in $\triangle BDE$ and $\triangle BDC$,

$$\varphi_2 = \arccos \frac{2l_6^2 - 2l_8l_6 \cos \theta_2}{2l_{BD}l_6} + \arccos \frac{l_8^2 + l_6^2 + l_5^2 - l_4^2 - 2l_8l_6 \cos \theta_2}{2l_{BD}l_5} \tag{18}$$

Thus joint angle 2 is:

$$\beta = 2\pi - \varphi_2 - \lambda \tag{19}$$

By substituting variable expressions and parameters into expression of β in turn, relation among joint angle 2 β, joint angle 1 α and motor output angle θ can be obtained. That is $\beta(\theta, \alpha)$, which is used to plot the graph by mathematical software. Figure 8(a) shows the relation between β and θ, with α varying from 70° to 130° with the interval of 10°. The range of motor output angle is selected as range of θ, and the range of joint 2 mechanical limiting angle is selected as range of β. In this range, these series curves have good straightness, which means that the motion in this phase is smooth and motion shock moment does not exist. Figure 8(b) shows the relation between $\frac{d\beta}{d\theta}$ and θ. The $\frac{d\beta}{d\theta}$ value keeps in range from -100 to -180 and changes slowly, indicating that distal phalanx speed changes slowly, without significant shock.

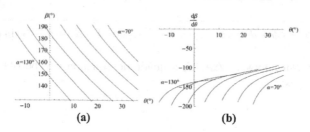

Fig. 8. Curves of joint angle 2 (a) and its rate of change (b) with motor output θ under different value of α

3 Adaption Degree of Mechanism

For an under-actuated mechanism, even if the motor outputs a certain angle, redundant DOFs exists to realize adjustment among variables of joints according to demands. Therefore, the finger adopts under-actuated mechanism to keep the adjustment range of joint angles to adapt different shapes, so as to fully contact with more kinds of objects. To measure the ability to adapt kinds of sizes and shapes, the definition of adaption degree is proposed. It is defined as the variation range of joint angle 1 while the motor output angle is constant. Adaption degree is represented by $\Delta\alpha$. The larger the parameter of $\Delta\alpha$, the stronger the adaption ability, which means the larger range of grasped object's style.

According to $\beta(\theta, \alpha)$ deduced in the previous chapter. The curve graph is plotted by mathematical software as Fig. 9(a). It shows the relation between joint angle 1 α and joint angle 2 β, with the value of motor output angle θ varying from $-10°$ to $40°$ with the interval of $10°$. Figure 9(a) illustrates that α increases with β reducing.

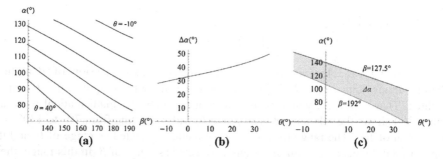

Fig. 9. Relation of joint angle 1 and joint angle 2 under different value of θ (a); Changes of joint angle 1 (b) and its adjustment range (c) with motor output angle θ

Constraint range of joint angle 2 is realized by joint limit structure. For a given motor output angle, joint angle 2 can vary from 127.5° to 192° continuously. So these two values are set as extreme angles of β. It is equivalent to locking the joint 2 by the extreme angles and plotting two α–θ curves in the same graph. Hatched area in Fig. 9(b) between two curves represents the achievable angle range of α, which is the adaption degree $\Delta\alpha$ defined by this manuscript.

It is indicated that adaption degree $\Delta\alpha$ grows with increasing of motor output angle θ under the given parameters above. That means that adaption ability increases with decreasing of grasped object's size. In general, adjustment range of joint angle 1 is approximately 40°, which can guarantee adaption ability under any output of motor.

4 Experiment Design and Effect of Grasp

According to different size and geometry shape of objects, experiments are designed to verify whether the grasp movement can be carried out as pre-planned sequence by the dexterous hand.

Object in small size has light mass in general. Small output force can be set to realize grasp, as is shown in Fig. 10(a). When accelerations in any directions are applied to objects, grasp stability can still be guaranteed. As is shown in Fig. 10(b), object in medium size is grasped by contacting with finger pulp, where the silicone pad was designed concave in shape to increase contact area. Figure 10(c) shows, all two joints bend to make the inside of whole finger fully contact with object when grasping objects in large size. If radius of object's cross-sectional is less than 100 mm, the wrapping angle can be larger than 180°, difficult to slide over in mechanism plane.

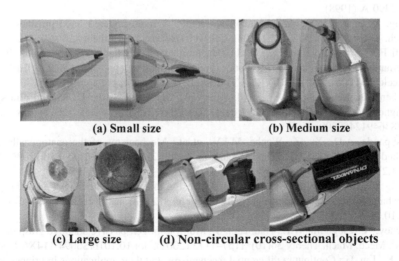

(a) Small size   (b) Medium size

(c) Large size   (d) Non-circular cross-sectional objects

Fig. 10. Grasp effect for different sizes and shapes

Experiments and analyses above are mainly for circular cross-sectional objects, the following analyses show grasp results for non-circular cross-sectional objects.

As shown in Fig. 10(d), the finger mechanism designed by this paper can realize self-adaption in mechanical kinematic plane for non-circular cross-sectional objects. However, its effect is not ideal as experiments for circular cross-sectional objects, concretely showed in slide in the direction along the finger in the mechanical plane. This phenomenon can be avoided by providing greater output torque to increase the friction. Experiments show that the dexterous hand can stably grasp non-circular cross-sectional objects which have relatively light mass.

The series experiments show that the dexterous hand designed by this paper can realize reliable grasp of objects in various sizes and shapes. It can adapt various shapes well and its joint movement sequence is consistent with pre-designed sequence.

5 Discussion and Conclusion

The under-actuated double fingered dexterous hand can realize grasp of objects in different sizes and geometry shapes. It effectively reduces number of actuators and complexity of control system. Thus it is achievable to integrate actuation motors and control circuitry into the palm, making mechanical-electrical integration come true. With the advantages of small volume, light weight, strong self-adaption, it is suitable for popularization and application.

References

1. Gosselin, C.M., Laliberte, T.: Underactuated mechanical finger with return actuation. US 5762390 A (1998)
2. Yang, J., Pitarch, E.P., Abdel-Malek, K., et al.: A multi-fingered hand prosthesis. Mech. Mach. Theory 39(6), 555–581 (2004). doi:10.1016/j.mechmachtheory.2004.01.002
3. Lotti F., Tiezzi P., Vassura G., et al.: Development of UB hand 3: early results. In: IEEE International Conference on Robotics and Automation, pp. 4488–4493. IEEE Press, Barcelona (2005). doi:10.1109/ROBOT.2005.1570811
4. Grebenstein, M., Chalon, M., Friedl, W., et al.: The hand of the DLR hand arm system: designed for interaction. Int. J. Robot. Res. 31(13), 1531–1555 (2012). doi:10.1177/0278364912459209
5. Deshpande, A.D., Xu, Z., Weghe, M.J.V., et al.: Mechanisms of the anatomically correct testbed hand. IEEE/ASME Trans. Mechatron. 18(1), 238–250 (2012). doi:10.1109/TMECH.2011.2166801
6. Xiong, C.H., Chen, W.R., Sun, B.Y., et al.: Design and implementation of an anthropomorphic hand for replicating human grasping functions. IEEE Trans. Rob. 32(3), 652–671 (2016). doi:10.1109/TRO.2016.2558193
7. Dechev, N., Cleghorn, W.L., Naumann, S.: Multiple finger, passive adaptive grasp prosthetic hand. Mech. Mach. Theory 36(10), 1157–1173 (2001). doi:10.1016/S0094-114X(01)00035-0
8. Xu, K., Liu, H.: Continuum differential mechanisms and their applications in gripper designs. IEEE Trans. Rob. 32(3), 754–762 (2016). doi:10.1109/TRO.2016.2561295

LIPSAY Hand: A Linear Parallel and Self-adaptive Hand with Y-Shaped Linkage Mechanisms

Jian Hu[1], Ke Li[1], Wenzeng Zhang[2], Xiangrong Xu[1(✉)], and Aleksandar Rodic[3]

[1] School of Mechanical Engineering, Anhui University of Technology,
Ma'anshan 243032, China
xuxr@ahut.edu.cn
[2] Department of Mechanical Engineering, Tsinghua University,
Beijing 100084, China
[3] University of Belgrade, Volgina 15, 11060 Belgrade, Serbia

Abstract. In order to overcome the shortcoming that traditional self-adaptive underactuated robot hands cannot perform linear parallel pinching, this paper presents a novel linear parallel and self-adaptive hand with y-shaped linkage mechanism, LIPSAY hand. The LIPSAY hand uses y-shaped linkage mechanisms to achieve linear parallel pinching, and the pulley-belt and double-slider mechanisms to perform self-adaptive grasping. Compared with traditional underactuated hands, the LIPSAY hand can realize the hybrid function of precise linear parallel pinching and self-adaptive grasping. The LIPSAY hand has strong grasping adaptability and high grasping stability, which is suitable for industrial applications.

Keywords: Underactuated hand · Linear parallel pinching · Self-adaptive grasping · Linkage mechanism

1 Introduction

Robot hand, as the end effector and one of the key parts in a robot system, has gained more attentions in all works of life during recent years. In order to meet the needs of industrial production, the robot hand must have high reliability, stability and accuracy of grasping.

During past decades of research, there have been four main types of the robot hands, including underactuated hands, industrial grippers, dexterous hands and special hands. Dexterous hands have more degrees of freedom, which can easily grasp objects. Generally, special hands barely have things in common with human hands. For some special areas, traditional robot hands cannot work well, which initiates the development of special hands. Due to the complexity and expensive costs, the application of dexterous hands and special hands are still relatively limited. There are many dexterous hands and special hands, including Stanford/JPL Hand [1], Robonaut Hand [2], KAWABUCHI Hand [3], BH Hand [4, 5], Vacuum Gripper [6], Magnetic Gripper [7, 8] and Soft Robot Hand [9].

Y. Huang et al. (Eds.): ICIRA 2017, Part II, LNAI 10463, pp. 739–751, 2017.
DOI: 10.1007/978-3-319-65292-4_64

The underactuated robot hands and industrial grippers have very broad application prospects. The number of actuators of an underactuated hand is less than its degrees of freedom, which makes the control system very simple. Compared with dexterous hands and special hands, underactuated hands have got less cost and more stable and reliable working status. Industrial grippers have fairly high grasping accuracy, which are suitable for industrial use. Up to now, many industrial hands and underactuated hands have been developed, including Dollar Hand [10], FRH-4 Hand [11], and Underactuated Elastic Hand [12]. Besides, Tsinghua University has been doing many researches on underactuated hands, developing a variety of underactuated hands like PASA finger [13], SCUH hand [14] and COSA hand [15].

The traditional underactuated robot hand cannot achieve linear parallel pinching. For instance, PASA finger [13] cannot grasp some small and thin objects. Although some industrial grippers solve the problem of linear parallel pinching [16], their grasping modes are simple. Thus, the object can only be pinched, not grasped in enveloping form.

A linear parallel pinching and self-adaptive grasping function was proposed in [17], which was realized in each LIPSA finger by a Chebyshev multi-linkage mechanism. The LIPSA finger just can perform a close linear motion based on reasonable parameter set.

This paper introduces a novel linear parallel and self-adaptive underactuated hand with link-wheel mechanism (the LIPSAY hand). The LIPSAY hand overcomes the disadvantages of the traditional underactuated hands and industrial grippers. It can not only achieve the linear parallel pinching, but also perform self-adaptive enveloping.

2 Design of LIPSAY Hand

2.1 The Concept of Linear Parallel and Self-adaptive Underactuated Hand

As we all known, linear parallel and self-adaptive underactuated hand contains two grasping modes: linear parallel pinching grasp and self-adaptive enveloping grasp. Figure 1 shows the concept of linear parallel pinching and self-adaptive grasping modes.

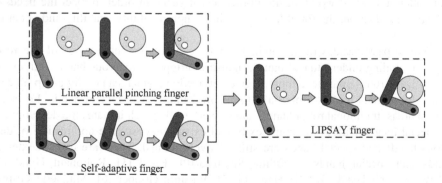

Fig. 1. Linear parallel pinching and self-adaptive finger.

Linear parallel pinching means using the two parallel distal phalanxes of two fingers to clamp the object by approaching it. In this pinching process, the ends of two distal phalanxes are always kept in the same straight line. This is mainly used to grip objects of small and thin size, of which the precision is highly required. Whereas self-adaptive enveloping means that if an external force is applying on the proximal phalanx, the robot finger will then execute an enveloping motion. And the process of enveloping grasp is more stable.

2.2 Design of the Structure of Linear Parallel Pinching

Linear parallel pinching can improve the grasping precision. In this paper, a special linkage mechanism is used to realize the function of linear parallel pinching of the robot finger. Figure 2 shows the principle of the linkage mechanism.

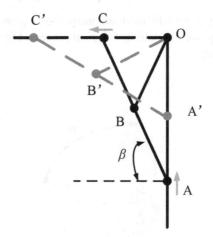

Fig. 2. The principle of linear parallel pinching of the y-shaped linkage mechanism.

The linkage mechanism must meet the following equation:

$$l_{AC} = 2l_{OB} \tag{1}$$

In this paper, set the coordinates of point A, C and O as:

$$A(x_A, 0), C(x_C, y_C), O(0, 0) \tag{2}$$

So the coordinates of point B is:

$$B\left(\frac{x_A + x_C}{2}, \frac{y_C}{2}\right) \tag{3}$$

Due to $l_{AC} = 2l_{OB}$, so:

$$\sqrt{(x_C - x_A)^2 + y_C^2} = 2\sqrt{\left(\frac{x_A + x_C}{2}\right)^2 + \left(\frac{y_C}{2}\right)^2} \tag{4}$$

Finally, we can get:

$$4x_A x_C = 0 \tag{5}$$

According to Eq. (5), we know if $x_A \neq 0$, then we get $x_C = 0$. So the linkage mechanism meets the design requirement.

Due to the dead point existing in linkage mechanism, we have $\beta \neq \pi/2$.

2.3 Design of the Structure of Self-adaptive Grasping

Self-adaptive grasping is mainly used to grasp the objects of big sizes. And this grasping mode can improve the stability of grasp process. Figure 3 shows the principle of self-adaptive grasping.

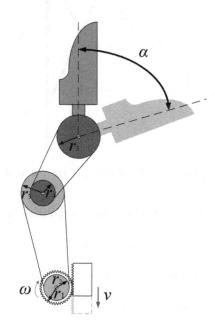

Fig. 3. The principle of self-adaptive grasping of the LIPSAY finger.

From the Fig. 3 we can know, the structure of self-adaptive grasping mode mainly contains two sets of pulleys, one set of gear and rack. When the rack moves down, the gear clockwise rotates. Then the distal phalanx can achieve clockwise rotation.

When the rack moves down, robot hand can achieve self-adaptive grasping. Assume some parameters of the finger as follows:

v-the speed of the rack
r_1-the radius of the gear
r_2, r_3, r_4, r_5-the radius of the pulleys
θ-the rotation angle of the gear
α-the rotation angle of the distal phalanx

One gets:

$$\theta = \frac{vt}{r_1} \tag{6}$$

And the angle of the distal phalanx:

$$\alpha = \frac{r_2 r_4}{r_1 r_3 r_5} vt \tag{7}$$

Considering the actual demand, we can set the rotation angle of the distal phalanx from 0 to $\pi/2$.

3 Overall Design of LIPSAY Hand

In order to meet the needs of practical use, the robot hand should have an appropriate size, large grasping force, high stability and more grasping modes. So the LIPSAY has 2 DOFs and 2 grasping modes. The linkage mechanism is for linear parallel pinching, and the pulleys and slider mechanism for self-adaptive grasping. The structure of LIPSAY hand is shown in Fig. 4.

From Fig. 4 we can see that when distal phalanx and proximal phalanx both haven't contact any objects, first slider and second slider, connected with each other by a spring, move down together. If there is only the distal phalanxes contacting object, LIPSAY hand will pinch it. Once the proximal phalanxes reach the object in first place, the second slider will then overcome the force of the spring to move down, and meanwhile the first slider keeps motionless. Being fixed to the second slider, the rack will move down synchronously. On the other hand, the power passes through the gear and the belts to the distal phalanx, helping the distal phalanx achieve self-adaptive grasping. It should be pointed out that the second slider has internal thread, while the first slider has none, so the power can be passed from motor to the second slider. The length of first link is equal to the second link. The length of the third link is half of the first link.

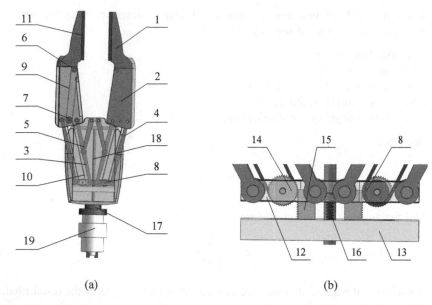

Fig. 4. The structure of the LIPSAY hand. 1-distal phalanx; 2-proximal phalanx; 3-first link;4-second link; 5-third link; 6-first pulley;7-second pulley; 8-third pulley; 9-first transmission belt; 10-second transmission belt;11-rubber mat; 12-first slider;13-second slider; 14-gear; 15-rack; 16-spring;17-transmission mechanism; 18-worm; 19- motor.

4   Analysis of the Grasping Force

The LIPSAY hand has two grasping modes, of which the forces are not the same. This part will analyze the forces of the two grasping modes respectively.

4.1   Analysis of the Pinching State

As is discussed already, pinching state occurs when only the distal phalanx contact the object. The force analysis diagram is shown in Fig. 5.

For the sake of simplification, the gravity of the object and the friction between phalanxes are neglected, and all contact forces are applied on points.

There are some parameters assumed as follow:

l_1-the length of first link and second link;
l_2-the length of proximal phalanx;
l_3-the length from the point of F_1 to proximal phalanx;
F_1-the force between distal phalanx and object;
F_3-the force between first link and third link;
F-the force acting on the second slider;
G_1-the contact point between object and distal phalanx;
G_3-the contact point between first link and third link;
δ-the angle between third link and vertical line;

β-the angle between second link and horizontal line;
M, N-the points on the first link.

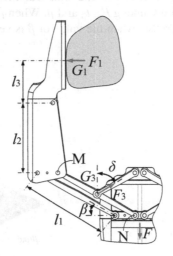

Fig. 5. The forces of pinching state.

When the LIPSAY hand is working, torques at any point of it should be balanced.
The torque equilibrium of the point G_3 is:

$$F_1\left(l_2 + l_3 + \frac{l_1}{2}\sin\beta\right) - F\frac{l_1}{2}\cos\beta = 0 \tag{8}$$

The torque equilibrium of the point M is:

$$Fl_1\cos\beta = F_1(l_2 + l_3) + F_3\cos\delta\frac{l_1}{2}\cos\beta + F_3\sin\delta\frac{l_1}{2}\sin\beta \tag{9}$$

Combining Eqs. (8) and (9), one gets:

$$F_1 = \frac{Fl_1\cos\beta}{2l_2 + 2l_3 + l_1\sin\beta} \tag{10}$$

$$F_3 = \frac{2Fl_1\cos\beta(l_2 + l_3 + l_1\sin\beta)}{\cos(\delta - \beta)(2l_2 + 2l_3 + l_1\sin\beta)} \tag{11}$$

Because l_3, δ and β are variables, F_1 and F_3 vary in different states.
The relation of β and δ is:

$$\sin\beta = \cos\delta \tag{12}$$

From Eqs. (10) and (11), we can get the relations among F_1, F_3, l_3 and β.

Figure 6(a) shows the trends of F_1 and F_3 with respect to β. With the increasing of β, F_1 and F_3 decrease. During this process, F_1 decreases fairly slower, where F_3 decrease rapidly when β is between 0 and 0.6. But then F_3 is almost a constant. Figure 6(b) shows the relations among F_n, l_3 and β. When β is of medium values, F_1 is usually larger than F_3. Otherwise, when the value of β is very small or very large, then F_3 is larger than F_1.

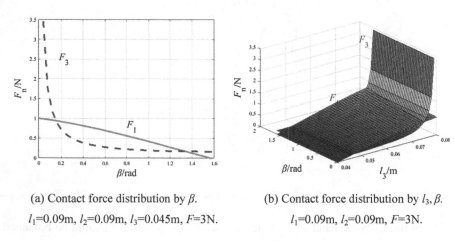

(a) Contact force distribution by β.

l_1=0.09m, l_2=0.09m, l_3=0.045m, F=3N.

(b) Contact force distribution by l_3, β.

l_1=0.09m, l_2=0.09m, F=3N.

Fig. 6. Contact-force analysis of pinching.

4.2   Analysis of the Self-adaptive Grasping State

In order to achieve self-adaptive grasping, the proximal phalanx should contact object in the first place, followed by distal phalanx. Figure 7 shows the force condition of self-adaptive grasping.

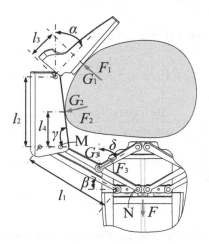

Fig. 7. The forces of self-adaptive state.

There are also some parameters assumed as follows:

l_1-the length of first link and second link;
l_2-the length of proximal phalanx;
l_3-the length from the point of F_1 to the proximal phalanx;
l_4-the length from the point of F_2 to point M;
F_1-the force between distal phalanx and the object;
F_2-the force between proximal phalanx and the object;
F_3-the force between first link and third link;
G_1-the contact point between object and distal phalanx;
G_2-the contact point between object and proximal phalanx;
G_3-the contact point between first link and third link;
β-the angle between second link and horizontal line;
γ-the angle of the proximal phalanx;
α-the angle between distal phalanx and vertical line;
δ-the angle between third link and vertical line;
M, N-two points of the first link;
F'-the force of the spring;
F''-the force of the rack.

When proximal phalanx contact object, the first slider stop moving down. Thus, the second slider has to overcome the force of spring to keep moving.

The value of $\cos\gamma$ is very small, so we can neglect it. Here is the torque equilibrium of the point M:

$$F'l_1 \cos\beta = F_2 \sin\gamma l_4 + F_3 \cos\delta \frac{l_1}{2}\cos\beta + F_3 \sin\delta \frac{l_1}{2}\sin\beta \tag{13}$$

And the torque equilibrium of the point G_3 is:

$$F_2 \sin\gamma(l_4 + \frac{l_1}{2}\sin\beta) - F'\frac{l_1}{2}\cos\beta = 0 \tag{14}$$

Combining Eqs. (12) and (13), one gets:

$$F_2 = \frac{F'l_1 \cos\beta}{2l_4 \sin\gamma + l_1 \sin\beta \sin\gamma} \tag{15}$$

$$F_3 = \frac{2F' \cos\beta(l_4 + l_1 \sin\beta)}{\cos(\delta - \beta)(2l_4 + l_1 \sin\beta)} \tag{16}$$

Figure 8 shows the relations among F_n, l_4 and β. We can know that F_2 is always smaller than F_3. The change rate of F_3 is very large when β is in the range of 0 to 0.6. And when β increases from 0.6 to 1.6, F_3 is almost a constant.

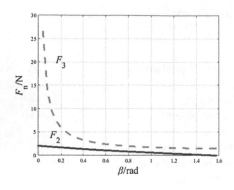
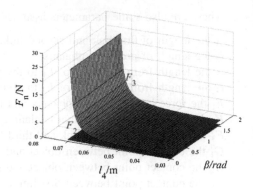

(a) Contact-force distribution by β.
l_1=0.09m, l_4=0.045m, γ=80°, F'=3N.

(b) Contact-force distribution by l_4, β.
l_1=0.09m, l_2=0.09m, F'=3N.

Fig. 8. Contact-force analysis of enveloping.

Obviously, F_1 is related to F''. We can know from the transmission law of belt pulley:

$$F'' r_1 r_3 r_5 = F_1 l_3 r_2 r_4 \tag{17}$$

So the force of second slider:

$$F = F' + F'' \tag{18}$$

Combing Eqs. (12), (13), (17) and (18), one gets:

$$F = \frac{2F_1 l_1 l_3 \cos\beta \frac{r_2 r_4}{r_1 r_3 r_5} + 2F_2 l_4 \sin\gamma + F_3 l_3 \sin 2\beta}{2l_1 \cos\beta} \tag{19}$$

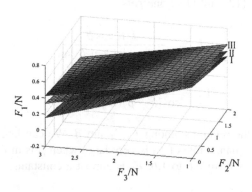

Fig. 9. Contact-force distribution among F_1, F_2 and F_3. $r_1 = r_2 = r_4 = 5$ mm, $r_3 = r_6 = 6$ mm, $l_1 = l_2 = 0.09$ m, $r_3 = r_4 = 0.045$ m, $\gamma = 80°$, $F = 3$ N.

Figure 9 shows the contact force distribution among F_1, F_2 and F_3. When $\beta = 30°$, the distribution situation is I; $\beta = 45°$, the distribution situation is II; Otherwise $\beta = 60°$, the distribution situation is III.

5   Experiment of Motion Process

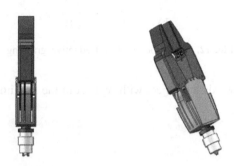

Fig. 10. The appearance of LIPSAY hand.

As is shown above, LIPSAY hand has two grasping modes: pinching and self-adaptive enveloping. So we will mainly discuss the motion process of pinching and self-adaptive in the next step.

Figure 10 shows the appearance of LIPSAY hand.

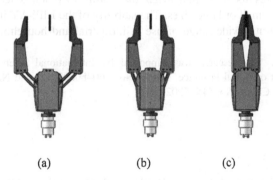

(a)   (b)   (c)

Fig. 11. The process of pinching.

Figure 11 shows the process of pinching. The mode of pinching is suitable for small and thin size of objects. And the end of distal phalanx keeps moving on linear in all motion processes.

Figure 12 shows the process of self-adaptive enveloping. Enveloping is suitable for objects in large sizes and complex shapes to achieve stable grasping. The whole process is in two steps: First, proximal phalanx reach the object; Then the distal

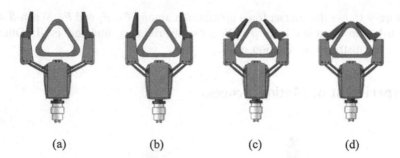

| (a) | (b) | (c) | (d) |

Fig. 12. The process of self-adaptive grasping.

phalanx will rotate towards the object with respect to the proximal one to achieve the rest of enveloping.

6 Conclusion

This paper proposes a novel underactuated robot hand (LIPSAY hand) for linear parallel and self-adaptive grasping. The LIPSAY hand has two grasping modes: linear parallel pinching and self-adaptive enveloping grasping. The key components of each finger of the LIPSAY hand are y-shaped linkage mechanism, pulley-belt and double-slider mechanism.

LIPSAY hand overcomes the shortcomings of traditional self-adaptive hand, which executes multiple grasping modes depending on dimensions, positions and shapes of objects. Analysis results show that it has the ability to pinch objects accurately and envelop objects of small or large sizes. The stability of the LIPSAY hand in the motion process guarantees it a wide range of use in industrial and household surroundings.

Acknowledgement. This research was supported by International Science and Technology Cooperation Project of Anhui Province, China (No. 1604b0602018) and National Natural Science Foundation of China (No. 51575302).

References

1. Mason, M.T., Salisbury, J.K.: Robot Hands and the Mechanics of Manipulation. The MIT Press, Cambridge (1985). 111(1), pp. 879–880
2. Lovchik, C.S., Diftler, M.A.: The robonaut hand: a dexterous robot hand for space. In: IEEE International Conference Robotics and Automation (ICRA), Detroit, Michigan, pp. 907–912 (1999)
3. Ichior, K.: A designing of humanoid robot hands in endoskeleton and exoskeleton and exoskeleton styles. Int. J. Adv. Robot. Syst. 401–426 (2007)
4. Li, J., Zhang, Y., Zhang, Q.: Analysis and evaluation of force manipulation dexterity of finger mechanisms. Robot **22**(2), 115–1211 (2000). (in Chinese)

5. Zhang, Y., Han, Z., Zhang, H., et al.: Design and control of the BUAA four-finger hand. In: IEEE International Conference on Robotics and Automation (ICRA), New York, America, pp. 2517–2522 (2001)
6. Callies, R., Fronz, S.: Recursive modeling and control of multi-link manipulators with vacuum grippers. Math. Comput. Simul. **79**(4), 906–916 (2008)
7. Xu, D., Chen, L., Zhu, C.: Semi-active control mechanical hand based on magnetorheological fluid. Patent CN204054061B (2014)
8. Zhu, X., He, G., Wu, J., et al.: Flexible clamping device and clamping method based on magnetorheological fluid. Patent CN103831837A (2014)
9. Xuance, Z., Carmel, M., Oliver, M.: Soft hands: an analysis of some gripping mechanism in soft robot design. Int. J. Solids Struct. 1–11 (2015)
10. Dollar, A.M., Howe, R.D.: SDM hand: a highly adaptive compliant grasper for unstructured environments. In: Khatib, O., Kumar, V., Pappas, G.J. (eds.) Experimental Robotics. Springer Tracts in Advanced Robotics, vol. 54, pp. 3–11. Springer, Heidelberg (2009)
11. Gaiser, I., Schule, S., Kargov, A., et al.: A new anthropomorphic robotic hand. In: IEEE International Conference on Humanoid Robots (ICHR), Daejeon, Korea, pp. 418–422 (2008)
12. Odhner, L.U., Dollar, A.M.: Dexterous manipulation with underactuated elastic hands. In: IEEE International Conference on Robotics and Automation (ICRA), pp. 5254–5260 (2011)
13. Liang, D., Zhang, W., et al.: PASA finger: a novel parallel and self-adaptive underactuated finger with pinching and enveloping grasp. IEEE International Conference on Robotics and Biomimetics (ROBIO), Zhuhai, China, pp. 1323–1328 (2015)
14. Yang, S., Zhang, W.: SCHU hand: a novel self-adaptive robot hand with single-column hybrid underactuated grasp. In: IEEE Conference on Robotics and Biomimetics (ROBIO), Zhuhai, China, pp. 1337–1342 (2015)
15. Yang, S., Song, J., Zhang, W., et al.: Development of the CA robot finger with a novel coupled and active grasping mode. Int. J. Humanoid Robot. **13**(3), 1650012 (2016)
16. Murakami, J., et al.: Robot hand and robot. Patent WO2016063314A1 (2016)
17. Yang, Y., Zhang, W., Xu, X., et al.: LIPSA hand: a novel underactuated hand with linearly parallel and self-adaptive grasp. In: Zhang, X., Wang, N., Huang, Y. (eds.) Mechanism and Machine Science. LNEE, vol. 408, pp. 111–119. Springer, Singapore (2017)

A Novel Parallel and Self-adaptive Robot Hand with Triple-Shaft Pulley-Belt Mechanisms

Qingyuan Jiang[1], Shuang Song[2], and Wenzeng Zhang[1(✉)]

[1] Department of Mechanical Engineering, Tsinghua University,
Beijing 100084, China
wenzeng@tsinghua.edu.cn
[2] Department of Mechanical Science and Engineering,
University of Illinois at Urbana-Champaign, Champaign, IL 61801, USA

Abstract. This paper proposes a novel parallel pinching and self-adaptive grasping hand with triple-shaft pulley-belt mechanisms, named TPM hand. It contains two TPM fingers, and each finger has two phalanges, and mainly consists of a driving shaft, triple pulleys, two belts, and two springs. The TPM hand is capable of grasping objects with parallel pinching and enveloping, and simultaneously adapting to the contours of the objects. Particularly, compared with other parallel pinching and self-adaptive grasping hands, the TPM hand separates the proximal shaft and the driving shaft so that the grasping range can be expanded. Besides, due to the elasticity of flexible devices, objects can be protected in emergencies. New mechanical parameters are added and are conformed to specific domain and function. This function is given in this paper through geometry model. Meanwhile, a feasible analysis is provided to discuss the influence of parameters brought by the departure above. Also, the process of grasping is discussed, and the kinematic analysis is provided.

Keywords: Robot hand · Underactuated finger · Parallel pinching · Self-adaptive grasp

1 Introduction

With the unprecedented development of modern industry, intelligent industrial robots and robot hands have been used more than ever in different circumstances. The robot hand, as one of the most important parts of industrial robots, its method and capability to grasp objects are among the core techniques, which have drawn a lot of attention around the world.

Most multi-fingered robot hands existed can be divided into two types after decades of research: the dexterous hand and the underactuated hand. Dexterous hand has been designed to imitate the human hand. The multiple grasping modes of the human hand include (1) Cylindrical grasping, (2) Tip grasping, (3) Hook or snap grasping, (4) Palmar grasping, (5) Spherical grasping, (6) Lateral grasping, etc. With the help of many actuators, the dexterous hand is able to accomplish part of multiple grasping tasks assemble to the human hand. For example, the Utah/MIT Dexterous Hand [1],

© Springer International Publishing AG 2017
Y. Huang et al. (Eds.): ICIRA 2017, Part II, LNAI 10463, pp. 752–763, 2017.
DOI: 10.1007/978-3-319-65292-4_65

Stanford/JPL Hand [2], DLR Hand [3] and Gifu Hand II [4] have provided an excellent sample with good effect.

Another pattern of the robot hand is called the underactuated hand. The concept of under-actuation was first proposed by Laval Univ. [5], along with an underactuated robot hand, SARAH Hand [5–7]. The concept is defined as: the number of actuators is less than the number of degrees of freedom (DOFs). The invention of underactuated robot hand perfectly solved the limitation in some working circumstances, in which less degree of movements are necessary, or its volume or degree of freedom are limited. Compared to the dexterous hand, fewer actuators and simpler grasping patterns are used. Afterward, many kinds of underactuated robot hand have been designed to match different grasping tasks, such as SDM Hand [8], and LARM Hand [9].

In this paper, a novel underactuated robot hand is proposed, named the TPM hand. In the TPM hand, the driving shaft and the proximal shaft are separated, so that the driving shaft and the driving wheel can be placed in accordance with the particular designing requirement. As a result, the volume of the robot hand can also change in a wider range depending on the requirement (Fig. 1).

 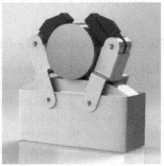

(a) Parallel grasping mode   (b) Self-adaptive grasping mode

Fig. 1. The parallel grasping mode and the self-adaptive grasping mode.

The second part of this paper introduces the concept of the parallel grasp and self-adaptive grasp. The third part provides details of the structure of the TPM hand, and the methods to achieve parallel grasp and enveloping grasp of the TPM hand. The fourth part analyzes the kinematics of the system and discusses the domain of parameters and their functional relationships.

2   Concept of Parallel and Self-adaptive Underactuated Hand

The grasping mode of the underactuated finger can be divided into three basic groups, parallel grasp (PA), coupled grasp (CO) and self-adaptive grasp (SA). Based on these three grasping modes, hybrid grasping modes are proposed. The following part will focus on the parallel grasping and self-adaptive grasping which was achieved on the TPM hand.

2.1 The Parallel Grasp (PA)

In parallel grasp mode, the distal phalanx keeps parallel to the original orientation. When the proximal phalanx rotates a certain angle forward relative to the base, the distal phalanx rotates an equal angle rear to the proximal phalanx (shown in Fig. 2a), thus the orientation of distal phalanx does not rotate relative to the base. The parallel grasp mode is very common in robot hand grasp as well as in human hand, for instance, when grasping a coin, a pamphlet, and other thinner objects.

2.2 The Self-adaptive Grasp (SA)

The definition of self-adaptive grasp means that the gesture of a robot hand varies according to the shape and the size of the objects. The function of self-adaptive grasp is required when the object is placed closely to the proximal phalanx, and it can be achieved in two steps (shown in Fig. 2b). In the first step, the proximal phalanx and distal phalanx both rotate around the proximal joint until the proximal phalanx reaches the objects. Then the rotation of proximal phalanx ceases while the distal phalanx continues to envelop the object. Objects in different sizes or different shape can be held tightly.

2.3 The Parallel and Self-adaptive Grasp (PASA)

The PASA hand combines both functions of parallel grasp and self-adaptive grasp, applying them to grasp objects in different shape and size (show in Fig. 3). In the first step, it functions the same as parallel grasp. The distal phalanx keeps parallel to its original orientation so that objects in a plane shape can be directly grasped in this step. If the object is round, the proximal phalanx will be blocked, and it will come into the second step. The distal phalanx rotates around the distal joint and envelops the objects similarly to the process in self-adaptive grasp.

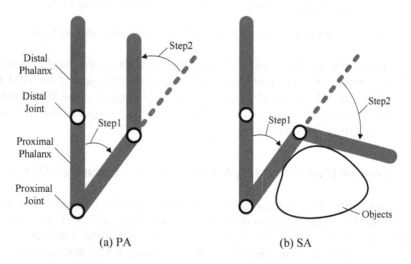

Fig. 2. Basic grasping modes of underactuated fingers.

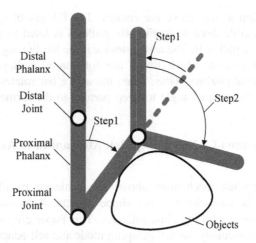

Distal
Phalanx

Distal
Joint

Step1

Step1

Proximal
Phalanx

Step2

Proximal
Joint

Objects

Fig. 3. The hybrid grasping mode (PASA) of underactuated fingers.

The TPM hand proposed in this paper has adopted the hybrid grasping mode that can simultaneously grasp in both parallel mode and self-adaptive mode (shown in Fig. 3). In Sect. 3, the structure of the hand will be introduced in detail.

3 Architecture of the TPM Hand

3.1 The Architecture of Parallel Grasping Mode

The four-bar mechanism makes up most traditional parallel systems, including the active bar, the base, bar-III, and bar-IV (shown in Fig. 4a). The distal phalanx is fixed to bar-IV. The four bars form a parallelogram, which limits the distal phalanx to keep

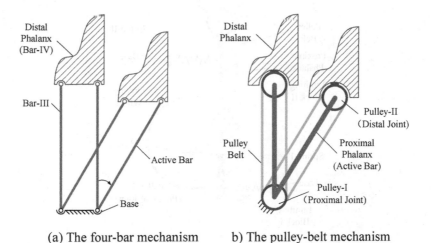

Distal
Phalanx
(Bar-IV)

Bar-III

Active Bar

Base

Distal
Phalanx

Pulley-II
(Distal Joint)

Pulley
Belt

Proximal
Phalanx
(Active Bar)

Pulley-I
(Proximal Joint)

(a) The four-bar mechanism   b) The pulley-belt mechanism

Fig. 4. Traditional mechanisms to achieve PA function.

the translation motion as the active bar rotates. The PA grasping mode can also be accomplished by flexible devices. In Fig. 4b, pulley-I is fixed to the base. The distal phalanx is fixed to the pulley-II. The active device is the bar linking two joints. The belt is set in a parallel pattern, which limits the rotation of pulley-II, as well as the movement of the distal phalanx. Thus, when the active bar rotates, the distal phalanx will turn backward to an equal angle to keep parallel to the former orientation.

3.2   The Architecture of a New Pulley Belt Mechanism to Achieve PA and SA

Although the pulley-belt mechanism above has achieved parallel grasping mode, self-adaptive grasp is necessary to other shape objects. A hybrid grasping mode is usually needed in many working circumstances. This paper proposes a novel mechanism to realize simultaneously parallel grasping mode and self-adaptive grasping mode.

As shown in Fig. 5, the TPM hand with flexible devices mainly consists of two phalanges, a driving wheel, pulleys I and II, a pulley belt, and two springs. The proximal phalanx rotates around the proximal joint, which is fixed to the base, and the distal phalanx is fixed to the pulley-II. (Shown in Fig. 5, the proximal phalanx is replaced by a link connecting two pulleys.) The driving shaft is placed rear to the pulley-I according to the requirements. The pulley belt is fixed to the pulley-I at one end, winding around the pulley-II, and is fixed on the driving wheel at the other end. Two springs are connected to pulley-I and proximal phalanx each, namely spring-I and spring-II, which separately limit the rotation angle of the pulley-I and proximal

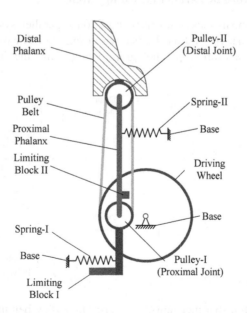

Fig. 5. The parallel grasp and self-adaptive finger.

phalanx. In addition, two bumps are connected to the pulley-I and the proximal phalanx separately, and two limiting blocks are fixed to the base.

As the primary gesture, the whole finger is vertical to the base. The rotation angle of pulley-I keeps zero because of tension from the spring-I. The proximal phalanx keeps vertical due to the tension from spring-II as well. When the driving wheel rotates, the pulley belt is coiled, driving pulley-II and proximal phalanx to rotate. However, the pulley-I do not rotate under tension from spring-I, which leads to the translation motion of the distal phalanx.

When the proximal phalanx contacts with the objects and is blocked to rotate, it will autonomously come into the self-adaptive grasp step. The force from driving wheel has transformed through pulley belt to pulley-I, which pulls the pulley-I rotate against tension from spring-I. The spring-I will be stretched. The enveloping step will be finished until distal phalanx has contacted the objects. When driving wheel resets after grasping, the spring-I and spring-II separately pull the pulley-I and proximal phalanx rotates backward to the primary gesture.

3.3 Details in the TPM Hand

The detailed mechanical design of the TPM hand is in Fig. 6. The TPM finger mainly consists of an actuator, a decelerator, a base, a transmission device, two pulleys, a driving wheel, a proximal shaft, triple shafts, two phalanxes, a tension spring, a torsion

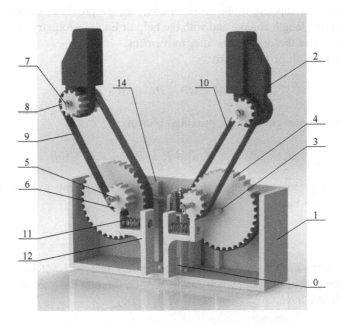

Fig. 6. The structure of TPM hand. 0 actuator, 1 base, 2 distal phalanx, 3 driving shaft, 4 driving wheel, 5 proximal joint, 6 pulley-I, 7 distal joint, 8 pulley-II, 9 pulley belt I, 10 pulley belt II, 11 spring-I, 12 limiting block I, 14 limiting block II.

spring, and two pairs of bumps and limit blocks. In the prototype, the pulley belt is divided into two parts. Each partial belt is fixed on the pulleys separately because of the interference between the driving wheel and pulley-I. This interference is required according to the analysis in Sect. 4.

4   Analysis of the TPM Hand

4.1   Kinematic Analysis and Feasible Field

According to the mechanical design in Sect. 3, a new parameter-distance between proximal shaft and driving shaft is introduced to the system. This distance, however, is influenced by other parameters, for example, the radius of driving wheel and pulleys, in some potential functional relationship. In this section, this functional relationship is discussed through the kinematic analysis.

4.1.1   The Kinematic Analysis

To describe the movement of the robot finger, an equilibrium containing all parameters is required. Define:

θ-the rotation angle of the proximal phalanx relative to the base, rad;
α-the rotation angle of the driving wheel relative to the base, rad;
δ-the rotation angle of the cut point between pulley belt and driving wheel, rad;
h-the length of the proximal phalanx, mm;

Meanwhile, we suppose in the initial orientation, the proximal keep vertical. Then by calculating the length change and with the help of Euclid geometry, we can describe the initial state and the orientation after movement.

The initial length of pulley part:

$$L = NM + MP - PA + AB + BB'$$ (1)

After a rotation angle θ of the proximal phalanx:

$$L' = NN' + N'M' + M'P' - P'A' + A'B' + roll$$ (2)

Here, *roll* represents the length that is twined on the driving wheel because of its rotation. Thus, we have:

$$L = L'$$ (3)

The geometry limitation is $\alpha \geq \delta$, so that pulley belt will not be fractured. The length of each part of the pulley belt can be written through the geometry relationship in the Fig. 7. We have:

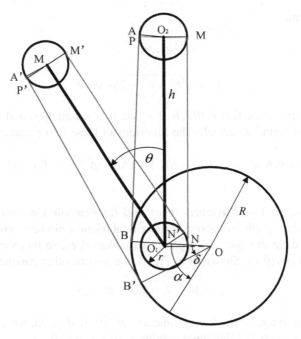

Fig. 7. The kinematic diagram of the mechanism.

$$\delta = \left[\arccos\left(\frac{R-r}{l_2}\right) - \arcsin\left(\frac{h\cos\theta}{l_2}\right) \right] - \left[\arccos\left(\frac{R-r}{l_1}\right) - \arctan\left(\frac{h}{e}\right) \right] \quad (4)$$

$$BB' = \delta \cdot R \quad (5)$$

$$NN' = \theta \cdot r \quad (6)$$

$$AB = \sqrt{l_1^2 - (R-r)^2} \quad (7)$$

$$A'B' = \sqrt{l_2^2 - (R-r)^2} \quad (8)$$

$$PA = \left[\arccos\left(\frac{R-r}{l_1}\right) - \arctan\left(\frac{h}{e}\right) \right] \cdot r \quad (9)$$

$$P'A' = \left[\arccos\left(\frac{R-r}{l_2}\right) - \arcsin\left(\frac{h\cos\theta}{l_2}\right) - \theta \right] \cdot r \quad (10)$$

Among them:

$$l_1 = \sqrt{h^2 + e^2} \tag{11}$$

$$l_2 = \sqrt{h^2 + e^2 + 2he \sin \theta} \tag{12}$$

Thus, we have a function $roll(\theta, h, R, e)$ on four parameters, and every part consisting the equilibrium are actually the functions of these four parameters as well:

$$roll(\theta, h, R, e) = (BB' - NN') + (AB - A'B') - (PA - P'A') \tag{13}$$

4.1.2   Analysis on Two Variables (θ, e) and the Feasible Condition

Once we have the equilibrium above, a functional relationship between parameters has been established. In this part, we separate variables (θ, e), so that we could obtain a binary function $roll(\theta, e)$. Shown as in Fig. 7, we assume other parameters as follows:

$$r = 10, \quad h = 40, \quad R = 25 \tag{14}$$

Through the image of the binary function $roll(\theta, e)$ (Fig. 8), we could obtain the feasible field to satisfy the limitation condition $roll(\theta, e) > 0$:

$$0 < e < 15, \quad 0 < \theta < \tfrac{\pi}{2} \tag{15}$$

When the radius of the pulley r, the radius of driving wheel R, and the length of proximal phalanx h keep constants. θ and e have to obey the following three groups of relationships:

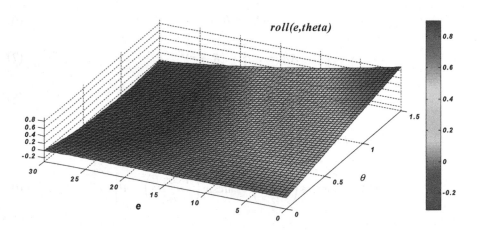

Fig. 8. Image of binary function: $roll(\theta, e)$.

1. In the initial state, only when the driving wheel rotates an infinite small angle, the proximal phalanx can rotate:

$$\alpha(\theta, e) = \frac{roll(\theta, e)}{R} + \delta, \quad \lim_{\theta \to 0} roll(\theta, e) = 0, \quad \lim_{\theta \to 0} \alpha(\theta, e) = 0 \qquad (16)$$

2. Along with the increasing rotation angle of the proximal phalanx, the length of twined pulley belt keeps to be positive, and increasing monotonously:

$$roll(\theta, e) > 0, \quad \frac{\partial roll(\theta, e)}{\partial \theta} > 0 \qquad (17)$$

3. Along with the increasing rotation angle of the proximal phalanx, the rotation angle of driving wheel keeps to be positive, and increasing monotonously:

$$\alpha(\theta, e) > 0, \quad \frac{\partial \alpha(\theta, e)}{\partial \theta} > 0 \qquad (18)$$

4.2   Feasible Field of Driving Wheel's Radius R

With three groups of equilibrium above, we could separate other parameters each, and explore their domain separately. Assume $r = 10$ is the only constant.

According to the analysis, $r = 10, h = 40, R = 25, 0 < e < 15$ is the feasible field. In this part, assume $\theta = 0.01, \quad r = 10, \quad h = 40, \quad R = 25$. Explore the relationship between R and e:

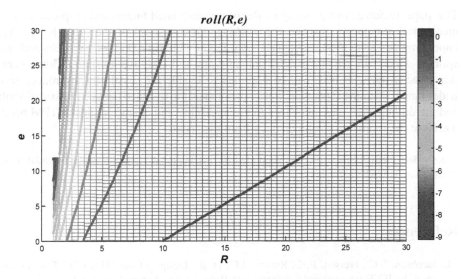

Fig. 9. Image of binary function: $roll(R, e)$.

Shown in Fig. 9 above, With the constant h, R and e approximately has a linear relationship:

$$R > 10 + e \qquad (19)$$

Bring the kinematic function (12) into the three group of inequality (15), (16), (17), and let θ approaches to 0, e, r, and R have to follow:

$$0 < e < R - r \qquad (20)$$

As long as e, r, and R follow the inequality (20) above, the parallel function can be achieved, as long as no interferences occur. Particularly, it is unrelated to the length of the proximal phalanx h.

4.3   Conclusion of the Limiting Condition

Based on the analysis above, we have obtained all required conditions. As a conclusion, three inequalities have been proposed to describe the relationships between parameters.

$$\begin{aligned} R &> r \\ 0 < e &< R - r \\ h &> 2r \end{aligned} \qquad (21)$$

5   Conclusion

This paper reviewed two grasping modes of underactuated fingers and proposes a novel mechanical structure to expand the working circumstances of the robot hand. This paper proposes a novel underactuated robot finger with triple pulleys, two belts and two springs, which can achieve parallel pinching and self-adaptive grasping. This separation is achieved with belts and two springs to enable the distal phalanx rotates in a wider range. With the separation, a new parameter, the distance between proximal shaft and the driving shaft, has influenced to the feasibility and characteristics of TPM hand, therefore the limitation functions have been discussed.

Acknowledgment. This paper was supported by the National Natural Science Foundation of China (No. 51575302).

References

1. Jacobsen, S.C., Iversen, E.K., Knutti, D.F., et al.: Design of the UTAH/M.I.T. dexterous hand. In: IEEE International Conference on Robotics and Automation, San Francisco, USA, April, pp. 1520–1532 (1986)

2. Loucks, C.S.: Modeling and control of the stanford/JPL hand. In: 1987 International Conference on Robotics and Automation, pp. 573–578 (1987)
3. Butterfass, J., Grebenstein, M., Liu, H., et al.: DLR-hand II: next generation of a dexterous robot hand. In: IEEE International Conference on Robotics and Automation (ICRA), vol. 1, pp. 109–114 (2001)
4. Kawasaki, H., Komatsu, T., Uchiyama, K., et al.: Dexterous anthropomorphic robot hand with distributed tactile sensor: gifu hand II. In: IEEE International Conference on Systems, Man, and Cybernetics (SMC), pp. 782–787 (1999)
5. Laliberte, T., Gosselin, C.: Simulation and design of underactuated mechanical hands. Mech. Mach. Theory 33(1/2), 39–57 (1998)
6. Birglen, L., Gosselin, C.M.: Kinetostatic analysis of underactuated fingers. IEEE Trans. Robot. Autom. 20(2), 211–221 (2004)
7. Salisbury, J.K., Craig, J.J.: Articulated hands: force control and kinematic issues. Int. J. Robot. Res. 1(4), 4–17 (1982)
8. Dollar, A.M., Howe, R.D.: The SDM hand as a prosthetic terminal device: a feasibility study. In: 2007 IEEE 10th International Conference on Rehabilitation Robotics, Noordwijk, Netherlands, June, pp. 978–983 (2007)
9. Carbone, G., Iannone, S., Ceccarelli, M.: Regulation and control of LARM Hand III. Robot. Comput.-Integr. Manuf. 24(2), 202–211 (2010)
10. Pons, J.L., Rocon, E., Ceres, R., et al.: The MANUS-hand dexterous robotics upper limb prosthesis: mechanical and Manipulation Aspects. Auton. Robots. 16, 143–163 (2004)
11. Mizusawa, S., Namiki, A., Ishikawa, M.: Tweezers type tool manipulation by a multifingered hand using a high-speed visual servoing. In: 2008 IEEE/RSJ International Conference on Intelligent Robots and Systems. Nice, France, September, pp. 2709–2714 (2008)
12. Zhang, W., Che, D., Liu, H., et al.: Super underactuated multi-fingered mechanical hand with modular self-adaptive gear-rack mechanism. Ind. Robot: Int. J. 36(3), 255–262 (2009)
13. Li, G., Li, B., Sun, J., et al.: The development of a directly self-adaptive robot hand with pulley-belt mechanism. Int. J. Precis. Eng. Manuf. 14(8), 1361–1368 (2013)
14. Li, G., Liu, H., Zhang, W.: Development of multi-fingered robot hand with coupled and directly self-adaptive grasp. Int. J. Humanoid Robot. 9(4) (2012)
15. Zhang, W., Zhao, D., Zhou, H., et al.: Two-DOF coupled and self-adaptive (COSA) finger: a novel underactuated mechanism. Int. J. Humanoid Robot. 10(2) (2013)

A Novel Robot Finger with a Rotating-Idle Stroke for Parallel Pinching and Self-adaptive Encompassing

Jingchen Qi[1], Linan Dang[1,2], and Wenzeng Zhang[1(✉)]

[1] Department of Mechanical Engineering, Tsinghua University,
Beijing 100084, China
wenzeng@tsinghua.edu.cn
[2] College of Electronic Information Engineering, Inner Mongolia University,
Hohehot 010021, China

Abstract. This paper presents the design of a novel underactuated robot finger with a rotating-idle stroke, called PASA-RIS finger. The PASA-RIS finger consists of multiple gears, a pair of rotating blocks, a pair of limiting blocks, and two springs. The finger has two joints driven by one reduced motor and a transmission mechanism. Grasping force analysis and experimental results show that the PASA-RIS finger has the parallel pinching and self-adaptive grasping function. The PASA-RIS finger can adapt objects of different shapes and sizes and automatically switch from parallel pinching to self-adaptive encompassing.

Keywords: Robot hand · Self-adaptive grasp · Parallel pinching · Underactuation

1 Introduction

In the 21th century, robots have been widely used in various fields to complete large tasks instead of human labors. Robotic hand, as the end-effector of the humanoid robots, has become the key point in the research of robotics.

One of these is the dexterous hand, in which actuators are equal to their degrees of freedom (DOFs), such as Gifu Hand II [1], DLR-HIT Hand [2], NAIST Hand [3], Robonaut 2 Hand [4], KIST Hand [5]. The dexterous hands are capable of complex tasks through precise control of their joints, but their control and sensing systems are too complicated as well as expensive, and their contact forces are not large enough.

Another one of those hands is the underactuated hand, in which actuators are less than their DOFs, such as the highly-underactuated robotic hand [6], Biomimetic Robot Hand [7], Underactuated adaptive hand [8], GCUA Hand [9], COSA Hand [10], TH-3R Hand [11], KIST Hand [12, 13], MAP Hand [14, 15], self-adaptive prosthetic finger [16].

The larger power motors of underactuated hands can be set in their palms to reach greater grasping forces. The grasping mode of a finger is generally fixed by its mechanism so that the control and the sensing system are simple. The underactuated hands are small, easy to control, and low in cost compared with dexterous hands. However, it is hard for underactuated hands to conduct precise grasping such as

Y. Huang et al. (Eds.): ICIRA 2017, Part II, LNAI 10463, pp. 764–775, 2017.
DOI: 10.1007/978-3-319-65292-4_66

pinching, which is easy for industrial grippers [17]. The Robotiq hand [18], combining the advantages of traditional underactuated hands and industrial grippers, adopts a linkage mechanism, which brings out the problem of dead zone in motion.

This paper presents the design of a novel underactuated robot finger with a rotating-idle stroke, called PASA-RIS finger. The PASA-RIS finger can perform parallel pinching as well as self-adaptive grasping according to the shapes, sizes, and locations of the objects grasped. The design and motion analysis of the PASA-RIS finger are given in detail. Simulation and grasping experimental results are given and discussed to demonstrate the feasibility of the PASA-RIS finger.

2 Concept of the PASA Grasping Mode

The underactuated robot fingers from the view of mechanism design can be classified into three kinds:

(1) Parallel finger with parallel pinching (PA) mode,
(2) Self-adaptive finger with self-adaptive (SA) mode,
(3) PASA finger with PASA mode (a hybrid of parallel and self-adaptive modes).

The PA mode is shown in Fig. 1a. The SA mode is shown in Fig. 1b.

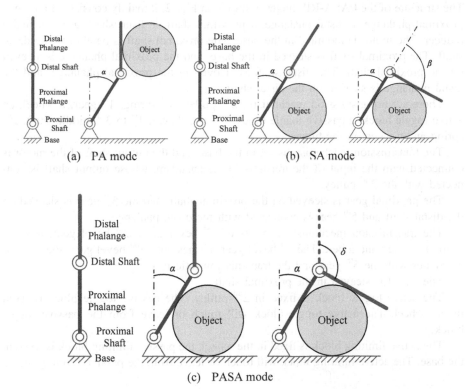

Fig. 1. Concepts of PA, SA and PASA modes.

Table 1. Comparison of coupled, self-adaptive, and PASA-RIS fingers.

Robot finger	Coupling	Self-adaptation	PASA-RIS
Pinching ability	10	0	10
Enveloping ability	0	10	10
Control difficulty	10	10	10
Adaptation	5	10	10
Plane grasp	5	0	10
Total score	30	30	50

The PASA grasping mode is shown in Fig. 1c. The PASA finger will conduct parallel pinching when the object is close to the top of the finger. The finger will conduct parallel action until the proximal phalange touches the object when the object is close to the bottom of the finger, and then it will switch to self-adaptive grasping mode (Table 1).

3 Design and Grasping Process of the PASA-RIS Finger

3.1 Design of the PASA-RIS Finger

The structure of the PASA-RIS finger is shown in Fig. 2. It mainly consists of a base, a proximal phalange, a distal phalange, a proximal shaft, a distal shaft, a motor and a reducer. The motor is mounted in the base. The proximal shaft is parallel to the distal shaft. The proximal shaft is sleeved in the base, and the proximal phalange is sleeved on the proximal shaft. The distal shaft is sleeved in the proximal phalange, and the distal phalange is sleeved on the distal shaft.

There are a transmission mechanism, 2nd pulley, 1st spring, 1st gear, a wheel, an active toggle block, a passive toggle block, 2nd to 5th gear, 1st to 3rd middle shaft, 2nd spring and a wheel in the finger.

The transmission mechanism is set in the base, and the output shaft of the motor is connected with the input of the transmission mechanism, whose output shaft is connected with the 2nd pulley.

The proximal gear is sleeved on the proximal shaft, and the 5th gear is sleeved on the distal shaft and 5th gear is connected with the distal phalange.

The transmission mechanism consists of 1st bevel gear, 2nd bevel gear, a lower shaft, 1st pulley and a belt. The 1st bevel gear meshes with 2nd bevel gear, the 1st gear co-rotates with the 5th gear and the transmission ratio is 1.

The wheel is sleeved in the proximal shaft.

The active toggle block is fixed in 2nd pulley. The passive toggle block is fixed in the wheel. The active toggle block will touch or leave from the passive toggle block.

The active limiting block is fixed in the wheel, the passive limiting block is fixed in the base. The active limiting block will touch or leave from the passive limiting block.

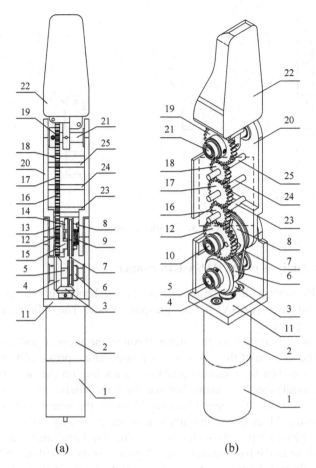

(a)   (b)

Fig. 2. Structure of the PASA-RIS finger. 1-motor; 2-reducer; 3-1st bevel gear; 4-2nd bevel gear; 5-lower shaft; 6-1st pulley; 7-belt; 8-2nd pulley; 9-1st spring; 10-proximal shaft; 11-base; 12-1st gear; 13-wheel; 14-active toggle; 15- passive toggle; 16-2nd gear; 17-3rd gear; 18-4th gear; 19-5th gear; 20-proximal phalange; 21-distal shaft; 22-distal phalange; 23-25-1st -3rd middle shaft.

1st spring connects 2nd pulley and the proximal phalange.

2nd spring connects the active limiting block and the base or the passive limiting block.

The active limiting block touches the passive limiting block in the initial status. Then the rotating direction of the active limiting block is set as same as one of the proximal phalange and the passive limiting block restricts the rotation angle of the active limiting block.

The 1st spring is at tension in the initial status. The active toggle block is away from the passive toggle block. The active toggle block is able to touch the passive toggle block in the motion range of the active toggle block.

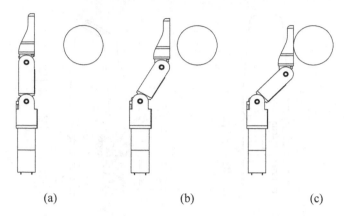

(a)   (b)   (c)

Fig. 3. Parallel grasping.

3.2   Motion Analysis of the PASA-RIS Finger

The PASA-RIS finger with a rotating idle-stroke mechanism is able to complete parallel and self-adaptive (PASA) hybrid grasping mode. The following is a specific implementation process:

The 2nd pulley is driven by the motor through the reducer and the transmission mechanism. The rotation of the 2nd pulley is passed to the proximal phalange through the 1st spring, and then the proximal phalange is rotating on the proximal shaft.

The active toggle block is rotating but not touch the passive toggle block for a short time, namely idle stroke. The active limiting block on the wheel is closely touched to the passive limiting block due to the function of the 2nd spring, so the 1st gear and the active limiting block keep in the initial status. The distal phalange realizes translation instead of rotating to the base because the 1st gear co-rotates with the 5th gear and the transmission ratio is 1. Therefore, the finger realizes the parallel pinching motion.

(a) The parallel pinching mode will be conducted until the distal phalange touches the object. The force is passed to the object through the proximal phalange, distal shaft and the distal phalange. The motor stops running if the contact force is big enough to grasp the object. The whole process is shown in Fig. 3.

(b) The self-adaptive mode will be conducted until the proximal phalange touches the object. The motor keeps running to push the passive toggle block by the rotation of the active toggle block. The active limiting block and the 1st gear is rotating when the contact force between the passive toggle and the active toggle is greater than the one conducted by the 2nd spring.

The stretching of the 2nd spring exceeds. The 5th gear and the distal phalange is rotating on the distal shaft till the distal phalange touching the object and exerting force upon it. Then the motor stops running and the enveloping grasp is completed. The motions of the active toggle block and the passive toggle block are shown in Fig. 4. The stretching of the 2nd spring is shown in Fig. 5. The whole grasping progress is shown in Fig. 6.

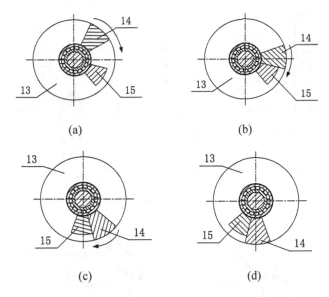

Fig. 4. Motion of the rotating block. 13-wheel; 14-active toggle; 15- passive toggle;

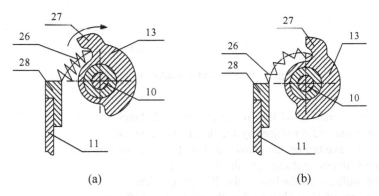

Fig. 5. Stretching of the first spring. 10-proximal shaft; 11-base; 13-wheel; 26-2nd spring; 27-28-active, passive limiting blocks.

4   Distribution Analysis of Grasping Forces

This part is going to introduce the force analysis in the grasping progress. The friction between the phalanges and the object is ignored as well as the gravity. Meaning of the variables is following:

θ_1 - the 1st joint angle of the proximal phalange, rad;
θ_2 - the 2nd joint angle of the distal phalange, rad;
β - the rotation angle of the 1st gear, rad;
α - the rotation angle of the 5th gear, rad;

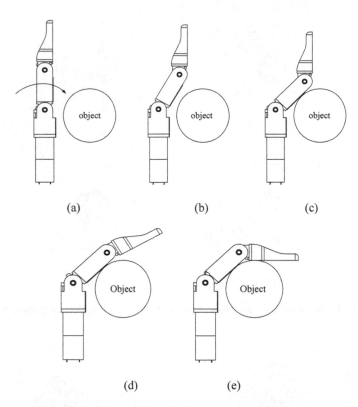

(a)   (b)   (c)

(d)   (e)

Fig. 6. Parallel and self-adaptive grasping.

F_1 - force exerted on object by the proximal phalange, N;
F_2 - force exerted on object by the distal phalange, N;
T - torque exerted on the proximal shaft from the motor, Nm;
k_1 - the stiffness coefficient of the 1st spring, N/m;
k_2 - the stiffness coefficient of the 2nd spring, N/m;
h_1 - the arm of F_1 with regard to the proximal shaft, m;
h_2 - the arm of F_2 with regard to the distal shaft, m;
l_1 - the length of the proximal phalange, m;
v - the potential energy of the finger, j (Fig. 7).

In the stage of parallel pinching, one can get:

$$\theta_1 = \beta \tag{1}$$

$$\theta_2 = 0 \tag{2}$$

Once the active toggle block pushes the passive toggle block going down, the grasping goes into self-adaptive stage. In this stage, the distal phalange starts rotating. Before the distal phalange touches the object, the F_2 is equal to 0 and the F_1 will be:

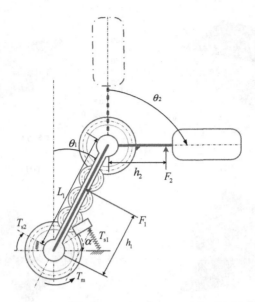

Fig. 7. Force analysis of entire grasping state of the PASA-RIS finger.

$$F_1 = \frac{k_2(\beta - \theta_{10}) - k_1\alpha}{h_1} \tag{3}$$

Once the distal phalange touches the object, the finger realizes enveloping grasp, the potential power of the finger at this moment is:

$$v = \frac{1}{2}k_2(\beta - \theta_1)^2 + \frac{1}{2}k_1(\beta - 2\theta_1)^2 - T\beta \tag{4}$$

Then take the derivative of Eq. (4),

$$\partial v = [F_1, F_2] \begin{bmatrix} -h_1 & 0 \\ -l_1\cos(\theta_2 - 3\theta_1) & -h_2 \end{bmatrix} \begin{bmatrix} \partial\theta_1 \\ \partial\theta_2 - 2\theta_1 \end{bmatrix} \tag{5}$$

Based on the previous formulas, one can easily get the final equation of F_1 and F_2:

$$
\begin{aligned}
F_1 = {} & \frac{(\theta_2 - 2\theta_1)(k_2 - 2k_1) - \theta_1(k_2 - 4k_1)}{h_1} \\
& - \frac{l_1\cos(3\theta_1 - \theta_2)(T - (\theta_2 - 2\theta_1)(k_2 + k_1) + \theta_1(k_2 + 2k_1))}{h_1 h_2}
\end{aligned} \tag{6}
$$

$$F_2 = \frac{T - (\theta_2 - 2\theta_1)(k_2 + k_1) + \theta_1(k_2 + 2k_1)}{h_2} \tag{7}$$

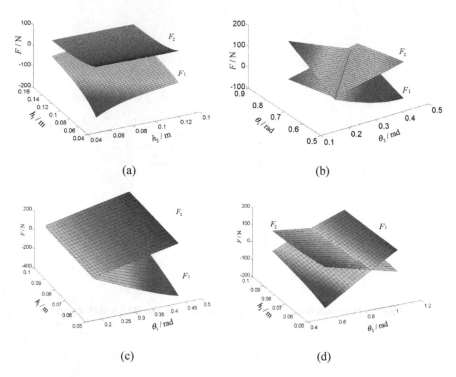

(a)　　　　　　　　　　　　　　　　(b)

(c)　　　　　　　　　　　　　　　　(d)

Fig. 8. Relation among the grasping force (N), rotation angle of the proximal phalange (rad) and the distal phalange (rad)

The relationships among F_1, F_2, h_1, h_2, and joints angle are shown in Fig. 8a, b. The relationships among F_1, F_2, h_1, θ_1, and F_1, F_2, h_1, θ_2 are shown in Fig. 8c, d. The simulation results show that the PASA-RIS finger is effective to grasp objects stably.

5 Experiments

In order to validate the feasibility of the design, we developed a prototype of the PASA-RIS finger for grasping experiment. The experimental process is shown in Fig. 9. Once the object is a spherical, if the distal phalange touches the object at first, then the grasping mode will be parallel pinching; if the proximal phalange touches the object at first, the grasping mode of the finger will switch from parallel pinching to self-adaptive grasping.

The finger is able to have a stable grasp whether the size of the spherical object changes from small to large, which is both phalanges contact the object and the grasping forces are bigger than zero. If the object needs friction forces to grasp, the grasping stability can be improved by rubber or silica gel increasing the friction coefficients of the surfaces of phalanges.

Experimental results show that the parallel pinching ability of the PASA-RIS finger is valid.

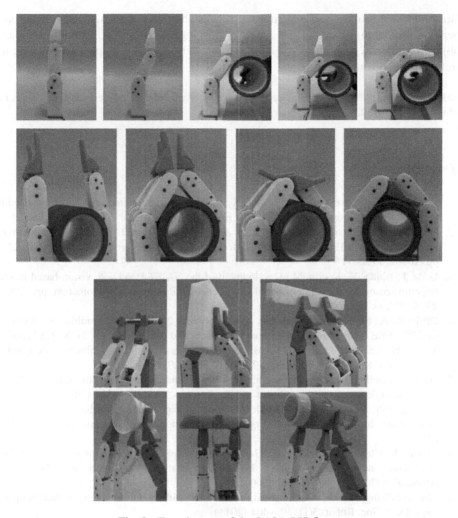

Fig. 9. Experiments of the PASA-RIS finger.

In the parallel pinching grasp stage, the grasping force of the distal phalange is large enough to achieve stable grasp. In the self-adaptive encompassing grasp stage, the proximal phalange and the distal phalange both have large grasping force to achieve stable grasp. After several experiments, we can conclude that the PASA-RIS finger is able to achieve the goal of original design, and automatically switch grasping modes from parallel pinching to self-adaptive encompassing mode.

6   Conclusion

This paper presents the design of a novel underactuated robot finger with a rotating-idle stroke, called PASA-RIS finger. The PASA-RIS finger has two joints driven by only one motor. Based on the finger, a PASA–RIS robot hand can be designed.

Grasping force analysis and experimental results show that the PASA-RIS finger can perform parallel pinching and self-adaptive grasping. The PASA-RIS finger can adapt objects of different shapes and sizes and automatically switch from parallel pinching to self-adaptive encompassing.

Acknowledgment. This paper was supported by the National Natural Science Foundation of China (No. 51575302).

References

1. Kawasaki, H., Komatsu, T., Uchiyama, K.: Dexterous anthropomorphic robot hand with distributed tactile sensor: gifu hand II. IEEE/ASME Trans. Mech. **7**(3), 296–303 (2002)
2. Butterfass, J., Grebenstein, M., Liu, H., et al.: DLR-hand II: next generation of a dextrous robot hand. In: IEEE International Conference on Robotics and Automation (ICRA), pp. 109–114 (2001)
3. Ueda, J., Ishida, Y., Kondo, M.: Development of the NAIST-Hand with vision-based tactile fingertip sensor. In: IEEE International Conference on Robotics and Automation, pp. 2343–2348 (2005)
4. Diftler, M.A., Mehling, J.S., Abdallah, M.E.: Robonaut 2 - the first humanoid robot in space. In: 2011 IEEE International Conference on Robotics and Automation, pp. 2178–2183 (2011)
5. Kim, E.H., Lee, S.W., Lee, K.Y.: A dexterous robot hand with a bio-mimetic mechanism. Int. J. Precis. Eng. Manuf. (IJPEM) **12**(2), 227–235 (2011)
6. Wang, L., DelPreto, J., Bhattacharyya, S.: A highly-underactuated robotic hand with force and joint angle sensors. In: Intelligent Robots Systems, pp. 1380–1385 (2011)
7. Lee, S., Noh, S., Lee, Y.K., et al.: Development of biomimetic robot hand using parallel mechanisms. In: IEEE International Conference on Robotics and Biomimetics, pp. 550–555 (2010)
8. Zhang, W., Chen, Q., Sun, Z., et al.: Under-actuated passive adaptive grasp humanoid robot hand with control of grasping force. In: IEEE International Conference on Robotics and Automation (ICRA), pp. 696–701 (2003)
9. Che, D., Zhang, W.: GCUA humanoid robotic hand with tendon mechanisms and its upper limb. Int. J. Soc. Robot. **3**(1), 395–404 (2011)
10. Li, G., Liu, H., Zhang, W.: Development of multi-fingered robotic hand with coupled and directly self-adaptive grasp. Int. J. Humanoid Robot. **9**(4), 1–18 (2012)
11. Zhang, W., Che, D., Liu, H.: Super underactuated multi-fingered mechanical hand with modular self-adaptive gear-rack mechanism. Ind. Robot **36**(3), 255–262 (2009)
12. Cho, C., Lee, Y., Kim, M.: Underactuated hand with passive adaptation. In: IEEE International Symposium in Industrial Electronics, pp. 995–1000 (2009)
13. Choi, B., Lee, S., Choi, H.R., et al.: Development of anthropomorphic robot hand with tactile sensor: SKKU Hand II. In: IEEE/RSJ International Conference on Intelligent Robots and Systems, pp. 3779–3784 (2009)
14. Dalley, S.A., Wiste, T.E., Withrow, T.J., et al.: Design of a multifunctional anthropomorphic prosthetic hand with extrinsic actuation. In: IEEE International Conference on Rehabilitation Robotics (ICORR), pp. 675–681 (2009)
15. Dalley, S.A., Wiste, T.E., Varol, H.A., et al.: A multigrasp hand prosthesis for transradial amputees. In: International Conference of the IEEE Engineering in Medicine & Biology Society, pp. 5062–5065 (2010)

16. Sheng, X., Hua, L., Zhang, D., et al.: Design and testing of a self-adaptive prosthetic finger with a compliant driving mechanism. Int. J. Humanoid Robot (IJHR) **11**(3), 1450026 (2014)
17. Ali, H., Hoi, L.H., Seng, T.C.: Design and development of smart gripper with vision sensor for industrial applications. In: IEEE International Conference on Computation, pp. 175–180 (2011)
18. Demers, L.A., Lefrancois, S., Jobin, J.: Gripper having a two degree of freedom underactuated mechanical finger for encompassing and pinch grasping, US Patent, US8973958 (2015)

Remote Live-Video Security Surveillance via Mobile Robot with Raspberry Pi IP Camera

Xiaolong Jing, Changyang Gong, Zhenyu Wang, Xudong Li,
Zhao Ma, and Liang Gong[✉]

Shanghai Jiao Tong University, 800 Dongchuan Road, Minhang District,
Shanghai, China
707929484@qq.com, gongliang_mi@sjtu.edu.cn

Abstract. This paper presents a robot prototype designed to overcome the lack of current security monitoring system: low flexibility, limited monitoring range, disability of copping with sudden special circumstances. It's designed and built for remote live-video security surveillance and object recognition and capture in highly structured small space environment in industrial field, involving remote controlling, live video taking, distance and position identifying of an object, trajectory planning and object capturing. The robot is controlled by means of human-machine interface. The operator drives the robotic device and performs the related tasks by means of a control user interface developed on android APP. The robot computer operates robot grasping and manipulation automatically, that is, performs the precise localization of a specific object, computes adequate capturing sequence and controls the motion (robot moving and object capturing) of all the mechanical components. Throughout this paper, the specific design of every module of the robotic device is presented. The device has been built. Related laboratory tests have been done to check the robot performance. Results show excellent robot controllability, targeted object finder and live field operations. Some figures and tables showing overall performance are given.

Keywords: Remote control · Live-video · Object recognition and capture · Live field operation

1 Introduction

Industrial application and disaster rescue background has been traditionally and normally the application field for robotics. Robots perform well in structured environments where working positions as well as obstacles are somehow predictable or visible. Robots are designed to cope with a specific problem such as auto parts assemblies, disaster relief, material processing and manufacturing, etc.

In early 1970s, professor Nilsson of the Stanford Institute developed a mobile robot, Shakey [1, 2], who could perceive the surroundings and make relevant reactions.

The original version of this chapter was revised: the acknowledgement section was added. The erratum to this chapter is available at https://doi.org/10.1007/978-3-319-65292-4_78

Y. Huang et al. (Eds.): ICIRA 2017, Part II, LNAI 10463, pp. 776–788, 2017.
DOI: 10.1007/978-3-319-65292-4_67

Prof. DN Green of Carleton University designed a planetary robot LARES-L [3, 4], which used the operation control algorithm to achieve path tracking and automatic obstacle avoidance. Professor Shigeo Hirose is one of the earliest scholars engaged in special field rescue and detection robot researching. He developed the "ACM" [5–8] series of snake-like robots from the perspective of bionic and the idea of ultra-mechanical system. The "ACM" robots can be used in snake amphibious detectors and disaster rescue, etc. For applications such as earthquake and fire rescue, he has also developed "GENBU" and "SORYU" series robots [9–12]. Based on the World Trade Center disaster in 2003, Prof. Jennifer presented a study method of human-robot interaction in fields such as disaster rescue [13]. [14] presented a ground rescue robot to the real disaster sites at Aomori Prefecture and Iwate Prefecture and discussed the effectiveness and problems of applying the rescue robot in the real disaster sites. [15] presented the improvements to the rescue robot quince toward future indoor surveillance missions.

As can be seen from the references above, security and rescue tasks have been an important application area of different kinds of technologies and robots to improve safety of people, equipment and machines, prevent fire and electric leakage in advance and do rescue work when disaster occurs. But all the robots or devices mentioned above mainly focus on the detection/investigation of the site or methodology of application when applied in special scenarios.

So, we hope to establish a robot that can both "see" and detect the site and operate live field operation, i.e., a remote live-video security surveillance mobile robot. Taking into account the complexity of the problems related to the industrial/disaster site environment and limitations of current approaches, the implementation of a fully automatic and real-time solution for these tasks seems far away.

We present a semi-automatic way of operation, with realistic goal, combining harmoniously the human and machine functions, which is what we call human-in-the-loop controlling method. We aim to devote the human skills into more complex and intellectual parts of the operation, i.e. robot movement control and task signal sending, which don't require much effort, while the robot itself will perform the hardest, most dangerous and precise job (in this project, we hope it can identify an specific object itself, localize where the object is, plan the movement trajectory and capture the identified object independently).

After a brief introduction, we will present the mechanical and electronic design of the whole remote live-video security surveillance robot including all subcomponents, that is, lower movable chassis, middle action execution robot, upper visual system and the whole controlling system. The robot is mainly applied to real-time security monitoring. Besides, it can be used in anti-terrorism in dangerous environment, fruit harvesting and human-computer interactive entertainment, traffic police, etc.

Our robot can realize four primary targets:

1. The robot have good adaptability and flexible movability. It can adapt to a variety of complex industrial conditions. Besides, the movement of the execution system can be remote controlled through an Android APP.
2. The robot can take live HD video through two Raspberry Pi Cameras located in the execution robot's eyes and upload the HD video to Tyco's server ExacqVision via WIFI so that people can monitor specific areas in case of emergency. Besides, the video can also be seen on the Android APP screen.

3. The robot can independently identify a specific object and accurately localize where the object is through binocular vision system using two the Kinect camera. The robot can intercept images from the video stream for the identification operation.
4. The robot can independently plan the best path to move the lower chassis and control the execution robot to capture the specific, identified object.

So, we use the open-sourced 3D printed InMoov and the Mecanum wheel platform as the Execution system to realize the high DOF and flexible movement, the Raspberry Pi and Kinect camera as the visual system to realize live-video transmission, object identification and localization, and develope an Android APP as the controlling terminal to realize remote control.

2 The Execution System

2.1 The Mecanum Wheel

The Mecanum wheel (as shown in Fig. 1) is an omnidirectional wheel having the ability of free planar 3DOF movement and moving in any direction. We use four wheels and a matched Aluminum platform as the lower moving platform.

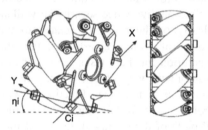

Fig. 1. The Mecanum Wheel. It's a conventional wheel with a sequence of small rollers attached to its circumference with a typical axis of rotation at 45° to the plane of the wheel and at 45° to a line through the center of the roller parallel to the axis of rotation of the wheel.

From the kinematics of the Mecanum wheel [16–18], the relationship between the speed of the Mecanum wheel platform and the rotation input speed of the four motors corresponding to the wheels can be written in:

$$\begin{bmatrix} \dot{\theta}_{1y} \\ \dot{\theta}_{2y} \\ \dot{\theta}_{3y} \\ \dot{\theta}_{4y} \end{bmatrix} = \frac{1}{R} \begin{bmatrix} 1 & -1 & -(a+b) \\ 1 & 1 & a+b \\ 1 & 1 & -(a+b) \\ 1 & -1 & a+b \end{bmatrix} \begin{bmatrix} \dot{x} \\ \dot{y} \\ \dot{\theta} \end{bmatrix} \tag{1}$$

Where $\dot{\theta}_{iy}$ is the rotational speed of the ith wheel counting from the right-front wheel, R is the radius of the wheel, a or b is the half of the distance between the left and right or front and back wheel, $\dot{x}$, $\dot{y}$ are translational speed and $\dot{\theta}$ is rotational speed of the whole platform.

We can conclude from the equation that running all four wheels in the same direction causes the platform forward or backward movement, running the wheels on one side in the opposite direction to those on the other side causes in-place rotation of the platform and running the wheels on one diagonal in the opposite direction to those on the other diagonal causes sideways movement. Any combinations of the four wheels' movement can result in various motions in any direction with any platform rotation you desire.

The characteristics of the Mecanum wheel platform helps a lot when operating in small space and complex industrial field if high flexibility and movability is required.

2.2   The Execution Robot

We use the open-sourced InMoov robot as the execution robot controlled by Arduino microcontroller. InMoov is able to perceive sound, see, speak and move independently. Because of its characteristics of open-source, high degree of freedom and high flexibility, we use it as the execution robot for the live field operation.

We construct the communication framework of each separate part of the execution robot based on ModBus RTU [19–21] protocol star communication architecture and RS485 [22] hardware port. The execution robot realizes full functionalities of RS485 star network at the bottom of InMoov ModBus communication architecture, including data transmission, command and error feedback between the hosts (Mega2560) located in the trunk of the robot and the slaves (Nano) located at the main joints.

Based on the robot, we developed a variety of platforms for the robot debugging and interfaces for functional calls, including personal computer serial port debugging, APP terminal command control, functional controls under ROS (robot operating system) [23, 24] in IPC (industrial personal computer), direct driven controlling of the head and neck monitoring camera part and so on. Besides, it completes the receiving, storage and scheduling of the trajectory planning information as an InMoov robot action management platform. Any user can easily control i through these platforms.

We hope the device is able to cope with sudden special emergencies and do some onsite work independently. To realize these functions, it should has the ability to plan appropriate movement trajectory work and grasp a targeted object such as red ball. The system uses IPC for the inverse kinematics solutions [25–27]. Taking into consideration the target position and current pose of the robot read from servo registers, Rviz running under the ROS will operate inverse kinematics to work out how the robot can get close to the target position by changing the joints from current configuration to a new configuration. Rviz solver will find out the required angle of each rotational joint that enables the palm of the hand to access the target position.

Because the robot's joints can't move freely in a large space, Rviz will determine whether there is a suitable solution according to the corresponding range of motion and the actual situation of the current position and orientation. If seeking out a suit-able solution, the ROS will break down the movement into multiple steps according to the current position or orientation and the targeted position or orientation. Then, ROS will send control commands to the master chip in the robot's trunk according to time series, which in turn drives the movement of each joint servo. Changing the angles of the rotational joints by modulating the PWM duty cycle step by step, the continuous movement of the robot is realized. Each servo drives its corresponding rotational joint to the required angle, thus resulting in the accomplishment of targeted pose. Then, the servos corresponding to five fingers drive fingers from fully extended to fully tightened state. Because of the characteristics of the servo, the five fingers will automatically grasp the specified object tightly with the strength of the servos kept the maximum torque, thus each finger remaining on the surface of the object.

2.3   Performance of the Execution System

Figure 2 shows an integration of the whole subsystems, including the execution robot and the Mecanum wheel platform and other mechanical parts.

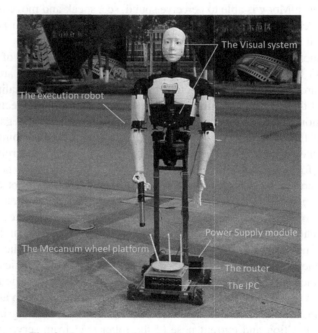

Fig. 2. Integration of the whole mechanical modules

After testing, we found that the Mecanum wheel platform had great flexibility and movability as expected. The platform has a maximum move speed of 1.00 m/s on industrial ordinary ground and 0.80 m/s on rugged ground and this will cause acceptable damage to the wheels. In-place rotation and lateral translation movement made the robot adapt to highly structured space environment in industrial field great.

The execution robot is expected to do live field operation as flexible and fast as it can. So, the motion range and motion velocity (Table 1) of each DOF are carefully

Table 1. Motion range/velocity/precision of each DOF of the execution robot. Bigger motion range indicates bigger workspace and dexterous workspace. This means that on the assumption that the lower platform keeps in the same place, the robot can operate on objects located in a larger space. Motion velocity means how fast the robot moves from an initial configuration to a desired configuration. The sign ">" indicates that the motion velocity of corresponding servo is so fast that the time it takes for the movement is less than one second.

DOF	Range(°)	Velocity(°/s)	Precision(°)
Head swing	−90−+90	>180	<2
Head nodding	−20−+30	>50	<2
Shoulder extension	0–70	20	<3
Shoulder swing	−70−+70	30	<3
Should rotation	−90−+90	30	<3
Forearm swing	0–80	40	<2
Wrist rotation	−30−+30	60	<2
Each finger	0–90	>90	

considered. The motion range is referred to the range of angle the servo can drive the corresponding links to. The bigger the motion range is, the bigger the workspace is. The faster the motion velocity of each DOF is, the faster the live operation operates.

Besides, to do live operation precisely and automatically, the kinematic accuracy of each DOF is required, too. So, we drove each servo to the same configuration several times to measure the actual angle the joint accessed and computed the kinematic accuracy. The results of the repeat precision showed that the desired configuration and the actual configuration wouldn't deviate much except for the fingers. That's because the fingers' extended or tightened state is controlled by wire. There exists a course without wire response when driving the finger in two directions, i.e., from fully extended to fully tightened state or the reverse direction, which is so-called return error. We reduced ruturn error by driving the finger in a single direction or establishing the two accurate relationships between them, which was what we have done.

Besides, workspace is an important index of the device. Workspace is the region described by the origin of the end-effector (hand palm) frame when all the joints execute all possible angles. Workspace has a crucial impact on the live field operation.

The workspace of a single arm is an ellipsoid envelope as shown in Fig. 3. A great number of points are located in the positions with a certain distance with respect to the origin of the coordinate. In case the platform and the robot waist are still, the execution robot can dexterously capture an object located in the space with a higher density of the point cloud. Figure 3 shows an isometric drawing of workspace of a single arm. In order to figure out the shape of it clearly, we draw it in 2D views in Fig. 4.

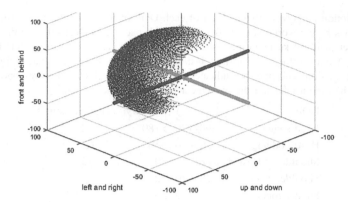

Fig. 3. Isometric drawing of workspace of the arm. In order to see the shape of the workspace, we set the z-axis as "front and behind" direction. Each point in the point cloud is a workspace point that the execution robot can access. The density of the point cloud reflects the number of possible solutions to the point. The greater the density of the point cloud, the more configurations of the arm to reach the corresponding point.

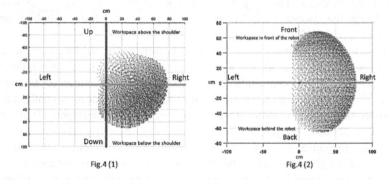

Fig. 4. Workspace of a single arm in 2D views. In Fig. 4 (1) we know that it's easier to capture an object below the shoulder of the robot because of a lager workspace. Due to a smaller dexterous workspace, the configuration to capture an object above the shoulder of the robot is relative less than that below the shoulder. In Fig. 4 (2), the density of the point cloud in front of the robot is slightly greater than that in the back, especially the position far away from the origin, which is the result of the fact that the motion range of forearm swing is 0–80°. So it's easier to capture an object in front of the robot. However, owing to the DOF of shoulder swing, the workspace behind the robot is big enough for the robot to capture an object.

3   The Visual System

In this project, we construct a robot used in industrial field security monitoring situation. So, the most indispensable function of the device is remote live-video shooting and video stream transmission [28]. The live-video is shot by two Raspberry Pi camera [29] modules located in the eyes of the InMoov robot and the Raspberry Pi will drive the two cameras and package the HD video stream into the file form suitable for transmission. The packaged video stream will uploaded to a RTSP(real time streaming protocol) [30] server built by Raspberry Pi. The remote computer running Tyco ExacaVision will be able to connect to the RTSP server and view the live video with the help of WIFI. Besides, the video stream can be viewed on the Android control terminal, i.e., an APP, screen we developed, too.

We also hope our device is able to cope with sudden special emergencies and do some on-site work independently. To realize it, it should have the ability of object identification and localization. Here we exploit Kinect type II [31] to accomplish this function, which consists of a RGB camera and a 3D depth sensor.

The 3D depth sensor of Kinect type II mainly contains two components: the infrared ray emitter and the infrared ray CMOS camera. This sensor is generally regarded as a depth sensor for structured light, mainly based on calculating the phase difference between the emitted and the received infrared ray. Such calculation method can be classified as ToF depth measurement method. Under ROS-based environment, we use *freenect* library to drive the Kinect cameras and apply *iai_kinect* package to transform the video stream into ROS image messages. The workflow can be illustrated in Fig. 5.

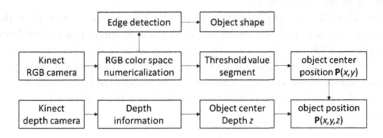

Fig. 5. Workflow of targeted object identification and localization

To classify the environment and the targeted object, a specific visual feature is required. When processing images captured by cameras based on content, color is usually considered as a crucial visual feature. Based on color difference, the visual system will firstly divide the image into "object" and "non-object" part. This process is called "binary classification". Given the threshold value of specific color, computer is able to segment the color space easily. After binary classification, we need to identify the shape of the target object. This process is accomplished with the help of edge detection method. By detecting the points whose grey value varies greatly, we can connect these point to form lines, which can be regarded as the object edge. The shape of the target object will be identified on the basis of specific edges. All these calculations are based on *OpenCV* library.

Bottom left image in Fig. 6 showed a successful identification of the targeted object. We carried the laboratory test under good lighting conditions. Results showed that identification success rate fluctuated at about 85% depending on the light conditions and other interference factor. So, we could assume that our visual system

Fig. 6. A successful identification of the targeted object (bottom left). The yellow and red circle indicate the comparison between the identified size of object and the real size of it. The red numbers in the image are the 3D coordinate of a point measured by Kinect (Color figure online)

performs well when identifying the object with regular shape and obvious color compared with the environment.

Having identified the target object and its shaped, the adjacent work is the localization of the target object. We make use of the image frame intercepted from video stream captured by Kinect RGB camera, so we can localize the specific object with the help of binocular vision. Figure 6 shows a successful localization of a point. The localization ability of the visual system is what we concerned best, which is defined by the ratio of the position identified by the system and the real position of the point, described by:

$$R_x = \frac{X_{kinect}}{X}$$
$$R_y = \frac{Y_{kincet}}{Y} \tag{2}$$
$$R_z = \frac{Z_{kincet}}{Z}$$

Where Ri is the localization ability in the i direction, X_{kinect}, Y_{kinect}, Z_{kinect} is the position identified in each direction, X, Y, Z is the real position of the point. As Ri grows closer to 1.00, the localization ability grows greatly.

Testing result showed that the localization ability in x, y, z direction is 1.029, 0.925, 1.149. So, we concluded that the localization ability of the visual system is great. But, when it comes to the localization ability of depth, the localization performs relatively bad. Here we mainly used the point clout built by the Kinect cameras, so the localization effect relies on the point cloud greatly. We think that by optimizing the process of building the point cloud will improve the depth localization effect.

4   The Controlling System

Our device is designed for remote live-video security surveillance, so we hope we can not only monitor the industrial site on a specific fixed screen in the monitoring room, but also check the current situation of the monitored area on a mobile terminal. Besides, we expected to use all these functions we developed easily and integratedly.

Based on the requirements, we developed an Android APP realizing two main functions. First, we can check real-time remote video of the monitored areas on the screen conveniently. Second, we can send commands to control the whole execution system, including the movement of the Mecanum wheel platform and the live-field operation such as object capturing and targeted action execution. The Android APP UI we developed is shown in Fig. 7. The communication between the APP and the execution robot is through UDP [32]. The UI contains three main areas: video stream display area, the Mecanum platform controlling area, function controlling area.

The video stream display area shows what the execution robot "sees". In order to display the spatial information such as depth of field, we use equidistant line that can be adjusted automatically and dynamically according to the visual angle of the cameras. Once the operator click on one point on the screen, the visual system will localize the spatial coordinate of the point automatically and then display it on the middle "waiting

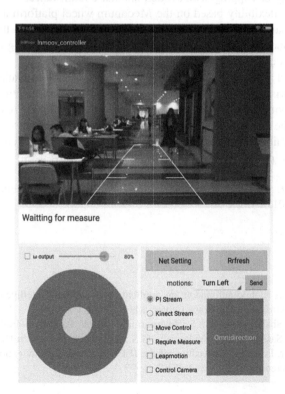

Fig. 7. Android APP UI that integrates all the functions we developed

for measure" blank area. If you send the "Capture" in the "Motions" tab, the execution system will do inverse kinematic solution and then capture it automatically.

The circular and rectangle dark grey color blocks are the Mecanum wheel platform controller, through which, the operator can control the movement of the platform easily. The "motions" tab include object capturing, robot turning left, pre-edited traffic police gestures and so on. Once the "send" tab clicked, the execution system or visual system will do corresponding operations according to different functions.

5   Conclusion

This paper presents the design of a robotic movable security surveillance system based on the co-operation between the human operator and the machine. The whole device can shoot HD video and upload it to the Tyco's server ExacqVision and the mobile APP terminal. It relies on an operator to guide the platform in the industrial site. Once receiving related capturing control order, the visual and execution systems will inter-cept frame image information from the video stream, identify the distance and position information of the specific object and plan the best trajectory to capture it. And the execution of other orders is similar. This new device is expected to overcome the lack of current security monitoring system: low flexibility, limited monitoring range and position, disability of copping with sudden special circumstances.

The mobility, flexibility based on the Mecanum wheel platform and the execution robot, InMoov, have shown a great movability and applicability to the industrial site. The execution robot InMoov has a high degree of freedom, enabling it to do live field operation such as object capturing can door opening freely.

The presented follows a new approach with remarkable advantages concerning industrial monitoring system. The robot system can identify and capture specific object, do some preset motions such as door opening and robot turning left, etc. In a long term, this product can be applied to real-time security monitoring, anti-terrorism in dangerous environment, fruit harvesting, human-computer interactive entertainment and mechanical traffic police, etc.

Acknowledgement. This work is supported by the National Key Technology R&D Program under Grant No. 2015BAF13B02.

References

1. Nilsson, N.J.: A mobile automaton: an application of artificial intelligence techniques. Sri International Artificial Intelligence Center, Menlo Park, CA (1969)
2. Nilsson, N.J.: Shakey the robot. Sri International, Menlo Park, CA (1984)
3. Green, D.N., Sasiadek, J.Z., Vukovich, G.S.: Guidance and control of an autonomous planetary rover. In: 1993, Proceedings of the IEEE-IEE Vehicle Navigation and Information Systems Conference, pp. 539–542. IEEE (1993)

4. Green, D.N., Sasiadek, J.Z., Vukovich, G.S.: Path tracking, obstacle avoidance and position estimation by an autonomous, wheeled planetary rover. In: 1994 Proceedings, IEEE International Conference on Robotics and Automation, pp. 1300–1305. IEEE (1994)
5. Hirose, S.: Snake, walking and group robots for super mechano-system. In: IEEE SMC 1999 Conference Proceedings. 1999 IEEE International Conference on Systems, Man, and Cybernetics, vol. 3, pp. 129–133. IEEE (1999)
6. Ohno, H., Hirose, S.: Design of slim slime robot and its gait of locomotion. In: Proceedings. 2001 IEEE/RSJ International Conference on Intelligent Robots and Systems, vol. 2, pp. 707–715. IEEE (2001)
7. Hirose, S., Fukushima, E.F.: Snakes and strings: new robotic components for rescue operations. Int. J. Robot. Res. 23(4–5), 341–349 (2004)
8. Hirose, S., Yamada, H.: Snake-like robots [tutorial]. IEEE Robot. Autom. Mag. 16(1), 88–98 (2009)
9. Hirose, S., Fukushima, E.F.: Development of mobile robots for rescue operations. Adv. Robot. 16(6), 509–512 (2002)
10. Kimura, H., Hirose, S.: Development of genbu: active wheel passive joint articulated mobile robot. In: 2002 IEEE/RSJ International Conference on Intelligent Robots and Systems, vol. 1, pp. 823–828. IEEE (2002)
11. Masayuki, A., Takayama, T., Hirose, S.: Development of "Souryu-III": connected crawler vehicle for inspection inside narrow and winding spaces. In: Proceedings. 2004 IEEE/RSJ International Conference on Intelligent Robots and Systems, 2004. (IROS 2004), vol. 1, pp. 52–57. IEEE (2004)
12. Arai, M., Tanaka, Y., Hirose, S., et al.: Development of "Souryu-IV" and "Souryu-V:" Serially connected crawler vehicles for in-rubble searching operations. J. Field Robot. 25(1–2), 31–65 (2008)
13. Casper, J., Murphy, R.R.: Human-robot interactions during the robot-assisted urban search and rescue response at the world trade center. IEEE Trans. Syst. Man Cybern. Part B (Cybern.) 33(3), 367–385 (2003)
14. Matsuno, F., Sato, N., Kon, K., Igarashi, H., Kimura, T., Murphy, R.: Utilization of robot systems in disaster sites of the great eastern Japan earthquake. In: Yoshida, K., Tadokoro, S. (eds.) Field and Service Robotics. Springer Tracts in Advanced Robotics, vol. 92, pp. 1–17. Springer, Berlin (2014). doi:10.1007/978-3-642-40686-7_1
15. Yoshida, T., Nagatani, K., Tadokoro, S., Nishimura, T., Koyanagi, E.: Improvements to the rescue robot quince toward future indoor surveillance missions in the fukushima daiichi nuclear power plant. In: Yoshida, K., Tadokoro, S. (eds.) Field and Service Robotics. Springer Tracts in Advanced Robotics, vol. 92, pp. 19–32. Springer, Berlin (2014). doi:10.1007/978-3-642-40686-7_2
16. Tsai, C.C., Tai, F.C., Lee, Y.R.: Motion controller design and embedded realization for Mecanum wheeled omnidirectional robots. In: 2011 9th World Congress on Intelligent Control and Automation (WCICA), pp. 546–551. IEEE (2011)
17. Diegel, O., Badve, A., Bright, G., et al.: Improved mecanum wheel design for omni-directional robots. In: Proceedings 2002 Australasian Conference on Robotics and Automation, Auckland, pp. 117–121 (2002)
18. Gfrerrer, A.: Geometry and kinematics of the Mecanum wheel. Comput. Aided Geom. Des. 25(9), 784–791 (2008)
19. Modbus, I.D.A.: Modbus application protocol specification v1. 1a. North Grafton, Massachusetts (www.modbus.org/specs.php) (2004)
20. Dao-gang, P., Hao, Z., Li, Y., Hui, L.: Design and realization of ModBus protocol based on embedded Linux system. In: The 2008 International Conference on Embedded Software and Systems Symposia (ICESS, 2008)

21. Peng, H., Zheng, Y., Lu, G., Luo, M.: The realization of serial communication between kingview 6.55 and MCU based on Modbus-RTU protocol. In: 6th International Conference on Machinery, Materials, Environment, Biotechnology and Computer (MMEBC 2016)
22. Jia, H.J., Guo, Z.H.: Research on the technology of RS485 over ethernet. In: 2010 International Conference on E-Product E-Service and E-Entertainment (ICEEE)
23. Quigley, M., Conley, K., Gerkey, B., et al.: ROS: an open-source Robot Operating System. In: ICRA Workshop on Open Source Software, vol. 3(3.2), pp. 5 (2009)
24. Gossow, D., Leeper, A., Hershberger, D., et al.: Interactive markers: 3-D user interfaces for ros applications [ros topics]. IEEE Robot. Autom. Mag. **18**(4), 14–15 (2011)
25. Siciliano, B., Sciavicco, L., Villani, L., et al.: Robotics: Modelling, Planning and Control. Springer Science & Business Media, London (2010). doi:10.1007/978-1-84628-642-1
26. Murray, R.M., Li, Z., Sastry, S.S., et al.: A Mathematical Introduction to Robotic Manipulation. CRC Press, Boca Raton (1994)
27. Springer handbook of robotics, Springer (2016)
28. Biedermann, D.H., Dietrich, F., Handel, O., et al.: Using raspberry pi for scientific video observation of pedestrians during a music festival. arXiv preprint arXiv:1511.00217 (2015)
29. RASPBERRY PI 2 MODEL B. Tillgänglig (2015). https://www.raspberrypi.org/products/raspberry-pi-2-model-b/. Använd 10 Feb 2016
30. Schulzrinne, H.: Real time streaming protocol (RTSP) (1998)
31. Khoshelham, K., Elberink, S.O.: Accuracy and resolution of kinect depth data for indoor mapping applications. Sensors **12**(2), 1437–1454 (2012)
32. Postel, J.: User Datagram Protocol, RFC768 (1980)

Dynamic Identification for Industrial Robot Manipulators Based on Glowworm Optimization Algorithm

Li Ding[1(✉)], Wentao Shan[1], Chuan Zhou[2], and Wanqiang Xi[3]

[1] College of Mechanical Engineering, Jiangsu University of Technology,
Changzhou 213001, China
nuaadli@163.com, shanwentao520@163.com
[2] School of Business, Jiangsu University of Technology,
Changzhou 213001, China
447030849@qq.com
[3] College of Mechanical and Electrical Engineering,
Nanjing University of Aeronautics and Astronautics, Nanjing 210016, China
1161330474@qq.com

Abstract. Dynamical identification methods for the industrial robot manipulators are widely and successfully applied to obtain a model that is suitable for controller design. In this paper, the dynamical model of robot was obtained by Newton-Euler method and linearized by a particular approach. A novel glowworm swarm optimization algorithm was introduced to estimate the unknown parameters. The algorithm had been coded in the popular Matlab environment and the procedure was tested in a practical case research to identify the dynamical model of a six degree-of-freedom industrial robot. The results of the identification experiment showed the efficiency of the proposed algorithm.

Keywords: Industrial robot manipulators · Dynamical identification · Glowworm swarm optimization · Identification experiment

1 Introduction

With the advantages that the robot manipulators can deal with some dangerous and tiring works in industries, especially in the automotive, shipbuilding and aerospace manufacturing industries, robot manipulators have been much attractive in the industry domain recently [1, 2]. Advanced control techniques for robots have become more and more affordable thanks to increasing power of computing resources and their dramatic cost reduction. However, the design and implementation of a model-based controller requires an accurate knowledge of robot's dynamic model. For those purpose, dynamical parameters identification method has importance for developing robot controllers.

Although several theoretical researches about the identification for the robot manipulators are available, the research still lack in presenting a detailed and reliable procedure in terms of giving users instructions to perform identification experiments for a generic industrial robot. This paper is aimed at providing a systematic procedure for

© Springer International Publishing AG 2017
Y. Huang et al. (Eds.): ICIRA 2017, Part II, LNAI 10463, pp. 789–799, 2017.
DOI: 10.1007/978-3-319-65292-4_68

the parameter identification of the robot. A standard robot identification procedure consists of six steps: dynamical modeling, excitation design, data collection, signal preprocess, parameter identification and model validation. And the parameter identification methods have drawing many researchers to take the challenge of realizing it fleetly and efficiently. Such as, Atkeson et al. [3] proposed the least square method to realize the estimation of dynamical parameters. Grotjahn et al. [4] used the two-step approach to perform the identification of robot dynamics. Gautier and Poignet [5] obtained a dynamical model of SCARA robot from experimental data with weighted least squares method. Behzad et al. [6] applied fractional subspace method to identify a robot model in simulation field. Recently, some intelligence computation algorithms have been reported as a useful tool in robot model identification. Ding et al. [7] use the artificial bee colony algorithm to solve the problem of parameters identification for a 6-DOF industrial robot. Liu et al. [8] introduced the improved genetic algorithm to obtain the space robot model.

Glowworm swarm optimization (GSO) was originally presented by Krishnanand and Ghose [9], under the inspiration of collective behavior on glowworms. Compared with usual algorithms, the major advantage of GSO lies in that it can capture all local search maxima of multi-modal function and is globally convergent, and we adopt this new algorithm to process the identification of robot.

The outline of this paper is organized as follows. At first, the linear robot dynamical model is given in Sect. 2. Then, Sect. 3 presents the detailed identification process based on GSO. Later on, the experiment and analysis are presented in Sect. 4. At last, the conclusions are given in Sect. 5.

2 Dynamic Modeling

The dynamic model of 6-DOF of the robot manipulators is derived from the Lagrangian or Newton-Euler method [10]. Both approaches yield the dynamical equation:

$$\tau_{dyn} = \mathbf{M}(q)\ddot{q} + \mathbf{C}(q,\dot{q}) + \mathbf{G}(q) \tag{1}$$

Where τ_{dyn} denotes the n-vector of the actuator torque as well as the joint positions q, velocities $\dot{q}$ and accelerations $\ddot{q}$. $\mathbf{M}(q)$ is the n × n inertia matrix, $\mathbf{C}(q,\dot{q})$ is the n-vector including Coriolis and centrifugal forces, and $\mathbf{G}(q)$ is the n-vector of gravity.

According to [11] or [12], the Eq. (1) can be rewritten as a linear form by using modified Newton-Euler parameters or the barycentric parameters:

$$\tau_{dyn} = \mathbf{\Phi}_{dyn}(q,\dot{q},\ddot{q})\mathbf{\theta}_{dyn} \tag{2}$$

Where $\mathbf{\Phi}_{dyn}$ is the n × 10n observation matrix or identification matrix, which depends only on the motion data. $\mathbf{\theta}_{dyn}$ is the barycentric parameters vector. This process vastly decreases the number of the unknown identified parameters.

Joint frictions contribute the additional dynamical parameters due to the geared actuator robot spinning at high velocity. In fact, the model of joint frictions is a complex nonlinear form, especially during motion reversal [13]. A simplified and linear friction model consisting of only Coulomb and viscous friction is given by:

$$\tau_{fric} = f_c sign(\dot{q}) + f_v \dot{q} \tag{3}$$

Where τ_{fric} is the friction torques, f_c and f_v respectively are the Coulomb and viscous friction parameters.

Integrating Eqs. (2) and (3), the whole dynamic model of the robot can be rewritten as:

$$\tau_s = \mathbf{\Phi}_s(q, \dot{q}, \ddot{q})\mathbf{\theta}_s \tag{4}$$

Where τ_s is the actuator torque including τ_{dyn} and τ_{fric}. $\mathbf{\Phi}_s$ is the n × 12n observation matrix, and $\mathbf{\theta}_s$ is 12n-vector unknown dynamical parameters. In addition, the dynamical parameters of link i are governed by the form:

$$\mathbf{\theta}_s^i = [I_{xxi}, I_{xyi}, I_{xzi}, I_{yyi}, I_{yzi}, I_{zzi}, m_i r_{xi}, m_i r_{yi}, m_i r_{zi}, m_i, f_{ci}, f_{vi}]^T \tag{5}$$

Where $I_{\zeta\zeta i}(\zeta = x, y, z)$ is the inertial of the link i. Similarly, $m_i r_{\zeta i}$ is the first-order mass moment and m_i is the mass of the link i.

However, the observation matrix $\mathbf{\Phi}_s$ in Eq. (4) is not a full rank, i.e., not all dynamical parameters have an influence on the robot model. In order to obtain a set of minimum base parameters, a case-by-case analysis method is adopted to handle this particular problem. The method uses the Singular Value Decomposition (SVD) approach [14] to obtain the basic dynamical parameters. Hence, the ultimate form of the dynamical model can be rewritten as:

$$\tau = \mathbf{\Phi}(q, \dot{q}, \ddot{q})\mathbf{\theta} \tag{6}$$

Where $\mathbf{\Phi}$ is the n × (p + 2n) observation matrix. $\mathbf{\theta}$ is (P + 2n)-vector of dynamic parameters, including the basic parameters and the friction parameters. p is the number of the minimum base parameters, 2n means the number of the friction parameters.

3   Dynamical Identification Based on GSO

In this section, GSO algorithm is employed to search the optimal value of the dynamical parameters. GSO is based on the following principles: initialization of the glowworm, update of the luciferin values and position of glowworm and update of the decision domain.

Consider the position and luciferin value of the particular glowworm i (i=1, 2, $\cdots$, M) is $x_i(t)$ and $l_i(t)$ at t iteration, and $x_i(t)$ is governed by:

$$x_i(t) = L + rand(0, 1) \cdot (U - L) \tag{7}$$

Where M denotes the number of glowworm, L and U mean the lower and upper bounds, respectively.

The objective function corresponding to each $x_i(t)$ can be described as follows:

$$F_i = \frac{1}{N} \sum_{i=1}^{N} (\delta_1 \|\tau_{1i} - \tau_{p1i}\| + \delta_2 \|\tau_{2i} - \tau_{p2i}\| + \delta_3 \|\tau_{3i} - \tau_{p3i}\|) \quad (8)$$

Where N is the data length, $\tau_{\xi i}(\xi = 1, 2, 3)$ is the vector of the actual torques data from the first three joints. Similarly, $\tau_{p\xi i}(\xi = 1, 2, 3)$ is the vector of the predicted data from the identified model. $\delta_\xi(\xi = 1, 2, 3)$ is a weight coefficient between 0 and 1.

The glowworms communicate with each other by releasing luciferin and each glowworm gives it response to the surrounding local environment that is determined by its dynamic decision domain $r_d^i(t)(0 < r_d^i < r_s)$. $r_d^i(t)$ is updated according to the following equation:

$$r_d^i(t+1) = \min\{r_s, \max[0, r_d^i(t) + \beta(n_t - |N_i(t)|)]\} \quad (9)$$

Where r_s is the circular sensor range. n_t is the neighborhood threshold, which controls the number of nearby each glowworm. β is the rate of change of the neighborhood range. $N_i(t)$ is the set of i-th nearby glowworm at the t-th iteration, which consists of those glowworms that have a relatively higher luciferin value and that are located within a dynamic decision domain. That is to say:

$$N_i(t) = \{j : \|x_j(t) - x_i(t)\| < r_d^i; l_i(t) < l_j(t)\} \quad (10)$$

Where the position and luciferin value of glowworm j are $x_j(t)$ and $l_j(t)$, respectively.

When the glowworm moving, it need decide direction of movement in accordance with luciferin values of glowworms in its set of neighbourhood. $p_{ij}(t)$ represents probability of movement of the i-th glowworm moving toward the j-th glowworm in its set of neighbourhood. And it is computed on the basis of the following equation:

$$p_{ij}(t) = \frac{l_j(t) - l_i(t)}{\sum_{k \in N_i(t)} l_k(t) - l_i(t)} \quad (11)$$

After moving, the new position of the glowworm i is calculated by:

$$x_i(t+1) = x_i(t) + s \left(\frac{x_j(t) - x_i(t)}{\|x_j(t) - x_i(t)\|} \right) \quad (12)$$

Where s is the step length.

The new luciferin value for glowworm i is updated by the equation:

$$l_i(t) = (1 - \rho)l_i(t - 1) + \gamma f(x_i(t)) \quad (13)$$

Where $\rho \in [0,1]$ is the ratio of luciferin vaporization. $(1-\rho)$ represents the reflection of the cumulative goodness of the path followed by the glowworm in their current luciferin values. γ scales the fitness values.

Generally speaking, in the set of neighbourhood, if the glowworm i find the glowworm j with higher luciferin value and the distance between the two glowworms is less than the circular sensor range, the glowworm i moves towards glowworm j according to the probability $p_{ij}(t)$. The glowworm i updates its luciferin values and position.

4 Experiment and Analysis

4.1 Excitation Trajectories

In order to ensure the accuracy of estimation, a finite Fourier series is adopted as excitation trajectories [15]. And the trajectories for joint i of a robot are designed as:

$$q_i(t) = q_{i,0} + \sum_{k=1}^{N} a_{i,k} \sin(k\omega_f t) + \sum_{k=1}^{N} b_{i,k} \cos(k\omega_f t) \tag{14}$$

Where $q_{i,0}$ denotes the offset term, ω_f is the fundamental pulsation of the Fourier series. This Fourier series specifies a periodic function with period $2\pi/\omega_f$. Each Fourier series contains $2N+1$ parameters, and $a_{i,k}$, $b_{i,k}$ are the amplitudes of the sine and cosine functions.

The noise immunity and convergence rate of an identification experiment depend directly upon the constraints of the excitation trajectories. It is important to emphasize that the configurations for which measurements are taken must correspond to a well-conditioned reduced observation matrix since the constraints represent some limits for input and output. According to [16], the constraints of the excitation trajectories can be described as:

$$\begin{cases} \min cond(\Phi) \\ q_{\min} \leq q(\beta) \leq q_{\max}, \ |\dot{q}(\beta)| \leq \dot{q}_{\max} \\ |\ddot{q}(\beta)| \leq \ddot{q}_{\max} \\ w(q(\beta)) \subset W_o \\ \tau_{\min} \leq \Phi(q(\beta), \dot{q}(\beta), \ddot{q}(\beta)\theta \leq \tau_{\max} \end{cases} \tag{15}$$

Where $q_{\min}$, $q_{\max}$ are the lower and upper values of the joint positions, $\dot{q}_{\max}$, $\ddot{q}_{\max}$ are the upper value of velocities and accelerations, β is optimal trajectory parameters, W_o is the available workspace of robot, $\tau_{\max}$ is the maximum joint torque.

The experimental torque data is obtained through the conversion of collected motor current. Before identification, in order to eliminate the effects of the trim and systematic noises, the collected experimental data is preprocessed by passing through a ten-point average FIR filter and executing the trim removal. The filtering equation is written as:

$$y_f(k) = \frac{1}{10} \sum_{i=0}^{9} y(k-i) \qquad (16)$$

It is well known that the above FIR filter will be trapped into delay. Hence, the above filter should be implemented via Matlab function 'filtfilt' [17] which process the input data in both the forward and reverse directions.

4.2   Experiment Results

As shown in Fig. 1, the identification experiment is conducted on the robot manipulator ER-50. The geometric parameters of the ER-50 are given in Table 1. The values of the dynamic parameters of the first three joints are much bigger than those of the last three joints, therefore only the first three joints are considered here. The commanded trajectories used in our experiment are five-term Fourier series, which is depicted in Fig. 2. A fundamental pulsation of 0.05 Hz is selected for the excitation trajectories, resulting in a period of 20 s. All the experimental data are sampled at 1 kHz and recorded in the encoder of actuating motors.

Fig. 1. ER-50 6-DOF robot manipulator

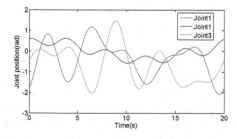

Fig. 2. Five-term excitation trajectories

Table 1. DH parameters of ER-50

Link i	α_{i-1} (rad)	a_{i-1} (m)	d_i (m)	θ_i (rad)
1	0	0	0.412	θ_1
2	$-\pi/2$	0.16	0	θ_2
3	0	0.68	0	θ_3
4	$\pi/2$	0.13	0.75	θ_4
5	$-\pi/2$	0	0	θ_5
6	$-\pi/2$	0	0	θ_6

The identification procedure is carried out with GSO in Matlab 2014b programming environment on an Intel Core i7-3770 PC running Windows 7. No commercial tools are used. The initial conditions of GSO are set as: $M = 40$, $\rho = 0.4$, $\gamma = 0.6$, $\beta = 0.1$, $n_t = 8$, $s = 0.03$. The algorithm is run 30 times and the maximum iteration time is 100. Figure 3 shows the objective value of optimization process for the parameters. The result shows that our algorithm becomes convergence when iterating about 18 times and the final calculated objective value is 0.3356.

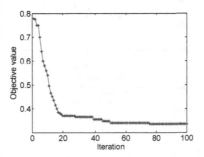

Fig. 3. Evolutionary curves of identification algorithm

Table 2. Dynamic parameters of robot

Parameter	Value	Parameter	Value
I_{zz1} (kg.m^2)	61.4671	I_{zz3} (kg.m^2)	−4.2246
I_{xx2} (kg.m^2)	48.6976	$m_3 r_{x3}$ (kg.m)	4.0345
I_{xy2} (kg.m^2)	28.6346	$m_3 r_{y3}$ (kg.m)	2.0367
I_{xz2} (kg.m^2)	−16.9477	f_{c1} (N.m)	0.1018
I_{yz2} (kg.m^2)	1.0183	f_{v1} (Nm.s/rad)	0.4332
I_{zz2} (kg.m^2)	4.3318	f_{c2} (N.m)	2.1801
$m_2 r_{x2}$ (kg.m)	−0.8854	f_{v2} (Nm.s/rad)	−0.3165
$m_2 r_{y2}$ (kg.m)	71.1972	f_{c3} (N.m)	5.8281
I_{xx3} (kg.m^2)	45.8305	f_{v3} (Nm.s/rad)	13.2567
I_{xy3} (kg.m^2)	−0.8354	I_{xz3} (kg.m^2)	5.9615
I_{yz3} (kg.m^2)	67.6530		

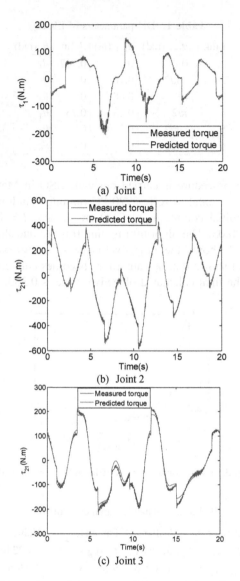

Fig. 4. Comparison of the measured torques and predicted torques

The robot dynamical model for the first three joints contains 21 parameters, 6 friction parameters and 15 base parameters. And the result of the identified parameters is listed in Table 2. Figure 4 shows the simulation of the identified model over a set of real experiment data. It shows that all the predicted torque can match the trends of the measured torque well. The results demonstrate that identified dynamical model based on GSO is effective.

The error between the measured torque and predicted torque is shown in Fig. 5 to investigate the validity of the dynamical model obtained from our proposed method.

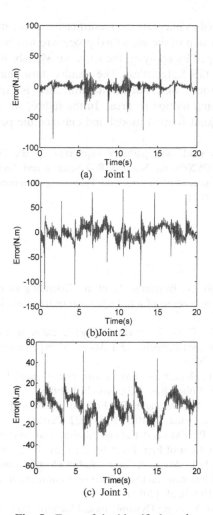

Fig. 5. Error of the identified results

From the pictures, we can see the tendency of the error is smooth and all the data are almost small. It should be noted that the error suddenly becomes greater when the joints turn to reverse direction. The dynamical model of joints might not be completely suitable for describing the friction model. In general, the above indicates that the GSO method can get a good identified model with its ability to discover the information contained in the collected data.

5 Conclusion

This paper gave a novel algorithm to identify the dynamical model of the industrial robot manipulators. The model of robot including simplified joint friction model is linear with respect to the unknown parameters. An optimal periodic excitation trajectories were designed to integrate the identification experiment, data collection and signal preprocess. Each step of the identified procedure had been deeply implemented as software way developed by applying the software Matlab. And the unknown model was well identified by GSO algorithm. The results show that the measured torques match the predicted torques well, but the identification is especially sensitive to the joint friction model during motion reversal. In the future, we will devote ourselves to research the nonlinear joint friction model and enhance the performance of GSO.

Acknowledgement. This work was partially supported by the National Natural Sciences Foundation of China (51405209), the National 863 Science and Technology Support program (2013AA041004) and the project of Science and Technology Support Plan of Jiangsu province (BE2013003-1, BE2013010-2).

References

1. Blomdell, A., Bolmsjö, G., Brogårdh, T., et al.: Extending an industrial robot controller-implementation and applications of a fast open sensor interface. IEEE Robot. Autom. Mag. **12**(3), 85–94 (2005)
2. Wu, Y., Klimchik, A., Caro, S., et al.: Geometric calibration of industrial robots using enhanced partial pose measurements and design of experiments. Robot. Comput.-Integr. Manuf. **35**, 151–168 (2015)
3. Atkeson, C., An, C.H., Hollerbach, J.M.: Estimation of inertial parameters of manipulator loads and links. Int. J. Robot. Res. **5**(3), 101–119 (1986)
4. Grotjahn, M., Daemi, M., Heimann, B.: Friction and rigid body identification of robot dynamics. Int. J. Solids Struct. **38**(10), 1889–1902 (2001)
5. Gautier, M., Poignet, P.: Extended Kalman filtering and weighted least squares dynamic identification of robot. Control Eng. Pract. **9**(12), 1361–1372 (2001)
6. Behzad, H., Shandiz, H.T., Noori, A., et al.: Robot identification using fractional subspace method. In: 2011 2nd International Conference on Control, Instrumentation and Automation (ICCIA), pp. 1193–1199. IEEE (2011)
7. Ding, L., Wu, H., Yao, Y., et al.: Dynamic model identification for 6-DOF industrial robots. J. Robot. **2015**, 1–9 (2015)
8. Liu, Y., Li, G.X., Xia, D., et al.: Identifying dynamic parameters of a space robot based on improved genetic algorithm. J. Harbin Inst. Technol. **42**, 1734–1739 (2010)
9. Krishnanand, K.N., Ghose, D.: Glowworm swarm optimisation: a new method for optimising multi-modal functions. Int. J. Comput. Intell. Stud. **1**(1), 93–119 (2009)
10. Swevers, J., Verdonck, W., Schutter, D.S.: Dynamic model identification for industrial robots. IEEE Control Syst. **27**(5), 58–71 (2007)
11. Calanca, A., Capisani, L.M., Ferrara, A., et al.: MIMO closed loop identification of an industrial robot. IEEE Trans. Control Syst. Technol. **19**(5), 1214–1224 (2011)
12. Vuong, N.D., Ang, M.H.: Dynamic model identification for industrial robots. Acta Plolytechnica Hungarica **6**(5), 51–68 (2009)
13. He, W., Ge, W., Li, Y., et al.: Model identification and control design for a humanoid robot. IEEE Trans. Syst. Man Cybern.: Syst. **47**(1), 45–57 (2017)
14. Antonelli, G., Caccavale, F., Chiacchio, P.: A systematic procedure for the identification of dynamic parameters of robot manipulators. Robotica **17**(04), 427–435 (1999)

15. Ganseman, C., Swevers, J., De Schutter, J., et al.: Experimental robot identification using optimized periodic trajectories. In: Proceedings of the International Conference on Noise and Vibration Engineering, pp. 585–595 (1994)
16. Wu, W., Zhu, S., Wang, X., et al.: Closed-loop dynamic parameter identification of robot manipulators using modified fourier series. Int. J. Adv. Robot. Syst. **9**, 29 (2012)
17. Wang, T., Chen, Y., Liang, J., et al.: Chaos-genetic algorithm for the system identification of a small unmanned helicopter. J. Intell. Robot. Syst. **67**(3–4), 323–338 (2012)

A Method of Computed-Torque Deviation Coupling Control Based on Friction Compensation Analysis

Yao Yan[1], Le Liang[1,2(✉)], Yanyan Chen[1,2], Yue Wang[1], and Yanjie Liu[2]

[1] No. 716 Research Institute, China Shipbuilding Industry Corporation,
Lianyungang 222061, Jiangsu, China
liangle2007@163.com
[2] State Key Laboratory of Robotics and System, Harbin Institute of Technology,
Harbin 150080, Heilongjiang, China

Abstract. In order to improve the trajectory tracking precision and reduce the synchronization error of a 6-DOF lightweight robot, computed-torque deviation coupling control strategy based on friction compensation analysis is presented. The mathematical models of the robot which include kinematic model and dynamic model are established. The single joint Lugre friction model is proposed, and the parameters of the friction model are identified by experiment. Since it is difficult to describe the real-time contour error of the robot for complex trajectory, the adjacent coupling error is analyzed to solve the problem. Combined with friction compensation and coupling performance of the robot, computed-torque deviation coupling controller is designed and validated by simulation analysis. A servo control experimental system is constructed, and verified that the synchronization error are significantly decreased and the trajectory error is reduced from $(-0.8° \sim 1°)$ to $(-0.230° \sim 0.587°)$ after the friction compensation is added. The effectiveness of the control algorithm is validated by the experimental results, thus the control strategy can improve the robot's trajectory tracking precision significantly.

Keywords: Computed-torque control · Lugre friction model · Friction compensation · Adjacent coupling error · Trajectory precision

1 Introduction

Compared with traditional industrial robots, the lightweight robot system is a type of robot which offers several advantages such as faster system response, lower energy consumption, better maneuverability and transportability, higher load-to-weight ratio and safer operation owing to reduced inertia. Since these kinds of lightweight robots are multi-input multi-output and strong nonlinear coupling system and applied in the complex trajectory conditions which need high real time capability, it is important to improve the control strategy for the purpose of enhance the robotic trajectory tracking precision rather than just repeat positioning accuracy.

© Springer International Publishing AG 2017
Y. Huang et al. (Eds.): ICIRA 2017, Part II, LNAI 10463, pp. 800–811, 2017.
DOI: 10.1007/978-3-319-65292-4_69

When the dynamics model of the robot is accurate, the computed-torque control method can decouple the system theoretically, which solves the requirement of track precision and system stability under high speed condition to a certain extent, but the disadvantage is that the calculation of the algorithm is too large. In order to improve the influence of the inaccuracy of the dynamic parameters of the robot, the researchers applied the intelligent algorithm such as adaptive control [1], synovial variable structure control [2], fuzzy control [3] and neural network control [4] into the computed-torque control method. However, these advanced intelligent algorithms still require a lot of on-line computing which limits its application.

The key factor of the imprecision of the robot dynamics is the uncertainty of the joint friction. The control of the multi-joint robot can be achieved by establishing the joint friction model or designing the friction disturbance observer [5]. Since adding the disturbance observer will increase the complexity of the controller, many scholars use the robot's friction model to compensate for the uncertainty of the dynamic model. ALONGE F proposed an adaptive control method combining computed-torque and static friction compensation to solve the trajectory tracking problem of tandem robot [6]. LIU adopted the Coulomb friction model to describe the joint friction of flexible space robot, and designed an active controller by the computed-torque control method for trajectory tracking control [7]. ZHU extended the traditional model-based adaptive control algorithms with adaptive friction compensation [8].

Since the compensation law needs large amount of online calculation while the system contains excessive number of motors, hard-real-time and high-performance communication bus are in greater demand. Real-time EtherCAT can achieve high performance of anti-interface and robustness in process, especially for high-speed, high-reliability, high-real-time data transfer applications [9]. GUO K introduced the EtherCAT fieldbus for the control system of quadruped robot [10]. CHOI T used EtherCAT in multi-robot system [11].

In this paper, a method of computed-torque deviation coupling control based on dynamic friction compensation analysis is used to improve the trajectory tracking accuracy of the lightweight robot. The dynamic model of the robot is improved by introducing the LuGre friction model, and the accuracy of the model is enhanced by the method of friction parameter identification. Then, combining the idea of deviation coupling and friction compensation, the computed-torque synchronization control algorithm is designed. Experiments conducted on a 6-DOF lightweight serial robot system under the EtherCAT protocol demonstrate that the proposed approach can improve the trajectory tracking precision effectively.

2 Mathematical Model of 6-DOF Lightweight Robot

The mechanical body of IRB120 robot and a reconstructive control system are used as experimental platform for verification of the control algorithm. The mechanical body and D-H coordinate system of the robot are shown in Fig. 1. Kinematic calibration method was used to identify the real values of the kinematic parameters and correct the robot kinematic model to improve the robot accuracy [12].

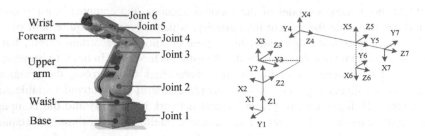

Fig. 1. Mechanical body and D-H coordinate system of IRB120 robot.

According to the geometrical characteristics of the robot, the forward kinematics model can be got using D-H method:

$$
\begin{cases}
p_x = -l_2 c\theta_1 s\theta_2 - l_3 c\theta_1 c\theta_2 s\theta_3 - l_3 c\theta_1 c\theta_3 s\theta_4 - d_4 c\theta_{23} c\theta_1 \\
p_y = -l_2 s\theta_1 s\theta_2 - l_3 c\theta_2 s\theta_1 s\theta_3 - l_3 c\theta_3 s\theta_1 s\theta_2 - d_4 c\theta_{23} s\theta_1 \\
p_z = d_1 - d_4 s\theta_{23} + l_2 c\theta_2 + l_3 c\theta_2 c\theta_3 - l_3 s\theta_2 s\theta_3
\end{cases}
\tag{1}
$$

where p_x, p_y, p_z denote the position coordinate of the end-effector, l_i is the length of robot link, d_i is the offset distance, θ_i is the joint angle, $c\theta_i$ is short for $\cos\theta_i$, $s\theta_i$ is short for $\sin\theta_i$, θ_{ij} is short for $\theta_i + \theta_j$.

The kinetic energy of all components can be calculated and the dynamics equation can be obtained using the second type Lagrange equation [13]:

$$
\begin{cases}
Q_i = \sum_{k=1}^{n} D_{ik}\ddot{\theta}_k + I_i\ddot{\theta}_i + \sum_{k=1}^{n}\sum_{m=1}^{n} D_{ikm}\dot{\theta}_k\dot{\theta}_m + D_i \\
D_{ik} = \sum_{j=\max(i,k)}^{n} Trace\left(\frac{\partial T_j}{\partial\theta_k} I_j \frac{\partial T_j^T}{\partial\theta_k}\right) \\
D_{ijk} = \sum_{j=\max(i,k,m)}^{n} Trace\left(\frac{\partial^2 T_j}{\partial\theta_k\partial\theta_m} I_j \frac{\partial T_j^T}{\partial\theta_i}\right) \\
D_i = \sum_{j=i}^{n} -m_j g^T \frac{\partial T_j}{\partial\theta_i} {}^j r_c
\end{cases}
\tag{2}
$$

where I_i is the rotational inertia of the ith transmission mechanism, T_j is the transformation matrix.

The dynamic model of the robot can be edited as follow:

$$
M(\theta)\ddot{\theta} + c(\theta,\dot{\theta})\dot{\theta} + G(\theta) = \tau
\tag{3}
$$

where $M(\theta)$ is the inertia matrix, $c(\theta,\dot{\theta})$ is coriolis force item and centrifugal force item, $G(\theta)$ is gravity item.

3 Computed-Torque Control Algorithm

The law of computed-torque control algorithm is as follows, and the control block diagram is shown in Fig. 2.

$$\tau = M(\theta)(\ddot{\theta}_d + T_d\dot{e} + K_p e) + c(\theta, \dot{\theta})\dot{\theta} + G(\theta) \tag{4}$$

where, e is velocity error, T_d, K_p are diagonal positive definite gain matrixes.

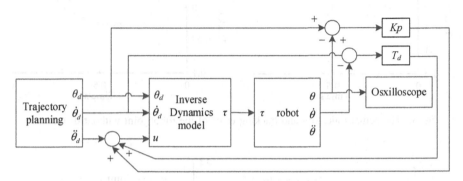

Fig. 2. Control block diagram of computed-torque algorithm.

However, because of the external interference and friction, the actual system dynamics model should be:

$$\tau = M(\theta)(\ddot{\theta}_d + T_d\dot{e} + K_p e) + c(\theta, \dot{\theta})\dot{\theta} + G(\theta) + \tau_l(\theta, \dot{\theta}, t) + f(\dot{\theta}) \tag{5}$$

where, $\tau_l(\theta, \dot{\theta}, t)$ is the external disturbance and other uncertainties, $f(\dot{\theta})$ is the friction matrix of each joint.

In order to analyze the influence of robot dynamics uncertainties on the control system, using virtual prototype simulation analysis, the dynamics of the uncertainties are simulated by the interference simulation, the simulation results are shown in Figs. 3 and 4. From the two figures can be seen, the tracking error of each axis is less than 0.03 rad without interference, and each axis can track the sinusoidal trajectory with high speed and accuracy. After adding the interference in 0.5 s, the tracking error of each joint is obviously increased, especially the joint 3, the error is 0.2 rad. The tracking effect of each axis after adding the interference is obviously worse. It can be seen that the computed-torque control method can not achieve complete accurate tracking, the tracking error of the system is related to the accuracy of the dynamics model of the robot.

The main reason for the inaccurate robot dynamics model is the uncertainty of the dimensional accuracy and inertia parameters and the influence of friction. In the actual control of the robot, it is found that the friction of each joint has a great influence on the position control precision. This is mainly because there is a complex friction between the motor and the reducer, as well as harmonic reducer itself. Therefore, the friction compensated controller should be designed by identifying the friction model of the joint to eliminate the influence of nonlinear factors on the dynamic model.

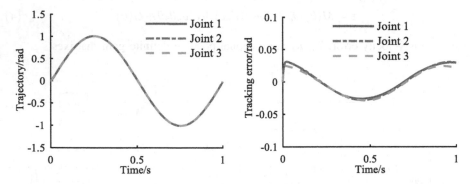

Fig. 3. The actual trajectory and tracking error of the robot joint without disturbance.

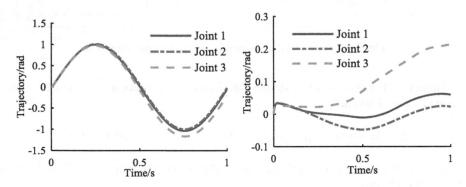

Fig. 4. The actual trajectory and tracking error of the robot joint with disturbance.

4   Computed-Torque Deviation Coupling Control Based on Friction Compensation

4.1   Single Joint LuGre Friction Model

LuGre model approximates the contact surface into a mane model microscopically, the elastic deformation of the mane is represented by the state variable x, and the mathematical expression is

$$
\begin{cases}
\frac{dx}{dt} = \dot{\theta} - \sigma_0 \frac{|\dot{\theta}|}{s(\dot{\theta})} x \\
F = \sigma_0 x + \sigma_1 \frac{dx}{dt} + \sigma_3 \dot{\theta}
\end{cases}
\tag{6}
$$

where, $\sigma_0, \sigma_1, \sigma_3$ is respectively stiffness coefficient, damping coefficient and viscous friction coefficient, F is the predicted friction force.

$S(\dot{\theta})$ is the stribeck effect of static friction and sliding friction, which is

$$S(\dot{\theta}) = f_v + (f_s - f_v)e^{-(\theta/v_s)^2} \tag{7}$$

where, f_s corresponds to the stiction force, f_v is the coulomb friction force, v_s determines how quickly $S(\dot{\theta})$ approaches f_v.

Assume that the deformation rate of mane in the slip zone is 0, when $dx/dt = 0$, there is

$$\sigma_0 x_{ss} = \frac{g(\dot{\theta})}{|\dot{\theta}|}\dot{\theta} = \text{sgn}(\dot{\theta})S(\dot{\theta}) \tag{8}$$

where, x_{ss} is the deformation value when the mane is in a stable state, sgn(.) is symbolic function. So the predicted friction force can be expressed as

$$F = \text{sgn}(\dot{\theta})(f_v + (f_s - f_v)e^{-(\theta/v_s)^2}) + \sigma_3\dot{\theta} \tag{9}$$

From the above analysis we can see that the dynamic LuGre friction model can reflect the static friction and stribeck phenomenon, and can accurately describe the friction behavior. When the mane deformation rate is 0, only four parameters need to be identified which are $(f_s, f_v, v_s, \sigma_3)$, so the LuGre model is selected for friction identification of each joint of the robot.

4.2   Parameter Identification of Friction Model

When the robot moves freely in the Cartesian space at low speed, the friction torque of the robot joint can be expressed as,

$$F(\dot{\theta}) = u - M(\theta)\ddot{\theta} - c(\theta, \dot{\theta})\dot{\theta} \tag{10}$$

where u is motor input torque.

When the joints in the case of low speed, the friction torque become the dominant factor, it can be considered that the motor drive torque equal to the friction torque. In order to obtain a more accurate friction model, the high-gain closed-loop control method is adopted to control the position of single joint, so as to obtain a more stable average velocity. According to the parameters of forward and reverse identification, the friction curves can be got as shown in Fig. 5.

The comparison between the fitting curve and the measured actual torque shows that the selected LuGre friction model has good fitting effect on the friction torque of the robot joint. In the case of joint 1, when joint 1 in the low speed of v < 2°/s, the robot movement is very noisy, there is a significant stribeck effect. In this range, the friction torque decreases with the increase of velocity. As the velocity increases, the friction torque of the joint is in good linear relationship with the velocity.

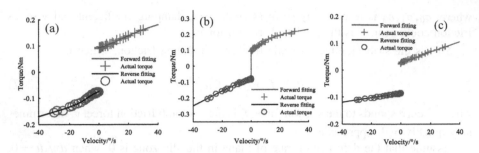

Fig. 5. Friction characteristic curve of each joint. (a) Joint 1, (b) Joint 2, (c) Joint 3.

4.3 Adjacent Coupling Error Analysis

For a multi-axis motion control system, the contouring errors intensely affect the quality of the machined workpieces [14], and the robot task trajectory are usually complex such as arc welding process. So it is difficult to establish proper coupling rules and there will be huge amounts of calculation on-line using the existed algorithms. Aimed at these problems, a synchronized control idea in 6-DOF robot systems with the minimum number of correlative axis was produced [15].

The objectives for synchronous control system of the 6-DOF robot are

$$\frac{\dot{\theta}_1}{a_1} = \frac{\dot{\theta}_2}{a_2} = \cdots = \frac{\dot{\theta}_6}{a_6} \tag{11}$$

where a_i is the synchronization proportional coefficient.

The tracking velocity error is the difference between desired velocity of the ith joint $\dot{\theta}_i$ and real velocity $\dot{\theta}_i^d$, which is as follows:

$$e_i = \dot{\theta}_i - \dot{\theta}_i^d \tag{12}$$

In order to obtain the proper synchronization motion, $e_i \to 0$ is required along with

$$e_1/a_1 = e_2/a_2 = \cdots = e_6/a_6 \tag{13}$$

ζ_i is the synchronous speed error of joint i which can be calculated as follows:

$$\xi_i = \frac{2}{a_i} e_i - \left[\frac{1}{a_{i+1}} e_{i+1} + \frac{1}{a_{i-1}} e_{i-1} \right] \tag{14}$$

The goal of synchronous control is to make $\zeta_i = 0$. Substituted ζ_i into Eq. (15) to obtain the adjacent coupling error E_i, which not only consider the tracking speed error about the ith joint but also its adjacent joints.

$$E_i = e_i + \sigma_i \xi_i \qquad (15)$$

where σ_i is a positive coupled error coefficient. And the larger that σ_i is, the bigger the weight of the synchronous speed error will be.

4.4 Controller Design

In order to achieve high-precision trajectory tracking under high-speed conditions, a computed-torque deviation coupling control algorithm based on friction compensation is proposed. Define that,

$$u' = \theta_d - \gamma E \qquad (16)$$

where, E is adjacent coupling error, γ is the coupling error coefficient.

The robot position control vector is

$$r' = u' - \theta = \Delta\theta + \gamma E \qquad (17)$$

The computed-torque deviation coupling control law is designed as

$$\tau = M(\theta)\ddot{u}' + c(\theta, \dot{\theta})\dot{u}' + G(\theta) - K_p r' - K_d \dot{r}' \qquad (18)$$

where K_p, K_d are diagonal positive definite gain matrixes.

After Eq. (18) is brought into Eq. (3), the closed-loop equation of the robot control system is obtained as follows

$$M(\theta)\ddot{r}' + c(\theta, \dot{\theta})\dot{r}' = -K_p r' - K_d \dot{r}' \qquad (19)$$

The Lyapunov function was designed as follows:

$$V = 0.5 * \dot{r}'^T M(\theta)\dot{r}' + 0.5 * r'^T K_p r' \qquad (20)$$

Since the bounded inertia matrix $M(\theta)$ is symmetric positive definite, V is a positive definite function. The derivative of function V is as follows:

$$\dot{V} = \dot{r}'^T \dot{M}(\theta)\dot{r}' + \dot{r}'^T M(\theta)\ddot{r}' + \dot{r}'^T K_p r' \qquad (21)$$

Combining Eq. (18) and antisymmetric matrix $\dot{M}(q) - 2c(q, \dot{q})$, it can be obtained as follows:

$$\dot{V} = -\dot{r}'^T K_d \dot{r}' \leq 0 \qquad (22)$$

Due to $\dot{V} \leq 0$, $\dot{r}_i'$ is bounded. Since $M(\theta)$ and K_p are positive definite, and $\dot{V} \leq 0$, so V (0) is the maximum value of the function $V(t)$. So the following equation can be obtained

$$0.5 * r'^T K_p r' \leq V(0) \tag{23}$$

So r_i' is bounded which can be concluded that $\ddot{V} = -2\dot{r}'^T K_d \ddot{r}'$ is bounded.

From Barbalat lemma it can be seen that when $t \to \infty$, $\dot{V} \to 0$, so $r' \to 0$. From Eq. (17), $\Delta\theta \to 0$ and $E \to 0$. That is, the proposed control law is globally asymptotically stable.

The friction compensated term after identification is introduced into the computed-torque synchronous controller, and the computed-torque deviation coupling control based on friction compensation is obtained as follows

$$\tau = M(\theta)\ddot{u}' + c(\theta, \dot{\theta})\dot{u}' + G(\theta) - K_p r' - K_d \dot{r}_i'$$
$$+ \operatorname{sgn}(\dot{\theta})(f_v + (f_s - f_v)e^{-(\theta/v_s)^2}) + \sigma_3 \dot{\theta} \tag{24}$$

5 Simulation Analysis and Experimental Verification

5.1 Simulation Analysis

Move the first three joints of the robot according to Sect. 3, using the controller shown in Eq. (24). Set the proportional coefficient $K_p = \operatorname{diag}(400)$, differential coefficient $K_d = \operatorname{diag}(200)$, and two sets of simulation were carried out. Adding the disturbance to each axis in 0.5 s for one group, simulation time is 1 s, the results shown in Fig. 6.

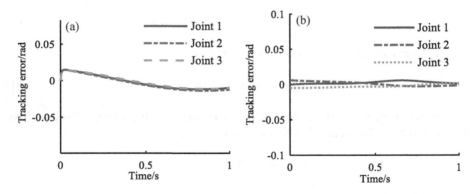

Fig. 6. Synchronization error of each joint. (a) Without disturbance, (b) With disturbance.

From Fig. 6(a) it can be seen that the tracking error of each axis is obviously reduced compared with the computed-torque control algorithm, and the maximum tracking error is 0.01 rad which quickly converges to 0. It can be seen from the comparison of the results in Figs. 6(b) and 4 that the maximum uniaxial tracking error is reduced from 0.2 rad to 0.02 rad, and the error is only slightly fluctuated after disturbance and then quickly converges. Accordingly, the controller can significantly improve the trajectory tracking accuracy and system's anti-jamming capacibility.

5.2 Experimental System

A servo control system based on EtherCAT bus and 6-DOF lightweight robot body experimental platform were built in this work. Figure 7 shows the entire experimental system which mainly consists of three components: laser tracking system, servo control system and the robot mechanical system.

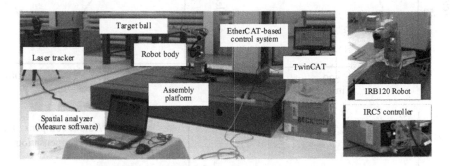

Fig. 7. Experimental system.

Beckhoff IPC was used as industrial computer which included TwinCAT3 programming software. The ADSA servo drives were applied in this work which supported EtherCAT bus communications. So real-time communication and motor control could be realized via high-speed EtherCAT bus.

5.3 Experimental Verification

In order to verify the accuracy of the friction model at high speed, the friction model compensation experiment is carried out as follows.

Set K_p = diag[12, 30, 30], K_d = diag [1, 5], diag[.] is a positive definite diagonal matrix. From point $A(-0.2, -0.2, 0.3)$ to point $B(-0.2, 0.2, 0.3)$ with a run time of 1 s. Taking axis 1 as an example, the tracking performance comparison of axis 1 before and after the compensation is shown in Fig. 8.

From Fig. 8(a) and (b) it can be seen that after adding the friction compensation, the acceleration of axis 1 has obvious fluctuation only at the start and stop time, and the speed and acceleration trajectory are very smooth after the robot is stable. And the maximum acceleration of joint 1 is reduced from $240°/s^2$ to $128°/s^2$ with the compensation.

From Fig. 8(c) it can be seen that considering the friction compensation, the control torque is more smooth, not only can reduce the noise of the robot, but also improve the service life of the motor. It can be seen from Fig. 8(d) that the trajectory error of the axis 1 is reduced from $(-0.8° \sim 1°)$ to $(-0.230° \sim 0.587°)$ after the friction compensation is added. In summary, the design of the computed-torque deviation coupling controller based on the friction compensation, can accurately compensate for the dynamic uncertainties of the robot, thereby significantly improving the robot trajectory tracking accuracy.

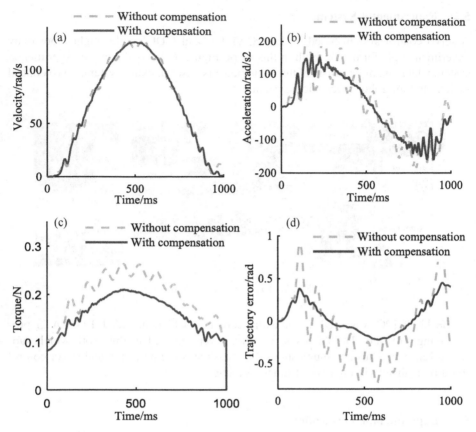

Fig. 8. Comparison of experimental results of axis 1. (a) Velocity comparison, (b) Acceleration comparison, (c) Torque comparison, (d) Trajectory error comparison.

6 Conclusion

(1) The mathematical model of the 6-DOF lightweight robot is deduced which can be used for robot controller design and trajectory planning. The single joint Lugre friction model is proposed, and the parameters of the friction model are identified.

(2) A control strategy based on friction compensation analysis is proposed which has considered adjacent coupling error analysis of the robot. And the stability of the control method is validated by Lyapunov function.

(3) The trajectory tracking ability of the robot in the case of disturbance and without disturbance are compared through simulation respectively. The simulation results show that the control strategy can significantly improve the system's anti-jamming capability.

(4) The experimental system is set up, which consists of laser tracking system, the robot mechanical system and servo control system. The synchronization error are significantly decreased and the trajectory error is reduced from $(-0.8° \sim 1°)$ to $(-0.230° \sim 0.587°)$ after the friction compensation is added.

Acknowledgments. The authors would like to thank the financial support from the National Natural Science Foundation of China (61305050) and the National High Technology Research and Development Program of China (2015AA043102) and (2015AA043003).

References

1. Li, Z.B., Wang, Z.L., Li, J.F.: A hybrid control scheme of adaptive and variable structure for flexible spacecraft. J. Aerosp. Sci. Technol. **8**(5), 423–430 (2004)
2. Hwang, C.L., Chao, S.F.: A fuzzy-model-based variable structure control for robot arms: theory and experiments. In: IEEE International Conference on Systems, Man and Cybernetics, vol. 6, No. 6, pp. 5252–5258. IEEE Xplore (2004)
3. Chen, Y., Ma, G., Lin, S., et al.: Computed-torque plus robust adaptive compensation control for robot manipulator with structured and unstructured uncertainties. J. Ima. J. Math. Control. Inf. **33**(1), 37–52 (2014)
4. Jankovics, V., Mátéfi-Tempfli, S., Manoonpong, P.: Artificial neural network based compliant control for robot arms. In: Tuci, E., Giagkos, A., Wilson, M., Hallam, J. (eds.) SAB 2016. LNCS, vol. 9825, pp. 91–100. Springer, Cham (2016). doi:10.1007/978-3-319-43488-9_9
5. Tien, L.L., Albu-Schaffer, A., Luca, A.D., et al.: Friction observer and compensation for control of robots with joint torque measurement. In: IEEE/RSJ International Conference on Intelligent Robots and Systems, pp. 3789–3795. IEEE Press (2008)
6. Alonge, F., D'Ippolito, F., Raimondi, F.M.: Globally convergent adaptive and robust control of robotic manipulators for trajectory tracking. J. Control Eng. Pract. **12**(9), 1091–1100 (2004)
7. Liu, X., Li, H., Wang, J., Cai, G.: Dynamics analysis of flexible space robot with joint friction. J. Aerosp. Sci. Technol. **47**, 164–176 (2015)
8. Zhu, Y., Pagilla, P.R.: Static and dynamic friction compensation in trajectory tracking control of robots. In: IEEE International Conference on Robotics and Automation (ICRA), vol. 3, pp. 2644–2649. IEEE Press (2002)
9. Gu, K., Cao, Q.: Control system design of 6-DOFs serial manipulator based on real-time ethernet. In: IEEE International Conference on Information and Automation, pp. 118–120. IEEE Press (2014)
10. Guo, K., Li, S., Huang, D.: Real-time quadruped robot control system based on Xenomai. In: Chinese Automation Congress, pp. 342–347. IEEE Press (2015)
11. Choi, T., Do, H., Park, D., Park, C.: Real-time synchronisation method in multi-robot system. J. Electron. Lett. **50**(24), 1824–1826 (2014)
12. Li, T., Sun, K., Liu, H., et al.: Optimal measurement configurations for kinematic calibration of six-dof serial robot. J. Cent. South. Univ. Technol. **18**(3), 618–626 (2011)
13. You, W., Kong, M.X., Sun, L.N., et al.: Optimal design of dynamic and control performance for planar manipulator. J. Cent. South. Univ. **19**(1), 108–116 (2012)
14. Zhang, D.L., Zhang, X., et al.: Contouring error modeling and simulation of a four-axis motion control system. J. Cent. South. Univ. **22**(1), 141–149 (2015)
15. Sun, D., Tong, M.C.: A synchronization approach for the minimization of contouring errors of cnc machine tools. J. IEEE. T. Autom. Sci. Eng. **6**(4), 720–729 (2009)

Fuzzy PD-Type Iterative Learning Control of a Single Pneumatic Muscle Actuator

Da Ke[1,2], Qingsong Ai[1,2(✉)], Wei Meng[1,2], Congsheng Zhang[1,2],
and Quan Liu[1,2]

[1] School of Information Engineering, Wuhan University of Technology,
Wuhan 430070, China
qingsongai@whut.edu.cn
[2] Key Laboratory of Fiber Optic Sensing Technology and Information
Processing (Wuhan University of Technology), Ministry of Education,
Wuhan 430070, China

Abstract. Pneumatic muscles actuator (PMA) is widely used in the field of rehabilitation robot for its good flexibility, light weight and high power/mass ratio as compared to traditional actuator. In this paper, a fuzzy logic-based PD-type iterative learning controller (ILC) is proposed to control the PMA to track a predefined trajectory more precisely during repetitive movements. In order to optimize the parameters of the learning law, fuzzy logic control is introduced into ILC to achieve smaller errors and faster convergence. A simulation experiment was first conducted by taking the PMA model fitted by support vector machine (SVM) as controlled target, which showed that the proposed method achieved a better tracking performance than traditional PD-type ILC. A satisfactory control effect was also obtained when fuzzy PD-type ILC was applied to actual PMA control experiment. Result showed that it takes 25 iterations for the maximum error of trajectory converges to a minimum of about 0.2.

Keywords: Iterative learning control · Pneumatic muscle actuator · Fuzzy logic

1 Introduction

Robot-assisted rehabilitation and therapy has been used to help the elderly and the disabled or movement disorders patients to perform exercise and training [1]. As a compliant actuator, pneumatic muscle actuator (PMA) is widely used in rehabilitation robot for its "force - shrink" characteristics, which is very similar to human muscle. And it has many advantages such as low price, good flexibility, high ratio of output power and weight and so on [2]. However, the strong non-linearity of the PMA, coupled with the non-linear effect of the aerodynamic dynamics and the pressure proportional valve in pneumatic control system, making the PMA difficult to be controlled with high precision. Traditional model-based control method is not suitable for PMA due to its difficulties in modeling. Therefore, it is particularly important to find a model-free and high-precision control algorithm for PMA.

© Springer International Publishing AG 2017
Y. Huang et al. (Eds.): ICIRA 2017, Part II, LNAI 10463, pp. 812–822, 2017.
DOI: 10.1007/978-3-319-65292-4_70

   Iterative learning control (ILC) algorithm is a kind of non-model-dependent control method which is suitable for the object with repetitive motion characteristics. It was proposed by Arimoto et al. in 1984 and has been applied to various fields of control theory and engineering recently. It focuses on the problems where the interaction between different durations is normally zero but the repetition of tracking the same trajectory creates a possibility of improving tracking performance [3]. When ILC was first proposed, only differential item of the error was used to correct the last input, i.e. D-type ILC [4]. After that, some other scholars take proportional and integral items into account using PID control for reference, and propose ILC methods of PI-type, PD-type and PID-type. With the development of intelligent control method in recent years, fuzzy control [5], neural network [6], adaptive control [7] and other methods have been introduced into ILC.

   The application of ILC algorithm to PMA control has several prominent advantages. Meng et al. proposed a robust iterative feedback tuning (IFT) technique for repetitive training control of a PMAs driven compliant parallel ankle rehabilitation robot, which achieves a better and better tracking performance during the robot repetitive control [8]. Schindele et al. studied a lot on ILC for PMA. First, they studied a fast linear axis driven by PMA [9, 10] and proposed a model free PID-type ILC and a model-based norm-optimal ILC for PMA control. Both methods have achieved good tracking results. Then, they designed a 2-DOF PMA-driven parallel robot and proposed a P-type ILC with leading phase compensation to iteratively compensate the uncertainties of remaining model, which makes the desired trajectories for the end-effector position to be tracked with high accuracy [11]. Balasubramanian et al. designed a 4-DOF upper limb repetitive therapy wearable robot named "RUPERT" driven by PMA, using for the shoulder, elbow and wrist rehabilitation training [12]. On the basis of RUPERT, the adaptive control strategy was designed. Using the closed-loop control method, the PID feedback controller and the ILC were combined, and fuzzy controller was introduced, leading to strong robustness of RUPERT [13, 14].

   In this paper, a fuzzy logic-based PD-type ILC method is proposed to control a PMA to track the desired trajectories precisely after several iterations. Support vector machine (SVM) is used to fit the relationship between shrinkage and air pressure of PMA. Comparative experiments between general PD-type ILC algorithm and fuzzy PD-type ILC algorithm are carried out. Results show the latter is proved to be better than the former in controlling the fitted model above. A good control effect is achieved when fuzzy PD-type ILC applied in the actual PMA control experiment.

2 Fuzzy Logic-Based ILC

2.1 Iterative Learning Control

A non-linear system is described as follows:

$$\begin{cases} \dot{x}(t) = f(x(t), u(t), t) \\ y(t) = g(x(t), u(t), t) \end{cases} \tag{1}$$

where $u(t) \in R^{n_u}$, $x(t) \in R^{n_x}$, $y(t) \in R^{n_y}$ are the control input, the state and the output of the system respectively, and $f(\cdot)$ and $g(\cdot)$ are the state functions and output functions of the system, whose structure and parameters are unknown. If $k = 1, 2, 3 \cdots$ and $t \in [0, T]$ represents the number of iterations and the sampling time points respectively, ILC can be described as: in the finite time interval $t \in [0, T]$, the expected response $y_d(t)$ and the corresponding expected initial state $x_d(0)$ of the controlled object are known, seeking for a control quantity $u_k(t)$ that makes $y_k(t)$ to be improved when compared to $y_{k-1}(t)$. If $k \to \infty$, $y_k(t) \to y_d(t)$, i.e.:

$$\lim_{k \to \infty} y_k(t) = y_d(t) \tag{2}$$

ILC method updates the input for the next iteration by the current iteration input $u_k(t)$ and the error $e_k(t)$. This method is reflected by the iterative learning law. The general form of learning law is PID-type:

$$u_{k+1}(t) = u_k(t) + Le_k(t) + \Gamma \dot{e}_k(t) + \Psi \int_0^t e_k(\tau) d\tau \tag{3}$$

where L, Γ, Ψ are the gain matrix. Different choices for Γ and Ψ will lead to P-type, PD-type or PI-type learning law. The algorithm flowchart is shown in Fig. 1.

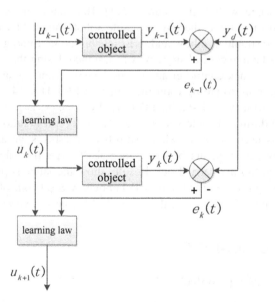

Fig. 1. ILC algorithm flow

2.2   Fuzzy Logic-Based Iterative Learning Control

Fuzzy logic control has been proved effective for complex, non-linear systems which are difficult to define. Moreover, it is used in various industrial applications nowadays because of prominent performance [15]. Fuzzy set theory is the basis of fuzzy logic control. In the theory, accurate quantity is described as fuzzy variables such as big, medium, small, etc. Linguistic variables and defined rules are utilized in fuzzy inference mechanism.

The choice of parameters in learning law has an important effect on the convergence of the algorithm. Inappropriate parameters often lead to the rise of controller input and output errors, even the situation in which the system does not converge. The introduction of fuzzy logic to avoid the blindness in selecting the parameters, the fuzzification of the input quantity can reduce the effect caused by special value and improve the convergence speed of the algorithm. A basic fuzzy controller mainly includes the following three procedures:

(a) Fuzzification: Convert the precise error e and the error change ratio ec into fuzzy quantity E and EC;
(b) Fuzzy reasoning: Define fuzzy rules in the form of "if-then", for example, if E is NB and EC is NM, then U is PB (NB, NM, PB are fuzzy level);
(c) Defuzzification: Convert the results of fuzzy reasoning into the precise quantity that can be used for actual control.

In fuzzy logic ILC algorithm, after appropriate membership functions and fuzzy rules determined, output error e and the differential of error ec are set to be the input of the fuzzy controller and the output is the corrected value of gain matrix. In the case of large output deviation, the fuzzy corrected value is used to increase the ILC convergence speed; In the case of small output deviation, the fuzzy compensation is used to ensure the stability of the system, achieving the effect of rapid convergence. Fuzzy logic iterative learning law is described as follows:

$$u_{k+1}(t) = u_k(t) + f(e_k(t), \dot{e}_k(t)) + g(e_k(t), \dot{e}_k(t)) \tag{4}$$

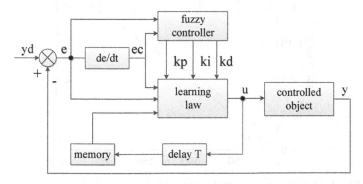

Fig. 2. Fuzzy ILC algorithm flow

where $f(\cdot)$ is the proportion, integral, and differential item of error in traditional iterative learning law, $g(\cdot)$ is fuzzy controller. Fuzzy ILC algorithm flowchart is shown in Fig. 2.

3 Experiments

3.1 Simulation

The control object is a PMA. Many researchers have investigated different modeling method of PMAs [16]. Among them, Yao took its internal friction and non-ideal cylinder and other characteristics caused by the hysteresis phenomenon into account, regarding it as the ideal model, hysteresis item, elastic item in parallel, where the hysteresis item consists of the position-dependent hysteresis and the velocity-dependent hysteresis [17]. The model with high precision and simple form reflects the physical properties of PMA. According to this model, the relationship between PMA pressure and shrinkage can be described as follows:

$$P = \frac{F}{a(1-\varepsilon)^2 - b} + P_{hys} + P_{vis} + P_e \tag{5}$$

where the first item is the ideal model (F is external load, a and b are parameters related to PMA diameter and braided mesh angle, ε is shrinkage), P_{hys} is the position-dependent hysteresis and P_{vis} is the velocity-dependent hysteresis, P_e is the elastic item. Equation (5) describes air pressure as a function of shrinkage. However, in the simulation process, a description of shrinkage by air pressure is more expected, i.e.:

$$\varepsilon = f(P, F) \tag{6}$$

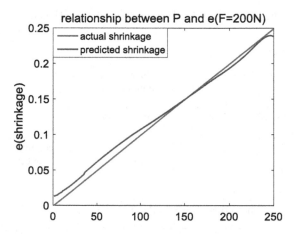

Fig. 3. Actual value and predict value of shrinkage (Color figure online)

Therefore, support vector machine (SVM) is used in re-fitting the above model in order to obtain the relationship between shrinkage and air pressure when external load is fixed. SVM is a supervised learning method, which presents many unique advantages in solving small sample, nonlinear and high dimensional pattern recognition, and can be applied to other machine learning problems such as function fitting. Figure 3 shows the result of real PMA and SVM method (external load F = 200 N), where the red line indicates that the shrinkage increases linearly by 0.01 as the air pressure increases in real PMA and the blue line indicates the shrinkage predicted by SVM under the same pressure. It shows that SVM method has a good effect on fitting the relationship between shrinkage and air pressure of PMA and the predict accuracy is 99.86% when the error is allowed.

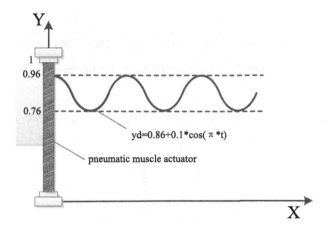

Fig. 4. Experimental method

The experimental method was designed and shown in Fig. 4. A PMA was fixed on the Y axis, whose bottom coincides with the origin and the top was the free movement end. The length was 1 m in the case of non-inflated. Inflate the PMA and make the top move in cosine trajectory between 0.76 and 0.96, i.e. make the shrinkage change between 0.04 and 0.24. Therefore, the desired trajectory is $y_d = 0.86 + 0.1 * \cos(\pi t)$ and the actual trajectory is $y = 1 - \varepsilon$, where ε is shrinkage.

Figure 5 shows that the output trajectory of PMA after iterations of 2, 5, 8, 15 times when general PD-type iterative learning law was used. It can be obviously seen that with the number of iterations grows, the actual trajectory is approaching the desired trajectory gradually.

In order to improve the convergence rate of the algorithm and the performance of the algorithm, fuzzy logic control was introduced. Fuzzy PD controller was designed as follow steps: the domain of input error e and differential of error ec was [−0.3, 0.3], outputs were corrected value Δp and Δd, whose domain was [0, 8]. Membership functions (MFs) of each variable was Gaussian-type shown in Fig. 6. Fuzzy levels of each variable were divided into seven grades: NB, NM, NS, ZO, PS, PM, PB. Fuzzy rules are shown in Table 1.

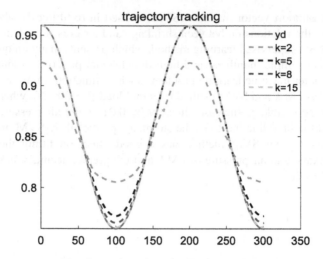

Fig. 5. Trajectories of different iterations

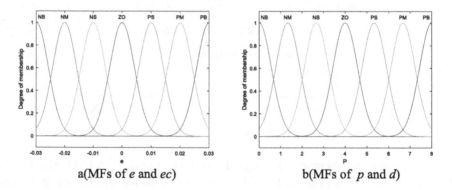

Fig. 6. Membership function curve of variables

Table 1. Fuzzy rules of Δp and Δd

$\Delta p, \Delta d$		e						
		NB	NM	NS	ZO	PS	PM	PB
ec	NB	PB,NS	PM,NS	NB,NS	PB,PB	NS,PS	PM,NB	ZO,NS
	NM	PB,NS	PM,NS	NB,NM	PM,PM	NS,PS	PM,NB	PS,NS
	NS	PM,NS	PS,NM	NM,NM	PM,PM	NS,ZO	PS,NM	PS,NS
	ZO	PM,NS	PS,NM	NM,ZO	PM,PS	NM,ZO	PS,NM	PM,NS
	PS	PS,NS	PS,NM	NS,ZO	PM,ZO	NM,NM	PS,NM	PM,NS
	PM	PS,NS	PS,NB	NS,PS	PM,PS	NB,NM	PM,NS	PB,NS

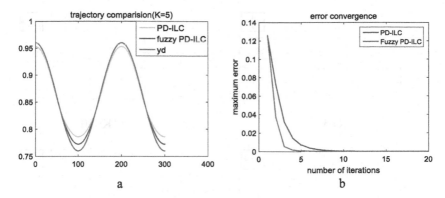

Fig. 7. Results Comparison

And iterative learning law is described as follows:

$$u_{k+1}(t) = u_k(t) + (p + \Delta p)e_k(t) + (d + \Delta d)\dot{e}_k(t) \tag{7}$$

Figure 7 shows the comparison results of general PD-type ILC and fuzzy PD-type ILC. In Fig. 7(a), three trajectory tracking effects at the 5th iteration are compared. It can be seen that compared to the general method, fuzzy PD-type ILC achieves a closer trajectory to the desired one. In Fig. 7(b), the error convergence curve are compared, where the blue line indicates that the maximum error varies with the number of iterations when fuzzy controller was not used, and the red line indicates the result after

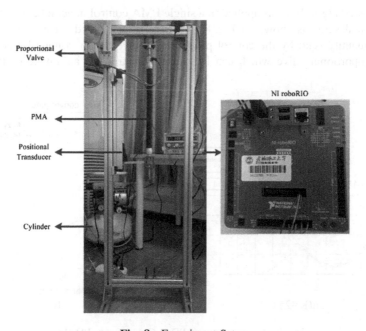

Fig. 8. Experiment Setup

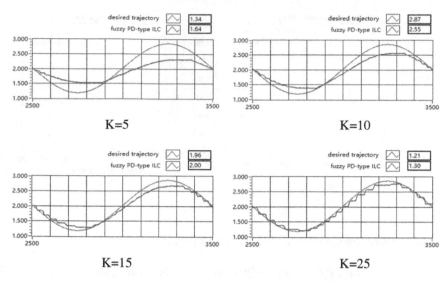

Fig. 9. Experimental trajectory of different iteration (Color figure online)

adding fuzzy controller. It can be seen that the error convergence is obviously accelerated in fuzzy PD-type ILC. The validity and superiority of the algorithm are proved.

3.2   PMA Control Experiment

The proposed algorithm was applied to a single PMA control in actual experiment. The experimental setup is shown in Fig. 8. A PMA is suspended on a stable stent, the control quantity send by the control program is converted to voltage by NIroboRIO to the proportional valve which can be used to control the air intake of the PMA.

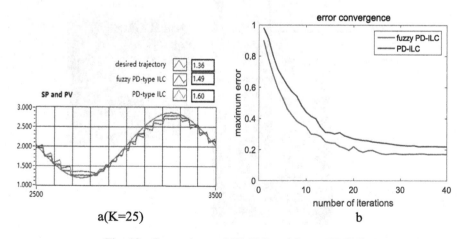

Fig. 10. Comparisons of PD-ILC and fuzzy PD-ILC

The displacement is measured by the positional transducer fixed at the bottom of the PMA and send to the control program via NI roboRIO.

The control program was built by LibVIEW, where the desired trajectory is $y_d = 0.85 * \sin(0.02 * \pi * t) + 2$ and the number of iterations was 40. Experimental trajectories of different iteration are shown in Fig. 9, where the red line indicates the desired trajectory and the blue line indicates the experimental trajectory. It can be seen that the experimental trajectory is getting closer to the desired trajectory with the number of iterations increasing. Figure 10 shows the comparison of PD-ILC and fuzzy PD-ILC in trajectory tracking and error convergence. It is easily observed in Fig. 10(a) that the actual trajectory is closer to the desired trajectory when fuzzy PD-ILC is applied in the same iteration ($K = 25$). Error convergence rate is shown in Fig. 10(b), fuzzy PD-ILC presents a faster convergence rate and a smaller error. The experiment shows that the application of the proposed algorithm has a good effect in PMA control.

The results of the simulation and the actual control experiment indicate that fuzzy PD-type ILC is an effective method for trajectory tracking in PMA repetitive movements. The fitting of PMA model by SVM makes simulation availabel so that the theoretical basis of the proposed method can be obtained. The introduction of fuzzy logic contributes a better tracking effect for the desired trajectory and a higher convergence rate. In the future work, the method to reduce the jitter of PMA during the iterative process will be studied in depth.

4 Conclusion

In this paper, a PD-type ILC algorithm based on fuzzy logic for PMA is proposed. A simulation experiment is carried out on computer. The controlled object is the model fitted by SVM and the result shows that the accuracy is 99.86%. General PD-type ILC performs well in controlling the model and the actual trajectory can almost completely track the desired trajectory after 10 iterations. Fuzzy logic controller is introduced into ILC algorithm in order to improve the convergence rate. Compared with the general PD-type ILC, fuzzy PD-type ILC can significantly reduce the number of iterations required for algorithm convergence. In actual PMA control experiment, the maximum error converges to about 0.2 cm after 25 iterations. Therefore, fuzzy ILC is an effective method of controlling the PMA.

Acknowledgments. Research supported by National Natural Science Foundation of China under grants Nos. 51675389, 51475342 and The Excellent Dissertation Cultivation Funds of Wuhan University of Technology with No. 2016-YS-062.

References

1. Meng, W., Liu, Q., Zhou, Z., et al.: Recent development of mechanisms and control strategies for robot-assisted lower limb rehabilitation. Mechatronics **31**, 132–145 (2015)
2. Liu, Y., Wang, T., Fan, W., et al.: Model-free adaptive control for the ball-joint robot driven by PMA group. Robot **35**(2), 129–134 (2013). (in Chinese)

3. Shao, Z., Xiang, Z.: Iterative learning control for non-linear switched discrete-time systems. IET Control Theory Appl. **11**(6), 883–889 (2017)
4. Arimoto, S., Kawamura, S., Miyazaki, F.: Better operation of robots by learning. J. Field Robot. **1**(2), 123–140 (1984)
5. Wang, Y., Chien, C.: A fuzzy iterative learning control for nonlinear discrete-time systems with unknown control directions. In: 54th IEEE Conference on Decision and Control, pp. 3081–3086. Institute of Electrical and Electronics Engineers Inc, Osaka (2015)
6. Czajkowski, A., Patan, M., Patan, K.: Design of iterative learning control by the means of state space neural networks. In: 3rd Conference on Control and Fault-Tolerant Systems, pp. 293–298. IEEE Computer Society, Barcelona (2016)
7. Liu, B., Zhou, W.: Adaptive iterative learning control for a class of MIMO discrete-time nonlinear systems. In: 11th International Conference on Computer Science and Education, pp. 832–838. Institute of Electrical and Electronics Engineers Inc, Nagoya (2016)
8. Meng, W., Xie, S., Liu, Q., et al.: Robust iterative feedback tuning control of a compliant rehabilitation robot for repetitive ankle training. IEEE/ASME Trans. Mechatron. **22**(1), 173–184 (2016)
9. Schindele, D., Aschemann, H.: Norm-optimal iterative learning control for a high-speed linear axis with pneumatic muscles. In: 8th IFAC Symposium on Nonlinear Control System, pp. 463–468. IFAC Secretariat, Bologna (2010)
10. Schindele, D., Aschemann, H.: ILC for a fast linear axis driven by pneumatic muscle actuators. In: 2011 International Conference on Mechatronics, pp. 967–972. IEEE Computer Society, Istanbul (2011)
11. Schindele, D., Aschemann, H.: P-type ILC with phase lead compensation for a pneumatically driven parallel robot. In: 2012 American Control Conference, pp. 5484–5489. Institute of Electrical and Electronics Engineers Inc, Montreal (2012)
12. Sugar, T., He, J., Koeneman, E., et al.: Design and control of RUPERT: a device for robotic upper extremity repetitive therapy. IEEE Trans. Neural Syst. Rehabil. Eng. **15**(3), 336–346 (2007)
13. Balasubramanian, S., Wei, R., He, J.: Rupert closed loop control design. In: 2008 Engineering in Medicine and Biology Society. 30th Annual International Conference of the IEEE Engineering in Medicine and Biology Society, pp. 3467–3470. Institute of Electrical and Electronic Engineers Computer Society, Vancouver (2008)
14. Wei, R., Balasubramanian, S., Xu, L., et al.: Adaptive iterative learning control design for RUPERT IV. In: 2nd Biennial IEEE/RAS-EMBS International Conference on Biomedical Robotics and Biomechatronics, pp. 647–652. Institute of Electrical and Electronic Engineers Computer Society, Scottsdale (2008)
15. Dursun, E., Durdu, A.: Position control by using PD type fuzzy logic: Experimental study on rotary servo system. In: 8th Conference on Electronics, Computers and Artificial Intelligence, pp. 1–6. Institute of Electrical and Electronics Engineers Inc. Ploiesti (2016)
16. Cao, J., Xie, S.Q., Zhang, M., Das, R.: A new dynamic modelling algorithm for pneumatic muscle actuators. In: Zhang, X., Liu, H., Chen, Z., Wang, N. (eds.) ICIRA 2014. LNCS, vol. 8918, pp. 432–440. Springer, Cham (2014). doi:10.1007/978-3-319-13963-0_44
17. Yao, B.: Research on the Position Control Strategy of Pneumatic Muscle Actuator (PMA) and the PMA-actuated Parallel Robot. Wuhan University of Technology, Wuhan (2012)

Cascade Control for SEAs and Its Performance Analysis

Yuancan Huang$^{(\boxtimes)}$, Yin Ke, Fangxing Li, and Shuai Li

School of Mechatronical Engineering, Beijing Institute of Technology,
Beijing, China
yuancanhuang@bit.edu.cn

Abstract. Serial Elastic Actuators (SEAs) have several superiorities
over conventional rigid actuators *e.g.*, greater shock tolerance, energy
storage, safety, and so on. However, there exist performance limitations
due to existence of elasticity in SEAs, *i.e.*, low accuracy in position control, low control bandwidth in force control as coupling stability is guaranteed, and impedance constraints in impedance control. Variable Stiffness Actuators (VSAs) are designed to address the performance degradation but increases complexity in mechanism. As a result, control in
SEAs plays a more important role than that in their rigid counterparts.
We address this challenge by cascade control severing as the same purpose as that of VSAs' without losing favorable advantages of SEAs to
imitate humanoid manipulation. Then stability and passivity constraints
is derived. Performance analysis is conducted in terms of quasi-rigid performance and apparent stiffness. Finally, experiments are carried out to
test performance of the suggested control scheme.

Keywords: SEAs · Cascade control · Stiffness · Performance analysis

1 Introduction

SEA [1] was first posed in 1995 by Pratt and Williamson. Nowadays, its superiority over rigid actuators has been generally recognized in robotics society, and
SEAs were used as building bricks for a lot of collaborative robots, for example, Rethink Robotics' Baxter and Sawyer [2], DLR hand-arm system [3], iit
CompAct arm and iCub [4,5], MIT Cog's humanoid robot [6], BIT soft massage
robotic arm [7–9], and so forth.

SEA primarily consists of a power unit, a speed reducer, an elastic element,
and a transmission mechanism, which can be chosen and configured in various
ways, resulting in a number of SEA variants with tradeoffs on power output, volumetric size, weight, efficiency, backdrivability, impact resistance, passive energy
storage, backlash, and torque ripple [10]. Nevertheless, SEA may be simply modeled as an ideal force/torque source in series with a mass/pure inertia through
an ideal fixed- or variable-stiffness spring.

© Springer International Publishing AG 2017
Y. Huang et al. (Eds.): ICIRA 2017, Part II, LNAI 10463, pp. 823–834, 2017.
DOI: 10.1007/978-3-319-65292-4_71

Numerous research works concerning SEA control exist in the literature, which may be classified into force, position, impedance, and admittance control schemes [11]. In regard to force control, Williamson [12] proposed a PID controller plus two feedforward terms and a noncollocated acceleration feedback that compensates the system dynamics; The linear controllers has been presented in [12], and [13] sometimes using an inner velocity (or position) loop to improve performance; Calanca and Fiorini [14] described an adaptive control for rehabilitative devices, which automatically varies force control gains in response to the changing environment; A passivity-based approach with motor-side position feedback and inner torque loop was adopted to provide a unified framework for force, position, and impedance control in [15], in which the inner torque loop is interpreted physically as the shaping of the actuator inertia.

Although various advantages have been widely exploited in robotics society, there exist performance limitations due to existence of elasticity in SEAs, *i.e.*, low accuracy in position control and low control bandwidth in force control as coupling stability is guaranteed, and variation range of impedance is restricted by its elasticity in impedance control. Variable Stiffness Actuators (VSAs) are designed to address the performance degradation, where stiffness of elastic element can be physically modulated during operation or pre-set relevant to a specific application such that high performance are recovered at the price of increasing design complexity. Here, we propose the cascade control with link-side position and torque loops serving as the same purpose as that of VSAs without losing favorable advantages of SEAs, which may serve as an universal inner position loop in position-based impedance control or position-based force control.

2   Cascade Control of SEAs

2.1   Modelling of SEAs

SEAs are naturally split into two cascade subsystems (link- and motor-side dynamics) by series elasticity, and then are simply modeled as an ideal force/torque source in series with a mass/pure inertia through an ideal fixed-stiffness spring in Fig. 1.

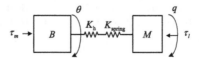

Fig. 1. Simplified model of SEAs

The dynamical equation of a rotary SEA is written as:

$$B\ddot{\theta} + \tau_s = \tau_m \tag{1}$$

$$M\ddot{q} = \tau_s + \tau_{ext} \tag{2}$$

$$\tau_s = K(\theta - q) \tag{3}$$

where θ and q represent motor- and link-side positions, respectively. K_h, K_{spring} are respective stiffness of speed reducer and elastic element, where $K_h \gg K_{spring}$ and thus effective stiffness $K \approx K_{spring}$. τ_m is motor torque, B is rotor inertia, M is link inertia and τ_{ext} is external torque applied on the link.

2.2 Implementation of Controllers

Cascade control for SEAs composed of the outer link-side position loop and the inner motor-side torque loop is shown in Fig. 2(a), where PD-type controller is used in both outer and inner loops for the sake of simplicity. The PD-type cascade control law is given by:

$$v = -k_P^{out}\tilde{q} - k_D^{out}\dot{\tilde{q}} \tag{4}$$

$$\tau_m = -k_{in}(k_P^{in}e + k_D^{in}\dot{e}) + \tau_s \tag{5}$$

where v is the link-side virtual input, which is reproduced in the spring torque through the inner torque loop, $\tilde{q} = q - q^d$ and $e = \tau_s - v$ are errors in the outer and inner loops. q^d is the desired link-side position. k_P^{out}, k_D^{out}, $k_{in}k_P^{in}$, $k_{in}k_D^{in}$ represent, respectively, proportional and derivative parameters in the outer and inner PD controllers, and k_{in} is scaled coefficient.

A target system is plotted in Fig. 2(b) as if there is no series elasticity in SEAs or driving torque were collocated with link-side position measurement. In addition, the actuator has zero effective inertia in the equivalent system as $k_{in} \to \infty$.

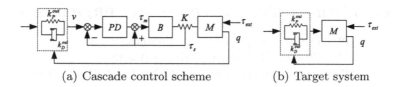

(a) Cascade control scheme   (b) Target system

Fig. 2. Cascade control structure and target system

3 Stability and Passivity Analysis

3.1 Stability Criterion

After simple derivation, we obtain the following polynomial description of overall closed-loop system:

$$A(s)Q = B(s)Q_d + C(s)T_{ext} \tag{6}$$

where

$$A(s) = BMs^4 + MKk_{in}k_d^{in}s^3 + K(B + Mk_{in}k_p^{in} + k_{in}k_d^{in}k_d^{out})s^2 +$$
$$Kk_{in}(k_p^{in}k_d^{out} + k_p^{out}k_d^{in})s + Kk_{in}k_p^{in}k_p^{out} \tag{7}$$

$$B(s) = Kk_{in}(k_d^{in}k_p^{out}s + k_p^{in}k_p^{out}) \tag{8}$$

$$C(s) = Bs^2 + Kk_{in}k_d^{in}s + Kk_{in}k_p^{in} \tag{9}$$

By Routh-Hurwitz Criterion, the closed-loop stability is guaranteed whenever the following inequalities are satisfied:

$$
\begin{cases}
BM > 0 \\
MKk_{in}k_d^{in} > 0 \\
k_P^{out} < \dfrac{K(Mk_P^{in}+k_D^{in}k_D^{out})}{B}k_{in} + \dfrac{K-k_D^{out}k_P^{in}}{k_D^{in}} \\
k_P^{out} < \dfrac{(Kk_D^{in}k_D^{out})k_{in}+(Kk_D^{in}(KB^2k_D^{in}+(2KBk_D^{in^2}k_D^{out}+4MBk_D^{out}k_P^{in^2})k_{in}+Kk_D^{in^3}k_D^{out^2}k_{in}^2))^{1/2}}{2B} \\
\qquad + \dfrac{K}{2} - \dfrac{k_D^{out}k_P^{in}}{k_D^{in}} \\
k_P^{out} > \dfrac{(Kk_D^{in}k_D^{out})k_{in}-(Kk_D^{in}(KB^2k_D^{in}+(2KBk_D^{in^2}k_D^{out}+4MBk_D^{out}k_P^{in^2})k_{in}+Kk_D^{in^3}k_D^{out^2}k_{in}^2))^{1/2}}{2B} \\
\qquad - (\dfrac{K}{2} - \dfrac{k_D^{out}k_P^{in}}{k_D^{in}}) \\
Kk_p^{in}k_p^{out} > 0
\end{cases}
$$

From the inequalities set, k_P^{out} is upper bounded by k_{in}. One conclusion can be drawn that k_P^{out} can exceed spring stiffness with high inner loop gain k_{in}, namely, apparent stiffness breaks through limitation of spring stiffness, which is depicted in Sect. 4.2.

3.2 Passivity and Coupled Stability

In physical interaction, passivity of the overall closed-loop system is desired, coupled stability is guaranteed when controlled SEAs contact passive environments [16].

Definition 1. *If $H(s) \in \mathbf{R}[s]$, $H(s)$ is Positive Real (PR) if and only is*

$$
\begin{aligned}
&1^o \quad (\forall s | Res > 0) : f(s) \quad is \quad analytic; \\
&2^o \quad (\forall \omega | \omega \in \mathbf{R}, j\omega \bar{e} r(d)) : Ref(j\omega) \geq 0; \\
&3^o \quad (\forall j\omega_0 \in r(d)) : \lim_{s \to j\omega_0} f(s) = k, k \in \mathbf{R}, k > 0
\end{aligned}
\tag{10}
$$

Consider transfer function $H(s)$ where external torque is taken as input and link end velocity as output. From (7) and (9), we have

$$
H(s) = \frac{V(s)}{T_{ext}(s)} = \frac{sQ(s)}{T_{ext}(s)} = \frac{sC(s)}{A(s)}
\tag{11}
$$

By Definition 1, $H(s)$ is positive real if the following condition is satisfied:

$$
(\forall \omega \in \mathbf{R}) : Re\{H(j\omega)\} \geq 0
\tag{12}
$$

Therefore, the overall closed-loop system is passive when the following conditions holds:

$$
k_P^{out} < \frac{Kk_D^{out}k_D^{in}}{B}k_{in} - \frac{k_D^{out}k_P^{in}}{k_D^{in}} + K
\tag{13}
$$

Above condition ensures passivity and coupled stability, obviously limitation of k_P^{out} can be broken with high inner loop gain k_{in}.

4   Performance Analysis

4.1   Quasi-Rigid Performance

The closed-loop transfer function can be rewritten as:

$$C_c(s) = \frac{k_P^{out}(k_D^{in}s + k_P^{in})}{\frac{Bs^2(Ms^2+K)}{Kk_{in}} + (k_D^{in}s + k_P^{in})(Ms^2 + k_D^{out}s + k_P^{out})} \tag{14}$$

Regard the equivalent system as the desired system, which is given as:

$$C_t(s) = \frac{k_D^{out}s + k_P^{out}}{Ms^2 + k_D^{out}s + k_P^{out}} \tag{15}$$

The fact that the dominant poles of (14) are exactly the same as those of (15) implies that the cascade control of SEAs achieves quasi-rigid performance as $k_{in} \to \infty$, i.e., there is high gain in the inner torque loop. Figure 3 plots the root locus of (14) as k_{in} varies from 0 to ∞. The simulation parameters are given in Table 1, where M, K, B are derived from actual measurements, and controller parameters are carefully chosen to achieving best performance.

Table 1. Simulation parameters

Symbol	Meaning	Value
B	Inertia of motor	$0.3789\,\mathrm{N\,m\,s^2/rad}$
M	Inertia of link	$0.2\,\mathrm{N\,m\,s^2/rad}$
K	Spring stiffness	$700\,\mathrm{N\,m}$
k_P^{out}	Position controller gain	1300
k_D^{out}	Position controller gain	20
k_P^{in}	Torque controller gain	100
k_D^{in}	Torque controller gain	1

4.2   Apparent Stiffness

Without loss of generality, assume that $q_d = 0$ in equilibrium position. Now, the input of system is τ_{ext}, with q as output. Then the system error is $E(s) = -Q(s)$, and from (6) yields

$$Q(s) = \frac{C(s)}{A(s)}T_{ext}(s) \tag{16}$$

Let $T_{ext}(s) = \frac{\tau_0}{s}$. The steady-state error in terms of final value theorem of Laplace transform is

$$e_{ss} = \lim_{s \to 0} sE(s) = \lim_{s \to 0} s\frac{C(s)}{A(s)}T_{ext}(s) = \lim_{s \to 0}\frac{C(s)}{A(s)}\tau_0 \tag{17}$$

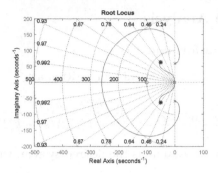

Fig. 3. Root locus with gain k_{in}

Thus, the apparent stiffness can be expressed as follows:

$$K_q = \frac{\tau_0}{e_{ss}} = \frac{\tau_0}{\lim_{s\to 0} \frac{C(s)}{A(s)}\tau_0} = \lim_{s\to 0} \frac{A(s)}{C(s)} = \frac{A_0}{C_0} \qquad (18)$$

Since $A(s)$ and $C(s)$ both are polynomials of s, the limit of (18) is $K_q = \frac{A_0}{C_0}$. As for motor-side position θ feedback, apparent stiffness has the same form, here we present directly,

$$K_q = \frac{Kk_{in}k_P^{in}k_P^{out}}{Kk_{in}k_P^{in}} = k_P^{out}, \quad K_\theta = \frac{Kk_P^{out}}{K + k_P^{out}} \qquad (19)$$

which shows that K_θ is upper bounded by spring physical stiffness K while K_q is not, or put this differently, $K_\theta \to K$ while $K_q \to \infty$ with $k_P^{out} \to \infty$.

5   Practical Consideration

Without velocity sensor in this design, velocity measurement is implemented by actual differentiator, which is presented in form of Laplace operator:

$$D = \frac{s}{1 + \frac{s}{N}} \qquad (20)$$

where N is filtering coefficient.

5.1   Effects on Apparent Stiffness and Passivity

As discussed above, stability and passivity (or coupled stability) are guaranteed with high inner loop gain k_{in} and apparent stiffness k_P^{out} is free from limitation of spring stiffness (Table 2).

In order to illustrate effects of actual differentiator on apparent stiffness, closed-loop transfer function is rewritten with actual differentiator:

$$A_N(s)Q = B_N(s)Q_d + C_N(s)T_{ext} \qquad (21)$$

and closed-loop system is $C(s) = \frac{B_N(s)}{A_N(s)}$, where

$$A_N(s) = BMs^6 + 2BMNs^5 + (BMN^2 + KMk_D^{in}k_{in}N + BK + KMk_{in}k_P^{in})s^4 +$$
$$(2BKN + 2KMNk_{in}k_P^{in} + KMN^2k_D^{in}k_{in})s^3 +$$
$$K(BN^2 + k_{in}k_P^{in}k_P^{out} + MN^2k_{in}k_P^{in} + N^2k_D^{in}k_D^{out}k_{in}$$
$$+Nk_D^{in}k_{in}k_P^{out} + Nk_D^{out}k_{in}k_P^{in})s^2 +$$
$$(2KNk_{in}k_P^{in}k_P^{out} + KN^2k_D^{in}k_{in}k_P^{out} + KN^2k_D^{out}k_{in}k_P^{in})s$$
$$Kk_{in}(k_p^{in}k_d^{out} + k_p^{out}k_d^{in})s + KN^2k_{in}k_P^{in}k_P^{out}$$
$$B_N(s) = K(k_{in}k_P^{in}k_P^{out} + Nk_D^{in}k_{in}k_P^{out}))s^2 +$$
$$(KN(k_{in}k_P^{in}k_P^{out} + Nk_D^{in}k_{in}k_P^{out}) + KNk_{in}k_P^{in}k_P^{out})s + KN^2k_{in}k_P^{in}k_P^{out}$$
$$C_N(s) = KN^2k_{in}k_P^{in}s^2 + (Kk_D^{in}k_{in}N^2 + 2Kk_{in}k_P^{in}N)s + Kk_{in}k_P^{in} + KNk_D^{in}k_{in}$$

Generalized root locus is plotted in term of k_P^{out} with different $N(10^2, 10^4, 10^6)$. From Fig. 4, one can tell that upper bound of apparent stiffness, namely k_P^{out}, increases as N raises. Simulation parameters is presented in Table 1.

Table 2. Simulation parameters

Symbol	Meaning	Value
k_D^{out}	Position controller gain	200
k_{in}	Inner loop gain	100
k_P^{in}	Torque controller gain	1000
k_D^{in}	Torque controller gain	100

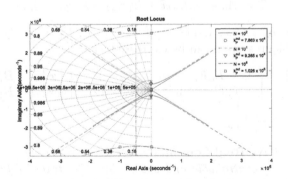

Fig. 4. Root locus with different N

Obviously, relative order of closed-loop system $H_N(s) = \frac{sC_N(s)}{A_N(s)}$ is bigger than 1, where external torque is taken as input and link end velocity as output. Which means passivity is deteriorated due to actual differentiator. A method base on state observer for recovering passivity is attempted in the following part.

5.2   Passivity Recovery by State Observer

Consider open-loop system in state-space form:

$$\begin{cases} \dot{x} = A_0 x + B_0 \tau_m + B_1 \tau_{ext} \\ y = C_0 x \end{cases} \tag{22}$$

where state variable $x = [q \quad \dot{q} \quad \theta \quad \dot{\theta}]^T$,

$$A_0 = \begin{bmatrix} 0 & 1 & 0 & 0 \\ -M^{-1}K & 0 & M^{-1}K & 0 \\ 0 & 0 & 0 & 1 \\ B^{-1}K & 0 & -B^{-1}K & 0 \end{bmatrix}, \quad B_0 = \begin{bmatrix} 0 \\ 0 \\ 0 \\ B^{-1} \end{bmatrix}, \quad B_1 = \begin{bmatrix} 0 \\ M^{-1} \\ 0 \\ 0 \end{bmatrix}, \quad C_0 = \begin{bmatrix} 0 \\ 1 \\ 0 \\ 0 \end{bmatrix}$$

We choose full-state feedback:

$$\tau_m = K_c x = [k_1 \quad k_2 \quad k_3 \quad k_4] x \tag{23}$$

where

$$\begin{cases} k_1 = -k_{in} k_P^{in} (k_P^{out} - K) - K + \frac{K}{M} k_{in} k_D^{in} k_D^{out} \\ k_2 = -k_{in} k_D^{in} (k_P^{out} - K) - k_{in} k_P^{in} k_D^{out} \\ k_3 = K(1 - k_{in} k_P^{in}) - \frac{K}{M} k_{in} k_D^{in} k_D^{out} \\ k_4 = -K k_{in} k_D^{in} \end{cases} \tag{24}$$

Then we have controlled system

$$\begin{cases} \dot{x} = (A_0 + B_0 K_c) x + B_1 \tau_{ext} \\ y = C_0 x \end{cases} \tag{25}$$

and state observer is given as

$$\dot{\hat{x}} = (A_0 + B_0 K_c)\hat{x} + B_1 \tau_{ext} + L_0 (y_r - \hat{y}_r) \tag{26}$$

where $y_r = C_r x$, $\hat{y}_r = C_r \hat{x}$, and $C_r = [0 \quad 1 \quad 0 \quad 1]$, $L_0 = [l_1 \quad l_2 \quad l_3 \quad l_4]$.

Final closed-loop system can be expressed as

$$\begin{cases} \begin{bmatrix} \dot{x} \\ \dot{\hat{x}} \end{bmatrix} = \begin{bmatrix} A_0 + B_0 K_c & 0 \\ L_0 C_r & A_0 + B_0 K_c - L_0 C_r \end{bmatrix} \begin{bmatrix} x \\ \hat{x} \end{bmatrix} + \begin{bmatrix} B_1 \\ B_1 \end{bmatrix} F \\ y = \begin{bmatrix} C_0 & 0 \end{bmatrix} \begin{bmatrix} x \\ \hat{x} \end{bmatrix} \end{cases} \tag{27}$$

Lemma 1 (Positive Real Lemma or KYP Lemma). *Let system in general state-space form be controllable and observable. The transfer function* $H(s) = C(sI_n - A)^{-1} B + D$, *with* $A \in \mathbb{R}^{n \times n}$, $B \in \mathbb{R}^{n \times n}$, $C \in \mathbb{R}^{n \times n}$, $D \in \mathbb{R}^{n \times n}$ *is PR with* $H(s) \in \mathbb{R}^{n \times n}$, $s \in \mathbb{C}$, *if and only if there exists matrices* $P = P^T > 0$, $P \in \mathbb{R}^{n \times n}$, $L \in \mathbb{R}^{n \times m}$ *and* $W \in \mathbb{R}^{m \times m}$ *such that:*

$$\begin{cases} PA + A^T P = -LL^T \\ PB - C^T = -LW \\ D + D^T = W^T W \end{cases} \tag{28}$$

In our case, $D = D^T = 0$, then $W = W^T = 0$, from above derivation, we have

$$A = \begin{bmatrix} A_0 + B_0 K_c & 0 \\ L_0 C_r & A_0 + B_0 K_c - L_0 C_r \end{bmatrix}, \quad B = \begin{bmatrix} B_1 \\ B_1 \end{bmatrix}, \quad C = \begin{bmatrix} C_0 & 0 \end{bmatrix} \quad (29)$$

As long as **KYP Lemma** is satisfied, passivity is recovered. Without losing generality, we present a numerical instance here. Consider a symmetric positive definite matrix $P \in \mathbb{R}^{8 \times 8}$. From (28), we have:

$$PB = C^T \quad (30)$$

where $B = \begin{bmatrix} 0 & M^{-1} & 0 & 0 & 0 & M^{-1} & 0 & 0 \end{bmatrix}^T$, $M = 0.2$, $C = \begin{bmatrix} 0 & 1 & 0 & 0 & 0 & 0 & 0 & 0 \end{bmatrix}$.

Then we have the following constraints of elements of P

$$\begin{array}{ll} P_{1,2} + P_{1,6} = 0 & P_{2,5} + P_{5,6} = 0 \\ P_{2,2} + P_{2,6} = 0.2 & P_{2,6} + P_{6,6} = 0 \\ P_{2,3} + P_{3,6} = 0 & P_{2,7} + P_{6,7} = 0 \\ P_{2,4} + P_{4,6} = 0 & P_{2,8} + P_{6,8} = 0 \end{array} \quad (31)$$

We present directly P that satisfies all constraints mentioned above. Passivity is recovered with state observer.

$$\begin{bmatrix} 1 & 0 & 0 & 0 & 0 & 0 & 0 & 0 \\ 0 & 1.2 & 0 & 0 & 0 & -1 & 0 & 0 \\ 0 & 0 & 1 & 0 & 0 & 0 & 0 & 0 \\ 0 & 0 & 0 & 1 & 0 & 0 & 0 & 0 \\ 0 & 0 & 0 & 0 & 1 & 0 & 0 & 0 \\ 0 & -1 & 0 & 0 & 0 & 1 & 0 & 0 \\ 0 & 0 & 0 & 0 & 0 & 0 & 1 & 0 \\ 0 & 0 & 0 & 0 & 0 & 0 & 0 & 1 \end{bmatrix} \quad (32)$$

6 Position-Based Impedance and Force Control

6.1 Impedance Control

For the goal of compliant behavior, an impedance controller is implemented based on the cascade control structure mention above. PD-type position controller have exhibited compliant behavior already, while with impedance controller implemented based on inner position loop, wider variation range of impedance can be achieved in form of $Z = \frac{1}{M_d s^2 + B_d s + K_d}$. Impedance control scheme is depicted in Fig. 5(a).

6.2 Force Control

As proved in [17] that force control stability can be assured only if exists a compliant interface between the robot and the environment and that compliance is helpful to ensure well defined force control dynamics, if combined with a low robot inertia. Here, we present a force control scheme, as shown in Fig. 5(b).

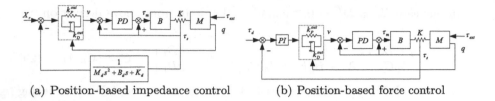

(a) Position-based impedance control   (b) Position-based force control

Fig. 5. Impedance and control scheme

7   Experimental Verification

7.1   Setup of Testbed

Experiments are carried out to verify effectiveness of the suggested control scheme in comparison with the conventional motor-side position control in terms of rise time, overshoot, and adaptability to load variation. As shown in Fig. 6, a SEA prototype, containing two Avago AEDT-9811-W00 photoelectric incremental encoders placed on both sides of the torsional spring, has been designed in [7]. A testing rig is built to evaluate performance of the suggested control scheme, where a flat bar is used as a lever loading the weighting mass to simulate load changes.

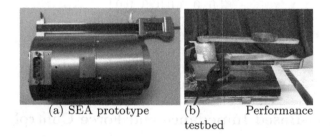

(a) SEA prototype   (b)   Performance
testbed

Fig. 6. Experimental devices

7.2   Performance Comparison

Rise time and overshoot are chosen as criteria for evaluating performance of link- and motor-side position feedback control strategies in overdamping and underdamping cases. As shown in Fig. 7(a), in the overdamping case, rise time is much shorter in q feedback than that in θ feedback, where is around $t = 5\,$s in q feedback and is around $15\,$s in θ feedback. In the underdamping case, overshoot is observed in both feedback strategies. However, thanks to higher effective stiffness, the overshoot in q feedback is less than that in θ feedback. To conclude, the suggested cascade control is superior to the conventional motor-side position control in terms of rise time and overshoot.

7.3   Adaptibility to Load Variation

In order to test adaptibility to load variation under link- and motor-side position feedback control strategies, several weighting masses are fixed on the link such that effective mass m_e increases and thus effective natural frequency ω_e lowers, which is $\propto \frac{K_e}{m_e}$ and K_e is apparent stiffness. In q feedback, loss of ω_e due to incremental of m_e may be compensated by increasing k_e, which cannot occur in θ feedback, for there is a upper bound of apparent stiffness in this case. In Fig. 7(b), red line and red dotted line denote standard performance with acceptable rapidity and overshoot under q and θ feedback, respectively. As load increases, overshoot increases as expected in θ feedback. On the contrary, the suggested cascade control may adapt to load variation as indicated by three solid lines.

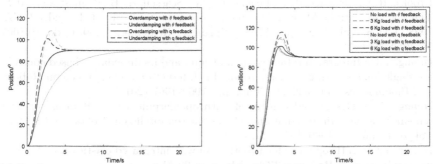

(a) Performance comparison between link- and motor-side position feedback control strategies: Desired position is set to be 90° in the experiment

(b) Adaptibility to load variation under link- and motor-side position feedback control strategies

Fig. 7. Performance test results

8   Conclusions

A cascade control with outer link-side position feedback and inner torque loops is proposed to improve performance for SEAs. The outer link-side position loop is responsible for electrically-adjusting apparent stiffness at the link side which is free from limitation of stiffness and quasi-rigid performance is achieved as if there were no series elasticity in SEAs. Stability is derived to expound constraints for apparent stiffness and passivity criterion is examined in terms of positive real lemma. High gain in torque loop produces fast actuation for reproducing desired torque needed by the outer link-side position through spring and shapes simultaneously the actuator inertia, which can add robustness to environment uncertainty in force control. Experiments are carried out to compare performance under link- and motor-side position feedback control strategies. Results show that the suggested cascade control is superior to the conventional motor-side position control, at least, in terms of rise time, overshoot and adaptibility to load variation.

References

1. Pratt, G.A., Williamson, M.M.: Series Elastic Actuators. In: International Conference on Intelligent Robots and Systems, pp. 399–406. IEEE Computer Society, Pittsburg (1995)
2. RETHINKROBOTICS Homepage: http://www.rethinkrobotics.com/. Accessed 01 June 2017
3. Wolf, S., Eiberger, O., Hirzinger, G.: The DLR FSJ: energy based design of a variable stiffness joint. In: International Conference on Robotics and Automation, Shanghai, China, pp. 5082–5089 (2011)
4. Kashiri, N.: Dynamic modeling and adaptable control of the CompAct^{T}M arm. In: International Conference on Mechatronics, Vicenza, Italy, pp. 477–482 (2013)
5. Tsagarakis, N.G.: iCub - the design and realization of an open humanoid platform for cognitive and neuroscience research. Adv. Robot. **21**(10), 1151–1175 (2007)
6. Brooks, R.A., Breazeal, C., Marjanović, M., Scassellati, B., Williamson, M.M.: The cog project: building a humanoid robot. In: Nehaniv, C.L. (ed.) CMAA 1998. LNCS, vol. 1562, pp. 52–87. Springer, Heidelberg (1999). doi:10.1007/3-540-48834-0_5
7. Huang, Y.C.: Integrated rotary compliant joint and its impedance-based controller for single-joint pressing massage robot. In: International Conference on Robotics and Biomimetics, Guangzhou, China, pp. 1962–1967 (2012)
8. Huang, Y.C.: Design and control of anthropomorphic BIT soft arms for TCM remedial massage. In: International Conference on Intelligent Robots and Systems, Tokyo, Japan, pp. 1960–1965 (2013)
9. Huang, Y.C.: Anthropomorphic robotic arm with integrated elastic joints for TCM remedial massage. Robotica **33**(2), 348–365 (2015)
10. Ham, R.V.: Compliant actuator designs. IEEE Robot. Autom. Mag. **16**(3), 81–94 (2009)
11. Calanca, A., Muradore, R., Fiorini, P.: A review of algorithm for compliant control of stiff and fixed-compliance robots. IEEE/ASME Trans. Mechatron. **21**(2), 613–624 (2016)
12. Williamson, M.: Series elastic actuators. Ph.D. dissertation, Massachussetts Institute of Technology, Cambridge, MA, USA (1995)
13. Pratt, J., Krupp, B., Morse, C.: Series elastic actuators for high fidelity force control. Ind. Robot **29**(3), 234–241 (2002)
14. Calanca, A., Fiorini, P.: Human-adaptive control of series elastic actuators. Robotica **2**(8), 1301–1316 (2014)
15. Albu-Schäffer, A., Ott, C., Hirzinger, G.: A unified passivity-based control framework for position, torque and impedance control of flexible joint. Int. J. Robot. Res. **26**(1), 23–39 (2007)
16. Colgate, J.E., Hogan, N.: Robust control of dynamically interacting systems. Int. J. Control **48**(1), 65–88 (1988)
17. Calanca, A., Fiorini, P.: On the role of compliance in force control. In: Menegatti, E., Michael, N., Berns, K., Yamaguchi, H. (eds.) Intelligent Autonomous Systems 13. AISC, vol. 302, pp. 1243–1255. Springer, Cham (2016). doi:10.1007/978-3-319-08338-4_90

One of the Gait Planning Algorithm
for Humanoid Robot Based on CPG Model

Liqing Wang[1,2]([✉]), Xun Li[1,2], and Yanduo Zhang[1,2]

[1] School of Computer Science and Engineering, Wuhan Institute of Technology,
Wuhan, China
1373642524@qq.com
[2] Hubei Provincial Key Laboratory of Intelligent Robot, Wuhan, China

Abstract. Stable walking is the most basic human behavior of humanoid robots and one of the most important research contents in the field of robots. However, reasonable gait planning is the basis for the stable walking of humanoid robots. Therefore, in this paper, we analyzes one of the CPG model and applies it to our own laboratory robot, aim at the problem that it is prone to shock forward and backward in the Robot Athletics Sprint, this paper increased centroid offset control and proposed an algorithm to optimize the walking control parameters for the improvement of the whole gait planning algorithm. Stability margin of ZMP and balanced oscillator amplitude were combined as an optimization target, genetic algorithm was used as a solution tool, the purpose is to get the optimal gait parameters under different input speed. The proposed algorithm was tested on the laboratory robot, the results of simulation and real robot experiments show the effectiveness of our algorithm.

Keywords: Gait planning · CPG · ZMP · Centroid offset · Optimization

1 Introduction

Central Pattern Generator (CPG) is a local oscillation network composed of neurons, it can produce a stable phase lock by mutual inhibition between neurons and generate rhythmic motion of the body-related parts by self-excited oscillation [1–3]. Inspired by it, some researchers has proposed bionics-based gait planning algorithm [4–6]. However, there are a lot of parameters in the CPG model without explicit physical meaning, it is difficult to determine the value, resulting in the difficulty of applying the CPG model directly to the robot gait planning. To this end, Endo simplified the CPG model [7], they used the oscillator, which have fewer parameters, to plan the humanoid robot gait. Ha proposed a Linear Coupled Oscillator model and applied it to small humanoid robots for affordable computational cost [8]. The proposed of linear coupled oscillator greatly simplifies the traditional CPG model, however, the choice of parameters is still an unavoidable problem when use it to plan the robot gait. Ha makes use of the ZMP criterion and the centroid trajectory to qualitatively analyze the influence of some parameters on the stability under the condition that other parameters are invariable [8]. Wang Liu Qing, from University of Science and Technology of China, combined the walking speed, centroid amplitude and the winding rate of the ZMP curve as an

© Springer International Publishing AG 2017
Y. Huang et al. (Eds.): ICIRA 2017, Part II, LNAI 10463, pp. 835–845, 2017.
DOI: 10.1007/978-3-319-65292-4_72

optimization target, took the ZMP stability margin as the constraint condition, used the constrained nonlinear solver as a solving tool, to optimize some parameters of the oscillator [9].

The research of this paper is developed from the sprint events of FIRA Roboworld Cup. The competition was first presented by Professor Jong-Hwan Kim of Korea Advanced Institute of Science and Technology in 1995, since then, FIRA Roboworld Cup was held in the world once a year, throughout China, South Korea, the United States, Germany and Britain. The sprint events requires the robot to move from the starting point to the target point and then back to the starting point as quickly as possible, the whole process is fast and the speed is variable. The robots of large group, their centroid is high, will be prone to shock forward and backward as the speed is fast or variable. When the shock amplitude exceeds a certain value, the divergent oscillation will be caused, then the robot will fall forward or backward. Therefore, it is very important to ensure that the robot can walk at a wide range of stability margin and small vibration amplitude. To this end, this paper analyzed the principle of linear coupling oscillator model for gait planning of Humanoid robot and increased centroid offset control to reduce the dependence of stability on oscillator amplitude when the walking speed is fast. Then, the stability of the most unstable points in a period and the centroid amplitude are combined as the optimization target. Finally, The genetic algorithm was used as a solution tool to optimize the parameters, which affect the stability of x-axis. Our algorithm reduces the shock amplitude of centroid and increases the stability margin of ZMP, which gives the robot a lower probability of falling forward or backward. The results of simulation and real robot experiments show the effectiveness of our algorithm.

2   Robot Gait Design

This chapter use one of the CPG model, linear coupling oscillator, for gait planning and analyzing the effect of the parameters on gait stability. The coupled oscillator model consists of two kinds of oscillators, one is the movement oscillator, represents the trajectory of the feet relative to the fixed centroid, and the other one is the balanced oscillator, represents the trajectory of the centroid relative to the fixed reference system. The two oscillators are used to describe the position transformation of the centroid and feet as shown in Fig. 1.

The mathematical expression of the coupled oscillator model was given by Literature [8].

$$osc_s(t) = osc_b(t) + osc_m(t) \tag{1}$$

In formula (1), $osc_s(t)$ represents the total motion trajectory of the robot's foot relative to the center of mass, $osc_b(t)$ represents the output of the balanced oscillator and $osc_m(t)$ represents the output of the motion oscillator. They are defined as follows.

$$osc_b(t) = \rho_b \, sin(\omega_b t + \Delta_b) + u_b \tag{2}$$

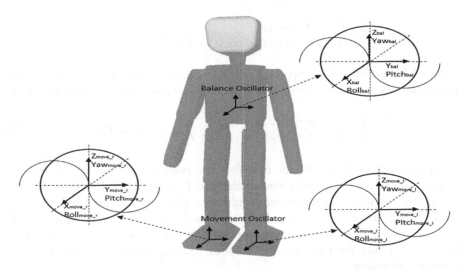

Fig. 1. Movement oscillator and balanced oscillator

$$
osc_m(t) = \begin{cases}
\rho_m & \left[0, \frac{rT}{4}\right) \\
\rho_m \sin(\omega_m t + \Delta_m) & \left[\frac{rT}{4}, \frac{T}{2} - \frac{rT}{4}\right) \\
-\rho_m & \left[\frac{T}{2} - \frac{rT}{4}, \frac{T}{2} + \frac{rT}{4}\right) \\
\rho_m \sin\left(\omega_m\left(t - \frac{rT}{2}\right) + \Delta_m\right) & \left[\frac{T}{2} + \frac{rT}{4}, T - \frac{rT}{4}\right) \\
\rho_m & \left[T - \frac{rT}{4}, T\right)
\end{cases}
\tag{3}
$$

In the above formula, the balanced and movement oscillator parameters are defined, $[\rho_b, \rho_m]$ are the amplitudes, $[\omega_b, \omega_m]$ are the frequencies, $[\Delta_b, \Delta_m]$ are the initial phase, u_b is the offset of the balanced oscillator. The movement oscillator can be expressed by five stages, T is the time of one walking cycle, r represents the proportion of the foot support time in the total cycle.

The relative positions of the centroid and the end of the legs can be obtained at each sampling time t using formula (2) to (3). In the actual gait planning, as to reduce the computational complexity, the center of the body is used as the center of mass. At time t, the relationship between the centroid position c (t), the right foot end position r (t) and the left foot end position l (t) is as follows.

$$
r(t) = c(t) + osc_s(t)
\tag{4}
$$

$$
l(t) = c(t) + osc_s(t + T/2)
\tag{5}
$$

In a walk cycle, both legs support half the time. In [0, T/2] the right leg as the support point, in [T/2, T] left leg as the support point. When the right leg as the support leg and the left foot as the swing leg, the right foot end coordinates unchanged, we set it to r0, then based on the above formula we can get

$$\begin{cases} c(t) = r0 - osc_s(t) \\ l(t) = r0 - osc_s(t) + osc_m(t + T/2) \end{cases} t \in [0, T/2]. \tag{6}$$

When the left leg as the support leg and the right foot as the swing leg, the left foot end coordinates unchanged, we set it to l0, then based on the above formula we can get

$$\begin{cases} c(t) = l0 - osc_s(t) \\ r(t) = l0 - osc_b(t) + osc_s(t + T/2) \end{cases} t \in [T/2, T]. \tag{7}$$

The centroid position c (t), the right foot end position r (t) and the left foot end position l (t) of each moment t can be obtained from the formulas (4) to (7) after determining the initial support point coordinates. Then, we can get the value of 12 motors of the robot legs by inverse kinematics. Finally, the motors rotate to the specified position each sampling time to realize the robot's walking. For the inverse kinematics solution, this paper uses the analytical method, because of the space, there is no more explanation.

3   Gait Trajectory Optimization

From the above gait planning based on coupled oscillator model, we can calculate the trajectory of the end of legs and centroid, simulate the human motion system to produce self-excited oscillations. Theoretically, the stability of the walking can be achieved by adjusting the parameters, but it is difficult to find the appropriate parameters, which can plan an ideal gait curve, by manual adjustment. In the experience of robot sprint race we found that when the robot accelerates or slows down, its centroid prone shock forward and backward. Therefore, this paper optimized the oscillator model for this phenomenon as below.

3.1   Stability Criterion

In this paper, we use ZMP stability criterion to judge whether the robot gait curve is stable. The calculation of ZMP can be divided into static walking and dynamic walking two cases. Static walking refers that the robot is at rest or low speed movement, under this case the robot inertia characteristics can be ignored, then the ZMP position is the projection point of the gravity on the support surface. When the robot is in the high-speed walking state, the inertia characteristic can't be ignored, the ZMP position is the projection point of the force, which is composed of gravity and inertia force, on the support surface. Robot sprint needs high speed, it belongs to dynamic walking, so its position of ZMP in the X axis can be obtained from formula (8).

$$X_{zmp} = \frac{\sum_{i=1}^{n} m_i \{ (\ddot{z}_i + g) x_i - \ddot{x}_i z_i \}}{\sum_{i=1}^{n} m_i (\ddot{z}_i + g)} \tag{8}$$

In the above formula, n is the number of link of the robot, m_i is the quality of link-i, g is the local gravitational acceleration, $[x_i, z_i]$ are the positions of the x-axis and z-axis of the link-I, $[\ddot{x}_i, \ddot{z}_i]$ are the corresponding acceleration, X_{zmp} is the ZMP point position of X-axis.

The speed of the robot can be expressed by the distance of centroid movement per unit time. According to the oscillator model, we can see that the robot's centroid moves $4\rho_m^x$ in a walking cycle, that is, the walking speed of the robot is $4\rho_m^x/T$. The speed change with the change of ρ_m^x as T is constant. Then from the formulas (4) to (7), we can see that the change of ρ_m^x affects The centroid position c (t), the right foot end position r (t) and the left foot end position l (t), which is acts on x_i, z_i, $\ddot{x}_i$ and $\ddot{z}_i$. Therefore, when the speed changes, the ZMP trajectory will change accordingly, the gait curve is Unknown stability. In order to meet the requirements of high-speed in sprint, the speed can't be used as a stable control parameters. We must find other parameters changing with the speed to adjust the ZMP trajectory. So that, at any speed the robot will has a greater ZMP stability margin.

3.2 Centroid Offset Control

From the formula (2)–(3), it can be known that the parameters affecting the trajectory of X-axis are $\rho_m^x, \omega_m^x, \Delta_m^x, \rho_b^x, \omega_b^x, \Delta_b^x$, T, r, u_b, ρ_m^x. T is given by the speed task, then $[\omega_m^x, \omega_b^x]$ will be obtained from T. $[\Delta_m^x, \Delta_b^x]$ and u_b can be defined by the robot's feet end position in the initial moment. The rest only is ρ_b^x, it is the amplitude of balance oscillator, which affects the balance of X axis to some extent. The results show that the amplitude of the centroid should be proportional to the walking speed for stability. However, when the amplitude increases, the shock amplitude of the robot will also increase, which becomes the unstable factor. So ρ_b^x may not be the best control output.

According to the law of human walking, it is found that the centroid is always pushed forward when the speed is accelerated and backward when the speed is decelerated, to overcome the change of inertia force caused by the change of speed. Inspired by this, this paper adds a centroid offset to the coupled oscillator model. The modified oscillator model is show as (9). $offect_x$ is the deviation of centroid at the initial position in X axis. When the coordinate of the support leg is X0, the coordinate $x_c(t)$ of centroid is as in (10).

$$osc_{total}^x = osc_{bal}^x + osc_{move}^x + offect_x \tag{9}$$

$$x_c(t) = x_0 - osc_{bal}^x(t) - osc_{move}^x(t) - offect_x \tag{10}$$

From (10), it can be known that the centroid position changes with the change of $offect_x$. By adjusting the value of $offect_x$, we can adjust the relative position of the robot's centroid in X-axis, and then control the walking balance.

3.3 Optimization Design of Genetic Algorithm

Genetic algorithm has only a few mathematical requirements for solving the optimization problem, so it is suitable for solving the problem of complex nonlinear optimization. What's more, because of its random variation and selective evolution, it is not easy to fall into the local optimal solution. The humanoid robot has many degrees of freedom and the coupling oscillator mathematical model is complicated, so this paper use the genetic algorithm to optimize parameters.

We set the population size M = 100, the crossover probability Pc = 0.5, the evolution probability Pe = 0.02, the evolutionary algebraic limit T = 1500, and use the roulette gambling method to select the excellent population. Genetic algorithm process shown in Fig. 2

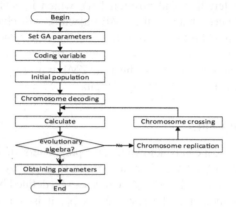

Fig. 2. Flow chart of genetic algorithm

Variables to be optimized: ρ_{bal}^{x} and offect$_x$

Optimization goal determination:

In the case of taking the projection point of the ankle of support leg on the ground as the origin. Coordinates of the center point of sole of the support leg in X axis

$$X_{fcenter} = \frac{X_{tip} - X_{heel}}{2}. \qquad (11)$$

X_{tip} is the distance between the tiptoe of the support leg and the origin, X_{heel} is the distance between the heel of the support leg and the origin. According to the ZMP stability constraint, the closer the X_{zmp} is to the $X_{fcenter}$ in a walking cycle the more stable the walking is. Therefore, in this paper, the stability of the most unstable point in a walking cycle is used as the optimization goal. The optimal expression is as in (12).

$$Obj : minimize \; f\left(\rho_{bal}^{x}, offect_{x}\right) = max\left|X_{zmp}(t) - X_{fcenter}\right| + \rho_{bal}^{x}, t \in [0, T] \qquad (12)$$

$max\left|X_{zmp}(t) - X_{fcenter}\right|$ is the longest distance between X_{zmp} and $X_{fcenter}$ in a walking cycle. In order to reduce the shoke amplitude of the robot, ρ_{bal}^{x} is also included in the optimization object.

4   Experimental Analysis

In this paper, the laboratory robot manufactured by imitating NimbRo-OP model are used as the experimental platform, which has 20 degrees of freedom, including the neck of the two, the two arms of each of the three, and the legs of each of the 6 degrees of freedom. The detailed parameters of each link are given in Table 1 and Fig. 3.

Table 1. Laboratory robot link weight

Link name	Mass(g)	Link name	Mass(g)	Link name	Mass(g)
Head	426.6	Left arm lower	160.1	Lower left leg	189.8
Neck	65.6	Upper right arm	455.6	Upper right leg	323.1
Body	2634.1	Lower right arm	160.1	Lower right le	1189.8
Shoulder	69.93	Hip joint	323.1	Ankle	323.1
Left arm upper	454.7	Upper left leg	323.2	Foot	214.38

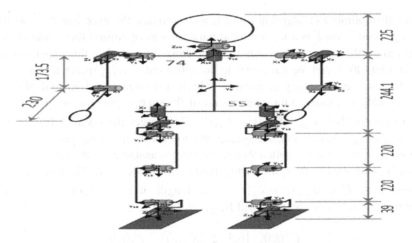

Fig. 3. Laboratory robot link dimensions

4.1   Simulation Experiment

The experiment is running on a Linux operating system with a version of 15.04. Data simulation experiment and optimization software are Matlab which version is 2013a. The visual simulation software platform is Robot Operating System. In this experiment, firstly, we use Matlab's genetic algorithm toolbox to realize the optimization algorithm proposed in this paper, then reproduce the algorithm [9] in the experimental environment, next we compared the optimization results of the two algorithms.

In the experiment, the oscillator model period T is 800 ms, the foot support period is the walking cycle ratio r = 0.2, the values of ρ_{move}^y and ρ_{move}^z are 0, by changing the value of ρ_{move}^x to change the robot walking speed, The critical parameter ρ_{bal}^x is in the range of [0, 5] in centimeters by the critical experiment. In this experiment, we optimize the parameters with Matlab—Genetic Algorithm for Function optimization designed by Christopher R. Houch, the optimization results is as in Fig. 4.

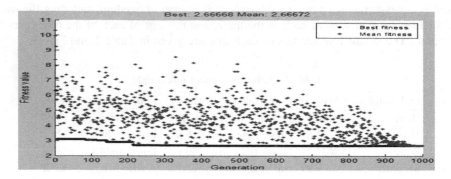

Fig. 4. Average fitness and optimal fitness evolution process

For the optimization algorithm in [9], we reproduce the experiment ② which the walking speed is used as a known input parameter to optimize other parameters. The algorithm in [9] has two parameters to be optimized (ρ_{bal}^x, ρ_{bal}^y), this paper introduced the centroid offset, and the parameter to be optimized is (ρ_{bal}^x, offect$_x$). We use instability to describe the quality of stability, which is defined as the integral of distance, from the ZMP trajectory to the center point of the support polygon, to time in one leg support period, that is $\int_0^T \|X_{zmp}(t) - C_x(t)\|$. The larger the distance from the ZMP point to the support polygon, the greater the impact on the robot gait, and when the support polygon is exceeded the robot gait becomes unstable. So the view of instability should be multiplied by the reward and punishment coefficient w. The instability can be ex-pressed as $\int_0^T w.\|X_{zmp}(t) - C_x(t)\|$. The length of robot' foot is 21 cm in the experimental and the value of w will be given by (13).

$$
w = \begin{cases} 1000, & 10.5 \ll \|X_{zmp}(t) - C_x(t)\| \\ 1, & 5.25 \ll \|X_{zmp}(t) - C_x(t)\| \ll 10.5 \\ 0.1, & \|X_{zmp}(t) - C_x(t)\| \ll 5.25 \end{cases} \tag{13}
$$

The experimental results of two different methods are shown in Table 2.

Table 2. The experimental results of two different methods

Experimental source	Optimized parameters	The value of instability	Walking speed (m/s)
This paper	(0.23, 12.43)	6.481	0.1
Literature [9]	(2.56, 23.42)	7.253	0.1
This paper	(0.42, 29.25)	8.264	0.3
Literature [9]	(5.32, 24.63)	12.235	0.3
This paper	(0.56, 36.43)	8.563	0.5
Literature [9]	(7.24, 25.46)	18.235	0.5

As shown in Table 2, the amplitude of balanced oscillator obtained by this paper is much smaller than that in literature [9], as the case of lower instability. The lower the instability, the greater the stability margin of the robot walking, and the smaller the amplitude of the balanced oscillator, the less the vibration of the robot's centroid. This advantage is even more pronounced when the speed increased, which proves the rationality greatly improving the walking stability of X-axial by the algorithm of this paper while the robot walking fast.

4.2   Real Robot Experiment

This experiment apply the above gait parameters into the laboratory real robot, letting the robot walking on the same road for some time, at the same time, the value of the twenty electric machineries on the robot were read and sent to the visual simulation program achieved by Rviz. So that, the simulation robot will synchronize with the real robot. In order to facilitate the analysis of the stability, the ZMP value was constantly calculated by the motor values read above and marked by the arrow on the simulation robot. The laboratory real robot experiment environment as in Fig. 5.

Because this gait optimization is used for robot sprint that the walking speed is fast, this experiment only compared the two sets of parameters with a speed of 0.5 m/s. Figure 6 shows the gait conditions for two different parameters in a walking cycle. In the figure, the intersection of the red arrow pointing to the support surface and the support surface is the real-time ZMP point. The upper row of pictures is controlled by the parameters obtained from literature [9], it can be seen that the ZMP point oscillated between the upper and lower bounds of support domain, what makes the gait unstable, a small disturbance will make the ZMP point out of the support domain. The lower row of pictures is controlled by this paper, the ZMP point is always near the center of the support domain on X axis. Its stability margin is great that in the case of small disturbance the robot can repair independent. From the picture, we can be very intuitive

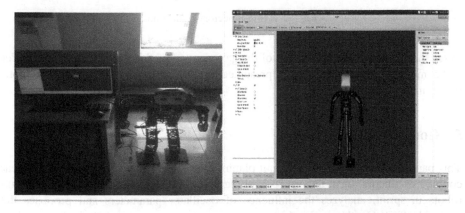

Fig. 5. Laboratory real robot experiment environment

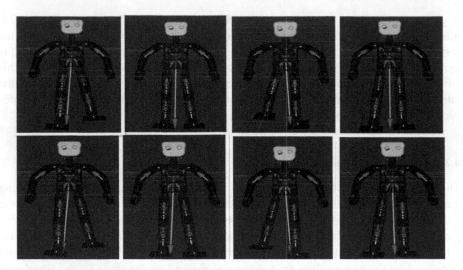

Fig. 6. Stability comparison of real robot experiment data in simulation platform (Color figure online)

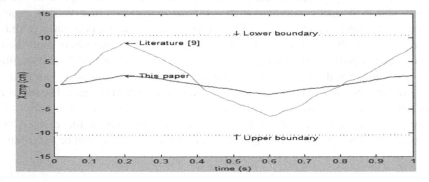

Fig. 7. Xzmp trajectory with time

to see that the ZMP stability margin of the gait planned by this paper is greater. In order to be more persuasive, this paper presented the Xzmp trajectories of the two optimization algorithms as in Fig. 7.

5 Conclusion

This paper first analyzes the robot gait planning model based on the principle of linear coupled oscillators, then, against rocking back and forth phenomenon of large group robots in sprint, analyzes the influence of oscillator parameters and proposes the centroid offset. On this basis, this paper proposes a parameter optimization method based on genetic algorithm. The rationality of this paper was verified through the results of the simulation experiment and the laboratory real robot experiment environment. What's more, this algorithm has been applied to the sprint for large group robot in 21th FIRA, second grade proved the validity of the algorithm. But only in the case of small or no disturbance, can this algorithm greatly improve the walking stability, the robot cannot repair independent when the disturbance large enough. Therefore, on the basis of the work by this paper, it is necessary to study that greater combining the feedback information of sensors with the CPG model, for better anti-interference ability.

References

1. Yamaguchi, T.: The central pattern generator for forelimb locomotion in the cat. J. Prog. Brain Res. **143**, 115–122 (2004)
2. Dickinson, P.S.: Neuromodulation of central pattern generators in invertebrates and vertebrates. J. Curr. Opin. Neurobiol. **16**(6), 604 (2006)
3. Wu, Q., Liu, C., Zhang, J., et al.: Survey of locomotion control of legged robots inspired by biological concept. J. Sci. China Inf. Sci. **52**(10), 1715–1729 (2009)
4. Son, Y., Kamano, T., Yasuno, T., et al.: Generation of adaptive gait patterns for quadruped robot with CPG network including motor dynamic model. J. Electr. Eng. Jpn. **155**(1), 35–43 (2006)
5. Kuwata, N.: Analysis of coupled Van Der Pol oscillators and implementation to a myriapod robot. J. IFAC Proc. **41**(2), 767–772 (2008)
6. Lu, Z., Ma, S., Li, B., et al.: Gaits-transferable CPG controller for a snake-like robot. J. Sci. China Inf. Sci. **51**(3), 293–305 (2008)
7. Endo, G., Nakanishi, J., Morimoto, J., et al.: Experimental studies of a neural oscillator for biped locomotion with QRIO. In: Proceedings of the 2005 IEEE International Conference on Robotics and Automation, pp. 596–602. IEEE, Barcelona (2005)
8. Ha, I., Tamura, Y., Asama, H.: Gait pattern generation and stabilization for humanoid robot based on coupled oscillators. In: 2011 IEEE/RSJ International Conference on Intelligent Robots and Systems, pp. 3207–3212. IEEE, San Francisco (2011)
9. Wang, L., Shang, W., Automation, D.: A gait pattern planning algorithm based on linear coupled oscillator model for humanoid robots. J. Univ. Sci. Technol. China **44**(10), 795–803 (2014). (in China)

Design of an Active Compliance Controller for a Bionic Hydraulic Quadruped Robot

Xiaoxing Zhang, Xiaoqiang Jiang, Xin Luo$^{(\boxtimes)}$, and Xuedong Chen

Huazhong University of Science and Technology, Wuhan 430074, China
mexinluo@hust.edu.cn

Abstract. Active compliance control is an effective method for a legged robot to decrease impact disturbances at the moment of foot-ground contact and to realize harmonic locomotion by actively adjusting the leg stiffness and damping in real-time, which will significantly improve the adaptability of the robot to the irregular terrain. However, the design of an active compliance controller remains to be a challenge when the nonlinearity of actuators must be taken into account. This paper presents an active compliance controller for a bionic hydraulic quadruped robot. The nonlinear dynamics of the hydraulic system, including the compressibility of the fluid, the friction of the hydraulic cylinders, and the flexibility of the tubing are well modeled and identified. The nonlinearity of the hydraulic actuators are compensated through friction compensation and feedback linearization. The proposed active compliance controller has been applied to a bionic hydraulic robot prototype. Experimental results indicate that the active controller can handle the impact disturbances from robot feet.

Keywords: Bionic hydraulic quadruped robot · Active compliance control · Hybrid force-position control

1 Introduction

Legged robots have promising advantages in mobility compared with wheeled and tracked robots in unstructured environments since they only need a discrete number of isolated footholds [1]. Unfortunately, discrete support legged locomotion will result in intermittent foot-ground contacts which may cause impact disturbances to the robot locomotion and even damaging the robot mechanical structure. It gets even worse when the contacts are rigid. Precise kinematic planning and execution of contacts may reduce the impact disturbances, but not a feasible solution for legged robots in unstructured environments since neither a perfect map of the environment nor a perfect estimation of the robot state will be obtained.

In the recent years, researchers proposed several possible ways to cope with impact disturbances in legged robots arising from intermittent foot-ground contacts. A simple and effective approach is to utilize passive compliant elements in the robot leg [2, 3]. A spring was mounted on the distal segment of robot legs and a rubber pad at the foot bottom to absorb impact peaks and temporarily store energy during legged locomotion. Hutter and Gehring developed a compliant quadruped robot named StarlETH driven by highly compliant series elastic actuator (SEA) [4, 5]. The spring in series with the

© Springer International Publishing AG 2017
Y. Huang et al. (Eds.): ICIRA 2017, Part II, LNAI 10463, pp. 846–855, 2017.
DOI: 10.1007/978-3-319-65292-4_73

motor has important functions: (a) reducing transmission stiffness and peak impact forces, (b) to turn the force control problem into a position control problem, greatly improving force accuracy, (c) to store energy, and (d) to protect the actuator or gearbox from damage due to impact forces [6]. These springs, however, should be appropriately chosen for a certain task to guarantee task performance. Semini made a breakthrough in active impedance control of legged robot and demonstrated their control strategy on their hydraulically actuated quadruped robot [7, 8]. However, the robustness of the low-level hydraulic force control and the accuracy of the friction model should be improved as they themselves mentioned. Although a lot of work have been done by former researchers, there remain many issues to be further studied about the compliance control of hydraulic robots, especially when the nonlinear dynamics of the hydraulic system, friction of the hydraulic cylinders, and impact forces on robot feet are considered.

This paper presents an active compliance controller and demonstrates it on a bionic hydraulic quadruped robot. The compliance controller considers the nonlinear dynamics of the hydraulic system, including the compressibility of the fluid, the friction of the hydraulic cylinders, and the flexibility of the tubing, so that high performance of compliance control can be accomplished. The compliance ability can help the robot to handle the impact disturbances from robot feet.

2 Modeling of the Hydraulic Actuators

2.1 Overview of the Bionic Quadruped Robot

A bionic hydraulic quadruped robot is built to study the motion planning and control issues. The bionic quadruped robot is hydraulically actuated and powered by an off-board hydraulic pump. The sensing system includes a 12-bit digital potentiometer at each joint, 2 pressure sensors for each cylinder which are located in chambers at piston and ring side, respectively, a 3-axis force sensor for each foot, and an IMU for the robot torso motion estimation. Table 1 lists the physical specifications of the robot. The prototype is controlled by an NI based controller from National Instrument, USA, as shown in Fig. 1.

Table 1. Physical specifications of the quadruped robot

Object	Parameter	Unit
Length/width/height	1000 × 660 × 800	mm
Total leg length	550	mm
Leg segment length	360/300	mm
Range of joint rotation	80(HAA)/100(HFE)/100(KFE)	deg
Joints of actuation	3 per leg	–
Segments of per leg	2	–
Total mass	75	kg

(a) NI-based controller   (b) The bionic quadruped robot prototype

Fig. 1. Overview of the bionic hydraulic robot

The main advantage of hydraulic actuation is the good power-to-weight ratio [9], which is extremely important for legged robots. A basic hydraulic actuation system consists of a tank, pump, filter, valves, cylinders, accumulator and tubing as is shown in Fig. 2. The pump generates high-pressure hydraulic flow, the pressure relief valve keeps the maximum pressure, the servo-valve controls the flow direction and magnitude and the cylinder generates force to move a load. To obtain stable pressure of the

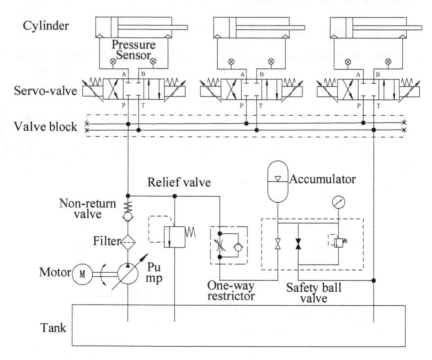

Fig. 2. Configuration of the hydraulic actuation system of the bionic hydraulic quadruped robot

hydraulic actuation systems, an accumulator and a safety ball valve along with a one-way restrictor are used between the inlet and outlet oil line.

2.2   Modeling of the Hydraulic Actuators

Figure 3 presents an overview of the valve-cylinder combination studied in this paper. The main nonlinearities in hydraulic systems arise from the compressibility of the hydraulic fluid, the complex flow properties of the servo-valve and the friction in the hydraulic actuators [10]. It is essential to model the nonlinear dynamics of the hydraulic system to obtain high control performance.

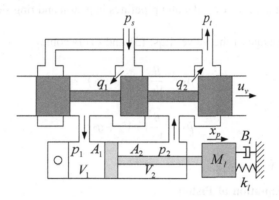

Fig. 3. Valve-cylinder combination

The nonlinear dynamics of the hydraulic system are mainly governed by the pressure-flow equations of valves, the continuity equation of fluid and the force balance equation.

Pressure-Flow Equations of Valves
Neglecting leakage in the valve, the flow into sides 1 and 2 of the cylinder is

$$q_1 = \begin{cases} c_1 u\sqrt{p_s - p_1}, u \geq 0 \\ c_2 u\sqrt{p_1 - p_r}, u < 0 \end{cases} \tag{1}$$

$$q_2 = \begin{cases} -c_3 u\sqrt{p_2 - p_r}, u \geq 0 \\ -c_4 u\sqrt{p_s - p_2}, u < 0 \end{cases} \tag{2}$$

where c_1, c_2, c_3, c_4 are the valve orifice coefficients, u represents the control voltage of servo value, p_1 and p_2 denote the hydraulic pressure on the two cylinder sides, p_s is the supply pressure, and p_r is the reservoir pressure. The pressures above are measured on real time by pressure sensors, and the flow is calculated by piston velocity. The valve orifice coefficients can be obtained using off-line testing.

Continuity Equation of Fluid

Neglecting pipe compressibility and pressure loss in pipelines, the continuity equation of fluid can be expressed as

$$q_1 - c_i(p_1 - p_2) - c_e p_1 = A_1 \frac{dx}{dt} + \frac{V_1}{\beta} \frac{dp_1}{dt} \tag{3}$$

$$q_2 - c_i(p_1 - p_2) + c_e p_2 = A_2 \frac{dx}{dt} - \frac{V_2}{\beta} \frac{dp_2}{dt} \tag{4}$$

where $\beta = -\frac{Vdp}{dV}$ is the fluid bulk modulus, c_i and c_e are the leakage flow coefficients (internal, external), A_1 and A_2 are the piston area in piston and ring side, V_1 and V_2 are the chamber volume of the cylinder and pipelines in piston and ring side, x is the piston displacement.

Neglecting leakage in the valve, Eqs. (3) and (4) become

$$\dot{p}_1 = \frac{\beta}{V_1}(-\dot{V}_1 + q_1) \tag{5}$$

$$\dot{p}_2 = \frac{\beta}{V_2}(-\dot{V}_2 + q_2) \tag{6}$$

where $\dot{V}_1 = A_1 \dot{x}$, $\dot{V}_2 = -A_2 \dot{x}$.

Force Balance Equation of Piston

The net fluid force on the piston is defined as the difference of forces applied on the two sides of piston, i.e.

$$F = p_1 A_1 - p_2 A_2 \tag{7}$$

which is determined by the loads of the cylinder. Differentiating Eq. (7) yields

$$\dot{F} = \dot{p}_1 A_1 - \dot{p}_2 A_2 \tag{8}$$

Substituting Eqs. (5) and (6) into (8) we have

$$\dot{F} = -\dot{x}\beta(\frac{A_2^2}{V_2} + \frac{A_1^2}{V_1}) + z(x, p_1, p_2)u \tag{9}$$

where

$$z = \begin{cases} \beta(\frac{A_2 c_3}{V_2} \sqrt{p_2 - p_r} + \frac{A_1 c_1}{V_1} \sqrt{p_s - p_1}) & u \geq 0 \\ \beta(\frac{A_2 c_4}{V_2} \sqrt{p_s - p_2} + \frac{A_1 c_2}{V_1} \sqrt{p_1 - p_r}) & u < 0 \end{cases} \tag{10}$$

Equation (9) expresses the intrinsic relationship between the fluid force and the control voltage, which can be used to design the control law of fluid force.

2.3   Friction Identification

The piston sealing of a hydraulic cylinder is usually quite tight to avoid internal leakages, which will cause significant friction force between the piston and cylinder body. Since the friction force has an influence on the control performance of the joint motion, especially in the case of hybrid force position based control method, it is necessary to find an accurate mathematical model of friction force. In order to observe the friction force in the hydraulic cylinder of our robot, we used a PID position control method to exert a sinusoidal motion of different frequency on the piston, and measured the pressure in piston and ring side of the cylinder to calculate the friction force. The red lines in Fig. 4 show the measured friction force versus time and cylinder velocity under sinusoidal motion of different frequency.

The friction force of the cylinders shows obvious hysteresis in the Fig. 4, which may due to the slip characteristics between the piston and cylinder body, the inertia load (e.g. the fluid between the serve-value and cylinder) and the structural flexibility of the hydraulic system [10]. Considering the hysteresis of the friction force, a composite arctangent function is used to estimate the friction force. The structure of the arctangent function can be expressed as

$$F_f = p_1\left(\frac{\arctan(-\frac{p_2 - \dot{x}}{p_3})}{\pi} + 0.5\right) + p_4\left(\frac{\arctan(-\frac{p_5 - \dot{x}}{p_6})}{\pi} + 0.5\right) \tag{11}$$

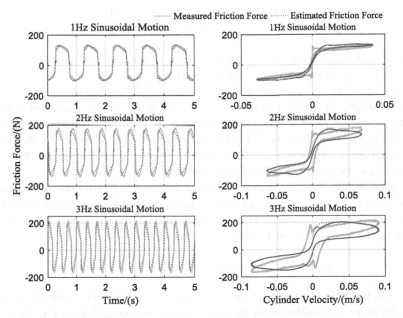

Fig. 4. Friction force versus time and cylinder velocity under sinusoidal motion of different frequency (Color figure online)

where $\dot{x}$ and $\ddot{x}$ denote the velocity and acceleration of the cylinder, and p_i are the constant parameters obtained by data fitting with the measured friction force and cylinder motion. The estimated friction force of the hydraulic cylinder is shown in Fig. 4 with blue lines.

3   Design of the Active Compliance Controller

For the translational actuation, the basic relationship between the force and position can be expressed as [8]

$$f = K_p(x_{ref} - x) + K_d(\dot{x}_{ref} - \dot{x}) + K_m(\ddot{x}_{ref} - \ddot{x}) \tag{12}$$

where f is the force, K_p, K_d and K_m denote the stiffness, damping and inertia parameters, respectively, x_{ref} is the generalized position reference, and x is the actual position.

Active compliance control means that the position and force of the hydraulic cylinder must be cooperatively controlled through adjusting the stiffness, damping and inertia. To realize the compliance control of the hydraulic robot, we use the position controller as the outer control loop to produce the desired fluid force and the force controller as the inner control loop to generate the corresponding control voltage [11].

As mentioned above, the Eq. (9) expresses the intrinsic relationship between the fluid force and the control voltage. A model-based force controller can be designed by solving the inverse of the system in Eq. (9) as

$$u = \frac{1}{z}(\dot{F}_d - k_L(F - F_d) + \dot{x}\beta(\frac{A_2^2}{V_2} + \frac{A_1^2}{V_1})) \tag{13}$$

where F_d is the desired fluid force of the hydraulic cylinder, k_L is a positive force error gain and the nonzero quantity z is defined in (10).

The model-based force controller expressed in (13) involves the nonlinear dynamics of the hydraulic system, including the nonlinear hydraulic pressure-flow dynamics of sever-valve and compressibility of the fluid, and make a compensation through feedback linearization. Using the real-time pressure measured by the pressure sensors and the piston motion calculated by the joint encoder measurement, the desired fluid force can be achieved by the force controller.

Inspired by the Eq. (12), velocity and position feedbacks are used to calculate the desired fluid force. Since the actual acceleration obtained by the second order difference of position is unreliable, a feedforward of acceleration is used instead of acceleration feedback. The desired force can be expressed as

$$f_d = m\ddot{x}_d(t) + k_v(\dot{x}_d(t) - \dot{x}(t)) + k_p(x_d(t) - x(t)) \tag{14}$$

where m is the equivalent mass of the load, $x_d(t)$ is the desired position of the piston, $x(t)$ is the measured position of the piston, k_p is the position feedback gain, and k_v is the velocity feedback gain. The first term of Eq. (14) represents the desired motion of the actuator, the second and third terms refer to a modification of the actuator motion through velocity feedback and position feedback respectively.

The calculated force above is the desired force applied on the moving load to produce the desired motion. Practically, the friction force between piston and cylinder body and the external force applied on the moving load, such as the reaction force of robot gravity on the foot, also make an important impact. Therefore, the actual desired fluid force is

$$F_d = m\ddot{x}_d(t) + k_v(\dot{x}_d(t) - \dot{x}(t)) + k_p(x_d(t) - x(t)) + \hat{F}_f(\dot{x}) + \hat{F}_{ext} \tag{15}$$

Where $\hat{F}_f$ is the estimated friction force of the hydraulic cylinder, and $\hat{F}_{ext}$ is the estimated external force applied on the piston. The friction force is estimated using identification result expressed as Eq. (11). The external force is zero when the foot is in the air and is simplified as a constant of the robot gravity when the foot is on the ground.

The Eqs. (13) and (15) represent the main scheme of the compliance controller for the bionic hydraulic robot. The k_v and k_p can be actively controlled to achieve desired compliance of the robot.

4 Experiments

To evaluate the performance of the proposed compliance controller, the position/force tracking and impact disturbance experiments were conducted on the quadruped robot system.

We hang the robot in the air with a 25 kg load on the foot to mimic the torso mass of the robot, so that we can use a single leg of the robot to carry out the experiment. The squatting gait of 50 mm height is designed using polynomial fitting method to generate the desired motion of the hip joint and knee joint. Figure 5 shows the joint force and position tracking control performance of the quadruped robot. Force tracking errors of the hip and knee joints are within range of ±100 N as shown in Fig. 5a), and position tracking errors are within ±0.5 mm as shown in Fig. 5b), which means the proposed hybrid force-position control method perform well and can achieve accurate force and position control at the same time.

The ideal control accuracy of joint force and position is not our ultimate goal. We expect appropriate compliance ability of the joints to deal with impact disturbances coming from contacts between the robot feet and the environment. To verify the compliance controller expressed as Eq. (12), impact forces were applied to the foot of our quadruped robot hanging in the air. The stiffness of the hydraulic cylinder is set as 500/mm. We measure the impact forces with the 3-axis force sensor mounted at the foot end and calculate the deformation of a virtual spring with the same stiffness to be the desired position response. The measured position response of the hydraulic cylinder is compared with the desired one as is shown in Fig. 6a. It can be seen that, the hydraulic cylinder behaves like a spring under the impact disturbances and has obvious compliance ability. Figure 6b shows the hydraulic cylinder position response of a simple PID controller under similar impact disturbance. In contrast with the proposed controller, obvious overshoot can be observed in the PID controlled position response, which will cause instability to the robot motion.

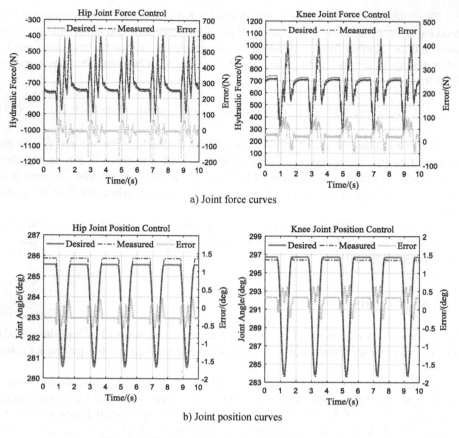

a) Joint force curves

b) Joint position curves

Fig. 5. Joint force and position control performance

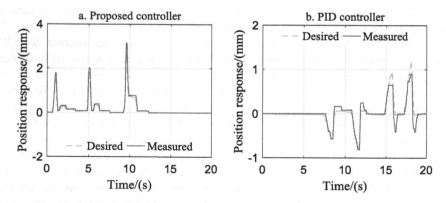

Fig. 6. Position response of the hydraulic cylinder under impact disturbances

5   Conclusions

This study presents an active compliance controller for hydraulic legged robots based on hybrid force-position control. The nonlinear dynamics of hydraulic system are modeled and the friction force of hydraulic cylinder is compensated to achieve accurate force and position tracking at the same time. Experiments verify the effectiveness of the proposed active compliance controller, and the force/position tracking accuracy is remarkable. Benefitting from the active compliance control, the robot can handle the impact disturbances from robot feet like a spring.

In the future work, we will extend this active compliance controller to the whole body motion control of our bionic quadruped robot. Also, we will pay attention to the intelligent algorithms of stiffness changing referring to gaits of the robot and environment parameters to achieve more efficient and stable legged locomotion.

Acknowledgment. This work is partially supported by the National Natural Science Foundation of China (NSFC) under grant numbers 61175097 and 51475177, and the Research Fund for the Doctoral Program of Higher Education of China (RFDP) under grant number 20130142110081.

References

1. Raibert, M.H.: Legged Robots that Balance. The MIT Press, Cambridge (1986)
2. Ding, L., Gao, H., Deng, Z., Song, J., Liu, Y., Liu, G., Iagnemma, K.: Foot-terrain interaction mechanics for legged robots: modeling and experimental validation. Int. J. Robot. Res. **32**(13), 1585–1606 (2013)
3. Wooden, D., Malchano, M., Blankespoor, K., Howardy, A., Rizzi, A.A., Raibert, M.: Autonomous navigation for BigDog. In: 2010 IEEE International Conference on Robotics and Automation (ICRA), pp. 4736–4741. IEEE, Anchorage (2010)
4. Hutter, M., Gehring, C., Hopflinger, M.A., Blosch, M., Siegwart, R.: Toward combining speed, efficiency, versatility, and robustness in an autonomous quadruped. IEEE Trans. Robot. **30**(6), 1427–1440 (2014)
5. Gehring, C., Coros, S., Hutler, M., Bellicoso, C.D., Heijnen, H., Diethelm, R., Bloesch, M., Fankhauser, P., Hwangbo, J., Hoepflinger, M., Siegwart, R.: Practice makes perfect: an optimization-based approach to controlling agile motions for a quadruped robot. IEEE Robot. Autom. Mag. **23**(1), 34–43 (2016)
6. Pratt, G.A., Williamson, M.M.: Series elastic actuators. In: Proceedings of the 1995 IEEE/RSJ International Conference on Intelligent Robots and Systems 1995, Human Robot Interaction and Cooperative Robots, pp. 399–406. IEEE, Pittsburgh (1995)
7. Boaventura, T., Buchli, J., Semini, C., Caldwell, D.G.: Model-based hydraulic impedance control for dynamic robots. IEEE Trans. Robot. **31**(6), 1324–1336 (2015)
8. Semini, C., Barasuol, V., Boaventura, T., Frigerio, M., Focchi, M., Caldwell, D.G., Buchli, J.: Towards versatile legged robots through active impedance control. Int. J. Robot. Res. **34**(7), 1003–1020 (2015)
9. Jelali, M., Kroll, A.: Hydraulic Servo-Systems. Springer, London (2003)
10. Jiang, W., Luo, X., Chen, X.D.: Influence of structural flexibility on the nonlinear stiffness of hydraulic system. Adv. Mech. Eng. **8**(8), 1–7 (2016)
11. Sohl, G.A., Bobrow, J.E.: Experiments and simulations on the nonlinear control of a hydraulic servosystem. IEEE Trans. Control Syst. Technol. **7**(2), 238–247 (1999)

Motion Control Strategy of Redundant Manipulators Based on Dynamic Task-Priority

Weiyao Bi[1], Xin-Jun Liu[1,2(✉)], Fugui Xie[1,2], and Wan Ding[3]

[1] Department of Mechanical Engineering,
The State Key Laboratory of Tribology & Institute of Manufacturing
Engineering, Tsinghua University, Beijing 100084, China
xinjunliu@mail.tsinghua.edu.cn
[2] Beijing Key Lab of Precision/Ultra-Precision Manufacturing Equipments
and Control, Tsinghua University, Beijing 100084, China
[3] Department of Mechanism Theory and Dynamics of Machines (IGM),
RWTH Aachen University, Aachen, Germany

Abstract. Kinematic redundancy has been widely used in manipulator design. Nevertheless, the control of redundancy is a tough task. This paper proposed a control method for serial kinematic redundant manipulators in consideration of multiple performance criteria. The method is based on the classical gradient projection method, and in order to handle the relations between different performance criteria, a combination of task priority strategy and fuzzy inference is used. The method is applied to a 10 DOFs (degrees of freedom) manipulator as an example to illustrate its efficiency.

Keywords: Multiple performance criteria · Redundancy · Fuzzy inference · Task priority method

1 Introduction

Redundancy is a popular concept which is often adopted in modern manipulator design process, especially humanoid robots [1]. It offers manipulators possibility of handling multiple tasks and constraints except for tracking desired trajectory (end-effector task). On the other hand, redundancy brings great difficulty to control, since it increases the complexity of kinematics and the solution of inverse kinematics is non-unique [2].

Kinematic redundancy occurs when a manipulator possesses more degrees of freedom than the required to execute a given task [3]. In this case an infinite number of solutions are yielded from the inverse kinematics and so the redundant manipulator, where infinite solutions may exist, forms a subset of the joint space. As a consequence, an approach is required in order to select one satisfying set of them.

Many selection approaches have been proposed and some have been applied in the control of redundant manipulators, for instance, gradient projection method [4], weighted least-norm solution [5], and task-space augmentation [6]. Among them, gradient projection method has been proved to be the most popular and influential optimal control strategy. The fundamentals of gradient projection method are like follows: on the basis of pseudo-inverse of the Jacobian matrix, a term proportional to

© Springer International Publishing AG 2017
Y. Huang et al. (Eds.): ICIRA 2017, Part II, LNAI 10463, pp. 856–868, 2017.
DOI: 10.1007/978-3-319-65292-4_74

the gradient of the criterion (a cost function H) is projected onto the null space of the Jacobian matrix so that the end-effector task is not affected [4]. With the help of pseudo-inverse, the solution of the least-squares problem can be found. Several modifications have been made to improve its efficiency. Most of the improvement focus on the optimization of Jacobian matrix like generalizing the matrix or adding damping.

Apart from selection of suitable optimal control methods, how to handle multiple optimization objectives in one certain method invites our consideration as well. In actual scenes, manipulator motion planning is often restrained by multiple criteria at the same time, e.g., maximization of joint range availability, obstacle avoidance or minimization of joint torques. Unlike single objective optimization, in multiple optimum all objectives should be taken into account at the same time.

The general approach to multi-objective optimization is to combine the individual objectives into a single composite function and assign them different weights [4]. In many applications. Fuzzy logic control (FLC) is used to assign the weights [7]. The methodology of the FLC appears very useful when the processes are too complex for analysis by conventional quantitative techniques or when the available sources of information are interpreted qualitatively, inexactly, or uncertainly. However, how to build reliably and effective fuzzy rules is a time-consuming task and for programmers and developers there is no existing regulation to consult.

In recent ten years a novel approach has been developed. It is called the task-priority strategy. Just as its name implies, conflicts between the end-effector task and multiple constraint tasks are handled by suitable assigning an order of priority to the given tasks and then satisfying the lower priority task [8]. In the typical case, the end-effector task is considered as the primary task. The task priority strategy becomes more and more popular these years.

In this paper we will focus our attention on the handling of kinematic redundancy under consideration of multiple criteria. The target is to develop a method that can execute a redundant robot action in real time and fulfil the multiple goals. A 10 degree-of-freedom manipulator is considered in numerical case study to demonstrate practical application of the method and its effectiveness.

The rest of this paper is organized as follows: Sect. 2 explains the kinematics of redundant manipulators and the gradient projection method. Section 3 introduces the way to handle the multiple criteria by the task-priority strategy and fussy inference. In Sect. 4 a real case of a 10 degree-of-freedom manipulator is taken as an example. Section 5 concludes the paper.

2 Gradient Projection Method Algorithm

2.1 Kinematics of Redundant Manipulators

We consider a manipulator with n degrees of freedom whose joint variables are denoted by $\mathbf{q} = [q_1, q_2, \ldots, q_n]^T$. Assuming that the class of tasks that we are interested in can be described by m variables, $\mathbf{x} = [x_1, x_2, \ldots, x_m]^T$, $m < n$, the relation between $\mathbf{q}$ and $\mathbf{x}$ is given by the direct kinematics:

$$\mathbf{x} = f(\mathbf{q}) \tag{1}$$

The differential kinematics was introduced by Whitney [10] that proposed the use of differential relationships to solve for the joint motion from the Cartesian trajectory of the end-effector. Differentiating (1) with respect to time yields:

$$\dot{\mathbf{x}} = \mathbf{J}(\mathbf{q})\dot{\mathbf{q}} \tag{2}$$

Where $\dot{\mathbf{x}} \in \mathbb{R}^m$, $\dot{\mathbf{q}} \in \mathbb{R}^n$, and $\mathbf{J}(\mathbf{q}) = \partial f(\mathbf{q})/\partial \mathbf{q} \in \mathbb{R}^{m \times n}$. Hence, it is possible to calculate a path $\mathbf{q}(t)$ in terms of a prescribed trajectory $\mathbf{x}(t)$ in the operational space.

Equation (2) can be inverted to provide a solution in terms of the joint velocities:

$$\dot{\mathbf{q}} = \mathbf{J}^+(\mathbf{x})\dot{\mathbf{x}} \tag{3}$$

In which $\mathbf{J}^+$ is pseudo-inverse Jacobian matrix. When the manipulator is redundant, $m < n$, the Jacobian $\mathbf{J}$ is not a square matrix. In order to obtain a square matrix, pseudo-inverse Jacobian matrix $\mathbf{J}^+$ is defined [9].

In the closed-loop pseudo-inverse (CLP) method the joint positions can be computed through the time integration of the expression:

$$\Delta \mathbf{q} = \mathbf{J}^+(\mathbf{x})\Delta \mathbf{x} \tag{4}$$

Where $\Delta \mathbf{x} = \mathbf{x}_r - \mathbf{x}$ and $\mathbf{x}_r$ is the vector of reference (desired) position in the operational space.

2.2   Gradient Projection Method

With the pseudo-inverse method, we find the solution of the least-squares problem. Nevertheless, when having redundant degrees of freedom, we can optimize according to another criterion. The fundamental of gradient projection method is to create a cost function H, calculate its gradient ∇H, and project it to the kernel of the matrix J [4].

Projecting a vector (often gradient of cost function) to the null space of Jacobian in Eq. (4), this method could be expressed as:

$$\Delta \mathbf{q} = \alpha \cdot \mathbf{J}^+ \cdot \Delta \mathbf{x} + \mu(\mathbf{I} - \mathbf{J}^+ \cdot \mathbf{J}) \cdot \nabla H \tag{5}$$

In which μ is a scalar indicating the magnitude of the projection, and ∇H is the vector to project.

The formula of fixed scaling factor gradient projection method with damped pseudo-inverse is shown below:

$$\dot{\theta} = \mathbf{J}^+\dot{\mathbf{x}} + \mu(\mathbf{I} - \mathbf{J}^+\mathbf{J})(k_1\nabla \mathbf{H}_1 + \ldots + k_n\nabla \mathbf{H}_n) \tag{6}$$

In which, $k_1, \ldots k_n$ is weighting factors of performance criteria.

3 Dynamic Task Priority Motion Control Strategy

From the foregoing, the general way to balance multiple criteria in gradient projection method is to give them different weights. However, the assignment of weights leans heavily on existing experience. In view of this, this paper proposes a fuzzy inference-task priority method combined approach. On the one hand, this combined method realizes real-time adjustment of control method according to the changing conditions during the movement of manipulators. On the other hand, it simplifies the fuzzy logic level so that the difficulties of establishment of fuzzy rule table are abated.

In this section, three performance criteria H_1, H_2, H_3 are taken as examples. 3.1 introduces the underlying algorithm of the proposed method, task priority method in detail. On this basis, 3.2 introduces the logic in the higher layer, which is fuzzy inference.

3.1 Task Priority Method

Instead of assigning diverse weights to criteria, the task priority method considers them successively according to their importance and urgency. The idea is to define a set of tasks $t_1, \ldots t_k$ and their errors $e_1, \ldots e_k$ and then compute a Jacobian matrix for each task. The procure is like follows [8]:

The basis task, which can be referred to the end-effect task, is to track desired trajectory. The equation can be written as follows:

$$\dot{\mathbf{q}} = \mathbf{J}_1^+ \dot{\mathbf{x}}_1 \tag{7}$$

On this basis, task. 1 can be brought in by using gradient projection method.

$$\dot{\mathbf{q}} = \mathbf{J}_1^+ \dot{\mathbf{x}}_1 + (\mathbf{I} - \mathbf{J}_1^+ \mathbf{J}_1)\boldsymbol{\omega}_1 \tag{8}$$

In which, $\boldsymbol{\omega}_1$ is an arbitrary vector. Normally $\boldsymbol{\omega}_1$ is a gradient vector of a certain performance function.

Taken task. 2 into consideration, a new relational function between joint velocity vector and end-effector speed vector is established.

$$\dot{\mathbf{x}}_2 = \mathbf{J}_2 \dot{\mathbf{q}} \tag{9}$$

By solving two simultaneous Eqs. (8) and (9), expression of $\boldsymbol{\omega}_1$ can be obtained

$$\boldsymbol{\omega}_1 = (\mathbf{J}_2 \mathbf{P}_1)^+ (\dot{\mathbf{x}}_2 - \mathbf{J}_2 \mathbf{J}_1^+ \dot{\mathbf{x}}_1) \tag{10}$$

By using gradient projection method, a modified $\boldsymbol{\omega}_1$ is yielded under consideration of the task. 2 criteria function gradient $\boldsymbol{\omega}_2$.

$$\boldsymbol{\omega}_1 = (\mathbf{J}_2 \mathbf{P}_1)^+ (\dot{\mathbf{x}}_2 - \mathbf{J}_2 \mathbf{J}_1^+ \dot{\mathbf{x}}_1) + (\mathbf{I} - (\mathbf{J}_2 \mathbf{P}_1)^+ (\mathbf{J}_2 \mathbf{P}_1))\boldsymbol{\omega}_2 \tag{11}$$

In which, $\mathbf{P}_1 = \mathbf{I} - \mathbf{J}_1^+ \mathbf{J}_1$.

To make it simple, use $\hat{\mathbf{J}}_2^+$ to signify $(\mathbf{J}_2\mathbf{P}_1)^+$, use $\hat{\dot{\mathbf{x}}}_2$ to signify $\dot{\mathbf{x}}_2 - \mathbf{J}_2\mathbf{J}_1^+\dot{\mathbf{x}}_1$, use $\mathbf{P}_2$ to signify $\mathbf{I} - (\mathbf{J}_2\mathbf{P}_1)^+(\mathbf{J}_2\mathbf{P}_1)$. A simplified equation can be obtained as follow:

$$\boldsymbol{\omega}_1 = \hat{\mathbf{J}}_2^+\hat{\dot{\mathbf{x}}}_2 + \mathbf{P}_2\boldsymbol{\omega}_2 \tag{12}$$

Then we substitute $\boldsymbol{\omega}_1$ expressed by Eq. (12) for $\boldsymbol{\omega}_1$ in Eq. (8). The vector of joint angle velocity can be yielded as following equation.

$$\dot{\mathbf{q}} = \mathbf{J}_1^+\dot{\mathbf{x}}_1 + \hat{\mathbf{J}}_2^+\hat{\dot{\mathbf{x}}}_2 + \mathbf{P}_1\mathbf{P}_2\boldsymbol{\omega}_2 \tag{13}$$

Above is process of managing the end-effector task, the first and second performance criteria tasks. The third performance criteria task can be imported a similar way.

$$\dot{\mathbf{q}} = \mathbf{J}_1^+\dot{\mathbf{x}}_1 + \hat{\mathbf{J}}_2^+\hat{\dot{\mathbf{x}}}_2 + \hat{\mathbf{J}}_3^+\hat{\dot{\mathbf{x}}}_3 + \mathbf{P}_1\mathbf{P}_2\mathbf{P}_3\boldsymbol{\omega}_3 \tag{14}$$

By that analogy, other performance criteria can be adopted. Assume that the number of performance criteria is n, (in other words, the number of tasks is $n+1$, from the end-effector task to task. n.), then the joint velocity vector can be yielded as follows.

$$\dot{\mathbf{q}} = \mathbf{J}_1^+\dot{\mathbf{x}}_1 + \hat{\mathbf{J}}_2^+\hat{\dot{\mathbf{x}}}_2 + \hat{\mathbf{J}}_3^+\hat{\dot{\mathbf{x}}}_3 + \ldots + \hat{\mathbf{J}}_n^+\hat{\dot{\mathbf{x}}}_n + \mathbf{P}_1\mathbf{P}_2\mathbf{P}_3\ldots\mathbf{P}_n\boldsymbol{\omega}_n \tag{15}$$

The schematic diagram is concluded in Fig. 1.

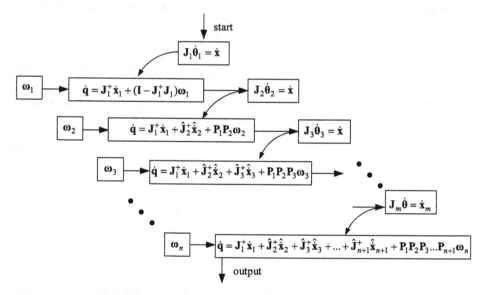

Fig. 1. Schematic diagram of inverse kinematics based on task priority

3.2 Dynamic Priority Sequence Based on Fuzzy Inference

The task priority algorithm handles the multiple performance criteria one after another in line with their priority sequence. Thus, a big concern is how to determine the priority sequence. The method proposed in this paper uses fuzzy inference to determine the real- time sequences instead of invariant sequence during the whole motion.

Firstly, we assume that the larger the value of performance function is, the more urgent the necessary to handle the requirement becomes. Thus the performance function of criterion with large value should be considered in a high priority. The priority queue of three arbitrary criteria, H_1, H_2, H_3, is determined by fuzzy inference. In fuzzy inference whose rules is shown in Table 1, H_1, H_2, H_3 are inputs and symbols: SM, MM and LM respectively denote that the numerical value of input is small, middle and large according to their each range size [10]. The output is shown in form of roman numerals which refer to different priority queues. The correspondence between these roman numerals and priority queues is shown in Table 2.

Besides, the original priority queue should be determined. The original priority queue can help handle a dilemma that two or three criteria are in same level (SM, MM or LM). In this case, their sequence is determined by the original priority queue.

Table 1. Fuzzy rules for sequences decided by three performance functions

H_1	SM								
H_2	SM			MM			LM		
H_3	SM	MM	LM	SM	MM	LM	SM	MM	LM
Sequence	I	III	III	II	II	III	II	II	II
H_1	MM								
H_2	SM			MM			LM		
H_3	SM	MM	LM	SM	MM	LM	SM	MM	LM
Sequence	I	I	III	I	I	III	II	II	II
H_1	LM								
H_2	SM			MM			LM		
H_3	SM	MM	LM	SM	MM	LM	SM	MM	LM
Sequence	I	I	I	I	I	I	I	I	I

Table 2. Mapping relations between sequence numbers and priority queues

Sequence number	Priority queue
Sequence I	$H_1 > H_2 > H_3$
Sequence II	$H_1 > H_3 > H_2$
Sequence III	$H_2 > H_1 > H_3$
Sequence IV	$H_2 > H_3 > H_1$
Sequence V	$H_3 > H_1 > H_2$
Sequence VI	$H_3 > H_2 > H_1$

4 Case Study of a 10-DOF Manipulator

In this section a 10-DOF redundant robot arm is taken as an example to illustrate the joint space planning algorithm and its efficiency. In the process, three criteria are taken into consideration: singularity avoidance, obstacle avoidance and joint limitation avoidance.

4.1 Illustration of the 10-DOF Manipulator

The 10-DOF manipulator is composed of two subsystems. The first subsystem is Portal3, a *xyz*-axis translation platform connected to the fixed base. The second subsystem is a scaled 7-DOF robot arm developed by the Mitsubishi Ltd., called PA-10 [11]. The geometrical parameters of PA-10 in this example is 100 times smaller than the real mechanism. The combined manipulator has 10 DOFs and therefore has 4 DORs (degree of redundancy). The large number of degrees of redundancy offers possibility of handling multiple optimization tasks. In fact, it is a simplified model of one kind of popular application where robot arms are placed on mobile platforms. The schematic diagram of the 10-DOF manipulator is shown in Fig. 2(a). The simulation model is built in Matlab. It is shown in Fig. 2(b).

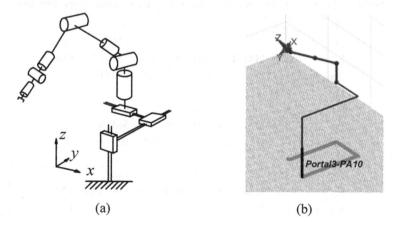

(a)   (b)

Fig. 2. Simulation model of 10-DOF manipulator: (a) schematic diagram; (b) model in Matlab

The Mitsubishi PA-10 is a 7-axis general purpose robot arm composed of a succession of roll joints and pitch joints. Its connecting rod coordinate department is shown in Fig. 3. What's more, the geometrical information of PA-10 is described by Denavit-Hartenberg (DH) convention. The corresponding DH-parameters are listed in Table 3.

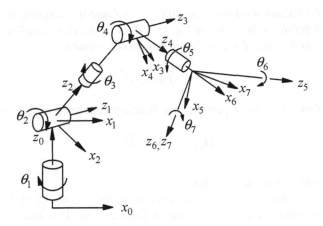

Fig. 3. Connecting rod coordinate department of PA-10

Table 3. DH-parameters of PA-10 robot arm (scale: 1:100)

Link i	a_i/mm	α_i	d_i/mm	θ_i
1	0	$-90°$	3.15	θ_1
2	0	$90°$	0	θ_2
3	0	$-90°$	4.50	θ_3
4	0	$90°$	0	θ_4
5	0	$-90°$	5.00	θ_5
6	0	$90°$	0	θ_6
7	0	0	0.80	θ_7

4.2 Explanation of Three Performance Criteria

Based on real needs in practice, singularity avoidance, obstacle avoidance and obstacle avoidance are adopted as three criteria in this case. Their performance functions are introduced to prepare for gradient projection method. The zero value of three performance functions all denote the failure of avoidance.

4.2.1 Singularity Avoidance Analysis

The first criterion is singularity avoidance. Obviously, the xyz-axis moving platform has no effect on the singular value. So our attention is mainly on PA-10.

For redundant manipulators, the pseudo-inverse Jacobian matrix is taken into account to study the singularity. The equation is shown in Eq. (16), whose solutions denote sets of joint angles when singularity occurs.

$$\det(JJ^T) = 0 \tag{16}$$

In 1985 Yoshikawa used manipulability to evaluate the distance from singularity Poses [12]. Its definition is shown in Eq. (17). It is a quantitative index and so can be used an index to describe the singularity avoidance distance.

$$W = \sqrt{|\det(JJ^T)|} \qquad (17)$$

The singularity avoidance performance function is defined as follows:

$$\mathbf{H}_{sin} = \sqrt{|\det(JJ^T)|} \qquad (18)$$

4.2.2 Obstacle Avoidance Analysis

In this section, a sphere is taken as an obstacle. In the process of obstacle avoidance analysis, the minimum distance from joints or links to obstacle is analysed and then the main task of optimization is to maximize the minimum distance [13]. So the criteria function $\mathbf{H}_{obs}(\mathbf{q})$ is defined as:

$$\mathbf{H}_{obs}(\mathbf{q}) = \min_{\substack{a \in \text{ obstacle} \\ b \in \text{ robot}}} \|a(\mathbf{q}) - b\|^2 \qquad (19)$$

4.2.3 Joint Limitation Avoidance Analysis

In the analysis of joint limitation avoidance, the 'distance' from the mid points of the joint ranges to the threshold values is taken into consideration [14].

In order to match up with the other two performance functions, the performance function $\mathbf{H}_{limit}(\mathbf{q})$ is defined as follows.

$$\mathbf{H}_{limit}(\mathbf{q}) = \sum_{i=1}^{n} 4 \frac{(q_{max}[i] - q[i])(q[i] - q_{min}[i])}{(q_{max}[i] - q_{min}[i])^2} \qquad (20)$$

When the joint angle equals the upper or lower joint range limit, the value of performance function equals zero. The larger the deviation from the median is, the higher the value of joint limitation avoidance performance function is.

4.3 Motion Control Process and Simulation Results

In last section, three performance functions are proposed. The initial conditions of this case study are set as follows. For the sake of simplicity, the three translation joint ranges of xyz-axis moving platform are [0, 30]. As for PA-10, the seven revolute joint angle ranges are set to [−120°, 120°].

To test the ability of obstacle avoidance, a spherical obstacle is placed in the workspace of manipulators. The coordinate of the ball obstacle is (2.5, 12.5, 12.5) and the radius is 2.5. The spherical obstacle is shown in Fig. 4.

In the analysis process, each task is based on the corresponding performance criterion function $\mathbf{H}_{sin}, \mathbf{H}_{obs}, \mathbf{H}_{lim}$. In last section, function $\mathbf{H}$ equals zero when the failure

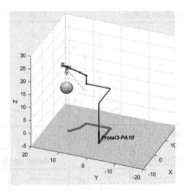

Fig. 4. Schematic diagram of simulation conditions

occurs, namely singular pose, collision and excess of joint angle range. Thus the mapping relations between function values and symbols in fuzzy rules can be set (see Table 4). The component of H_{lim} equals 0.382 when the respective joint angle equals $\pm 100°$. And it equals 0.555 when the respective joint angle equals $\pm 80°$. Of note is that, the range partitioning of each criterion function is determined by practical experience. And the priority queue is determined by fuzzy logic rules shown in Tables 1 and 2.

Table 4. Mapping relations between sequence numbers and priority queue

	LM	MM	SM
H_{sin}	[0, 1]	(1, 3]	$(3, \infty)$
H_{obs}	[0, 3]	(3, 10]	$(10, \infty)$
Any component of H_{lim}	[0, 0.382]	(0.0.555]	(0.555, 1]

In this case study the 10-DOF manipulator is supposed to tracking a specified trajectory. Coordinate of the start point is $(2.28, 13.55, 20.89)^T$ and coordinate of the end point is $(1.42, 5.49, 13.43)^T$. The calculation of trajectory tracking is down in 50 steps. Values of different performance indexes change along with the increase of step number, which means that the indexes values change in real time and so does the priority queue. Schematic diagram of simulation conditions is shown in Fig. 4.

To get it more convincing a comparing experiment irrespective of any performance optimization is done firstly. The results are shown in Fig. 5. It is not hard to find that for this specific trajectory the joint limitation angles of the 5th and 7th joints exceed the boundary.

According to the algorithm presented in Sect. 3, new set of joint angle can be yielded. The performance indexes are shown in Fig. 6.

Of note is that, the sequence of task priority changes due to the fuzzy inference. At the 27th step the seventh joint angle is over 100°. As a result, Sequence IV $(H_{obs} > H_{lim} > H_{sin})$ changes into Sequence VI $(H_{lim} > H_{obs} > H_{sin})$ according to

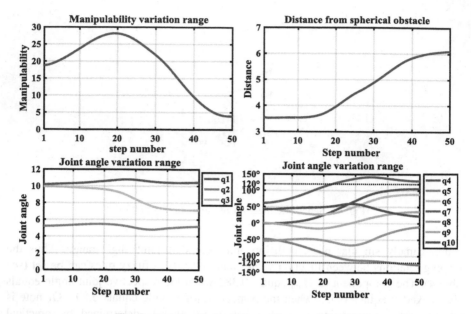

Fig. 5. Results of simulation without optimization

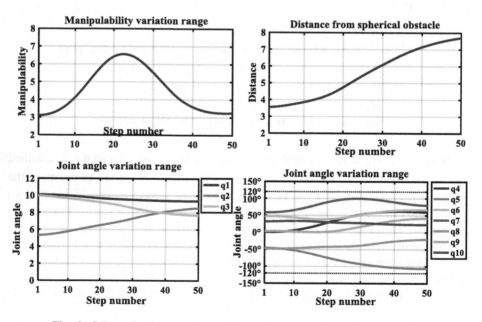

Fig. 6. Schematic diagram of simulation with proposed optimization method

Table 2 and the joint limitation avoidance becomes the primary task. As a result, the growth rate of the seventh joint angle decreases and the joint limitation avoidance is guaranteed.

5 Conclusions

In this paper a motion control strategy is proposed for purpose of multiple performance criteria optimization on kinematic redundant manipulators. In order to handle the relationship of multiple criteria, fuzzy inference and task priority method are used in combination. To be exact, Fuzzy inference yields different sequences according to the changing urgency and importance of different performance criteria. And each sequence decides a priority queues for the following task priority method. Therefore, the task priority method becomes real-time and dynamic and the resulting joint angle space is qualified to handle the changing working conditions during the manipulator motions. The method is applied to a 10 degree-of-freedom manipulator in numerical case study to illustrate its efficiency. In the case study, three performance criteria are taken into account, which are singularity avoidance, obstacle avoidance and joint limitation avoidance. Concluded from its better behavior than the case without optimization, the method is proved to be effective for multiple criteria control.

Acknowledgment. This project is supported by the National Key Scientific and Technological Project (Grant No. 2015ZX04005006) and Sino-German (CSC-DAAD) Postdoc Scholarship Program 2014 (Grant No. 57165010).

References

1. Nishiguchi, S., Ogawa, K., Yoshikawa, Y.: Theatrical approach: designing human-like behaviour in humanoid robots. Robot. Auton. Syst. **89**, 158–166 (2017)
2. Nenchev, D.N., Tsumaki, Y.: Motion analysis of a knematically redundant seven-DOF manipulator under the singularity-consistent method. In: IEEE Conference on Robotics and Automation, pp. 2760–2765 (2003)
3. Baillieul, J.: Kinematic programming alternatives for redundant manipulators. In: IEEE Conference on Robot and Automation, St. Louis, MI, Mar, pp. 722–728 (1985)
4. Figueiredo, M.A.T., Nowak, R.D., Wright, S.J.: Gradient projection for sparse reconstruction: application to compressed sensing and other inverse problems. IEEE J. Sel. Top. Sign. Process. **1**, 586–597 (2007)
5. Xiang, J., Zhong, C., Wei, W.: General-weighted least-norm control for redundant manipulators. IEEE Trans. Robot. **26**, 660–669 (2010)
6. Feng, B., Ma, G., Wen, Q.: Adaptive robust control of space robot in task space. In: IEEE International Conference on Mechatronics and Automation, Henan University of Science and Technology, Luoyang, China, June, 2006
7. Jang, J.S.R.: Anfis-adaptive-network-based fuzzy inference system. IEEE Trans. Syst. Man Cybern. **23**, 665–685 (1993)
8. Nakamura, Y., Hanafusa, H., Yoshikawa, T.: Task-priority based redundancy control of robot manipulators. Int. J. Robot. Res. **6**, 3–15 (1987)
9. Keith, L., Doty, C., Bonivento, M.: A theory of generalized inverses applied to robotics. Int. J. Robot. Res. **12**, 1–19 (1993)
10. Liegeois, A.: Automatic supervisory control of the configuration and behavior of multibody mechanisms. IEEE Trans. Syst., Man Cybern. **7**, 868–871 (1977)

11. Llama, M.A., Flores, A., Santibanez, V., Campa, R.: Global convergence of a decentralized adaptive fuzzy control for the motion of robot manipulators: application to the Mitsubishi PA10-7CE as a case of study. J. Intell. Robot. Syst. **82**, 363–377 (2016)
12. Yoshikawa, T.: Manipulators of robotic mechanisms. Int. J. Robot. Res. **4**, 3–9 (1985)
13. Maciejewski, A.A., Klein, C.A.: Obstacle avoidance for kinematically redundant manipulators in dynamically varying environment. Int. J. Robot. Res. **4**, 109–117 (1985)
14. Dubey, R.V., Euler, J.A., Babock, S.M.: An efficient gradient projection optimization scheme for a seven degree-of-freedom redundant robot with spherical wrist. In: Proceedings of the IEEE International Conference on Robotics and Automation, pp. 28–36 (1988)

A Boundary Control Method for Suppressing Flexible Wings Vibration of the FMAV

Yunan Chen[1], Wei He[1(✉)], Xiuyu He[1], Yao Yu[1], and Changyin Sun[2]

[1] School of Automation and Electrical Engineering,
University of Science and Technology Beijing, Beijing 100083, China
hewei.ac@gmail.com
[2] School of Automation, Southeast University, Nanjing 210096, China

Abstract. In this paper, we propose a boundary control strategy for vibration suppression of two flexible wings and a rigid body. As a basic approach, Hamilton's principle is used to ascertain the system dynamic model, which includes governing equations and boundary conditions. Considering the coupled bending and torsional deformations of flexible wings, boundary control force and torque act on the rigid body to regulate unexpected deformations of flexible wings that caused by air flow. Then, we present stability analysis of the closed-loop system through Lyapunov's direct method. Simulations are carried out by using finite difference method. The results illustrate the significant effect of the developed control strategies.

Keywords: Boundary control · Flexible wings · FMAV · Robot

1 Introduction

The advent of advanced high-strength flexible materials and the development of Micro Electro Mechanical Systems (MEMS) make it possible to invent sophisticated flapping-wing micro air vehicles (FMAVs) with flexible wings. FMAVs using flexible wings can reduce the impact of atmospheric turbulence so that improve its wind resistance ability [1]. Moreover, comparing with the rigid wing, the flexible wing has higher lift coefficient, larger stall angle. This leads that FMAVs with flexible wings are considered as promising robotic aircrafts while unmanned aerial vehicles (UAV) toward miniaturization. FMAV's lift and thrust are produced via flapping and twisting wings continuously. Besides, it can realize

This work was supported by the National Basic Research Program of China (973 Program) under Grant 2014CB744206, the National Natural Science Foundation of China under Grant 61522302, 61761130080, 61520106009, 61533008, the Newton Advanced Fellowship from The Royal Society, UK, under Grant NA160436, the Beijing Natural Science Foundation under Grant 4172041, and the Fundamental Research Funds for the China Central Universities of USTB under Grant FRF-BD-16-005A and FRF-TP-15-005C1.

© Springer International Publishing AG 2017
Y. Huang et al. (Eds.): ICIRA 2017, Part II, LNAI 10463, pp. 869–878, 2017.
DOI: 10.1007/978-3-319-65292-4_75

hovering with a quick transition to forward flight and side flight through adjusting the flap angle or twist angle [2]. In [3], the authors studied partial differential equations (PDE) control method for a robotic aircraft with flexible wings that can be actuated actively for control. However, the flight mechanisms of insects and birds are not yet clear. That pose an obstacle to design the FMAV in a simple way of imitating the flying animals.

From an extensive review, many of the FMAV dynamic models presented in the literatures focus on the standard six-degree of freedom model in Lagrangian form [4]. Afshin and Neda presented an adaptive attitude and position control of insect bionic aircraft based on the standard aircraft model in [5]. They transformed the time-varying equation of motion into the time-invariant system in order to design the adaptive controller. The wing flexibility is believed to play a key role in flight performance and agility for FMAV. However, undesired deformations occur in the nonlinear aeroelastic flexible wings system when the FMAV flap its wings. This could result in undamped vibrations. Such vibrations are unacceptable since the antisymmetric motion of wings will exert adverse impacts on FMAVs [6].

Various vibration control methods have been introduced to regulate the attitude of flexible satellites [7], spacecraft, and beams. In theory, flexible systems require an infinite number of elastic modes to thoroughly describe their dynamics. Nguyen and Tuzc used an equivalent beam-rod model to describe an integrated flexible aircraft [8]. Similar analyses of the flexible wing dynamics can be found in [9] and [10]. We develop control strategies for full flexible wings that are modeled as a set of PDEs and associated boundary conditions by using Hamilton's principle. In order to achieve autonomous flight, the control calculations need to be minimal as possible. Boundary control has several merits for suppressing vibration of flexible wings [11,12]. By installing sensors and actuators only at the boundary, like the wing root or tip, dynamic characteristics of the system will not be affected. Furthermore, the boundary control strategy can be developed with the help of a specific Lyapunov candidate function, which is designed according to the kinematic and potential energies of the system [13,14].

2   Problem Formulation

The dynamics of two flexible wings and a rigid body system is derived under distributed air loads $F_{bL}(x,t)$ and $F_{bR}(x,t)$ in this section.

Considering an FMAV with two flexible wings and a fuselage that is modeled as a point load, as depicted in Fig. 1. Herein, the origin is located in the point load body. We assume that cross-section areas of flexible wings are uniform and they are enduring bending and torsional deformations while flapping flight. Let $w_{L(R)}(x,t)$ and $\theta_{L(R)}(t)$ be the bending and torsional deformations of the elastic axis with respect to its initial equilibrium position at the position x, time t. The subscript L and R represent the left and the right wing, respectively. Each wingspan is $L \in \mathbb{R}$, the chord length is $c \in \mathbb{R}$, and the mass per unit span is $\rho \in \mathbb{R}$. The polar moment of inertia per unit span is $I_p \in \mathbb{R}$. The bending rigidity and torsional stiffness of wings are $EI \in \mathbb{R}$ and $GJ \in \mathbb{R}$, respectively. The mass of the point load fuselage is $m_b \in \mathbb{R}$. Let $x_e c$ denote the distance between the wing centroid (GC) and the shear center (EA). $x_a c$ is the distance between the aerodynamic center (AC) and the shear center of the wing, as shown in Fig. 2.

$u(t)$ and $\tau(t)$ are the control input force and torque. The study of undesired vibrations of flexible wings is based on Timoshenko and Euler-Bernoulli beam theory [15]. In the meantime, we introduce Kelvin-Voigt viscoelastic law to establish the damping mechanism [16]. η is the Kelvin-Voigt damping coefficient. $F_{bL(R)}(x,t)$ is unknown time-varying distributed air loads exert along the flexible wings.

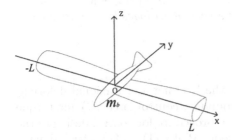
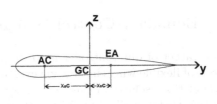

Fig. 1. Diagram of the flexible-wings FMAV.

Fig. 2. Sectional view of the flexible wing.

Remark 1. To simplify formulae, we use notations $(*)' = \frac{\partial(*)}{\partial x}$ and $\dot{(*)} = \frac{\partial(*)}{\partial t}$ in the whole article.

We use Hamilton's principle $\int_{t_1}^{t_2} \delta[E_k(t) - E_p(t) + W(t)]dt = 0$ to acquire the system equations of motion from energy analysis in a variational form. It is proven that the variation of the kinetic energy subtract potential energy plus total virtual work done during time interval $[t_1, t_2]$ equal to zero. We get the dynamics equations [17] of flexible wings and the body as follows:

$$\rho \ddot{w}_L(x,t) + EI w_L''''(x,t) - \rho x_e c \ddot{\theta}_L(x,t) + \eta EI \dot{w}_L''''(x,t) = F_{bL}(x,t) \quad (1)$$

$$-I_p \ddot{\theta}_L(x,t) + GJ \theta_L''(x,t) + \rho x_e c \ddot{w}_L(x,t) + \eta GJ \dot{\theta}_L''(x,t) = x_a c F_{bL}(x,t) \quad (2)$$

$$\rho \ddot{w}_R(x,t) + EI w_R''''(x,t) - \rho x_e c \ddot{\theta}_R(x,t) + \eta EI \dot{w}_R''''(x,t) = F_{bR}(x,t) \quad (3)$$

$$- I_p \ddot{\theta}_R(x,t) + GJ \theta_R''(x,t) + \rho x_e c \ddot{w}_R(x,t) + \eta GJ \dot{\theta}_R''(x,t) = x_a c F_{bR}(x,t) \quad (4)$$

and corresponding ordinary differential equations (ODEs) boundary conditions are

$$w_L(0,t) = w_R(0,t) = w(0,t) \quad (5)$$

$$\theta_L(0,t) = \theta_R(0,t) = \theta(0,t) \quad (6)$$

$$w'_L(0,t) = w'_R(0,t) = w''_L(-L,t) = w''_R(L,t) = 0 \tag{7}$$

$$w'''_L(-L,t) + \eta\dot{w}'''_L(-L,t) = 0, w'''_R(L,t) + \eta\dot{w}'''_R(L,t) = 0 \tag{8}$$

$$\theta'_L(-L,t) + \eta\dot{\theta}'_L(-L,t) = 0, \theta'_R(L,t) + \eta\dot{\theta}'_R(L,t) = 0 \tag{9}$$

$$m_b\ddot{w}(0,t) - EI[w'''_L(0,t) - \eta\dot{w}'''_L(0,t) + w'''_R(0,t) + \eta\dot{w}'''_R(0,t)] = u(t) \tag{10}$$

$$I_p\ddot{\theta}(0,t) + GJ[\theta'_L(0,t) + \eta\dot{\theta}'_L(0,t) - \theta'_R(0,t) - \eta\dot{\theta}'_R(0,t)] = \tau(t) \tag{11}$$

Assumption 1 [17]. *The spatiotemporal-varying distributed airloads $F_{bL}(x,t)$ and $F_{bR}(x,t)$ are assumed to be bounded. That is, constants $\bar{F}_{bL} \in \mathbb{R}^+$ and $\bar{F}_{bR} \in \mathbb{R}^+$ make $|F_{bL}(x,t)| \leq \bar{F}_{bL}$ and $|F_{bR}(x,t)| \leq \bar{F}_{bR}$, $\forall(x,t) \in [0,L] \times [0,\infty)$. This is a reasonable assumption, since the distributed air loads have finite energy.*

3   Boundary Control Design

The target of our controller is to suppress the bending and torsional deformations of flexible wings in time via exerting boundary control force $u(t)$ and torque $\tau(t)$ at the fuselage. In this section, we give a procedure for control design of the system that described by four nonhomogeneous PDEs (1) – (4) under unknown distributed air loads $F_{bL}(x,t)$ and $F_{bR}(x,t)$. Boundary control design is illustrated according the Lyapunov's direct method [18]. We also give a rigorous mathematical proof of the stability of the closed-loop FMAV system. Under the condition that system parameters are known, we design boundary control laws as

$$u(t) = -k_1\dot{w}(0,t) \tag{12}$$

$$\tau(t) = -k_2\dot{\theta}(0,t) \tag{13}$$

where k_1 and k_2 are two positive control gains.

Considering the Lyapunov candidate function we designed as follows

$$V(t) = V_1(t) + V_2(t) + \Delta(t) \tag{14}$$

In detail

$$V_1(t) = \frac{1}{2}\int_{-L}^{0}\left[a\rho\dot{w}_L^2(x,t) + aEIw''^2_L(x,t) + aI_p\dot{\theta}_L^2(x,t) + aGJ\theta'^2_L(x,t)\right]dx$$

$$+ \frac{1}{2}\int_{0}^{L}\left[a\rho\dot{w}_R^2(x,t) + aEIw''^2_R(x,t) + aI_p\dot{\theta}_R^2(x,t) + aGJ\theta'^2_R(x,t)\right]dx \tag{15}$$

$$V_2(t) = \frac{a}{2}m_b[\dot{w}(0,t)]^2 + \frac{a}{2}I_p[\dot{\theta}(0,t)]^2 \tag{16}$$

$$\Delta(t) = -a\rho x_e c\int_{-L}^{0}\dot{w}_L(x,t)\dot{\theta}_L(x,t)dx - a\rho x_e c\int_{0}^{L}\dot{w}_R(x,t)\dot{\theta}_R(x,t)dx \tag{17}$$

where a is a positive constant.

Theorem 1. *The Lyapunov candidate function $V(t)$ has upper and lower bounds*

$$0 \le \xi_1[V_1(t) + V_2(t)] \le V(t) \le \xi_2[V_1(t) + V_2(t)] \tag{18}$$

where $\xi_1 > 0$ and $\xi_2 > 0$ are two positive numbers.

Proof: Defining an auxiliary function $\nu(t)$

$$\nu(t) = \int_{-L}^{0} \{[\dot{w}_L(x,t)]^2 + [\dot{\theta}_L(x,t)]^2 + [w_L''(x,t)]^2 dx + [\theta_L'(x,t)]^2\} dx$$

$$+ \int_{0}^{L} \{[\dot{w}_R(x,t)]^2 + [\dot{\theta}_R(x,t)]^2 + [w_R''(x,t)]^2 + [\theta_R'(x,t)]^2\} dx \tag{19}$$

From the definition of $V_1(t)$, its lower bound can be expressed with $\nu(t)$ as

$$\gamma_1 \nu(t) \le V_1(t) \tag{20}$$

where $\gamma_1 = \frac{a}{2} \min\{\rho, I_p, EI, GJ\}$.

Furthermore, we have

$$|\Delta(t)| \le a\rho x_e c \nu(t) \le \gamma_2 V_1(t) \tag{21}$$

where $\gamma_2 = \frac{a\rho x_e c}{\gamma_1}$, that is

$$-\gamma_2 V_1(t) \le \Delta(t) \le \gamma_2 V_1(t) \tag{22}$$

Then, we add $V_1(t)$ to the both sides of (22)

$$(1 - \gamma_2)V_1(t) \le V_1(t) + \Delta(t) \le (1 + \gamma_2)V_1(t) \tag{23}$$

We can choose proper parameters to guarantee that $0 < \gamma_2 < 1$.

Further more, we have

$$0 \le \xi_1[V_1(t) + V_2(t)] \le V_1(t) + V_2(t) + \Delta(t) \le \xi_2[V_1(t) + V_2(t)] \tag{24}$$

where $\xi_1 = \min\{1 - \gamma_2, 1\} = 1 - \gamma_2$, $\xi_2 = \max\{1 + \gamma_2, 1\} = 1 + \gamma_2$.   ∎

Theorem 2. *If the boundary control laws (12) and (13) are utilized, the time derivative of $V(t)$ has upper bound*

$$\dot{V}(t) \le -\xi V(t) + \varepsilon \tag{25}$$

where $\xi > 0$ is a constant.

Proof: The derivative of the Lyapunov candidate (14) with respect to time as follows

$$\dot{V}(t) = \dot{V}_1(t) + \dot{V}_2(t) + \dot{\Delta}(t) \tag{26}$$

Furthermore, we have

$$
\dot{V}_1(t) = \int_{-L}^{0} \big[a\rho\dot{w}_L(x,t)\ddot{w}_L(x,t) + aEIw_L''(x,t)\dot{w}_L''(x,t) + aI_p\dot{\theta}_L(x,t)\ddot{\theta}_L(x,t)
$$
$$
+ aGJ\theta_L'(x,t)\dot{\theta}_L'(x,t) \big]\mathrm{d}x + \int_{0}^{L} \big[a\rho\dot{w}_R(x,t)\ddot{w}_R(x,t) + aI_p\dot{\theta}_R(x,t)\ddot{\theta}_R(x,t)
$$
$$
+ aEIw_R''(x,t)\dot{w}_R''(x,t) + aGJ\theta_R'(x,t)\dot{\theta}_R'(x,t) \big]\mathrm{d}x \tag{27}
$$

Differentiating $V_2(t)$ respect to time t, then substituting the boundary conditions (10) and (11) into it, $\dot{V}_2(t)$ is calculated as

$$
\dot{V}_2(t) = a\dot{w}(0,t)\{u(t) + EI[w_L'''(0,t) + \eta\dot{w}_L'''(0,t) - w_R'''(0,t) - \eta\dot{w}_R'''(0,t)]\}
$$
$$
+ a\dot{\theta}(0,t)\{\tau(t) - GJ[\theta_L'(0,t) + \eta\dot{\theta}_L'(0,t) - \theta_R'(0,t) - \eta\dot{\theta}_R'(0,t)]\} \tag{28}
$$

Then, differentiating $\Delta(t)$ with respect to t, it yields

$$
\dot{\Delta}(t) = - a\rho x_e c \int_{-L}^{0} [\ddot{w}_L(x,t)\dot{\theta}_L(x,t) + \dot{w}_L(x,t)\ddot{\theta}_L(x,t)]\mathrm{d}x
$$
$$
- a\rho x_e c \int_{0}^{L} [\ddot{w}_R(x,t)\dot{\theta}_R(x,t) + \dot{w}_R(x,t)\ddot{\theta}_R(x,t)]\mathrm{d}x \tag{29}
$$

From (27), (28) and (29), substituting the proposed control laws (12), (13) and Assumption 1, we further obtain the time derivative of Lyapunov function (26) as

$$
\dot{V}(t) \leqslant - ak_1[\dot{w}(0,t)]^2 - ak_2[\dot{\theta}(0,t)]^2 - \left(\frac{a\eta EI}{L^4} - a\delta_1\right) \int_{-L}^{0} [\dot{w}_L(x,t)]^2\mathrm{d}x
$$
$$
- \left(\frac{a\eta EI}{L^4} - a\delta_3\right) \int_{0}^{L} [\dot{w}_R(x,t)]^2\mathrm{d}x - \left(\frac{a\eta GJ}{L^2} - ax_a c\delta_2\right) \int_{-L}^{0} [\dot{\theta}_L(x,t)]^2\mathrm{d}x
$$
$$
- \left(\frac{a\eta GJ}{L^2} - ax_a c\delta_4\right) \int_{0}^{L} [\dot{\theta}_R(x,t)]^2\mathrm{d}x + \left(\frac{a}{\delta_1} + \frac{ax_a c}{\delta_2}\right) \int_{-L}^{0} [F_{bL}(x,t)]^2\mathrm{d}x
$$
$$
+ \left(\frac{a}{\delta_3} + \frac{ax_a c}{\delta_4}\right) \int_{0}^{L} [F_{bR}(x,t)]^2\mathrm{d}x \leqslant -\xi_3[V_1(t) + V_2(t)] + \varepsilon \tag{30}
$$

The positive constants a and $\delta_1 - \delta_4$ are chosen to satisfy $\sigma_1 = \left(\frac{a\eta EI}{L^4} - a\delta_1\right) \geqslant 0, \sigma_2 = \left(\frac{a\eta EI}{L^4} - a\delta_3\right) \geqslant 0, \sigma_3 = \left(\frac{a\eta GJ}{L^2} - ax_a c\delta_2\right) \geqslant 0, \sigma_4 = \left(\frac{a\eta GJ}{L^2} - ax_a c\delta_4\right) \geqslant 0.$ Meanwhile, $\varepsilon = \left(\frac{a}{\delta_1} + \frac{ax_a c}{\delta_2}\right)\bar{F}_{bL}^2 + \left(\frac{a}{\delta_3} + \frac{ax_a c}{\delta_4}\right)\bar{F}_{bR}^2, \xi_3 = \min \frac{2}{a}\{\frac{\sigma_1}{\rho}, \frac{\sigma_2}{\rho}, \frac{\sigma_3}{I_p}, \frac{\sigma_4}{I_p}\}.$ Combining Theorem 1 and (30), we can prove that

$$
\dot{V}(t) \leqslant -\xi V(t) + \varepsilon \tag{31}
$$

where $\xi = \frac{\xi_3}{\xi_2} > 0$. ∎

Theorem 3. *For the dynamical system of a pair of flexible wings and the fuselage, which is described by governing Eqs. (1) – (4) as well as boundary conditions (5) – (11). Under the effect of proposed control schemes (12) and (13), giving that initial conditions are bounded, we can obtain that: all the system signals, especially boundary outputs $w_L(-L,t)$, $w_R(L,t)$ and $\theta_L(-L,t)$, $\theta_R(L,t)$ can realize uniformly ultimate bounded (UUB), i.e., as t tends to infinity, they will eventually converge to small compact sets.*

Proof: Integrating of the inequality (31) by multiplying $e^{\xi t}$

$$V(t) \leqslant V(0)e^{-\xi t} - \frac{\varepsilon}{\xi}e^{-\xi t} + \frac{\varepsilon}{\xi} \leqslant V(0)e^{-\xi t} + \frac{\varepsilon}{\xi} \in \mathcal{L}_\infty \qquad (32)$$

which indicates that $V(t)$ has its upper bound. From the introduced Lemma A.12 in [19], we can further obtain

$$\frac{1}{L^3}w_L^2(x,t) \leqslant \int_{-L}^{0} [w_L''(x,t)]^2 dx \leqslant \nu(t) \leqslant \frac{1}{\gamma_1\xi_1}V(t) \in \mathcal{L}_\infty \qquad (33)$$

$$\frac{1}{L}\theta_L^2(x,t) \leqslant \int_{-L}^{0} [\theta_L'(x,t)]^2 dx \leqslant \nu(t) \leqslant \frac{1}{\gamma_1\xi_1}V(t) \in \mathcal{L}_\infty \qquad (34)$$

where γ_1 and ξ_1 are dimensionless positive constants.

The boundedness of deformations of the right wing side $w_R(x,t)$ and $\theta_R(x,t)$ can be obtained in the same way. Appropriately rearranging the formulae (32) – (34), two DOF deformations of the flexible wings $w_{L(R)}(x,t)$ and $\theta_{L(R)}(x,t)$ have upper bounds as follows

$$|w_{L(R)}(x,t)| \leqslant \sqrt{\frac{L^3}{\gamma_1\xi_1}(V(0)e^{-\xi t} + \frac{\varepsilon}{\xi})} \qquad (35)$$

$$|\theta_{L(R)}(x,t)| \leqslant \sqrt{\frac{L}{\gamma_1\xi_1}(V(0)e^{-\xi t} + \frac{\varepsilon}{\xi})} \qquad (36)$$

Based on (35) and (36), we can state that when t intends to infinity, $w_{L(R)}(x,t)$ and $\theta_{L(R)}(x,t)$ will remain in the small domains, that are

$$|w_{L(R)}(x,t)| \leqslant \sqrt{\frac{L^3\varepsilon}{\gamma_1\xi_1\xi}}, \forall x \in [-L,L] \qquad (37)$$

$$|\theta_{L(R)}(x,t)| \leqslant \sqrt{\frac{L\varepsilon}{\gamma_1\xi_1\xi}}, \forall x \in [-L,L] \qquad (38)$$

Namely, $\exists T_0$, when $t > T_0$, the system states $w_{L(R)}(x,t)$ and $\theta_{L(R)}(x,t)$ will approach to small vicinities of above domains. According to the above analysis, it is indicated that the boundary control laws (12) and (13) can guarantee that all system states are uniformly ultimately bounded (UUB), $\forall (x,t) \in [0,L] \times [0,\infty)$. ∎

4   Numerical Simulation

In this section, the performance of control schemes (12) and (13) is demonstrated by numerical simulations. We introduce finite difference approximation method to approximate the nonlinear PDE system which is described by (1) – (4). Corresponding initial conditions of the flexible wings are set as $w_L(x,0) = -xm, w_R(x,0) = xm, \theta_L(x,0) = \frac{-x}{2L}\deg, \theta_R(x,0) = \frac{x}{2L}\deg$, and flexible wings are affected by the spatiotemporally varying distributed loads $F_{bL}(x,t) = F_{bR}(x,t) = [1 + \sin(t) + \cos(3t)]xN$.

Figures 3 and 4 indicate that the bounded initial condition result in undesired deformations of flexible wings of FMAV. From Fig. 3, the bending deformations of the flexible wings are increasing as time going. From Fig. 4, the torsional deformations get larger and larger. During the whole simulation process, the flexible wings vibrate unexpectedly, even beyond the maximum range of mechanical structures can bear.

By utilizing boundary control force (12) and torque (13) with the control gains are chose as $k_1 = 80, k_2 = 5$, the spatial time representation of bending and torsion deformations are shown in Figs. 5 and 6, respectively. The comparison

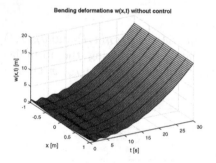

Fig. 3. Bend deformations $w_{L(R)}(x,t)$ of flexible wings without control.

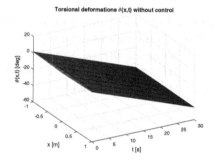

Fig. 4. Torsional deformations $\theta_{L(R)}(x,t)$ of flexible wings without control.

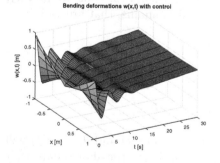

Fig. 5. Bending deformations $w_{L(R)}(x,t)$ of flexible wings with control.

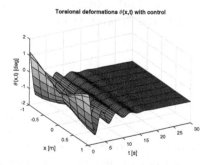

Fig. 6. Torsional deformations $\theta_{L(R)}(x,t)$ of flexible wings with control.

of the open-loop and closed-loop deformations at the left wing tip and the wing root are shown in Figs. 7 and 8. We can notice that the errors of wing deformations between wing tip and wing root approach to zero within a few seconds after applying control inputs. Based on above results, the developed controls are capable of suppressing irregular vibrations of flexible wings effectively.

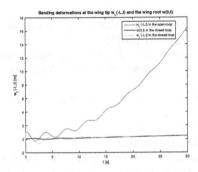

Fig. 7. Bending deformations at wing tip and root: the blue dot-dash line is the open-loop $w_L(-L,t)$; the black line is the closed-loop $w(0,t)$; the red dot line is the closed-loop $w_L(-L,t)$. (Color figure online)

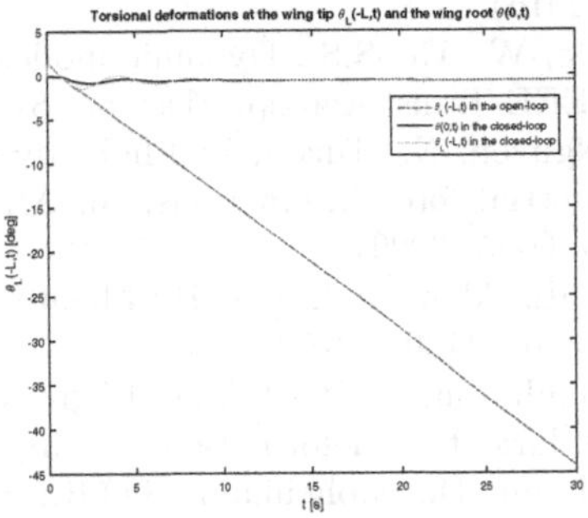

Fig. 8. Torsional deformations at wing tip and root: the blue dot-dash line is the open-loop $\theta_L(-L,t)$; the black line is the closed-loop $\theta(0,t)$; the red dot line is the closed-loop $\theta_L(-L,t)$. (Color figure online)

5   Conclusion

This article puts forward a boundary control design to suppress the undesired deformations of two flexible wings and a body of the FMAV system. Using boundary inputs to deal with this problem is more practical because there are only a small number of actuators. In this paper, two boundary control inputs are needed: a force and a torque. A limitation also exists in this paper. Numerical simulations are carried out by using finite difference method to approximate derivative terms, thereby noises are introduced. In future, we will focus on disturbance observers design to reduce influences of external noises.

References

1. Zhang, F., Zhu, R., Zhou, Z.: Experiment research on aerodynamics of flexible wing MAV. Chin. J. Aeronaut. **29**(6), 1440–1446 (2008)
2. Orlowski, C.T., Girard, A.R.: Dynamics, stability, and control analyses of flapping wing micro-air vehicles. Prog. Aerosp. Sci. **51**, 18–30 (2012)
3. Paranjape, A.A., Guan, J.Y., Chuang, S.J., Krstic, M.: PDE boundary control for flexible articulated wings on a robotic aircraft. IEEE Trans. Robot. **29**(3), 625–640 (2013)

4. He, W., Huang, H., Chen, Y., Xie, W., Feng, F., Kang, Y., Sun, C.: Development of an autonomous flapping-wing aerial vehicle. Sci. China Inf. Sci. **60**(6), 063201 (2017)

5. Banazadeh, A., Taymourtash, N.: Adaptive attitude and position control of an insect-like flapping wing air vehicle. Nonlinear Dyn. **85**(1), 47–66 (2016)

6. Bialy, B.J., Chakraborty, I., Cekic, S.C., Dixon, W.E.: Adaptive boundary control of store induced oscillations in a flexible aircraft wing. Automatica **70**, 230–238 (2016)

7. He, W., Ge, S.S.: Dynamic modeling and vibration control of a flexible satellite. IEEE Trans. Aerosp. Electron. Syst. **51**(2), 1422–1431 (2015)

8. Nguyen, N., Tuzcu, I.: Flight dynamics of flexible aircraft with aeroelastic and inertial force interactions. In: AIAA Atmospheric Flight Mechanics Conference, p. 6045 (2009)

9. Patil, M.J., Hodges, D.H.: Flight dynamics of highly flexible flying wings. J. Aircr. **43**(6), 1790–1798 (2006)

10. Raghavan, B., Patil, M.J.: Flight dynamics of high-aspect-ratio flying wings: effect of large trim deformation. J. Aircr. **46**(5), 1808–1812 (2009)

11. Halim, D., Moheimani, S.O.R.: Spatial resonant control of flexible structures-application to a piezoelectric laminate beam. IEEE Trans. Control Syst. Technol. **9**(1), 37–53 (2001)

12. Guo, B.Z., Jin, F.F.: The active disturbance rejection and sliding mode control approach to the stabilization of the Euler-Bernoulli beam equation with boundary input disturbance. Automatica **49**(9), 2911–2918 (2013)

13. Nguyen, Q.C., Hong, K.S.: Asymptotic stabilization of a nonlinear axially moving string by adaptive boundary control. J. Sound Vib. **329**(22), 4588–4603 (2010)

14. He, W., Ge, S.S.: Robust adaptive boundary control of a vibrating string under unknown time-varying disturbance. IEEE Trans. Control Syst. Technol. **20**(1), 48–58 (2012)

15. Wu, H.N., Wang, J.W.: Static output feedback control via PDE boundary and ODE measurements in linear cascaded ODE-beam systems. Automatica **50**(11), 2787–2798 (2014)

16. Mozaffari-Jovin, S., Firouz-Abadi, R.D., Roshanian, J.: Flutter of wings involving a locally distributed flexible control surface. J. Sound Vib. **357**, 377–408 (2015)

17. He, W., Zhang, S.: Control design for nonlinear flexible wings of a robotic aircraft. IEEE Trans. Control Syst. Technol. **25**(1), 351–357 (2017)

18. Cai, X., Krstic, M.: Nonlinear stabilization through wave PDE dynamics with a moving uncontrolled boundary. Automatica **68**, 27–38 (2016)

19. Queiroz, M.D., Dawson, D.M., Nagarkatti, S.P., Zhang, F., Bentsman, J.: Lyapunov-based control of mechanical systems. Appl. Mech. Rev. **54**(5), B81 (2001)

Proxy Based Sliding Mode Control for a Class of Second-Order Nonlinear Systems

Guangzheng Ding, Jian Huang$^{(\boxtimes)}$, and Yu Cao

Huazhong University of Science and Technology,
Wuhan 430074, People's Republic of China
huang_jan@mail.hust.edu.cn

Abstract. In this paper, an extended Proxy-based Sliding Mode Control (PSMC) which combines sliding mode control with conventional stiff position control, is proposed and applied for the tracking control of a class of second-order nonlinear systems. The well-known chattering problem can be solved by using the proposed control scheme. As a result, the overdamped dynamics which is extremely important for system safety can be achieved without losing the advantage of accurate control performance in the presence of external disturbances and parameter uncertainties. Based on Lyapunov theory, the stability analysis of the proposed scheme is presented and the passivity of the system is also proven. Experiments are carried out to verify the proposed method.

Keywords: Proxy based sliding mode control · Safety · Chattering · Nonlinear control

1 Introduction

Many physical systems can be represented by a class of second-order nonlinear model, including inertia wheel pendulums (Chen and Chen 2010; Iriarte et al. 2017), robotic manipulators (Eksin et al. 2017), magnetic levitation systems (Yang et al. 2017), and pneumatic muscle actuators (Lilly 2017). Various model-based control methods have been proposed for the second-order nonlinear model by researchers. The main difficulty is that the available mathematical models are usually imprecise due to parameter uncertainties and the external disturbances, which are usually unmeasurable, commonly exists in practice. To deal with the problem, many strategies have been proposed, for example, robust fuzzy method (Xiang et al. 2017). Besides the above problem, what is specially considered in this paper is, in case the actuator force becomes excessively large, which may do harm to the environment, even to the human body in some special cases, overdamped dynamics of the system is desired. This is the so-called safety problem in (Kikuuwe and Fujimoto 2006), which can not be solved by conventional stiff positional control when the robustness and tracking accuracy are required simultaneously. For example, the PID control can possess an overdamped and smooth recovery without overshoot from large positional errors by using a very

© Springer International Publishing AG 2017
Y. Huang et al. (Eds.): ICIRA 2017, Part II, LNAI 10463, pp. 879–888, 2017.
DOI: 10.1007/978-3-319-65292-4_76

high velocity feedback gain, which guarantees the safety of the actuator, but this can magnify the noise, which will deteriorate the tracking accuracy.

Sliding mode control (SMC) has been applied in many non-linear systems due to its design simplicity and robustness against matched disturbances and parameter uncertainties. Theoretically, the SMC can produce the control signal which will force the state variables to the sliding surface, and the system state can be robustly constrained to the surface and convergent to zero in finite time. Besides, the magnitude limit of the control signal can be set arbitrarily by the controller. Therefore, in an ideal condition, the tracking accuracy and overdamped dynamics can be both obtained by using SMC. However, it is just due to the purpose to account for the presence of modelling imprecision and of disturbances, the discontinuous function (e.g. signum-type function) need to be introduced across the sliding surface. The implementation of the proposed control method is imperfect, because the switching of signum function can not be infinite fast in practice, and the value of sliding mode is not known with infinitely precision. This will lead to undesired phenomenon, chattering. Chattering can excite high frequency dynamics to the system and may produce excessively large control signal when big positional error occurs, which can do damage to physical systems. To reduce the chattering phenomenon in SMC, two main approaches are commonly used. One method is introducing a boundary layer so that the control law is linear in the vicinity of the sliding surface (Slotine and Li 1991). In this method, the selection of an appropriate value for the boundary thickness is very important. The chattering can not be solved with too narrow boundary layer while the tracking performance and robustness will be deteriorated if the value is too large. Therefore, it is obvious that this method can not guarantee both the safety and the accuracy with large position error in the system. The other is using higher order sliding mode control (John 2005) in which the derivative of the state variable can not be measured. Thus, the observer is needed to be designed to estimate that variable. Whereas, the high order SMC requires more complex calculation than normal SMC.

Proxy-based Sliding Mode Control (PSMC), introduced by Kikuuwe and Fujimoto in 2006, can be viewed as a modified version of SMC, and also an extension of the force limited PID control, which achieves the chattering free target from another perspective. Thus, both the safety and the tracking accuracy can be guaranteed by using the PSMC. We can notice from the above that the root of the chattering problem is the existence of the signum-type function in the control law. An imaginary proxy object is introduced in PSMC to remove the signum-type function out of the closed loop of the controller. By this way, the signum-type function can be algebraically transferred to the saturation function, which eliminates the chattering in essence. Therefore, this control method can achieve overdamped recovery when big positional error occurs and accurate tracking performance during normal operation.

At the same time, it should be noted that the stability problem of PSMC is not well addressed so far. The stability analysis of PSMC in (Kikuuwe et al. 2010) is based on a conjecture of the local passivity of the well-tuned PID control, which, however, remains an issue to study in the future. In this paper, we proposed a model-based PSMC for a class of second-order nonlinear systems, and its stability analysis based on Lyapunov stability theory is also given.

The rest of the paper is organized as follows. A class of second-order nonlinear system is described in Sect. 2. Based on the PSMC, a newly designed model-based controller is also derived for the system in this section. The stability analysis is given in Sect. 3. Experiments on pneumatic muscle system are carried out to demonstrate the validity of the designed controller in Sect. 4. Finally a conclusion is given in Sect. 5.

2   System Formulation and Controller Design

2.1   System Description

Consider a single-input second-order nonlinear system defined as following differential equation:
$$\ddot{q} = f(q, \dot{q}, t) + b(q, \dot{q}, t)u(t) + d(t) \tag{1}$$
where q denotes the actuator position. u is the control input signal. f is the nonlinear dynamic. b is the control gain and $d(t)$ is the lumped disturbance containing the model uncertainties and external disturbances.

2.2   Design of Model-Based PSMC

To generate overdamped response when big tracking error occurs, without sacrificing accurate, responsive tracking ability during normal operations, chattering need to be eliminated. From the above, we know that chattering problem is caused by the existence of discontinuous function sgn($\cdot$). When the sgn($\cdot$) is enclosed in the feedback loop which passes through the physical devices, time delay is inevitable. To solve the problem, we need to put closed loop containing the sgn($\cdot$) in the control software. For this purpose, an imaginary object called "proxy" is introduced to the control scheme. A proxy is a virtual object which is supposed to be connected to the physical actuator, and the connection between actuator and proxy is the so-called virtual coupling. Then, the ideal SMC can be thought to control the proxy which does not contain any physical object, so that the chattering can be eliminated since no time delay exists. On the other hand, the virtual coupling is supposed to maintain its length to be zero, which means the error between actuator position and proxy position is going to be convergent to zero. The physical model of PSMC is illustrated in Fig. 1. For the stability of the system, in this paper, we proposed a new control method to control the proxy which can be seen as an extension of the classical SMC. The sliding manifold is defined as:

$$\sigma_p = \dot{q}_d - \dot{p} + H(q_d - p) \tag{2}$$

$$\sigma_q = \dot{q}_d - \dot{q} + H(q_d - q) \tag{3}$$

where H is a positive constant. p denotes the proxy position. q_d denotes the desired position. The newly designed sliding mode controller produces the force f_p, which is applied to the proxy, as follows

$$
\begin{aligned}
f_p = \Gamma\mathrm{sgn}\,(\sigma_p) &- m_p H\dot{p} - K_P\,(p - q) \\
&- K_D\,(\dot{p} - \dot{q}) + m_p\ddot{q}_d + m_p H\dot{q}_d \\
&+ \tfrac{1}{b}(-f + K_P\,(p - q) + K_D\,(\dot{p} - \dot{q}) \\
&- H\dot{q} + H\dot{q}_d + \ddot{q}_d)
\end{aligned}
\tag{4}
$$

where q denotes the actuator position. m_p denotes the proxy mass. K_P and K_D are positive constants which represent the proportional and differential gains. $\mathrm{sgn}(\cdot)$ is the signum function which is defined by

$$
\mathrm{sgn}(x) = \begin{cases} 1, & x > 1 \\ [-1, 1], & x = 1 \\ -1, & x < 1 \end{cases}
\tag{5}
$$

The virtual coupling, which is designed to make convergent to, can be selected in many ways. In (Kikuuwe and Fujimoto 2006), a PID-type virtual coupling is adopted; in (Kashiri and Jinoh 2016), a PD-type is used. In this paper, the virtual coupling is defined as:

$$
\begin{aligned}
f_c = \tfrac{1}{b}(K_P\,(p - q) &+ K_D\,(\dot{p} - \dot{q}) \\
&- f - H\dot{q} + H\dot{q}_d + \ddot{q}_d)
\end{aligned}
\tag{6}
$$

The force f_c is directly applied to the physical controlled actuator, and its reaction force is applied to the proxy. Thus, the dynamics of the proxy can be described as:

$$
\begin{aligned}
m_p\ddot{p} = f_p - f_c \\
= \Gamma\mathrm{sgn}\,(\sigma_p) &- m_p H\dot{p} - K_P\,(p - q) \\
&- K_D\,(\dot{p} - \dot{q}) + m_p\ddot{q}_d + m_p H\dot{q}_d
\end{aligned}
\tag{7}
$$

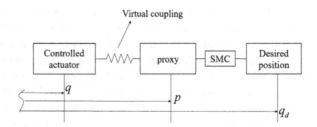

Fig. 1. Principle of PSMC

Figure 2 is a block-diagram representation of the controller. From Fig. 2, we know that the feedback loop containing the signum function does not pass through the physical controlled object. It should be noticed that the proxy is a virtual object whose mass can be set to 0. Then (7) can be rewritten as:

$$
\dot{p} + Xw = Y\mathrm{sgn}(z - Z\dot{p})
\tag{8}
$$

where

$$\begin{cases} Xw = \frac{K_P(p-q)-K_D\dot{q}}{K_D} \\ Y = \frac{\Gamma}{K_D} \\ z = \dot{q}_d + H(q_d - p) \\ Z = 1 \end{cases} \tag{9}$$

$$y + Xw = Y\,\mathrm{sgn}(z - Zy) \\ \Leftrightarrow y = -Xw + Y\,sat(\tfrac{z/Z+Xw}{Y}) \tag{10}$$

By (10), in a similar way to that in (Kikuuwe et al. 2010), (8) can be rewritten as:

$$\dot{p} = -Xw + Y\,sat(\frac{z/Z + Xw}{Y}) \tag{11}$$

In this way, the sgn($\cdot$) function is rolled out as a sat function. (8) and (11) are algebraically equivalent, which implies that (8) is actually a saturated controller. This is really different from simply substituting the sgn($\cdot$) by sat($\cdot$) approximately. Therefore, the proposed control scheme doesn't cause chattering.

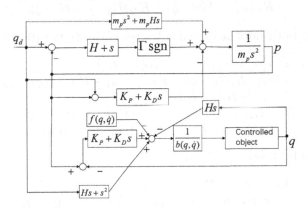

Fig. 2. Block diagram of PSMC

3 Stability Analysis

The stability analysis in Theorem 1 is based on the case of free motion, i.e. $d(t) = 0$. The passivity of the system in the presence of the lumped disturbance, i.e. $d(t) \neq 0$ is considered in Theorem 2.

Theorem 1. *Consider the second-order nonlinear system (1) with $d(t) = 0$. The stability of the system can be guaranteed by using the controller (6), whose dynamics is defined by (7).*

Proof. We choose a Lyapunov candidate as

$$V = V_1 + V_2 + V_3 \tag{12}$$

where

$$V_1 = \frac{1}{2} m_p \sigma_p^2 \tag{13}$$

$$V_2 = \frac{1}{2} \sigma_q^2 \tag{14}$$

$$V_3 = \frac{1}{2} (K_P + H K_D)(p - q)^2 \tag{15}$$

Apparently,

$$\begin{aligned} V &> 0, x_e \neq 0 \\ V &= 0, x_e = 0 \end{aligned} \tag{16}$$

is satisfied, where $x_e = [\sigma_p, \sigma_q, p - q]$. Differentiating both sides of V_1 and substitute (7) and (2) into it, we have:

$$\begin{aligned} \dot{V}_1 &= \sigma_p m_p \dot{\sigma}_p = \sigma_p \left(m_p \ddot{q}_d - m_p \ddot{p} + m_p H \dot{q}_d - m_p H \dot{p} \right) \\ &= \sigma_p \left(-\Gamma \operatorname{sgn}(\sigma_p) + K_P (p - q) + K_D (\dot{p} - \dot{q}) \right) \\ &= -\Gamma |\sigma_p| + K_P (\dot{q}_d - \dot{p})(p - q) + K_P H (q_d - p)(p - q) \\ &\quad + K_D (\dot{q}_d - \dot{p})(\dot{p} - \dot{q}) + K_D H (q_d - p)(\dot{p} - \dot{q}) \end{aligned} \tag{17}$$

Differentiating both sides of V_2 and substitute (1), (3) and (6) into it, we have:

$$\begin{aligned} \dot{V}_2 &= \sigma_q \dot{\sigma}_q = \sigma_q \left(\ddot{q}_d - \ddot{q} + H \dot{q}_d - H \dot{q} \right) \\ &= \sigma_q (-K_P (p - q) - K_D (\dot{p} - \dot{q}) + H \dot{q} - H \dot{q}_d - \ddot{q}_d \\ &\quad + \ddot{q}_d + H \dot{q}_d - H \dot{q}) \\ &= -K_P (\dot{q}_d - \dot{q})(p - q) - H K_P (q_d - q)(p - q) \\ &\quad - K_D (\dot{q}_d - \dot{q})(\dot{p} - \dot{q}) - H K_D (q_d - q)(\dot{p} - \dot{q}) \end{aligned} \tag{18}$$

Differentiating both sides of V_3, we have:

$$\begin{aligned} \dot{V}_3 &= (K_P + H K_D)(p - q)(\dot{p} - \dot{q}) \\ &= K_P (p - q)(\dot{p} - \dot{q}) + H K_D (p - q)(\dot{p} - \dot{q}) \end{aligned} \tag{19}$$

Thus, the derivation of the Lyapunov candidate is given by:

$$\begin{aligned} \dot{V} &= \dot{V}_1 + \dot{V}_2 + \dot{V}_3 \\ &= -\Gamma |\sigma_p| - K_D (\dot{p} - \dot{q})(\dot{p} - \dot{q}) \\ &\quad - H K_P (p - q)(p - q) \\ &\leq 0 \end{aligned} \tag{20}$$

From (16) and (20), we know that V is a positive-definite function and $\dot{V}$ is a negative semi-definite function. Thus, The stability of the system around the equilibrium $[\sigma_p, \sigma_q, p - q] = 0$ is proven in the sense of Lyapunov.

Theorem 2. *The second-order nonlinear system*

$$\begin{aligned} \ddot{q} &= f(q, \dot{q}, t) + b(q, \dot{q}, t) u(t) + d(t) \\ y &= -\sigma_q \end{aligned} \tag{21}$$

mapping $d(t)$ to y is passive.

Proof. We choose the storage function as same as (12). With $d(t) \neq 0$, $\dot{V}_1$ and $\dot{V}_3$ are calculated as the same as (17) and (19), while $\dot{V}_2$ is calculated as follows:

$$
\begin{aligned}
\dot{V}_2 &= \sigma_q \dot{\sigma}_q = \sigma_q \left(\ddot{q}_d - \ddot{q} + H\dot{q}_d - H\dot{q} \right) \\
&= \sigma_q (-K_P \left(p - q \right) - K_D \left(\dot{p} - \dot{q} \right) + H\dot{q} - H\dot{q}_d \\
&\quad -\ddot{q}_d - d + \ddot{q}_d + H\dot{q}_d - H\dot{q}) \\
&= -K_P(\dot{q}_d - \dot{q})(p - q) - HK_P(q_d - q)(p - q) \\
&\quad - K_D(\dot{q}_d - \dot{q})(\dot{p} - \dot{q}) - HK_D(q_d - q)(\dot{p} - \dot{q}) + yd
\end{aligned}
\tag{22}
$$

Therefore, we have:

$$
\begin{aligned}
\dot{V} &= \dot{V}_1 + \dot{V}_2 + \dot{V}_3 \\
&= -\Gamma |\sigma_p| - K_D \left(\dot{p} - \dot{q} \right) \left(\dot{p} - \dot{q} \right) \\
&\quad -HK_P \left(p - q \right) \left(p - q \right) + yd \\
&\leq yd
\end{aligned}
\tag{23}
$$

which satisfied the necessary condition for the passivity of the system. This completes the stability proof with the condition $d(t) \neq 0$.

4 Example

To verify the proposed control method, the experiments on Pneumatic Muscle Actuators (PMAs) system, which is a typical second order nonlinear system, are carried out.

PMA is widely used as the actuator in human muscle-like application due to its similarity with human muscle in size, weight and power output. The dynamic equation is based on the three elements model proposed by Reynolds (Reynolds et al. 2003). In this model, as shown in Fig. 3, the dynamics of the PMA is regarded as being composed of a contraction element, a spring element and a damping element in parallel, and the coefficients of the three elements depend on the magnitude of the input air pressure. The dynamic model of PMA is given by:

$$
M\ddot{x} - B(P)\dot{x} - K(P)x = F(P) - Mg
\tag{24}
$$

$$
\begin{aligned}
K(P) &= K_0 + K_1 P \\
F(P) &= F_0 + F_1 P \\
B(P) &= \begin{cases} B_{0i} + B_{1i}P \text{(inflation)} \\ B_{0d} + B_{1d}P \text{(deflation)} \end{cases}
\end{aligned}
\tag{25}
$$

where M is the mass. g is the acceleration of gravity. x is the axial contraction length and P is the system input, the air pressure. $B(P)$ and $K(P)$ denote the damping coefficient and spring coefficient respectively. $F(P)$ is the effective force generated by the contractile element. According to (1), the dynamics of PMA system is actually described as:

$$
\begin{aligned}
\ddot{x} &= f + bP \\
f &= \frac{1}{M}(F_0 - Mg - B_0\dot{x} - K_0 x) \\
b &= \frac{1}{M}(F_1 - B_1\dot{x} - K_1 x)
\end{aligned}
\tag{26}
$$

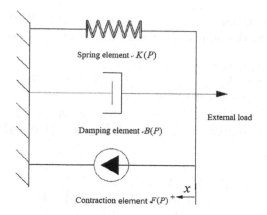

Fig. 3. Three elements model.

Then we can apply the propose control method on PMA conveniently.

The PMA system experiment platform is shown in Fig. 4. It consists of a pneumatic muscle, a displacement sensor, a force sensor, a baroceptor, an air compressor and an xPC target system. After getting the real time sensory data of the PMA system, the computer computes the control signal according to the control algorithm. The output is then fed into the electromagnetic proportion valve to adjust the air pressure. Then the pneumatic muscle can be stretched to drive the load. The xPC Target from MATLAB is used in our experiment. In this environment, a desktop PC is used as a host computer to run SIMULINK models and the second PC compiler system is used as a target computer to generate the code which can executes in real-time.

Based on three elements model, the extended PSMC control strategy is carried out on real PMA system to evaluate its control performance. The coefficients are chosen as $K_P = 31000, K_D = 0.1, H = 0.0005, F = 100000$. To make a comparison, the experiment using SMC is also carried out. The position tracking

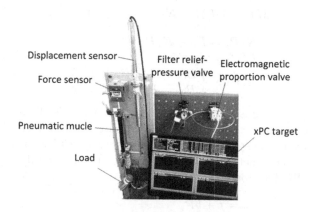

Fig. 4. PMA system experiment platform.

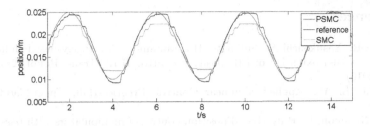

Fig. 5. Trajectory tracking control result of the model.

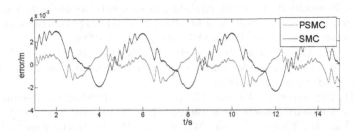

Fig. 6. Comparison of the tracking error using PSMC and SMC.

result of two control strategy is shown in Fig. 5. The corresponding tracking error result is shown in Fig. 6. The SMC can hardly achieve in tracking reference when the peak value occurs, and the chattering is obvious. Besides, the tracking error of PSMC is smaller.

5   Conclusion

The PSMC is proposed to guarantee both safety and tracking accuracy. It has already been verified experimentally (e.g. (Van Damme et al. 2009; Kikuuwe et al. 2008)), while the theoretical stability analysis is not well addressed. In this paper, we proposed an extended PSMC where a new type of virtual coupling between proxy and actuator is adopted. In this way, stability of the system is proven based on Lyapunov theory. Experiments on PMA system are also carried out to show the validity of the proposed scheme. The future study will focus on model-free PSMC by using neural network.

Acknowledgement. This work was supported by the National Natural Science Foundation of China under Grant 61473130, the Science Fund for Distinguished Young Scholars of Hubei Province (2015CFA047), the Fundamental Research Funds for the Central Universities (HUST: 2015TS028) and the Program for New Century Excellent Talents in University (NCET- 12-0214).

References

Kikuuwe, R., Yasukouchi, S., Fujimoto, H., Yamamoto, M.: Proxy-based sliding mode control: a safer extension of PID position control. IEEE Trans. Robot. **26**(4), 670–683 (2010)

Slotine, J., Li, W.: Applied Nonlinear Control. Prentice-Hall, Upper Saddle River (1991)

Kashiri, N., Jinoh, L.: Proxy-based position control of manipulators with passive compliant actuators: stability analysis and experiments. Robot. Auton. Syst. **75**, 398–408 (2016)

Kikuuwe, R., Fujimoto, H.: Proxy-based sliding mode control for accurate and safe position control. In: Proceedings of the 2006 IEEE International Conference on Robotics and Automation, pp. 25–30. IEEE, Orlando, FL, USA (2006)

Van Damme, M., Vanderborght, B., Verrelst, B., VanHam, R., Daerden, F., Lefeber, D.: Proxy-based sliding mode control of a planar pneumatic manipulator. Int. J. Robot. Res. **28**(2), 266–284 (2009)

Reynolds, D., Repperger, D., Phillips, C., Bandry, G.: Modeling the dynamic characteristics of pneumatic muscle. Ann. Biomed. Eng. **31**(3), 310–317 (2003)

Kexin, X., Jian, H., Yongji, W., Jun, W., Xu, Q., He, J.: Tracking control of pneumatic artificial muscle actuators based on sliding mode and non-linear disturbance observer. IET Control Theory Appl. **4**(10), 2058–2070 (2010)

Kikuuwe, R., Yamamoto, T., Fujimoto, H.: A guideline for low-force robotic guidance for enhancing human performance of positioning and trajectory tracking: it should be stiff and appropriately slow. IEEE Trans. Syst. Man Cybern. - Part A Syst. Hum. **38**(4), 945–957 (2008)

Jun, W., Jian, H., Yongji, W., Kexin, X.: Nonlinear disturbance observer-based dynamic surface control for trajectory tracking of pneumatic muscle system. IEEE Trans. Control Syst. Technol. **22**(2), 440–455 (2014)

Lilly, J.H., Liang, Y.: Trajectory tracking of a pneumatically driven parallel robot using higher-order SMC. IEEE Trans. Control Syst. Technol. **23**(3), 387–392 (2010)

Jian, H., Zhihong, G., Matsuno, T., Fukuda, T., Sekiyama, K.: Sliding-mode velocity control of mobile-wheeled inverted-pendulum systems. IEEE Trans. Robot. **26**(4), 750–758 (2010)

Xiang, X., Caoyang, Y., Qin, Z.: Robust fuzzy 3d path following for autonomous underwater vehicle subject to uncertainties. Comput. Oper. Res. **84**, 165–177 (2017)

Chen, M., Chen, W.H.: Sliding mode control for a class of uncertain nonlinear system based on disturbance observer. Int. J. Adapt. Control Signal Process. **24**(1), 51–64 (2010)

Yang, Z., Tsubakihara, H., Kanae, S., Wada, K., Su, C.: A novel robust non-linear motion controller with disturbance observer. IEEE Trans. Control Syst. Technol. **16**(1), 137–147 (2008)

Eksin, B., Tokat, S., Guzelkaya, M., Soylemez, M.: Design of a sliding mode controller with a nonlinear time-varying sliding surface. Trans. Inst. Meas. Control **25**(2), 145–162 (2003)

Lilly, J.H., Yang, L.: Sliding mode tracking for pneumatic muscle actuators in opposing pair configuration. IEEE Trans. Control Syst. Technol. **13**(4), 550–558 (2005)

Iriarte, R., Aguilar, L.T., Fridman, L.: Second order sliding mode tracking controller for inertia wheel pendulum. J. Franklin Inst. **350**(1), 92–106 (2013)

Numerical Methods for Cooperative Control of Double Mobile Manipulators

Víctor H. Andaluz$^{(\boxtimes)}$, María F. Molina, Yaritza P. Erazo, and Jessica S. Ortiz

Universidad de las Fuerzas Armadas ESPE, Sangolquí, Ecuador
{vhandaluz1,mfmolina1,yperazo,jsortiz4}@espe.edu.ec

Abstract. In this paper proposes a numerical methods based control algorithm for tracking trajectories applied in two anthropomorphic robotic arms mounted on an omnidirectional platform, which allows the transport of an object in common. For this, the kinematic modelation of the entire coupled system is performed, considering as the position of interest the midpoint of the operative ends of each manipulator and its formation characteristics. The stability of the proposed controller for tracking trajectories using numerical methods is demonstrated analytically, obtaining that the control is asymptotically stable. Finally the results obtained are evaluated in a virtual reality simulation environment.

Keywords: Double mobile manipulator · Kinematic modeling · Numerical methods · Virtual reality

1 Introduction

Robots nowadays are considered as a precision tool well they perform productive tasks in modern industry with a high level of complexity [1]; this reality has given rise to the programming and implementation of advanced control algorithms using different tools and simulation techniques the same serve as a fundamental basis for their real implementation [2].

A mobile manipulator robot refers to robots that are made up of a robotic arm mounted in a movable platform with legs or wheels [3]. This type of system has the ability to combine the skills and dexterities of *manipulation* of a fixed base manipulator with the *locomotion* of a mobile platform, granting the robot manipulator a larger workspace [4, 5], thus allowing the more usual missions of robotic systems that require locomotion and manipulation skills, increasing accuracy and getting better the quality of tasks [5, 6]. They offer multiple applications in different industrial and productive areas such as mining and construction or for the assistance of people [5].

There are areas where individual robots are limited in terms of manipulability, flexibility, accessibility and maneuverability, so that cooperative robots are implemented to execute the work, giving greater efficiency in industrial processes [7]. Transportation and handling are typical assignments in the industry, where large-scale use of autonomous robotic arms is being introduced [8, 9]. On the other hand, multiple robots that simultaneously manipulate a single object, called cooperative manipulation

© Springer International Publishing AG 2017
Y. Huang et al. (Eds.): ICIRA 2017, Part II, LNAI 10463, pp. 889–898, 2017.
DOI: 10.1007/978-3-319-65292-4_77

of the robot, have been an increasingly popular focus of research during recent decades, since it is recognized that many tasks can not be Performed by a single robot arm [10]. The most obvious case is that the load is too large for the robot to handle alone, so it is important to control both the absolute movement of the object and the internal stresses [11].

There are different works in the literature in which different control techniques, *e.g.*, sliding mode based on PSO (particle optimization in-jambre) and ANFIS (adaptive system of neuro-fuzzy inference) for trajectory follow-up, PSO serves for to obtain discrete articulation angles and ANFIS to convert the angles of discrete joints into angles of continuous joints [12]. In [13], they present a system of N robots of a manipulator arm and a mobile platform that carry an object to follow a desired trajectory, position and orientation, they develop a control algorithm of commutation for the manipulation of distributed cooperation of rigid bodies in a flat environment and test the stability of control by Lyapunov. Multiple mobile manipulators are considered, which hold a common object in a cooperative way to follow a desired trajectory, the control of coordinated systems allows control of the object in non limited directions [14, 15]. In [16, 17], they present the transport of an object in cooperative form of multiple omnidirectional mobile manipulators through a centralized and decentralized architecture for tracking of trajectory with out collisions in order to capture the object and place it In the destination position.

In this work, it is considered two anthropomorphic robotic arms of omnidirectional type mounted on a mobile platform, for which the kinematic modeling is performed, considering as the point of interest the midpoint of the two operative ends. The design of the coordinated cooperative controller for tracking trajectories allows the transport of a common object, for this system the control is developed by the numerical methods, which eliminates the inherent errors of kinematics. The experiments through a virtual structure allow to demonstrate that the controller is appropriate for the solution of movement problems.

The article is organized in 5 Sections including the Introduction. Section 2 presents the kinematic modeling of the anthropomorphic arms mounted on an omnidirectional platform. The design and stability analysis of the control algorithm are presented in Sect. 3. The results and discussion are shown in Sect. 4. Finally, the conclusions are presented in Sect. 5.

2 Kinematic Models

The particular case of a omnidirectional mobile manipulator with 11DOF, composed by two robotic arms –of 4DOF each arm- mounted on a omnidirectional mobile platform. The mobile platform has four driven wheels which are controlled independently by four DC motors. The robotic arms are placed at a distance a of the center of mass G of the mobile platform and at a height h_{alt} in relation to the absolute coordinate system $<R>$. Each of the arms links is driven by independent DC motors.

For a mobile platform of omnidirectional wheels with location $\mathbf{q}_p = \begin{bmatrix} x & y & \psi \end{bmatrix}^T$ in which two robotic arms are mounted with generalized coordinates $\mathbf{q}_{a1}$, $\mathbf{q}_{a2}$, and with positions of the operative end $\mathbf{h}_1$ and $\mathbf{h}_2$ corresponding to each one of the robotic arms, is given by

$$
\begin{cases}
h_{xi} = x \pm aC_\psi + l_{2i}C_{q2i}C_{q1i,\psi} + l_{3i}C_{q2i,q3i}C_{q1i,\psi} + l_{4i}C_{q2i,q3i,q4i}C_{q1i,\psi} \\
h_{yi} = y \pm aS_\psi + l_{2i}C_{q2i}S_{q1i,\psi} + l_{3i}C_{q2i,q3i}S_{q1i,\psi} + l_{4i}C_{q2i,q3i,q4i}S_{q1i,\psi} \\
h_{zi} = h_{alt} + l_{1i} + l_{2i}S_{q2i} + l_{3i}S_{q2i,q3i} + l_{4i}S_{q2i,q3i,q4i}
\end{cases}
\tag{1}
$$

where, $i = 1, 2$ represents each robotic arm mounted on the omnidirectional mobile platform; ψ is the orientation of the mobile platform; q_j $j = 1, 2, .., n_a$ ($n_a = 4$) are the joint positions of the robotic arm; $C_{q2i,q3i} = \cos(q_{2i} + q_{3i})$; $S_{q2i,q3i} = \sin(q_{2i} + q_{3i})$; $C_{q1i,\psi} = \cos(q_{1i} + \psi)$; $S_{q1i,\psi} = \sin(q_{1i} + \psi)$; $C_{q2i,q3i,q4i} = \cos(q_{2i} + q_{3i} + q_{4i})$; and $S_{q2i,q3i,q4i} = \sin(q_{2i} + q_{3i} + q_{4i})$.

To determine the system modeling of a two-arm mobile platform, the virtual point in the X-Y-Z plane between the midpoint of each end effector of the robotic arms is fixed; The virtual point is defined by $\mathbf{P}_F = \begin{bmatrix} h_x & h_y & h_z \end{bmatrix}$ that represents the position of its centroid on the inertial frame $<\mathsf{R}>$,

$$
\begin{cases}
h_x = \dfrac{h_{x1} + h_{x2}}{2} \\[2mm]
h_y = \dfrac{h_{y1} + h_{y2}}{2} \\[2mm]
h_z = \dfrac{h_{z1} + h_{z2}}{2}
\end{cases}
\tag{2}
$$

While, the vector structure of the virtual shape is defined by $\mathbf{S}_F = \begin{bmatrix} d_F & \theta_F & \phi_F \end{bmatrix}$, where, d represents the distance between the position of the end-effector $\mathbf{h}_1$ and $\mathbf{h}_2$, θ and ϕ represents its orientation with respect to the global Y-axis and X-axis, respectability on the inertial frame $<\mathsf{R}>$, Fig. 1.

$$
\begin{cases}
d_F = \sqrt{(h_{x2} - h_{x1})^2 + (h_{y2} - h_{y1})^2 + (h_{z2} - h_{z1})^2} \\[2mm]
\theta_F = \arctan\left(\frac{h_z}{h_x}\right) \\[2mm]
\phi_F = \arctan\left(\frac{h_y}{h_x}\right)
\end{cases}
\tag{3}
$$

The point of interest of the system is represented in a simplified way as $\mathbf{h} = \begin{bmatrix} \mathbf{P}_F & \mathbf{S}_F \end{bmatrix}$, i.e.,

$$
\mathbf{h} = \begin{bmatrix} hx & hy & hz & d & \theta & \phi \end{bmatrix}^T.
\tag{4}
$$

The other hand, the kinematic model of the omnidirectional mobile platform is represented as,

$$
\begin{bmatrix} \dot{x} \\ \dot{y} \\ \dot{\psi} \end{bmatrix} =
\begin{bmatrix} \cos\psi & -\sin\psi & 0 \\ \sin\psi & \cos\psi & 0 \\ 0 & 0 & 1 \end{bmatrix}
\begin{bmatrix} u_l \\ u_m \\ \omega \end{bmatrix}
\tag{5}
$$

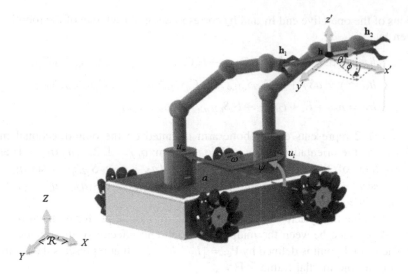

Fig. 1. Mobile manipulator with two robotic arms of 4 DOF for the experiment

where, each linear velocity is directed as one of the axes of the frame $<P>$ attached to the center of gravity of the mobile platform: u_l points to the frontal direction; u_m points to the left-lateral direction, and the angular velocity ω rotates the referential system $<P>$ counterclockwise, around the axis P_Z (considering the top view).

Now, substituting (1), (2), (3) in (4) and deriving (4), we obtain the kinematic model of the point of interest of the mobile manipulators. Also the kinematic model can be written in compact form as $\dot{\mathbf{h}} = f(\mathbf{q_p}, \mathbf{q_a})\mathbf{v}$, *i.e.*,

$$\dot{\mathbf{h}}(t) = \mathbf{J}(\mathbf{q_p}, \mathbf{q_{ai}})\mathbf{v}(t) \tag{6}$$

where, $\dot{\mathbf{h}}(t) = \begin{bmatrix} \dot{h}_x & \dot{h}_y & \dot{h}_z & \dot{d}_F & \dot{\theta}_F & \dot{\phi}_F \end{bmatrix}^T$ is the velocity vector of the interest point; $\mathbf{J}(\mathbf{q})$ represents the Jacobian matrix, $\mathbf{v}(t) = \begin{bmatrix} u_l & u_m & \omega & \dot{q}_{11} & \cdots \end{bmatrix}$ $\dot{q}_{na1} \quad \dot{q}_{12} \quad \cdots \quad \dot{q}_{na2} \end{bmatrix}^T$ is the control vector of mobility of the interest point.

3 Control

Is proposed a control system based on the system kinematics (platform, arms and object), as shown in Fig. 2.

Considering the differential equation

$$\dot{y} = f(y, u, t) \text{ con } y(0) = y_0$$

where, y represents the output of the system to be controlled, $\dot{y}$ first derivative u the control action and t, the time. The values of $y(t)$ in the discret time $t = kT_0$, are called, $y(k)$ where T_0 represents the sampling period, and $k \in \{0, 1, 2, 3, 4, 5 \ldots\}$.

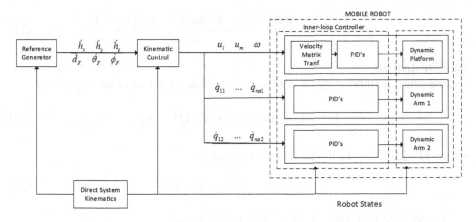

Fig. 2. Trajectory tracking control

The use of numerical methods for the calculus of the system evolution is based mainly on the possibility to approximation the sistema state on the instant time $k + 1$, if the state and the control action on the time instant k are known, this approximation is called the Euler method.

$$y(k+1) = y(k) + T_0 f(y, u, t) \tag{7}$$

On this way the discrete model for the robot could be expressed by

$$\mathbf{h}(k+1) = \mathbf{h}(k) + T_0 \mathbf{J}(\mathbf{q}(k))\mathbf{v}(k) \tag{8}$$

In order that the tracking error tends to zero the following expresition is used

$$\mathbf{h}(k+1) = \mathbf{hd}(k+1) - \mathbf{W}(\mathbf{hd}(k) - \mathbf{h}(k)) \tag{9}$$

where, $\mathbf{W}$ is a diagonal matrix $0 < \mathbf{diag}(w_x, w_y, w_z) < 1$ and $\mathbf{hd}$ is the desired trajectory. From (7) and (8) is possible to get (9) using $\mathbf{Au} = \mathbf{b}$.

$$\underbrace{\mathbf{J}(\mathbf{q}(k))}_{\mathbf{A}} \underbrace{\mathbf{v}(k)}_{\mathbf{u}} = \underbrace{\frac{\mathbf{hd}(k+1) - \mathbf{W}(\mathbf{hd}(k) - \mathbf{h}(k)) - \mathbf{h}(k)}{T_0}}_{\mathbf{b}} \tag{10}$$

Hence, viable solution method is to formulate the problem as a constrained linear optimization problem.

$$\frac{1}{2}\|\mathbf{v}\|_2^2 = \min.$$

Then yields the particular solution.

$$v(t) = J^T (JJ^T)^{-1} \frac{hd(k+1) - W(hd(k) - h(k)) - h(k)}{T_0} \tag{11}$$

Stability Analysis

In kinematics is fulfilled $v_{ref} = v$, therefore the closed-loop equation is given by,

$$h(k+1) - h(k) = T_0 J \left(J^T (JJ^T)^{-1} b \right) \tag{12}$$

where $JJ^T (JJ^T)^{-1} = I_m$ the Eq. (11) is reduced to,

$$h(k+1) - h(k) = T_0 I_m b \tag{13}$$

Through the properties of the identity matrix have

$$h(k+1) - h(k) = T_0 \left(\frac{hd(k+1) - W(hd(k) - h(k)) - h(k)}{T_0} \right) \tag{14}$$

Reducing terms and grouping them you have that the error in the following state $h_d(k+1) - h(k+1)$ depends only on the previous error by a gain $W(hd(k) - h(k))$

$$\begin{bmatrix} h_x(k+1) \\ h_y(k+1) \\ h_z(k+1) \end{bmatrix} = \begin{bmatrix} h_{xd}(k+1) - w_x(e_x) \\ h_{yd}(k+1) - w_y(e_y) \\ h_{zd}(k+1) - w_z(e_z) \end{bmatrix}$$

$$\begin{bmatrix} e_x(k+1) \\ e_y(k+1) \\ e_z(k+1) \end{bmatrix} = \begin{bmatrix} w_x(e_x(k)) \\ w_y(e_y(k)) \\ w_z(e_z(k)) \end{bmatrix}$$

The error on the following states comes by

$$e_i(k+1) = w_i e_i(k)$$
$$e_i(k+2) = w_i e_i(k+1) = w_i^2 e(k)$$
$$e_i(k+3) = w_i e_i(k+2) = w_i^3 e(k)$$
$$\vdots$$
$$e_i(k+n) = w_i e_i(k+n-1) = w_i^n e(k)$$

Them the error approaches asymptotically to cero when $0 < w_i < 1$ y $n \to \infty$. This implies that the equilibrium point of the closed loop is asymptotically stable, so that the position error of the *i-th* final effector verifies $\tilde{h}_i(t) \to 0$ asymptotically with $t \to \infty$.

4   Results

In order to evaluate the performance of the proposed controller, was developed in Matlab a 3D simulator in which the KUKA Youtbot robot is considered, which is made up of an omnidirectional mobile platform with mecanum wheels and two anthropomorphic robotic arms [18].

The simulation experiments implemented recreate an application of cooperation and coordination between two robotic arms and a mobile platform in order to perform tasks of handling a load. Figure 3 shows the stroboscopic movement in the X-Y-Z space of the reference system $<R>$, which allows to verify that the proposed controller works correctly in cooperation tasks when moving an object in common. The Fig. 4 shows the trajectories described by the operative ends of the robotic arms when performing the task; In the graph you can see the trajectories described by the robot.

Figure 5 show that the control errors of the position of the point of interest formed by the ends of the robotic arms; While Fig. 6 illustrates shape and orientation errors, I.e. the distance between the operative ends and the angles forming the object with the

Fig. 3. Stroboscopic movement of the mobile manipulator

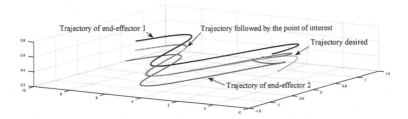

Fig. 4. Path followed by the point of interest.

arms over the XY and YZ planes with respect to the reference system $<R>$; in the
two graphs it can be seen that control errors tend to zero asymptotically when $t \to \infty$.

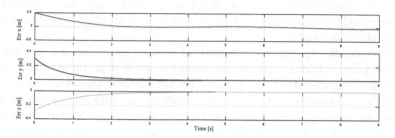

Fig. 5. Errors of position of the point of interest or midpoint of the operative ends.

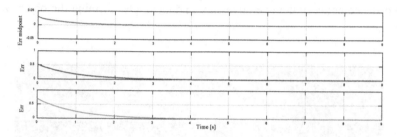

Fig. 6. Errors of shape of the object to be transported

Finally Figs. 7 and 8 its show the maneuverability commands applied to the
omnidirectional platform and the anthropomorphic arms respectively in order to
accomplish the task objective.

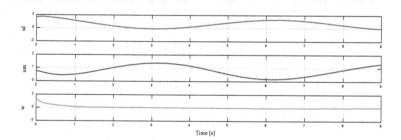

Fig. 7. Omnidirectional platform speeds

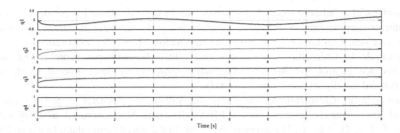

Fig. 8. Robotic arm speeds.

5 Conclusions

In this work the design of a coordinated cooperative controller for the trajectory tracking that allows to transport an object in common was presented. The design of the controller is based on a kinematic control that meets the purpose of movement, reference velocities were defined for the omnidirectional platform and the robotic arms. Stability and robustness are tested by the numerical methods. Experiments carried out trough a virtual reality structure confirm the scope of the controller to solve different problems of movement through a strong choice of control references.

References

1. Markus, E., Agee, J., Jimoh, A., Ceccarelli, M.: Flat control of industrial robotic manipulators. Robot. Auton. Syst. **87**, 226–236 (2017)
2. Materna, Z., Kapinus, M., Spanel, M., Beran, Z., Smrz, P.: Simplified industrial robot programming: effects of errors on multimodal interaction in WoZ experiment. In: 25th IEEE Internacional Symposium on Robot and Human Communication (RO-MAN), pp. 200–205, USA (2016)
3. Meng, Z., Liang, X., Andersen, H., Ang, M.: Modelling and control of a 2-link mobile manipulator with virtual prototyping. In: 13th Internacional Conference on Ubiquitous and Ambiente Intelligence (URAI), pp. 363–368, China (2016)
4. Andaluz, V., Roberti, F., Toibero, J., Carelli, R.: Passivity-based visual feedback control with dynamic compensation of mobile manipulators: stability and L2-gain performance analysis. Robot. Auton. Syst. **66**, 64–74 (2015)
5. Andaluz, V., Roberti, F., Toibero, J., Carelli, R.: Adaptive unified motion of mobile manipulators. Control Eng. Pract. **20**, 1337–1352 (2012)
6. Khatib, O.: Mobile manipulator: the robotic assistant. Robot. Auton. Syst. **26**, 175–183 (1999)
7. Markus, E., Yskander, H., Agee, J., Jimoh, A.: Coordination control of robot manipulators using flat outputs. Robot. Auton. Syst. **83**, 169–176 (2016)
8. Szabó, R., Gontean, A.: Robotic arm autonomous movement in 3D space using stereo image recognition in Linux. In: 11th International Symposium on Electronics and Telecommunications (ISETEC) (2014)

9. Iossifidis, I., Schöner, G.: Autonomous reaching and obstacle avoidance with the anthropomorphic arm of a robotic assistant using the attractor dynamics approach. In: Internacional Conference on Robotics & Automation, pp. 4295– 4300 (2004)

10. Caccavale, F., Chiacchio, A., De Santis, A., Villani, L.: An experimental investigation on impedance control for dual-arm cooperative systems. In: Internacional Conference in Advanced Intelligent Mechatronics (2007)

11. Caccavale, F., Chiacchio, A., De Santis, A., Villani, L.: Six-DOF impedance control of dual-arm cooperative manipulators. IEEE/ASME Trans. Mechatron. **13**, 576–586 (2008)

12. Ke, Z., Lou, G., Lai, X.: Trajectory tracking for 3-DOF robot manipulator based on PSO and adaptive neuro-fuzzy inference system. In: Control Conference (CCC), pp. 1934–1768, Chinese (2016)

13. Markdalh, J., Karayiannidis, Y., Hu, X.: Cooperative object path following control by means of mobile manipulators: a switched systems approach. In: 10th IFAC Symposium on Robot Control, vol. 45, pp. 773–778 (2012)

14. Galicki, M.: Finite-time trajectory tracking control in a task space of robotic manipulators. Automatica **67**, 165–170 (2016)

15. Abbaspour, A., Alipour, K., Jafari, H., Moosavian, A.: Optimal formation and control of cooperative wheeled mobile robots. C. R. Méc. **343**, 307–321 (2015)

16. Hekmatfar, T., Masehian, E., Maousavi, S.: Cooperative object transportation by multiple mobile manipulators though a hierarchical planning architecture. In: 2nd RSI/ISM International Conference on Robotics and Mechatronics, pp. 503–508 (2014)

17. Ponce, A., Castro, J., Guerrero, H., Parra, V., Olguín, E.: Cooperative redundant omnidirectional mobile manipulators: modelfree decentralized integral sliding modes and passive velocity fields. In: IEEE International Conference on Robotics and Automation (ICRA), pp. 2375–2380 (2016)

18. Kuka youBot store. http://www.youbot-store.com/G

Erratum to: Intelligent Robotics and Applications

YongAn Huang[1]([⊠]), Hao Wu[1], Honghai Liu[2], and Zhouping Yin[1]

[1] School of Mechanical Science and Engineering,
Huazhong University of Science and Technology, Wuhan, China
yahuang@hust.edu.cn
[2] Institute of Industrial Research, University of Portsmouth, Portsmouth, UK

Erratum to:
Chapter "Real-Time Normal Measurement
and Error Compensation of Curved Aircraft Surface Based
on On-line Thickness Measurement" in: Y. Huang et al. (Eds.):
Intelligent Robotics and Applications, Part II, LNAI 10463,
https://doi.org/10.1007/978-3-319-65292-4_15

In an older version of this paper, the acknowledgement section was missing. This has been corrected.

Erratum to:
Chapter "Remote Live-Video Security
Surveillance via Mobile Robot with Raspberry Pi IP Camera"
in: Y. Huang et al. (Eds.): Intelligent Robotics
and Applications, Part II, LNAI 10463,
https://doi.org/10.1007/978-3-319-65292-4_67

In an older version of this paper, the acknowledgement section was missing. This has been corrected.

The updated online version of this book can be found at
https://doi.org/10.1007/978-3-319-65292-4
https://doi.org/10.1007/978-3-319-65292-4_15
https://doi.org/10.1007/978-3-319-65292-4_67

Author Index

Printed in the United States
By Bookmasters